Progress in Botany 65

Springer

Berlin
Heidelberg
New York
Hong Kong
London
Milan
Paris
Tokyo

65 PROGRESS IN BOTANY

Genetics
Physiology
Systematics
Ecology

Edited by

K. Esser, Bochum
U. Lüttge, Darmstadt
W. Beyschlag, Bielefeld
J. Murata, Tokyo

Springer

With 48 Figures

ISSN 0340-4773
ISBN 3-540-40721-9 Springer-Verlag Berlin Heidelberg New York

The Library of Congress Card Number 33-15850

Springer-Verlag is a part of Springer Science+Business Media

springeronline.com

Cover design: Design & Production, Heidelberg
Typesetting: M. Masson-Scheurer, Neckargemünd
31/3150 - 5 4 3 2 1 0 - Printed on acid-free paper

Contents

Review

Genetics

Physiology

Systematics

Ecology

Plants and Geothermal CO_2 Exhalations – Survival in and Adaptation to a High CO_2 Environment 499
Hardy Pfanz, Dominik Vodnik, Christiane Wittmann, Guido Aschan, and Antonio Raschi (With 4 Figures)

List of Editors

Professor Dr. Dr. h.c. mult. K. Esser
Lehrstuhl für Allgemeine Botanik, Ruhr Universität
Postfach 10 21 48
44780 Bochum, Germany

Phone: +49-234-32-22211, Fax: +49-234-32-14211
e-mail: karl.esser@ruhr-uni-bochum.de

Professor Dr. U. Lüttge
TU Darmstadt, Institut für Botanik, FB Biologie (10)
Schnittsphahnstraße 3–5
64287 Darmstadt, Germany

Phone: +49-6151-163200, Fax: +49-6151-164630
e-mail: luettge@bio.tu-darmstadt.de

Professor Dr. W. Beyschlag
Fakultät für Biologie,
Lehrstuhl für Experimentelle Ökologie
und Ökosystembiologie
Universität Bielefeld
Universitätsstraße 25
33615 Bielefeld, Germany

Phone: +49-521-106-5573, Fax: +49-521-106-6038
e-mail: w.beyschlag@biologie.uni-bielefeld.de

Professor Dr. J. Murata
Botanical Gardens, Graduate School of Science
University of Tokyo
3-7-1 Hakusan, Bunkyo-ku, Tokyo 112-0001, Japan

Phone: +81-3-38142625, Fax: +81-3-38140139
e-mail: murata@ns.bg.s.u-tokyo.ac.jp

Eberhard Schnepf was born on April 4, 1931, in Nürnberg, Germany. After his Abitur (1950, Grammar School in Wolfenbüttel) and apprenticeship as gardener, he studied biology, plant pathology and chemistry at the universities of Munich (1952–1954) and Bonn (1964–1958) where he received his doctoral degree (Dr. rer. nat.) in 1958. From 1958–1964, he was assistant lecturer (Wissenschaftlicher Assistent) at the Botanical Institute, University of Marburg, and habilitated 1963 with studies on structure and function of plant glands. In 1959, he married Rosemarie Langbein. They have two children. From 1964–1966 he was research associate (Wissenschaftlicher Rat) and professor at the Institute of Plant Physiology, University of Göttingen. In 1966, he became Außerordentlicher Professor (associate professor) and in 1970 Ordentlicher Professor (full professor) at the University of Heidelberg (Chair of Cytology) until his retirement in 1996. In 1969 and 1979, he declined professorships offered at the Universities of Kiel and Hamburg, respectively.

The main research activities of Eberhard Schnepf included: secretory processes in plant glands, plant cell ultrastructure and its dynamics, cell compartmentation, structure and formation of cell walls, morphogenesis of plant cells, tip growth and cell polarity, ultrastructure and taxonomy of algae, parasites of algae and the symbiogenesis of plastids.

The scientific work of Eberhard Schnepf comprises more than 320 publications. He was editor or co-editor of several journals (*Protoplasma*, *Botanica Acta*, *Planta*, *Biochemie* and *Physiologie der Pflanzen*) and the book series, *Progress in Botany*. He was engaged for many years in the Deutsche Forschungsgemeinschaft as referee and was active in the Deutsche Botanische Gesellschaft (member of the Board of Directors) and in the Deutsche Gesellschaft für Zellbiologie (Vice-President).

Since 1974 he has been a member of the Deutsche Akademie der Naturforscher Leopoldina, since 1982 corresponding member of the Akademie der Wissenschaften Göttingen. He was awarded the title Dr. h. c. (honorary doctorate) in 1984 at the University of Thessaloniki (Greece).

Protoctists and Microalgae: Antagonistic and Mutualistic Associations and the Symbiogenesis of Plastids

Eberhard Schnepf

1 Introduction

Microscopical studies of associations of protoctists and microalgae have provided new insight into phycopathology and symbiogenesis of plastids also allowing a better understanding of fundamental problems of cell biology and cell evolution. Ecological problems and molecular biological findings are discussed marginally, lichens are not considered.

The discovery of intracellular membrane systems has been of prime importance for cell biology and biochemistry. Endomembranes completely enclose various kinds of compartments in eukaryotic cells (Sitte 1998). Initially it was supposed that each compartment contained a special kind of cell plasma (Ruska 1960). Secretory processes in plant glands, based on membrane flow mechanisms (Schnepf 1961, 1969a) suggested a different interpretation of cell compartmentation. Electron microscopical investigations in *Geosiphon pyriforme* (Schnepf 1964) and *Glaucocystis* (Schnepf et al. 1966) helped to establish a new concept which took into consideration also the two-membrane envelopes of chloroplasts and mitochondria: There are "plasmatic" compartments (cytoplasm s.str. and the nuclear matrix, the mitochondrial matrix, and the plastid stroma) and "nonplasmatic" compartments (vacuoles, ER, Golgi cisternae and vesicles, microbodies as well as the spaces of mitochondrial and plastidal envelopes, of the mitochondrial cristae and the thylakoid lumen). Biomembranes separate a plasmatic from a nonplasmatic compartment (Schnepf 1966). This compartmentation rule has a few exceptions (Schnepf 1984, see also Cavalier-Smith 1993, 2000, and his interpretation of the two-membrane envelopes) which will be discussed below, but a general meaning, as demonstrated, e.g., in the nomenclature of freeze-fractured membranes.

The electron microscopy of *Geosiphon* and *Glaucocystis* initiated an understanding of the compartmentation of plastids, while also opening a door to understanding the phylogeny and evolution of chloroplasts and mitochondria. The idea of an endosymbiotic origin, first suggested by Schimper (1883) and Mereschkowsky (1905), largely disregarded at the

Progress in Botany, Vol. 65

time, was more recently renewed (Schnepf 1966). The serial endosymbiosis theory is now widely accepted (Margulis 1981; Maier et al. 1996), supported by molecular biological data and by electron microscopical studies of existing associations between prokaryotic (cyanobacteria) or eukaryotic algae and protoctists. Genetic changes arising from endocytosymbiosis are by far greater than those arising from other intrinsic changes such as mutations or hybridization (Jeong 1983).

Antagonistic and mutualistic associations are living models showing how chloroplasts can arise, and in fact evolve. The electron microscopy of antagonistic and mutualistic associations between microalgae and protoctists in respect to the interface between the two partners and its transformation into chloroplast envelopes in conjunction with the various strategies of food uptake are the central issues of this article.

2 Antagonistic Associations

The terms predator/prey and parasite/host have been coined for antagonistic associations of multicellular animals and plants. A typical predator kills the prey, a typical parasite feeds on the host keeping it alive. It consumes the interest, while the predator consumes the capital. A clear distinction is difficult, even in multicellular organisms. A unicellular alga is usually, but not always, killed when attacked by a protoctist. Sommer (1994) therefore preferred the term "parasitoid" instead of "parasite". Canter-Lund and Lund (1995) discriminated between grazers, organisms which digest the algae within themselves, and parasites, which digest the algae externally.

Following largely Gaines and Elbrächter (1987) "parasite" is applied here for organisms which divide repeatedly after each feeding period whereas in "predators" feeding and cell division are not connected (see also Schnepf and Elbrächter 1992). A parasite is moreover frequently smaller than its host and its feeding process takes a long time (often in the range of hours or even days) whereas a predator is usually larger than the prey and it swallows the complete algal cell quickly and at once to digest it in a food vacuole.

Parasitic eumycota and oomycetes generally digest the algae extracellularly and take up the nutrients by resorbing the molecules, a process which is not quite correctly named "osmotrophy". Most algivorous protoctists engulf portions of the alga or whole cells by endocytosis and digest them in food vacuoles. Only those associations comprise models of chloroplast evolution.

2.1 Parasites Feeding by Resorption

2.1.1 Endocytic Parasites

Most endocytic "fungal" parasites of microalgae belong to oomycetes or chytrids (Sparrow 1960, Canter-Lund and Lund 1995). *Lagenisma coscinodisci* is perhaps the most extensively studied example among the oomycetes. It was detected by Drebes (1966) parasitizing the centric marine diatom *Coscinodiscus. Lagenisma* is diplanetic. A primary zoospore forms a primary cyst. It releases an isomorphic secondary zoospore which attaches to a host cell and encysts there (Schnepf and Drebes 1977; Schnepf et al. 1978c,e). It germinates with a thin infection tube which penetrates the diatom frustule between the overlapping cingula and grows out into a thick, irregularly branched, non-septated, wall-less, multinucleate hypha (Drebes 1966; Schnepf et al. 1978a). The host plasmalemma is pushed back, not pierced. The narrow space around the parasite thallus remains in contact with the periplasmic space of the diatom so that *Lagenisma* is endocytic but extrabiotic. The host plasmalemma is in close contact with the parasite plasmalemma. It disintegrates when the *Coscinodiscus* protoplast breaks down.

Being an obligate, biotrophic parasite, *Lagenisma* then changes from the trophic into the reproductive phase.The thallus becomes covered by a cell wall and develops into a holocarp zoosporangium (Drebes 1966; Schnepf et al. 1978b, 1978e). The (primary) zoospores are released through a single discharge tube which opens apically (Drebes 1966; Schnepf et al. 1978e). They seem to be driven out by the swelling of a mucilaginous material within the zoosporangium. The life cycle takes a few days.

The sexual reproduction of *Lagenisma* differs considerably from that of other oomycetes (Schnepf and Drebes 1977; Schnepf et al. 1978b,d). In old, overpopulated cultures the last nuclear division is meiotic, not mitotic. The arising haploid swarmers resemble morphologically the diploid zoospores. Female-determined swarmers settle down near the host cell and encyst, forming an oogonium. Male-determined swarmers encyst close to an oogonium to form an antheridium. That drives a thin fertilization tube toward and into the oogonium. During plasmogamy the oogonium develops a thick, short hypha into which the fused cytoplasms with the two nuclei migrate. It is surrounded by a thick wall and becomes a resting spore (oospore) in which karyogamy takes place. The germination of a resting spore could not be observed as yet.

An *Olpidium* sp., parasitizing the marine diatom *Pseudonitzschia* seems to be truly endobiotic (Elbrächter and Schnepf 1998), perhaps with a host-parasite interface of a single membrane, the parasite plasmalemma, as in *Brassica* cells infected with *Olpidium brassicae* (Lesemann and Fuchs 1970)

Some endobiotic parasites belong to the dinoflagellates (Schnepf and Elbrächter 1992). *Amoebophrya ceratii*, a complex of host specific taxa (Coats and Park 2002), is an obligate parasite of dinoflagellates (Fritz and

Nass 1992; Maranda 2001). It has a very peculiar life cycle. A flagellate (dinospore) penetrates into a host cell and is then included in a periparasitic vacuole, situated within the nucleus or within the cytoplasm of the host. It becomes a round trophont and begins to rapidly grow. The girdle lengthens and the hypocone forms a cup-shaped circumvallation which finally encloses the epicone. The inner surface of the parasite is covered by a typical amphiesma (plasmalemma plus a subsurface layer of flattened amphiesmal vesicles) and bears many flagella. The trophont becomes multinucleate. When at the end of the feeding phase the trophont largely fills the host cell, it everts into a vermiform stage, thereby enclosing portions of host cytoplasm. The host-parasite interface consists hitherto of generally three membranes, the homology of which is unclear (Fritz and Nass 1992). During the main growth phase food vacuoles are absent. The nutrients are obviously taken up by resorption. The vermiform leaves the host cell and divides into hundreds to thousand new dinospores.

2.1.2 Epicytic Parasites

Many epicytic parasites (often named "epibiotic") of microalgae feed by resorption. Most of them belong to the chytrids. A *Chytridium* sp., parasitising the chlorococcalean alga *Scenedesmus armatus* has been investigated by Schnepf et al. (1971b). A *Chytridium* zoospore attaches to a *Scenedesmus* cell, retracts the flagellum, encysts and drives a haustorium through the wall of the alga. The host deposits electron dense material around the perforation site. Occasionally a "callosity" blocks the invasion. It consists of irregular host wall material and includes remnants of host cytoplasm. It inhibits the further development of the parasite that consequently dies.

A successful haustorium pushes the host plasmalemma back but does not pierce it. It is surrounded by the fungal wall and does not contain mitochondria. Accompanied by a small layer of host cytoplasm it may even penetrate through a chloroplast, the envelope of which is likewise invaginated but not pierced.

The main body of the chytrid remains epicytic. It grows while the host cytoplasm degenerates. At the end of the feeding period a plug separates the then multinuclear cytoplasm of the developing epicytic zoosporangium from the haustorium. Subsequently the sporangial wall is thickened. A second, inner wall layer is deposited, and the zoospores are formed.

A marine chytrid on the chain-forming diatom *Bellerochea malleus* resembles the *Scenedesmus* parasite in essential details (Schweikert 1997). Deposition of host material around the invading haustorium has not been observed here. The host protoplast does not retract locally, in contrast to the freshwater diatom *Asterionella formosa* when attacked by the chytrid *Rhizophydium planktonicum* where the fungus becomes deprived of food and dies (Canter and Jaworsky 1979).

2.2 Phagocytotic Predators

Most predatory algivorous protoctists phagocytose complete algal cells by means of pseudopodia or ingest them through a cytostome.

Amoeboid, plasmodial and heliozoan organisms surround a prey cell with a pseudopodium and enclose it in a food vacuole. This process depends on membrane flow mechanisms (Hausmann and Radek 1993). When the heliozoan *Actinophrys sol* catches the green flagellate *Chlorogonium elongatum*, the flagellate is attached to the surface of axopodia by means of the content of extrusomes (Sakaguchi et al. 1998). By the fusion of the extrusomes, the membranes are now transformed into the membranes of the food vacuoles (Hausmann and Patterson 1982).

In the vampirellid amoebae, membrane stores seem to pre-exist which can be used rapidly for the formation of food vacuoles. The vampirellids open the cell wall of filamentous (*Oedogonium*) and unicellular (*Closterium*) green algae obviously by local enzymatic activity. Due to the turgor of the algal cells, part of the prey cytoplasm is explosively released when the cell is opened. The material is enclosed within a few seconds in the arising food vacuole which has a membrane area of more than 4000 μm^2. Further portions of the algal cytoplasm are ingested by a pseudopodium which invades the cell while the main body of the predator is epicellular (Hausmann and Radek 1993).

Various heterotrophic and mixotrophic heterokont flagellates and haptophytes likewise use pseudopodia to ingest algal cells. Surveys are given in the book of Patterson and Larsen (1991). They are supplemented here by a few further examples. The heterokont marine alga *Reticulosphaera socialis* consists of spherical cells which are connected by a common reticulopodium which is used to capture and digest diatoms. They can survive only for a restricted time period by photosynthesis alone (Grell 1989). *Parapedinella reticulata* is a heterokont, apoplastidic flagellate with long, thin filopodia. They adhere to small diatoms and are then retracted so that the prey can be enveloped and digested (Schnepf 1999).

More sophisticated with very complex structures are the cytopharyngeal apparatuses in ciliates and euglenoids. Only a few examples can be presented here. The ciliate *Pseudomicrothorax dubius* preys specifically on filamentous cyanobacteria. The algae are ingested through a cytopharyngeal basket consisting of hundreds of microtubules forming nematodesmata and nematodesmal lamellae (Hausmann and Peck 1978). The algal filaments are ingested at rates of up to 15 µm per second. The force for phagocytosis seems to be generated by actin filaments whereas the microtubules serve as skeletal elements. During food uptake the membrane of the food vacuole increases rapidly. Vacuole growth results from the fusion of vesicles containing hydrolytic enzymes which digest the wall of the cyanobacterium in a matter of seconds (Hausmann and Peck 1979). Other ciliates ingesting cyanobacteria are less specific. They often require several hours to accomplish digestion of a single filament (Hausmann and Hülsmann 1996).

The ciliate, *Climacostomum virens*, uses oral membranelles to drive suspended food particles into its buccal cavity and ingests them periodically at the end of this tube. Up to 140 *Chlorogonium elongatum* cells can be taken up in 2–5 min (Fischer-Defoy and Hausmann 1981).

Only a few euglenids are predators of unicellular algae though many species are heterotrophs. Their feeding apparatuses have been reviewed by Triemer and Farmer (1991). *Peranema trichophorum* employs two different modes of feeding. It ingests a wide array of particles, also living cells of the green euglenid *Lepocinclis buetschlii* or it uses the feeding apparatus to take up the prey protoplast in a myzocytosis-like process (Sect. 2.4) (Triemer 1997).

Dinoflagellates have developed diverse nutritional strategies (Gaines and Elbrächter 1987; Schnepf and Elbrächter 1992). About half of the extant species are heterotrophs and many photosynthetic species are mixotrophic.

Noctiluca scintillans (synon. *N. miliaris*) is an omnivorous dinoflagellate (Elbrächter and Qi 1998) which ingests even large pieces of glass but also small algae. The prey is trapped at the tip of a long, mobile tentacle by means of mucilage. The tentacle then places the prey into the cytostome where a food vacuole, bounded by a single membrane, is pinched off (Nawata and Sibaoka 1983). Membrane hyperpolarization and transmembrane ion currents through the cytostome region precede the feeding process (Nawata and Sibaoka 1987).

The athecate dinoflagellate, *Oxyrrhis marina*, feeds on a wide range of algae and is even cannibalistic (Dodge and Crawford 1974). The phagocytosis of a cell of *Dunaliella bioculata* takes not more than 10–15 s. A well-elaborated cytostome is lacking. During the ingestion the ventral peripheral microtubules are reversibly rearranged (Höhfeld and Melkonian 1998).

Even thecate dinoflagellates are able to take up other algae by the whole. The mixotrophic marine *Fragilidium subglobossum* feeds exclusively on *Ceratium* spp. (Skovgaard 1996). The sulcus plates move away from one another creating a longitudinal feeding gap along the sulcus plates. The theca of the *Ceratium* cell is dissolved early, in part even when the prey is not yet ingested completely. As the ingestion proceeds and the body of *Fragelidium* increases, the thecal plates become widely separated and the shape of the cell alters drastically. A *Ceratium tripos* cell is engulfed within 15 min, the smaller *C. lineatum* within 5 min.

2.3 Pallium Feeding

Several dinoflagellates have overcome the problem to feed on cells much larger than themselves, by pallium feeding, first observed in *Protoperidinium* by Gaines and Taylor (1984). Prey species belong to the dinoflagellates, chlorophycean, cryptophycean and prymnesiophycean flagellates and diatoms (Naustvoll 2000a).

Jacobson and Anderson (1986) describe the feeding process as follows. The dinoflagellate swims about with a cytoplasmic peduncle, extruded near the flagellar grooves. If it encounters a prey it changes its swimming behaviour to move in tight circles. Eventually it makes contact with the tip of the peduncle. A pseudopodium then extends from the flagellar grooves along the peduncle which shortens pulling the prey closer to the predator. The pseudopodium enlarges and encloses the prey, usually single cells or chains of diatoms, as a feeding veil (pallium). Even large chains of cells with long spines are captured, e.g. a 600-μm-long chain of the diatom *Chaetoceros* within 15 min by the pallium of *Protoperidinium spinulosum*, the diameter of which was 50 μm. Eventually the prey is enclosed in a large food vacuole outside the theca and digested there, generally within less than 30 min. Only digested material is transported into the cell body. At the end of the feeding phase the pallium is retracted and undigested material is released.

The pallium is a vesiculate, membraneous sac containing several microtubular ribbons but neither mitochondria nor ER. The microtubules originate from an internal microtubular basket and pass through a sphincter-like ring located inside the posterior flagellar groove (Jacobson and Anderson 1992). The organization of the pallium strongly suggests that it is homologous with the feeding tube of other dinoflagellates (Sect. 2.4) albeit morphologically and physiologically very different.

2.4 Peduncle Feeding

Myzocytosis (Schnepf and Deichgräber 1984) is a special form of endocytosis observed in various epicytic parasitic dinoflagellates (reviewed in Schnepf and Elbrächter 1992). The prey protoplast is pierced by a cytoplasmic feeding tube, the peduncle, which is homologous with the pallium (Sect. 2.3) or by a phagopod. Peduncle and phagopod have not always been clearly distinguished in previous publications (Drebes and Schnepf 1998). The host cytoplasm is sucked up gradually and included in a food vacuole. The host plasmalemma is not taken up so that the ingested cytoplasm is separated from the parasite cytoplasm by a single membrane, the membrane of the food vacuole.

Peduncle feeding and the structure of the peduncle were first described in detail for *Gymnodinium fungiforme* (synon. *Katodinium f.*) (Spero and Morée 1981) and for *Paulsenella* (Drebes and Schnepf 1982; Schnepf et al. 1985).

The most extensively studied species is *Paulsenella vonstoschii* (in former publications as *P.*cf. *chaetoceratis* or *P.* sp.), feeding on *Streptotheca thamesis* (newer name: *Helicotheca t.*). The *P. vonstoschii* dinospores (= flagellates) are chemotactically attracted by the host, presumably by short-lived components of its mucilage. These components are not released by egg cells, sperm cells and auxozygotes. The appetence of *Paulsenella* is light dependent (Schnepf and Drebes 1986). The zoospore attaches to the girdle region and drives a cytoplasmic peduncle through the overlap of the cingula into the host cell. The peduncle emerges near the flagellar grooves through a sphincter of striated filaments. Its basis is surrounded by a cellulosic sheath, the distal part is mobile. It pierces the host plasmalemma the rim of which becomes closely attached to the apical opening of the peduncle. It retains its integrity elsewhere though it shrinks rapidly during the uptake process. The host vacuoles contract and the protoplast is sucked out within less than one hour (Drebes and Schnepf 1982; Schnepf et al. 1985). When an extremely large *Streptotheca* cell is attacked, the parasite is unable to ingest the whole cell contents. The cell recovers if the host nucleus is left intact – a true case of parasitism.

The peduncle is formed by the emergence of a preformed microtubular basket consisting of bands of microtubules. It is a thin cytoplasmic tube bounded by a single membrane, the plasmalemma, which continues into the inner tube membrane and the membrane of the developing food vacuole. It is stiffened by the bands of microtubules. The elongation of the peduncle and its movement within the host cell seems to be driven by the sliding of the bands and of the microtubules within the bands. Before its protrusion the microtubular basket is characteristically associated with long electron-transparent and long electron-dense vesicles, as in the microtubular basket that forms a pallium (Sect. 2.4).

Paulsenella chaetoceratis invades the interior of *Chaetoceros* cells through a yet non-silicified tip of the setae. The feeding tube may here reach a length of more than 100 µm (the dinospores are 12–15 µm long). *Chaetoceros* chloroplasts pass the narrow feeding tube of *P. chaetoceratis*, within a few seconds being deformed drastically from a lense-shaped organelle 3.5 µm in diameter and 1.8 µm thick to a cylinder 8 µm long and 1.8 µm thick. Within the food vacuole the chloroplasts regain their previous shape and internal structure. Their envelope is not ruptured. Even the periplastidal ER ("chloroplast ER", Sect. 6.2) is partially preserved (Schnepf et al. 1988). At the end of the feeding process the peduncle shortens, collapsing apically, and is retracted within a few seconds. Digestion does not begin before food uptake has been concluded completely. The ingested cytoplasm and the chloroplasts therein appear to be structurally intact (perhaps also active) for more than an hour. The trophont encysts during digestion and divides repeatedly to eventually give rise to dinospores. The life cycle is concluded within about a day. Sexual processes have been observed but the knowledge on the details is still incomplete (Drebes and Schnepf 1988).

Myzocytosis by means of a typical peduncle was observed also in a *Cypthecodinium cohnii*-like dinoflagellate preying on the small unicellular red alga *Porphyridium* sp. (Ucko et al. 1997). The cell is sucked out within 10–30 s, leaving an empty, soon disintegrating cell wall. The peduncle is retracted after each uptake process. Actin seems to be involved in the suction and the retraction of the peduncle. Up to 20 *Porphyridium* cells are taken up until the dinoflagellate encysts and divides.

A 66-kDa glycoprotein in the *Porphyridium* cell wall is the recognition site for the dinoflagellate which contains enzymes that degrade the cell wall complex of this alga but not that of other rhodophytes. The dinoflagellate is prevented to prey when the algal cells are blocked with antiserum specific to the 66-kDa glycoprotein or the lectin conA (Ucko et al. 1999).

Peridiniopsis berolinensis feeds on cryptophytes but also on nematodes and rotifers, with a typical peduncle which is also used to take up particulate food and whole algal cells and is, therefore, not a myzocytotic apparatus in the narrow sense (Calado and Moestrup 1997).

2.5 Phagopod Feeding

Phagopod feeding was first described in detail for the dinoflagellate *Amphidinium cryophilum* (Wilcox and Wedemayer 1991). The phagopod is a hollow cylinder that extends from the antapex. In *A. cryophilum* it consists of electron-opaque material that is possibly deposited on a membrane but is no longer cytoplasmic when in function. In contrast to a peduncle it does not pass through a sphincter. The prey cells, other dinoflagellates, are sucked out within 10 min or more. The food is deposited in a nascent vacuole. After feeding the phagopod is left in the prey cell. The peduncle of *A. cryophilum* is not involved in food uptake.

The phagopod of *A. cryophilum* has a wide apical opening, in contrast to that of *Gyrodinium undulans*, a dinoflagellate that feeds on the marine diatom *Odontella aurita*. No other diatoms are accepted but copepod and rotifer eggs are also sucked out (Drebes and Schnepf 1998). A *G. undulans* flagellate may attach to any site of the diatom frustule. It sticks with its antapex and protrudes a short phagopodium which broadens to a flat appressorium exactly over an areolar chamber. The frustule is pierced through the areole which has a diameter of only 0.1–0.25 μm. Within the diatom the phagopod is a 2-μm-wide and up to 60-μm-long cellulosic tube.

The *Odontella* protoplast contracts but is eventually pierced by the tip of the phagopod which then opens to form a narrow funnel. The cytoplasmic fluid and perhaps small organelles are sucked up. The chloroplasts are too large to enter the funnel. They thus cannot plug the bottle neck where the phagopod passes through the areola. The feeding phase takes 1.5–3 h and changes from myzocytosis to phagocytosis when the host plasmalemma is broken down. The phagopodium and an aggregation of discoloured chloroplasts are left behind when the ingestion is finished and the dinoflagellate detaches and divides.

Myzocytosis-like processes do not seem to be restricted to dinoflagellates. The kinetoplastid flagellate *Bodo bacillariophagus* parasitizes on the diatom *Navicula* extending a delicate "haustorium" through the raphe into the cell. It restricts itself to sucking up the nucleoplasm (Bursa 1963). The euglenoid *Peranema trichophorum* feeds phagotrophically and myzocytotically (Triemer 1997). Some ciliates belonging to the Colpodea have a feeding apparatus that structurally and functionally resembles a dinoflagellate peduncle (Foissner and Didier 1984) as do the suctorian tentacles (Bardele 1972).

2.6 Epicytic Parasites Feeding with Pseudopodia

Diatoms are well protected by their frustule against the attack of small parasites or predators. Peduncle feeding, phagopod feeding and pallium feeding are ways to overcome this barrier, special pseudopodia may also serve towards these means.

An epicytic amoeba with an endocytic, phagocytic pseudopodium is *Rhizamoeba schnepfii* (Kühn 1996/97). It is obligately algivorous and feeds on various marine diatoms. The pseudopodium penetrates into the frustule between the girdle bands and gradually phagocytizes portions of the host protoplast. The food vacuoles are transported into the main body which remains outside the frustule. Within a few minutes up to a few hours most of the diatom protoplast is consumed. Only life diatoms are fed upon. Intact frustules of *Coscinodiscus granii* and *Odontella sinensis* cannot be penetrated but phagocytosis on damaged cells is possible.

The apoplastidic nanoflagellate *Pirsonia guinardiae* is an epicytic parasite of the marine diatom *Guinardia flaccida*. It feeds in a unique mode (Schnepf et al. 1990). Further recently detected *Pirsonia* species are distinguished by their morphology, their development and the host range (Kühn et al. 1996; Schweikert and Schnepf 1997b). Electron microscopy demonstrates that *Pirsonia* belongs to the heterokonts (stramenopiles) (Schnepf and Schweikert 1996/97).

The mobile stages are small (generally about 5×10 µm), naked flagellates. They attach to a host cell with the posterior cell pole, guided by chemotaxis as well as by the topography of the frustule surface to host-parasite specific sites (Schnepf and Schweikert 1996/97; Kühn 1997). The attachment depends on an intact actomyosin system. An antapical pseudopodium penetrates into the frustule. The main body remains epicytic and becomes the "auxosome". The flagella wind around the cell apex and are retracted in most species. The auxosome does not encyst. Inside the diatom the pseudopodium develops into a "trophosome", consisting of a single, enlarging proximal digestion vacuole and distal pseudopodia which phagocytize portions of host cytoplasm and transport them into the digestion vacuole. The thin connection between trophosome and auxosome is not passed by particulate host material. The diatom reacts by shrinking of the vacuoles and/or systrophical movement of cytoplasm and chloroplasts toward the infection site. The feeding phase takes many hours.

Electron microscopy revealed that the host plasmalemma remains uninterrupted during the feeding process. Freshly phagocytized portions of host cytoplasm are still bounded by their plasmalemma when included in a food vacuole. The trophosome cytoplasm does not contain larger organelles such as mitochondria, only some fibrillar structures and a few vesicles (Schnepf and Schweikert 1996/97).

Under culture conditions an already attacked but still living cell is usually more attractive than an uninjured one. It seems to release more chemotactically effective, specific substances than healthy cells. The trophosomes of adjacent auxosomes may fuse to form a common digestion vacuole (Kühn et al. 1996). Adjacent sister cells are obviously recognized as "self".

The auxosome begins to divide already during phagocytosis and digstion. The arising daughter cells continue to divide until the direct or indirect contact with the trophosome is lost. They then become flagellate mother cells which divide again one or two times to give rise to new flagellates. Up to about 60 offspring may result from a single infection (Schweikert and Schnepf 1997b).

A *Guinardia flaccida* cell may survive the attack of a single *P. guinardiae* because the nucleus is often not ingested. In plankton samples only about one third of the infested cells were found to be killed (Schnepf et al. 1990).

The genus *Cryothecomonas* comprises nanoflagellates with two smooth, heterodynamic, apically inserting flagella of unequal length. The cell is surrounded by a delicate, close-fitting theca. The theca is multilayered and contains a fribrillar, presumably non-cellulosic polysaccharide. Some species have extrusomes (Schnepf and Gold 2000; Schnepf and Kühn 2000). Analyses of the small subunit rRNA indicate that *Cryothecomonas* belongs to the Cercozoa and is related with the sarcomonad flagellate *Heteromita globosa* (Kühn et al. 2000). *Cryothecomonas aestivalis* (Drebes et al. 1996) (Sect. 2.7) and *C. longipes* (Schnepf and Kühn 2000) are parasites of marine diatoms.

An attacking *C. longipes* protrudes a broad pseudopodium through a ventral slit of its theca. It penetrates the frustule between overlapping cingula, pinches off portions of the diatom protoplast and transports the food vacuoles into the epicytic trophont. The pseudopodium may branch and reach a length of more than 30 μm when the host protoplast shrinks and retracts. It contains a largely ribosome-free mass of fine fibrils, some vesicles and a few microtubules in its proximal part. When it emerges special flat vesicles are incorporated into the plasmalemma at the basis of the pseudopodium to enlarge it (Schnepf and Kühn 2000). The food uptake can take more than 3 h and is finished when the diatom protoplast begins to disintegrate. The trophonts divide after feeding.

2.7 Endocytic Phagotrophic Parasites

Cryothecomonas aestivalis belongs to a group of parasitic protoctists which completely invade the host cell gradually phagocytosing the host cytoplasm but remaining extrabiotic. The organism was detected in 1993 feeding on the marine diatom *Guinardia delicatula* (Drebes et al. 1996). When a flagellate has traced a *G. delicatula* chain, it slides along its surface and scans it with the tip of the anterior flagellum. Eventually it attaches with its basal pole, generally close to the overlap between valva and cingulum. Becoming slightly amoeboid it squeezes through the frustule, inserting the posterior first. Inside the diatom the flagella shorten. The invasion takes 5–10 min.

The trophic phase begins with the emergence of a pseudopodium from the basal part of the flagellate. The theca has a gap here. The pseudopodium gradually phagocytizes portions of the diatom protoplast, contracting rhythmically. At the contact zone with the diatom protoplast, the pseudopodium has a fibrillar-granular appearance. Many vesicles and invaginations of the plasmalemma indicate a high rate of membrane flow. Mitochondria are not found here. The trophont grows considerably. It fills a large part of the host cell when the protoplast has been completely ingested. The food uptake comes to an end when the *Guinardia* plasmalemma is ruptured. The feeding phase takes about 4 h and is followed by multiple binary divisions. Usually 8 (but up to 32) flagellates arise.

Examples for endocytotic phagotrophic flagellates that feed with a well-elaborated cytostome are the euglenoid *Rhynchopus coscinodiscivorus* and the bodonoid (kinetoplastid)

Hemistasia phaeocysticola. *R. coscinodiscivorus* feeds on the large, marine diatom *Coscinodiscus concinnus* (Schnepf 1994a). It has a subapical flagellar pocket with two concealed, rudimentary flagella and a subapical cytostome. Cell apex, cytostome, cytopharynx and flagellar pocket are reinforced by microtubules and a band of rods. The mitochondria have conspicuously poorly developed cristae.

The flagellate tears off pieces of the retracted host protoplast with its cytostome by jerking contractions of the cell apex. They are transported through the cytopharynx and added to one of the digestion vacuoles. Cell debris is eventually also ingested. *R. coscinodiscivorous* divides as long as food is available. A single host cell can provide food for up to 100 offspring. It takes several days until a *Coscinodiscus* cell is consumed.

Sometimes confused with *R. coscinodiscivorus* (both are also incorrectly named "*Pronoctiluca*") was *Hemistasia phaeocysticola*. The latter has a typical mitochondrion of the polykinetoplastic type (Elbrächter et al. 1996). The two flagella insert in a deep, subapical pocket. The apical rostrum, flagellar pocket, cytostome and cytopharynx are reinforced by various cytoskeletal elements. Food is ingested phagocytotically mainly from weakened or decaying organisms, preferentially diatoms, and accumulated in a single, large digestion vacuole. Feeding and division are not separated.

Amoeboaphelidium and *Aphelidium* are endocytic, extrabiotic parasites of chlorococcalean algae. Diverse species of *Amoeboaphelidium* have been studied with respect to their host range, their mode of food uptake, their development and their fine structure (Gromov and Mamkaeva 1968, 1970). A similar parasite is *Aphelidium* (Schnepf et al. 1971a; Schnepf 1972). *Amoeboaphelidium* propagates itself by amoebae, *Aphelidium* by zoospores with a single opisthokont flagellum. Their taxonomic position is unclear.

Aphelidium cf. *chlorococcarum* is parasitic on *Scenedesmus*. A zoospore attaches to a host cell, encysts, and penetrates the cell wall. The alga reacts with the formation of a tube-like channel of cell wall material around the invasion site. The naked *Aphelidium* invades into the algal protoplast the plasmalemma of which is not pierced but structurally modified at the host/parasite interface. The periparasitic space remains in connection with the infection channel.

The food vacuoles are initially bounded by two membranes closely sticking together. The outer one belongs to the parasite, the inner one to the alga. The latter eventually disintegrates. Closely sticking parasite and host membranes ("compound membranes") are found also elsewhere at the host/parasite interface, especially in the region around the infection channel (Schnepf 1972). The amoeba develops into a multinucleate plasmodium which eventually cleaves into 10–20 zoospores. A sporangial wall is not formed. Under certain conditions a thick-walled resting spore arises instead.

Another, somewhat similar organism is *Pseudaphelidium drebesii*, a rare parasite of the marine diatom *Thalassiosira punctigera* (Schweikert and Schnepf 1996, 1997a). Its development differs from that of *Aphelidium*.

Pseudaphelidium has a unique form of mitosis. The nuclear envelope remains closed and the spindle microtubules are extranuclear. In *Aphelidium* and *Amoeboaphelidium* the details of mitosis are unknown as yet.

The motile, infective stages of *P. drebesii* are small zoospores with a single, opisthokont flagellum. They attach to a host cell, retract the flagellum and encyst. The cyst contains a long, invaginated infection tube. It is internally bounded by a continuation of the plasmalemma and opens over a gap in the cyst wall, positioned at overlaps in the diatom frustule. The infection tube everts, penetrates through the frustule and creates an opening through which the *P. drebesii* protoplast enters the cell lumen. The *Thalassiosira* plasmalemma is pushed back. The parasite is endocytic but extrabiotic. Portions of host cytoplasm are included in food vacuoles which fuse to form a large, central digestion vacuole, surrounded by the then multinucleate parasite plasmodium. Host and parasite are separated by their plasma membranes. The phagocytotic areas of the plasmodium contain bundles of microfilaments.

At the end of the trophic phase the plasmodium is a hollow sphere and fills the frustule completely. It cleaves to form amoeboid cells which soon encyst. The cyst releases four, sometimes less zoospores.

Phagomyxa algarum was found by Karling (1944) as an endoparasite of filamentous brown algae. He placed it in the Plasmodiophorida though it feeds phagocytotically, in contrast to the other members of this taxon which take up their nutrients by resorption. A similar organism, *Phagomyxa bellerocheae* (Schnepf et al. 2000), is a frequent endoparasite of the marine diatom *Bellerochea malleus* (Schnepf 1994b). A related species is *P. odontellae*, feeding on the marine diatom *Odontella sinensis* (Schnepf et al. 2000).

Electron microscopical investigations of the zoospores and their flagellar apparatus confirmed that *Phagomyxa* is related with the other Plasmodiophorida (Schnepf 1994b). Analyses of the SSUrRNA revealed that *Phagomyxa* and Plasmodiophorida form a monophyletic clade. It clusters most closely with a rhizopod assemblage consisting of sarcomonads and chlorarachniophytes (Bulman et al. 2001). The *Phagomyxa* sequences appeared distant enough from those of the Plasmodiophorida to justify a separate order, the Phagomyxidae.

The first infection stages of *Phagomyxa* are not yet known because the two species could not be cultivated. *P. bellerocheae* forms a plasmodium, a hollow sphere surrounding a large digestion vacuole, within the host protoplast. The structure of the host/parasite interface is not yet known.

The plasmodia of *P. bellerocheae* cleave when the host material is consumed. A single sporangiosorus is formed, consisting of numerous zoosporangia. The zoosporangia are surrounded by a thin cell wall and release tiny zoospores (2.5×4 μm). *P. odontellae* plasmodia form peripheral, three-dimensional networks within the host cell, with numerous small food vacuoles and develop several sporangiosori (Schnepf et al. 2000).

2.8 General Aspects

In spite of the great diversity of predators and parasites some general conclusions are possible. Parasites are usually host specific, rather than predators, especially those predators which are much larger than their prey. *Noctiluca,* a big cell, is an example for omnivory; ciliates acquiring their food by filter feeding are another example. In those cases (Sherr et al. 1991) as also, e.g., in some naked dinoflagellates (Naustvoll 2000b) the size of the particles is the determinative factor. There are, however, also highly selective predators. The ciliate *Pseudomicrothorax dubius* can ingest its prey, filamentous cyanobacteria, only by immediate digestion of the cell walls by specific enzymes (Hausmann and Peck 1979). An omnivorous feeder has a better chance of acquiring food than a specialist but it has to be equipped with a broader array of digestive enzymes. Most predators therefore possess sensory prey-detecting mechanisms (Sibbald et al. 1988).

Pallium feeders have a restricted choice of prey organisms (Buskey et al. 1994), the food preference being influenced by the size of the prey (Naustvoll 2000a).

Small parasites need to be clever. They have to find access to their hosts. The host range may indicate taxonomic relationships (Cook 1963) and can be extremely narrow. There are chytrids which discriminate between morphologically similar strains ("demes") of what has traditionally been considered single diatom species. Even individual clones exhibit different degrees of susceptibility (Mann 1999). – An interesting curiosity is *Gyrodinium undulans.* It feeds on a single diatom species, *Odontella aurita,* but also on eggs of rotifers and copepods (Drebes and Schnepf 1998).

The infectivity of certain parasites can depend on light (Schnepf and Drebes 1986; Bruning 1991). The incidence of preying and of infections may be influenced by the degree of turbulence (Kühn and Hofmann 1999) which favours random collisions on the one side but destroys gradients of chemical signals around the prey or host on the other side.

Chemotaxis is the most important mechanism for long-distance attraction (Spero 1985). Chemosensory stimulation by algal exudates can change the swimming pattern of flagellates and zoospores (Schnepf and Drebes 1986; Kühn 1997). Already attacked cells are especially attractive, indicating that they release more attractants than uninjured ones (Kühn 1998).

In contact recognition, nanoflagellates frequently use the tip of the anterior flagellum (Drebes et al. 1996; Kühn et al. 1996). Cannibalism is generally avoided but possible. The chemical as well as the morphological topography of the host cell surface guides the parasite to find a site for invading (Schnepf and Schweikert 1996/1997; Kühn 1997). Surface charges also seem to play a role in food recognition (Hammer et al. 1999).

Gap-free host cell walls can be digested (Ucko et al. 1999); sometimes even locally (Hausmann and Radek 1993). Diatom frustules are usually invaded through overlaps of the shell or through openings such as rimoportulae. If a parasite fails to attach and to encyst at a proper site it will perish (Schweikert and Schnepf 1997a).

An invading obligatory biotrophic parasite does not penetrate through the host plasmalemma but pushes it back. The feeding phase comes to an end when the host plasmalemma disintegrates. The interface between a host and an extrabiotic parasite thus consists always of the two plasma membranes. Haustoria are additionally surrounded by the fungal cell wall which is not present in intracellular, extrabiotic oomycetes like *Lagenisma* until the thallus becomes a zoosporangium.

Parasites which feed myzocytotically pierce the host plasmalemma the rim of which becomes closely attached to the opening of the feeding tube or phagopod. When the host cell is sucked out, the surface of the protoplast decreases rapidly. It is still unexplained how the plasmalemma is reduced, for it is not included in the food vacuole, where host and parasite cytoplasm are separated only by the membrane of the food vacuole.

Curiously the pallium, the dinoflagellate peduncle, the *Pirsonia* trophosome, and the feeding parts of *Cryothecomonas* pseudopodia do not contain mitochondria though the uptake processes surely require ATP and are, at least in part, driven by the actomyosin system.

The host/parasite interface of some endobiotic parasites consists of a single membrane, the parasite plasmalemma. The host membrane is lost in these cases.

Eumycotes and oomycetes digest their food extracellularly. It is unknown how the digestive enzymes of biotrophic parasites pass the two membranes at the host/parasite interface and how the nutrients are resorbed. Digestion vacuoles outside the cell body proper are formed by *Pirsonia* and by pallium feeding dinoflagellates.

Only a few defense mechanisms are known as yet (Verity and Smetacek 1996; Wolfe 2000). Ingestion by filter feeders like ciliates is impeded by spikes (*Chrysochromulina*), the formation of coenobia (*Scenedesmus*) or colonies (*Phaeocystis*) (Sournia 1982; P.J. Hansen et al. 1995; Heckmann 1995). These defense mechanisms are ineffective against infections by parasites. The frustule of a diatom protects against the attacks of invasive parasites (Kühn 1996/97) as does the skin of the *Phaeocystis* colony (Smetacek 1999). In the diatom *Thalassiosira rotula* a wound-activated chemical defense has been detected (Pohnert 2002; further examples see:Verity and Smetacek 1996). Effective against the propagation of biotrophic parasites are hypersensitive reactions (Canter and Jaworsky 1979).

2.9 Ecological Impact and Epidemiology

Protoctists feeding on plankton algae play an important role in the biotic control of primary production and of carbon flux through marine and freshwater food webs, a role that was underestimated for a long time. It was commonly believed that "big ingest small" (e.g., copepods/euphasids feeding on diatoms) (Smetacek 1999). The detection of many algivorous protoctists has changed this inaccurate perception, the more because it is now recognized that many photosynthetic protoctists also feed on plankton algae (Stoecker 1998). Bockstahler and Coats (1993a) found up to 30% of the pigmented dinoflagellates with food vacuoles containing ciliates, diatoms and dinoflagellates. The mixotrophic *Gymnodinium sanguineum* was estimated to remove ciliate biomass by up to 67% of the standing crop, daily (Bockstahler and Coats 1993b).

A few selected data may illustrate the ecological impact of algivorous protoctists. Heterotrophic dinoflagellates feeding by peduncle or pallium were most abundant in the Kattegat during periods characterized by large phytoplankton blooms (diatoms and dinoflagellates). During these periods the heterotrophic biomass corresponded to between 13–37% of the phytoplankton biomass (P.J. Hansen 1991). In the Oslofjord the *Protoperidinium* biomass may reach nearly 40% of the algal carbon biomass in the form of diatoms and phototrophic dinoflagellates (Kjaeret et al. 2000). Heterotrophic protoctists grazed off 20–100% of the daily primary production in antarctic coastal waters (Archer et al. 1996).

Paulsenella on *Streptotheca* reached infection rates of up to 20% (Drebes and Schnepf 1982), *Gyrodinium undulans* on *Odontella aurita* of up to 85% (Drebes and Schnepf 1998). *Guinardia delicatula* and *G. flaccida* were infected of up to 35% and 65%, respectively, by *Pirsonia* and *Cryothecomonas* (Tillmann et al. 1999). During a bloom of *Palmeria hardmania* in Jamaica, up to 90% of the diatom cells became parasitized by *Lagenisma* (Grahame 1976).

In the Schöhsee, a mesotrophic lake near Plön (Germany), epidemics of chytrids on plankton algae occurred throughout the year. Parasites appeared at population densities as low as 1 cell ml^{-1} in some species, with infection rates sometimes exceeding 80%. The proportion of the total phytoplankton biovolume infected by fungi was usually less than 1% but occasionally reached 10%. Parasitism was highly species specific with one parasite species usually infecting one host species (Holfeld 1998). Because of the specificity, infection of one algal species favours successional development of other, competing species (Tillmann et al. 1999).

The dynamics of parasite populations depends on the population of possible hosts. *Lagenisma coscinodisci* was common in the German Bight up to the 1980s but is rare now (H. Halliger, List, pers. comm.). The decrease coincides with the spreading of a newly introduced *Coscinodiscus* species, *C. wailesii.* The latter is by far less accepted as host as the other species.

During a host algal bloom its parasite finds optimal conditions. Following the outbreak of a red tide of the photosynthetic dinoflagellate *Gymnodinium mikimotoi* (in European waters misidentified as *Gyrodinium aureolum= Gymnodinium aureolum* is now also called *Keramia mikimotoi*, Schnepf and Elbrächter 1999; G. Hansen et al. 2000), abundance and growth rates of heterotrophic dinoflagellates increased rapidly. Grazing by dinoflagellates effectively contributed to the disappearance of the red tide. The ciliate population did not respond (Nakamura et al. 1995). Due to their longer generation times multicellular zooplankters can "miss" blooms and are less successful in reducing them than protoctists. – The ecological impact of algal parasites is influenced also by predators of the parasites and by hyperparasites (example of hyperparasitism, see Kühn and Schnepf 2002).

3 Kleptochloroplasts

Kleptochloroplasts (Schnepf et al. 1989) result from an endocytic uptake of alien cytoplasm containing chloroplasts which are ultimately retained and used as photosynthetic organelles for a restricted period of time. The other components of the ingested cell inclusive of the nucleus are usually digested much prior to the digestion of the chloroplasts. Kleptoplastidy takes an intermediate position between antagonistic and mutualistic, symbiotic associations. The enslaved chloroplasts deliver photosynthetic products into the feeder which in turn support the chloroplasts to allow their long survival. The acquisition of kleptochloroplasts converts a heterotroph into a mixotroph. This occurrence is widespread among protoctists (Stoecker 1998) and has been known for a long time from marine molluscs (opisthobranchs) (recently reviewed by Rumpho et al. 2000).

Dinoflagellates take the kleptochloroplasts generally from cryptophytes. Investigations of *Gymnodinium aeruginosum* (Schnepf et al. 1989) and the perhaps conspecific *G. acidotum* (Farmer and Roberts 1990, Fields and Rhodes 1991) as well as of *Amphidinium poecilochroum* (Larsen 1988) revealed the following course of events:

The dinoflagellates in question are initially colourless and contain a single dinokaryotic nucleus. They can survive in this state for only a restricted period of time. Various cryptophytes are ingested myzocytotically. Horiguchi and Pienaar (1992) found three types of cryptophytes in *Amphidinium latum*. The naked cryptophycean cytoplasm without plasmalemma, periplast and flagellar apparatus is included in a vacuole. The algiphorous vacuole is irregular in shape, the chloroplasts are highly lobed. After some time, presumably in the range of days, the cryptophycean nucleus and the nucleomorph (see Sect. 5.2.2) disappear. Presumably they are selectively digested. In wild populations of *G. acidotum* only 30% contained a cryptophycean nucleus (Farmer and Roberts 1990), 10% in *G. aeruginosum* (Schnepf et al. 1989).

G. acidotum becomes colourless in pure culture 10–14 days after feeding on cryptophytes and dies. In co-cultivation with the cryptophyte *Chroomonas* it has been maintained over 9 months. Starch grains in the dinophycean cytoplasm indicate that the kleptochloroplasts are functional and that carbohydrates are exported into the dinoflagellate. The photosynthetic activity of the kleptochloroplasts of *Gymnodinium gracilentum* is lost within a few days (Skovgaard 1998).

In *Amphidinium vigrense* cryptophycean-like chloroplasts were found to be bounded by three membranes (Wilcox and Wedemayer 1985); it was suggested that the chloroplasts had become permanent residents. *A. vigrense* was, however, described as colourless, feeding on cryptophytes (Woloszynska 1925). If chloroplasts are present, their number is highly variable. Kleptoplastidy is probable also here. Variable chloroplast numbers are known also from several other dinoflagellates. Kleptoplastidy seems to be much more common in dinoflagellates than initially thought.

Ciliates receive their kleptochloroplasts from a variety of algae. Even a single cell may contain a mixture of plastid types (Stoecker 1993, 1998). In the Strombidiidae (Oligotricha) plastid retention is common. In marine waters about 40% of the ciliate cells in the euphotic zone are "plastidic" strombidiids (Stoecker et al. 1987). Mixotrophic ciliates comprised 88% of the total ciliate biovolume during summer and fall (Holen 2000).

Plastid retention is a stable feature in some oligotrich ciliates. *Laboea strobila* is obligately mixotrophic and does not grow in the dark even when suitable food is present (Stoecker et al. 1988). Photosynthesis has been demonstrated (Stoeker 1993). The half life of *Strombidium* kleptochloroplasts is between 12 h and some days.

Kleptochloroplasts in foraminiferans are received only from chromophytes (Lopez 1979). In centrohelid heliozoans (Patterson and Dürrschmidt 1987) they are derived from different algae. Chlorophytic and chromophytic chloroplasts can be found in one and the same cell, together with digestion vacuoles, either lying free in the heliozoan cytoplasm, surrounded by only the two membranes of the chloroplast envelope, even in the case of chromophytic origin, or additionally by a membrane provided by the heliozoon.

The ciliate *Mesodinium rubrum* (= *Myrionecta rubra*) has cryptophyte-derived chloroplasts representing a special kind of kleptoplastidy. The details of this association are not yet well understood. *M. rubrum* has neither a cytostome nor a cytoproct. It contains numerous vacuoles with usually a chloroplast, mitochondria, not always a microbody, but neither dictyosomes nor elaborated ER. They are bounded by a single membrane. The chloroplast-mitochondrial complexes are in general not connected with a portion of endosymbiont cytoplasm containing a cryptophyte nucleus. Digestion vacuoles have not been found (Hibberd 1977, Lindholm et al. 1988). It is believed that *M. rubrum* is permanently associated with an endosymbiotic cryptophyte.

Only recently has it become possible to cultivate *M. rubrum* (Gustavson et al. 2000). Surprisingly it turned out that, when cultivated unfed for 28 days, the ciliate became colourless and ceased to grow. When supplied with free-living cryptophytes, the biomass of the culture and pigment contents increased. The number of ciliates with cryptophyte nuclei increased considerably within minutes after feeding. Chlorophyll *a* was synthesized then also within the ciliates. Details of food uptake could not be observed. It remained unclear whether non-functional chloroplasts and the nuclei are egested or digested.

Mesodinium rubrum may occur in dense blooms. Its non-toxic red tides have extremely high rates of primary production, with some of the highest values for chlorophyll *a* and primary production in the marine environment (Gustavson et al. 2000) – a convincing example for the efficiency of symbiotic cooperation – here at the expense of the cryptophytes.

Is kleptoplastidy a model for a first step in the evolution of secondary chloroplasts? In protoctists the plastids remain active for only some days. In marine slugs (reviewed by Rumpho et al. 2000) chloroplasts of mainly siphonaceous green algae, red algae or *Vaucheria* are still functional after months (*Vaucheria* chloroplasts in *Elysia chlorotica*: nine months) in the absence of any algal nucleo-cytoplasmatic influence. The *Vaucheria* chloroplasts lie "naked" in the *Elysia* cytosol, surrounded only by two membranes, i. e., without the two membranes of the chloroplast ER and a vacuole membrane. Plastid protein synthesis has nevertheless been demonstrated. The question whether nuclear encoded proteins are synthesized in the slug cytosol and targeted to the chloroplasts has not been conclusively answered yet. It is also discussed whether the long-term activity of the kleptochloroplasts depends on an unusually high level of chloroplast gene autonomy, on extremely stable proteins, whether the minimal protein composition needed to support chloroplast activity is less than expected, whether mitochondrial or animal encoded proteins with related functions are redirected or whether lateral gene transfer has taken place here (Rumpho et al. 2000). It should be added that the isolation of the kleptochloroplasts may abolish the "normal" turn-over of chloroplast proteins.

4 Endosymbiotic Associations with Prokaryotic Algae

An association between cyanobacteria and the apoplastidic euglenoid flagellate, *Petalomonas sphagnophila,* has recently been detected (Schnepf et al. 2002). It resembles kleptoplastidy in some respect. *P. sphagnophila* always contains *Synechococcus*-like cyanobacteria. Their number varies between 6 and 20 per cell. They lie individually in perisymbiont vacuoles, remain alive for at least several weeks, and are occasionally seen to divide, indicating a mutualistic relationship. Cyanobacteria in digestion indicate that they directly serve as food, too. The association is specific and seems to be a transient, relatively undeveloped form of endosymbiosis.

P. sphagnophila lives under hypoxic conditions in floating *Sphagnum* mats of bog lakes. The production of oxygen by the symbiont may be as important for the host as the supply with carbohydrates or nitrogen compounds.

A somewhat similar association has been found by Hargraves (2002). *Hermesium adriaticum,* an apoplastidic flagellate, has become mixotrophic, harbouring numerous endobiotic *Synechococcus*-like cyanobacteria. They are not merely ingested, for dividing cells have been observed.

The heterotrophic dinophysoid dinoflagellates *Ornithocercus, Histioneis* and *Citharistes* often are associated with *Synechococcus* or *Synechocystis*-like cyanobacteria, previously described as "phaeosomes" (Lucas 1991; Schnepf and Elbrächter 1992). In *Citharistes* and *Histioneis* they live extrathecally in special chambers, in *Ornithocercus* intracellularly in the region of the adcingular collar. In the Gulf of Aqaba they occur mainly in the fall, at a time of extended nitrogen limitation (Gordon et al. 1994). It is believed that the cyanobacteria supply the dinoflagellates with nitrogen compounds and the hosts provide the symbionts with the anaerobic microenvironment necessary for efficient nitrogen fixation.

Associations between endosymbiotic cyanobacteria and hosts which have their own chloroplasts likewise use (primarily?) the potential for nitrogen fixation of the prokaryotic partner (Floener and Bothe 1980; Janson et al. 1995). The filamentous cyanobacterium *Richelia intracellularis* is a temporary, intracellular, extrabiotic symbiont in the marine diatoms *Hemiaulus* spp. and *Rhizosolenia* spp. (Villareal 1994; Jansson et al. 1995). The filaments have terminal heterocysts. They may fix a considerable amount of nitrogen (Werner 1992).

The coccoid cyanobacteria in *Rhopalodia* are endobiotic. A diatom contains between two and five endosymbionts. They have a normal cell wall, a somewhat peculiar thylakoid system, lie in perialgal vacuoles and fix nitrogen (Floener and Bothe 1980).

Geosiphon pyriforme is a consortium between prokaryotic alga and fungus. It initiated my own research in this field. A recent review of *Geosiphon* is given by Schüßler and Kluge (2001). The host is a coenocytic fungus related to the arbuscular mycorrhizal fungi (Schüßler et al. 1994) and is now included in an new phylum, the Glomeromycota (Schüssler et al. 2001) - the algal partner is *Nostoc punctiforme*. Resting spores of the fungus do not contain cyanobacteria. When a resting spore germinates and the apical region of a hypha meets a *Nostoc* filament in a special developmental stage, the "early primordium", it bulges to surround the alga and encloses it excepting heterocysts (Mollenhauer et al. 1966). After a shock period the then endobiotic *Nostoc* grows and forms filaments with heterocysts. They are enclosed in a single, peripheral compartment, the symbiosome. The symbiosome membrane is a derivative of the host plasmalemma. Its inner surface is covered by a rudimentary fungal wall (Schüßler et al. 1996). The *Nostoc* cells and the heterocysts have normal cell walls and a normal internal structure. The endosymbiotic cyanobacteria photosynthesize (Kluge et al. 1991) and show nitrogenase activity (Kluge et al. 1992).

Geosiphon houses also endobiotic bacteria. They lie free in the cytoplasm, not in a perisymbiont vacuole (Schüßler et al. 1994). Associations between bacteria and unicellular algae are common and reported from nearly all major groups (Chesnick and Cox 1986). Cases of parasitism are known (Schnepf et al. 1974). The endobiotic bacteria live in peribacterial vacuoles or free in the cytoplasm. The peribacterial membrane of the host is lost then. In a *Cryptomonas* sp. cell only bacteria within peribacterial vacuoles were seen to divide. Free bacteria in the same cell contained partly bacteriophages which were subsequently digested in autolysosomes (Schnepf and Melkonian 1990).

5 Endosymbiotic Associations with Eukaryotic Algae

5.1 Potentially Autonomous Endosymbionts

The genetically autonomous partners discussed in this chapter can live independently, at least for some time. In nature they are metabolically interdependent. The algae (phycobionts) belong mainly to the chlorococcales (zoochlorellae) or to the dinoflagellates (zooxanthellae). The hosts are diverse protoctists or invertebrates. The zooxanthellae of marine hydrozoa and the zoochlorellae in freshwater hydrozoa, sponges and turbellaria are not within the scope of this review (see Werner 1992).

The phycobionts are generally enclosed in a special microhabitat, the perialgal vacuole or symbiosome. One of the exceptions is *Noctiluca*. In Southeast Asian waters a prasinophyte flagellate, *Pedinomonas noctilucae*, lives in the large central vacuole (Sweeney 1976).

Diverse amoebae, thecamoebae, heliozoa and, especially, ciliates (holotrichs, spirotrichs and peritrichs) have zoochlorellae (Reisser 1994; Görtz 1996). The algae remain permanently in their hosts and are distributed to the daughter cells when the host cell divides. Both partners of the association between *Paramecium bursaria* and *Chlorella* can easily be cultivated separately but are in nature found only in symbiosis. A *P. bursaria* cell contains several hundred zoochlorelles. They are individually enclosed in a perialgal vacuole, the membrane of which surrounds tight-fitting the alga, even during and after cell division. Its structure and function differs from the membranes of food vacuoles (Meier et al. 1984).

Symbiotic *Chlorella* strains secrete various sugars, some of them mainly maltose, much more than non-symbiotic strains. Isolated *Chlorella* of *P. bursaria* releases up to 90% of the photosynthetically fixed carbon (Reisser 1994). The secretion is promoted by an acidic environment. During photosynthesis the perialgal vacuole becomes acidified, suggesting that the sugar transport to the host is driven by proton antiport (Schüßler and Schnepf 1992). The host receives also O_2 from the phycobiont, in exchange for CO_2 and inorganic ions. *P. bursaria* reacts phototactically (Niess et al. 1982) but not when aposymbiotic. It is a good shepherd, not only because it provides an optimal environment for photosynthesis but also because the algae in the symbiosomes are protected against the attacks of *Chlorella* viruses. Free in nature they are highly susceptible to viral infections (Reisser 1993).

P. bursaria depends not exclusively on its zoochlorellae. It feeds phagotrophically on bacteria and algae. In prolonged darkness the zoochlorella are also digested, perhaps because sugar secretion has ceased then. If they are then re-fed with suitable *Chlorella*, some of the algae are not passed into digestion vacuoles but are enclosed in perisymbiont vacuoles. Reisser (1993) suggests that the ciliates recognize suitable algae by carbohydrates on the surface of their cell wall which interact zipper-like with lektin-like proteins of the phagocytotic membrane to form the perialgal membrane that then does not fuse with the membranes of primary lysosomes.

Zoochlorellae occur also in a few marine protoctists, such as foraminiferans, which may house also other algae (Gastrich 1987). Symbiotic chlamydomonal flagellates loose transiently their flagella, and diatoms do no longer form a frustule (Lee 1983). In culture, after their isolation they produce silica shells again, demonstrating that they belong to the pennate genera *Fragilaria* and *Nitzschia* (Lee et al. 1979).

5.2 Reduced Endosymbionts

5.2.1 Endosymbionts with Nucleus

Several dinoflagellates contain eukaryotic endosymbionts with chloroplasts (reviewed in Schnepf 1993). The chloroplasts are not lost in permanent unialgal culture so that kleptoplastidy can be excluded. Mandelli (1968) found that *Glenodinium* (synon. *Peridinium, Kryptoperidinium*) *foliaceum* has fucoxanthin as the main carotene instead of the characteristic dinoflagellate pigment, peridinin. Dodge (1971) detected that *G. foliaceum* has not only a dinokaryon but also a typical eukaryotic nucleus, belonging to a reduced cell with chloroplasts (but see Sect. 5.2.3!). *Peridinium balticum* (Tomas and Cox 1973), *Peridinium quinquecorne* (Horiguchi and Pienaar 1991) and *Gymnodinium quadrilobum* (Horiguchi and Pienaar 1994) are similar chimeras.

The dinoflagellates are quite normal in shape and in their internal structure. They have genuine dinoflagellate plastids with a characteristic envelope of three membranes (Sect. 6.3.2). These are small and are without stacks of thylakoids and chlorophyll and are, therefore, not chloroplasts, functioning in photosynthesis. They contain many carotene droplets and function as stigmata. In "normal" phototrophic dinoflagellates the stigma is likewise a part of the chloroplast.

The photosynthetically active chloroplasts have thylakoids in stacks of three and a girdle lamella. They are surrounded by two pairs of membranes, the chloroplast envelope and a peripheral envelope ("chloroplast ER") (see Sect. 6.2). Their structure as well as their pigment composition suggests that the endosymbionts are derived from chromophytes (Withers et al. 1977; Horiguchi and Pienaar 1994). DNA sequence analyses revealed that the ancestor of the endosymbionts of *P. balticum* and *G. foliaceum* is a *Navicula*-like pennate diatom (Chesnick et al. 1996, 1997). Dinoflagellate and endosymbiont cytoplasms are separated by a single membrane, the host membrane because it is likely that the symbiont has been taken up by myzocytosis (Sect. 2.4).

The two nuclei of *P. balticum* divide synchronously, the endosymbiont nucleus amitotically (Tippit and Pickett-Heaps 1976). Its (consequent?) variance in appearance resembles a ciliate macronucleus, suggesting that it has become genetically incomplete. During sexual reproduction the karyogamy of the dinoflagellate is followed by that of the endosymbiont (Chesnick and Cox 1989).

Peridinium and *Gymnodinium* belong to different orders. It is probable that the associations arose, at least in part, from separate endosymbiotic events, while there are also alternative explanations (Morris et al. 1993; Horiguchi and Pienaar 1994).

5.2.2 Endosymbionts with Nucleomorph

Cryptomonads and chlorarachniophytes are chimeras, with reduced nuclei, "nucleomorphs", in the photosynthetic endosymbiont. The cryptomonad chloroplasts contain phycobiliproteins. They are not assembled in phycobilisomes at the outside of the thylakoids but are localized inside the thylakoid lumen, in contrast to most other algae with phycobilins (an exception is *Dinophysis*, Sect. 6.4.2). The chloroplasts are surrounded by two pairs of membranes. The inner pair is the chloroplast envelope. The outer pair is the "chloroplast ER". Its outer membrane is studded with ribosomes and is continuous with the nuclear envelope (Gibbs 1981).

A "periplastidal cytoplasm" between the two membrane pairs contains ribosomes of the eukaryotic type, starch grains, a few vesicles and the nucleomorph. This special structure is surrounded by a nuclear envelope-like membrane pair with a few pores. It is undoubtedly the vestigial nucleus of a reduced eukaryotic endosymbiont (Gillott and Gibbs 1980). It divides amitotically just prior to the host cell nucleus (McKerracher and Gibbs 1982). The nucleomorph contains DNA (Hansmann et al. 1985; Ludwig and Gibbs 1985) as well as RNA, a part of which is concentrated in a fibrillo-granular region, a structural and functional equivalent of a nucleolus (Hansmann 1988). The nucleomorph DNA is organized in three linear, tiny chromosomes. They encode a mere 511 genes, 30 of which are for proteins required by the plastids (Douglas et al. 2001).

The nucleomorph and its genes share a common ancestry with the nucleus of red algae (Leitsch et al. 1999; Douglas et al. 2001). The starch grains in the periplast cytoplasm occur in the same compartment as floridean starch. It is generally accepted now that the endosymbiont is a reduced red algal cell. The inner membrane of the periplastidal envelope is derived from the plasmalemma of the red alga, the outer one represents the membrane of the perialgal vacuole which is connected now with the ER of the host cell and the nuclear envelope.

The chlorarachniophytes are a small phylum comprising amoeboid, coccoid and flagellate species (Moestrup and Sengco 2001). *Chlorarachnion reptans* forms large plasmodia and is mixotrophic, phagocytizing small algae and flagellates. Its chimeric nature has been revealed by Hibberd and Norris (1984). Further details of the chlorarachniophyte fine structure have been described by Moestrup and Sengco (2001).

The chloroplasts contain chlorophyll *a* and *b* (Wilhelm et al. 1991). They have a stalked pyrenoid but do not contain starch. As in cryptomonads, their envelope consists of two membranes and is surrounded by a thin layer of periplastidal cytoplasm which is delimited against the host cytoplasm by the two membranes of the periplastidal envelope. In contrast to cryptomonads the outer membrane of this pair is not studded with ribosomes and not connected with the host ER, thus demonstrating more clearly the character of a perisymbiont membrane. The pyrenoid is capped by a vesicle which is located in the host cytoplasm but in close contact with the periplast envelope. It encloses a storage carbohydrate, presumably a β-1,3-linked glucan.

The periplastidal cytoplasm contains ribosomes of the eukaryote type and a nucleomorph resembling a cryptomonad nucleomorph. It divides by infolding of the envelope membranes, microtubules are not involved (Ludwig and Gibbs 1989). The DNA is arranged in three tiny, linear chromosomes, as in cryptomonad nucleomorphs, but organization of the genome differs considerably (Douglas et al. 2001, McFadden 2001). It contains about 300 genes packed into a DNA of 30 kb (Gilson and McFadden 1997). During the evolution of the chlorarachniophytes genes have been transferred from the endosymbiont nucleus into the host nucleus (Deane et al. 2000).

The pigments as well as the molecular data (Van De Peer et al. 1996) indicate that the endosymbiont is a vestigial green alga. The host is related to the filose amoebae with *Euglypha* and *Paulinella* (Cavalier-Smith and Chao 1996/97). Curiously one species of *Paulinella* has cyanelles (Sect. 6.1.1).

5.2.3 Enigmatic Chimeras

The dinoflagellates *Lepidodinium viride* (= "strain Y 100") (Watanabe et al. 1987, 1990) and *Gymnodinium chlorophorum* (Schnepf 1993, Elbrächter and Schnepf 1996) can be maintained in unialgal cultures. They are similar in essential details. The dinoflagellates are apoplastidic and harbour a green, vestigial, eukaryotic endosymbiont. The partners are separated by a pair of ribosome-free membranes, the plasmalemma of the endosymbiont and the symbiosome membrane. Obviously the endosymbiont has been taken up by phagocytosis, not by myzocytosis. The chloroplasts have the chlorophylls *a* and *b*, the carotene prasinoxanthin but lacks chlorophyll *c* and the typical dinophycean carotene peridinin. They are enveloped by two membranes. The thylakoids are arranged in stacks of three. A pyrenoid is present. Starch grains are found only in the dinoflagellate cytoplasm.

Chloroplast ultrastructure and pigmentation suggest that the endosymbiont is derived from a prasinophyte, the cytoplasm of which is highly reduced. Mitochondria, dictyosomes, an elaborated ER and cytoskeletal structures are absent, only abundant ribosomes and a few vesicles found in the narrow periplastidal cytoplasm. Some of these "vesicles" are enveloped by a pair of membranes which is interrupted by pores and has the appearance of a nuclear envelope. These "vesicles" thus resemble nucleomorphs. In contrast to a typical nucleomorph their interior is loose and homogenous. It has to be confirmed whether they are indeed relicts of the prasinophyte nucleus and contain DNA, which could not yet be detected. A DNA is required for the synthesis of the ribosomal constituents in the vestigial prasinophyte cytoplasm.

It was a great surprise when recently it turned out that there are strains of *Glenodinium foliaceum* in which an endosymbiont nucleus could not be detected (Kempton et al. 2002). Morphological, biochemical and DNA analyses clearly showed that the mononucleate *G. foliaceum* strain from South Carolina is identical with the previously investigated binucleate strains and has chloroplasts with the same pigment composition. Further mononucleate strains (with a dinokaryon only) have been found in Florida and in the Baltic Sea off Hiddensee (Elbrächter, pers. comm.). To make the curious situation even more complex: There are strains with dinokaryon and endosymbiont nucleus but lack a stigma. Obviously the reduced dinophycean plastid (see Sect. 5.2.1) is lost here. Some of the strains are not yet in culture so that kleptoplastidy could be possible, but certain strains have been in culture for a long time, in which kleptoplastidy can be excluded (Elbrächter, pers. comm.). These unexpected organisms wait to be investigated further. They demonstrate the incredible plasticity of dinoflagellate-endosymbiont associations.

6 Chloroplasts

The idea of a symbiotic origin of plastids is generally accepted now. Chloroplasts evolved from endocytobiotic symbionts, either directly from a prokaryotic alga (primary plastids) or from a eukaryotic alga (secondary endosymbiosis, complex plastids).

6.1 Primary Chloroplasts: Two-Membrane Envelopes

The glaucocystophytes and rhodophytes have chloroplasts with phycobilins located in phycobilisomes at the outside of the unstacked thylakoids.

They store carbohydrates in the form of starch grains free in the cytosol. The glaucocystophytes have a vestigial cell wall between the two membranes of the cyanelle (chloroplast) envelope which is lacking in rhodophyte chloroplasts. The chloroplasts of chlorophytes/embryophytes contain chlorophylls *a* and *b*, stacked thylakoids and deposit starch grains in their matrix.

In the three taxa with primary plastids apoplastidic species are not known though apochlorotic ones can be found in rhodophytes and chlorophytes/embryophytes. The plastids, and in apochlorotic organisms the leucoplasts, are the sole site for the synthesis of fatty acids, certain amino acids and haem (Howe and Smith 1991).

6.1.1 Cyanelles

The glaucocystophytes are a small group of algae that includes flagellated, palmelloid and coccoid cells (Bhattacharya and Medlin 1995). The cyanelles as well as the presence of peripheral lacunae and a cruciate flagellar root system with multilayered structures characterize the Glaucocystophytes (Kies and Kremer 1990). The different genera had an uncertain, often changing position before (Schnepf et al. 1966; Schnepf and Brown 1971).

Cyanelles have frequently been considered as models for the symbiogenesis of primary chloroplasts. Indeed they have both features of free-living cyanaobacteria and cell organelles (McFadden 2001). They retain their shape when isolated, due to the presence of a vestigial but rigid cell wall, the existence of which has been demonstrated electron microscopically in *Glaucocystis* (Schnepf et al. 1966). It is dissolved by lysozyme (Schenk 1970: *Cyanophora*, Scott et al. 1984: *Glaucocystis*), like a typical peptidoglycan layer. Its synthesis is blocked by penicillin (Kies 1988). The cyanelles of *Cyanophora* have hydrogenases with similar properties as in free-living cyanobacteria (Bothe and Floener 1978) but they lack respiration (Floener and Bothe 1982) and the enzymes of nitrate assimilation are distributed as in cells with chloroplasts (Floener et al. 1982).

The complete sequence of the *Cyanophora paradoxa* cyanelle genome is now available (Löffelhardt et al. 1997). With about 135 kb it is one order of magnitude smaller than that of free-living cyanobacteria and is comparable to that of plastids. It comprises 191 genes, including the standard set of chloroplast-encoded genes. A certain number of cyanelle genes are absent in higher plants. Among them is the gene for the small subunit of ribulose bisphosphate carboxylase (rubisco), but also in rhodophytes it is in the plastid DNA. There are further similarities between cyanelle and rhodophyte chloroplast genomes. Curiously, the cyanelle genome codes only for one of the proteins needed to synthesize the peptidoglycan wall layer.

The most conspicuous feature of the gross organization in the circular cyanelle genome is the inverted repeat, typical for chloroplast DNA. Cyanelles are, therefore, true chloroplasts with a vestigial prokaryotic cell wall as distinct character. Schenk (1994) consequently uses the term "cyanoplast" instead of "cyanelle".

The wall remnant lies between two membranes. The inner one is undoubtedly derived from the plasmalemma of the endosymbiotic cyanobacterium. The nature of the outer one is controversial (Whatley 1993). It was initially thought that it represents the membrane of the symbiosome vesicle (Schnepf 1966; Schnepf et al. 1966) as in *Geosiphon* (see Sect. 4). Cavalier-Smith (1993) argues on the contrary that it is homologous with the outer membrane of the symbiotic cyanobacterium.

Other controversial issues are the age of the glaucocystophytes and whether they are descendants of a link in the monophyletic evolution of primary chloroplasts (Kowallik 1997) or are a relatively young group and an analogous model for (polyphyletic) plastid evolution (Schenk 1994). Because of the small size of the cyanelle genome and because the partners are highly adapted to each other, it is believed that the glaucocystophytes are a relatively old taxon (Kies and Kremer 1990). Most molecular biological data suggest a monophyletic origin of the three algal lineages with primary chloroplasts (Bhattacharya and Schmidt 1997).

Quite another organism with cyanelles is the thecamoeba *Paulinella chromatophora* (Kies 1974). It contains two sausage-shaped cyanelles and is related to the chlorarachniophytes (see Sect. 5.2.2). The other species of the genus *Paulinella* are cyanelle-free. The cyanelles have a thin (6–13 nm) vestigial cell wall and are lying in perisymbiont vesicles. It is discussed whether *P. chromatophora* and the glaucocystophytes represent analogous evolutionary branches, from independent primary endosymbiotic events, or whether the cyanelles of *P. chromatophora* are gained from a secondary symbiosis of a glaucocystophyte (Bhattacharya and Medlin 1995; Bhattacharya et al. 1995).

6.1.2 Rhodophyte and Chlorophyte Plastids

The two membranes of the chloroplast envelope of rhodophytes and chlorophytes/embryophytes have likewise be taken as evidence for the symbiogenetic origin of the organelles directly from cyanobacteria, with the inner membrane homologous to the plasmalemma of the symbiont and the outer one homologous to the membrane of the perisymbiont vesicle, a derivative of the phagosomal membrane (Schnepf 1964, 1966; Whatley 1993). The chemical composition of the outer membrane differs, however, from that of a phagosomal membrane (Joyard et al. 1991) and resembles somehow that of the outer membrane of cyanobacteria and Gram-negative bacteria (Cavalier-Smith 1993, 2000). Both prokaryotes are enveloped by two "membranes" (plus cell wall proper). Their major glycolipids resemble that of the two membranes of the plastidal envelope. An exception is phosphatidyl choline, present only in the outer leaflet of the outer membrane of the plastidal envelope and common in eukaryotic membranes but absent from prokaryotic membranes and the inner plastid envelope membranes (Cavalier-Smith 1993; Whatley 1993). It has also to be remembered that a perisymbiont membrane is frequently lost around bacteria (Sect. 4). The transfer of nuclear gene-encoded proteins into chloroplasts would perhaps be simpler if the accepting (outer) organelle membrane is of eukaryotic origin and faced as other membranes in the endomembrane system.

The problem whether primary chloroplasts have a monophyletic or a polyphyletic origin is to be resolved by molecular rather than by microscopical methods and is discussed here but briefly. The prochlorophytes, prokaryotic algae distinguished from cyanobacteria s. str. by the presence of chlorophylls *a* and *b* as photosynthetic pigments, the lack of phycobilins and the closely stacked thylakoids have initially been considered as a missing link in the genesis of green algal chloroplasts (Swift and Palenik 1993). Analyses of various gene sequences suggest, however, that none of the known prochlorophytes can be an ancestor of the chlorophyte chloroplasts. They even showed that the prochlorophytes are polyphyletic.

It is generally assumed now that the three lines of primary plastids (in glaucocystophytes, rhodophytes and chlorophytes) are of monophyletic origin, from a single endosymbiosis of a heterotrophic phagotrophic flagellate with a cyanobacterium (Melkonian 1996; Delwiche and Palmer 1997; Kowallik 1997; McFadden 2001). This opinion is not always accepted (Gueneau et al. 1998).

A gene transfer from the prokaryote into the eukaryotic nucleus and the therewith necessarily connected evolution of protein import mechanisms from the cytosol into the plastid (Bodyl 2002) would characterize the first step of the transition from symbiont to organelle.

6.2 Complex Plastids: Four-Membrane Envelopes

6.2.1 Plastids of "Golden Algae"

The chloroplasts of various groups of algae were shown to be additionally surrounded by a pair of membranes, the periplastidal ER or chloroplast ER (Gibbs 1981) . Here, in the "golden algae" (Medlin et al. 1997), namely the photosynthetic heterokontophytes (chromophytes) and haptophytes (prymnesiophytes) the vestigial eukaryotic cytoplasm between the periplast ER and the chloroplast, still found in cryptophytes and chlorarachniophytes, is further reduced. The chloroplasts contain chlorophylls *a* and *c* and β-1,3-glucans as storage polysaccharide, deposited in vacuoles outside the chloroplasts. The space between the periplast ER and the plastids contains a periplast reticulum, consisting of a few smooth vesicles and tubules. Nucleomorphs and ribosomes are lost. Leucoplasts of golden algae have a periplast ER and a periplast reticulum, as well (Schnepf 1969b: the diatom *Nitzschia alba*).

The outer membrane of the periplastidal ER belongs to the host. It is studded with ribosomes and is continuous with the nuclear envelope when feasible by spatial relationships. The inner membrane of the periplastidal ER is interpreted to be likewise a part of the "host" ER. Alternatively it has been suggested that the inner membrane is homologous to the plasmalemma of the former eukaryotic endosymbiont while the outer one is derived from the membrane of the perisymbiont vesicle (Whatley et al. 1979).

DNA sequence analyses indicate that the endobiotic ancestors of chromophyte and haptophyte chloroplasts were red algal-like (Melkonian 1996; Kowallik 1997) and arose polyphyletically from unrelated symbiotic events (Daugbjerg and Andersen 1997).

6.2.2 Plastids of Apicomplexans

It was a great surprise when plastids were detected in apicomplexan parasites like *Toxoplasma gondii*, *Eimeria tenella* and *Plasmodium falciparum*. It has been shown recently that they have an envelope of four membranes (Köhler et al. 1997) In *Psalteriomonas* even traces of chlorophylls *a* and *b* and carotenoids have been detected (Hackstein et al. 1997). The coccidial plastid DNA has a length of 35 kb and contains 66 genes in *P. falciparum*. Most of them are involved in transcription and translation. The ribosome genes are arranged as inverted repeats, as in other plastids (Hackstein et al. 1997)

The apicomplexans (alveolates) could thus be taken as algae with extremely reduced plastids. Probably the plastids have been acquired from a chlorophyte by secondary endosymbiosis (Köhler et al. 1997). Red algae, perhaps via a chromophyte-like organism (McFadden 2001), or dinoflagellates (Melkonian 1996) are also suspected as plastid source. DNA sequence analyses indicate that the apicomplexans evolved from a single, free-living, photosynthetic progenitor (Denny et al. 1998; McFadden et al. 1997).

6.3 Complex Plastids: Three-Membrane Envelopes

The photosynthetic euglenophytes and the peridinin-containing dinoflagellates have chloroplasts with an envelope of three membranes. Cavalier-Smith (1993) suggested that the three membranes result from a primary endosymbiosis, derived from the phagosomal membrane and the outer and inner membrane of a cyanobacterium. It is, however, largely believed now that the chloroplasts arose from a secondary endosymbiosis as already suggested by Gibbs (1978). The initially four membranes (Sects. 5.2.2 and 6.2) are reduced. The outermost membrane does not bear ribosomes. It may be homologous to the plasmalemma of the eukaryotic symbiont and the two inner ones with the envelope of the primary chloroplast while the phagosomal membrane is lost (Gibbs 1978). Others have proposed that the outer membrane is derived from the phagosomal membrane (Whatley et al. 1979). An uptake of the later endosymbiont by myzocytosis would explain the presence of only three envelope membranes (Bodyl 2002; Shin et al. 2002).

6.3.1 Euglenophyte Chloroplasts

Only about 30% of the euglenoids have plastids (Whatley 1993). The others are obligate heterotrophs. The euglenophyte chloroplasts contain the chlorophylls *a* and *b*, like chlorophyte chloroplasts. The spectrum of the xanthophylls differs from that of chlorophytes. The thylakoids are in stacks of three or two. The storage carbohydrate is paramylum, a β-1,3 glucan, deposited in the form of crystalline granules in vacuoles.

The euglenoid plastid DNA has a peculiarity. The genes for rRNA are arranged in tandem, not in inverted repeats as in other plastids. The genome organization and its gene content is otherwise similar to that of chlorophyte chloroplasts (Hallick et al. 1993), suggesting that a chlorophyte was the ancestor of the euglenophyte chloroplasts (Bhattacharya and Medlin 1998).

The euglenoid "hosts" are a sister group of the kinetoplastids (Triemer and Farmer 1991), a group of apoplastidic protoctists. The comparison of kinetoplastid (bodonid) cytostomes, colourless euglenoid cytostomes and the reservoir pockets of photosynthetic euglenophytes suggests that the photosynthetic euglenophytes arose from phagotrophic or myzocytotic ancestors (Shin et al. 2002).

6.3.2 Dinoflagellate Chloroplasts with Peridinin

The dinoflagellates are most diverse with respect to their plastids. Only 50% are photosynthetic. Many families and even genera contain taxa with or without chloroplasts or with different types of plastids (Schnepf and Elbrächter 1999; G. Hansen 2001). Even parasitic dinoflagellates may have chloroplasts. The "typical" dinoflagellate chloroplasts are characterized by chlorophylls *a* and c_2 and the carotenoid peridinin. The other types are discussed in Sect. 6.4. Only the peridinin-chloroplasts have a three-membrane envelope. The thylakoids are arranged in stacks of three, occasionally of four or two. The storage carbohydrates are deposited as starch grains within the cytosol.

The outermost membrane is ribosome-free. The middle membrane shows sometimes an unusual structure in thin sections (Schnepf 1993: *Prorocentrum micans*) as well as after freeze-fracturing (Sweeney 1981: *Gonyaulax polyedra*). In *Prorocentrum micans* and some other dinoflagellates, mostly Prorocentrales, it is even locally absent, especially close to the pyrenoid (Schnepf and Elbrächter 1999). The question consequently arises whether it is actually a "biomembrane" in the narrow sense. In uninjured cells each true biomembrane surrounds a compartment completely, without gap. The molecular architecture of biomembranes does hardly allow free edges.

Cavalier-Smith (1993) discussed whether the peridinin chloroplasts are directly derived from cyanobacteria. The middle membrane would then be homologous with the outer membrane of the prokaryote and represent an example of membrane loss, an indication that the outer membrane of Gram-negative bacteria and cyanobacteria are not "true" biomembranes, but membrane-like cell wall layers. This interpretation is suppported by their "aberrant" position (they do not separate a cytoplasmic from a non-cytoplasmic compartment, see Introduction) and the lack of active transport processes.

The original idea of Gibbs (1978) that the peridinin chloroplasts are the result of a secondary endosymbiosis is favoured now (Medlin et al. 1997).The endosymbiont was perhaps red algal-like (Melkonian 1996; Zhang et al. 1999; McFadden 2001). If the ingestion was by myzocytosis the outer envelope membrane would represent the phagosomal membrane, the two inner ones the chloroplast membrane. Hypotheses about protein import into the chloroplast have to consider the presence or absence of the middle membrane. It might be extremely permeable.

The DNA of peridinin chloroplasts is unique in consisting of minicircles each containing a single gene (Zhang et al. 1999; Barbrook and Howe 2000). Another peculiarity concerns rubisco. All organisms with oxygenic photosynthesis have form-I rubisco, with eight catalytic large and eight small subunits, but peridinin dinoflagellates have form-II rubisco with only two large subunits, an enzyme which was previously known only from certain anaerobic proteobacteria such as *Rhodospirillum rubrum* (Morse et al. 1995; Rowan et al. 1996). In contrast to form-I rubisco the dinoflagellate rubisco is encoded as a polyprotein by nuclear genes. Palmer (1995) discussed several scenarios how rubisco II genes might be incorporated into the dinoflagellate nucleus.

6.4 Aberrant Dinoflagellate Chloroplasts

Aberrant dinoflagellate chloroplasts have either fucoxanthin or fucoxanthin derivatives instead of peridinin as main carotenoid or they contain phycobilins (reviewed by Schnepf and Elbrächter 1999).

6.4.1 Fucoxanthin Chloroplasts

The most extensively investigated dinoflagellate with fucoxanthin-chloroplasts is *Gymnodinium mikimotoi.* The characteristic carotenoid is 18'-hexanoylfucoxanthin (Tangen and Björnland 1981), a pigment known otherwise only from *Emiliana huxlei* and some other prymnesiophytes. The *G. mikimotoi* chloroplasts have thylakoids in stacks of three and a prominent stalked pyrenoid. It is thylakoid-free and connected with the chloroplast by a small bridge. The chloroplast envelope is not well enough preserved in electron microscopical studies (Kite and Dodge 1985; Schnepf and Elbrächter 1999) to state definitively the number of membranes. Probably it consists of two membranes, and a chloroplast ER, as present in *Emiliana,* is lacking here. The further organelle-cytoplasm interface is likewise not yet fully explored. *G. mikimotoi* is grown in unialgal culture. Kleptoplastidy is thus excluded. One may speculate that the chloroplasts are derived from a prasinophyte, taken up by myzocytosis. The periplastidal ER and other prasinophyte cell components have apparently been lost subsequently.

6.4.2 Phycobilin-Chloroplasts

Phycobilin-containing chloroplasts are known from *Dinophysis* and some related dinoflagellates. There are also apoplastidic *Dinophysis* species (Hallegraeff and Lucas 1988; Schnepf and Elbrächter 1988). Some of the photosynthetic species are mixotrophic and feed by myzocytosis (Jacobson and Andersen 1994).

The chloroplasts have thylakoids in stacks of two. Their lumen is relatively wide and filled with an electron-dense material. They resemble highly the thylakoids of cryptophytes (Schnepf and Elbrächter 1988) and contain likewise phycoerythrin in their lumen (Vesk et al. 1996). The major carotenoid is alloxanthin, as in cryptophytes. In contrast to previous statements peridinin is absent (Meyer-Harms and Pollehne 1998) as supposed earlier (Vesk et al. 1996). The chloroplast envelope consists of two membranes (Schnepf and Elbrächter 1988). Starch grains are deposited in the cytoplasm, as in other dinoflagellates.

Photosynthetic *Dinophysis* species could as yet not be grown in culture. The idea of Melkonian (1996) that the chloroplasts may be kleptochloroplasts is nevertheless unlikely because chloroplast-free cells have never been found in photosynthetic species. Schnepf and Elbrächter (1988) suggested that the chloroplasts were derived from a cryptophyte, thus representing a tertiary endosymbiosis, with subsequent loss of cryptophyte membranes and cytoplasm. Cavalier-Smith (1993) assumed on the contrary that *Dinophysis* chloroplasts result from a primary symbiosis directly from a cyanobacterium and represent an early stage of dinophyte chloroplast evolution. His idea was based on the wrong assumption that *Dinophysis* chloroplasts contain phycobilins as well as peridinin. Molecular studies indicate moreover that the Dinophysales are not primitive dinoflagellates but a sister group of the Prorocentrales and that the photosynthetic species have evolved rather recently (Rehnstam-Holm et al. 2002).

7 Conclusions and Perspectives

This review was initiated by the question of the general validity of the "rules of cell compartmentation" (Sect. 1), especially for associations between protoctists and microalgae. It was shown that there are, indeed, some exceptions (Sects. 2.1.1, 4, 5.2.1). The host plasmalemma-derived membrane around endobiotic parasites and symbionts is lost sometimes. A myzocytotic uptake of host cytoplasm results likewise in a single membrane, a host membrane, between host/predator and symbiont/prey (Sect. 2.4). In most parasitic and symbiotic associations the rules are not violated (Sects. 2, 3, 4, 5) In a variety of associations the interface between the two partners is not yet clearly explored (Sects. 2.1.1, 2.7, 5.1).

The further search for protoctists feeding on or symbiotic with algae will result in new insights in their structural and cell biological relationships as well as in the biological role of those associations which has been underestimated for a long time. They will furthermore give new hints on the symbiogenesis of plastids.

The endosymbiotic origin of plastids from a previously free-living cyanobacterium, first deduced from structural evidence (Sect. 1) is well established by analyses of plastid DNA phylogeny. It is generally accepted now that the plastids with two envelope membranes trace back to a single endosymbiosis with a heterotrophic, phagotrophic flagellate as host (Sect. 6.1). Glaucocystophytes, rhodophytes and chlorophytes/embryophytes are thus sister groups, the glaucophytes being the least evolutionarily different from cyanobacteria. Further research on DNA phylogeny will refine our image of the evolutionary scenario and will perhaps present further surprises as was the detection that in rhodophytes (and in algae with secondarily gained rhodophyte chloroplasts) the genes encoding rubisco are not of cyanobacterial but proteobacterial provenance (Delwiche and Palmer 1997).

Biochemical and molecular biological studies will perhaps resolve the controversy around the homology of the outer membrane of the plastid envelope (Sect. 6.1.2). Is it derived from the host plasmalemma (via the membrane of the symbiosome) or does it represent the "outer membrane" of the symbiotic cyanobacterium? In this context it is necessary to discriminate between biomembranes s. str. and membrane-like cell wall layers which do not participate in active transport processes.

There is also general agreement about the origin of plastids with an envelope of two membrane pairs from a secondary endosymbiosis between (unrelated) eukaryotic hosts and red or green algae, respectively (Sect. 6.3). The algae with nucleomorphs represent stages on the way from endosymbiont to organelle (Sect. 5.2.2). Not yet fully clear is the number of secondary symbioses having led to complex chloroplasts.

An interesting idea about the evolutionary advantages of secondary chloroplasts over primary ones was recently proposed (Lee and Kugrens 2000). The latter authors noticed that the (successful) secondary symbioses arose before 275 Ma, at a time when the atmospheric CO_2 level had decreased drastically. An acidification of the perisymbiotic space, which is homologous to a digestion vacuole, would remarkably improve the supply of rubisco with CO_2. In the alkaline sea water inorganic carbon is dissolved in the form of HCO_3 – which is not accepted by rubisco. Back then, but not before that time, algae with secondary plastids would have had a selective advantage over other existing algae, as perhaps in the case of kleptochloroplasts (Sect. 3).

This discussion includes also the peridinin-chloroplasts of dinoflagellates and euglenophyte chloroplasts, both with three-membrane envelopes (Lee and Kugrens 2000).

They are most probably the result of secondary endosymbioses though an origin via primary symbioses is also discussed (Sect. 6.3). It is perhaps important to note in this context that some dinoflagellates and a few euglenoids ingest host cytoplasm by myzocytosis, leaving out the plasmalemma (Sect. 2.4), and that both groups comprise many apoplastidic species. While some heterokont taxa, e.g., diatoms (Sect. 6.2.1), comprise apochlorotic but no apoplastic species, many larger taxa of heterokonts, e.g., the oomycetes, are apoplastidic. Secondary chloroplasts seem to be less firmly integrated into the host than primary ones which are never lost but transformed into leucoplasts in non-photosynthetic species.

Were the ancestors of apoplastidic heterokonts, dinoflagellates, euglenoids and algae with secondary chloroplasts originally phagotrophic mixotrophs with primary chloroplasts which were then lost? If so, the incorporation of a secondary endosymbiont would be facilitated by the presence of genes for photosynthetic activities already transferred into the nucleus, and of protein import mechanisms (Häuber et al. 1994). A coexistence of two different types of photosynthetic organelles would, however, result in competition which is less likely and not represented by living examples. Endosymbiotic cyanobacteria in cells with chloroplasts fix primarily nitrogen (Sect. 4).

The photosynthetic endosymbiont of *Glenodinium foliaceum* and similar dinoflagellates coexist with a genuine achlorophyllous plastid which is reduced to an eyespot (Sect. 5.2.1). The dinoflagellates with their different types of secondary and tertiary chloroplasts and their different strategies of food uptake (Sects. 2.2, 2.3, 3, 4, 5.1, 5.2.1, 5.2.3, 6.3, 6.4) comprise some not yet fully understood symbiotic associations (Sect. 5.2.3) and will surely present further surprising examples. Microscopical studies of extraordinary algae are especially useful for understanding the symbiogenetic origin of chloroplasts. In the extremes the borders of general laws can be traced better than in "normal" organisms. The main progress will come by molecular phylogenetic analyses, the analyses of lateral gene transfer and of the mechanisms of protein import into the organelles, considering kleptochloroplasts, as well.

Many of my own studies have been carried out at Wattenmeerstation List/Sylt, Biologische Anstalt Helgoland (now Alfred Wegener Institute). I gratefully acknowledge the fruitful cooporation with Dr. G. Drebes, Dr. M. Elbrächter, H. Halliger, G. Deichgräber and Dr. S. Kühn. Dr. M. Elbrächter and Prof. Dr. P. Sitte provided valuable information and helped to condense the text.

References

Archer SD, Leakey RJG, Burkill PH, Sleigh MA (1996) Microbial dynamics in coastal waters of East Antactica: herbivory by heterotrophic dinoflagellates. Mar Ecol Progr Ser 139:239–255

Bardele CF (1972) A microtubule model for ingestion and transport in the suctorian tentacle. Z Zellforsch 126:116–134

Barbrook AC, Howe CJ (2000) Minicircular plastid DNA in the dinoflagellate *Amphidinium operculatum*. Mol Gen Genet 263:152–158

Bhattacharya D, Medlin L (1995) The phylogeny of plastids: a review based on comparisons of small-subunit ribosomal RNA coding regions. J Phycol 31:489–498

Bhattacharya D, Medlin L (1998) Algal phylogeny and the origin of land plants. Plant Physiol 116:9–15

Bhattacharya D, Schmidt HA (1997) Division Glaucocystophyta. In: Bhattacharya D (ed) Origin of algae and their plastids. Springer, Vienna New York, pp 139–148

Bhattacharya D, Helmchen T, Melkonian M (1995) Molecular evolutionary analyses of nuclear-encoded small subunit ribosomal RNA identify an independent rhizopod lineage containing Euglyphina and Chlorarachniophyta. J Eukar Microbiol 42:65–69

Bockstahler KR, Coats DW (1993a) Spatial and temporal aspects of mixotrophy in Chesapeake Bay dinoflagellates. J Eukar Microbiol 40:49–60

Bockstahler KR, Coats DW (1993b) Grazing of the mixotrophic dinoflagellate *Gymnodinium sanguineum* on ciliate populations of Chesapeake Bay. Mar Biol 116:477–487

Bodyl A (2002) Is protein import into plastids with four membrane envelopes dependent on two Toc systems operating in tandem? Plant Biol 4:423–431

Bothe H, Floener L (1978) Physiological characterization of *Cyanophora paradoxa*, a flagellate containing cyanelles in endosymbiosis. Z Naturforsch 33c:1030–1035

Bruning K (1991) Infection of the diatom *Asterionella* by a chytrid. II Effects of light on survival and epidemic development of the parasite. J Plankton Res 13:119–129

Bulman SR, Kühn SF, Marshall JW, Schnepf E. (2001) A phylogenetic analysis of the SSUrRNA from members of the Plasmodiophorida and Phagomyxida. Protist 152:43–51

Bursa A (1963) Phytoplankton in coastal waters of the Arctic Ocean at Point Barrow, Alaska. Artic 4:239–262

Buskey EJ, Coulter CJ, Brown SL (1994) Feeding, growth and bioluminescence of the heterotrophic dinoflagellate *Protoperidinium huberi*. Mar Biol 212:373–380

Calado AJ, Moestup O (1997) Feeding in *Peridiniopsis berolinensis* (Dinophyceae):new observations on tube feeding by an omnivorous, heterotrophic dinoflagellate. Phycologia 36:47–59

Canter HM, Jaworsky GHM (1979) The occurrence of hypersensitive reaction in the planktonic diatom *Asterionella formosa* Hassal parasitized by the chytrid *Rhizophydium planktonicum* Canter emend. in culture. New Phytol 82:187–206

Canter-Lund H, Lund JW (1995) Fresh water algae. Biopress, Bristol

Cavalier-Smith T (1993) The origin, losses and gains of chloroplasts. In: Lewin RA (ed) Origin of plastids. Chapman and Hall, New York, pp 291–348

Cavalier-Smith T (2000) Membrane heredity and early chloroplast evolution. Trends Plant Sci 5:174–182

Cavalier-Smith T, Chao EE (1996/97) Sarcomonad ribosomal RNA sequences, rhizopod phylogeny, and the origin of euglyphid amoebae. Arch Prostistenkd 147:227–236

Chesnik JM, Cox ER (1986) Specialization of endoplasmic reticulum architecture in response to a bacterial symbiosis in *Peridinium balticum* (Pyrrhophyta). J Phycol 22:291–298

Chesnick JM, Cox ER (1989) Fertilization and zygote development in the binucleate dinoflagellate *Peridinium balticum* (Pyrrhophyta). Am J Bot 76:1060–1072

Chesnick JM, Morden CW, Schmieg AM (1996) Identidy of the endosymbiont of *Peridinium foliaceum* (Pyrrophyta). Analysis of the rbcLS operon. J Phycol 32:850–857

Chesnick JM, Kooistra WHCF, Wellbrock U, Medlin LA (1997) Ribosomal RNA analysis indicates a benthic pennate diatom ancestry for the endosymbiont of the dinoflagellates *Peridinium foliaceum* and *Peridinium balticum* (Pyrrhophyta). J Eukar Microbiol 44:314–320

Coats DM, Park MG (2002) Parasitism of photosynthetic dinoflagellates by three strains of *Amoebophrya* (Dinophyta): parasite survival, infectivity, generation time, and host specificity. J Phycol 38:520–528

Cook PW (1963) Host range studies of certain phycomycetes parasitic on desmids. Am J Bot 50:580–588

Daugbjerg N, Andersen RA (1997) Phylogenetic analyses of the rbcL sequences from haptophytes and heterokont algae suggest their chloroplasts are unrelated. Mol Biol Evol 14:1242–1251

Deane J, Fraunholz M, Su V, Maier U-G, Martin W, Durnford D, McFadden G (2000) Evidence for nucleomorph to host nucleus gene transfer: light harvesting complex proteins from cryptomonads and chlorarachniophytes. Protist 151:239–252

Delwiche CF, Palmer JD (1997) The origin of plastids and their spread via secondary symbiosis. In: Bhattacharya D (ed) Origins of algae and their plastids. Springer, Vienna New York, pp 53–86

Denny P, Preiser P, Williamson D, Wilson I (1998) Evidence for a single origin of the 35 kb plastid DNA in apicomplexans. Protist 149:51–59

Dodge JD (1971) A dinoflagellate with both a mesocaryotic and a eukaryotic nucleus. I. Fine structure of the nuclei. Protoplasma 73:145–157

Dodge JD, Crawford RM (1974) Fine structure of the dinoflagellate *Oxyrrhis marina* III. Phagotrophy. Protistologica 10:239–244

Douglas S, Zauner S, Fraunholz M, Beaton M, Penny S, L-TXW Deng, Reith M, Cavalier-Smith T, Maier U-G (2001) The highly reduced genome of an enslaved algal nucleus. Nature 410:1091–1096

Drebes G (1966) Ein parasitischer Phycomycet (Lagenidiales) in *Coscinodiscus*. Helgol Meeresunters 13:426–435

Drebes G, Schnepf E (1982) Phagotrophy and development of *Paulsenella* cf. *chaetoceratis* (Dinophyta), an ectoparasite of the diatom *Streptotheca thamesis*. Helgol Meeresunters 35:501–515

Drebes G, Schnepf E (1988) *Paulsenella* (Chatton (Dinophyta), ectoparasites of marine diatoms: development and taxonomy. Helgol Meeresunters 42:563–581

Drebes G, Schnepf E (1998) *Gyrodinium undulans* Hulburt, a marine dinoflagellate feeding on the bloom-forming diatom *Odontella aurita*, and on copepod and rotifer eggs. Helgol Meeresunters 52:1–14

Drebes G, Kühn SF, Gmelch A, Schnepf E (1996) *Cryothecomonas aestivalis* sp. nov., a colourless nanoflagellate feeding on the marine centric diatom *Guinardia denticulata* (Cleve) Hasle. Helgol Meeresunters 50:497–515

Elbrächter M, Qi Y-Z (1998) Aspects of *Noctiluca* (Dinophyceae) population dynamics. In: Anderson DM, Cembella AD, Hallegraeff GM (eds) Physiological ecology of harmful algal blooms. NATO ASI Series, vol G 41. Springer, Berlin Heidelberg New York, pp 315–335

Elbrächter M, Schnepf E (1996) *Gymnodinium chlorophorum*, a new, green, bloom-forming dinoflagellate (Gymnodiniales, Dinophyceae) with a vestigial prasinophyte endosymbiont. Phycologia 35:381–393

Elbrächter M, Schnepf E (1998) Parasites of harmful algae. In: Anderson DM, Cembella AD, Hallegraeff GM (eds) Physiological ecology of harmful algal blooms. NATO ASI Series, vol G 41. Springer, Berlin Heidelberg New York, pp 351–365

Elbrächter M, Schnepf E., Balzer I (1996) *Hemistasia phaeocysticola* (Scherffel) comb. nov., redescription of a free-living, marine, phagotrophic kinetoplastid flagellate. Arch Protistenkd 147:125–136

Farmer MA, Roberts KR (1990) Organelle loss in the endosymbiont of *Gymnodinium acidotum* (Dinophyceae). Protoplasma 153:178–185

Fields SD, Rhodes RG (1991) Ingestion and retention of *Chroomonas* spp (Cryptophyceae) by *Gymnodinium acidotum* (Dinophyceae). J Phycol 27:525–529

Fischer-Defoy D, Hausmann K (1981) Microtubules, microfilaments and membranes in phagocytosis:Structure and function of the oral apparatus of the ciliate *Climacostomum virens*. Differentiation 20:141–151

Floener L, Bothe H (1980) Nitrogen fixation in *Rhopalodia gibba*, a diatom containing blue-greenish inclusions symbiotically. In: Schwemmler W, Schenk HEA (eds) Endocytobiology:endosymbiosis and cell biology, a synthesis of recent research, vol 1. Walter de Gruyter, Berlin, pp 541–552

Floener L, Bothe H (1982) Metabolic activities in *Cyanophora paradoxa* and its cyanelles. II. Photosynthesis and respiration. Planta 156:78–83

Floener L, Danneberg G, Bothe H (1982) Metabolic activities in *Cyanophora paradoxa* and its cyanelles. I. The enzymes of assimilatory nitrate reduction. Planta 156:70–77

Foissner W, Didier P (1984) Nahrungsaufnahme, Lebenszyklus und Morphogenese von *Pseudoplatyophrya nana* (Kahl, 19226) (Ciliophora, Colpodida). Protistologica 19:103–109

Fritz L, Nass M (1992) Development of the endoparasitic dinoflagellate *Amoebophrya ceratii* within different host species. J Phycol 28:312–320

Gaines G, Elbrächter M (1987) Heterotrophic nutrition. In: Taylor FJR (ed) The biology of dinoflagellates. Blackwell, Oxford, pp 224-268

Gaines G, Taylor FJR (1984) Extracellular digestion in marine dinoflagellates. J Plank Res 6:1057–1061

Gastrich MD (1987) Ultrastructure of a new intracellular symbiotic alga found within planktonic foraminifera. J Phycol 23:623–632

Gibbs SP (1978) The chloroplasts of *Euglena* may have evolved from symbiotic green algae. Can J Bot 56:2883–2889

Gibbs SP (1981) The chloroplast endoplasmic reticulum: structure, function, and evolutionary significance. Int Rev Cytol 72:49–99

Gillott MA, Gibbs SP (1980) The cryptomonad nucleomorph: its ultrastructure and evolutionary significance. J Phycol 16:558–568

Gilson P, McFadden G (1996) The miniturized nuclear genome of a eukaryotic endosymbiont contains genes that overlap, genes that are contrascribed, and smallest known spliceosomal introns. Proc Natl Acad Sci USA 93:7737–7742

Görtz H-D (1996) Symbiosis in ciliates. In: Hausmann K, Bradbury PC (eds) Ciliates. Cells as organisms. Fischer, Stuttgart, pp 441–462

Gordon N, Angel DL, Neori A, Kress N, Kimor B (1994) Heterotrophic dinoflagellates with symbiotic cyanobacteria and nitrogen limitation in the Gulf of Aquba. Mar Ecol Progr Ser 107:83–88

Grahame ES (1976) The occurrence of *Lagenisma coscinodisci* in *Palmeria hardmaniana* from Kingston harbour, Jamaica. Br Phycol J 11:57–61

Grell KG (1989) *Reticulosphaera socialis* n. gen. n. sp., ein plasmodialer und phagotropher Vertreter der heterokonten Algen. Z Naturforsch 44c:330–332

Gromov BV, Mamkaeva KA (1968) *Amoeboaaphelidium protococcarum* sp. n. and *Amoeboaphelidium chlorellavorum* sp. n. – endoparasites of protococcous algae. Acta Protozool 6:221–225

Gromov BV, Mamkaeva (1970) The fine structure of *Amoeboaphelidium protococcarum* Gromov et Mamkaeva – an endoparasite of green alga *Scenedesmus*. Arch Hydrobiol 67:452–459

Gueneau P, Morel F, LarocheJ, Erdner D (1998) The petF region of the chloroplast genome from the diatom *Thalassiosira weissflogii*:sequence, organization and phylogeny. Eur J Phycol 33:2033–211

Gustavson DE Jr, Stoecker DK, Johnson, MD, Van Heukelen WF, Sneider K (2000) Cryptophyte algae are robbed of their organelles by the marine ciliate *Mesodinium rubrum.* Nature 405:1049–1052

Hackstein JPH, Schubert H, Rosenberg J, Mackenstedt U, van den Berg M, Brul, S, Derksen J, Matthijs HCP (1997) Plastid-like organelles in anaerobic mastigotes and parasitic apicomplexans. In: Schenk, HEA, Herrmann RG, Jeon KW, Müller NE, Schwemmler W (eds) Eukaryotism and symbiosis. Springer, Berlin Heidelberg New York, pp 49–56

Häuber MM, Müller SB, Speth V, Maier UG (1994) How to evolve a complex plastid? – A hypothesis. Bot Acta 107:383–386

Hallegraeff GN, Lucas IAN (1988) The marine dinoflagellate genus *Dinophysis* (Dinophyceae):Photosynthetic, neritic and non-photosynthetic oceanic species. Phycologia 27:25–42

Hallick RB, Hong L, Droger RG, Favreau MR, Montfort A, Orsat B, Spielmann A, Stutz E (1993) Complete sequence of *Euglena gracilis* Chloroplast DNA. Nucleic Acid Res 21:3537–3544

Hammer A, Grüttner C, Schumann R (1999) The effect of electrostatic charge of food particles on capture efficiency by *Oxyrrhis marina* Dujardin (Dinoflagellata). Protist 150:375–382

Hansen G (2001) Ultrastructure of *Gymnodinium aureolum* (Dinophyceae): toward a further redefinition of *Gymnodinium* sensu stricto. J Phycol 37:612–623

Hansen G, Daugbjerg N, Henriksen P (2000) Comparative study of *Gymnodinium mikimotoi* and *Gymnodinium aureolum*, comb. nov. (= *Gyrodinium aureolum*) based on morphology, pigment composition, and molecular data. J Phycol 36:394–410

Hansen PJ (1991) Quantitative importance and trophic role of heterotrophic dinoflagellates in a coastal pelagic food web. Mar Ecol Progr Ser 73:253–261

Hansen PJ, Nielsen TG, Kaas H (1995) Distribution and growth of protists and mesozooplankton during a bloom of *Chrysochromulina* spp (Prymnesiophyceae, Prymnesiales). Phycologia 34:409–416

Hansmann P (1988) Ultrastructural localization of RNA in cryptomonads. Protoplasma 146:81–88

Hansmann P, Falk H, Sitte P (1985) DNA in the nucleomomorph of *Cryptomonas* demonstrated by DAPI fluorescence. Z Naturforsch 40c:933–935

Hargraves PE (2002) The ebridian flagellates *Ebria* and *Hermesium.* Plankton Biol Ecol 49:9–16

Hausmann K, Hülsmann N (1996) Protozoology, 2nd edn. Thieme, Stuttgart

Hausmann K, Patterson DJ (1982) Pseudopod formation and membrane production during prey capture by a heliozoon (Feeding by *Actinophrys* II). Cell Motility 2:9–24

Hausmann K, Peck RK (1978) Microtubules and microfilaments as major components of a phagocytic apparatus:the cytopharyngeal basket of the ciliate *Pseudomicrothorax dubius.* Differentiation 11:157–167

Hausmann K, Peck RK (1979) The mode of function of the cytopharyngeal basket of the ciliate *Pseudomicrothorax dubius.* Differentation 14:147–158

Hausmann K, Radek R (1993) A comparative survey on phagosome formation in protozoa. In: Plattner H (ed) Advances in cell and molecular biology of membranes. IIB Membrane traffic in Protozoa. JAI Press, Greenwich, CT, pp 259–282

Heckmann K (1995) Räuber-induzierte Feindabwehr bei Protozoen. Naturwissenschaften 82:107–116

Hibberd DJ (1977) Observations on the ultrastructure of the cryptomonad endosymbiont of the red-water ciliate *Mesodinium rubrum.* J Mar Biol Ass UK 57:45–61

Hibberd DJ, Norris RE (1984) Cytology and ultrastructure of *Chlorarachnion reptans* (Chlorarachniophyta, Divisio nova, Chlorarachniophyceae Classis nova). J Phycol 20:310–330

Höhfeld I, Melkonian M (1998) Lifting the curtain? The microtubular cytoskeleton of *Oxyrrhis marina* (Dinophyceae) and its rearrangement during phagocytosis. Protist 149:75–88

Holen DA (2000) The relative abundance of mixotropohic and heterotrophic ciliates in an oligotrophic lake. Arch Hydrobiol 150:1–15

Holfeld H (1998) Fungal infections of the phytoplankton seasonality, minimal host density, and specificity in a mesotrophic lake. New Phytol 138:507–517

Horiguchi T, Pienaar RN (1991) Ultrastructure of a marine dinoflagellate, *Peridinium quinquecorne* Abé (Peridiniales) from South Africa with particular reference to its chrysophyte endosymbiont. Bot Mar 34:123–131

Horiguchi T, Pienaar RN (1992) *Amphidinium latum* (Dinophyceae), a sand-dwelling dinoflagellate feeding on cryptomonads. Jpn J Phycol (Sôrni) 40:353–363

Horiguchi T, Pienaar RN (1994) Ultrastructure of a new marine sand-dwelling dinoflagellate, *Gymnodinium quadrilobum* sp nov. (Dinophyceae) with special reference to its endosymbiotic alga. Eur J Phycol 29:237–245

Howe CH, Smith AG (1991) Plants without chlorophyll. Nature 349:109

Jacobson DM, Anderson DM (1986) Thecate heterotrophic dinoflagellates: Feeding behavior and mechanisms. J Phycol 22:249–258

Jacobson DM, Anderson DM (1992) Ultrastructure of the feeding apparatus and myonemal system of the heterotrophic dinoflagellate *Protoperidinium spinulosum*. J Phycol 28:69–82

Jacobson DM, Andersen RA (1994) The discovery of mixotrophy in photosynthetic species of *Dinophysis* (Dinophyta):light and electron microscopical observations of food vacuoles in *Dinophysis acuminate, D. norvegica* and two heterotrophic dinophysoid dinoflagellates. Phycologia 33:97–110

Janson S, Rai SAN, Bergman B (1995) Intracellular cyanobiont *Richelia intracellularis* – ultrastructure and immuno-localization of phycoerythrin, nitrogenase, Rubisco and glutamine synthase. Mar Biol 124:1–8

Jeong KW (1983) Preface. Intracellular symbiosis. Int Rev Cytol (Suppl 14):XIII–XIV

Joyard J, Block MA, Douce R (1991) Molecular aspects of plastid envelope biochemistry. Eur J Biochem 191:489–509

Karling JS (1944) *Phagomyxa algarum*, n. g., n. sp., an unusual parasite with plasmodiophoralean and proteomyxan characteristics. Am J Bot 31:38–52

Kempton JW, Wolny J, Tengs T, Rizzo P, Morris R, Tunnell J, Scott P, Steidinger K, Hymel SN, Lewitus AJ (2002) *Kryptoperidinium foliaceum* blooms in South carolina:a multi-analytical approach to identification. Harmful Algae 1:383–392

Kies L (1974) Elektronenmikroskopische Untersuchungen an *Paulinella chromatophora* Lauterborn, einer Thekamöbe mit blau-grünen Endosymbionten (Cyanellen). Protoplasma 80:69–89

Kies L (1988) The effect of penicillin and the morphology and ultrastructure of *Cyanophora, Gloeochaete* and *Glaucocystis* (Glaucocystophyceae) and their cyanelles. Endocytobios Cell Res 5:361–372

Kies L, Kremer BP (1990) Phylum Glaucocystophyta. In: Margulis, L, Corliss JO, Melkonian M, Chapman DJ (eds) Handbook of protoctista. Jones and Bartlett, Boston, pp 152–166

Kite GC, Dodge JD (1985) Structural organization of plastid DNA in two anomalously pigmented dinoflagellates, *Glenodinium foliaceum, Gyrodinium aureolum*. J. Phycol 21:50–56

Kjaeret, AH, Naustvoll LJ, Paasche E (2000) Ecology of the heterotrophic dinoflagellate genus *Protoperidinium* in the inner Oslofjord (Norway). Sarsia 85:453–460

Kluge M, Mollenhauer D, Mollenhauer R (1991) Photosynthetic carbon assimilation in *Geosiphon pyriforme* (Kützing) F. v. Wettstein, an endosymbiotic association of fungus and cyanobacterium. Planta 185:311–315

Kluge M, Mollenhauer D, Mollenhauer R, Kape R (1992) *Geosiphon pyriforme*, an endosymbiotic consortium of a fungus and a cyanobacterium (*Nostoc*), fixes nitrogen. Bot Acta 105:343–344

Köhler S, Delwiche CF, Denny PW, Tilney LG, Webster P, Wilson RJM, Palmer JD, Roos DS (1997) A plastid of probable green algal origin in apicomplexan parasites. Science 275:1485–1488

Kowallik KV (1997) Origin and evolution of chloroplasts: current status and future perspectives. In: Schenk HEA, Herrmann RG, Jeon KW, Müller NE, Schwemmler W (eds) Eukaryotism and symbiosis. Springer, Berlin Heidelberg New York, pp 3–23

Kühn SF (1996/97) *Rhizamoeba schnepfii* sp. nov., a naked amoeba feeding on marine diatoms/North Sea, German Bight. Arch Protistenkd 147:277–282

Kühn SF (1997) Infection of *Coscinodiscus* spp. by the parasitoid nanoflagellate *Pirsonia diadema*:I. Behavioural studies on the infection process. J Plankton Res 19:791–804

Kühn SF (1998) Infection of *Coscinodiscus* spp. by the parasitoid nanoflagellate *Pirsonia diadema*. II. Selective infection behaviour for host species and individual host cells. J Plankton Res 20:443–453

Kühn SF, Hofmann M (1999) Infection of *Coscinodiscus granii* by the parasitoid nanoflagellate *Pirsonia diadema*. III. Effects of turbulence on the incidence of infection. J Plankton Res 21:2323–2340

Kühn SF, Schnepf E (2002) Infection of *Glaucocystis nostochinearum* (Glaucophyta) by *Lagenidium* sp. (Oomycota) and its hyperparasite *Pythiella* sp. (Oomycota). Hydrobiologia 481:165–171

Kühn SF, Drebes G, Schnepf E (1996) Fife new species of the nanoflagellate *Pirsonia* in the German Bight, North Sea, feeding on planktic diatoms. Helgol Meeresuntersuch 50:205–222

Kühn S, Lange M, Medlin LK (2000) Phylogenetic position of *Cryothecomonas* inferred from nuclear-encoded small subunit ribosomal RNA. Protist 151:337–345

Larsen J (1988) An ultrastructural study of *Amphidinium poecilochroum* (Dinophyceae), a phagotrophjic dinoflagellate feeding on small species of cryptophytes. Phycologia 27:366–377

Lee JJ (1983) Perspective on algal endosymbionts in larger foraminifera. Int Rev Cytol Suppl 14:49–77

Lee JJ, McEnery ME, Shilo M, Reiss Z (1979) Isolation and cultivation of diatom symbionts from larger Foraminifera (Protozoa). Nature 280:57–58

Lee RE, Kugrens P (2000) Ancient atmospheric CO_2 and the timing of evolution of secondary endosymbioses (Commentary). Phycologia 39:167–172

Leitsch CEW, Kowallik KV, Douglas S (1999) The atpA gene cluster of *Guillardia theta* (Cryptophyta): a piece in the puzzle of chloroplast genome evolution. J Phycol 35:128–135

Lesemann DE, Fuchs WH (1970) Die Ultrastruktur des Penetrationsvorganges von *Olpidium brassicae* an Kohlrabi-Wurzeln. Arch Mikrobiol 71:20–30

Lindholm T, Lindroos P, Mörk A-C (1988) Ultrastructure of the photosynthetic ciliate *Mesodinium rubrum*. BioSystems 21:141–149

Löffelhardt W, Bohnert HJ, Bryant DA (1997) The complete sequence of the *Cyanophora paradoxa* cyanelle genome. In: Bhattacharya D (ed) Origins of algae and their plastids. Springer, Berlin Heidelberg New York, pp 149–162

Lopez E (1979) Algal chloroplasts in the protoplasm of three species of benthic foraminifera: taxonomic affinity, viability and persistence. Mar Biol 53:201–211

Lucas IAN (1991) Symbiosis of the tropical Dinophysiales (Dinophyceae). Ophelia 33:213–224

Ludwig M, Gibbs SP (1985) DNA is present in the nucleopmorph of *Cryptomonas*. Further evidence that the chloroplast evolved from a eukaryotic endosymbiont. Protoplasma 127:9–20

Ludwig M, Gibbs SP (1989) Evidence that the nucleomorphs of *Chlorarachnion reptans* (Chlorarachniophyceae) are vestigial nuclei: morphology, division and DNA-DAPI fluorescence. J Phycol 25:385–394

Maier UG, Hofmann CJB, Sitte P (1996) Die Evolution von Zellen. Naturwissenschaften 83:103–112

Mandelli EF (1968) Carotenoid pigments of the dinoflagellate *Glenodinium foliaceum*. J Phycol 4:347–348

Mann DG (1999) The species concept in diatoms. Phycologia 38:437–495

Maranda L (2001) Infection of *Prorocentrum minimum* (Dinophyceae) by the parasite *Amoebophrya* sp. (Dinoflagellea). J Phycol 37:245–248

Margulis L (1981) Symbiosis in cell evolution: life and its environment on the early earth. Freeman, San Francisco

McFadden GI (2001) Primary and secondary endosymbiosis and the origin of plastids. J Phycol 37:951–959

McFadden GI, Walter RF, Reith M, Munholland J, Lang-Unnasch N (1997) Plastids in apicomplexan parasites (1997) In: Bhattacharya D (ed) Origins of algae and their plastids. Springer, Berlin Heidelberg New York, pp 263–287

McKerracher L, Gibbs SP (1982) Cell and nucleomorph division in the alga *Cryptomonas*. Can J Bot 60:2440–2452

Medlin LK, Kooistra WHCF, Potter D, Saunders GW, Andersen RA (1997) Phylyogenetic relationships of the 'golden algae` (haptophytes, heterokont chromophytes) and their plastids. In: Bhattacharya D (ed) Origin of algae and their plastids. Springer, Berlin Heidelberg New York, pp 187–219

Meier R, Lefort-Tran M, Pouphile M, Reisser W, Wiessner W (1984) Comparative freeze-fracture study of perialgal and digestive vacuoles in *Paramecium bursaria*. J Cell Sci 71:121–140

Melkonian M (1996) Phylogeny of photosynthetic protists and their plastids. Verh Dtsch Zool Ges 89.2:71–96

Mereschkowsky C (1905) Über Natur und Ursprung der Chromatophoren im Pflanzenreiche. Biol Ctrbl 25:593–604

Meyer-Harms B, Pollehne F (1998) Alloxanthin in *Dinophysis norvegica* (Dinophysales, Dinophyceae) from the Baltic Sea. J Phycol 34:280–285

Moestrup O, Sengco M (2001) Ultrastructural studies on *Bigelowiella natans*, gen. et sp. nov., a chlorarachniophyte flagellate. J Phycol 37:624–646

Mollenhauer D, Mollenhauer R, Kluge M (1996) Studies on initiation and development of the partner association in *Geosiphon pyriforme* (Kütz.) v. Wettstein, a unique endocytobiotic system of a fungus (Glomales) and the cyanobacterium *Nostoc punctiforme* (Kütz.) Hariot. Protoplasma 193:3–9

Morris RL, Fuller CB, Rizzo PJ (1993) Nuclear basic proteins from the binucleate dinoflagellate *Peridinium foliaceum* (Pyrrhophyta). J Phycol 29:342–347

Morse D, Salois P, Markovic P, Hastings JW (1995) A nuclear-encoded form II RuBisCo in dinoflagellates. Science 268:1622–1624

Nakamura Y, Suzuki S, Hiromi J (1995) Population dynamics of heterotrophic dinoflagellates during a *Gymnodinium mikimotoiu* red tide in the Seto Inland Sea. Mar Ecol Progr Ser 125:269–277

Naustvoll L-J (2000a) Prey size spectra and food preference in thecate heterotrophic dinoflagellates. Phycologia 39:187–198

Naustvoll L-J (2000b) Prey size spectra in naked heterotrophic dinoflagellates. Phycologia 39:448–455

Nawata T, Sibaoka T (1983) Experimental induction of feeding behaviour in *Noctiluca miliaris*. Protoplasma 115:34–42

Nawata T, Sibaoka T (1987) Local ion currents controlling the localized cytoplasmic movement associated with feeding initiation of *Noctiluca*. Protoplasma 137:125–133

Niess D, Reisser W, Wiessner W (1982) Photobehaviour of *Paramecium bursaria* infected with different symbiotic and aposymbiotic species of *Chlorella.* Planta 156:475–480
Palmer J (1995) Rubisco rules fall; gene transfer triumphs. BioEssays 17:1005–1008
Pattersson DJ, Dürrschmidt M (1987) Selective retention of chloroplasts by algivorous Heliozoa:Fortuitous chloroplast symbiosis? Eur J Protistol 23:51–55
Patterson DJ, Larsen J (eds) (1991) The biology of free-living heterotrophic flagellates. Clarendon Press Oxford
Pohnert G (2002) Phospholipase A_2 triggers the wound-activated chemical defence in the diatom *Thalassiosira rotula.* Plant Phys 129:103–111
Rehnstam-Holm A-S, Goelke A, Anderson DM (2002) Molecular studies of *Dinophysis* (Dinophyceae) species from Sweden and North America. Phycologia 41:348–357
Reisser W (1993) Green ciliates:principles of symbiosis formation between autotrophic and heterotrophic partners. In: Lewin RA (ed) Origin of plastids. Chapman and Hall, New York, pp 27–43
Reisser W (1994) Enigmatic chlorophycean algae forming symbiotic associations with ciliates. In: Seckbach J (ed) Evolutionary pathways and enigmatic algae:*Cyanidium caldarium* (Rhodophyta) and related cells. Kluver, Dordrecht, pp 87–95
Rowan R, Whitney SM, Fowler A, Yellowiees D (1996) Rubisco in marine symbiotic dinoflagellates: form II enzymes in eukaryotic oxygenic phototrophs encoded by a nuclear multigene family. Plant Cell 8:539–553
Rumpho ME, Summer EJ, Manhart JR (2000) Solar-powered sea slugs. Mollusk/algal chloroplast symbiosis. Plant Physiol 123:29–38
Ruska H (1960) Der Einfluß der Elektronenmikroskopie auf die biologische Forschung. Sitzber Ges Beförd ges Naturwiss Marburg 82:3–38
Sakaguchi M, Hausmann K, Suzaki T (1998) Food capture and adhesion by the heliozoon *Actinophrys sol.* Protoplasma 203:130–137
Schenk HEA (1970) Nachweis einer lysozymempfindlichen Stützmembran der Endocyanellen von *Cyanophora Paradoxa* Korschikoff. Z Naturforsch 25b:656
Schenk HEA (1994) Glaucocystophyta model for symbiogenous evolution of new eukaryotic species. In: Seckbach J (ed) Evolutionary pathways and enigmatic algae: *Cyanidium caldarium* (Rhodophyta) and related cells. Kluver, Dordrecht, pp 19–52
Schimper AFW (1883) Über die Entwicklung der Chlorophyllkörner und Farbkörper (1. Teil). Bot Z 41:105–114
Schnepf E (1961) Quantitative Zusammenhänge zwischen der Sekretion des Fangschleimes und den Golgi-Strukturen bei *Drosophyllum lusitanicum.* Z Naturforsch 16b:605–610
Schnepf E (1964) Zur Feinstruktur von *Geosiphon pyriforme.* Ein Versuch zur Deutung cytoplasmatischer Membranen und Kompartimente. Arch Mikrobiol 49:112–131
Schnepf E (1966) Organellen-Reduplikation und Zellkompartimentierung. In: Sitte P (Herausg) Funktionelle und morphologische Organisation der Zelle. Probleme der biologischen Reduplikation. 3. Wiss Konf Ges Dtsch Naturf u Ärzte. Springer, Berlin Heidelberg New York, pp 372–393
Schnepf E (1969a) Membranfluß und Membrantransformation. Ber Dtsch Bot Ges 82:407–413
Schnepf E (1969b) Leukoplasten bei *Nitzschia alba.* Österr Bot Z 116:65–69
Schnepf E (1972) Strukturveränderungen am Plasmalemma *Aphelidium*-infizierter *Scenedesmus*-Zellen. Protoplasma 75:155–165
Schnepf E (1984) The cytological viewpoint of functional compartmentation. In: Wiessner W, Robinson D, Starr R (eds) Compartments in algal cells and their interaction. Springer, Berlin Heidelberg New York, pp 1–10
Schnepf E (1993) From prey via endosymbiosis to plastid: comparative studies in dinoflagellates. In: Lewin RA (ed) Origin of plastids. Chapman and Hall, New York, pp 53–76
Schnepf E (1994a) Light and electron microscopical observations in *Rhynchopus coscinodiscivorus* sp. nov., a colorless, phagotrophic euglenozoon with concealed flagella. Arch Protistenkd 144:63–74

Schnepf E (1994b) A *Phagomyxa*-like endoparasite of the centric marine diatom *Bellerochea malleus*: a phagotrophic plasmodiophoromycete. Bot Acta 107:374–382

Schnepf E (1999) Beobachtungen an Protisten, die sich von Algen ernähren. V. Teil Amöboide und plasmodiale Protisten. Mikrokosmos 88:305–312

Schnepf E, Brown RM Jr (1971) On relationships between endosymbiosis and the origin of plastids and mitochondria. In: Reinert J, Ursprung H (ed) Resuslts and problems in cell differentiation. Vol 2. Origin and continuity of cell organelles. Springer, Berlin Heidelberg New York, pp 299–320

Schnepf E, Deichgräber G (1984) "Myzocytosis", a kind of endocytosis with implications to compartmentation in endosymbiosis. Observations in *Paulsenella* (Dinophyta). Naturwissenschaften 71:218–219

Schnepf E, Drebes G (1977) Über die Entwicklung des marinen parasitischen Phycomyceten *Lagenisma coscinodiscici* (Lagenidiales). Helgol Meeresunters 29:291–301

Schnepf E, Drebes G (1986) Chemotaxis and appetence of *Paulsenella* sp. (Dinophyta), an ectoparasite of the marine diatom *Streptotheca thamesis*. Planta 167:337–343

Schnepf E, Elbrächter M (1988) Cryptophycean-like double membrane-bound chloroplast in the dinoflagellate, *Dinophysis* Ehrenb.: evolutionary, phylogenetic and toxicological implications. Bot Acta 101:196–203

Schnepf E, Elbrächter M (1992) Nutritional strategies in dinoflagellates. A review with emphasis on cell biological aspects. Eur J Protistol 28:3–24

Schnepf E, Elbrächter M (1999) Dinophyte chloroplasts and phylogeny – a review. Grana 38:81–97

Schnepf E, Gold S (2000) The optical brightener Calcofluor White disorders thecal architecture and inhibits cell division of *Cryothecomonas longipes* Schnepf et Kühn, a marine nanoflagellate incertae sedis. Protoplasma 210:133–137

Schnepf E, Kühn SF (2000) Food uptake and fine structure of *Cryothecomonas longipes* sp. nov., a marine nanoflagellate incertae sedis feeding phagotrophically on large diatoms. Helgol Mar Res 54:18–32

Schnepf E, Melkonian M (1990) Bacteriophage-like particles in endocytic bacteria of *Cryptomonas* (Cryptophyceae). Phycologia 29:338–343

Schnepf E., Schweikert M (1996/97) *Pirsonia*, phagotrophic nanoflagellates incertae sedis, feeding on marine diatoms: attachment, fine structure and taxonomy. Arch Protistenkd 147:361–371

Schnepf E, Koch W, Deichgräber G (1966) Zur Cytologie und taxonomischen Einordnung von *Glaucocystis*. Arch Mikrobiol 55:149–174

Schnepf E, Hegewald E, Soeder C-J (1971a) Elektronenmikroskopische Beobachtungen an Parasiten aus *Scenedesmus*-Massenkulturen. 2. Über Entwicklung und Parasit-Wirt-Kontakt von *Aphelidium* und virusartige Partikel im Cytoplasma infizierter *Scenedesmus*-Zellen. Arch Mikrobiol 75:209–229

Schnepf E, Deichgräber G, Hegewald E, Soeder C-J (1971b) Elektronenmikroskopische Beobachtungen an Parasiten aus *Scenedesmus*-Massenkulturen. 3. *Chytridium* sp. Arch Mikrobiol 75:230–245

Schnepf E, Hegewald E, Soeder CJ (1974) Elektronenmikroskopische Beobachtungen an Parasiten aus *Scenedesmus*-Massenkulturen. 4. Bakterien. Arch Microbiol 98:133–145

Schnepf E, Deichgräber G, Drebes G (1978a) Development and ultrastructure of the marine, parasitic oomycete, *Lagenisma coscinodisci* Drebes (Lagenidiales): the infection. Arch Microbiol 116:133–139

Schnepf E., Deichgräber G, Drebes G (1978b) Development and ultrastructure of the marine, parasitic oomycete, *Lagenisma coscinodisci* Drebes (Lagenidiales): Thallus, zoosporangium, mitosis, and meiosis. Arch Microbiol 116:141–150

Schnepf E, Deichgräber G, Drebes G (1978c) Development and ultrastructure of the marine, parasitic oomycete, *Lagenisma coscinodisci* (Lagenidiales):encystment of primary zoospores. Can J Bot 56:1309–1314

Schnepf E, Deichgräber G, Drebes G (1978d) Development and ultrastructure of the marine, parasitic oomycete, *Lagenisma coscinodisci* (Lagenidiales): sexual reproduction. Can J Bot 56:1315–1325

Schnepf E, Deichgräber G, Drebes G (1978e) Development and ultrastructure of the marine, parasitic oomycete, *Lagenisma coscinodisci* Drebes (Lagenidiales): formation of the primary zoospores and their release. Protoplasma 94:263–280

Schnepf E, Deichgräber G, Drebes G (1985) Food uptake and the fine structure of the dinophyte *Paulsenella* sp., an ectoparasite of marine diatoms. Protoplasma 124:188–204

Schnepf E, Meier R, Drebes G (1988) Stability and deformation of diatom chloroplasts during food uptake of the parasitic dinoflagellate, *Paulsenella* (Dinophyta). Phycologia 27:283–290

Schnepf E, Winter S, Mollenhauer D (1989) *Gymnodinium aeruginosum* (Dinophyta):a blue-green dinoflagellate with a vestigial, anucleate, cryptophycean endosymbiont. Pl Syst Evol 164:75–91

Schnepf E, Drebes G, Elbrächter M (1990) *Pirsonia guinardiae*, gen. et spec. nov.: a parasitic flagellate on the marine diatom *Guinardia flaccida* with an unusual mode of food uptake. Helgol Meeresuntersuch 44:275–293

Schnepf E, Kühn SF, Bulman S (2000) *Phagomyxa bellerocheae* sp. nov. and *Phagomyxa odontella* sp nov., Plasmodiophoromycetes feeding on marine diatoms. Helgol Mar Res 54:237–242

Schnepf E, Schlegel I, Hepperle D (2002) *Petalomonas sphagnophila* (Euglenophyta) and its endocytobiotic cyanobacteria:a unique form of symbiosis. Phycologia 41:153–157

Schüßler A, Kluge M (2001) *Geosiphon pyriforme*, an endocytosymbiosis between fungus and Cyanobacteria, and its meaning as a model system for arbuscular mycorrhizal research. In: Hock B (ed) The Mycota IX. Fungal associations. Springer, Berlin Heidelberg New York, pp 151–161

Schüßler A, Schnepf E (1992) Photosynthesis dependent acidification of perialgal vacuoles in the *Paramecium bursaria/Chlorella* symbiosis. Visualization by monensin. Protoplasma 166:218–222

Schüßler A, Mollenhauer D, Schnepf E, Kluge M (1994) *Geosiphon pyyriforme*, an endosymbiotic association of fungus and cyanobacterium; the spore structure resembles that of arbuscular mycorrhizal /AM) fungi. Bot Acta 107:36–45

Schüßler A, Bonfante P, Schnepf E, Mollenhauer D, Kluge M (1996) Characterization of the *Geosiphon pyriforme* symbiosome by affinity techniques: confocal laser scanning microscopy (CLSM) and electron microscopy. Protoplasma 190:53–67

Schüssler A, Schwarzott D, Walker C (2001) A new fungal phylum, the Glomeromycota:phylogeny and evolution. Mycol Res 105:1413–1421

Schweikert M (1997) Lichtmikroskopische und elektronenmikroskopische Untersuchungen an Parasiten von marinen Planktondiatomeen. Diss Fak Biol, Universität Heidelberg

Schweikert M, Schnepf E (1996) *Pseudaphelidium drebesii*, gen. et spec. nov. (incerta sedis), a parasite of the marine centric diatom *Thalassiosita punctigera*. Arch Protistenkd 147:11–17

Schweikert M, Schnepf E (1997a) Electron microscopical observations on *Pseudaphelidium drebesii* Schweikert and Schnepf, a parasite of the centric diatom *Thalassiosira punctigera*. Protoplasma 199:113–123

Schweikert M, Schnepf E (1997b) Light and electron microscopical observations on *Pirsonia punctigerae* spec. nov., a nanoflagellate feeding on the marine centric diatom *Thalassiosira punctigerae*. Eur J Protistol 33:168–177

Scott OT, Castenholz RW, Bonnett HT (1984) Evidence for a peptidoplycan envelope in the cyanelles of *Glaucocystis nostochinearum* Itsigsohn. Arch Microbiol 139:130–138

Sherr EB, Sherr BF, McDaniel J (1991) Clearance rates of 6 nm fluorescently labelled algae (FLA) by estuarine protozoa. Potential grazing impact of flagellates and ciliates. Mar Ecol Progr Ser 69:81–92

Shin W, Brosman S, Triemer RE (2002) Are cytoplasmic pockets (MTR/pocket) present in all photosynthetic euglenoid genera? J Phycol 38:790–799
Sibbald MJ, Sibbald PR, Albright LJ (1988) How advantageous is a sensory prey detection mechanisms to predatory microflagellates? J Plankton Res 10:455–464
Sitte P (1988) Facts and concepts in cell compartmentation. Prog Bot 59:3–45
Skovgaard A (1996) Engulfment of *Ceratium* spp. (Dinophyceae) by the thecate photosynthetic dinoflagellate *Fragilidium subglobosum*. Phycologia 35:490–499
Skovgaard A (1998) Role of chloroplast retention in a marine dinoflagellate. Aquat Microb Ecol 15:293–301
Smetacek V (1999) Diatoms and the oceanic carbon cycle. Protist 150:25–32
Sommer U (1994) Planktologie. Springer, Berlin Heidelberg New York
Sparrow FK (1960) Aquatic phycomycetes, 2nd edn. Univ Michigan Press, Ann Arbor
Spero HJ (1985) Chemosensory capabilities in the phagotrophic dinoflagellate *Gymnodinium fungiforme*. J Phycol 21:181–184
Spero HJ, Morée MD (1981) Phagotrophic feeding and its importance to the life cycle of the holozoic dinoflagellate, *Gymnodinium fungiforme*. J Phycol 17:43–51
Sournia A (1982) Form and function in marine phytoplankton. Biol Rev 57:347–394
Stoecker DK (1993) Chloroplast-retention in ciliated protozoa. In: Lewin RA (ed) Origin of plastids. Chapman and Hill, New York, pp 9–26
Stoecker DK (1998) Conceptual models of mixotrophy in planktonic protists and some ecological and evolutionary implications. Eur J Protistol 34:281–290
Stoecker DK, Michaels AE, Davis LH (1987) Large proportion of marine planktonic ciliates found to contain functional chloroplasts. Nature 326:790–792
Stoecker DK, Silver MW, Michaels AE, Davis LH (1988) Obligate mixotrophy in *Laboea strobila*, a ciliate which retains chloroplasts. Mar Biol 99:415–423
Sweeney BM (1976) *Pedinomonas noctilucae* (Prasinophyceae), the flagellate symbiotic in *Noctiluca* in Southeast Asia. J Phycol 12:460–464
Sweeney BM (1981) Freeze-fractures chloroplast membranes of *Gonyaulax polyedra* (Pyrrhophyta). J Phycol 17:95–101
Swift H, Palenik B (1993) Prochlorophyte evolution and the origin of chloroplasts: morphological and molecular evidence. In: Lewin RAS (ed) Origin of plastids. Chapman and Hill, New York, pp 123–139
Tangen K, Björnland T (1981) Observations on pigments and morphology of *Gyrodinium aureolum* Hulburt, a marine dinoflagellate containing 19'-hexanoyloxyfucoxanthin as the main carotenoid. J Plankton Res 3:389–401
Tillmann U, Hesse, KJ, Tillmann A (1999) Large-scale parasitic infection of diatoms in the Northfrisian Wadden Sea. J Sea Res 42:255–261
Tippit DH, Pickett-Heaps JD (1976) Apparent amitosis in the binucleate dinoflagellate *Peridinium balticum*. J Cell Sci 21:273–289
Tomas RN, Cox ER (1973) Observations on the symbiosis of *Peridinium balticum* and its intracellular alga. I Ultrastructure, J Phycol 9:304–323
Triemer RE (1997) Feeding in *Peranema trichophorum* revisited (Euglenophyta). J Phycol 33:649–654
Triemer RE, Farmer MA (1991) An ultrastructural comparison of the mitotic apparatus, feeding apparatus, flagellar apparatus and cytoskeleton in euglenoids and kinetoplastids. Protoplasma 164:91–104
Ucko M, Elbrächter M, Schnepf E (1997) A *Crypthecoodinium cohnii*-like dinoflagellate feeding myzocytotically on the unicellular red alga *Porphyridium* sp. Eur J Phycol 32:133–140
Ucko M, Prakash Shrestha R, Mesika P, Bar-Zvi D, Arad (Malis) S (1999) Glycoprotein moiety in the cell wall of the red microalga *Porphyridium* sp. (Rhodophyta) as the recognition site for the *Crypthecodinium cohnii*-like dinoflagellate. J Phycol 35:1276–1281

Van de Peer Y, De Wachter R (1997) Evolutionary relationships among the eukaryotic crown taxa taking into account site-to-site rate variation in the 18S rRNA. J Mol Evol 54:619–630
Van De Peer Y, Rensing S, Maier U-G, De Wachter R (1996) Substitution rate calibration of small subunit rRNA identifies chlorarachniophyte endosymbionts as remnants of green algae. Proc Natl Acad Sci USA 93:7732–7736
Verity PG, Smetacek V (1996) Organism life cycles, predation, and the structure of marine pelagic ecosystems. Mar Ecol Progr Ser 130:277–293
Vesk M, Dibbayavan TP, Vesk PA (1996) Immunogold localization of phycoerythrin in chloroplasts of *Dinophysis acuminata* and *D. fortii* (Dinophysales, Dinophyta). Phycologia 35:234–238
Villareal TA (1994) Widespread occurrence of the *Hemiaulus*-cyanobacterial symbiosis in the southwest North Atlantic Ocean. Bull Mar Sci 54:1–7
Watanabe MM, Takeda Y, Sasa T, Inouye I, Suda S, Sawaguchi T, Chihara M (1987) A green dinoflagellate with chlorophylls *a* and *b*: morphology, fine structure of the chloroplast and chlorophyll composition. J Phycol 23:382–389
Watanabe MM, Suda S, Inouye I, Sawaguchi T, Chihara M (1990) *Lepidodinium viride* gen. et sp. nov. (Gymnodiniales, Dinophyta), a green dinoflagellate with a chlorophyll *a*- and b-containing endosymbiont. J Phycol 26:741–751
Werner D (1992) Symbiosis of plants and microbes. Chapman and Hall, London
Whatley JM (1993) Membranes and plastid origins. In: Lewin RA (ed) Origin of plastids. Chapman and Hall, New York, pp 77–106
Whatley JM, John P, Whatley FR (1979) From extracellular to intracellular. The establishment of mitochondria and chloroplasts. Proc R Soc London B 204:165–187
Wilcox LW, Wedemayer GJ (1985) Dinoflagellate with blue-green chloroplasts derived from an endosymbiotic eukaryote. Science 227:192–194
Wilcox LW, Wedemayer GJ (1991) Phagotrophy in the freshwater, photosynthetic dinoflagellate *Amphidinium cryophilum*. J Phycol 27:600–609
Wilhelm C, Müller AM, Borstelmann B, Schnetter R (1991) The pigment composition of *Chlorarachnion reptans* and of *Cryptochlora perforans* (Chlorarachniophyta). Crypt Bot 2:201–204
Withers NW, Cox ER, Tomas R, Haxo FT (1977) Pigments of the dinoflagellate *Peridinium balticum* and its photosynthetic endosymbiont. J Phycol 13:354–358
Wolfe GV (2000) The chemical defense ecology of marine unicellular plankton: constraints, mechanisms, and impacts. Biol Bull 198:225–244
Woloszynska J (1925) Notatki algologiczne (Algologische Notizen) 1. *Amphidinium vigrense* n. sp. 2. *Peridinium* sp. Sprow Stac Hydrobiol Wigrach 1:1–9
Zhang Z, Green BR, Cavalier-Smith T (1999) Single gene circles in dinoflagellate chloroplast genome. Nature 400:155–159

Prof. Dr. E. Schnepf
Dürerweg 11
69168 Wiesloch, Germany

Genetics

Recombination: Implications of Single Nucleotide Polymorphisms for Plant Breeding

Gisela Neuhaus and Renate Horn

1 Introduction

Single-nucleotide polymorphisms (SNPs) are the most abundant type of DNA sequence variation in humans, animals and plants (Landegren et al. 1998; Schafer and Hawkins 1998; Wang et al. 1998; Gilles et al. 1999; See et al. 2002). SNPs, which are in the narrow sense single-base substitutions (transitions or transversions) in a DNA sequence, as well as small insertions and deletions (InDels) represent the most promising markers for plant breeding, so far. Although SNPs are biallelic, large numbers can be combined to form haplotypes making SNPs highly informative in comparison to other types of molecular markers such as RFLPs or SSRs (Meyer et al. 2000; Rafalski 2002a). Recent advances in SNP technologies, especially in the field of high-throughput applications (Gupta et al. 2001, Jander et al. 2002), allow SNP scoring and screening in plants on a large scale or even genome-wide (Buckler and Thornsberry 2002; Rafalski 2002a,b; Cho et al. 1999).

For plant breeding, SNPs are interesting for a wide range of applications, like e.g. estimation of genetic diversity or determination of allele distributions and frequencies within a large number of genotypes. Association mapping with SNPs based on candidate genes or genome-wide approaches also offers the possibility to resolve complex traits and the involved genes. Here, we will give an overview of SNP technologies and applications and will highlight potential future uses of SNPs in plant breeding.

2 Single Nucleotide Polymorphisms in Higher Plants

2.1 Identification of Single Nucleotide Polymorphisms

Several strategies have been employed to discover SNPs. Sequence data from direct sequencing or after cloning amplicons or fragments represent the major source for detecting SNPs. In small, trait-oriented approaches,

Progress in Botany, Vol. 65

sequences of candidate genes or markers linked to traits of interest have been successfully used to identify SNPs.

AFLP markers all linked to one of two loci, *rhg1* and *Rhg4*, which confer resistance to soybean cyst nematode, have proven to be a good source in soybean to develop SNP-based markers (Meksem et al. 2001a). In maize, DNA sequence data were used from coding regions and flanking sequences of four genes, *indeterminate1* (*id1*), *teosinte branched1*(*tb1*), *dwarf8* (*d8*) and *dwarf3*(*d3*), to identify SNPs (Remington et al. 2001). To determine the genetic diversity based on SNPs, six key genes involved in starch production in maize: *amylose extender1* (*ae1*), *brittle2* (*bt2*), *shrunken1* (*sh1*), *shrunken2* (*sh2*), *sugary1* (*su1*), and *waxy1* (*wx1*) were analysed for both coding and non-coding regions in 30 maize inbred lines (Whitt et al. 2002). Also, in maize, resistance gene analogues (RGA) mapped in relation to sugarcane mosaic virus (SCMV) resistance genes (*Scmv1* and *Scmv2*) were used to discover SNPs, which allowed mapping of the RGA (Quint et al. 2002). In barley, cytochrome P450 genes extracted from the International Triticeae EST Cooperative database were used to identify SNPs in barley varieties (Bundock et al. 2003).

For genome-wide discovery of SNPs, extensive databases are available that allow electronic SNP (eSNP) discovery in shotgun genomic libraries, and in expressed sequence tag (EST) libraries.

Tools for single nucleotide polymorphism detection, like e.g. Polybayes (http://genome.wustl.edu/groups/informatics/software/polybayes) or SNPFINDER (http://www.ktl.fi/bioinfo/maps.html) are available for public research in the Internet. These programmes allow one to cluster overlapping ESTs, filter possible pseudogenes or paralogues, and check trace files, which otherwise must be done by hand. Brett et al. (2000) recommend following three rules to locate SNPs with high accuracy: ESTs should be aligned using BLASTN with an 'expected value'of 1e-30. Then, only ESTs with over 95% identity over 100 bp should be considered to avoid processing pseudogenes and close homologues. In addition, the at least two different ESTs rule' should be employed, meaning that in a case where several ESTs match a sequence, at least two of them must have the same nucleotide change that is different from the majority of ESTs matching the sequence at that point.

Current genome sequencing projects in rice, maize, wheat and other crop species are providing increasing amounts of sequencing data, which will accelerate SNP discovery by comparative alignment using databases.

2.2 Distribution of SNPs in Plant Genomes

In plant species, sequence variation is widespread (Table 1). In *Arabidopsis*, 37,344 SNPs and 18,579 small InDels (http://www.arabidopsis.org/cereon; Jander et al. 2002) have been identified. In rice, investigations of a 2.3-Mb homologous region between the two subspecies *japonica* and *indica* identified 9,056 SNPs in each and 63 InDels in *indica* and 138 in *japonica* (Han and Xue 2003). These data demonstrate that SNPs can provide a much larger number of markers for plant breeding than any other known marker

Table 1. Distribution of single nucleotide polymorphisms (SNPs) in higher plants. Frequencies of SNPs per base pair (bp) are given based on sequence analysis. If data were available, frequencies are separately presented for coding and non-coding regions, as SNPs seem to be more abundant in non-coding regions of most plants

Plant species	Population type	Frequency	Reference
Arabidopsis thaliana	Ecotypes		Jander et al. (2002)
	Coding regions	1 SNP/3,100 bp	
	Non-coding regions	1 SNP/2,200 bp	
Beta vulgaris	Breeding lines	1 SNP/130 bp	Schneider et al. (2001)
Glycine max	Various genotypes		Zhu et al. (2001)
	Coding regions	1 SNP/610 bp	
	Non-coding regions	1 SNP/206 bp	
Hordeum vulgare	Barley varieties	1 SNP/189 bp	Kanazin et al. (2002)
	Barley lines (cytochrome P450 genes)	1 SNP/131 bp	Bundock et al. (2003)
	Barley varieties	1 SNP/58 bp	Neuhaus et al. (2003)
Oryza sativa	*Oryza sativa* ssp. *japonica* (three cultivars) *Oryza sativa* ssp. *indica* (two cultivars) *Oryza rufipogon*	1 SNP/89 bp	Nasu et al. (2002)
	Oryza sativa accessions	1 SNP/100 bp	Garris et al. (2003)
	Oryza sativa ssp. *japonica*		Han and Xue (2003)
	Coding regions	1 SNP/304 bp	
	Non-coding regions	1 SNP/272 bp	
	Oryza sativa ssp. *indica*		
	Coding regions	1 SNP/379 bp	
	Non-coding regions	1 SNP/315 bp	
Saccharum officinarum	Coding region	1 SNP/122 bp	Grivet et al. (2003)
Zea mays	Maize exotic land-races and inbred lines	1 SNP/104 bp	Tenaillon et al. (2001)
	Maize inbred lines:		Ching et al. (2002)
	Coding regions	1 SNP/124 bp	
	Non-coding region	1 SNP 31 bp	
	Maize inbred lines:		Quint et al. (2002)
	Resistance gene analogs	1 SNP/ 33 bp	
	Maize inbred lines:		Bhattramakki et al. (2002)
	3'Untranslated region	1 SNP/ 48 bp	
	Coding regions	1 SNP/ 130 bp	

system, so far. In theory, several SNPs should be detectable within each gene of interest and should allow allele differentiation. The lack of polymorphisms linked to valuable traits had been a limiting factor in marker-assisted breeding, which can be overcome by using SNPs as markers.

3 Overview of Different Methods for SNP Detection in Plant Breeding

Especially, in marker-assisted plant breeding, where thousands of genotypes have to be screened every year, novel and suitable assays must be fast, inexpensive, multiplexable and amenable to automatisation to allow high throughput (Bormans et al. 2002). Recent advances in SNP technologies provide screening methods that fulfil these criteria. Table 2 gives an overview of SNP technologies and their applications in plants.

Table 2. Mutation analysis in crop cultivars by different SNP detection methods. (BESS-T/G, base excision sequence scanning; CAPS, cleaved-amplified polymorphic sequence; DHPLC, denaturing high-performance liquid chromatography; EST, expressed sequence tag, MALDI-ToF MS, matrix assisted laser desorption/ionisation time-of-flight mass spectrometry: PCR, polymerase chain reaction; SnuPE, single nucleotide primer extension; SSCP, single-strand conformation polymorphism; SSR. simple sequence repeats)

Crop cultivar/ plants	SNP detection method	Application	Reference
Alopecurus myosuroides Huds.	Sequencing, allele-specific PCR	Genotyping of resistance to graminicidal herbicides	Délye et al. (2002)
Arabidopsis ecotypes	Heteroduplex analysis	Generation of co-dominant PCR-based markers	Hauser et al. (1998)
	Oligonucleotide microarray	Positional Cloning	Cho et al. (1999); Spiegelmann et al. (2000)
Arachis hypogae L., wild species	Sequencing	Phylogenetic relationship	Jung et al. (2003)
Beta vulgaris	SSCP	Segregation analysis of functional gene homologues	Schneider et al. (1999, 2001)
Eucalyptus urophylla, E. grandis	SSCP	Characterisation of QTL for wood quality	Gion et al. (2000)
Glycine max	TaqMan^{TM}	SNP screening of inbred lines	Meksem et al. (2001b)

Table 2. *Continued*

Crop cultivar/ plants	SNP detection method	Application	Reference
Hordeum vulgare L.	DHPLC	Genome-wide SNP analysis	Kota et al. (2001)
	BESS-T/G	Resistance against barley yellow mosaic virus	Neuhaus et al. (2003)
	SNuPE/MALDI-ToF MS	SNP typing for *mlo* powdery mildew resistance	Paris et al. (2001)
	Sequencing	Genome-wide haplotypes	Kanazin et al. (2002)
Oryza sativa	Sequencing	Differentiation of amylose content in cultivars	Ayres et al. (1997)
	READIT^{TM}	SNP genotyping of granule bound starch synthase	Bormans et al. (2002)
Saccharum officinarum, *S. spontaneum*	Sequencing, Alignment of EST sequences	SNP screening in the *Adh* gene family	Grivet et al. (2003)
Solanum tubersum	Pyrosequencing	SNP genotyping in polyploid species	Rickert et al. (2002)
Zea mays L.	Sequencing	Genetic diversity	Tenaillon et al. (2001)
	EST-Mapping/ Pyrosequencing^{TM}	SNP screening of three different loci	Ching and Rafalski (2002)
	Hybridization assay	Genotyping of different inbred lines	Mogg et al. (2002)
	CAPS	Resistance gene analogs	Quint et al. (2002)
	SNuPE assay	Genotyping SNPs in the flanking regions of SSRs	Batley et al. (2003)

3.1 SNP Detection Methods for Pre-Screening

Technologies to identify unknown SNPs are mostly gel-based methods characterised by a moderate throughput but relatively low start-up costs (Jander et al. 2002). Pre-screening of amplicons for SNPs prior to verification of a SNP by sequence analyses proved to be most economic for development of SNP assays.

Single-Strand Conformation Polymorphism (SSCP) analysis and Temperature Gradient Gel Electrophoresis (TGGE) represent simple and inexpensive techniques (White et al. 1992) based on the principle that an altered DNA conformation, caused by a single point mutation, affects the mobility of these fragments during electrophoretic separation. Modified sequencing techniques, such as Base Excision Sequence Scanning (BESS-T/G sequencing), are also available tools for SNP pre-screening (Hawkins and Hoffman 1997; Karp et al. 1997; Yu et al. 2001).

These gel-based, technically simple and well-established assays are the most suitable tools for pre-screening of a number of germplasms or genotypes to get an indication of SNP occurrences or frequencies within unknown sequences. However, only short fragments ranging from 100–400 bp can be analysed for SNP frequencies by these technologies and the capability for multiplex assays is limited.

3.2 Present Technologies for Allelic Selection in Plant Breeding

Allele differentiation including discrimination of homozygotes and heterozygotes is essential for marker-assisted selection in plant breeding. Various methods based on SNPs have been developed, which allow distinction of different alleles.

Markers identifying alleles based on SNPs that alter restriction sites are called Cleaved-Amplified Polymorphic Sequence (CAPS) markers (Konieczny and Ausubel 1993). After PCR amplification, PCR products are cleaved with the appropriate restriction enzyme, and separated on agarose gels. If no useful restriction site is affected, allele-specific primers containing one of the bases of the SNP can be designed to genotype respective polymorphisms. However, allele-specific PCR requires sequence information regarding the SNP and a detailed optimisation of the amplification conditions is necessary to exclude false-positives. In a modified allele-specific amplification procedure (Single Nucleotide Amplified Polymorphism – SNAP), one additional mismatch, apart from the one separating the alleles, is introduced, which considerably enhances the selectivity of the primers and thereby maximizes the efficiency to differentiate alleles of the SNP (Drenkard et al. 2000).

A newly developed assay for SNP genotyping is the READIT system. In the presence of a high concentration of pyrophosphate, the DNA polymerase catalyses depolymerisation of the template DNA when there is a perfect match at the 3'end between the probe and the target sequence, while the mismatched probe is not depolymerised. Released dNTPs are converted enzymatically to ATP, which activates a luciferin/luciferase detection system. Using a probe specific for each allele, homozygotes and heterozygotes can be distinguished. The TaqMan assay also uses allele-specific oligonucleotides, but conjugated with a pair of fluorophores. In an intact unhybridised TaqMan probe, the fluorescence is quenched by fluorescence resonance energy transfer (FRET), where upon cleavage of the hybridised TaqMan probe by 5'exonuclease activity of Taq polymerase the reporter dye is separated from the quencher. The TaqMan-assay is designed to detect SNPs in a high-throughput manner. By using more than one reporter dye, it is also possible to detect different alleles of a SNP in a single reaction. Alternative approaches for screening single sequence variations are assays based on single nucleotide primer extension (SNuPE), including mini-sequencing approaches that are well established techniques for screening polymorphisms (Pastinen et al. 1997; Syvänen 1999; Shapero et al. 2001).

A large number of technologies based on hybridisation and primer extension assays has been developed to genotype SNPs (Jander et al. 2002). These types of assays characterised by a high potential for automatisation and increased throughput, are rapid and sensitive systems and therefore promising technologies in future plant breeding, but often require special equipment, which might only be affordable for big breeding companies.

3.3 Advanced Technologies for SNP-Screening

SNP analysis becomes most powerful if used in genome-wide approaches and on a high-throughput level. For large-scale screening of SNPs, methods based on primer extension as described above can be combined with highly automated techniques such as Matrix Assisted Laser Desorption/Ionisation Time-of-Flight Mass Spectrometry (MALDI-ToF MS) that analyses exact physical characteristics, such as molecular mass of a nucleotide (Kim et al. 2002). High-throughput analysis can also be performed by using DHPLC, which allows the detection of SNPs through different retention time of heteroduplex and homoduplex DNA in reverse-phase HPLC under partially denaturing conditions (Spiegelman et al. 2000). Pyrosequencing (Ronaghi et al. 1998; Ahmadian et al. 2000; Alderborn et al. 2000), which utilizes an enzyme cascade system to analyse large numbers of DNA sequences and to determine allele frequencies, can be automated as well. A luminometric detection system is used to measure the pyrophosphate that is released upon nucleotide incorporation. Because 20 or more nucleotides are determined by this method, it is possible to detect several closely linked SNPs at once.

Microarrays offer the highest potential for large-scale detection of polymorphisms and for high throughput (Schafer and Hawkins 1998; Aharoni and Vorst 2001). DNA microarrays are currently produced and assayed by two main approaches involving either in situ synthesis of oligonucleotides ('oligonucleotide microarrays') or deposition of pre-amplified DNA fragments ('cDNA microarrays') on solid surfaces (reviewed by Aharoni and Vorst 2001).

A wide field of applications in plant breeding using these novel, non-gel based screening methods is imaginable in the future for determination of genetic diversity in higher plants, especially crops, or gene discovery and expression analysis (Aharoni and Vorst 2001). Especially, the development of high density arrays enabling a fast and precise detection of many markers at the same time (multiplex application) will provide great possibilities for effective large-scale genotyping or genome-wide screening of genotypes.

4 Application of Single Nucleotide Polymorphisms

4.1 Assessment of Genetic Diversity by SNPs

Nucleotide diversity is the average number of nucleotide differences per site between two sequences. A number of factors, e.g. mutation rate, population size, mating type, balancing selection, background selection, population structure, have an impact on nucleotide diversity (Buckler and Thornsberry 2002). In maize (*Zea mays* ssp. *mays*), nucleotide diversity at silent sites averages 1.6% for genes that appear to behave neutrally, while the diversity in maize's wild relative *Z. mays* ssp. *parviglumis* is about 2% (White and Doebley 1999; Gaut et al. 2000). At individual loci, the diversity ranged from 0.2 to 5% (White and Doebley 1999). There has been roughly a drop of 30% in diversity at the average locus from maize's relatives. Compared to other grass species, maize and its wild relatives seem to have a high level of genetic diversity, which is probably the result of high levels of outcrossing (Buckler et al. 2001).

Whitt et al. (2002) analysed six key genes, *amylose extender1* (*ae1*), *brittle2* (*bt2*), *shrunken1* (*sh1*), *shrunken2* (*sh2*), *sugary1* (*su1*), and *waxy1* (*wx1*), known to play a major role in starch production in maize. Although the starch loci showed a wide range in diversity, average diversity in the starch loci was 2.3-fold lower than in 29 random maize loci at silent sites (Whitt et al. 2002), and 4.8-fold lower at non-synonymous sites (Tenaillon et al. 2001). For chromosome I in maize, Tenaillon et al. (2001) related nucleotide diversity to chromosome structure, recombination and various types of selection. Population effects in the production of the breeding material in maize are also discussed.

In general, the domesticated relatives of the grasses (maize, sorghum, rice, oats, pearl millet) have two-thirds of the diversity found in wild relatives (Buckler et al. 2001). However, there has probably been a greater loss in terms of alleles for agronomic use, as nucleotide diversity estimates are relatively insensitive to the loss of rare alleles (Buckler et al. 2001).

4.2 Association Mapping

4.2.1 Association and Population Structure

Association of a SNP with a trait may be the result from at least three possible causes (Risch and Merikangas 1996; Przeworski et al. 2000): (1) the locus carrying the SNP is the cause of the phenotype, (2) the locus is in linkage disequilibrium (LD) with the cause of the phenotype, meaning linked and highly correlated with the phenotype but not representing the gene itself, which would be still perfect for plant breeding purposes, and (3) the population structure produces the association. To avoid any false association due to the population structure, populations used for association mapping should at the same time be characterised by random distributed markers, e.g. SSR markers. In maize, Remington et al. (2001) used SSR markers representing 47 highly polymorphic loci with a mean of 6.85 alleles per locus to define their maize population structure. Garris et al. (2003) employed 22 SSR markers to characterise the population structure of 114 rice accessions. Association approaches can be used to rapidly evaluate candidate genes with high resolution but it is essential to determine and understand the underlying population structure and to correct for it (Pritchard et al. 2000; Nordborg and Innan 2002).

4.2.2 Candidate Gene Approach Versus Genome-Wide Mapping

Linkage Disequilibrium (LD) is the non-random association of alleles. Values for LD index D' range from 0 (equilibrium) to 1 (disequilibrium). The effectiveness of association mapping depends on the extent of LD around the gene of interest. Islands of linkage equilibrium are large segments (up to the order of tens of kb) that are characterized by low recombination, high LD and a small number of haplotypes that are interspersed with regions in which recombination is frequent (Daly et al. 2001; Goldstein 2001; Rafalski 2002a). Factors that influence the linkage disequilibrium are the rate of outcrossing, the degree of selection, recombination rate, chromosomal location and the sampling, which determines the population structure (Garris et al. submitted). Investigating six genes in maize, Remington et al. (2001) detected that intragenic LD generally declined rapidly with distance ($r^2<0.1$ within 1,500 bp) but rates of decline were highly variable among genes. This extension of LD is much less than the 50 kb predicted for humans (Koch et al. 2000) where it is even assumed to extend much further (Moffatt et al. 2000; Reich et al. 2001). Only for *sugary1* (*su1*) evidence was found that LD might persist in maize at anywhere near these distances (Remington et al. 2001). This study of the local and genome-wide LD pattern based on six genes included 102 maize inbred lines representing much of the worldwide genetic diversity used in maize breeding. As there exists a clear relationship between the extent of LD and the resolution of association studies (Rafalski 2002a) these results allow to estimate future prospects of association mapping in maize. If LD declines slowly with increasing distance from the gene of interest, a low density of markers is sufficient to identify the associated markers. However, if LD very rapidly declines around the gene of interest a much higher density of markers is required to identify an associated marker (Rafalski 2002a).

The rapid breakdown of LD in maize will be favourable for association mapping of candidate genes that are located near mapped Quantitative Trait Loci (QTL) but the decay of LD is probably too rapid to allow a genome-wide association testing with SNPs as has been proposed for human populations (Reich et al. 2001; Remington et al. 2001). According to Tenaillon et al. (2001), intragenic LD even declines within 100–200 bp in maize. Whereas Remington et al. (2001) looked at maize inbred lines, Tenaillon et al. (2001) compared US landraces and inbred lines. In *Arabidopsis*, LD is more extensive, decaying within perhaps 250 kb (Nordborg and Innan 2002). This is consistent with the difference in breeding system between these two species: maize undergoes outcrossing, whereas *Arabidopsis* is highly selfing, and selfing is expected to increase LD greatly (Nordborg 2000). In sugarcane, LD extends for several cM (Jannoo et al. 1999).

As *A. thaliana* shows extensive LD, a genome-wide mapping of SNPs is favourable (Rafalski 2002a). Cho et al. (1999) constructed a biallelic genetic map in *Arabidopsis* with a resolution of 3.5 cM and investigated its use in mapping of *Eds16*, a gene involved in the defence response to the fungal pathogen *Erysiphe orontii*. *Eds16* could be placed to a 7-cM interval on chromosome 1. The results demonstrate that a whole-genome mapping using array-based genotyping of SNPs can accelerate the process of positional cloning (Cho et al. 1999).

Based on sequence information present in *Arabidopsis* 1,800 PCR products, averaging 280 bp, were amplified from both the Columbia and Landsberg *erecta* ecotypes. Variation was analysed by investigating the heteroduplex formation of Columbia-Landsberg PCR products by DHPLC. Using this method, 487 loci carrying SNPs were identified and sequenced afterwards to confirm the mutations between the two ecotypes. After screening for optimal melting temperatures and absence of AT-rich regions in the vicinity of the polymorphic base 412 polymorphisms were used to generate an oligonucleotide array. A total of 237 SNP markers were placed on the map in *Arabidopsis thaliana* (Cho et al. 1999).

According to Drenkard et al. (2000) map-based cloning is greatly facilitated in *Arabidopsis* by the possibility to use SNP markers as demonstrated in mapping the mutation *edr5*–1, which causes enhanced disease resistance to *Pseudomonas syringae* pv *maculicola* ES4326 and to *Erysiphe orontii*.

In maize, a two-step strategy of QTL mapping followed by association testing of positional candidate genes shows substantial potential for localising quantitative trait effects to individual genes or even subgenic regions (Thornsberry et al. 2001). In a candidate gene approach, the *dwarf8* gene could be significantly associated with flowering time in maize (Thornsberry et al. 2001). The rapid decay of LD in maize provides an opportunity to map quantitative trait loci with up to 5,000-fold greater resolution than current mapping with F_2 or recombinant inbred lines (Remington et al. 2001).

Association studies in rice showed that linkage disequilibirium in the 70-kb *xa5* region, conferring resistance to *Xanthomonas oryzae* pv. *oryzae*, was extensive but still informative in eliminating three of eight candidate genes for *xa5* (Blair et al. 2003).

The application of association studies in plant breeding has the potential to be extremely efficient for studies of agronomically important traits.

4.2.3 Haplotype-Specific SNP

The fluidity of genomic parameters, such as linkage disequilibrium, questions the applicability of genome-wide studies, which assume that SNPs, randomly distributed over the genome, will be sufficient to detect an association with a phenotype. However, screening of SNPs on a genome-wide scale is interesting because haplotype analyses of multiple SNPs are especially powerful for investigating alleles responsible for genetically complex diseases (Stephens et al. 1998; Tishkoff et al. 2000). A haplotype is the specific combination of nucleotides, one from each polymorphic sites that are present on an individual chromosome (J. C. Stephens et al. 2001). Computer programmes using Clarks algorithm (Clark 1990) can be used to analyse the combination of alleles present at each site of polymorphism in each individual that assign a specific pair of haplotypes to each individual, as well as a score reflecting the confidence in that assignment. SNPHAP (EM algorithm) developed by David Clayton and PHASE (Bayesian statistics) developed by M. Stephens et al. (2001) are also available (Zhang et al. 2002). As more SNP marker data become available in humans, it is becoming preferable to use haplotypes of markers for association analysis of candidate genes rather than individual polymorphisms (Clark et al. 2001). For example, J.C. Stephens et al. (2001) investigated SNPs present within 313 genes in humans and organised these SNPs into 4,304 haplotypes.

Haplotypes can correlate a specific phenotype with a specific gene in a small population sample even when individual SNPs cannot (Drysdale et al. 2000). Thus, attempts to draw associations between phenotypes and genomic variation are more likely to succeed when the SNPs used in such studies have been confirmed to be in linkage disequilibrium by methods such as haplotyping (J.C. Stephens et al. 2001). Current haplotype mapping focuses on haplotypes with ≥ 5% population frequency (Zhang et al. 2002).

Existing genome viewers, such as NCBI Map Viewer and the UCSC Genome Viewer (Kent et al. 2002) as well as static genotype or haplotype viewers such as Visual Genotype or Visual Haloptype (Nickerson et al. 1998; Rieder et al. 1999) are capable of presenting donor genotype or haplotype information but lack more sophisticated features, such as SNP functional classification, haplotype phase probability assessment, display of population haplotype structure and haplotype blocks (Zhang et al. 2002). The software system HapScope represents a tool for analysing the genetic and functional correlation of haplotypes in a population (Zhang et al. 2002). The minimum SNP set selection algorithms define the minimum SNP set, which is the smallest possible subset of SNPs in a haplotype block that represents the diversity of haplotypes within the block: such SNPs have also been referred to as 'haplotype tagging SNPs' or htSNPs' (Johnson et al. 2001).

5 Conclusions

In plants, SNP technology has not yet been exploited in any comparable form to humans or animals. However, the results obtained in plants allow the following conclusions:

1. The high frequency of SNPs in plant genomes makes SNPs a very interesting universal marker systems, also for plants. Using gel-based methods, which have relatively low start-up costs, SNPs can be applied in a moderate throughput.
2. Applications of SNPs become the most powerful tool in genome-wide approaches on high-throughput levels. This has been extremely facilitated by whole genome sequencing of model species like *Arabidopsis* and rice and by the availability of large EST and SNP databases for important crop species.
3. For high-throughput techniques, the initial high equipment costs will be a problem for laboratories that do not need to perform large numbers of genotyping reactions on routine basis (Jander et al. 2002).
4. One big challenge in plant breeding is the handling of quantitative traits. A two-step strategy of QTL mapping, combining marker-assisted QTL localisation with association testing of positional candidate genes, shows substantial promise for addressing quantitative trait effects to individual genes or even subgenic regions (Remington et al. 2001; Thornsberry et al. 2001).
5. However, the power to detect association between an SNP and a quantitative trait ultimately depends on having a sufficient density of SNP markers to ensure that some SNPs will be in LD with the allele that contributes to the phenotypic variation (Tenaillon et al. 2001). Associa-

tion mapping of SNPs from candidate genes will be more promising in plants than whole genome-wide association studies.

6. In plants, genome-wide mapping approaches as proposed for humans are less attractive due to the very high number of SNP markers needed. In plant breeding, haplotype-specific SNP markers for candidate genes represent a very promising marker tool with regard to marker-assisted selection of different traits, e.g. disease resistances.
7. The rapid development of SNPs technologies for high-throughput applications will make SNP the most interesting molecular marker system for plant research and plant breeding in the near future. SNP technology will help to handle complex traits and to resolve the genes involved in these traits.

References

Aharoni A, Vorst O (2001) DNA microarrays for functional plant genomics. Plant Mol Biol 48:99–118

Ahmadian A, Gharizadeh B, Gustafsson AC, Sterky F, Nyren P, Uhlen M, Lundeberg J (2000) Single-nucleotide polymorphism analysis by pyrosequencing. Anal Biochem 280:103–110

Alderborn A, Kristofferson A, Hammerling U (2000) Determination of single-nucleotide polymorphisms by real-time pyrophosphate DNA sequencing. Genome Res 10:1249–1258

Ayres NM, McClung AM, Larkin PD, Bligh HFJ, Jones CA, Park WD (1997) Microsatellites and a single-nucleotide polymorphism differentiate apparent amylose classes in an extended pedigree of US rice germplasm. Theor Appl Genet 94:773–781

Batley J, Mogg R, Edwards D, OSullivan H, Edwards KJ (2003) A high-throughput SNuPE assay for genotyping SNPs in the flanking regions of *Zea mays* sequence tagged simple sequence repeats. Mol Breed 11:111–120

Bhattramakki D, Dolan M, Hnanfey M, Wineland R, Vaske D et al. (2002) Insertion-deletion polymorphisms in the 3 regions of maize genes occur frequently and can be used as highly informative markers. Plant Mol Biol 48:539–547

Blair M, Garris AJ, Ayer AS, Chapman B, Kresovich S et al. (2003) High resolution genetic mapping candidate gene identification at the *xa5* locus for bacterial blight resistance in rice. Theor Appl Genet 107:62–73

Bormans CA, Rhodes RB, Kephart DD, McClung AM, Park WD (2002) Analysis of a single nucleotide polymorphism that controls the cooking quality of rice using a non-gel based assay. Euphytica 128:261–267

Brett D, Lehmann G, Hanke J, Gross S, Reich J, Bork P (2000) EST analysis online: WWW tools for detection of SNPs and alternative splicing. TIG 16:416–418

Brookes AJ (1999) The essence of SNPs. Gene 234:177–186

Buckler IV ES, Thornsberry JM (2002) Plant molecular diversity and applications to genomics. Curr Opin Plant Biol 5:107–111

Buckler IV ES, Thornsberry JM, Kresovich S (2001) Molecular diversity, structure and domestication of grasses. Genet Res 77:213–218

Bundock PC, Christopher JT, Eggler P, Ablett G, Henry RJ, Holton TA (2003) Single nucleotide polymorphisms in cytochrome P450 genes from barley. Theor Appl Genet 106:676–682

Ching A, Rafalski A (2002) Rapid genetic mapping of ESTs using SNP pyrosequencing and indel analysis. Cell Mol Biol Lett 7(2B):803–810

Ching A, Caldwell KS, Jung M, Dolan M, Smith OS, Tingey S, Morgante M, Rafalski AJ (2002) SNP frequency, haplotype structure, and linkage disequilibrium in elite maize inbred lines. BMC Genetics 3:19

Cho RJ, Mindrinos M, Richards DR, Sapolsky RJ, Anderson M, Drenkard E, Dewdney J, Reuber TL, Stammers M, Federspiel N, Theologis A, Yang WH, Hubbell E, Au M, Chung EY, Lashkari D, Lemiuex B, Dean C, Lipshutz RJ, Ausubel FM, Davis RW, Oefner PJ (1999) Genome-wide mapping with biallelic markers in *Arabidopsis thaliana*. Nat Genet 23:203–207

Clark A (1990) Inference of haplotypes from PCR-amplified samples of diploid populations. Mol Biol Evol 7:111–122

Clark VJ, Metheny N, Dean M, Peterson RJ (2001) Statistical estimation and pedigree analysis of CCR2-CCR5 haplotypes. Hum Genet 108:484–493

Daly MJ, Rioux JD, Schaffner SF, Hudson TJ, Lander ES (2001) High-resolution haplotype structure in the human genome. Nat Genet 29:229–232

Délye C, Calmès E, Matéjicek A (2002) SNP marker for black-grass (*Alopecurus myosuroides* Huds.) genotypes resistant to acetyl CoA-carboxylase inhibiting herbicides. Theor Appl Genet 104:1114–1120

Drenkard E, Richter BG, Rozen S, Stutius LM, Angell NA, Mindrinos M, Cho RJ, Oefner PJ, Davis RW, Ausubel FM (2000) A simple procedure for the analysis of single nucleotide polymorphisms facilitates map-based cloning in *Arabidopsis*. Plant Physiol 124:1483–1492

Drysdale CM, McGraw DW, Stack CB, Stephens JC, Judson RS, Nandabalan K, Arnold K, Ruano G, Liggett SB (2000) Complex promoter and coding region beta(2)-adrenergic receptor haplotypes alter receptor expression and predict in vivo responsiveness. Proc Natl Acad Sci USA 97:10483–10488

Gaut BS, Le Tierry d'Ennequin M, Peek AS, Sawkins MC (2000). Maize as a model for the evolution of plant nuclear genomes. Proc Natl Acad Sci USA 97:7008–7015

Gilles PN, Wu DJ, Foster CB, Dillon PJ, Chanock SJ (1999) Single nucleotide polymorphic discrimination by an electronic dot blot assay on semiconductor microchips. Nat Biotech 17:365–370

Gion JM, Rech P, Grima-Pettenati J, Verhaegen D, Plomion C (2000) Mapping candidate genes in *Eucalyptus* with emphasis on lignification genes. Mol Breed 6:441–449

Goldstein DB (2001) Islands of linkage disequilibrium. Nat Genet 29:109–111

Grivet L, Glaszmann JC, Vincentz M, da Silva F, Arruda P (2003) ESTs as a source for sequence polymorphism discovery in sugarcane: example of the *Adh* genes. Theor Appl Genet 106:190–197

Gupta PK, Roy JK, Prasad M (2001) Single nucleotide polymorphisms: a new paradigm for molecular technology and DNA polymorphism detection with emphasis on their use in plants. Curr Sci 80:524–535

Han B, Xue Y (2003) Genome-wide intraspecific DNA-sequence variation in rice. Curr Opin Plant Biol 6:134–138

Hauser MT, Adhami F, Dorner M, Fuchs E, Glössl J (1998) Generation of co-dominant PCR-based markers by duplex analysis on high resolution gels. Plant J 16:117–125

Hawkins G, Hoffman LM (1997) Base excision sequence scanning. Nat Biotech 15:803804

Jander G, Norris SR, Rounsley SD, Bush DF, Levin IM, Last RL (2002) *Arabidopsis* map-based cloning in the post-genome era. Plant Physiol 129:440–450

Jannoo N, Grivet L, Dookun A, DHont A, Glaszman JC (1999) Linkage disequilibrium among modern sugarcane cultivars. Theor Appl Genet 99:1053–1060

Johnson GC, Esposito L, Barratt BJ, Smith AN, Heward J, Di Genova G, Ueda H, Cordell HJ, Eaves IA, Dudbridge F et al. (2001) Haplotype tagging for the identification of common disease genes. Nat Genet 29:233–237

Kanazin V, Talbert H, See D, DeCamp P, Nevo E, Blake T (2002) Discovery and assay of single-nucleotide polymorphisms in barley (*Hordeum vulgare*). Plant Mol Biol 48:529–537

Karp A, Edwards KJ, Bruford M, Funk S, Vosman B, Morgante M, Seberg O, Kremer A, Boursot P, Arctander P, Tautz D, Hewitt G (1997) Molecular technologies for biodiversity evaluation: opportunities and challenges. Nat Biotech 15:625–628

Kent WJ, Sugnet CW, Furey TS, Roskin KM, Pringle TH, Zahler AM, Haussler D (2002) The human genome browser at UCSC. Genome Res 12:996–1006

Kim S, Edwards JR, Deng L, Chung W, Ju J (2002) Solid phase capturable dideoxynucleotides for multiplexing genotyping using mass spectrometry. Nucleic Acids Res 30 (16):e85

Koch HG, McClay J, Loh EW, Higuchi S, Zao JH, Sham P, Ball D, Craig IW (2000) Allele association studies with SSR and SNP markers at known physical distances within a 1 Mb region embracing the ALDH2 locus in the Japanese, demonstrates linkage disequilibrium extending up to 400 kb. Hum Mol Genet 9:2993–2999

Konieczny A, Ausubel FM (1993) A procedure for mapping *Arabidopsis* mutations using co-dominant ecotype-specific PCR-based markers. Plant J 4:403–410

Kota R, Wolf M, Michalek W, Graner A (2001) Application of denaturing high-performance liquid chromatography for mapping of single nucleotide polymorphisms in barley (*Hordeum vulgare* L.). Genome 44:523–528

Landegren U, Nilsson M, Kwok PY (1998) Reading bits of genetic information: methods for single-nucleotide polymorphism analysis. Genome Res 8:769–776

Meksem K, Ruben E, Hyten D, Triwitayakorn K, Lightfoot DA (2001a) Conversion of AFLP bands into high-throughput DNA markers. Mol Genet Genomics 265:207–214

Meksem K, Ruben E, Hyten DL, Schmidt ME, Lightfoot DA (2001b) High-throughput genotyping for a polymorphism linked to soybean cyst nematode resistance gene *Rhg4* by using TaqMan probes. Mol Breed 7:63–71

Meyer RC, Lawrence PE, Young GR, Thomas WTB, Powell W (2000) SSRs and SNPs diagnostic tools for barley yellow mosaic virus. Scottish Crop Research Institute, Invergowrie, Dundee DD2 5DA,UK

Moffatt MF, Traherne JA, Abecasis GR, Cookson WOCM (2000) Single nucleotide polymorphism and linkage disequilibrium within the TCR alpha/ delta locus. Hum Mol Genet 9:1011–1019

Mogg R, Batley J, Hanley S, Edwards D, O'Sullivan HO, Edwards KJ (2002) Characterization of the flanking regions of *Zea mays* microsatellites reveals a large number of useful sequence polymorphisms. Theor Appl Genet 105:532–543

Nasu S, Suzuki J, Ohta R, Hasegawa K, Yui R, Kitazawa N, Monna L, Minobe Y (2002) Search for and analysis of single nucleotide polymorphisms (SNPs) in rice (*Oryza sativa*, *Oryza rufipogon*) and establishment of SNP markers. DNA Res 9:163–171

Neuhaus G, Werner K, Weyen J, Friedt W, Ordon F (2003) First results on SNP scanning in fragments linked to resistance genes against the barley yellow mosaic virus complex. J Plant Dis Protect 110:296–303

Nickerson DA, Taylor SL, Weiss KM, Clark AG, Hutchinson RG, Stengard J, Salomaa V, Vartiainen E, Boerwinkle E, Sing CF (1998) DNA sequence diversity in a 9.7-kb region of the human lipoprotein lipase gene. Nat Genet 19:233–240

Nordborg M (2000) Linkage disequilibrium, gene trees and selfing: an ancestral recombination graph with partial self-fertilization. Genetics 154:923–929

Nordborg M, Innan H (2002) Molecular population genetics. Curr Opin Plant Biol 5:69–73

Paris M, Lance R, Jones MGK (2001) Single nucleotide primer extensions to type SNPs in barley. Proceedings of the 10th Australian Barley Technical Symposium, Rydges (Lakeside), Canberra ACT, Australia

Pastinen T, Kurg A, Metspalu, Peltonen L, Syvänen AC (1997) Minisequencing: a specific tool for DNA analysis and diagnostics on oligonucleotide arrays. Genome Res 7:606–614

Pritchard J, Stephens M, Donnelly P (2000) Inference of population structure using multilocus genotype data. Genetics 155:945–959

Przeworski M, Hudson RR, Rienzo A (2000) Adjusting the focus on human variation. Trends Genet 16:296–302

Quint M,. Mihaljevic R, Dussle CM, Xu ML, Melchinger AE, Lübberstedt T (2002) Development of RGA-CAPS markers and genetic mapping of candidate genes for sugarcane mosaic virus resistance in maize. Theor Appl Genet 105:355–363

Rafalski JA (2002a) Applications of single nucleotide polymorphisms in crop genetics. Curr Opin Plant Biol 5:94–100

Rafalski JA (2002b) Novel genetic mapping tools in plants: SNPs and LD-based approaches. Plant Sci 162:329–333

Reich DE, Cargill M, Bolk S, Ireland J, Sabeti PC, Richter DJ, Lavery T, Kouyoumjian R, Farhadian SF, Ward R, Lander ES (2001) Linkage disequilibrium in the human genome. Nature 411:199–204

Remington DL, Thornsberry JM, Matsuoka Y, Wilson LM, Whitt SR, Doebley J, Kresovich S, Goodman MM, Buckler IV ES (2001) Structure of linkage disequilibrium and phenotypic associations in the maize genome. Proc Natl Acad Sci USA 98:11479–11484

Rickert AM, Premstaller A, Gebhardt C, Oefner PJ (2002) Genotyping of SNPs in a polyploid genome by pyrosequencing™. BioTechniques 32:592–598

Rieder MJ, Taylor SL, Clark AG, Nickerson DA (1999) Sequence variation in the human angiotensin converting enzyme. Nat Genet 22:59–62

Risch N, Merikangas K (1996) The future of genetic studies of complex human diseases. Science 273:1516–1517

Ronaghi M, Uhlén M, Nyrén P (1998) A sequencing method based on real-time pyrophosphate. Science 281:363–365

Schafer AJ, Hawkins JR (1998) DNA variation and the future of human genetics. Nat Biotech 16:33–39

Schneider K, Borchardt DC, Schäfer-Pregel R, Nagl N, Glass C, Jeppsson A, Gebhardt C, Salamini F (1999) PCR-based cloning and segregation analysis of functional gene homologues in *Beta vulgaris*. Mol Gen Genet 262:515–524

Schneider K, Weisshaar B, Borchardt DC, Salamini F (2001) SNP frequency and allelic haplotype structure of *Beta vulgaris* expressed genes. Mol Breed 8:63–74

See D, Kanazin V, Talbert H, Blake T (2002) Gel-based detection of single nucleotide polymorphisms. http://hordeum.oscs. montana.edu/pubs/supfinal.PDF

Shapero MH, Leuther KK, Nguyen A, Scott M, Jones KW (2001) SNP genotyping by multiplexed solid-phase amplification and fluorescent minisequencing. Genome Res 11:1926–1934

Spiegelmann JI, Mindrinos MN, Fankhauser C, Richards D, Lutes J, Chory J, Oefner PJ (2000) Cloning of the *Arabidopsis RSF1* gene using a mapping strategy based on high-density DNA arrays and denaturing high-performance liquid chromatography. Plant Cell 12:2485–2498

Stephens JC, Reich DE, Goldstein DB, Shin HD, Smith MW, Carrington M, Winkler C et al. (1998) Dating the origin of the CCR5-Delta32 AIDS-resistance allele by coalescence of haplotypes. Am J Hum Genet 62:1507–1515

Stephens JC, Schneider JA, Tanguay DA, Choi J, Acharya T, Stanley SE, Jiang R, Messer CJ, Chew A, Han JH, Duan J, Carr JL, Lee MS, Koshy B, Kumar AM, Zhang G, Newell WR, Windemuth A, Xu C, Kalbfleisch TS, Shaner SL, Arnold K, Schulz V, Drysdale CM, Nandabalan K, Judson RS, Ruano G, Vovis GF (2001) Haplotype variation and linkage disequilibrium in 313 human genes. Science 293:489–493

Stephens M, Smith NJ, Donnelly P (2001) A new statistical method for haplotype reconstruction from population data. Am J Hum Genet 68:978–989

Syvänen AC (1999) From gels to chips: 'Minisequencing' primer extension for analysis of point mutations and single nucleotide polymorphisms. Hum Mutat 13:1–10

Tenaillon MI, Sawkins MC, Long AD, Gaut RL, Doebley JF, Gaut BS (2001) Patterns of DNA sequence polymorphism along chromosome 1 of maize (*Zea mays* ssp. *mays* L.) Proc Natl Acad Sci USA 98:9161–9166

Thornsberry JM, Goodman MM, Doebley J, Kresovich S, Nielsen D, Buckler IV ES (2001) *Dwarf8* polymorphisms associate with variation in flowering time. Nat Genet 28:286–289

Tishkoff SA, Pakstis AJ, Ruano G, Kidd KK (2000) The accuracy of statistical methods for estimation of haplotype frequencies: an example from the CD4 locus. Am J Hum Genet 67:518–522

Wang DG, Fan JB, Siao CJ, Berno A, Young P, Sapolsky R, Ghandour G, Perkins N, Winchester E, Spencer J, Kruglyak L, Stein L, Hsie L, Topaloglou T, Hubbell E, Robinson E, Mittmann M, Morris MS, Shen N, Kilburn D, Rioux J, Nusbaum C, Rozen S, Hudson Th J, Lipshutz R, Chee M, Lander ES (1998) Large-scale identification, mapping, and genotyping of single-nucleotide polymorphisms in the human genome. Science 280:1007–1082

White SE, Doebley (1999) The molecular evolution of *terminal ear1*, a regulatory gene in the genus *Zea*. Genetics 153:1455–1462

White MB, Carvallo M, Derse D, O'Brien SJ, Dean M (1992) Detecting single base substitutions as heteroduplex polymorphisms. Genomics 12:301–306

Whitt SR, Wilson LM, Tenaillon MI, Gaut BS, Buckler IV ES (2002) Genetic diversity and selection in the maize starch pathway. Proc Nat Acad Sci USA 99:12959–12962

Yu K, Haffner M, Poysa V (2001) Tailed Primer Base Excision Sequence Scanning (TP-BESS) for detection of single nucleotide polymorphism. Plant Mol Biol Rep 19:49–54

Zhang J, Rowe WL, Struewing JP, Buetow KH (2002) HapScope: a software system for automated and visual analysis of functionally annotated haplotypes. Nucl Acids Res 30:5213–5221

Zhu Y, Hyatt S, Quigley C, Song Q, Grimm D, Young N, Cregan P (2001) Single nucleotide polymorphisms (SNPs) in soybean genes, cDNAs, and random genomic sequences. Plant and Animal Genome IX Conference, San Diego, USA

PD Dr. Renate Horn
Department of Genetics and Biochemistry
Clemson University
100 Jordan Hall
Box 340324
Clemson, South Carolina 29634–0324, USA
Tel.: +1 864 656 3595
Fax: +1 864 656 6879
e-mail: renate.horn@agrar.uni-giessen.de
or rhorn@clemson.edu

Dipl.-Biol. Gisela Neuhaus
Institut für Pflanzenbau
und Pflanzenzüchtung I
Heinrich-Buff-Ring 26–32
35392 Giessen, Germany
Tel.: +49 641 99 37423
Fax : +49 641 99 37429
e-mail: gisela.neuhaus@agrar.uni-giessen.de

Function of Genetic Material: Contribution of Molecular Markers in Improving Crop Plants

Volker Mohler and Gerhard Wenzel

1 Introduction

Large-scale genome analysis and related technologies provide access to a refined understanding of the genome. More and more, the systematic molecular analysis of structure, function, and regulation of plant genomes opens new perspectives for future developments and their application in the plant sciences, i.e. production of new and better varieties. The increasing amount of information on DNA documented in molecular marker collections and in dense gene maps, together with an excellent bioinformation system, increasingly allows the development of engineering strategies towards superior cultivars. Common principles, e.g. in defence reactions to biotic and abiotic stress, offering strategies like the candidate gene approach, have been set out (Lübberstedt et al. 2002). These developments help in the two phases of plant breeding: the production and use of variability called the 'evolutionary phase', as well as in selection, called the 'evaluation phase' (Khush 2002). Molecular markers are a powerful tool for: (1) higher efficiency in parental selection, allowing a controlled combination of better combining and thus heterotic parents and (2) pyramiding of single traits to result in more complex characters (Ranjekar et al. 2002). Additionally, they are presently the most efficient system for uncovering a rare but desired genotype in a large, segregating population.

The tools used today not only for investigation of the model plant *Arabidopsis thaliana* but increasingly in genotype building programs by breeding companies are: (1) bioinformatics, i.e. database comparisons of functional domains of protein families; (2) expression analysis, i.e. analysis of the transcriptome, including inventory of expressed transcripts, global expression analysis using cDNA arrays, straightforward identification of genes of interest by differential techniques, and (3) the use of mutant collections (knock-out mutants up to complete gene machines).

Molecular markers are virtually unlimited in their number, detectable at all plant developmental stages showing no pleiotropic effects. However, one of their greatest advantages is that they can be used to dissect quantitatively

Progress in Botany, Vol. 65

inherited traits into single Mendelian factors. Thus, molecular markers are of great value in applying genomics to crop improvement (Tanksley 1993; Mohan et al. 1997; Wenzel 1998; Snowdon et al. 2002).

2 Range of Molecular Marker Types

Marker types are numerous; comparisons and evaluations of their usefulness, reproducibility and costs have been well documented (Powell et al. 1996; Jones et al. 1997; Virk et al. 2000). A choice for a particular marker system needs to be made for each specific purpose. For selection of the most suitable system, it is important to consider that a high degree of polymorphism is revealed and the entire genome is covered (Bohn et al. 1999). Historically, the development of molecular markers started with restriction fragment length polymorphism (RFLPs, Botstein et al. 1980), followed by the first PCR-based system: random amplified polymorphic DNA (RAPD, Welsh and McClelland 1990; Williams et al. 1990). PCR-based methods offer the striking advantage of being rapid and requiring only tiny bits of plant material for DNA extraction (McGregor et al. 2000). When entire genome sequences of major crops become known, as e.g. the rice genome in 2002 (Goff et al. 2002; Yu et al. 2002), single nucleotide polymorphism (SNP) markers are likely to become most important in research aimed at crop improvement.

2.1 Restriction Fragment Length Polymorphism

In complex plant genomes, restriction fragment length polymorphisms (RFLPs) are detected upon hybridization of (mainly) single-copy DNA probes to restriction enzyme-hydrolyzed, agarose gel electrophoresis-separated and nylon membrane-bound genomic DNA. For the detection of polymorphisms, the DNAs of the genotypes to be surveyed are digested usually with various restriction enzymes, and marker alleles are identified by size differences of the restriction fragments to which the probes hybridize. Thus, RFLP markers are specified by a single clone/restriction enzyme combination. Major sources of RFLP probes are species-specific genomic DNA and cDNA sequences. Heterologous probes can be used to link genomes from species of the same family (Paterson et al. 2000), and even of different plant families (Fulton et al. 2002). Using RFLPs, detailed genetic maps of many crop plants have been constructed, e.g. tomato and potato (Tanksley et al. 1992), rice (Kurata et al. 1994), maize (Davis et al. 1999) or barley (Kleinhofs and Graner 2001). The principle and technique of RFLP production have often been described (Gupta et al. 2002b).

RFLP analysis is a time-consuming and labour-intensive task and, therefore, not suitable for the rapid genotyping of large segregating populations used in a breeding program. An additional disadvantage is the large quantity of genomic DNA (10 µg or more) needed for target fragment detection. **Due to superiority and the cost-effective nature of PCR-based marker systems, RFLPs lost their importance in genomics without becoming applied in practice.** RFLPs represent still, however, a highly reproducible and reliable marker system that can be used to obtain anchor points for consensus maps. Moreover, **the development of simple PCR**, also termed sequence-tagged site (STS) markers, **or SNP (single nucleotide polymorphism) markers from sequenced RFLP probes may provide an opportunity to determine useful polymorphisms found in previous gene mapping studies.**

2.2 Random Amplified Polymorphic DNA and DNA Amplification Fingerprinting

The use of a single, short oligonucleotide primer of random sequence (usually ten-base oligomers of varying GC content, ranging from 40 to 100%) in a low-stringency PCR results in random amplified polymorphic DNA (RAPD) which can be directly visualized in conventional agarose gel electrophoresis (Williams et al. 1990). The condition precedent to amplification is the occurrence of short inverted repeats of the same random sequence on opposite DNA strands within an amplifiable distance. Longer primers of up to 30 nucleotides were proposed by Welsh and McClelland (1990), shorter ones of 5–8 nucleotides by Caetano-Anollés et al. (1991; Caetano-Anollés 2001). With the reduction of the primer length, the number of amplified products increases, making this technique useful for fingerprinting approaches and thus being named DNA amplification fingerprinting (DAF). Polymorphisms between individuals result primarily from sequence differences in one or both primer sites or from indels (insertions–deletions) exceeding the PCR-amplifiable distance, and are recognizable by the presence or absence of a particular RAPD band.

Though RAPDs have some striking advantages such as low cost, the small amount of DNA needed (10–100 ng) and the use of universal primers, the DNA profiles often are prone to non-reproducibility between laboratories due to variations in DNA quality and PCR conditions, and different models of thermo-cyclers. In replicability studies by Pérez et al. (1998), mispriming errors amounted to up to 60%. To overcome this problem,

Paran and Michelmore (1993) converted RAPD fragments to simple and robust PCR markers termed sequence-characterized amplified regions (SCARs). Primer pairs are deduced from cloned RAPD fragments usually by extending the original decamer primer sequence by 10–15 bases.

2.3 Amplified Fragment Length Polymorphism

Amplified fragment length polymorphism (AFLP, Vos et al. 1995) is based on PCR amplification of a set of restriction fragments under high stringency conditions from a pool of fragments that is generated through digestion with a pair of restriction endonucleases, one of them being a frequent, four-base cutter (e.g., *Mse*I, *Taq*I), and the other being a rare, six-base cutter (e.g., *Eco*RI, *Pst*I, *Hind*III). Another widespread enzyme system substitutes six- for eight-base cutting enzymes, such as *Sse*8387I or its isoschizomer *Sda*I (Law et al. 1998; Hartl et al. 1999). To allow for PCR tagging of restriction fragments, oligonucleotide adapters are ligated to the ends of the DNA fragments. The number of DNA fragments to be amplified can be restricted by adding two to three selective nucleotides to the core sequence of the primers which is composed of adapter- and restriction-site homologous sequences. The option to permute the order of these selective bases and to recombine the primers with each other theoretically offers the possibility of the gradual collection of all restriction fragments of a particular enzyme combination showing a suitable size for PCR fragment analysis from a genotype. Thus, this method generates many bands, but not too many to create difficulties in scoring: a well-balanced number of amplified restriction fragments ranges from 50–150. The AFLP process may be improved by using a multifluorophore detection technique (Schwarz et al. 2000). In comparison to conventional ^{33}P-based AFLP analysis the multifluorophore technique allows multimixing as well as simultaneous amplification of AFLP fragments with different primer combinations in one reaction (multiplexing). Today, AFLP is the most efficient technique for detecting polymorphisms. The polymorphic bands from AFLP patterns may also be converted into simple PCR markers (Shan et al. 1999) suitable for marker-assisted selection procedures.

A cDNA-AFLP technique (Bachem et al. 1996), with the standard AFLP protocol on a cDNA template, was used for the construction of genome-wide transcriptome maps (Brugmans et al. 2002). The AFLP technique can be modified so that one primer is obtained from a known multi-copy sequence to detect sequence-specific amplification polymorphisms (S-SAP). This approach has been used successfully to generate genome-wide *Bare-1* retrotransposon-like markers in barley (Waugh et al. 1997) and *Avena* (Yu and Wise 2000) as well as in alfalfa by making use of consensus sequences from long terminal repeats (LTRs) of *Tms1* retrotransposon (Porceddu et al. 2002). Based on the LTR regions of retrotransposable elements *Tps12* and *Tps19* of pea (Pearce et al. 2000) and *Ty1-copia* of sweet potato (Berenyi et al. 2002) S-SAP systems were established for their use in genetic diversity studies.

2.4 Simple Sequence Repeat

Simple sequence repeats (SSRs) or microsatellites are tandem repeats of short nucleotide sequence motifs (1–6 bp) and ubiquitous in eukaryotic genomes (Tautz and Renz 1984). SSR loci are extremely variable in the number of repeat units among individuals of a given species and can be easily typed via STS-PCR (Weber and May 1989). **The opportunity to tag highly variable DNA sequence motifs, which are evenly spaced throughout the genome as simple PCR markers, represented an ideal resource for developing a marker class to be applied in practical breeding.** As a consequence, for most important crops, a huge number of SSR markers are publicly available for research, such as potato (Milbourne et al. 1998), wheat (Röder et al. 1998; Gupta et al. 2002a; Song et al. 2002), soybean (Cregan et al. 1999), barley (Liu et al. 1996; Ramsay et al. 2000), rice (Temnykh et al. 2000), and maize (Sharopova et al. 2002). Furthermore, **these short repetitive sequence elements also occur within genes** as demonstrated by searching EST (expressed sequence tag) databases for the presence of microsatellites (wheat: Eujayl et al. 2002; rye: Hackauf and Wehling 2002; barley: Thiel et al. 2003). EST-derived SSRs are expected to be slightly less polymorphic than SSRs derived from genomic libraries, as there is pressure for sequence conservation in coding regions.

2.5 Single Nucleotide Polymorphism

The primary focus for the development of the next generation of genetic markers is on single nucleotide polymorphism (SNP). An SNP is a single nucleotide position in a defined DNA stretch at which there is variation between different individuals within a species. Thus, single base insertion/deletion (indels) variants would not formally be considered to be SNPs (Brookes 1999). SNPs are of particular interest for their utilization in crop improvement, since they (1) represent the most frequent variations in the genome of any organism, thus offering the potential of finding informative markers for a distinct genomic region in any genetic background and (2) can be simply treated as diallelic markers and, therefore, make them amenable to automated high-throughput genotyping and data handling.

The determination of SNP genotypes can be preferentially performed using non-gel-based technology platforms such as denaturing high-performance liquid chromatography (DHPLC, Oefner and Underhill 1998), which is based on the detection of heteroduplex formation between mismatched nucleotides of mixed double-stranded PCR samples by ion-paired reversed phase chromatography under partially heat-denaturing conditions. Kota et al. (2001) previously reported on the DHPLC mapping of EST-SNPs in the barley genome. Another biochemical reaction principle for SNP genotype calling is mini-sequencing which can be used either as single-base primer extension (Syvänen 1999) or in the pyrosequencing format (Ahmadian et al. 2000).

3 Use of Genetic Markers in Plant Breeding

Breeders use three kinds of markers for a successful selection: morphological, biochemical, and molecular. The clearest and up till now most successful ones are the morphological markers, since they have been the characters on which Mendel elaborated the Mendelian laws. They enabled the establishment of genetic maps for numerous important crop plants (e.g. barley: Tsuchiya 1986). The biochemical markers are very limited in number, however, a few are tightly coupled to important genes and thus very useful in selection (e.g. the endopeptidase isozyme allele *Ep-D1b* for detection of the *Pch1* gene conferring resistance to *Pseudocercosporella herpotrichoides* in wheat, or the correlation of bread-making quality and the presence of particular high-molecular weight glutenin subunits among the seed storage protein complex of wheat).

Although a DNA marker for *Wx-B1b*, the null allele important for starch quality, is available (McLauchlan et al. 2001), an ELISA assay that has now been developed for *Wx-B1* (Gale et al. 2001) is likely to be much more cost-effective than the DNA marker currently being used.

For molecular markers a severe restraint exists in application. The cost for this process in plants is as high as in human medicine. Since even the best new cultivar will have difficulties to make so much money that all molecular work can be paid by this gain, we are faced with the following dilemma. One of the biggest challenges in the future will be to understand how genetics and biology work together and it is not only scientific interest by which this development is driven, it is also the need for more food and feed. The annual increase in plant productivity has become less than in previous years (James 2002). The three pillars: genetics, chemistry, and mechanization, on which the increased yield in agricultural production is based, are no longer contributing equally; progress in the latter two is mainly improving environmental conditions rather than productivity. Thus, we urgently need further understanding of gene function in order to manipulate it for the benefit of mankind. On the other hand, the projected costs are unlikely to be covered by seed consumers nor by environmental revenues. The dispute whether MAS selection is a fund-adsorbing system (Melchinger and Utz 2002) or whether it is the most advanced screening remains unresolved. The answer will be neither a clear yes or no. It will depend very much on the type of trait in question (qualitative or quantitative) and the applicability of the respective marker.

4 Present Status of Validated Molecular Markers for Molecular Breeding of Important Crops

A huge number of agronomically important genes have been tagged with the aid of molecular markers. However, most of them, although not the case for 'perfect' (gene sequence-derived) markers, do not meet conditions for marker-assisted selection (MAS), which are as follows:

1. **Markers should co-segregate or map as closely as possible to the target gene** (2 cM or less) to have low recombination frequency between the target gene and the marker. A better estimate of map distance between the target gene and the marker will be obtained by analysing further mapping populations which have genotypes in common with those used in the initial mapping population. Accuracy of MAS will be improved if, rather than a single marker, two markers flanking the target gene are used (Peng et al. 2000).
2. For unlimited use in MAS, the marker should display polymorphism between genotypes that have and do not have the target gene.
3. Cost-effective, simple PCR markers are required to ensure genotyping power needed for the fast screening of large populations.

Markers that have been elaborated and validated for the monitoring of agronomically important traits and which are used in current breeding programs are listed in Table 1. Most of them are detectable using simple PCR. However, some need to be tested for polymorphism between parental lines using breeding programs. In cases where RFLP markers have been used, such as for monitoring the cereal cyst nematode resistance gene *Cre1* in wheat (Ogbonnaya et al. 2001b), the gain fitted the effort due to the importance of the disease, high costs and unreliability of the bioassays.

Resistance gene analogs (RGAs) become more and more interesting as molecular markers (Table 1). RGAs are primarily located close to resistance

Table 1. Validated markers for the monitoring of agronomically important traits in marker-assisted selection programs

Crop	Trait	Gene(s)	Marker type(s)	References
Wheat	Reduced height	*Rht-B1b* *Rht-D1b*	Gene-specific SNP markers (competitive AS[a]-PCR)	Ellis et al. (2003)
		Rht8	SSR	Korzun et al. (1998)
	Starch quality	*Wx-B1b*	Gene-specific AS	McLauchlan et al. (2001)
	Powdery mildew resistance	*Pm1c*	AFLP	Hartl et al. (1999)
		Pm17	RFLP-derived STS	Mohler et al. (2001)
		Pm24	SSR	Huang et al. (2000)
	Cereal cyst nematode resistance	*Cre1* *Cre3*	RGA-RFLP RGA-based STS	Ogbonnaya et al. (2001b)
	Leaf rust resistance	*Lr21*	RFLP-derived STS	Huang and Gill (2001)
		Lr47	RFLP-derived CAPS[b]	Helguera et al. (2000)
	VPM rust *resistance*	*Sr38/Lr38/ Yr17*	RGA-based STS	Seah et al. (2001)
Barley	Barley yellow mosaic virus resistance	*rym4* *rym5*	SSR	Graner et al. (1999)
		rym9	RAPD-derived STS	Werner et al. (2000)
Rice	Bacterial blight	*xa-5*	Trait-flanking RFLP-derived CAPS and STS	Huang et al. (1997) Sanchez et al. (2000)
		xa-13	RFLP-derived CAPS	Huang et al. (1997)
		Xa-21	RAPD-derived STS	Chunwongse et al. (1993)
	Blast resistance	*Pi1* *Piz-5* *Pita*	Trait-flanking RFLPs RFLP-derived CAPS Trait-flanking RFLPs	Hittalmani et al. (2000)
Potato	Potato virus Y resistance	Ry_{adg}	RGA-based CAPS	Sorri et al. (1999)
			RGA-based STS	Kasai et al. (2000)
Soybean	Soybean cyst nematode	*Rhg4*	TaqMan assay	Meksem et al. (2001)

[a] AS, allele-specific.
[b] CAPS, cleaved amplified polymorphic sequence.

genes and are often members of local multigene families (Lübberstedt et al. 2002). Due to their clustered organization, the identification of the active gene copy remains a major challenge, but a high probability is given to find an informative SNP associated with the resistance phenotype. Homologous sequences from the common wheat *Cre3* locus isolated from *Aegilops ventricosa*, and shown to be of chromosome 5Nv origin, were successfully used as perfect markers for the presence of the *Cre6* gene (Ogbonnaya et al. 2001a). Costs of marker development may also be reduced since there is a chance of RGA markers to be used across the genomes of species from the same plant family (Mohler et al. 2002). Such broader markers are urgently needed.

In this context, the availability of SSR markers from the expressed portion of the genome might facilitate their transferability across genera, compared to the low efficiency of SSR markers retrieved from gene-poor areas in cross-generic amplification (Peakall et al. 1998). This approach would also benefit plant species with minimal resource and research expenditure.

5 Marker-Assisted Selection (MAS)

MAS offers the opportunity to select desirable lines based on genotype rather than phenotype, especially in the case of **combining different resistance genes for a given pathosystem in a single genotype (gene pyramiding)**, since it is difficult to select plants with multiple resistance genes based on phenotype alone as the action of one gene may mask the action of another. Successful examples are the combination of genes *xa4+xa5+xa13+Xa21* against bacterial blight disease in rice (Huang et al. 1997; Sanchez et al. 2000; Singh et al. 2001), *Pi1+Piz-5+Pita* conferring blast resistance in rice (Hittalmani et al. 2000) and *Pm2+Pm4a*, *Pm2+Pm21*, *Pm4a+Pm21* against powdery mildew disease in common wheat (Liu et al. 2000). Novel approaches deal with the implementation of transgenes in breeding programs such as the pyramiding of *Bt* genes *cry1Ac* and *cry1C* conferring resistance to diamondback moths in broccoli (Cao et al. 2002).

Molecular markers also contribute to improvement in efficiency of backcrossing, allowing at each backcross cycle the identification of carriers of the target trait (foreground selection) with the closest fit to the recurrent parent genotype (background selection). Two studies (Chen et al. 2000, 2001) reported on the improvement of restorer lines widely used in Chinese hybrid rice production, to bacterial blight resistance, caused by *Xanthomonas oryzae* pv. *oryzae* (*Xoo*), through introgression of *Xa21*, a broad-spectrum bacterial blight resistance gene. The PCR-based foreground selection system consisted of a marker that was part of gene *Xa21*, while RFLP and AFLP markers, respectively, were applied for background selection.

6 Strategies to Handle Complex Characters

It is still beyond practical use to apply MAS routinely for QTLs. Only a few reports are available on the successful use of QTL-MAS in backcross-assisted breeding strategies (Shen et al. 2001; Willcox et al. 2002). **Advanced backcross QTL** (AB-QTL, Tanksley and Nelson 1996) analysis was proposed as a general strategy for the simultaneous detection of QTL qualified for breeding purposes and variety development. The use of QTL analysis will be delayed until an advanced backcross generation offers advantages for QTL characterization such that the probability is reduced for the detection of QTL displaying epistatic interactions among donor alleles due to the overall lower frequency of donor alleles. In fact, there will be a higher probability of detecting additive QTL which will still function in a nearly isogenic background.

Efforts in indexing and sequencing of EST collections allow high-throughput scanning of SNPs among EST alleles from different genotypes. Along with SNP technology, which principally makes comprehensive haplotype analyses possible, **association mapping** (Risch 2000) has the potential to play a central role in the task of uncovering which candidate gene alleles cause complex traits in crop plants.

7 Outlook from a Breeder's Perspective

The number of molecular marker techniques and the applicability of dominant and co-dominant DNA markers made molecular plant breeding in principle a reality. The speed by which a new marker technique was used in laboratories was, however, such that a new and more efficient system became available before the latest marker system could be fully optimized or applied (Gupta et al. 2002b). Thus, breeders became reluctant to invest in a new technique which was already outdated when ready for use. Another drawback has been the high costs for a single test, resulting in limited genotype numbers that could be screened at a reasonable price. Consequently, the most advanced SNP allele calling procedures are not gel based and, therefore, readily automated for high-throughput genotyping, allowing the cost-effective screening of large populations. They are probably the most attractive tool for marker-assisted selection in future plant breeding programs.

Presently, only under specific applications, has MAS become a practical tool. MAS is applied in the selection for traits controlled by major genes responsible for important characters which are difficult or expensive to phenotype, e.g. resistance to soil-borne viruses (Friedt et al. 2001). Also, pyramiding of monogenically controlled resistances by MAS that result in an oligogenic trait causing more comprehensive resistance has been adopted by the breeding industry (Wenzel et al. 2001). Before genes can be combined in such a programmed way, it is necessary to determine whether or not they are allelic. Of course, pyramiding works for additive genes only and a serious problem is the specificity of the majority of available markers. They work only for those parents for which they have been precisely characterized. **Breeders still need more universal markers.**

The situation for traits controlled polygenically is even less satisfactory. Although numerous papers have been published describing and localizing QTL, understanding their interaction is very difficult. The impact of factors like temperature, day length, humidity on gene expression and metabolism is multifaceted and will require much more understanding of general gene functions. Moreover, a clear strategy as to how to make use of QTL via gene transformation is not yet feasible. The ideas go via master or major genes, via regulatory elements to a complete slicing and subsequent stepwise combination (Khush 2002). Because of multigenic responses, the question of additivity of gene products needs to be explored, normally not an easy task.

The over-expression of phytochrome B in potato resulted, e.g. under greenhouse conditions, in an increased yield (Thiele et al. 1999). Running the experiment under field conditions could not verify the greenhouse results (Braun et al. 2002; Schittenhelm 2002). To optimize allele complexes with high stability and a minimum of side effects requires an understanding of factors that are limiting in their expression as close to the end product in signalling and transcription pathway as possible (Winicov 2002). Successful identification of genes being able to accomplish this would provide reliable markers for plant breeding. A more global understanding of the action of QTL might allow one to treat them in slices as Mendelian characters in a monogenic manner.

Marker-based approaches are reaching a limit, opening up the newer field of genomics, dealing with the genes/enzymes directly rather than with indirect markers. Without using map-based cloning, chances are improving to find precise gene functions, e.g. via differential display techniques (Thümmler and Wenzel 2000). Genomics refers to application of global (in contrast to gene by gene) experimental approaches using DNA sequence data to assess gene functions by parallel methods such as expression profiling and screening for insertion mutants (Bouchez and Höfte 1998). Fortunately, information gathered in one species might be transferred to another species by conserved map location or via orthologous genes in related species (e.g. in the Triticeae: Mohler et al. 2002) and detraction of sequence databases (*in silico* genetics).

Most mapping and marker isolation has been done at the diploid level. In tetraploids like potato the dosage effect by doubling the ploidy level might have a striking influence (Luo et al. 2001). Markers are also useful for organelle DNA where mitochondrial DNA is more variable than plastid DNA. It should become possible to select with the help of markers for mitochondria which fit our demands better than mitochondria developed through natural evolution (Frei et al. 2003). The necessary variability has to be produced via somatic fusion. Thus, this approach works quite efficiently in vegetatively propagated tetraploid crops like potato. An optimal combination of plastome, chondriome and genome, using somatic hybrids (Lössl et al. 2000), could increase starch production.

Fortunately, molecular markers are an accepted tool of molecular genetics even by the public, which is generally critical of gene technology. The combined application of molecular markers and genome-wide, global approaches will contribute to better genotype descriptions and securer selections. As soon as validated universal markers reach the breeding stations, new concepts for population sizes and selection intensities will support condensed breeding programs. Perhaps the size reduction in a conventional program can counterbalance the higher cost for laboratory assistance. There is also a strong tendency to believe that classical combination breeding together with MAS is more universal in achieving breeding goals than gene transfer strategies.

References

Ahmadian A, Gharizadeh B, Gustafsson AC, Sterky F, Nyren P, Uhlen M, Lundeberg J (2000) Single nucleotide polymorphism analysis by pyrosequencing. Anal Biochem 280:103–110

Bachem CWB, van der Hoeven RS, de Bruijn SM, Vreugdenhil D, Zabeau M, Visser RGF (1996) Visualization of differential gene expression using a novel method of RNA fingerprinting based on AFLP: analysis of gene expression during potato tuber development. Plant J 9:745–753

Berenyi M, Gichuki ST, Schmidt J, Burg K (2002) *Ty1-copia* retrotransposon-based S-SAP (sequence-specific amplified polymorphism) for genetic analysis of sweetpotato. Theor Appl Genet 105:862–869

Bohn M, Utz HF, Melchinger AE (1999) Genetic similarities among winter wheat cultivars determined on the basis of RFLPs, AFLPs, and SSRs and their use for predicting progeny variance. Crop Sci 39:228–237

Botstein D, White RL, Skolnick M, Davis RW (1980) Construction of a genetic linkage map in man using restriction fragment length polymorphisms. Am J Hum Genet 32:314–331

Bouchez D, Höfte H (1998) Functional genomics in plants. Plant Physiol 118:725–732

Braun A, Gatz C, Frei U, Wenzel G (2002) Field experiments with transgenic potato clones. In: Wenzel G, Wulfert I (eds) Potatoes today and tomorrow. EAPR, WPR Communication, Königswinter, p 210

Brookes AJ (1999) The essence of SNPs. Gene 234:177–186

Brugmans B, del Carmen AF, Bachem CWB, van Os H, van Eck HJ, Visser RGF (2002) A novel method for the construction of genome wide transcriptome maps. Plant J 31:211–222

Caetano-Anollés G (2001) Plant genotyping using arbitrarily amplified DNA. In: Henry RJ (ed) Plant genotyping: the DNA fingerprinting of plants. CABI, Oxford, pp 29–46

Caetano-Anollés G, Bassam BJ, Gresshoff PM (1991) DNA amplification fingerprinting using very short arbitrary oligonucleotide primers. Biotechnology 9:553–557

Cao J, Zhao J-Z, Tang JD, Shelton AM, Earle ED (2002) Broccoli plants with pyramided *cry1Ac* and *cry1C* Bt genes control diamondback moths resistant to Cry1A and Cry1C proteins. Theor Appl Genet 105:258–264

Chen S, Lin XH, Xu CG, Zhang Q (2000) Improvement of bacterial blight resistance of 'Minghui 63', an elite restorer line of hybrid rice, by molecular marker-assisted selection. Crop Sci 40:239–244

Chen S, Xu CG, Lin XH, Zhang Q (2001) Improving bacterial blight resistance of 6078, an elite restorer line of hybrid rice, by molecular marker-assisted selection. Plant Breed 120:133–137

Chunwongse J, Martin GB, Tanksley SD (1993) Pregermination genotypic screening using PCR amplification of half seeds. Theor Appl Genet 86:694–698

Cregan PB, Jarvik T, Bush AL, Shoemaker RC, Lark KG, Kahler AL, Kaya N, VanToai TT, Lohnes DG, Chung J, Specht JE (1999) An integrated genetic linkage map of the soybean genome. Crop Sci 39:1464–1490

Davis GL, Baysdorfer C, Musket T, Grant D, Staebell M, Xu G, Polacco M, Koster L, Melia-Hancock S, Houchins K, Chao S, Coe EH Jr (1999) A maize map standard with sequenced core markers, grass genome reference points and 932 expressed sequence tagged sites (ESTs) in a 1736-locus map. Genetics 152:1137–1172

Ellis MH, Spielmeyer W, Gale KR, Rebetzke GJ, Richards RA (2003) "Perfect" markers for the *Rht-B1b* and *Rht-D1b* dwarfing genes in wheat. Theor Appl Genet DOI 10.1007/s00122-002-1048-4

Eujayl I, Sorrells ME, Baum M, Wolters P, Powell W (2002) Isolation of EST-derived microsatellite markers for genotyping the A and B genomes of wheat. Theor Appl Genet 104:399–407

Frei U, Peiretti EG, Wenzel G (2003) Significance of cytoplasmic DNA in plant breeding programs. Plant Breed Rev 23 175–210

Friedt W, Weikorn C, Scheurer KS, Habekuss A, Huth W, Ordon F (2001) Barley Yellow Dwarf Virus-Toleranz und markergestützte Züchtungsansätze bei Gerste. Bericht über die 52. Tagung der Vereinigung der Pflanzenzüchter und Saatgutkaufleute Österreichs 2001, BAL Gumpenstein, pp 117–120

Fulton TM, Van der Hoeven R, Eannetta NT, Tanksley SD (2002) Identification, analysis, and utilization of conserved ortholog set markers for comparative genomics in higher plants. Plant Cell 14:1457–1467

Gale KR, Panozzo JF, Eagles HA, Blundell M, Olsen H, Appels R (2001) Application of a high-throughput antibody-based assay for identification of the granule-bound starch synthase *Wx-B1b* allele in Australian wheat lines. Aust J Agric Res 52:1417–1423

Goff SA, Ricke D, Lan TH, Presting G, Wang RL, Dunn M, Glazebrook J, Sessions A, Oeller P, Varma H, Hadley D, Hutchinson D, Martin C, Katagiri F, Lange BM, Moughamer T, Xia Y, Budworth P, Zhong JP, Miguel T, Paszkowski U, Zhang SP, Colbert M, Sun WL, Chen LL, Cooper B, Park S, Wood TC, Mao L, Quail P, Wing R, Dean R, Yu YS, Zharkikh A, Shen R, Sahasrabudhe S, Thomas A, Cannings R, Gutin A, Pruss D, Reid J, Tavtigian S, Mitchell J, Eldredge G, Scholl T, Miller RM, Bhatnagar S, Adey N, Rubano T, Tusneem N, Robinson R, Feldhaus J, Macalma T, Oliphant A, Briggs S (2002) A draft sequence of the rice genome (*Oryza sativa* L. ssp. *japonica*). Science 296:92–100

Graner A, Streng S, Kellermann A, Schiemann A, Bauer E, Waugh R, Pellio B, Ordon F (1999) Molecular mapping and genetic fine-structure of the *rym5* locus encoding resistance to different strains of the barley yellow mosaic virus complex. Theor Appl Genet 98:285–290

Gupta PK, Balyan HS, Edwards KJ, Isaac P, Korzun V, Röder M, Gautier M-F, Joudrier P, Schlatter AR, Dubcovsky J, De la Pena RC, Khairallah M, Penner G, Hayden MJ, Sharp P, Keller B, Wang RCC, Hardouin JP, Jack P, Leroy P (2002a) Genetic mapping of 66 new microsatellite (SSR) loci in bread wheat. Theor Appl Genet 105:413–422

Gupta PK, Varshney RK, Prasad M (2002b) Molecular markers: principles and methodology. In: Jain SM, Brar DS, Ahloowalia (eds) Molecular techniques in crop improvement. Kluwer, Dordrecht, pp 9–54

Hackauf B, Wehling P (2002) Identification of microsatellite polymorphisms in an expressed portion of the rye genome. Plant Breed 121:17–25

Hartl L, Mohler V, Zeller FJ, Hsam SLK, Schweizer G (1999) Identification of AFLP markers closely linked to the powdery mildew resistance genes *Pm1c* and *Pm4a* in common wheat *(Triticum aestivum* L.). Genome 42:322–329

Helguera M, Khan IA, Dubcovsky J (2000) Development of PCR markers for wheat leaf rust resistance gene *Lr47*. Theor Appl Genet 101:625–631

Hittalmani S,·Parco A, Mew TV, Zeigler RS, Huang N (2000) Fine mapping and DNA marker-assisted pyramiding of the three major genes for blast resistance in rice. Theor Appl Genet 100:1121–1128

Huang L, Gill BS (2001) An RGA-like marker detects all known *Lr21* leaf rust resistance gene family members in *Aegilops tauschii* and wheat. Theor Appl Genet 103:1007–1013

Huang N, Angeles ER, Domingo J, Magpantay G, Singh S, Zhang G, Kumaravadivel N, Bennett J, Khush GS (1997) Pyramiding of bacterial blight resistance genes in rice: marker-assisted selection using RFLP and PCR. Theor Appl Genet 95:313–320

Huang XQ, Hsam SLK, Zeller FJ, Wenzel G, Mohler V (2000) Molecular mapping of the wheat powdery mildew resistance gene *Pm24* and marker validation for molecular breeding. Theor Appl Genet 101:407–414

James C (2002) World wide deployment of GM crops; aims and results – state of the art. www.gruenegentechnik.de/dgg/Doku_Fachtagung/james_engl.pdf

Jones CJ, Edwards KJ, Castaglione S, Winfiled MO, Sala F, van de Wiel C, Bredemeijer G, Vosman B, Matthes M, Daly A, Brettschneider R, Bettini P, Buiatti M, Maestri E, Malcevschi A, Marmiroli N, Aert R, Volckaert G, Rueda J, Linacero R, Vazquez A, Karp A (1997) Reproducibility testing of RAPD, AFLP and SSR markers in plants by a network of European laboratories. Mol Breed 3:381–390

Kasai K, Morikawa Y, Sorri VA, Valkonen JPT, Gebhardt C, Watanabe KN (2000) Development of SCAR markers to the PVY resistance gene Ry_{adg} based on a common feature of plant disease resistance genes. Genome 43:1–8

Khush GS (2002) Molecular genetics – plant breeder's perspective. In: Jain SM, Brar DS, Ahloowalia (eds) Molecular techniques in crop improvement. Kluwer, Dordrecht, pp 1–8

Kleinhofs A, Graner A (2001) An integrated map of the barley genome. In: Phillips RL, Vasil IK (eds) DNA-based markers in plants, 2nd edn. Kluwer, Dordrecht, pp 187–199

Korzun V, Röder MS, Ganal MW, Worland AJ, Law CN (1998) Genetic analysis of the dwarfing gene (*Rht8*) in wheat. Part I. Molecular mapping of *Rht8* on the short arm of chromosome 2D of bread wheat (*Triticum aestivum* L.). Theor Appl Genet 96:1104–1109

Kota R, Wolf M, Michalek W, Graner A (2001) Application of denaturing high-performance liquid chromatography for mapping of single nucleotide polymorphisms in barley (*Hordeum vulgare* L.). Genome 44:523–528

Kurata N, Nagamura Y, Yamamoto K, Harushima Y, Sue N, Wu J, Antonio BA, Shomura A, Shimizu T, Lin S-Y, Inoue T, Fukuda A, Shimano T, Kuboki Y, Toyama T, Miyamoto Y, Kirihara T, Hayasaka K, Miyao A, Monna L, Zhong HS, Tamura Y, Wang Z-X, Momma T, Umehara Y, Yano M, Sasaki T, Minobe Y (1994) A 300 kilobase interval genetic map of rice including 883 expressed sequences. Nat Genet 8:365–372

Law JR, Donini P, Koebner RMD, Reeves JC, RJ Cooke (1998) DNA profiling and plant variety registration III: the statistical assessment of distinctness in wheat using amplified fragment length polymorphisms. Euphytica 102:335–342

Liu J, Liu D, Tao W, Li W, Wang S, Chen P, Cheng S, Gao D (2000) Molecular marker-facilitated pyramiding of different genes for powdery mildew resistance in wheat. Plant Breed 119:21–24

Liu ZW, Biyashev RM, Saghai Maroof MA (1996) Development of simple sequence repeat DNA markers and their integration into a barley linkage map. Theor Appl Genet 93:869–876

Lössl A, Götz M, Braun A, Wenzel G (2000) Molecular markers for cytoplasm in potato: male sterility and contribution of different plastid-mitochondrial configurations to starch production. Euphytica 116:221–230

Lübberstedt T, Mohler V, Wenzel G (2002) Function of genetic material – genes involved in qualitative and quantitative resistance. Progress in botany 63. Springer, Berlin Heidelberg New York, pp 80–105

Luo ZW, Hackett CA, Bradshaw JE, McNichol JW, Milbourne D (2001) Construction of a genetic linkage map in tetraploid species using molecular markers. Genetics 157:1369–1385

McGregor CE, Lambert CA, Greyling MM, Louw JH, Warnich L (2000) A comparative assessment of DNA fingerprinting techniques (RAPD, ISSR, AFLP and SSR) in tetraploid potato (*Solanum tuberosum* L.) germplasm. Euphytica 113:135–144

McLauchlan A, Ogbonnaya FC, Hollingsworth B, Carter M, Gale KR, Henry RJ, Holton TA, Morell MK, Rampling LR, Sharp PJ, Shariflou MR, Jones MGK, Appels R (2001) Development of robust PCR-based DNA markers for each homoeo-allele of granule-bound starch synthase and their application in wheat breeding programs. Aust J Agric Res 52:1409–1416

Meksem K, Ruben E, Hyten DL, Schmidt ME, Lightfoot DA (2001) High-throughput genotyping for a polymorphism linked to soybean cyst nematode resistance gene *Rhg4* by using Taqman probes. Mol Breed 7:63–71

Melchinger AE, Utz HF (2002) Einsatz von Markern für die Selektion in der Pflanzenzüchtung: Theorie und praktische Beispiele. Vort Pflanzenzüchtg 54:11–21

Milbourne D, Meyer RC, Collins AJ, Ramsay LD, Gebhardt C, Waugh R (1998) Isolation, characterisation and mapping of simple sequence repeat loci in potato. Mol Gen Genet 259:233–245

Mohan M, Nair S, Bhagwat A, Krishna TG, Yano M, Bhatia CR, Sasaki T (1997) Genome mapping, molecular markers and marker-assisted selection in crop plants. Mol Breed 3:87–103

Mohler V, Hsam SLK, Zeller FJ, G Wenzel (2001) An STS marker distinguishing the rye-derived powdery mildew resistance alleles at the *Pm8*/*Pm17* locus of common wheat (*T. aestivum* L. em Thell.). Plant Breed 120:448–450

Mohler V, Klahr A, Wenzel G, Schwarz G (2002) A resistance gene analog useful for targeting disease resistance genes against different pathogens on group 1S chromosomes of barley, wheat and rye. Theor Appl Genet 105:365–368

Oefner PJ, Underhill PA (1998) DNA mutation detection using denaturing high-performance liquid chromatography (DHPLC). In: Dracopoli NC, Haines JL, Korf BR, Moir DT, Morton CC, Seidman CE (eds) Current protocols in human genetics (Suppl 19). Wiley, New York, pp 7.10.1–7.10.12

Ogbonnaya FC, Seah S, Delibes A, Jahier J, López-Braña I, Eastwood RF, Lagudah ES (2001a) Molecular-genetic characterisation of a new nematode resistance gene in wheat. Theor Appl Genet 102:623–629

Ogbonnaya FC, Subrahmanyam NC, Moullet O, de Majnik J, Eagles HA, Brown JS, Eastwood RF, Kollmorgen J, Appels R, Lagudah ES (2001b) Diagnostic DNA markers for cereal cyst nematode resistance in bread wheat. Aust J Agric Res 52:1367–1374

Paran I, Michelmore RW (1993) Development of reliable PCR-based markers linked to downy mildew resistance genes in lettuce. Theor Appl Genet 85:985–993

Paterson AH, Bowers JE, Burow MD, Draye X, Elsik CG, Jiang C-X, Katsar CS, Lan T-H, Lin Y-R, Ming R, Wright RJ (2000). Comparative genomics of plant chromosomes. Plant Cell 12:1523–1540

Peakall R, Gilmore S, Keys W, Morgante M, Rafalski A (1998) Cross-species amplification of soybean (*Glycine max*) simple sequence repeats (SSRs) within the genus and other legume genera: implications for the transferability of SSRs in plants. Mol Biol Evol 15:1275–1287

Pearce SR, Knox M, Ellis THN, Flavel AJ, Kumar A (2000) Pea *Ty1-copia* group retrotransposons: transpositional activity and use as markers to study genetic diversity in *Pisum*. Mol Gen Genet 263:898–907

Peng JH, Fahima T, Röder MS, Li YC, Grama A, Nevo E (2000) Microsatellite high-density mapping of the stripe rust resistance gene *YrH52* region on chromosome 1B and evaluation of its marker-assisted selection in the F_2 generation in wild emmer wheat. New Phytol 146:141–154

Pérez T, Albornoz J, Domínguez A (1998) An evaluation of RAPD fragment reproducibility and nature. Mol Ecol 7:1347–1358

Porceddu A, Albertini E, Barcaccia G, Marconi G, Bertoli F, Veronesi F (2002) Development of S-SAP markers based on an LTR-like sequence from *Medicago sativa* L. Mol Genet Genomics 267:107–114

Powell W, Morgante M, Andre C, Hanafey M, Vogel J, Tingey S, Rafalski A (1996) The comparison of RFLP, RAPD, AFLP and SSR (microsatellite) markers for germplasm analysis. Mol Breed 2:225–238

Ramsay L, Macaulay M, Ivanissevich SD, MacLean K, Cardle L, Fuller J, Edwards KJ, Tuvesson S, Morgante M, Massari A, Maestri E, Marmiroli N, Sjakste T, Ganal M, Powell W, Waugh R (2000) A simple sequence repeat-based linkage map of barley. Genetics 156:1997–2005

Ranjekar PK, Davierwala AP, Gupta VS (2002) DNA markers and heterosis. In: Jain SM, Brar DS, Ahloowalia (eds) Molecular techniques in crop improvement. Kluwer, Dordrecht, pp 161–201

Risch NJ (2000) Searching for genetic determinants in the new millennium. Nature 405:847–856

Röder MS, Korzun V, Wendehake K, Plaschke J, Tixier MH, Leroy P, Ganal MW (1998) A microsatellite map of wheat. Genetics 149:2007–2023

Sanchez AC, Brar DS, Huang N, Li Z, Khush GS (2000) Sequence tagged site marker-assisted selection for three bacterial blight resistance genes in rice. Crop Sci 40:792–797

Schittenhelm S, Menge-Hartmann U, Oldenburg E (2002) Growth anlysis of transgenic potato overexpressing *Arabidopsis thaliana* phytochrome B. In: Wenzel G, Wulfert I (eds) Potatoes today and tomorrow. EAPR, WPR Communication, Königswinter, 289 pp

Schwarz G, Herz M, Huang XQ, Michalek W, Jahoor A, Wenzel G, Mohler V (2000) Application of fluorescent-based semi-automated AFLP analysis in barley and wheat. Theor Appl Genet 100:545–551

Seah S, Bariana H, Jahier J, Sivasithamparam K, Lagudah ES (2001) The introgressed segment carrying rust resistance genes *Yr17*, *Lr37* and *Sr38* in wheat can be assayed by a cloned disease resistance gene-like sequence. Theor Appl Genet 102:600–605

Shan X, Blake TK, Talbert LE (1999) Conversion of AFLP markers to sequence-specific PCR markers in barley and wheat. Theor Appl Genet 98:1072–1078

Sharopova N, McMullen MD, Schultz L, Schroeder S, Sanchez-Villeda H, Gardiner J, Bergstrom D, Houchins K, Melia-Hancock S, Musket T, Duru N, Polacco M, Edwards K, Ruff T, Register JC, Brouwer C, Thompson R, Velasco R, Chin E, Lee M, Woodman-Clikeman W, Long MJ, Liscum E, Cone K, Davis G, Coe EH Jr (2002) Development and mapping of SSR markers for maize. Plant Mol Biol 48:483–499

Shen L, Courtois B, McNally KL, Robin S, Li Z (2001) Evaluation of near-isogenic lines of rice introgressed with QTLs for root depth through marker-aided selection. Theor Appl Genet 103:75–83

Singh S, Sidhu JS, Huang N, Vikal Y, Li Z, Brar DS, Dhaliwal HS, Khush GS (2001) Pyramiding three bacterial blight resistance genes (*xa5*, *xa13* and *Xa21*) using marker-assisted selection into indica rice cultivar PR106. Theor Appl Genet 102:1011–1015

Snowdon R, Kusterer B, Horn R (2002) Structural genome analysis using molecular cytogenetic techniques. Progress in botany 63. Springer, Berlin Heidelberg New York, pp 55–79

Song QJ, Fickus EW, Cregan PB (2002) Characterization of trinucleotide SSR motifs in wheat. Theor Appl Genet 104:286–293

Sorri VA, Watanabe KN, Valkonen JPT (1999) Predicted kinase-3a motif of a resistance gene-like fragment as a unique marker for PVY resistance. Theor Appl Genet 99:164–70

Syvänen AC (1999) From gels to chips: "minisequencing" primer extension for analysis of point mutations and single nucleotide polymorphisms. Hum Mutat 13:1–10

Tanksley SD (1993) Mapping polygenes. Annu Rev Genet 27:205–233

Tanksley SD, Nelson JC (1996) Advanced backcross QTL analysis: a method for the simultaneous discovery and transfer of valuable QTLs from unadapted germplasm into elite breeding lines. Theor Appl Genet 92:191–203

Tanksley SD, Ganal MW, Prince JP, De Vicente MC, Bonierbale MW, Broun P, Fulton TM, Giovannoni JJ, Grandillo S, Martin GB, Messeguer R, Miller JC, Miller L, Paterson AH, Pineda O, Roder MS, Wing RA, Wu W, Young ND (1992) High-density molecular linkage maps of the tomato and potato genomes Genetics. 132:1141–1160

Tautz D, Renz M (1984) Simple sequences are ubiquitous repetitive components of eukaryotic genomes. Nucleic Acids Res 12:4127–4138

Temnykh S, Park WD, Ayres N, Cartinhour S, Hauck N, Lipovich L, Cho YG, Ishii T, McCouch SR (2000) Mapping and genome organization of microsatellite sequences in rice (*Oryza sativa* L.). Theor Appl Genet 100:697–712

Thiel T, Michalek W, Varshney RK, Graner A (2003) Exploiting EST databases for the development and characterization of gene-derived SSR-markers in barley (*Hordeum vulgare* L.). Theor Appl Genet DOI 10.1007/s00122-002-1031-0

Thiele A, Herold M, Lenk I, Quail PH, Gatz C (1999) Heterologous expression of *Arabidopsis* phytochrome B in transgenic potato influences photosynthetic performance and tuber development. Plant Physiol 120:73–82

Thümmler F, Wenzel G (2000) Function of genetic material – approaches to understanding the action of genes in higher plants. Progress in botany 61. Springer, Berlin Heidelberg New York, pp 54–75

Tsuchiya T (1986) Gene analysis and linkage studies in barley. In: Yasuda S, Konishi T (eds) Barley genetics V. Sanyo Press, Okoyama, pp 175–187

Virk PS, Newbury HJ, Jackson MT, Ford-Lloyd BV (2000) Are mapped markers more useful for assessing genetic diversity? Theor Appl Genet 100:607–613

Vos P, Hogers R, Bleeker R, Reijans M, van de Lee T, Hornes M, Frijters A, Pot J, Peleman J, Kupier M, Zabeau M (1995) AFLP: a new technique for DNA fingerprinting. Nucleic Acids Res 23:4407–4414

Waugh R, Mclean K, Flavell AJ, Pearce SR, Kumar A, Thomas BBT (1997) Genetic distribution of *Bare-1*-like retrotransposable elements in the barley genome revealed by sequence-specific amplification polymorphism (S-SAP). Mol Gen Genet 253:687–694

Weber JL, May PE (1989) Abundant class of human DNA polymorphism which can be typed using the polymerase chain reaction. Am J Hum Genet 44:388–396

Welsh J, McClelland M (1990) Fingerprinting genomes using PCR with arbitrary primers. Nucleic Acids Res 18:7213–7218

Wenzel G (1998) Function of genetic material responsible for disease resistance in plants. Progress in botany 59. Springer, Berlin Heidelberg New York, pp 80–107

Wenzel G, Lübberstedt T, El-Badawy M, Mohler V (2001) Verteilung von Resistenzgenen im Genom von Gerste und Mais und die daraus folgenden Konsequenzen für die Züchtung. Bericht über die 52. Tagung der Vereinigung der Pflanzenzüchter und Saatgutkaufleute Österreichs 2001, BAL Gumpenstein, pp 121–125

Werner K, Pellio B, Ordon F, Friedt W (2000) Development of an STS marker and SSRs suitable for marker-assisted selection for the BaMMV resistance gene *rym9* in barley. Plant Breed 119:517–519

Willcox MC, Khairallah MM, Bergvinson D, Crossa J, Deutsch JA, Edmeades GO, González-de-León D, Jiang C, Jewell DC, Mihm JA, Williams WP, Hoisington D (2002) Selection for resistance to southwestern corn borer using marker-assisted and conventional backcrossing. Crop Sci 2002 42:1516–1528

Williams JGK, Kubelik ARK, Livak JL, Rafalski JA, Tingey SV (1990) DNA polymorphisms amplified by random primers are useful genetic markers. Nucleic Acids Res 18:6531–6535

Winicov I (2002) Molecular markers and abiotic stresses. In: Jain SM, Brar DS, Ahloowalia (eds) Molecular techniques in crop improvement. Kluwer, Dordrecht, pp 203–237

Yu G-X, Wise RP (2000) An anchored AFLP- and retrotransposon-based map of diploid *Avena*. Genome 43:736–749

Yu J, Hu SN, Wang J et al. (2002) A draft sequence of the rice genome (*Oryza sativa* L. ssp. *indica*). Science 296:79–92

Dr. Volker Mohler
LS Pflanzenbau und Pflanzenzüchtung
Wissenschaftszentrum Weihenstephan
Technische Universität München, Germany

Prof. Dr. Gerhard Wenzel
LS Pflanzenbau und Pflanzenzüchtung
Wissenschaftszentrum Weihenstephan
Technische Universität München, Germany
e-mail: gwenzel@wzw.tum.de

Extranuclear Inheritance: Chloroplast Proteomics

Michael Hippler and Ralph Bock

1 Introduction: The Choroplast as a Model System in Proteomics

Biology has arrived in the "omics" age. Currently, there is no better justification for the importance of one's own research field than adding the suffix "omics" to it (Fig. 1). This somewhat unfortunate tendency was initiated with the systematic and high-throughput sequencing of entire genomes for which the term "genomics" was coined. Soon, researchers using systematic approaches to elucidate gene functions felt it important to distinguish between "structural genomics" (i.e. genome sequencing) and "functional genomics" (i.e. elucidation of gene functions; see e.g. Bock and Hippler 2002). What are the criteria for "omics"? Certainly, any "omics" should (1) take a systematic approach and (2) use high-throughput techniques with the ultimate goal of achieving completeness (complete sequence, complete set of RNAs, proteins, metabolites, etc.). From this viewpoint, clearly, some fields are relatively far (e.g. "metabolomics") or even very far (e.g. "structuromics") from meeting these criteria and, here, enthusiastic addition of the ending "omics" appears premature (Fig. 1).

In this review, we aim to cover a field for which the "omics" is very well justified: proteomics. **Proteomics** can be defined as **the systematic analysis of proteins produced in a single organism, tissue, cell or subcellular compartment at a certain time** and covers three main aspects: protein identity, quantity and function(s). The central concept of proteomics is the simultaneous study of all proteins in a given protein population, rather than analysis of one protein at a time, as in traditional biochemistry.

Recent technological progress has made it feasible to separate and analyze the full protein set of subcellular compartments within a reasonable time frame. Cell organelles provide excellent model systems for proteomics, because (1) they contain a relatively small, manageable set of proteins, (2) they can be easily isolated in large amounts and at high purity and (3) they can be further fractionated into suborganellar compartments ("subproteomes"), such as outer and inner membranes, the intermembrane space,

Progress in Botany, Vol. 65

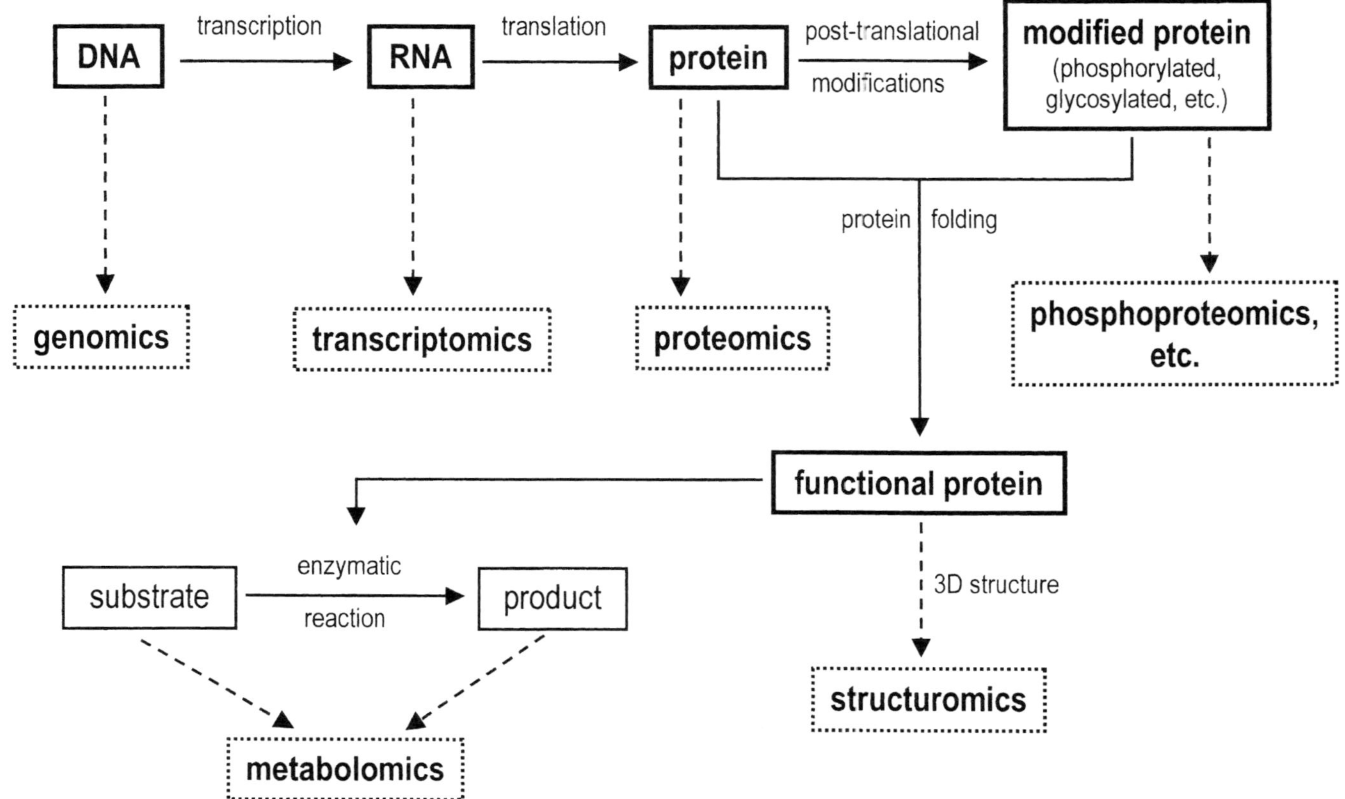

Fig. 1. The world of "omics" in modern biology

chloroplast thylakoids, etc. In this chapter, we shall summarize bioinformatic attempts to estimate the size of the chloroplast proteome, highlight the state of the art in proteome research and describe the results of recent studies on chloroplast subproteomes.

2 The Complexity of the Chloroplast Proteome and Its Evolutionary Origin

Plastids are derived from formerly free-living prokaryotes: More than a billion years ago, a cyanobacterium was engulfed by a eukaryotic host cell (already possessing mitochondria) and, different from endocytosis, was not digested but entered a symbiotic relationship. Subsequently, the cyanobacterial endosymbiont was gradually integrated into the metabolism of the host cell. Together, host and endosymbiont established a sophisticated division of labor by inventing efficient regulatory networks to coordinate the gene expression in the endosymbiont's genome with that in the host's nuclear genome. The evolutionary shaping of the endosymbiosis involved the (1) loss of dispensable genetic information (e.g., genes for cyanobacterial cell wall biosynthesis), (2) elimination of redundant genetic information (e.g., biosynthetic pathways for carbohydrate, amino acid and lipid biosyntheses), and (3) the massive translocation of genetic information from the endosymbiont's DNA to the host cell's nuclear genome. Consequently, present-day organellar genomes are greatly reduced and contain only a small proportion of the genes that their free-living ancestors had possessed: Higher plant plastid genomes harbor only about 130 genes in approximately 150 kbp whereas the cyanobacterium *Synechocystis* has more than 3100 genes (and open reading frames) in a genome of 3.57 Mbp. It has long been known that the vast majority of chloroplast proteins is not encoded by the plastid genome but is encoded by nuclear-localized genes, synthesized on cytosolic ribosomes and post-translationally imported into plastids. Protein targeting to plastids is directed by an N-terminal transit peptide which usually is cleaved off during the import process.

The availability of complete genome sequences for the cyanobacterium *Synechocystis* (Kaneko et al. 1996; Kaneko and Tabata 1997; Kotani and Tabata 1998) and the model plant *Arabidopsis thaliana* (The Arabidopsis Genome Initiative 2000) has facilitated the systematic reconstruction of the evolutionary origin of plant nuclear genes as well as the calculation of proteome sizes for different compartments of the plant cell. Analyzing a sample of 3961 *Arabidopsis* nuclear protein-coding genes (out of a total of ~25,000 genes in the *Arabidopsis* genome) using a number of bioinformatics tools (homology searches, protein targeting prediction, construction of phylogenetic trees; Table 1), Rujan and Martin calculated that between 400 (1.6%) and 2200 (9.2%) of the *Arabidopsis* nuclear genes stem from the cyanobacterial endosymbiont (Rujan and Martin 2001) and have ended up in the nuclear genome through gene transfer events out of the plastid. These numbers are in good agreement with another study which estimated that between 1150 and 1550 genes of cyanobacterial origin reside in the *Arabidopsis* nuclear genome (Abdallah et al. 2000). [A very recent new estimate incorporating data from 19 fully sequenced prokaryotic genomes suggests an even higher number of *Arabidopsis* genes of cyanobacterial origin (Martin et al. 2002).] The gene products of how many of these cyanobacterially derived nuclear genes are reimported into plastids and contribute to the size of the plastid proteome? Searches for putative N-terminal transit peptide sequences directing chloroplast protein import (see Table 1) suggest that between 650 and 900 *Arabidopsis* nuclear genes of cyanobacterial origin encode plastid-targeted proteins (Abdallah et al. 2000). The rest of the chloroplast proteome (estimated to be in the range of 1200 to 1600 proteins) comprises three categories: (1) genes coming from the genome of the eukaryotic host cell which, during evolution, have acquired sequences for chloroplast transit peptides, (2) new plant-specific genes which evolved after endosymbiosis and (3) genes of currently unknown evolutionary origin (some of which may be additional cyanobacterially derived genes). Thus, in total, the plastid proteome consists of 1900 to 2500 proteins of which:

- approximately 80 are encoded by the plastid genome,
- 650 to 900 are encoded by nuclear genes of cyanobacterial origin,
- 1200 to 1600 are encoded by nuclear genes of non-cyanobacterial (or uncertain) evolutionary origin.

All these calculations exclude additional protein diversity generated, for example, by processes like alternative splicing, differential protein processing or post-translational modifications. As it is currently unknown how widespread such protein isoforms made from one and the same gene are in chloroplasts, the total size of the plastid proteome may currently be underestimated.

Improved proteomics technologies (van Wijk 2000) nowadays make it possible to manage a set of 1900 to 2500 proteins as calculated to be present in higher plant chloroplasts, all the more since this set of proteins is further subcompartmentalized (into intermembrane space, stroma, thylakoid lumen and the respective membrane fractions) facilitating its convenient experimental splitting into subproteomes.

Table 1. Selected WWW tools for chloroplast genomics, proteomics and bioinformatics

Tool	Functions	Internet address
NCBI Genome	Database of complete organelle genome assemblies (plastid and mitochondrial genomes)	http://www.ncbi.nlm.nih.gov:80/PMGifs/Genomes/euk_o.html
GOBASE	Organelle genome database (plastid and mitochondrial genomes)	http://megasun.BCH.UMontreal.CA/gobase/
CyanoBase	Genome database for cyanobacteria	http://www.kazusa.or.jp/cyano/cyano.html
ChloroP	Predicts presence of plastid transit peptides in protein sequences and location of potential cleavage sites	http://www.cbs.dtu.dk/services/ChloroP/
PCLR	Predicts plastid localization of proteins	http://apicoplast.cis.upenn.edu/pclr/
Predotar	Predicts putative plastid and mitochondrial targeting sequences	http://www.inra.fr/Internet/Produits/Predotar/
PSORT	Predicts protein localization in cells	http://psort.nibb.ac.jp/
TargetP	Predicts protein localization in cells	http://www.cbs.dtu.dk/services/TargetP/
SignalP	Predicts presence and location of signal peptide cleavage sites in amino acid sequences	http://www.cbs.dtu.dk/services/SignalP/
MIPS AtDB	List of *Arabidopsis thaliana* proteins potentially targeted to chloroplasts (including scores, reliability classes and potential transit peptide lengths)	http://www.mips.biochem.mpg.de/proj/thal/target/chloro.html
ExPASy	Various proteomic tools for the analysis of protein sequences and structures as well as 2-D PAGE	http://www.expasy.ch/
MOTIF	Searches for protein (and nucleic acid) sequence motifs	http://www.motif.genome.ad.jp/

3 Methods in Chloroplast Proteomics

The analysis of complex protein profiles requires methods that allow for high resolution protein separation combined with very sensitive methods for protein identification (Fig. 2). The standard technique for quantitative proteome analysis is protein separation by high-resolution (isoelectric focusing/SDS-PAGE) two-dimensional gel electrophoresis (2-DE). This method separates proteins according to their isoelectric points in the first dimension and according to their molecular masses in the second. Al-

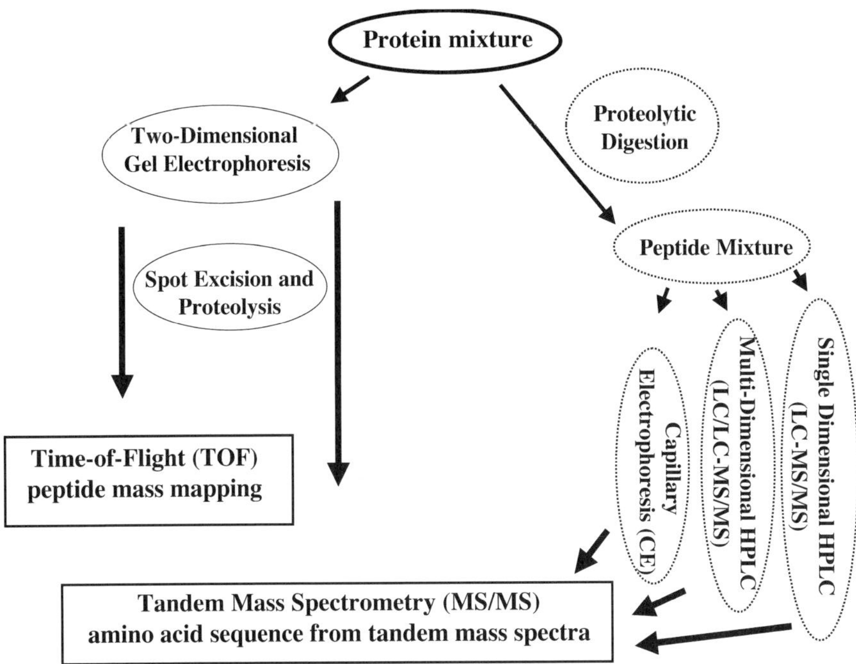

Fig. 2. High-resolution protein separation and mass spectrometric identification strategies. This simplified schematic diagram outlines strategies that depend on 2-DE (*solid ellipses*) or alternatives that do not require protein separation by 2-DE (*dotted ellipses*). In 2-DE-based approaches, a mixture of proteins is separated by 2-DE and visualized by staining. Individual protein spots are then excised and subjected to proteolysis. The resulting peptide mixture can subsequently be analyzed either by time-of-flight mass spectrometry (MALDI-TOF) to determine the masses of individual peptides (by mass mapping) or tandem mass spectrometry to determine (partial) amino acid sequences of the respective peptides. Protein separation strategies alternative to 2-DE utilize peptide mixtures resulting from proteolysis of a mixture of intact proteins. Peptide mixture can be separated by single (LC-MS/MS), multidimensional (LC/LC-MS/MS) HPLC or capillary electrophoresis (CE) and then analyzed on-line by tandem mass spectrometry. It should be noted that peptide mixtures obtained from proteolytic digestion of 2-DE separated spots can also be subjected to liquid chromatography prior to mass spectrometric analysis

though 2-DE was introduced already almost 30 years ago (O'Farrell 1975), new developments, like the introduction of immobilized pH gradients (IPG) or in-gel sample application, have greatly increased the reproducibility of 2-DE. This is important not only for comparative analyses of analytical 2-D protein maps, but also for their methodological standardization. The resolving power of this technique can be rationalized by the fact that over 10,000 proteins can be separated on a single gel (Klose 1999). To obtain sequence information from individual protein spots, the spots are cut out of the gel and digested in-gel with an appropriate protease, mostly trypsin. The resulting peptides are eluted and analyzed by mass spectrometry (MS). Based on sensitivity, mass spectrometry is the method of choice for identification of proteins. To perform mass spectrometry, several possible instrumentations are available. Matrix-assisted laser desorption ionization (MALDI) time-of-flight (TOF) instruments are commonly used for large-scale protein identification by the peptide mass mapping technique and are also used in chloroplast proteomics (Peltier et al. 2000, 2001, 2002; Yamaguchi and Subramanian 2000; Yamaguchi et al. 2000; Schubert et al. 2002). Peptide masses are determined for specific spots by MALDI-MS and these mass maps are then compared with predicted mass maps in the database to identify the respective protein. This peptide mass mapping strategy is a powerful tool for protein identification. However, one of its limitations is that the database must contain sufficient protein sequences from the organism under investigation, which is often a problem with plants since so far, only the *Arabidopsis* genome is completely sequenced. Another limitation of the MALDI-TOF mass mapping technique is that it is very sensitive to contamination with other proteins or protein isoforms arising, for example, from splice variants, RNA editing or post-translational modifications, which increase the ambiguity in protein identification. Besides the MALDI ionization technique, electrospray ionization (ESI) is frequently used and coupled with triple quadropole, ion-trap and hybrid quadropole-time-of-flight (Q-TOF) tandem MS. ESI-MS instrumentations are regularly utilized in combination with liquid chromatography (LC) or static nanospray techniques. Tandem MS can be employed to generate peptide fragment ion spectra by collision induced fragmentation (CID). CID predominantly causes fragmentation of peptide bonds. Therefore, it is possible to search for mass differences between peaks in the mass spectrum that differ by the mass of a single amino acid (which was deleted at a specific position within the peptide), thereby facilitating the deduction of short amino acid sequences. Software tools are available for the interpretation of amino acid sequence information from CID fragmentation spectra. These tools compare the experimental data with predicted spectra generated from database sequences. Comparisons between an experimental spectrum and spectra from the database can be done with Mascot (Perkins et al. 1999), an Internet-accessible program, or Sequest (Eng et al. 1994), a program that is commercially available. The extraction of short sequence "tags" from MS data requires partial amino acid sequence interpretation and can be carried out by PeptideSearch (Mann and Wilm 1994), another Internet-accessible program. In chloroplast proteomics, both ESI-MS and tandem MS have been used in a number of studies (Whitelegge et al. 1998; Peltier et al. 2000, 2001, 2002; Yamaguchi and Subramanian 2000; Yamaguchi et al. 2000; Hippler et al. 2001; Ferro et al. 2002; Gomez et al. 2002; Koller et al. 2002).

Recently, new methods in proteomic research have been developed which promise to be of enormous usefulness also for chloroplast proteomics. Here, we will briefly describe some of these new techniques (Fig. 2). Several of the new approaches take advantage of the coupling between liquid chromatography and tandem MS (LC/MS/MS) (Peng and Gygi 2001). Multi-dimensional liquid chromatography coupled to MS/MS allowed, for example, the identification of more than 70 proteins present in the yeast ribosome in a single analysis. This study was performed by analyzing tryptic peptides derived from digestion of the whole complex (Link et al. 1999). Multidimensional liquid chromatography coupled to tandem MS can also be used for large-scale proteome analysis. Using this combination of techniques, 1484 proteins could be identified from *Saccharomyces cerevisiae* including lowly abundant proteins like transcription factors and kinases which probably would have escaped detection by 2-DE (Washburn et al. 2001). The power of this experimental approach was also demonstrated by the systematic proteomic analysis of rice (*Oryza sativa*) leaf, root and seed tissues (Koller et al. 2002). This comparative study resulted in detection and identification of 2528 unique proteins. Among these proteins, 622 were leaf-, 862 root- and 512 seed-specific proteins, demonstrating that the overall protein expression profile is highly tissue-specific.

For quantifying differential protein expression, the isotope-coded affinity (ICAT) strategy is another approach that can be applied to LC/MS/MS (Gygi et al. 1999). Also, a robust gene tagging method is now available that facilitates the systematic analysis of components of multiprotein complexes (Gavin et al. 2002). Here, a gene encoding a subunit of the complex is tagged with a cassette coding for protein A and a calmodulin-binding protein (separated by a TEV protease cleavage site). Upon expression of the tagged gene in vivo, the respective gene product incorporates into its native complex allowing subsequent tandem-affinity purification (TAP) of the multiprotein complex by a two-step affinity chromatography procedure: The first step involves binding of the tagged complex via protein A to an

IgG column, followed by cleavage with the TEV protease releasing the purified complex. The second purification step employs calcium-calmodulin affinity chromatography to eliminate residual contaminants. Finally, individual components of the complex are analyzed by MS/MS (Puig et al. 2001).

4 Proteomics of Chloroplast Subcompartments

4.1 Proteomics of Soluble Proteins

As mentioned above, possibly more than 2000 nuclear-encoded gene products, in addition to the about 80 chloroplast-encoded proteins, are present in the chloroplast. To establish number, identity and abundance of these proteins, proteomics has become the most important tool. Especially the large-scale analysis of soluble proteins is straightforward since 2-DE based approaches are directly applicable to high-resolution separation of these proteins combined with mass spectrometric identification techniques. An impressive proteomics study of a soluble multiprotein complex succeeded in the identification of all ribosomal proteins (RPs) in spinach chloroplasts (Yamaguchi and Subramanian 2000; Yamaguchi et al. 2000). The spinach plastid ribosome is of the prokaryotic 70S type and comprises 59 proteins: 33 in the 50S and 25 in the 30S subunit. Plastid ribosomes are typical organellar multiprotein complexes in that they consist of both nuclear and organelle genome-encoded proteins (25 of the plastid 50S and 13 of the 30S RPs are nuclear encoded, whereas 8 of the 50S and 12 of the 30S RPs are plastid encoded). The plastid RPs also serve as good examples for the expression of a single gene resulting in a protein population that displays micro-heterogeneity. Such heterogeneity is observed with several plastid RPs and is caused by post-translational modifications, such as, α-*N*-acetylation of S9, differential N-terminal processing leading to five mature forms of S6 and two mature forms of S10 (30S subunit) and N-terminal/internal modifications in L2, L11 and L16 (50S subunit). The functional implications of these modifications are still unclear but it is tempting to speculate that they may serve regulatory functions in the plastid translational apparatus.

Another chloroplast multi-protein complex that was dissected by proteomics is the ClpP protease complex from *Arabidopsis* (Peltier et al. 2001). The 350-kDa protease complex was isolated by blue-native gel electrophoresis and SDS-PAGE and ten different Clp isoforms were identified by mass spectrometry.

In a first systematic analysis of the proteome of a chloroplast subcompartment, 2-DE based approaches were taken to analyze soluble lumenal and peripheral thylakoid proteins from pea (Peltier et al. 2000). From this study, it was estimated that at least 200 to 230 different lumenal and peripheral proteins exist. The proteome of the thylakoid lumen was investigated in greater detail and a 2-D map of *Arabidopsis* lumenal proteins has been established (Peltier et al. 2002; Schubert et al. 2002). In both studies, 30 to 36 lumenal proteins were identified and it was estimated that the *Arabidopsis* thylakoid lumen may contain a total of about 80 proteins. The presence of multiple isomerases, peroxiredoxins, m-type thioredoxins and a putative ascorbate peroxidase indicates that the lumenal proteome contains protein folding activities and is involved in the antioxidative defense network. It is noteworthy that among these proteins, a family of novel PsbP domain proteins was found. The extrinsic proteins PsbO, PsbP and PsbQ participate in the regulation of oxygen evolution and are present in the photosystem II (PSII) water-splitting complex in a stoichiometric ratio of 1:1:1. These novel PsbP domain proteins are possibly interchangeable at the lumenal side of PSII and could thereby provide a functional tool to modulate oxygen evolution activity of the PSII protein complex in response to environmental cues and/or developmental programs (Peltier et al. 2002; Schubert et al. 2002). These studies nicely demonstrate that proteomics cannot only identify protein sets, but also provides new functional implications and hypotheses. To convert these hypotheses into real understanding of biological function, new experiments can now be designed.

4.2 Proteomics of Membrane Proteins

The separation of hydrophobic intrinsic membrane proteins by 2-DE has long been a difficult task (Santoni et al. 2000). However, with the recent development of procedures for the analysis of transmembrane thylakoid proteins by 2-DE, proteomics of membrane proteins has become feasible. The successful separation of transmembrane proteins by 2-DE was achieved by combining extraction of hydrophobic proteins with organic solvents and subsequent solubilization of the precipitated proteins in a mixture of detergents together with urea and thiourea. 2-DE separation was combined with identification of protein spots by MS/MS and immunobiochemical techniques (Hippler et al. 2001). This procedure allowed the separation and identification of hydrophobic transmembrane proteins, such as the light-harvesting proteins, and was used to create 2-DE protein maps of thylakoid membrane proteins from wild type and mutant strains of *Chlamydomonas reinhardtii* (Hippler et al. 2001). The technique was also used to monitor changes in the protein composition of the thylakoid membrane during adaptation to iron deficiency (Moseley et al. 2002; see

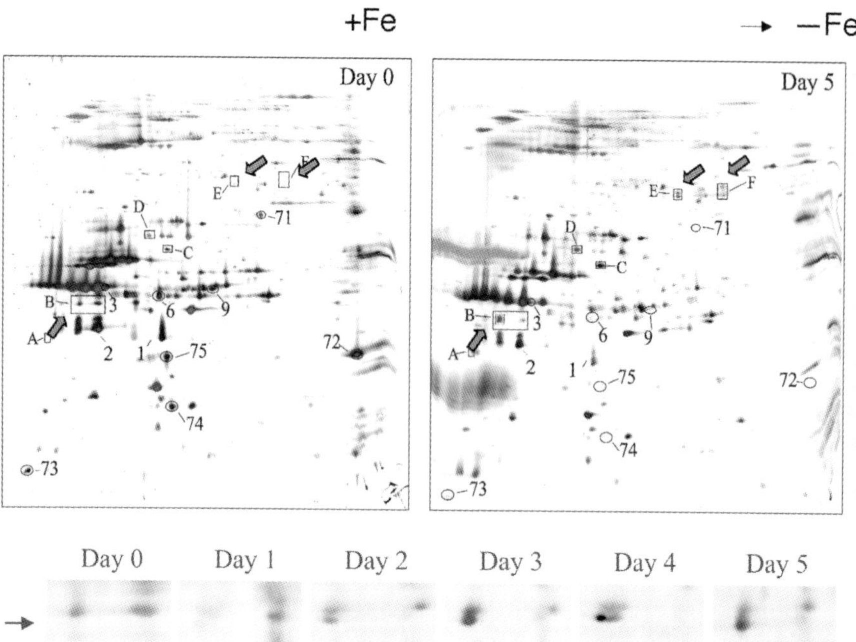

Fig. 3. Example of a comparative proteomics study in *Chlamydomonas* chloroplasts. The *top panel* (day 0 and day 5) represents silver-stained 2-DE gels of thylakoid membranes from wild-type cells before and 5 days after growth in iron-deficient (0 μM Fe) medium, respectively. *Circles* indicate spots that disappear and *squares* indicate spots that newly appear (see also *arrows*) or increase in intensity under iron deficiency. The *lower panel* shows the evolution of a new putative LHCI protein during the adaptation to iron deficiency. Note that spots 3, 6 and 9 in the *upper panel* represent LHCI polypeptides. The *lower panel* shows the enlarged box B (*upper panel*) from silver-stained 2-DE gels of thylakoid membranes from wild-type cells before and after 1 to 5 days of growth in iron-deficient medium. (Adapted from Moseley et al. 2002)

Fig. 3). This adaptation process is rather complex and highly dynamic since several protein spots disappear and a number of new protein spots appear as compared to 2-DE maps from iron repletion conditions (Fig. 3). Interestingly, some of the photosystem I light-harvesting complex (LHCI) subunits are degraded during adaptation to iron deficiency whereas new LHCI protein spots are induced under these conditions (Fig. 3). It is currently unclear whether these iron deficiency-induced LHCI spots represent newly synthesized isoforms or arise from post-translational modification or processing of already existing proteins.

In another study, a detailed 2-D protein map of Lhcb proteins from *C. reinhardtii* was established (A. Fink, E. Stauber, U. Johanningmeier, and M. Hippler, in prep.). Lhcb1 (CabII-1), LhcII-1.3, CabII-2 and LhcII-4 as well as CP29 and CP26 protein spots were identified. Interestingly, two Lhcb1 proteins with different molecular masses and different isoelectric points were found. Previous mass spectrometric analysis of 2-DE separated Lhcb1 had already indicated that this LHCII protein may contain two alternative N-terminal transit peptide cleavage sites (Hippler et al. 2001).

An HA-tag was introduced into the *lhcb1* gene in such a way that it is localized behind the second suggested N-terminal processing site in the expressed gene product (Imbault et al. 1988). Anti-HA monoclonal antibodies recognized two protein bands with different molecular masses (mass difference: 1–2 kDa) after SDS-PAGE fractionation of thylakoids from the Lhcb1-HA-tagged algal strain. Immunoblot experiments with 2-DE separated Lhcb1-HA-tagged thylakoids using anti-Lhcb1 peptide antibodies (directed against a sequence upstream of the second processing site) and anti-HA antibodies revealed that the lower molecular weight spots recognized by the HA antibody are not recognized by the anti-Lhcb1 antibodies, demonstrating that processing occurs at the N-terminus. Thus, the Lhcb 2-D mapping results and the tagging experiment independently confirmed that two differentially N-terminally processed Lhcb1 forms exist in vivo (A. Fink, E. Stauber, U. Johanningmeier, and M. Hippler, in prep.).

A non-2-DE based proteomic analysis of integral membrane proteins of the chloroplast envelope has been established by combining extraction of hydrophobic proteins with organic solvents, SDS-PAGE separation and MS/MS (Ferro et al. 2002). This approach has led to the identification of altogether 54 proteins. 27 of them were newly identified envelope proteins, most of them having several F128a-helical transmembrane domains. Some of these integral proteins are most likely transporter proteins, which have not been predicted to be localized in the chloroplast envelope before. The identification of these new proteins together with proteins that were already known for their localization in the chloroplast inner envelope allowed to define features shared by these proteins. These common features were then used to generate a virtual database of integral membrane proteins in the *Arabidopsis* plastid envelope. This database will be an important tool for the further characterization of chloroplast envelope membrane proteins by biochemical and/or reverse genetic approaches.

In addition to protein separation by 2-DE or SDS-PAGE, intact membrane proteins can also be extracted with organic solvents and separated by reverse-phase HPLC. Identification of chloroplast transmembrane proteins has been achieved by coupling liquid chromatography with electrospray-ionisation MS and MS/MS. This combination of techniques also succeeded in the characterization and identification of intact intrinsic thylakoid membrane proteins (Whitelegge et al. 1998; Corradini et al. 2000; Huber et al. 2001; Gomez et al. 2002).

All these proteomics analyses of chloroplast intrinsic membrane proteins demonstrate that, contrary to earlier belief, the hydrophobic protein core of a membrane system can be dissected by systematic proteomics approaches. The technical progress that has come with these studies is likely to have general implications for research on biological membranes.

5 Summary and Outlook

With recent fascinating progress in protein separation techniques and mass spectrometry methods, the global analysis of prokaryotic proteomes and eukaryotic subproteomes has now become feasible. Chloroplast subproteomes have provided excellent model systems to improve proteomics methods and refine bioinformatics tools for large-scale protein data analysis. Clearly, it will be important to combine the structural analysis of proteomes with appropriate functional studies. At the same time, the increasing identification of novel protein isoforms and post-translational modifications by proteomics approaches raises new questions about the functional significance of this protein heterogeneity.

What comes next? It seems clear that a complete picture of the chloroplast proteome (and any other cellular or compartmental proteome) cannot be achieved in the near future, especially since the limited sensitivity of protein detection after electrophoretic separation at present prevents the identification of lowly abundant proteins. It is for this reason that, for example, most regulator proteins of key processes in gene expression and metabolic control currently escape detection in proteomics studies. Novel and more sensitive methods in protein separation and mass spectrometry will have to be applied to plastid proteomics (see above). However, with the recent acceleration of proteomics research, it can be envisioned that even low abundant regulatory proteins can be identified in the near future facilitating the dissection of regulatory protein networks with proteomics tools.

Very obviously, the proteome of a given organism or compartment is more than the sum of the parts and by no means a static state of affairs. Although some information has been acquired on the protein set of the chloroplast, virtually nothing is known about the proteome of chromoplasts, proplastids or amyloplasts. Also, how does the chloroplast proteome change during plant development or in response to changing environmental conditions (e.g. upon exposure of the plant to biotic or abiotic stresses)? These questions lead into a field which may be called comparative proteomics. Here, plastids again provide an excellent model system since a large number of differentiation forms, developmental stages and environmental conditions can be compared for a reasonably small proteome. Moreover, all the many chloroplast mutants in photosynthesis and plastid gene expression that have been isolated or experimentally generated (Bock and Hippler 2002) over the past decades are likely to become attractive targets of future comparative proteomics studies.

In summary, we are still a long way from understanding the full complexity and dynamics of even a relatively small subproteome as the one of the chloroplast. Last, but not least, it needs to be borne in mind that proteomics, just like any other type of large-scale data collection, ultimately must lead to the generation of new knowledge: Huge amounts of collected data alone do not mean much. They must be converted into understanding – a challenge which goes far beyond collecting information.

References

Abdallah F, Salamini F, Leister D (2000) A prediction of the size and evolutionary origin of the proteome of chloroplasts of *Arabidopsis*. Trends Plant Sci 5:141–142

Bock R, Hippler M (2002) Extranuclear inheritance: functional genomics in chloroplasts. Progress in botany 63. Springer, Berlin Heidelberg New York, pp 106–131

Corradini D, Huber CG, Timperio AM, Zolla L (2000) Resolution and identification of the protein components of the photosystem II antenna system of higher plants by reversed-phase liquid chromatography with electrospray-mass spectrometric detection. J Chromatogr A 886:111–121

Eng J, McCormack AL, Yates JR (1994) An approach to correlate tandem mass spectral data of peptides with amino acid sequences in a protein database. J Am Soc Mass Spectrom 5:976–989

Ferro M, Salvi D, Riviere-Rolland H, Vermat T, Seigneurin-Berny D, Grunwald D, Garin J, Joyard J, Rolland N (2002) Integral membrane proteins of the chloroplast envelope: identification and subcellular localization of new transporters. Proc Natl Acad Sci USA 99:11487–11492

Gavin AC, Bosche M, Krause R, Grandi P et al. (2002) Functional organization of the yeast proteome by systematic analysis of protein complexes. Nature 415:141–147

Gomez SM, Nishio JN, Faull KF, Whitelegge JP (2002) The chloroplast grana proteome defined by intact mass measurements from liquid chromatography mass spectrometry. Mol Cell Proteomics 1:46–59

Gygi SP, Rist B, Gerber SA, Turecek F, Gelb MH, Aebersold R (1999) Quantitative analysis of complex protein mixtures using isotope-coded affinity tags. Nat Biotechnol 17:994–999

Hippler M, Klein J, Fink A, Allinger T, Hoerth P (2001) Towards functional proteomics of membrane protein complexes: analysis of thylakoid membranes from *Chlamydomonas reinhardtii*. Plant J. 28:595–606

Huber CG, Timperio AM, Zolla L (2001) Isoforms of photosystem II antenna proteins in different plant species revealed by liquid chromatography-electrospray ionization mass spectrometry. J Biol Chem 276:45755–45761

Imbault P, Wittemer C, Johanningmeier U, Jacobs JD, Howell SH (1988) Structure of the *Chlamydomonas reinhardtii* cabII-1 gene encoding a chlorophyll-*a*/*b*-binding protein. Gene 73:397–407

Kaneko T, Tabata S (1997) Complete genome structure of the unicellular cyanobacterium Synechocystis sp. PCC6803. Plant Cell Physiol 38:1171–1176

Kaneko T, Sato S, Kotani H, Tanaka A, Asamizu E, Nakamura Y, Miyajima N, Hirosawa M, Sugiura M, Sasamoto S, Kimura T, Hosouchi T, Matsuno A, Muraki A, Nakazaki N, Naruo K, Okumura S, Shimpo S, Takeuchi C, Wada T, Watanabe A, Yamada M, Yasuda M, Tabata S (1996) Sequence analysis of the genome of the unicellular cyanobacterium *Synechocystis* sp. strain PCC6803. II. Sequence determination of the entire genome and assignment of potential protein-coding regions. DNA Res 3:109–136

Klose J (1999) Genotypes and phenotypes. Electrophoresis 20:643–652

Koller A, Washburn MP, Lange BM, Andon NL, Deciu C, Haynes PA, Hays L, Schieltz D, Ulaszek R, Wei J, Wolters D, Yates JR 3rd (2002) Proteomic survey of metabolic pathways in rice. Proc Natl Acad Sci USA 99:11969–11974

Kotani H, Tabata S (1998) Lessons from sequencing of the genome of a unicellular cyanobacterium, *Synechocystis* SP. PCC6803. Annu Rev Plant Physiol Plant Mol Biol 49:151–171

Link AJ, Eng J, Schieltz DM, Carmack E, Mize GJ, Morris DR, Garvik BM, Yates JR 3rd (1999) Direct analysis of protein complexes using mass spectrometry. Nat Biotechnol 17:676–682

Mann M, Wilm M (1994) Error-tolerant identification of peptides in sequence databases by peptide sequence tags. Anal Chem 66:4390–4399

Martin W, Rujan T, Richly E, Hansen A, Cornelsen S, Lins T, Leister D, Stoebe B, Hasegawa M, Penny D (2002) Evolutionary analysis of *Arabidopsis*, cyanobacterial, and chloroplast genomes reveals plastid phylogeny and thousands of cyanobacterial genes in the nucleus. Proc Natl Acad Sci USA 99:12246–12251

Moseley JL, Allinger T, Herzog S, Hoerth P, Wehinger E, Merchant S, Hippler M (2002) Adaptation to Fe-deficiency requires re-modelling of the photosynthetic apparatus. EMBO J 21:6709-6720

O'Farrell P (1975) High resolution two-dimensional electrophoresis of proteins. J Biol Chem 250:4007–4021

Peltier JB, Friso G, Kalume DE, Roepstorff P, Nilsson F, Adamska I, van Wijk KJ (2000) Proteomics of the chloroplast: systematic identification and targeting analysis of lumenal and peripheral thylakoid proteins. Plant Cell 12:319–341

Peltier JB, Ytterberg J, Liberles DA, Roepstorff P, van Wijk KJ (2001) Identification of a 350-kDa ClpP protease complex with 10 different Clp isoforms in chloroplasts of *Arabidopsis thaliana*. J Biol Chem 276:16318–16327

Peltier JB, Emanuelsson O, Kalume DE, Ytterberg J, Friso G, Rudella A, Liberles DA, Söderberg L, Roepstorff P, von Heijne G, van Wijk KJ (2002) Central functions of the lumenal and peripheral thylakoid proteome of Arabidopsis determined by experimentation and genome-wide prediction. Plant Cell 14:211–236

Peng J, Gygi SP (2001) Proteomics: the move to mixtures. J Mass Spectrom 36:1083–1091

Perkins DN, Pappin DJ, Creasy DM, Cottrell JS (1999) Probability-based protein identification by searching sequence databases using mass spectrometry data. Electrophoresis 20:3551–3367

Puig O, Caspary F, Rigaut G, Rutz B, Bouveret E, Bragado-Nilsson E, Wilm M, Seraphin B (2001) The tandem affinity purification (TAP) method: a general procedure of protein complex purification. Methods 24:218–229

Rujan T, Martin W (2001) How many genes in Arabidopsis come from cyanobacteria ? An estimate from 386 protein phylogenies. Trends Genet 17:113–121

Santoni V, Molloy M, Rabilloud T (2000) Membrane proteins and proteomics: un amour impossible? Electrophoresis 21:1054–1070

Schubert M, Petersson UA, Haas BJ, Funk C, Schroder WP, Kieselbach T (2002) Proteome map of the chloroplast lumen of *Arabidopsis thaliana*. J Biol Chem 277:8354–8365

The Arabidopsis Genome Initiative (2000) Analysis of the genome sequence of the flowering plant *Arabidopsis thaliana*. Nature 408:796–815

van Wijk KJ (2000) Proteomics of the chloroplast: experimentation and prediction. Trends Plant Sci 5:420–425

Washburn MP, Wolters D, Yates JR 3rd (2001) Large-scale analysis of the yeast proteome by multidimensional protein identification technology. Nat Biotechnol 19:242–247

Whitelegge JP, Gundersen CB, Faull KF (1998) Electrospray-ionization mass spectrometry of intact intrinsic membrane proteins. Protein Sci 7:1423–1430

Yamaguchi K, Subramanian AR (2000) The plastid ribosomal proteins. Identification of all the proteins in the 50S subunit of an organelle ribosome (chloroplast). J Biol Chem 275:28466–28482

Yamaguchi K, von Knoblauch K, Subramanian AR (2000) The plastid ribosomal proteins. Identification of all the proteins in the 30S subunit of an organelle ribosome (chloroplast). J Biol Chem 275:28455–28465

PD Dr. Michael Hippler
Friedrich-Schiller-Universität Jena
Institut für Allgemeine Botanik
Dornburger Strasse 159
07743 Jena, Germany
Tel.: +49-3641-949237
Fax: +49-3641-949232
e-mail: m.hippler@uni-jena.de

Prof. Dr. Ralph Bock
Westfälische Wilhelms-Universität Münster
Institut für Biochemie und
Biotechnologie der Pflanzen
Hindenburgplatz 55
48143 Münster, Germany
Tel.: +49-251-83-24790/91
Fax: +49-251-8328371
e-mail: rbock@uni-muenster.de

Communicated by
K. Esser

Molecular Cell Biology: Organization and Molecular Evolution of rDNA, Nucleolar Dominance, and Nucleolus Structure

Roman A. Volkov, Francisco Javier Medina, Ulrike Zentgraf, and Vera Hemleben

1 Introduction

Recently, the ribosomes – molecularly well-characterized cell organelles where the process of translation of messenger RNA into polypeptides occurs – have attracted new interest since they can be considered as a huge ribozyme-like complex consisting of different proteins and RNA – the RNAs mainly fulfilling functional tasks in translation, the proteins serving more structural functions (Moore and Steitz 2002). Within the eukaryotic cell, two (animals and fungi) or three (plants) sites of a translational machinery are present, the cytoplasm, mitochondria and chloroplasts, and, respectively, the components are mostly encoded in the cell nucleus or the mitochondrial or chloroplast genome. Since new insights are gained into the production and the assembly of the ribosome components, ribosomal RNA (rRNA) and ribosomal proteins (r-proteins), in this chapter, we focus on the nucleolus, the site for cytoplasmic ribosome biogenesis and assembly occurring in the light-microscopically visible structure of the cell nucleus. Similarly as in prokaryotes, for all higher organisms two ribosome subunits are preformed in the nucleolus, which are evolutionary conserved: the small 40S subunit (SSU) containing 18S rRNA, and the large 60S subunit containing 25/28S plus 5.8S and 5S rRNA. After export of the ribosome subunits (LSU) into the cytoplasm, the functionally active ribosome is associated with translatable mRNA to an 80S complex corresponding to the 70S structure in chloroplasts and prokaryotic cells.

Following the ribozyme concept, ribosome subunits are built up by rRNA and r-proteins. The nuclear encoded rRNA genes (rDNA) are present as a multigene family in the genome, mostly located at the NOR (nucleolus-organizing region). In the transition area from fibrillar centers to the dense fibrillar component (FC/DFC) of the nucleolus, the 18S, 5.8S and 25S/28S rRNA are commonly transcribed by RNA polymerase I (pol I) as a large precursor, which in plants is approx. 35S in size (35S pre-rRNA); this 35S pre-rRNA is then stepwisely processed into the respective rRNAs. RNA polymerase III (pol III) transcribes the 5S rRNA component encoded by 5S

Progress in Botany, Vol. 65

rDNA located at other regions of the genome for most higher organisms. For processing of the pre-rRNA, small nucleolar RNAs (snoRNAs) genes are transcribed by either RNA polymerase II (pol II) or pol III. For methylation of rRNA and structural functions, the mRNA for enzymatic or structural proteins supporting the processing as snoRNPs (small nucleolar RNA/protein complexes) is produced by pol II. During the process of ribosome biogenesis, the rRNAs are associated with r-proteins to pre-ribosomes. Ribosomal proteins are encoded in the nuclear genome and transcribed by pol II into mRNA; this mRNA is translated into r-proteins on cytoplasmic ribosomes, which are mostly retransported to the nucleolus, where assembly of pre-ribosomes takes place, thus forming the granular component (GC). Therefore, we can expect a coordinated cross-talk not only between the different transcriptional and translational machineries between the cell nucleus and the cytoplasm, but also among the different cell organelles.

From extensive molecular studies of plant rRNA genes, we now have information on the structural organization of the multigene families, often comprising between 500 and more than 30,000 members, depending on the plant species (Hemleben et al. 1988), and their use for evolutionary and phylogenetic studies became evident. However, our knowledge on the regulation of transcription by specific factors and processing for animal and yeast cells is by far more detailed than for plant ribosomal genes.

Ribosomes can be considered as basic organelles of the cell; their components have been generally termed as fulfilling "house-keeping" functions and the biosynthesis of the respective components as occurring constitutively. Recently, more and more information is obtained on their highly regulated and finely tuned expression realized by transcriptional enhancement or by "epigenetic" processes like gene silencing. Especially for higher plants, where interspecific hybridization accompanied by polyploidization occurs in nature, the phenomenon of "nucleolar dominance", defined as active transcription of one partner rDNA and silencing of the respective other partner rDNA (Pikaard 2000), allows more insight into the regulation processes.

The new field of proteomics revealed more than 230 nucleolar proteins for humans and yeast, but up to now fewer than 100 nucleolar proteins have been identified in functional terms. Therefore, a significant proportion of these proteins are suggested to be involved in functions beyond the known role of the nucleolus in ribosome biogenesis (Andersen et al. 2002; Scherl et al. 2002), supporting the idea of a "plurifunctional nucleolus" (Pederson 1998).

Especially for higher plants, in the last decade, specific proteins gained much interest as therapeutic agents acting as anticancer or antiviral compounds: Ribosome-inactivating proteins (RIPs) are potent plant toxins with RNA *N*-glycosidase activity. They are strong inhibitors of eukaryotic protein synthesis that inactivate eukaryotic ribosomes by cleaving the *N*-glycosidic bond of a specific adenylate (A) residue in the 25/28S rRNA at the α-sarcin/ricin loop. Type 1 and type 2 RIPs were characterized for various plant species.

Here, we review and summarize results on the organization and transcriptional regulation of the nuclear-encoded rRNA genes and their use for evolutionary and phylogenetic studies, and emphasize nucleolar proteins, the nucleolus structure, and ribosome assembly, in order to supplement recent reviews on rDNA regulation and function (Hemleben and Zentgraf 1994; Zentgraf et al. 1998).

2 Structural Organization and Molecular Evolution of rDNA

2.1 Organization of 35S rDNA

The nuclear genes coding for the 18S, 5.8S, and 25/28S rRNA, belonging to the class of repeated sequences, are arranged in head-to-tail tandem arrays (with only rare exceptions) and located in one or more places in the genome. The transcribed genes, which form the NOR, can be cytologically distinguished as secondary constrictions (SC) in mitosis/meiosis or as nucleolus in interphase nuclei (Leweke and Hemleben 1982; Hemleben and Zentgraf 1994; Moss and Stefanovsky 1995). A rDNA repeat unit is composed of the rRNA coding regions, external and internal transcribed spacers (ETS and ITS), and non-transcribed spacer (NTS) regions (Fig. 1). The order of the coding regions is universal in all eukaryotes. Additionally, in rRNA-coding regions inserted sequences were described for insects (England et al. 1988; Bigot et al. 1992) and green algae (Wilcox et al. 1992).

Different portions of the rDNA repeat unit evolve with different rates. The coding regions represent one of the most conservative sequences in eukaryotes (Lipscomb et al. 1998), which is obviously a result of a strong selection against any loss-of-function mutations in components of the ribosome subunits (Caetano-Anolles 2002; Moore and Steitz 2002). The most conservative part appears to be the 3' end of 25S rDNA representing the α-sarcin/ricin (S/R) loop (see below). Similarly, a conservation of some sequence motifs and/or secondary structures in other subregions of rDNA/rRNA indicates a functional significance. For instance, sequence motifs conserved in the ITS seem to be necessary for interaction with the proteins involved in processing of the 35S pre-rRNA (Torres et al. 1990; Liu and Schardl 1994; Mai and Coleman 1997).

In contrast to the coding regions, the intergenic spacer (IGS) region evolves rather quickly; nearly no sequence similarity can be found in this region for members of different plant families (Hemleben and Zentgraf 1994). An exception represents a sequence motif for the transcription initiation site (TIS), TATATA(A/G)GGG in dicots and TATAGTAGGG(A/G) in monocots (Perry and Palukaitis 1990; Zentgraf et al. 1990; Doelling and Pikaard 1995; Volkov et al. 1996; Akhunov et al. 2001). Another comparatively conservative portion of the IGS is a region adjacent to the 18S rDNA in different families of dicots (King et al. 1993; Volkov et al. 1999b). These motifs could play a role in pre-rRNA processing.

A specific feature of the IGS is not only a higher rate of base substitutions, but also a remarkable length variability (Rogers and Bendich 1987a; Hemleben et al. 1988). Several length variants of IGS can be present in the same genome (Hemleben et al. 1988; Zentgraf et al. 1990). Species of the same genus often differ significantly in the length of the IGS: e.g. from 1.7 to 6.4 kb in *Trillium* (Yakura et al. 1983), from 2.6 to 6.5 kb in *Prunus* (Volkov et al. 1993) or from 3.4 to 5.9 kb in *Nicotiana* (Borisjuk et al. 1997). Moreover, even for individuals of the same population, the length of the IGS can differ (Learn et al. 1987; Rogers and Bendich 1987a). For instance, high variability of IGS was found for several species of family Fabaceae: *Vicia faba* (Kato et al. 1985; Rogers and Bendich 1987b), *Pisum sativum* (Polans et al. 1986), and *Phaseolus* (Gerstner et al. 1988; Schiebel et al. 1989; Maggini et al. 1992). However, there are also examples of low variability or even uniformity of IGS length among species of the same genus (Choumane and Heizmann 1988; Santoni and Bervielle 1992; Borisjuk et al. 1994). The degree of IGS variability could be connected with character of pollination: In the genus *Hordeum*, cross-pollinated species mainly possess one length variant of the IGS, whereas self-pollinated species show intragenomic polymorphisms (Molnar et al 1989); however, such a correlation was not found for *Clematis* or *Rudbeckia* (Learn et al. 1987; King and Schaal 1989).

The length variability of IGS is mainly due to amplification of different subregions that results in the appearance of subrepeated elements within the IGS (see Fig. 1; Hemleben and Zentgraf 1994).

Normally, several classes of short subrepeated elements are present upstream of TIS (e.g. Kato et al. 1985; Kelly and Siegel 1989; Delseny et al. 1990; Perry and Palukaitis 1990; Gründler et al. 1991; Borisjuk and Hemleben 1993; King et al. 1993; Volkov et al. 1996; Borisjuk et al. 1997). This portion of IGS accumulates base substitutions with a high rate: obvious sequence similarity can be found here only among members of the same genus or closely related genera. In contrast, downstream of TIS, subrepeats are not always present. Downstream subrepeats accumulate base substitutions more slowly. Interestingly, different sequences were independently amplified downstream of TIS in *Nicotiana* and *Lycopersicon*, producing subrepeats of nearly the same size, 140–145 bp (Volkov et al. 1996; 2003; Borisjuk et al. 1997). Amplification of subrepeats in IGS occurred several times during evolution. For instance, in the genus *Nicotiana*, original duplication of the ancestral A-subrepeat sequence in the ETS took place before the divergence of subgenera, and produced two subvariants, A1 and A2. Additional amplification during later speciation of *Nicotiana* formed longer stretches of A1/A2-subrepeats independently in different species (Volkov et al. 1996). Similarly, several rounds of amplification/deletions generated a "chaotic mix" of C-subrepeats upstream of TIS (Volkov et al. 1999a).

In spite of length and sequence variability, the general arrangement of IGS often remained unaltered among members of the same family. Comparison of IGS organization in distantly related Solanaceae, *Lycopersicon esculentum* (Perry and Palukaitis 1990), *Solanum tuberosum* (Borisjuk and Hemleben 1993), *Atropa belladonna* (Volkov et al. 1997), and three species of *Nicotiana* (Borisjuk et al. 1997; Volkov et al. 1999a), shows that all these species possess (1) a unique AT-rich sequence in the central part of IGS, immediately upstream of TIS, (2) subrepeated elements upstream of TIS, and (3) similar localization of stem-loop structures (Fig. 1).

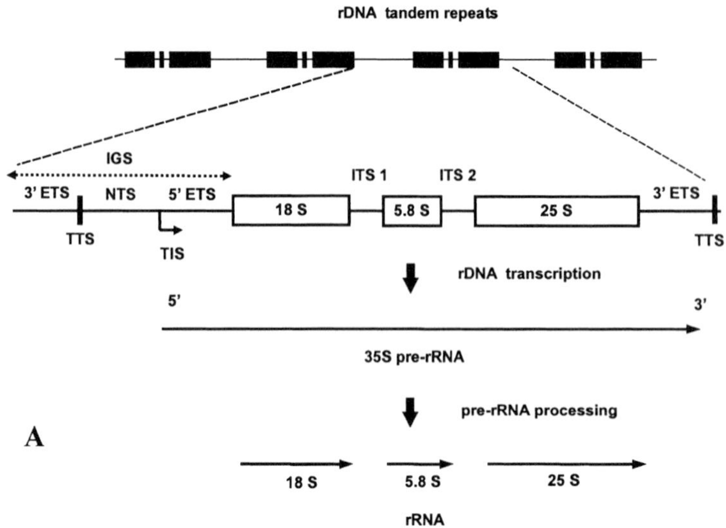

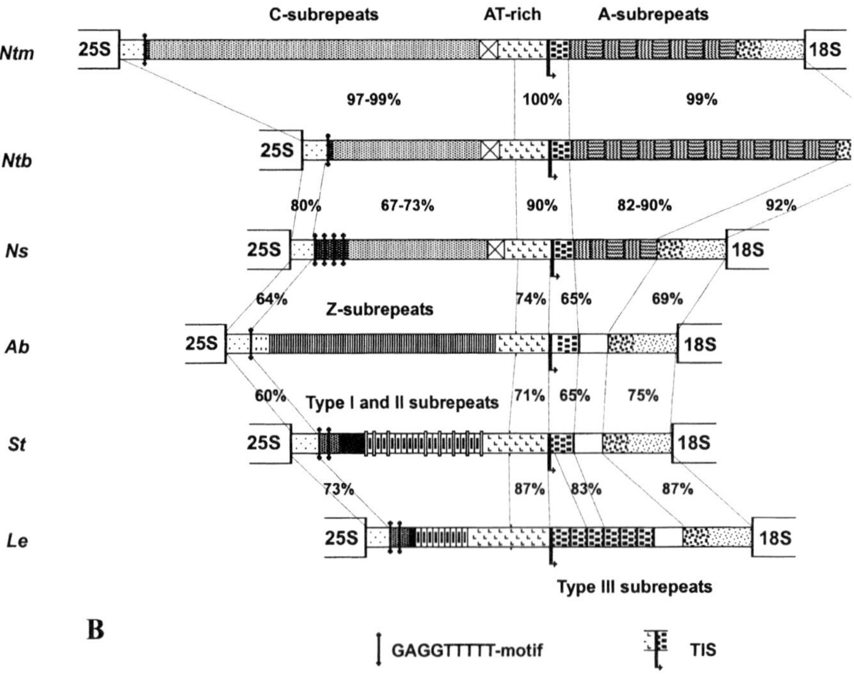

Fig. 1. Structural organization of 35S rDNA. **A** Schematic presentation of structural organization of rDNA, expression and processing. **B** Organization and sequence similarity of the intergenic spacers (IGS) of rDNA of different Solanaceae: *Ntm*, *N. tomentosiformis*; *Nbt*, *N. tabacum*; *Ns*, *N. sylvestris*; *Ab*, *Atropa belladonna*; *St*, *S. tuberosum*; *Le*, *L. esculentum*. Percent of similarity for different IGS subregiones are given. *TIS* Transcription initiation site; *TTS* transcription termination site; *ETS* external transcribed spacer; *ITS* internal transcribed spacer; *NTS* non-transcribed spacer

2.2 Organization of 5S rDNA

In eukaryotes, the 5S rDNA also belongs to the repeated sequence classes and is arranged in head-to-tail tandem arrays. Usually, 5S rDNA and 18–25S rDNA are located at different locations of the same chromosome or often at different chromosomes. The internal structural organization of 5S rDNA repeated unit is comparatively simple: Each repeat is composed of a conserved, approx. 120-bp-long coding region, and of a more variable intergenic spacer region (5S SR) (Hemleben and Werts 1988; Ellis et al. 1988; Scoles et al. 1988; Lapitan et al. 1991).

The high conservation of the 5S rRNA coding region exhibits some family-specificity, especially at the 5'end: 5S rRNA genes begin with GGA in Solanaceae (Frasch et al. 1989; Barciszewska et al. 1994; Volkov et al. 2001), with AGG in Fabaceae, and GGG in Brassicaceae (Hemleben and Werts 1988; Barciszewska et al. 1994) and in Poaceae (Van Campenhout et al. 1998; Röser et al. 2001). The 5S SR can be subdivided into three subregions: 3' and 5' flanking sequences (3' and 5' FS) and middle variable region (VR). In comparison to VR, both 3' and 5' FS are more conservative, suggesting presumptive functional significance (Hemleben and Werts 1988; Van Campenhout et al. 1998; Crisp et al. 1999; Trontin et al. 1999; Röser et al. 2001; Volkov et al. 2001). In the VR, two further portions can be distinguished: (1) the AT-rich and (2) the subrepeated regions. Rearrangements in VR are preferentially associated with the repeated motifs.

2.3 Molecular Evolution and Taxonomic Applications of rDNA

Concerted evolution, which enhances sequence similarity between members of multigene families, is a specific feature of repeated genome elements, including both 5S and 35S rDNA (Coen et al. 1982; Dover and Flavell 1984; Dvorak et al. 1987; Hemleben et al. 2000; Ganley and Scott 2002). A new mutation appearing in a single repeat unit has to be either quickly eliminated from the genome or distributed through all other repeats of the family. Although numerous copies of rDNA are present in the genome, concerted evolution often results nearly in identity of individual repeats of 35S (Volkov et al. 1996, 1999a; Linder et al. 2000) and 5S rDNA (Volkov et al. 2001).

Detailed comparison of numerous ETS clones isolated for 30 *Solanum* species (Volkov et al. 2003) demonstrated an unusually high level of identity: More than 99% sequence similarity was found for ten diploid species. For two allopolyploids, intergenomic sequence similarity was lower (93.5–97.7%), demonstrating that homogenization of rDNA inherited from parental species may take time. Remarkably, in two other polyploids, the rDNA repeats appear to be very homogeneous: 99.1–99.6%, considering the following reasons: (1) the polyploids appeared as a result of hybridization between closely related diploids; (2) the rDNA introduced by one of the parents was eliminated from the genome, similar to the situation described for the allotetraploid *N. tabacum* (Volkov et al. 1999a; see also below).

Efficiency of homogenization by concerted evolution depends on the local distance between repeated units (Dvorak et al. 1987). Particularly, 35S rDNA repeats within the same NOR are more similar to one another than repeats from different NORs, as it was originally shown for *Drosophila* (Schlotter and Tautz 1994). In *Arabidopsis thaliana*, a common IGS length variant was found among 35S rDNA repeats at NOR2 located on chromosome 2, whereas NOR4 on chromosome 4 contains three length variants (Copenhaver and Pikaard 1996). The NOR4 variants are not intermingled with one another, but highly clustered, suggesting that in the concerted evolution of rRNA genes homogenization is a consequence of local spreading of new rDNA variants. On the other hand, interspersion of 5S repeats of different length was observed in pea (Ellis et al. 1988) and in *Larix* (Trontin et al. 1999), suggesting different efficiency of concerted evolution of different repeated sequences.

Homogenization takes place not only between the complete copies of 35S rDNA repeats, but also between the subrepeats within the same repeat unit. Remarkably, the border subrepeat units are more divergent from the corresponding consensus sequence than the central ones, indicating that the homogenization process is more effective in the central portion of subrepeated region (Barker et al. 1989; Schiebel et al. 1989; Volkov et al. 1996). These observations agree with calculations derived from computer modeling of tandem arrays (Dvorak et al. 1987) and support the assumption that unequal crossing over is involved in the homogenization of repeated sequences.

Due to the existence of regions evolving with different rates, 35S rDNA represents a very attractive tool for molecular taxonomy. In addition, due to the higher similarity of individual repeat units found not only in coding, but also in spacer regions, it is often not necessary to sequence many individual clones; even application of direct sequencing could produce correct results. Coding regions with the lowest rate of evolution were successfully used for phylogenetic studies of distantly related taxa (Lipscomb et al. 1998), whereas comparison of more rapidly evolving ITS 1 and 2 was widely applied for taxonomic reconstructions among members of the same or closely related genera, e.g. of Compositae (Baldwin 1992), Cucurbitaceae (Jobst et al. 1998), Aveneae (Grebenstein et al. 1998), *Phaseolus, Vigna* (Goel et al. 2002), and many others. Again, in some groups, ITS does not appear to be phylogenetically informative. In such cases, comparison of ETS (Volkov et al. 1996, 2003; Baldwin and Markos 1998; Bena et al. 1998; Linder et al. 2000) can be used. In many plants, both ITS and ETS evolve preferentially by stepwise base substitutions allowing correct sequence alignments and successful phylogenetic reconstructions even for closely related species (Denk et al. 2002). However, specific indels in the 35S 5' ETS can be used also as taxonomically relevant characters for separation of major taxonomic groups (Volkov et al. 2003).

Furthermore, 5S rDNA can be a valuable molecular marker for phylogenetic reconstructions at different taxonomic levels, as widely used already for the coding region; but also the spacer region contains phylogenetic information for low-level taxonomy (Röser et al. 2001; Volkov et al. 2001; and above).

3 Transcriptional Regulation of rDNA

3.1 Promoter Structure and Transcription Initiation of 35S rDNA

Transcription of 35S rDNA is specifically carried out by pol I, starting from the TIS in the IGS, synthesizing the 35S pre-rRNA, and finishing at the transcription termination site (TTS). Hence, the rRNA precursor contains not only sequences of 18S, 5.8S, and 25–28S rRNA, but also of 5' and 3' ETS and of two ITS, which are removed in several subsequent steps during the processing (see Fig. 1; reviewed by Zentgraf et al. 1998).

In situ immunolocalization methods detected pol I in the nucleolar FC, and, in the onion nucleolus, in the transition area between FC and the DFC as well as in the mitotic secondary constriction of chromosomes, which corresponds to the NOR (Scheer and Rose 1984; Martín and Medina 1991; Gilbert et al. 1995). Interestingly, although the enzyme was unequivocally found to be rDNA-bound, this fact was not necessarily associated with its current involvement in active transcription (Scheer and Rose 1984; Weisenberger and Scheer 1995).

In different eukaryotic species, pol I was purified as a multimeric complex formed by many subunits (Sentenac 1985). In the yeast *Saccharomyces cerevisiae*, 14 proteins ranging in mass from 8.3 to 186 kDa were found, 7 of which are specific for pol I, whereas the other 7 represent components common also for pol II and/or pol III (Carles and Riva 1998). Phosphorylation of several subunits and binding of zinc seem to be involved in activity regulation and stabilization of pol I. Probably, three core subunits, A190, A135 and AC40, which are structurally and functionally related to the eubacterial RNA polymerase core enzyme, provide the main basal functions, i.e. template-dependent polymerization itself, whereas regulatory and/or species-specific aspects are connected with the remaining non-core subunits. In contrast to yeast and animals, much less is known about pol I in plants: Earlier, the chromatin-bound enzyme was isolated from cauliflower inflorescence (Guilfoyle 1980), and a putative pol I holoenzyme complex was purified from broccoli, *Brassica oleracea*. This cell-free system was able to support promoter-dependent pol I transcription in vitro (Saez-Vasquez and Pikaard 1997). The complex is composed of more than 30 polypeptides, including protein kinase, histone acetyltransferase and topoisomerase (Albert et al. 1999; Hannan et al. 1999; Saez-Vasquez et al. 2001).

It is known for animals and yeast that pol I requires additional transcription initiation factors (TIFs) for its function and several transcription-associated factors (TAFs) for interaction with the promoter. The rDNA promoter region consists of two functional subregions: core promoter (CP, from about −40 to +10 bp) and upstream element (UE, up to −150 bp). The only well-studied species in which UE has not been identified is *Acanthamoeba castellanii*, but this may be because the CP in this species is too strong to allow detection of UE in vitro (Paule 1998a,b). Although UE is not absolutely required for transcription initiation, it stimulates the intensity and provides species-specificity of expression (Heix and Grummt 1995; for review see: Moss and Stefanowsky 1995; Paule 1998a; Zentgraf et al. 1998; Grummt 1999). Direct binding of protein factors to UE seems to be the initial step of preinitiation complex formation for pol I, whereas several other TAFs are subsequently recruited to the promoter (Schnapp and Grummt 1991; Comai et al. 1992; Moss and Stefanowsky 1995; Steffan et al. 1996; Grummt 1999). Interestingly, in spite of a similar function, TAFs differ essentially in different taxonomic groups. In yeast, several proteins interact with UE, being substituted in vertebrates by a single large upstream binding factor (UBF), which contains multiple DNA binding domains known as HMG boxes due to their similarity to HMG non-histone chromosomal proteins (Reeder 1990; Schnapp and Grummt 1991; Zentgraf et al. 1998; Grummt 1999).

In contrast, little is known about the rDNA promoter structure and TAFs in plants (for review see Hemleben and Zentgraf 1994; Zentgraf et al 1998). In *A. thaliana*, a sequence between –55 and –33 bp upstream and +6 bp downstream of TIS was found to be sufficient for accurate pol I transcription; no domains equivalent to UE were found (Doelling and Pikaard 1995, 1996; Saez-Vasquez and Pikaard 1997). The authors speculate that plants may have simple rRNA promoters equivalent to the core promoters in non-plant systems (Pikaard 2002). Pol I could exist as a multiprotein holoenzyme, which can associate with the promoter in a single DNA binding event (Saez-Vasquez and Pikaard 2000). However, a longer promoter sequence, between –113 and +15 bp, was necessary to direct pol I-dependent transcription in *Triticum* species (Akhunov et al. 2001), suggesting the presence of UE also in plant pol I promoters.

Similarly, it remains unclear if any homologues of yeast or animal TAFs are present in plants. Remarkably, autoantiserum against human UBF is capable of detecting the homologous protein in onion cells. The protein appeared to be preferentially located in FCs and on the transition area FC/DFC, but a significant amount was found in the DFC (Rodrigo et al. 1992; De Cárcer and Medina 1999). However, it should be additionally clarified, if the respective plant protein represents a real homologue of mammalian UBF, or, alternatively, some other HMG protein, which cross-reacts with the antibody used. According to our preliminary results, at least five proteins, ranging in size from 34 to 120 kDa, interact with the promoter sequence from –165 to –97 bp upstream of TIS in tobacco (R.A. Volkov, V. Hemleben, unpubl.).

3.2 Modulation of rDNA Transcription

Different subregions of IGS other than the promoter can be involved in the regulation of rDNA transcription. AT-rich regions found in the IGS of different plants upstream of TIS represent presumptive enhancers (Borisjuk et al. 2000); similar sequences also improve the transcription of protein coding genes (Raschke et al. 1988; Schöffl et al. 1993; Sandhu et al. 1998). Besides, similarly to *Xenopus* (Moss 1983; Reeder 1984; Pape et al. 1989; Mougey et al. 1996), upstream subrepeats in plants could also represent enhancers (Flavell 1986; Zentgraf et al. 1998). In transgenic *Arabidopsis* plants, 35S rDNA upstream subrepeats increased transcription of a reporter gene up to four times if cloned upstream of the 35S minimal promoter (Schlogelhofer et al. 2002). However, the question still remains unanswered because later experiments with *Xenopus* did not confirm the enhancer function of upstream subrepeats (Caudy and Pikaard 2002). The functional role of downstream subrepeats also remains mostly unclear, however, new data demonstrate a role in the control of nucleolar dominance in potato-tomato hybrids (N.Y. Komarova, T. Grabe, V. Hemleben, R.A. Volkov, unpubl.).

3.3 Transcription Termination

Termination of pol I requires the interaction of a specific DNA binding factor with a terminator element downstream of the 3'end of 26S rDNA. These elements and corresponding protein factors are different in yeast and vertebrates, but the transcription termination factor of yeast, Reb1p, can function in vitro in combination with murine release factor, suggesting that the mechanism of pol I termination is highly conserved from yeast to mammals (Mason et al. 1997). In vivo analyses showed that in *Schizosaccharomyces pombe* the termination process strongly resembles the single element-mediated mechanism initially reported for mouse and is not dependent on an additional upstream sequence as first reported in yeast (Melekhovets et al. 1997). In the latter species, a stem-loop structure immediately downstream of the 25S rRNA region is necessary and sufficient for processing of the 3'ETS. Remarkably, the same stem loop is required for processing of ITS 1. This could be necessary to prevent processing of rRNA transcripts which are incomplete due to premature termination (Allmang and Tollervey 1998).

Still, termination of transcription of 35S rDNA in plants is poorly described. Position of transcription termination downstream of the pre-rRNA coding region was determined for several plants (for review see Hemleben and Zentgraf 1994). Although a region with 70% similarity to a terminator sequence in *Xenopus* rDNA was found near the termination signal in *Vigna radiata* (Schiebel et al. 1989), a consensus sequence for plants representing different families cannot be defined supposing a species specificity of rDNA transcription termination. However, in all studied genera of the family Solanaceae, a motif GAGGTTTTT appeared to be strongly conserved, indicating a presumptive involvement in transcription termination (Fig. 1). Additionally, termination of rRNA transcription could be controlled by a higher order stem-loop structure, as found in the IGS of *Vigna radiata* (Schiebel et al. 1989), *Cucumis sativus* (Zentgraf et al. 1990), and also in species of Solanaceae (Volkov et al. 1999b).

3.4 Replication of 35S Ribosomal DNA and Amplification Promoting Sequences

In eukaryotes, AT-rich sequence elements from IGS are not only involved in transcription initiation, but also participate in initiation of replication (Hernandez et al. 1988, 1993; Vant'Hof and Lamm 1992) and in amplification of rDNA repeats (Wegner et al. 1989; Hemann et al. 1994). Similarly, in AT-rich regions of potato and tobacco, a sequence element was identified which demonstrated a similarity to the muNTS1, an amplification-promoting sequence (APS) from the AT-rich region of mouse rDNA (Wegner et al. 1989). A similar function was suggested for plants, because in transgenic tobacco such an element was able to increase the copy number of adjacent reporter genes (Borisjuk et al. 2000). Short sequence motifs (A/T)TTTAT(A/G)TTT(A/T) appeared to be conserved in the APS of tobacco, potato, and mouse, and also in autonomous replication sequences (ARS) of yeast (Linkens and Hubermann 1988; Hernandez et al. 1993), indicating that amplification of rDNA repeats could function via repeated cycles of local replication (Hernandez et al. 1988; Borisjuk et al. 2000).

Another function of AT-rich regions may be an interaction with the nuclear matrix (Cockerill and Garrard 1986; Gasser and Laemmli 1986; Boulikas 1993). Remarkably, the same function could have TG-rich regions, present in the NTS of *Cucurbita maxima* (Kelly and Siegel 1989; King et al. 1993) and *Nicotiana* species (Borisjuk et al. 1997; Volkov et al. 1999b) because GT-rich stretches, usually 6–12 bp long, were found in nuclear matrix- or scafford-attached regions (SARs or MARs) of several genes (Boulikas 1993). Interestingly, GT-rich regions are completely absent in the NTS of *Solanum* and *Atropa* (Fig. 1). Recently, electron microscopic studies using anti-DNA antibody confirm that the NTS represents the anchoring site of *Allium cepa* rDNA (Yano and Sato 2002). Hence, AT-rich regions in rDNA probably provide (1) transcription enhancement, (2) rDNA replication and amplification, and (3) interaction with nuclear matrix, three functions which could be at least partially facilitated by topoisomerase II, one of the major scaffold proteins.

3.5 Functional Organization of 5S rDNA

In eukaryotes, 5S rDNA is transcribed by pol III which, as shown for animals and yeast, requires transcription factors TF IIIA, IIIB, and IIIC. TF IIIA, a zinc-finger protein, binds specifically to the internal control region (C-box) of 5S rRNA genes, which is followed by the sequential binding of TF IIIC, IIIB and pol III (Engelke et al. 1980; Pelham and Brown 1980; Ryan and Darby 1998). For plants, TF IIIA was purified and characterized from *Tulipa whittaili*, showing functional similarity of plant and amphibian factors (Wyszko and Barciszewska 1997); the factor-binding region appears also to be rather conserved (Hemleben and Werts 1988). In the 5S rDNA spacer some signals involved in transcription regulation are present as well: At the position –28 to –24 bp a conserved "TATA" box was found in distantly related *Vigna radiata* and *Matthiola incana* (Hemleben and Werts 1988). In *Solanum*, a similar TAATA motif was described at the same position, and in the 3' FS of 5S SR a pyrimidine-rich stretch probably functioning as termination site was found (Volkov et al. 2001). In the VR of 5S SR two portions can be distinguished: (1) the AT-rich and (2) the subrepeated regions. The AT-rich region shows similarity to the amplification-promoting sequences (Borisjuk et al. 2000) and may be involved in amplification of 5S rDNA repeats. Similarly, AT-rich segments were also observed in the 5S SR of Fabaceae (Hemleben and Werts 1988) and Poaceae (Röser et al. 2001). A function for the subrepeated region is not clear yet; however, repeated motifs present here show obvious homology to the box C, an internal control region of 5S rRNA genes (Pieler et al. 1987; Hemleben and Werts 1988).

4 Ribosomal DNA in Interspecific Hybrids

4.1 Nucleolar Dominance

A fascinating epigenetic phenomenon connected with the nuclear-encoded 35S rDNA, termed "nucleolar dominance" (ND; Honjo and Reeder 1973), was originally described as "differential amphiplasty" by Navashin (1928, 1934), who showed with cytological methods that in interspecific hybrids of *Crepis* only chromosomes of one crossing partner carry secondary constrictions in metaphase and produce active nucleoli in interphase. Obviously, the respective chromosomal regions were not lost in hybrids, but could be reactivated to produce normal nucleoli in hybrids with a different crossing partner. Based on these data, McClintock (1934) proposed that if the NOR of species A dominates over NOR of species B, and if B dominates over C, then NOR of species A should dominate over C. The existence of such a hierarchy for *Crepis* species was later confirmed experimentally (Wallace and Langbridge 1971). ND was demonstrated also for many other organisms, both plants and animals (for review see: Reeder 1985; Pikaard 2000).

With respect to current knowledge about the molecular functions of NOR, a differential transcription/silencing of parental 35S rDNA in hybrids could be anticipated, as it was demonstrated for hybrids of *Xenopus* (Honjo and Reeder 1973). In order to explain ND at the molecular level, several hypotheses were proposed. According to the species-specific transcription factor hypothesis (Reeder 1974, 1985; Dover and Flavell 1984; Saghai-Maroof et al. 1984), a rapid molecular evolution of the rDNA promoter region should be accompanied by the co-evolution of protein transcription factors interacting with the promoter and/or one with another. If so, inactivation of a species-specific transcription factor in hybrids should result in ND. This idea was confirmed for human and murine 35S rDNA promoters, which are transcriptionally active only in the cell-free extracts prepared from the same species, a phenomenon provided by the species-specific transcription factor SL1 (Grummt et al. 1982; Mishima et al. 1982; Learnd et al. 1985). Similar specificity was also observed for plants: A tobacco rDNA promoter was not recognized in cell-free transcription extracts from bean (Fan et al. 1995), and tomato promoter remained inactive after transfection into protoplasts of *Arabidopsis thaliana* (Doelling and Pikaard 1996). However, the hypothesis can not explain ND in hybrids of species which are sufficiently related to be crossable because in such species not only TIS, but also promoter regions upstream and downstream of it are as a rule highly homologous (King et al. 1993; Volkov et al. 1996, 2003; Chen and Pikaard 1997a,b; Akhunov et al. 2001). Similarly, protein factors interacting with the promoter region of *Solanum* species that exhibit differential expression of parental rDNA by crossing seem to be nearly identical (N.Y. Komarova, V. Hemleben, R.A.Volkov, unpubl.). Furthermore, 35S rDNA promoters of *Brassica* and *Arabidopsis* species, demonstrating ND when hybridized, are fully functional if transfected into protoplasts of the other species (Chen et al. 1998; Frieman et al. 1999).

Extensive investigation of ND in hybrids of wheat, rye and related genera of Triticeae revealed that rDNA repeats with longer IGS containing more subrepeated elements upstream of TIS, dominate over those with shorter IGS with less upstream subrepeats (Flavell and O'Dell 1979; Martini et al. 1982; Gustafson et al. 1988; Vakhitov et al. 1989; Dvorak 1993; Houchins et al. 1997). These data agree well with the enhancer imbalance hypothesis originally proposed for hybrids of *Xenopus*, where upstream subrepeats act as enhancers stimulating transcription if cloned adjacent to the gene promoter (Busby and Reeder 1983; Moss 1983; Reeder 1984; Pikaard and Reeder 1988; Pape et al. 1989). Assuming that enhancers bind transcription factors present in limited amounts, ND could be due to the competition for the factors between IGS of parental species (Reeder 1985; Flavell 1986). Selective methylation of under-dominant rDNA probably abolish the binding of transcription factors and results in differential NOR inactivation by chromatin condensation that is connected with histone deacetylation (Jasencakova et al. 2001) as shown for *Triticum-Secale* (Sardana et al. 1993; Houchins et al. 1997) and *Brassica* (Chen and Pikaard 1997a) hybrids. Nevertheless, in barley, differential expression of NORs present in wild type on chromosomes 6 and 7 does not always correlate with rDNA methylation in translocation/deletion lines (Papazova et al. 2001). Furthermore, in contrast to Triticeae, data obtained for interspecific hybrids of *Brassica* and *Arabidopsis* (Chen and Pikaard 1997a,b; Chen et al. 1998) were inconsistent with the enhancer imbalance hypothesis, because rDNA with longer spacers do not appear to be dominant in these species. However, for *Brassica* it is difficult to discuss the presumptive enhancer function of upstream subrepeats, because in the species studied upstream repetitive elements differ not only in number, but also in size and they belong to different types (Da Rocha and Bertrand 1995; Bhatia et al. 1996).

Several observations demonstrate that ND is a phenomenon regulated at the chromosomal level. In wild type of barley, NORs on both chromosomes 6 and 7 are active. If these NORs are placed together on the same chromosome via translocation, NOR 6 dominates over NOR 7, whereas two copies of NOR 6 or two parts of splitted NOR 7 on the same chromosome are co-dominant (Nicoloff 1979; Georgiev et al. 2001). In hybrids of wheat and rye, *Triticale*, the rye NOR on the short arm of chromosome 1R, is normally suppressed. However, translocation of rye NOR on wheat chromosome 1, deletion of the long arm of chromosome 1R (Viera et al. 1990a,b) or substitution of chromosome 2R by the wheat chromosome 2D (Neves et al. 1997a,b) result in the de-repression and co-dominance of rye NOR. Therefore, these cytological data obviously demonstrate that at least in Triticeae the chromosomal context of NOR can modulate ND. Application of molecular methods for *Brassica* species showing ND (Chen and Pikaard 1997b) revealed no difference in transcription efficiency of respective rDNA, either in protoplast transfection assays or by using an in vitro transcription system (Saez-Vasquez and Pikaard 1997; Frieman et al. 1999), i.e. for appropriate regulation, the rDNA promoter should be integrated into a chromosome.

A possible explanation of how the chromosomal context can modulate NOR activity could be that NORs are passively inactivated as a by-product of silencing of much larger chromosomal segments, with silencing spreading to the NOR (Lewis and Pikaard 2001). In this case, genes adjacent to the NOR should be also silenced when ND occurs. To test this idea, *Arabidopsis suecica*, an allotraploid species originated by crossing diploids *A. arenosa* with *A. thaliana*, was used. In this species, only NORs inherited from *A. arenosa* are active (Chen et al. 1998). However, protein-coding genes adjacent to the inactive NOR 4 inherited from *A. thaliana* remain active in *A. suecica* (Lewis and Pikaard 2001), demonstrating the local nature of silencing and arguing against the possibility that silencing spreads to the NOR from the adjacent region. Hence, some structural features of rDNA, perhaps specific sequences in the IGS, probably control differential expression/silencing of rDNA by ND.

4.2 Ribosomal DNA Rearrangement in Interspecific Hybrids

Genomes of alloploid hybrids contain rDNA of both or of one of the parental species (Borisjuk and Miroshnichenko 1989; Delseny et al. 1990; Volkov et al. 1993). However, the structure of rDNA repeats of allotetraploid *Nicotiana tabacum* differs from that of the diploid parents, *N. sylvestris* and *N. tomentosiformis* (Borisjuk et al. 1989, 1997; Volkov et al. 1991, 1996), although chromosomal localization of rDNA arrays seems to be similar in *N. tabacum* and in parental diploids (Kenton et al. 1993). Comparative sequence analysis showed that rDNA repeats of *N. tabacum* originate from *N. tomentosiformis,* but partial deletion of upstream subrepeats and additional amplification of downstream subrepeats occurred (Volkov et al. 1999a; see Fig. 1). Only about 8% of *N. tabacum* rDNA repeats remained from *N. sylvestris,* but they are located in a hypermethylated, inactive region of chromosome S12 (Lim et al. 2000).

Several steps of molecular evolution of *N. tabacum* rDNA following allopolyploidization were proposed. (1) Nucleolar dominance: After hybridization of the parental diploid species, *N. tomentosiformis* rDNA, with its longer IGS, dominated over the *N. sylvestris* rDNA, with a shorter IGS. (2) Differential elimination/replacement: Functional inactivation of *N. sylvestris* rDNA was accompanied by the structural inactivation via stepwise elimination. The remaining arrays of *sylvestris*-like rDNA were substituted by *tomentosiformis*-like repeats via interloci conversion or/and translocation. (3) rDNA rearrangement: In *tomentosiformis*-originating rDNA, the subrepeated region upstream of TIS endured substantial reduction due to partial deletion of C-subrepeats (formation of the short variant of *N. tabacum* rDNA) and subsequently the subrepeated region downstream of TIS became longer due to amplification of the A-subrepeats (formation of the long variant of *N. tabacum* rDNA). Thus, 35S rDNA in allopolyploid genomes appears to be a target for essential molecular reconstruction, obviously confirming the idea of dynamic nature of polyploid genomes (Miroshnichenko et al. 1988; Soltis and Soltis 1995; Comai 2000; Wendel 2000). Remarkably, the higher rate of rearangement/interlocal homogenization seems to be a specific feature of 35S rDNA. So, different 5S rDNA classes originating from diploid progenitors can still be distinguished in allopolyploid *Nicotiana tabacum* (Fulnecek et al. 1998) and in hexaploid *T. aestivum* (Van Campenhout et al. 1998).

5 Nucleolar Proteins and Structural-Functional Organization of the Nucleolus

5.1 Nucleolar Proteins

Under the group of "nucleolar proteins", we usually refer to proteins which control different steps of ribosome biogenesis (rDNA transcription and pre-rRNA processing), but which are not present in the mature ribosome (Olson 1991). These proteins are, in turn, the targets of other factors involved in the control and regulation of other activities of the cell; thus the role of nucleolar proteins is also to integrate ribosome biogenesis in the context of the concerted pattern of cellular activities. This is especially significant in proliferating cells, in which the nucleolar activity is recognized to be one of the best markers of cell proliferation and cell cycle progression by means of dynamic changes in the nucleolar structure, detectable in the different periods of interphase and in mitosis, e.g. by silver staining of the nucleolar organizer (Ag-NOR; Srivastava and Pollard 1999; Derenzini 2000; Medina et al. 2000).

Recent studies using modern powerful proteomic techniques have resolved the full catalogue of nucleolar proteins in the yeast (Pederson 2002) and human nucleolus. Two studies for human cells carried out on the same biological material did not arrive at the same results: The number of proteins detected was 271 in one case (Andersen et al. 2002), and 213 in another (Scherl et al. 2002). Probably, some proteins have been missed in each of the analyses, so that the true number of nucleolar proteins should be estimated at approx. 350. Less than 100 nucleolar proteins are identified in functional terms, but a significant proportion of these proteins are involved in functions beyond the known role of the nucleolus in ribosome biogenesis. This reinforces the idea of the "plurifunctional nucleolus" (Pederson 1998, 2002) and also indicates that the nucleolus is a dynamic organelle, not only from a morpho-functional point of view, but also regarding its protein composition, which might change according to the function and activity of the cell. Equivalent proteomic analyses have not yet been performed for the plant cell nucleolus, although the completion of the genomic sequence of *Arabidopsis* and rice and the current technological advancement (Thiellement et al. 1999) support expectations for a near achievement of such work.

In plants, the isolation and characterization of plant homologues of nucleolar proteins, e.g. AtNAP57a homologue to rat NAP57 or yeast Cbf5p pseudouridine synthase, indicate a conservation of many nucleolar functions (Maceluch et al. 2001). By complementation of the L25 mutation in yeast, At RPL23A-1, the first plant member of the L23/L25 r-protein family, demonstrates functional equivalence of the two ribosomal proteins (McIntosh and Bonham-Smith 2001). The pea homologue to the DNA helicase I from human (PDH65) is recognized by the antibody against HDH I which was localized within the dense fibrillar component of the nucleolus; therefore, it is most likely involved in rDNA transcription and in the early stages of pre-RNA processing. In addition, it is stimulated by phosphorylation with CK2 and cdc2 protein kinases (Tuteja et al. 2001).

Plants exhibit unique nucleolar features with respect to the snoRNAs (small nucleolar RNA) supporting processing of the 35S-rRNA. In contrast to vertebrates and yeast intron-encoded snoRNAs, which are processed from debranched introns by exonuclease activity, plants have a unique organization of the snoRNA genes (Brown et al. 2001). The maize gene clusters include genes from the two major classes of snoRNA: (1) box C/D snoRNA, which is considered to guide rRNA O-ribose methylations, and (2) box H/ACA snoRNA, thought to guide an rRNA pseudo-uridylation. These snoRNA gene clusters are transcribed from an upstream promoter as a polycistronic pre-snoRNA, which is then processed by a splicing-independent mechanism and gives rise to snoRNAs with distinct nucleolar localization; e.g. snoR2 and snoR3 are closely colocalized with the rDNA transcription sites, whereas snoR1 shows a similar distribution to that of 7–2/MRP, a snoRNA involved in the later pre-rRNA cleavage (Shaw et al. 1998; Leader et al. 1999). Furthermore, a novel histone H1 variant (p35) was isolated from the nucleoli of monocot as well as dicot plants. It strongly binds rDNA in vitro and is considered to be involved in the organization of nucleolar chromatin (Tanaka et al. 1999).

Considering the nucleolar proteins which are functionally conserved in higher organisms and also well studied in plant cells, we will concentrate on fibrillarin and nucleolin. Fibrillarin was originally discovered through the study of the target of a human autoantiserum (Ochs et al. 1985). It is a basic protein, 34–38 kDa in size, highly conserved between species throughout evolution. The sequence of fibrillarin in yeast, plants, and vertebrates contains two motifs which are also present in other nucleolar and nuclear proteins. The first of them is the RNA recognition motif (RRM), also called consensus RNA-binding domain (CS-RBD), which is a stretch of 80–90 amino acid residues containing two highly conserved sequences, the RNP-1 octapeptide and the RNP-2 hexapeptide; the second is the GAR (glycine- and arginine-rich) domain, a sequence formed by around 50% glycines and a substantial proportion of arginines, which are frequently in the form of dimethylarginine (Schimmang et al. 1989; Lapeyre et al. 1990; Pih et al. 2000). This GAR domain seems to be responsible for the targeting of fibrillarin to the nucleolus (Pih et al. 2000). Interestingly, the fibrillarin gene characterized for *Arabidopsis thaliana* also encodes a novel small nucleolar RNA involved in rRNA methylation (Barneche et al. 2000).

The most important functional feature of fibrillarin is the binding to snoRNAs of the C/D box group (U3, U8 and U13 snoRNAs) to form a snoRNP complex active in pre-rRNA processing (Baserga et al. 1991). Interaction of fibrillarin with U3 snoRNA is decisive for the first step of pre-rRNA processing, the cleavage of the ETS from the pre-rRNA (Kass et al. 1990). This interaction and the participation of fibrillarin in the earliest steps of pre-rRNA processing have been visualized by immunocytochemical localization of the protein in the terminal balls of "Christmas trees", the small globular structures located at the end of nascent pre-rRNA transcripts (Scheer and Benavente 1990; Mougey et al. 1993). Fibrillarin was named according to the localization in the nucleolar DFC (Ochs et al. 1985; Testillano et al. 1991). A detailed biochemical and quantitative immunolocalization study on plant fibrillarin in actively proliferating cells showed that two fractions of the protein could be identified, one soluble, active in pre-rRNA processing, and another insoluble form, associated with the nucleolar matrix; furthermore, the in situ distribution of fibrillarin was shown to be unequal throughout the massive DFC typical of onion root meristematic cells, being more concentrated in the transition area FC-DFC and in the proximal part of the DFC than in the zones of DFC more distant from FCs. Finally, the levels of fibrillarin appeared to be clearly dependent on the transcriptional activity of the nucleolus, and were shown to change during the progression of the interphase of the cell cycle (Cerdido and Medina 1995; Medina et al. 2000). Recent studies on the fibrillarin gene, characterized for *Arabidopsis*, have demonstrated that its expression is regulated by hormones, e.g. abscisic acid, and it is modulated by aging (Pih et al. 2000).

Nucleolin, a characteristic and prominent nucleolar protein (100 kD) that functions as a shuttle protein between nucleus and cytoplasm and is also found on the cell surface, was first identified for mammalian cells, where it is also called "C23" (Orrick et al. 1973). It appears evolutionarily conserved in animals (Lapeyre et al. 1987; Tuteja and Tuteja 1998), yeast (Lee et al. 1991; Gulli et al. 1995) and plants (Martín et al. 1992; Didier and Klee 1992; Bögre et al. 1996; De Cárcer et al. 1997; Tong et al. 1997). Nucleolin was initially found to be the target of the cytochemical silver staining of the nucleolar organizer (Ag-NOR staining; Lischwe et al. 1979); this is now commonly accepted, and it was shown that silver staining occurs at the N-terminal domain of the protein (Roussel et al. 1992).

Like fibrillarin, the structure of nucleolin is modular, containing molecular motifs which can also be found in other nucleolar and nuclear proteins. In particular, nucleolin shows a tripartite structure, formed by an N-terminal domain, containing a variable number of highly charged acidic sequence repeats interspersed with basic segments, a globular central domain, containing between two and four RRMs, and a C-terminal domain consisting in a GAR region. Variations in the numbers of acidic stretches, as well as in RRMs, are found in different taxa (Olson 1991; Tuteja and Tuteja 1998; Ginisty et al. 1999). Sequences in the N-terminal domain show homologies to the high-mobility group (HMG) of nonhistone chromatin proteins. This domain also contains target sites for phosphorylation by CK2 and cdc2 kinase (Caizergues-Ferrer et al. 1987; Belenguer et al. 1990). Homologies of the RRM motif and of the GAR region are present, as mentioned for fibrillarin.

The multipartite structure of nucleolin is reflected in the multiplicity of functions, defining it as a "multifunctional" protein. Nucleolin is involved in the control of ribosomal RNA transcription (Bouche et al. 1987), in the structural organization of rDNA chromatin, including the condensation of this chromatin in mitosis (Olson et al. 1983; Erard et al. 1988), in promoting secondary structures in pre-rRNA necessary for processing (Bugler et al. 1987), in preribosome maturation (Herrera and Olson 1986), and in the transport of preribosomal particles to the cytoplasm (Borer et al. 1989; reviewed by Tuteja and Tuteja 1998). The best-studied mechanism is the interaction of nucleolin with pre-rRNA in the transcript processing. Mouse nucleolin has been shown to bind with high affinity and specificity to a sequence of 18 nucleotides in pre-rRNA which adopts a stem-loop structure in the ETS, whose cleavage constitutes the first step in pre-rRNA processing

(Ghisolfi-Nieto et al. 1996). Nucleolin binding to pre-rRNA is required for the processing reaction in vitro, probably by promoting the association of other factors such as U3 snoRNP, thus contributing to the formation of the processing complex (Ginisty et al. 1998). A similar stem-loop structure has also been found in other positions in the sequence of mammalian pre-rRNA, which could indicate that the mechanism by which nucleolin acts in the first step of processing is repeated in further steps (Serin et al. 1996). On the contrary, the interaction of nucleolin with rDNA is still poorly understood. The affinity of nucleolin for rDNA sequences located in the IGS has been reported and, in some cases, these sequences have been identified as matrix-attachment regions (Olson et al. 1983; Dickinson and Kohwishigematsu 1995; McGrath et al. 1997). It has been demonstrated that nucleolin displaces the chromatin-binding domain of histone H1, thus being capable of modifying the basic structure of chromatin for transcription (Erard et al. 1988), and a further insight in the relationship of nucleolin with the transcription complex has recently evidenced that increasing amounts of nucleolin drastically reduces in vitro the number of transcription complexes formed by rDNA and RNA pol I (Roger et al. 2002).

Many of the functions of nucleolin are phosphorylation dependent (Ballal et al. 1975; Bourbon et al. 1983). The primary structure of nucleolin, in animal and plant cells, contains sequences corresponding to sites of phosphorylation by CK2 and cdc2 kinase (Lapeyre et al. 1987; Bögre et al. 1996; Tong et al. 1997); actually, nucleolin is phosphorylated on serine by CK2 during interphase of proliferating cells (Caizergues-Ferrer et al. 1987), and on threonine by the cyclin-dependent cdc2 kinase during mitosis (Belenguer et al. 1990).

In plants, the two proteins with homologies to nucleolin identified in onion root meristematic cells, NopA64 and NopA100, are phosphorylated in vitro by CK2 and cdc2 kinase, and the two kinases were shown to colocalize with the nucleolin homologues. NopA100 is the most phosphorylated nuclear protein of actively proliferating onion root cells (De Cárcer et al. 1997; Medina et al. 2001, and unpublished results). The exact mechanism of the functional modulation of nucleolin by phosphorylation is not well understood. Unphosphorylated nucleolin is a potent transcription inhibitor in vitro, and this form is recovered associated with chromatin in resting cells (Bourbon et al. 1983; Lapeyre et al. 1987). On the other hand, phosphorylated nucleolin is cleaved during transcription of rRNA genes (Warrener and Petryshyn 1991). It has been speculated that unphosphorylated nucleolin binds to rDNA and represses transcription; phosphorylation induces proteolysis, and consequently releases nucleolin from DNA, thus activating transcription. The physiological proteolysis of nucleolin is also related to the proliferative state of the cell, since the nucleolin self-cleaving activity was inhibited by a nucleolar extract prepared from proliferating cells (Chen et al. 1991

The levels of nucleolin have been shown to be highly dependent on cell proliferation, and on the phase of the cell cycle (De Cárcer et al. 1997; Sirri et al. 1997). In alfalfa, the nucleolin gene is not expressed in mitotic-arrested cells, but, upon mitogenic stimulation and cell cycle reentry, the gene is induced very early, exactly at the same time as a marker cyclin gene, indicating that nucleolin gene expression and cyclin expression could be regulated by the same factors. Otherwise, the expression of nucleolin is restricted to proliferating tissues of the plant (Bögre et al. 1996). In pea, the expression of nucleolin is regulated by light, following a close correlation with the effects of light in increasing cell proliferation and mitotic activity (Reichler et al. 2001). In fact, the modulation of the nucleolin function by phosphorylation by CK2 and cdc2 kinase is related to cell proliferation, since these two kinases are known to phosphorylate a battery of different targets as a response to proliferative signals, and as a modulation of the different periods of the cell cycle (Pinna 1990; Dorée and Galas 1994). During cell division, onion nucleolin, as a component of the pre-rRNA processing complex, is found first at the periphery of chromosomes, and then in prenucleolar bodies, playing a part in nucleologenesis (Medina et al. 1995, 2001). Immunolocalization of nucleolin in onion cells, either by a heterologous antibody (Martín et al. 1992) or by homologous anti-NopA64 (Medina et al. 2001), allowed the definition of different functional subdomains in the structurally homogeneous DFC; the same subdomains could also be defined after in situ localization of fibrillarin (Cerdido and Medina 1995). These are examples showing the contribution of the localization of nucleolar proteins to the elucidation of the functional architecture of the nucleolus.

5.2 Structural-Functional Organization of the Nucleolus

Structural domains in the nucleolus with functional significance could be defined by the localization in situ of transcription of rRNA genes (see Busch and Rothblum 1983). However, there was and is still an ongoing debate, whether transcription of rRNA takes place in nucleolar FCs or in the DFCs (Fakan and Puvion 1980; Scheer and Rose 1984; Thiry and Thiry-Blaise 1989; Wachtler et al. 1989). In situ run-on experiments showed the site of transcription at the boundary between FCs and the DFC (Dundr and Raska 1993; Hozák et al. 1994). For plants, these questions had been summarized in an earlier review on the structural and functional organization of the nucleolus (Leweke and Hemleben 1982). Here, we will focus on the recent state of the problem regarding plant cell nucleoli.

Quantitative immunolocalization of DNA and RNA pol I in the onion nucleolus showed the transition area between FCs and the DFC as a significant nucleolar domain for rDNA transcription (Martín et al. 1989; Medina et al. 1990; Martín and Medina 1991). Further studies, using different experimental methods, consistently showed the transcription marker in a restricted nucleolar area comprising the outer periphery of FCs and the inner periphery of the DFC (Testillano et al. 1994; Shaw et al. 1995; Melcák et al. 1996). The use of Br-UTP incorporation to isolated onion nuclei evidenced a focal arrangement of the transcription sites in the nucleolus; most of these foci were located at the periphery of FCs, but the core of every FC appeared devoid of labelling; otherwise, transcription foci did not totally surround FCs, so that transcription occurred at discrete points at the boundary domain between FCs and the DFC. Colocalization between transcription and fibrillarin, the latter used as a marker of early processing, was only partial in transcription foci, since fibrillarin was not found in the zones of these foci immediately bordering FCs. Finally, other transcription foci were found deep in the DFC, in zones located between FCs (De Cárcer and Medina 1999; Medina et al. 2001). The focal organization of nucleolar transcription had also been shown in other plant models (Thompson et al. 1997), but the relationship of these foci with FCs was questioned. Recently, Br-UTP incorporation has been detected in pea nucleoli using pre-embedding labeling and three-dimensional electron microscopy of entire nucleoli (González-Melendi et al. 2001). Labeling, observed as large irregular silver particles resulting from enhancement of nanogold, forms relatively large clusters of particles in panoramic views of thin serial sections of the nucleolus (which could correspond to transcription foci of the confocal microscope); these large clusters can be resolved in smaller clusters at higher magnifications, possibly representing single transcription units ("Christmas trees"). The structure underlying the clusters is supposed to be the DFC, but no indication is given as to where FCs could be situated. According to the authors' interpretation, actively transcribed rDNA would be an extended loop, emanating from the condensed rDNA chromatin adjacent to the nucleolus and continuously running through the DFC (González-Melendi et al. 2001).

Earlier work had shown the conversion of the heterogeneous FCs, undoubtedly containing condensed rDNA chromatin, into homogeneous FCs, the small type of FCs present in active plant nucleoli containing decondensed chromatin, in the course of germination (Deltour 1985; Risueño and Medina 1986) as well as during postmeiotic nucleolar reactivation in young microspores (Esponda and Giménez-Martín 1974; Medina et al. 1983). The existence of a nucleolar matrix or nucleolar proteinaceous skeleton with the function, among others, of gathering rDNA in FCs and structurally organizing it depending on the nucleolar activity and the requirements of the cells (Moreno Díaz de la Espina 1995) is now supported by the fact that MAR or SAR elements are present in the AT-rich region of every rDNA repeat (see above).

Therefore, it can be summarized that rRNA transcription complexes are assembled within FCs, but actual transcription only occurs at certain discrete points at their periphery in the DFC (Fig. 2). Furthermore, transcriptionally active chromatin can also be found deep in the DFC, in the stretches of rDNA chromatin connecting contiguous FCs. Certainly, some additional work is still needed for a full understanding of the molecular architecture of the transcription of rRNA genes in the nucleolus (see Huang 2002).

6 Ribosome Biogenesis

Ribosome biogenesis is a complex process which takes mainly place in the nucleolus starting with the transcription of the rRNA genes by pol I and III

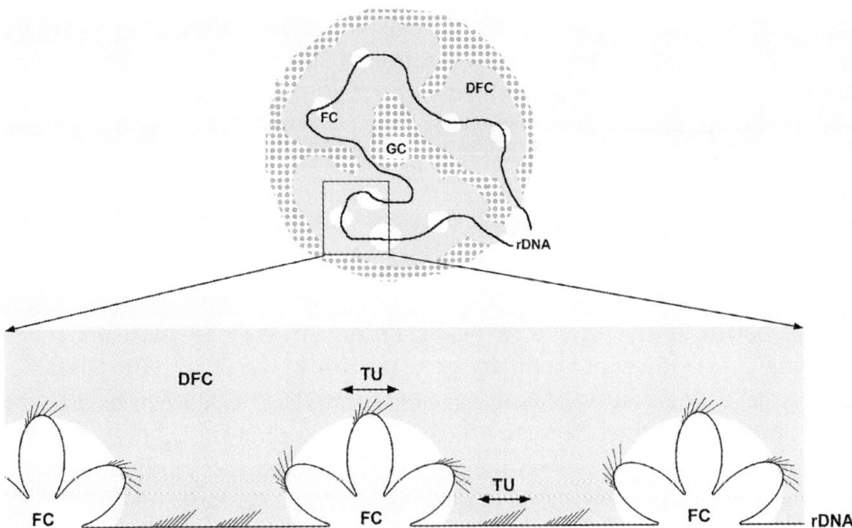

Fig. 2. Schematic presentation of the organization of 35S rDNA and transcription of rRNA genes in the structural components of the nucleolus, according to available experimental data. The drawing at the *top* represents the typical structure of the plant cell nucleolus, as observed in an ultrathin section under the electron microscope. The dense fibrillar component (*DFC*) is massive and appears surrounded by the granular component (*GC*). Lighter small areas inside the DFC are fibrillar centers (*FC*). The nucleolus is organized around rDNA, in such a way that the fiber of DNA containing rRNA genes runs across the nucleolus following a meandering pathway, which previously received the name of "nucleolar organizer track" by Godward (1950). This track passes through FCs and is embedded in the DFC. The *bottom drawing* represents the area inside the *square*, ideally stretched and magnified. Fibrillar centers represent foci of accumulation of rDNA which is ready for transcription, with bound pol I and factors (except for condensed chromatin inclusions of heterogeneous FCs), but actual transcription only occurs in loops emerging from the concentration of rDNA. The organization of these rDNA foci is probably assured by structural proteins of the nucleolar matrix. As a result of transcription, pre-rRNA is formed, which is located in the DFC. Therefore, the site of transcription is the transition area FC-DFC, also called the FC-DFC boundary. The organization of transcription in loops causes that the entire boundary of FCs is not occupied by transcription units, or "Christmas trees" (*TU*), which are only detected at certain places. These TUs also appear in the rDNA connecting contiguous FCs, which structurally corresponds to the bulk of DFC

transcribing the small 5S rRNA genes outside of the nucleolus, whereas RNA polymerase I generates a precursor molecule directly in the nucleolus which is rapidly processed by endonucleases and exonucleases into the 25S, 5.8S and 18S rRNAs and modified by pseudouridylation and methylation (Venema and Tollervey 1999). During these processing reactions, a large number of non-ribosomal proteins associate with the pre-RNAs (see above) in parallel with the starting assembly of the approximately 80 ribosomal proteins onto the RNA (Warner 1989, 1999, 2001). For the animal system, the view of the pathway of ribosome biogenesis is becoming more and more detailed (see Fig. 3). Early analyses in the 1970s already reported that the 45S pre-rRNA is initially assembled with ribosomal proteins to form a 90S particle (Trapman et al. 1975; Kruiswijk et al. 1978), which – as now known – also includes many non-ribosomal proteins (Dragon et al. 2002; Grandi et al. 2002). Subsequently, the pre-RNA undergoes three rapid cleaving reactions leading to the separation of the 90S particle into the large and the small subunits (Trapman et al. 1975, Venema and Tollervey 1999). Analyses of the large pre-60S particle revealed that they show dynamic changes during maturation and that a whole family of pre-60S particles exists carrying many different accessory proteins (Milkereit et al. 2001; Nissan et al. 2002). During maturation, the pre-60S particle is first released into the nucleoplasm and then exported to the cytoplasm via the Xpo1/Crm1/exportin-1 export pathway with the help of Nmd3 and Mtr2 and most likely other pre-60S associated proteins (Ho et al. 2000; Gadal et al. 2001). The export mechanism of the 40S subunit is still unclear, but it seems that it undergoes a comparatively simpler maturation (Moy and Silver 1999; Venema and Tollervey 1999). A model for the pathway of the 60S pre-ribosome maturation and export of 40S and 60S particles is presented in Fig. 3.

The complexity of ribosome biogenesis gives rise to many steps and opportunities where regulation could take place (reviewed by Leary and Huang 2001).

7 Ribosome-Inactivating Proteins (RIPs)

Ribosome-inactivating proteins (RIPs) consist of toxins that catalytically inactivate ribosomes at a universally conserved region of the large ribosomal RNA, the α-sarcin/ricin loop (S/R loop). Previously, it was found that a fungal toxin, α-sarcin, from *Aspergillus giganteus*, inactivates ribosomes by hydrolyzing a single phosphodiester bond on the 3' side of G4325 in the α-sarcin/ricin (S/R) loop of rat 28S ribosomal RNA or its equivalent residue G2661 in *E. coli* 23S ribosomal RNA (Endo and Wool 1982). Ribosome-inactivating proteins are also found in plants as toxic compounds that are remarkably potent inactivators of the eukaryotic protein synthesis system (for reviews, see Stirpe and Barbieri 1986; Stirpe et al. 1992; Parente et al. 1993). In contrast to fungal and bacterial translation inhibitors, RIPs are proposed to play a defence role in plants. RIPs belong to the group of RNA *N*-glycosidases, because they carry out a single *N*-glycosidation event at a specific and conserved A residue of the 25/28S rRNA of the eukaryotic 60S ribosome subunit that alters the binding site of the translational elonga-

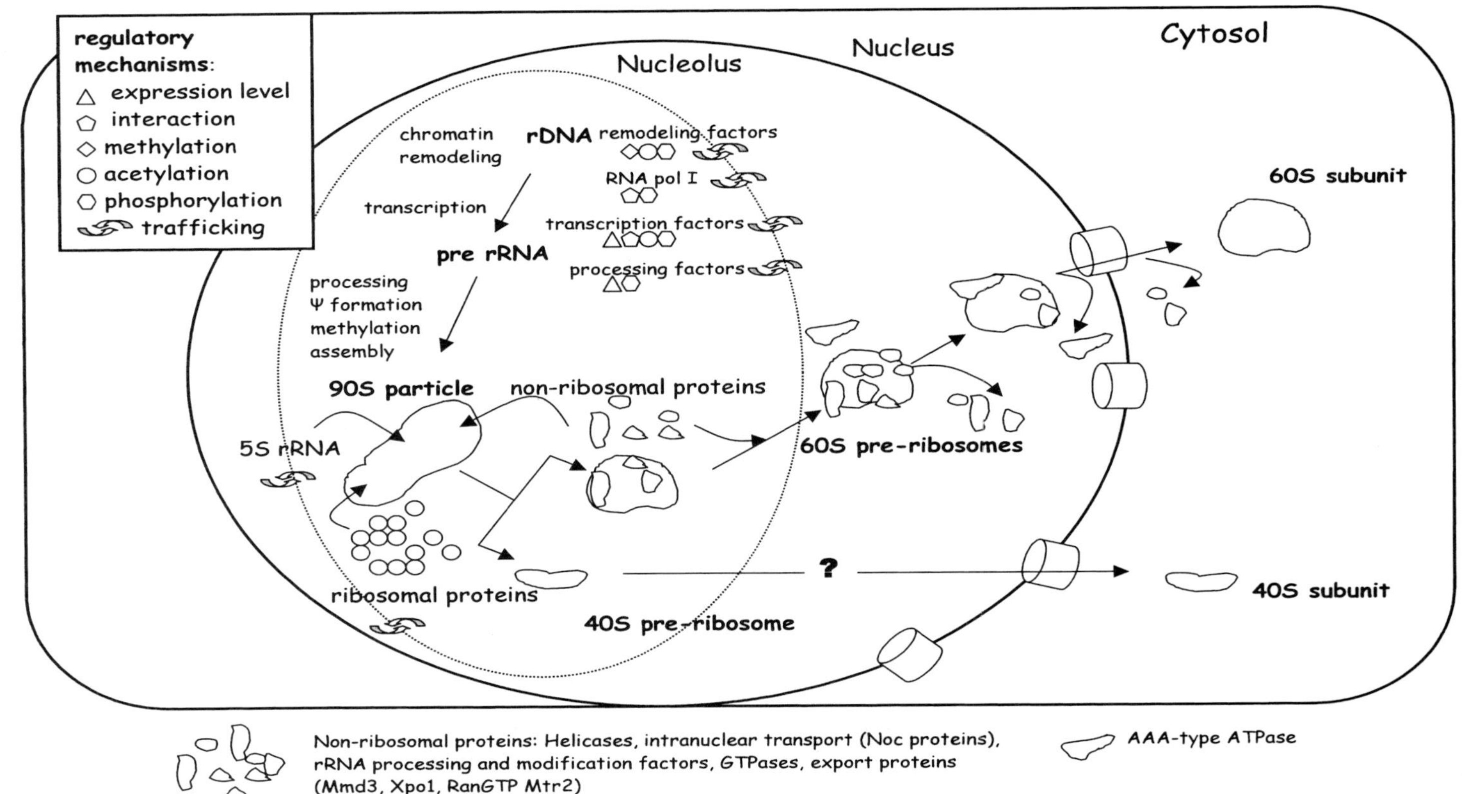

Fig. 3. Model of the pathway of the 60S and 40S pre-ribosome maturation, possible points of regulation and export. (Modified according to Leary and Huang 2001; Nissan et al. 2002)

tional factor eEF1A. Therewith they prevent protein synthesis leading to subsequent cell death.

Different forms of RIPs have been identified in plants. Most commonly RIPs are single-chain proteins (type-1 RIPs) with a molecular weight of 25–30 kDa. Type 2 RIPs consist of a toxin-part chain (A-chain), and a lectin-part chain (B-chain), bound by a disulfide bond, having molecular weights of 60–65 kDa and approximately 120 kDa, respectively, that binds to cell surfaces. The type 2 RIPs are potent toxins, the best known of which is ricin from *Ricinus communis*. Type 1 RIPs and the A-chains of type 2 RIPs possess sequence and structural homology. For a type 2 RIP, the B-chain of the molecule first binds to cell surface with its lectin facility, then the A-chain enters the cell through endocytosis to kill the cell by inactivating ribosomes. Such biomolecules causing cytotoxicity are being exploited for designing immunotoxins/hormonotoxins using heterobifunctional conjugates. These carriers conjugated with the RIPs can influence cellular trafficking and inhibit protein synthesis. It is assumed that these proteins from plants can be utilized for various therapeutical treatments including cancer, AIDS and other viral diseases of present times (Stirpe et al. 1992). RIPS have been tested extensively for use as plant defence molecules, particularly against fungal diseases (Logemann et al. 1992; Stirpe et al. 1992; Nielsen et al. 2001). Ribosome-inactivating proteins are found in different parts of plants, in concentrations ranging from a few micrograms to several hundred milligrams per 100 g of plant tissues (Singh and Singh 2000).

In particular, the abundant RIP I from maize (*Zea mays*) kernels was tested for antifungal activity (Nielsen et al. 2001). The substrate specificity of the maize RIP I differs across diverse taxa (Krawetz and Boston 2000). The maize RIP I was found to be produced as a zymogen, proRIP I. ProRIP I accumulates during seed development and becomes active during germination when cellular proteases remove acidic residues from a central domain and both termini. A wide spectrum of RIPs was isolated from different cucurbits (family Cucurbitaceae). The liposome-mediated delivery of the type 1 RIP luffin to human melanoma cells in vitro was described (Poma et al. 1999). The exposure of melanoma cells to two types of liposomes resulted in the inhibition of protein synthesis and cell growth. A new RIP (δ-momorcharin) and a candidate RIP (ε-momorcharin) were isolated from the seeds and fruits of the bitter gourd *Momordica charantia* (Cucurbitaceae; Paul et al. 1999). Recent studies have shown that some RIPs possess strong anti-human immunodeficiency virus (HIV) activity, and several common plant RIPs including agrostin, gelonin, luffin, lant RIPs including agrostin, gelonin, luffin, ss strong anti-human immunodeficiency virus (HIV) activity, and several common pype 1 RIP luffin to human melanoma cells in vitro was described (Poma et al. 1999)ed RIPs, MAP30 and GAP 31, isolated from *Momordica charantica* and *Gelonium multiflorum*, respectively, inhibited the replication of human immunodeficency virus type 1 (HIV-1) by inhibiting the viral integrase activity (Bourinbaiar and Lee-Huang 1996), indicating a different mechanism of these compounds.

Pokeweed (*Phytolacca americana*) antiviral protein (PAP), another type 1 RIP, was purified from *Agrobacterium rhizogenes*-transformed hairy roots of *Phytolacca americana* (Park et al. 2002a,b) from which it was constitutively secreted as part of the root exudates. Obviously, PAP-R depurinates fungal ribosomes only in vitro, suggesting a synergistic mechanism that enables PAP-R to penetrate fungal cells. PAP could also inhibit translation of mRNAs and viral RNAs that are capped by binding to the cap structure and depurinating the RNAs downstream of the cap (Parikh et al. 2002). Possibly, this single-chain RIP targets not only the large rRNA, but also its own mRNA, supporting the idea of a multifunctional protein.

The type 2 RIPs (RIP II), firstly represented by ricin, the toxic RIP of *Ricinus communis*, recently have attracted enormous interest (Di Cola et al. 2001). Another type 2 RIP, the mistletoe lectin I (ML I) from *Viscum album*, which consists also of two chains, is even more toxic than ricin: The A-chain fulfills the proper RIP function of an RNA *N*-glycosidase which acts at the specific A residue of the large rRNA and inhibits translation at the ribosome; the B-chain has lectin activity interacting with specific sugar residues and mediating the transport of the dimeric complex through the plasma membrane. Since it is proposed that this ML I will be used therapeutically in cancer therapy, this protein has been studied in detail. Also, cinnamomin was characterized as a new type 2 RIP (Xie et al. 2001; Liu et al. 2002). It is expressed specifically in cotyledons and accumulates in large amounts as a storage protein at later stages of seed development.

Thus RIPs comprise plant compounds which may be naturally used in plant defence. However, a clear biological significance of RIPs in plant physiology remains largely elusive. Nevertheless, they attract enormous interest as antiviral and possibly as anti-tumor therapeutically active substances.

8 Conclusions and Perspectives

Ribosome biogenesis and function of ribosomes are supposed to represent basic functions of the cell, and the components were often grouped into house-keeping functions which are constitutively provided. Now, it becomes more and more clear that expression is highly regulated on various levels and by multiple exogenous (light, different stress conditions etc.) and endogenous cellular stimuli which influence cell activity and proliferation status of the cells, like phytohormones and others. Similarly, the genes coding for rRNA and for nucleolar and r-proteins are expected to be highly regulated and the gene products to be delivered corresponding to the cell stage. Thus, the idea of the "plurifunctional nucleolus" now stimulates further research, and it reminds us that the nucleolus is a highly dynamic nuclear structure. Extensive investigations on the structural organization and molecular evolution of the rRNA genes are now available and can be applied for various aspects, such as following speciation processes and tracing phylogenetic relationships. In contrast, regulation of transcription of the ribosomal genes is still poorly understood and needs further research approaches in plants. Interesting phenomena, like nucleolar dominance or gene silencing, can be observed depending on the phylogenetic relationship of hybrid partners. Processing of pre-rRNA has been mainly studied for yeast, but similar snoRNPs may be involved in plant pre-rRNA processing which is supposed to be also a regulating step. Proteomic studies in animals and yeast slowly offer an overview on the functional nucleolar proteins. For the plant cell nucleolus, respective proteomic analyses have not yet consequently been performed and can be expected only in the future. In addition, plants provide various ribosome inactivating proteins which gain increasing importance as therapeutically useful substances.

Acknowledgement. The help of Dipl. Biol. Nataliya Komarova in assembling the literature is gratefully acknowledged.

References

Akhunov ED, Chemeris AV, Kulikov AM, Vakhitov VA (2001) Functional analysis of diploid wheat rRNA promoter by transient expression. Biochem Biophys Acta 1522:226–229

Albert AC, Denton M, Kermekchiev M, Pikaard CS (1999) Histone acetyltransferase and protein kinase activities copurify with the putative *Xenopus* RNA polymerase I holoenzyme self-sufficient for promotor-dependent transcription. Mol Cell Biol 19:796–806

Allmang C, Tollervey D (1998) The role of the 3' external transcribed spacer in yeast pre-rRNA processing. J Mol Biol 278:67–78

Andersen JS, Lyon CE, Fox AH, Leung AK, Lam YW, Steen HB, Mann M, Lamond AI (2002) Directed proteomic analysis of the human nucleolus. Curr Biol 12:1–11

Au TK, Collins RA, Lam TL, Ng TB, Fong WP, Wan DC (2000) The plant ribosome inactivating proteins luffin and saporin are potent inhibitors of HIV-1 integrase. FEBS Lett 471:169–172

Baldwin BG (1992) Phylogenetic utility of the internal transcribed spacers of nuclear ribosomal DNA in plants: an example from the *Compositae*. Mol Phylogenet Evol 1:3–16

Baldwin BG, Marcos S (1998) Phylogenetic utility of the external transcribed spacer (ETS) of 18S-26S rDNA: congruence of ETS and ITS trees of *Calycadenia* (Compositae). Mol Phylogenet Evol 10:449–463

Ballal NR, Kang YJ, Olson MOJ, Busch H (1975) Changes in nucleolar proteins and their phosphorylation patterns during liver regeneration. J Biol Chem 250:5921–5925

Barciszewska M, Erdman VA, Barciszewski J (1994) A new model for the tertiary structure of 5S ribonucleic acid in plants. Plant Mol Biol Rep 12:116–131

Barker RF, Harberd NP, Jarvis MG, Flavell RB (1989) Structure and evolution of the intergenic region in a ribosomal DNA repeat unit of wheat. J Mol Biol 201:1–17

Barneche F, Steinmetz F, Echeverria M (2000) Fibrillarin genes encode both a conserved nucleolar protein and a novel small nucleolar RNA involved in ribosomal RNA methylation in *Arabidopsis thaliana*. J Biol Chem 275:27212–27220

Baserga SJ, Yang XW, Steitz JA (1991) An intact Box C sequence in the U3 snRNA is required for binding of fibrillarin, the protein common to the major family of nucleolar snRNPs. EMBO J 10:2645–2651

Belenguer P, Caizergues-Ferrer M, Labbé JC, Dorée M, Amalric F (1990) Mitosis-specific phosphorylation of nucleolin by p34^{cdc2} protein kinase. Mol Cell Biol 10:3607–3618

Bena G, Jubier MF, Olivieri I, Lejeune B (1998) Ribosomal external and internal transcribed spacers: combined use in the phylogenetic analysis of *Medicago* (Leguminosae). J Mol Evol 46:299–306

Bhatia S, Singth Negi M, Lakshmikumaran M (1996) Structural analysis of the rDNA intergenic spacer of *Brassica nigra*: evolutionary divergence of the spacers of the three diploid *Brassica* species. J Mol Evol 43:460–468

Bigot Y, Lutcher F, Hamelin M-H, Periquet G (1992) The 28S ribosomal RNA-encoding gene of *Hymenoptera*: insreted sequences in the retrotransposon-rich region. Gene 121:347–352

Bögre L, Jonak C, Mink M, Meskiene I, Traas J, Ha DTC, Swoboda I, Plank C, Wagner E, Heberle-Bors E, Hirt H (1996) Developmental and cell cycle regulation of alfalfa *nucMs1*, a plant homolog of the yeast Nsr1 and mammalian nucleolin. Plant Cell 8:417–428

Borer RA, Lehner CF, Eppenberger HM, Nigg EA (1989) Major nucleolar proteins shuttle between nucleus and cytoplasm. Cell 56:379–390

Borisjuk N, Hemleben V (1993) Nucleotide sequence of potato rDNA intergenic spacer. Plant Mol Biol 21:381–384

Borisjuk NV, Miroshnichenko GP (1989) Organization of ribosomal RNA genes in *Brassica oleracea*, *Brassica campestris* and allotetraploid hybrid *Brassica napus*. Genetika (Moscow) 25:417–423

Borisjuk NV, Kostishin SS, Volkov RA, Miroshnichenko GP (1989) Ribosomal RNA gene organisation in higher plants from genus *Nicotiana*. Mol Biol (Moscow) 23:1067–1074

Borisjuk N, Borisjuk L, Petjuch G, Hemleben V (1994) Comparison of nuclear ribosomal RNA genes among *Solanum* species and other Solanaceae. Genome 37:271–279

Borisjuk NV, Davidjuk YM, Kostishin SS, Miroshnichenko GP, Velasko R, Hemleben V, Volkov RA (1997) Structural analysis of rDNA in the genus *Nicotiana*. Plant Mol Biol 35:655–660

Borisjuk N, Borisjuk L, Komarnytsky S, Timeva S, Hemleben V, Gleba Yu, Raskin I (2000) Tobacco ribosomal DNA spacer element stimulates amplification and expression of heterologous genes. Nat Biotechnol 18:1303–1306

Bouche G, Gas N, Prats H, Baldin V, Tauber JP, Teissie J, Amalric F (1987) Basic fibroblast growth factor enters the nucleolus and stimulates the transcription of ribosomal genes in ABAE cells undergoing G0-G1 transition. Proc Natl Acad Sci USA 84:6770–6774

Boulikas T (1993) Nature of DNA sequences at the attachment regions to the nuclear matrix. J Cell Biochem 52:14–22

Bourbon HM, Bugler B, Caizergues-Ferrer M, Amalric F (1983) Role of phosphorylation on the maturation pathways of 100 kD nucleolar protein. FEBS Lett 155:218–222

Bourinbaiar AS, Lee-Huang S (1996) The activity of plant-derived antiretroviral proteins MAP30 and GAP31 against herpes simplex virus in vitro. Biochem Biophys Res Commun 19:923–929

Brown JW, Clark GP, Leader DJ, Simpson CG, Lowe T (2001) Multiple snoRNA gene clusters from *Arabidopsis*. RNA 7:1817–1832

Bugler B, Bourbon HM, Lapeyre B, Wallace MO, Chang JH, Amalric F, Olson MOJ (1987) RNA binding fragments from nucleolin contain the ribonucleoprotein consensus sequence. J Biol Chem 262:10922–10925

Busby SJ, Reeder RH (1983) Spacer sequences regulate transcription of ribosomal gene plasmids injected into *Xenopus* embryos. Cell 34:989–996

Busch H, Rothblum L (1983) The cell nucleus: rDNA. Part B, Academic PressLondon

Caetano-Anolles G (2002) Tracing the evolution of RNA structure in ribosomes. Nucleic Acids Res 30:2575–2587

Caizergues-Ferrer M, Belenguer P, Lapeyre B, Amalric F, Wallace MO, Olson MOJ (1987) Phosphorylation of nucleolin by a nuclear type NII protein kinase. Biochemistry 26:7876–7883

Carles C, Riva M (1998) Yeast RNA polymerase I subunits and genes. In: Paule MR (ed) Transcription of ribosomal RNA genes by eukaryotic RNA polymerase I. Springer, Berlin Heidelberg New York, pp 9-38

Caudy AA, Pikaard CS (2002) *Xenopus* ribosomal RNA gene intergenic spacer elements conferring transcriptional enhancement and nucleolar dominance-like competition in oocytes. J Biol Chem 35:31577–31584

Cerdido A, Medina FJ (1995) Subnucleolar location of fibrillarin and variation in its levels during the cell cycle and during differentiation of plant cells. Chromosoma 103:625–634

Chen ZJ, Pikaard CS (1997a) Epigenetic silencing of RNA polymerase I transcription: a role for DNA methylation and histone modification in nucleolar dominance. Genes Dev 11:2124–2136

Chen ZJ, Pikaard CS (1997b) Transcriptional analysis of nucleolar domimamce in polyploid plants: biased expression/silencing of progenitor rRNA genes is developmentally regulated in *Brassica*. Proc Natl Acad Sci USA 94:3442–2447

Chen CM, Chiang SY, Yeh NH (1991) Increased stability of nucleolin in proliferating cells by inhibition of its self-cleaving activity. J Biol Chem 266:7754–7758

Chen ZJ, Comai L, Pikaard CS (1998) Gene dosage and stochastic effects determine the severity and direction of uniparental ribosomal RNA gene silencing (nucleolar dominance) in *Arabidopsis* allopolyploids. Proc Natl Acad Sci USA 95:14891–14896

Choumane W, Heizmann P (1988) Structure and variability of nuclear ribosomal genes in the genus *Helianthus*. Theor Appl Genet 76:481–489

Cockerill PN, Garrad WT (1986) Chromosomal loop anchorage of the kappa immunoglobulin gene occurs next to the enhancer in a region containing topoisomerase II sites. Cell 44:273–282

Coen ES, Strachan T, Dover G (1982) Dynamics of concerted evolution of ribosomal DNA and hystone gene families in the melanogaster species subgroup of *Drosophila*. J Mol Biol 158:17–35

Comai L (2000) Genetic and epigenetic interactions in allopolyploid plants. Plant Mol Biol 43:387–399

Comai L, Tanese N, Tjian R (1992) The TATA-binding protein and associated factors are integral components of the RNA polymerase I transcription factor, SL1. Cell 68:965–976

Copenhaver GP, Pikaard CS (1996) Two-demensional RFLP analyses reveal megabase-sized clusters of rRNA gene variants in *Arabidopsis thaliana*, suggesting local spreading of variants as the mode for gene homogenization during concerted evolution. Plant J 9:273–282

Crisp MD, Appels R, Smith FM, Keys WMS (1999) Phylogenetic evaluation of 5S ribosomal RNA gene and spacer in the *Callistachys* group (Fabaceae: Mirbelieae). Plant Syst Evol 218:33–42

Da Rocha PSCF, Bertrand H (1995) Structure and comparative analysis of the rDNA intergenic spacer of *Brassica rapa*: implications for the function and evolution of the Cruciferae spacer. Eur J Biochem 229:550–557

De Cárcer G, Medina FJ (1999) Simultaneous localization of transcription and early processing markers allows dissection of functional domains in the plant cell nucleolus. J Struct Biol 128:139–151

De Cárcer G, Cerdido A, Medina FJ (1997) NopA64, a novel nucleolar phosphoprotein from proliferating onion cells, sharing immunological determinants with mammalian nucleolin. Planta 201:487–495

Delseny M, McGrath JM, This P, Chevre AM, Quiros CF (1990) Ribosomal RNA genes in diploid and amphidiploid *Brassica* and related species: organisation, polymorphism and evolution. Genome 33:733–744

Deltour R (1985) Nuclear activation during early germination of the higher plant embryo. J Cell Sci 75:43–83

Denk T, Grimm G, Stögerer K, Langer M, Hemleben V (2002) The evolutionary history of *Fagus* in western Eurasia: Genes, morphology and the fossil record as evidence of reticulate evolution and suppressed speciation. Plant Syst Evol 232:213–236

Derenzini M (2000) The AgNORs. Micron 31:117–120

Dickinson LA, Kohwishigematsu T (1995) Nucleolin is a matrix attachment region DNA-binding protein that specifically recognizes a region with high base-unpairing potential. Mol Cell Biol 15:456–465

Di Cola A, Frigerio L, Lord JM, Ceriotti A, Roberts LM (2001) Ricin A chain without its partner B chain is degraded after retrotranslocation from the endoplasmic reticulum to the cytosol in plant cells. Proc Natl Acad Sci USA 98:14726–14731

Didier DK, Klee HJ (1992) Identification of an *Arabidopsis* DNA-binding protein with homology to nucleolin. Plant Mol Biol 18:977–979

Doelling JH, Pikaard CS (1995) The minimal ribosomal RNA gene promoter of *Arabidopsis thaliana* includes a critical element at the transcription initiation site. Plant J 8:683–692

Doelling JH, Pikaard CS (1996) Species-specificity of rRNA gene transcription in plants manifested as a switch in polymerase-specificity. Nucleic Acids Res 24:4725–4732

Dorée M, Galas S (1994) The cyclin-dependent protein kinases and the control of cell division. FASEB J 8:1114–1121

Dover GA, Flavell RV (1984) Molecular co-evolution: rDNA divergence and the maintenance of function. Cell 38:622–623

Dragon F, Gallagher JEG, Compagnone-Post PA, Mitchell BM, Porwancher KA, Wehner KA, Wormsley S, Settlage RE, Shabanowitz J, Osheim Y, Beyer AL, Hunt DF, Baserga SJ (2002) A large nucleolar U3 ribonucleoprotein required for 18S ribosomal RNA biogenesis. Nature 417:967–970

Dundr M, Raska I (1993) Nonisotopic ultrastructural mapping of transcription sites within the nucleolus. Exp Cell Res 208:275–281

Dvorak J, Jue D, Lassner M (1987) Homogenization of tandemly repeated nucleotide sequences by distance-dependent nucleotide sequence conversion. Genetics 116:487–498

Ellis THN, Lee D, Thomas CM, Simpson PR, Cleary WG, Newmann M-A, Burcham KWG (1988) 5S rRNA genes in *Pisum*: Sequence, long range und chromosomal organization. Mol Gen Genet 214:333–342

Endo Y, Wool IG (1982) The site of action of alpha-sarcin on eukaryotic ribosomes. The sequence at the alpha-sarcin cleavage site in 28 S ribosomal ribonucleic acid. J Biol Chem 257:9054–9060

Engelke DR, Ng SY, Shastry BS, Roeder RG (1980) Specific interaction of purified transcription factor with an internal control region of 5S RNA genes. Cell 19:717–728

England PR, Stokes HW, Frankham R (1988) Clustering of rDNA containing type 1 insertion sequence in the distal nucleolus organizer of *Drosophila melanogaster*: implications for the evolution of X and Y rDNA arrays. Genet Res Camb 51:209–215

Erard M, Belenguer P, Caizergues-Ferrer M, Pantaloni A, Amalric F (1988) A major nucleolar protein, nucleolin, induces chromatin decondensation by binding to histone H1. Eur J Biochem 175:525–530

Esponda P, Giménez-Martín G (1974) Cytochemical aspects of the nucleolar organizer in *Allium cepa* microspores. Chromosoma 45:203–213

Fakan S, Puvion E (1980) The ultrastructural visualization of nucleolar and extranucleolar RNA synthesis and distribution. Int Rev Cytol 65:255–295

Fan H, Kimitaka Y, Miyanishi M, Sugita M, Sugiura M (1995) In vitro transcription of plant RNA polymerase I-dependent rRNA genes is species-specific. Plant J 8:295–298

Flavell RB (1986) The structure and control of expression of ribosomal RNA genes. Ox Surv Plant Mol Cell Biol 3:252–274

Flavell RB, O'Dell M (1979) The genetic control of nucleolus formation in wheat. Chromosoma 71:135–152

Frasch M, Wenzel W, Hess D (1989) The nucleotide sequences of nuclear 5S rRNA genes and spacer regions of *Petunia hybrida*. Nucleic Acids Res 17:2857

Frieman M, Chen ZJ, Saez-Vasquez J, Shen LA, Pikaard CS (1999) RNA polymerase I transcription in a *Brassica* Interspecific hybrid and its progenitors: tests of transcription factor involvement in nucleolar dominance. Genetics 152:451–460

Fulnecek J, Matyasek R, Kovarik A, Bezdek M (1998) Mapping of 5-methylcytosine residues in *Nicotiana tabacum* 5S rRNA genes by genomic sequencing. Mol Gen Genet 259:133–141

Gadal O, Strauß D, Kessl J, Trumpower B, Tollervey D, Hurt E (2001) Nuclear export of 60S ribosomal subunits depends an Xpo1p and requires a NES-containing factor Nmd3p that associates with the large subunit protein RP110p. Mol Cell Biol 21:3405–3415

Ganley AR, Scott B (2002) Concerted evolution in the ribosomal RNA genes of an *Epichloe* endophyte hybrid comparison between tandemly arranged rDNA and dispersed 5S rrn genes. Fungal Genet Biol 35:39–51

Gasser SM, Laemmli UK (1986) Cohabitation of scaffold binding regions with upstream/enhancer elements of three developmentally regulated genes of *D. melanogaster*. Cell 46:521–530

Georgiev S, Papazova N, Gecheff K (2001) Transcriptional activity of an inversion split NOR in barley (*Hordeum vulgare* L.). Chromosome Res 9:507–514

Gerstner J, Schiebel K, v Waldburg G, Hemleben V (1988) Complex organisation of the length heterogeneous 5' external spacer of mung bean (*Vigna radiata*) ribosomal DNA. Genome 30:723–733

Ghisolfi-Nieto L, Joseph G, Puvion-Dutilleul F, Amalric F, Bouvet P (1996) Nucleolin is a sequence-specific RNA-binding protein: characterization of targets on pre-ribosomal RNA. J Mol Biol 260:34–53

Gilbert N, Lucas L, Klein C, Menager M, Bonnet N, Ploton D (1995) Three-dimensional co-location of RNA polymerase I and DNA during interphase and mitosis by confocal microscopy. J Cell Sci 108:115–125

Ginisty H, Amalric F, Bouvet P (1998) Nucleolin functions in the first step of ribosomal RNA processing. EMBO J 17:1476–1486

Ginisty H, Sicard H, Roger B, Bouvet P (1999) Structure and functions of nucleolin. J Cell Sci 112:761–772

Godward MBE (1950) On the nucleolus and nucleolar organizing chromosomes of *Spirogyra*. Ann Bot 14:39-53

Goel S, Raina SN, Ogihara Y (2002) Molecular evolution and phylogenetic implications of internal transcribed spacer sequences of nuclear ribosomal DNA in the *Phaseolus-Vigna* complex. Mol Phylogenet Evol 22:1–19

González-Melendi P, Wells B, Beven AF, Shaw PJ (2001) Single ribosomal transcription units are linear, compacted Christmas trees in plant nucleoli. Plant J 27:223–233

Grandi P, Rybin V, Bassler J, Petfalski E, Strauss D, Marzioch M, Schafer T, Kuster B, Tschochner H, Tollervey D, Gavin AC, Hurt E (2002) 90S pre-ribosomes include the 35S pre-rRNA, theU3 snoRNP and 40S subunit processing factors but predominantly lack 60S synthesis factors. Mol Cell 10:105–115

Grebenstein B, Röser M, Sauer W, Hemleben V (1998) Molecular phylogenetic relationships in Aveneae (Poaceae) species and other grasses as inferred from ITS1 and ITS2 sequences. Plant Syst Evol 213:233–250

Gründler P, Unfried I, Pascher K, Schweizer D (1991) rDNA intergenic region from *Arabidopsis thaliana*: structural analysis, intraspecific variation and functional implications. J Mol Biol 221:1209–1222

Grummt I (1999) Regulation of mammalian ribosomal gene transcription by RNA polymerase I. Progr Nucleic Acid Res Mol Biol 62:109–115

Grummt I, Roth E, Paule MR (1982) rRNA transcription in vitro is species-specific. Nature 296:173–174

Guilfoyle TJ (1980) Purification, subunit structure and immunological properties of chromatin-bound ribonucleic acid polymerase I from cauliflower inflorescence. Biochemistry 19:5966–5972

Gulli MP, Girard J-P, Zabetakis D, Lapeyre B, Melese T, Caizergues-Ferrer M (1995) gar2 is a nucleolar protein from *Schizosaccharomyces pombe* required for 18S rRNA and 40S ribosomal subunit accumulation. Nucleic Acids Res 23:1912–1918

Gustafson JP, Dera AR, Petrovic S (1988) Expression of modified rye ribosomal RNA genes in wheat. Proc Natl Acad Sci USA 85:3943–3945

Hannan RD, Cavanaugh A, Hempel WM, Moss T, Rothblum L (1999) Identification of a mammalian RNA polymerase I holoenzyme containing components of the DNA repair/replication system. Nucleic Acids Res 27:3720–3727

Heix J, Grummt I (1995) Species specificity of transcription by RNA polymerase I. Curr Opin Genet Dev 5:652–656

Hemann C, Gartner E, Weidle UK, Grummt F (1994) High-copy expression vector based on amplification-promoting sequences. DNA Cell Biol 13:437–445

Hemleben V, Werts D (1988) Sequence organization and putative regulatory elements in the 5S rRNA genes of two higher plants (*Vigna radiata* and *Matthiola incana*). Gene 62:165–169

Hemleben V, Zentgraf U (1994) Structural organization and regulation of transcription by RNA polymerase I of plant nuclear ribosomal RNA genes. In: Nover L (ed) Plant promoters and transcription factors. Results and problems in cell differentiation, vol 20. Springer, Berlin Heidelberg New York, pp 3–24

Hemleben V, Ganal M, Gerstner J, Schiebel K, Torres RA (1988) Organization and length heterogeneity of plant ribosomal RNA genes. In: Kahl G (ed) The architecture of eukaryotic gene. VHC, Weinheim, pp 371–384

Hemleben V, Schmidt T, Torres-Ruiz RA, Zentgraf U (2000) Molecular cell biology: role of repetitive DNA in nuclear architecture and chromosome structure. In: Esser K, Kadereit JW, Lüttge U, Runge M (eds) Progress in botany 61. Springer, Berlin, Heidelberg, New York, pp 91–117

Hernandez P, Bjerknes CA, Lamm SS, van't Hof J (1988) Proximity of an ARS consensus sequence to a replication origion of pea (*Pisum sativum*). Plant Mol Biol 10:413–422

Hernandez P, Martin-Parras L, Martinez-Robles ML, Schvartzman JB (1993) Conserved features in the mode of replication of eukaryotic ribosomal RNA genes. EMBO J 12:1475–1485

Herrera AH, Olson MOJ (1986) Association of protein C23 with rapidly labeled nucleolar RNA. Biochemistry 25:6258–6264

Hirschler-Laszkiewicz I, Cavanaugh A, Hu Q, Catania J, Avantaggiati ML, Rothblum LI (2001) The role of acetylation in rDNA transcription. Nucleic Acids Res 29:4114–4124

Ho JHN, Kallstrom G, Johnson AW (2000) Nmd3p is a Crm1p-dependent adapter protein for nuclear export of the large ribosomal subunit. J Cell Biol 151:1057–1066

Honjo T, Reeder RH (1973) Preferential transcription of *Xenopus laevis* ribosomal RNA in interspecific hybrids between *Xenopus laevis* and *Xenopus mulleri*. J Mol Biol 80:217–228

Houchins K, O'Dell M, Flavell RB, Gustafson JP (1997) Cytosine methylation and nucleolar dominance in cereal hybrids. Mol Gen Genet 255:294–301

Hozák P, Cook PR, Schöfer C, Mosgöller W, Wachtler F (1994) Site of transcription of ribosomal RNA and intranucleolar structure in HeLa cells. J Cell Sci 107:639–648

Huang S (2002) Building an efficient factory: where is pre-rRNA synthesized in the nucleolus? J Cell Biol 157:739–741

Jasencakova Z, Meister A, Schubert (2001) Chromatin organization and its relation to replication and histone acetylation during the cell cycle in barley. Chromosoma 110:83–92

Jobst J, King K, Hemleben V (1998) Molecular evolution of the internal transcribed spacers (ITS1 and ITS2) and phylogenetic relationships among species of Cucurbitaceae. Mol Phyl Evol 9:204–219

Kato A, Yakura K, Tanifuji S (1985) Repeated DNA sequences found in the large spacer of *Vicia faba* rDNA. Biochem Biophys Acta 825:411–415

Kass S, Tyc K, Steitz JA, Sollner-Webb B (1990) The U3 small nucleolar ribonucleoprotein functions in the first step of pre- ribosomal RNA processing. Cell 60:897–908

Kelly RJ, Siegel A (1989) The *Cucurbita maxima* ribosomal DNA intergenic spacer has a complex structure.Gene 80:239–248

Kenton A, Parakonny AS, Gleba YY, Bennett MD (1993) Characterization of the *Nicotiana tabacum* L. genome by molecular cytogenetics. Mol Gen Genet 240:159–169

King K, Torres RA, Zentgraf U, Hemleben V (1993) Moleular evolution of the intergenic specer in the nuclear ribosomal RNA genes of Cucurbitaceae. J Mol Evol 36:144–152

King LM, Schaal BA (1989) Ribosomal-DNA variation and distribution in *Rubeckia missouriensis.* Evolution 43:1117–1119

Krawetz JE, Boston RS (2000) Substrate specificity of a maize ribosome-inactivating protein differs across diverse taxa. Eur J Biochem 267:1966–1974

Kruiswijk T, Planta RJ, Knop JM (1978) The course of the assembly of ribosomal subunits. Biochem Biophys Acta 517:378–389

Lapeyre B, Bourbon HM, Amalric F (1987) Nucleolin, the major nucleolar protein of growing eukaryotic cells: an unusual protein structure revealed by the nucleotide sequence. Proc Natl Acad Sci USA 84:1472–1476

Lapeyre B, Mariottini P, Mathieu C, Ferrer P, Amaldi F, Amalric F, Caizergues-Ferrer M (1990) Molecular cloning of *Xenopus* fibrillarin, a conserved U3 small nuclear ribonucleoprotein recognized by antisera from humans with autoimmune disease. Mol Cell Biol 10:430–434

Lapitan NLV, Ganal MW, Tanksley SD (1991) Organization of the 5S ribosomal RNA genes in the genome of tomato. Genome 34:509–514

Leader DJ, Clark GP, Watters J, Beven AF, Shaw PJ, Brown JWS (1999) Splicing-independent processing of plant box C/D and box H/ACA small nucleolar RNAs. Plant Mol Biol 39:1091–1100

Learn GH, Schaal J, Schaal B (1987) Population subdivision for ribosomal DNA repeat variants in *Clematis fremontii.* Evolution 41:433–438

Learned RM, Cordes S, Tjian R (1985) Purification and characterization of a transcription factor that confers promotor specificity to human RNA polymerase I. Mol Cell Biol 5:1358–1369

Leary DJ, Huang S (2001) Regulation of the ribosome biogenesis within the nucleolus. FEBS Lett 509:145–150

Lee WC, Xue Z, Mélèse T (1991) The NSR1 gene encodes a protein that specifically binds nuclear localization sequences and has two RNA recognition motifs. J Cell Biol 113:1–12

Leweke B, Hemleben V (1982) Organization of rDNA in Chromatin: Plants. In: Busch H, Rothblum L (eds) The cell nucleus, Part B, vol XI. Academic Press, New York, pp 225–253

Lewis MS, Pikaard CS (2001) Restricted chromosome silencing in nucleolar dominance. Proc Natl Acad Sci USA 98:14536–14540

Lim KY, Kovarik A, Matýasek R, Bezdeek M, Lichtenstein CP, Leitch AR (2000) Gene conversion of ribosomal DNA in *Nicotiana tabacum* is associated with undermethylated, decondensed and probably active gene units. Chromosoma 109:161–172

Linder CR, Goertzen LR, Heuvel BV, Francisco-Ortega J, Jansen RK (2000) The complete external transcribed spacer of 18S-26S rDNA: amplification and phylogenetic utility at low taxonomic levels in asteraceae and closely allied families. Mol Phylogenet Evol 14:285–303

Linskens MH, Huberman JA (1988) Organization of replication of ribosomal DNA in *Saccharomyces cerevisiae.* Mol Cell Biol 8:4927–4935

Lipscomb DL, Farris JS, Kaellersjoe M, Tehler A (1998) Support, ribosomal sequences and the phylogeny of the eukaryotes. Cladistics 14:303–338

Lischwe MA, Smetana K, Olson MOJ, Busch H (1979) Proteins C23 and B23 are the major nucleolar silver staining proteins. Life Sci 25:701–708

Liu JS, Schardl CL (1994) A conserved sequence in internal transcribed spacer 1 of plant nuclear rRNA genes. Plant Mol Biol 26:775–778

Liu RS, Wei GQ, Yang Q, He WJ, Liu WY (2002) Cinnamomin, a type II ribosome-inactivating protein, is a storage protein in the seed of the camphor tree (*Cinnamomum camphora*). Biochem J 362:659–663

Logemann J, Jach G, Tommerup H, Mundy J, Schell J (1992) Expression of a barley ribosome-inactivating protein leads to increased fungal protection in transgenic tobacco plants. Biotechnology 10:305–308

Maceluch J, Kmieciak M, Szweykowska-Kulinska Z, Jarmolowski A (2001) Cloning and characterization of AtNAP57- a homologue of yeast pseudouridine synthase Cbf5p. Acta Biochem Pol 48:699–709

Martín M, Medina FJ (1991) A *Drosophila* anti-RNA polymerase II antibody recognizes a plant nucleolar antigen, RNA polymerase I, which is mostly localized in fibrillar centres. J Cell Sci 100:99–107

Martín M, Moreno Díaz de la Espina S, Medina FJ (1989) Immunolocalization of DNA at nucleolar structural components in onion cells. Chromosoma 98:368–377

Martín M, García-Fernández LF, Moreno Díaz de la Espina S, Noaillac-Depeyre J, Gas N, Medina FJ (1992) Identification and localization of a nucleolin homologue in onion nucleoli. Exp Cell Res 199:74–84

Martini G, O'Dell, Flavell RB (1982) Partial inactivation of wheat nucleolus organizer chromosomes from *Aegilops umbellulata*. Chromosoma 84:687–700

Maggini F, Tucci G, Demartis A, Gelati MT, Avanzi S (1992) Ribosomal RNA genes of *Phaseolus coccineus*. Plant Mol Biol 18:1073–1082

Mai JC, Coleman AW (1997) The internal transcribed spacer 2 exhibits a common secondary structure in green algae and flower plants. J Mol Evol 44:258–257

Mason SW, Wallisch M, Grummt I (1997) RNA polymerase I transcription termination: similar mechanisms are employed by yeast and mammals. J Mol Biol 268:229–234

McClintock B (1934) The relationship of a particular chromosomal element to the development of the nucleoli in *Zea mays*. Zeit Zellforsch Mik Anat 21:294–328

McGrath KE, Smothers JF, Dadd CA, Madireddi MT, Gorovsky MA, Allis CD (1997) An abundant nucleolar phosphoprotein is associated with ribosomal DNA in *Tetrahymena* macronuclei. Mol Biol Cell 8:97–108

McIntosh KB, Bonham-Smith PC (2001) Establishment of *Arabidopsis thaliana* ribosomal protein RPL23A-1 as a functional homologue of *Saccharomyces cerevisiae* ribosomal protein L25. Plant Mol Biol 46:673–682

Medina FJ, Risueño MC, Rodríguez-García MI, Sánchez-Pina MA (1983) The nucleolar organizer (NOR) and fibrillar centers during plant gametogenesis. J Ultrastruct Res 85:300–310

Medina FJ, Martín M, Moreno Díaz de la Espina S (1990) Implications for the function-structure relationship in the nucleolus after immunolocalization of DNA in onion cells. In: Harris JR, Zbarsky IB (eds) Nuclear structure and function. Plenum, London, pp 231–235

Medina FJ, Cerdido A, Fernández-Gómez ME (1995) Components of the nucleolar processing complex (pre-rRNA, fibrillarin and nucleolin) colocalize during mitosis and are incorporated to daughter cell nucleoli. Exp Cell Res 221:111–125

Medina FJ, Cerdido A, De Cárcer G (2000) The functional organization of the nucleolus in proliferating plant cells. Eur J Histochem 44:117–131

Medina FJ, González-Camacho F, Cerdido A, De Cárcer G (2001) In situ localization of the onion nucleolar protein NopA64 is dependent on cell proliferation mechanisms and cell cycle phases. In: Dini L, Catalano M (eds) Proc 5th Multinational Congr Electron Microscopy. Rinton Press, Princeton, New Jersey, pp 1976207

Melcák I, Risueño MC, Raska I (1996) Ultrastructural nonisotopic mapping of nucleolar transcription sites in onion protoplasts. J Struct Biol 116:253–263

Melekhovets YF, Shwed PS, Nazar RN (1997) In vivo analyses of RNA polymerase I termination in *Schizosaccharomyces pombe*. Nucleic Acids Res 25:103–5109

Milkereit P, Gadal O, Podtelejnikov A, Trumtel S, Gas N, Petfalski E, Tollervey D, Mann M, Hurt E, Tschochner H (2001) Maturation and intranuclear transport of pre-ribosomes requires Noc proteins. Cell 105:499–509

Miroshnichenko GP, Volkov RA, Kostishin SS (1988) Polynucleotide sequence divergence in DNAs of interspecific Solanaceae hybrids. Biochemistry (Moscow) 53:565–572

Mishima Y, Financsek I, Kominami R, Muramatsu M (1982) Fractionation and reconstitution of factors required for accurate transcription of mammalian ribosomal genes: identification of species-dependent initiation factor. Nucleic Acids Res 10:6659–6670

Molnar S, Gupta P, Fedak G, Wheatcroft R (1989) Ribosomal DNA repeat unit polymorphism in 25 *Hordeum* species. Theor Appl Genet 78:387–392

Moreno Díaz de la Espina S (1995) Nuclear matrix isolated from plant cells. Int Rev Cytol 16:75–139

Moore PB, Steitz TA (2002) The involvement of RNA in ribosome function. Nature 418:229–35

Moss T (1983) A transcription function for the repetitive ribosomal spacer in *Xenopus laevis*. Nature 302:223–228

Moss T, Stefanowsky VY (1995) Promotion and regulation of ribosomal transcription in eukaryotes by RNA polymerase I. Progr Nucleic Acid Res Mol Biol 50:25–66

Mougey EB, O'Reilly M, Osheim Y, Miller OL, Beyer A, Sollner-Webb B (1993) The terminal balls characteristic of eukaryotic rRNA transcription units in chromatin spreads are rRNA processing complexes. Genes Dev 7:1609–1619

Mougey EB, Pape LK, Sollner-Webb B (1996) Virtually the entire *Xenopus laevis* rDNA multikilobase intergenic spacer serves to stimulate polymerase I transcription. J Biol Chem 271:27138–27145

Moy TI, Silver PA (1999) Nuclear export of the small ribosomal subunits requires RanGTPase cycle and certain nucleoporins. Genes Dev 13:2118–2133

Navashin M (1927) Changes in number and form of chromosomes as a result of hybridization. Z Zellforsch Mikrosk Anat 6:195–223

Navashin M (1934) Chromosomal alterations caused by hybridization and their bearing upon certain general genetic problems. Cytologia 5:169–203

Neves N, Castilho A, Silva M, Heslop-Harrison JS, Viegas W (1997a) Genomic interactions: gene expression, DNA methylation and nuclear architecture. In: Henriques-Gil N, Parker JS, Puertas MJ (eds) Chromosomes today. Chapman and Hall, London, pp 182–200

Neves N, Silva M, Heslop-Harrison JS, Viegas W (1997b) Nucleolar dominance in triticales: control by unlinked genes. Chromosome Res 5:125–131

Nicoloff H (1979) Nucleolar dominance is observed in barley translocation lines with specifically reconstructed SAT chromosomes. Theor Appl Genet 55:247–251

Nielsen K, Payne GA, Boston RS (2001) Maize ribosome-inactivating protein inhibits normal development of *Aspergillus nidulans* and *Aspergillus flavus*. Mol Plant Microbe Interact 14:164–72

Nissan TA, Baßler J, Petfalski E, Tollervey D, Hurt E (2002) 60S pre-ribosome formation viewed from assembly in the nucleolus until export to the cytoplasm. EMBO J 21:5539–5547

Ochs RL, Lischwe MA, Spohn WH, Busch H (1985) Fibrillarin: a new protein of the nucleolus identified by autoimmune sera. Biol Cell 54:123–134

Olson MOJ (1991) The role of proteins in nucleolar structure and function. In: Strauss PR, Wilson SH (eds) The eukaryotic nucleus. Molecular biochemistry and macromolecular assemblies. The Telford Press, Caldwell, New Jersey, pp 519–559

Olson MOJ, Rivers ZM, Thompson BA, Kao WY, Case ST (1983) Interaction of nucleolar phosphoprotein C23 with cloned segments of rat ribosomal deoxyribonucleic acid. Biochemistry 22:3345–3351

Orrick L, Olson MOJ, Busch H (1973) Comparison of nucleolar proteins of normal rat liver and Novikoff hepatoma ascites cells by two dimensional polyacrylamide gel electrophoresis. Proc Natl Acad Sci USA 70:1316–1320

Papazova N, Hvarleva T, Atanassov A, Gecheff K (2001) The role of cytosine methylation for rRNA gene expression in reconstructed karyotypes of barley. Biotechnol Equipment 15:35–44

Pape LK, Windile JJ, Mougey EB and Sollner-Webb B (1989) The *Xenopus* ribosomal DNA 60- and 81-base-pair repeats are position-dependent enhancers that function at the establishment of the preinitiation complex: analysis in vivo and in an enhancer-responsive in vitro system. Mol Cell Biol 9:5093–5104

Parente A, De Luca P, Bolognesi A, Barbieri L, Battelli MG, Abbondanza A, Sande MJ, Gigliano GS, Tazzari PL, Stirpe F (1993) Purification and partial characterization of single-chain ribosome-inactivating proteins from the seeds of *Phytolacca dioica* L. Biochim Biophys Acta 1216:43–49

Parikh BA, Coetzer C, Tumer NE (2002) Pokeweed antiviral protein regulates the stability of its own mRNA by a mechanism that requires depurination but can be separated from depurination of the alpha-sarcin/ricin loop of rRNA. J Biol Chem 277:41428–41437

Park SW, Stevens NM, Vivanco JM (2002a) Enzymatic specificity of three ribosome-inactivating proteins against fungal ribosomes, and correlation with antifungal activity. Planta 216:227–234

Park SW, Lawrence CB, Linden JC, Vivanco JM. (2002b) Isolation and characterization of a novel ribosome-inactivating protein from root cultures of pokeweed and its mechanism of secretion from roots. Plant Physiol 130:164–178

Paul MF, Tse TBN, Fong WP, Wong RNS, Wan CC, Mak NK, Yeung HW (1999) New ribosome-inactivating proteins from seeds and fruits of the bitter gourd *Momordica charantia*. Int. J Biochem Cell Biol 31:895–901

Paule MR (1998a) Promoter structure of class I genes. In: Paule MR (ed) Transcription of ribosomal RNA genes by eukaryotic RNA polymerase I. Springer, Berlin Heidelberg New York, pp39-50

Paule MR (1998b) Structure of the fundamental transcription initiation factor from a simple eukaryote, *Acanthamoeba castellanii*. In: Paule MR (ed) Transcription of ribosomal RNA genes by eukaryotic RNA polymerase I. Springer, Berlin Heidelberg New York, pp 59-66

Pederson T (1998) The plurifunctional nucleolus. Nucleic Acids Res 26:3871–3876

Pederson T (2002) Proteomics of the nucleolus: more proteins, more functions? Trends Biochem Sci 27:111–112

Pelham HR, Brown DD (1980) A specific transcription factor that can bind either the 5S RNA gene or 5S RNA. Proc Natl Acad Sci USA 77:4170–4174

Perry KL, Palukaitis P (1990) Transcription of tomato ribosomal DNA and the organization of the intergenic spacer. Mol Gen Genet 221:102–112

Pieler T, Hamm J, Roeder RG (1987) The 5S gene internal control region is composed of three distinct sequence elements, organized as two functional domains with variable spacing. Cell 48:91–100

Pih KT, Yi MJ, Liang YS, Shin BJ, Cho MJ, Hwang I, Son D (2000) Molecular cloning and targeting of a fibrillarin homolog from *Arabidopsis*. Plant Physiol 123:51–58

Pikaard CS (2000) Nucleolar dominance: uniparental gene silencing on a multi-megabase scale in genetic hybrids. Plant Mol Biol 43:163–177

Pikaard CS (2002) Transcription and tyranny in the nucleolus: the organization, activation, dominance and repression of ribosomal RNA genes. The Arabidopsis Book, American Society of Plant Biologists, Rockville, MD

Pikaard CS, Reeder RH (1988) Sequence elements essential for function of the *Xenopus laevis* ribosomal DNA enhancers. Mol Cell Biol 8:4282–4288

Pinna LA (1990) Casein kinase 2: an 'eminence grise' in cellular regulation? Biochim Biophys Acta 1054:267–284

Polans NO, Weeden NF, Thompson WF (1986) Distribution, inheritance and linkage relationships of ribosomal DNA spacer length variants in pea. Theor Appl Genet 72:289–295

Poma A, Marcozzi G, Cesare P, Carmignani M, Spanò L (1999) Antiproliferative effect and apoptotic response in vitro of human melanoma cells to liposomes containing the ribosome-inactivating protein luffin. Biochim Biophys Acta (General Subjects) 1472:197–205

Raschke E, Baumann G, Schoeffl F (1988) Nucleotide sequence analysis of soybean small heat shock protein genes belonging to two different multigene families. J Mol Biol 199:549–557

Reeder RH (1974) Ribosomes from eukariyotes: genetics. In: Nomura M (ed) Ribosomes. Cold Spring Harbor Laboratory Press, Cold Spring Harbor, NY, pp 489–519

Reeder RH (1984) Enhancers and ribosomal gene spacer. Cell 38:349–351

Reeder RH (1985) Mechanisms of nucleolar dominance in animals and plants. J Cell Biol 101:2013–2016

Reeder RH (1990) rRNA synthesis in the nucleolus. Trends Genet 6:390–395

Reichler SA, Balk J, Brown ME, Woodruff K, Clark GB, Roux SJ (2001) Light differentially regulates cell division and the mRNA abundance of pea nucleolin during de-etiolation. Plant Physiol 125:339–350

Risueño MC, Medina FJ (1986) The nucleolar structure in plant cells. Cell Biol Rev (RBC) 7:1–154

Rodrigo RM, Rendón MC, Torreblanca J, García-Herdugo G, Moreno FJ (1992) Characterization and immunolocalization of RNA polymerase I transcription factor UBF with anti-NOR serum in protozoa, higher plant and vertebrate cells. J Cell Sci 103:1053–1063

Röser M, Winterfeld G, Grebenstein B, Hemleben V (2001) Molecular diversity and physical mapping of 5S rDNA in wild and cultivated oat grasses (Poaceae: *Aveneae*). Mol Phyl Evol 21:198–217

Roger B, Moisand A, Amalric F, Bouvet P (2002) Repression of RNA polymerase I transcription by nucleolin is independent of the RNA sequence that is transcribed. J Biol Chem 277:10209–10219

Rogers SO, Bendich AJ (1987a) Ribosomal DNA genes in plants: variability in copy number and in the intergenic spacer. Plant Mol Biol 9:509–520

Rogers SO, Bendich AJ (1987b) Heretability and variability in ribosomal RNA genes of *Vicia faba*. Genetics 117:285–295

Roth SY, Denu JM, Allis CD (2001) Histone acetyltransferases. Ann Rev Biochem 70:81–120

Roussel P, Belenguer P, Amalric F, Hernandez-Verdun D (1992) Nucleolin is an Ag-NOR protein; this property is determined by its amino-terminal domain independently of its phosphorylation state. Exp Cell Res 203:259–269

Ryan RF, Darby MK (1998) The role of zinc finger linkers in p43 and TFIIIA binding to 5S rRNA and DNA. Nucleic Acids Res 26:703–709

Saez-Vasquez J, Pikaard CS (1997) Extensive purification of a putative RNA polymerase I holoenzyme from plants that accurately initiates rRNA gene transcription in vitro. Proc Natl Acad Sci USA 94:11869–11874

Saez-Vasquez J, Pikaard CS (2000) RNA polymerase I holoenzyme-promoter interactions. J Biol Chem 275:37173–37180

Saez-Vasques J, Meissner M, Pikaard CS (2001) RNA polymerase I holoenzyme-promoter complexes include an associated CK2-like protein kinase. Plant Mol Biol 47:449–459

Saghai-Maroof MA, Soliman KM, Gorgensen RA, Allard RW (1984) Ribosomal DNA spacer-lengh polymorphisms in barley: nuclelian inheritance, chromosomal location, population dynamics. Proc Natl Acad Sci USA 81:8014–8018

Sandhu JS, Webster CI, Gray JC (1998) A/T-rich sequences act as quantitative enhancers of gene expression in transgenic tobacco and potato plants. Plant Mol Biol 37:885–896

Sandmeier JJ, French S, Osheim Y, Cheung WL, Gallo CM, Beyer AL, Smith JS (2002) RPD3 is required for the inactivation of yeast ribosomal DNA genes in stationary phase. EMBO J 21:4959–4968

Santoni S, Berville A (1992) Characterization of the nuclear ribosomal DNA units and phylogeny of *Beta* L. wild forms and cultivated beets. Theor Appl Genet 83:533–542

Sardana R, O'Dell A, Flavell R (1993) Correlation between the size of the intergenic regulatory region, the status of cytosine methylation of rRNA genes and nucleolar expression in wheat. Mol Gen Genet 236:155–162

Scheer U, Rose KM (1984) Localization of RNA polymerase I in interphase cells and mitotic chromosomes by light and electron microscopic immunocytochemistry. Proc Natl Acad Sci USA 81:1431–1435

Scheer U, Benavente R (1990) Functional and dynamic aspects of the mammalian nucleolus. BioEssays 12:14–21

Scherl A, Couté Y, Déon C, Callé A, Kindbeiter K, Sánchez JC, Greco A, Hochstrasser D, Díaz JJ (2002) Functional proteomic analysis of human nucleolus. Mol Biol Cell 13:4100–4109

Schiebel K, Waldburg G, Gerstner J, Hemleben V (1989) Termination of transcription of ribosomal RNA genes of mung bean occurs within a 175 bp repetitive element of the spacer region. Mol Gen Genet 218:302–307

Schimmang T, Tollervey D, Kern H, Frank R, Hurt EC (1989) A yeast nucleolar protein related to mammalian fibrillarin is associated with small nucleolar RNA and is essential for viability. EMBO J 8:4015–4024

Schlogelhofer P, Nizhynska V, Feik N, Chambon C, Potuschak T, Wanzenbock EM, Schweizer D, Bachmair A. (2002) The upstream Sal repeat-containing segment of *Arabidopsis thaliana* ribosomal DNA intergenic region (IGR) enhances the activity of adjacent protein-coding genes. Plant Mol Biol 49:655–667

Schlotter C, Tautz D (1994) Chromosomal homogeneity of Drosophila ribosomal DNA arrays suggests intrachromosomal exchanges drive concerted evolution. Curr Biol 4:777–783

Schnapp A, Grummt I (1991) Transcription complex formation at the mouse rDNA promoter involves the stepwise association of four transcription factors and RNA polymerase I. J Biol Chem 266:24588–24595

Schöffl F, Schröder G, Kliem M, Rieping M (1993) An SAR sequence containing 395 bp DNA fragment mediates enhanced, gene-dosage-correlated expression of a chimaeric heat shock gene in transgenic tobacco plants. Transgenic Res 2:93–100

Schubert I, Kunzel C (1990) Position-dependent NOR activity in barley. Chromosoma 99:352–359

Scoles CJ, Gill GS, Xin Z-Y, Clarke BC, McIntyre CL, Chapman C, Appels R (1988) Frequent duplication and deletion events in the 5S RNA genes and the associated spacer region of the Triticeae. Plant Syst Evol 160:105–122

Sentenac A (1985) Eukaryotic RNA polymerases. CRC Crit Rev Biochem 18:31–91

Serin G, Joseph G, Faucher C, Ghisolfi L, Bouche G, Amalric F, Bouvet P (1996) Localization of nucleolin binding sites on human and mouse pre-ribosomal RNA. Biochimie 78:530–538

Shaw PJ, Highett MI, Beven AF, Jordan EG (1995) The nucleolar architecture of polymerase I transcription and processing. EMBO J 14:2896–2906

Shaw PJ, Beven AF, Leader DJ, Brown JWS (1998) Localization and processing from a polycistronic precursor of novel snoRNAs in maize. J Cell Sci 111:2121–2128

Singh RC, Singh V (2000) Current status of ribosome inactivating proteins. Indian J Biochem Biophys 37:1–5

Sirri V, Roussel P, Gendron MC, Hernandez-Verdun D (1997) Amount of the two major AgNOR proteins, nucleolin and protein B23 is cell-cycle dependent. Cytometry 28:147–156

Soltis DE, Soltis PS (1995) The dynamic nature of polyploid genomes. Proc Natl Acad Sci USA 92:8089–8091

Srivastava M, Pollard HB (1999) Molecular dissection of nucleolin's role in growth and cell proliferation: new insights. FASEB J 13:1911–1922

Steffan JS, Keys DA, Dodd JA, Nomura M (1996) The role of TBP in rDNA transcription by RNA polymerase I in *Saccharomyces cerevisiae*: TBP is required for upstream activation factor-dependent recruitment of core factor. Genes Dev 10:2551–2563

Sterner DE, Berger SL (2000) Acetylation of histones and transcription-related factors. Microbiol Mol Biol Rev 64:435–459

Stirpe F, Barbieri L (1986) Ribosome-inactivating proteins up to date. FEBS Lett 195:1–8

Stirpe F, Barbieri L, Battelli MG, Soria M, Lappi DA (1992) Ribosome-inactivating proteins from plants: present status and future prospects. Bio Technology 10:405–412

Strahl BD, Allis CD (2000) The language of covalent histone modifications. Nature 403:41–45

Tanaka I, Akahori Y, Gomi K, Suzuki T, Ueda K (1999) A novel histone variant localized in the nucleoli of higher plant cells. Chromsoma 108:190–199

Testillano PS, Sánchez-Pina MA, Olmedilla A, Ollacarizqueta MA, Tandler CJ, Risueño MC (1991) A specific ultrastructural method to reveal DNA: the NAMA-Ur. J Histochem Cytochem 39:1427–1438

Testillano PS, Gorab E, Risueño MC (1994) A new approach to map transcription sites at the ultrastructural level. J Histochem Cytochem 42:1–10

Thiellement H, Bahrman N, Damerval C, Plomion C, Rossignol M, Santoni V, De Vienne D, Zivy M (1999) Proteomics for genetic and physiological studies in plants. Electrophoresis 20:2013–2026

Thiry M, Thiry-Blaise L (1989) In situ hybridization at the electron-microscope level: an improved method for precise localization of ribosomal DNA and RNA. Eur J Cell Biol 50:235–243

Thompson WF, Beven AF, Wells B, Shaw PJ (1997) Sites of rDNA transcription are widely dispersed through the nucleolus in *Pisum sativum* and can comprise single genes. Plant J 12:571–581

Tong CG, Reichler S, Blumenthal S, Balk J, Hsieh HL, Roux SJ (1997) Light regulation of the abundance of mRNA encoding a nucleolin-like protein localized in the nucleoli of pea nuclei. Plant Physiol 114:643–652

Torres RA, Ganal M, Hemleben V (1990) GC balance in the internal transcribed spacers ITS 1 and ITS 2 of nuclear ribosomal RNA genes. J Mol Evol 30:170–181

Trapman J, Retèl J, Planta RJ (1975) Ribosomal precursor particles from yeast. Exp Cell Res 90:95–104

Trontin J-F, Grandemange C, Favre J-M (1999) Two highly divergent 5S rDNA unit size classes occur in composite tandem array in European larch (*Larix deciduas* Mill.) and Japanese larch (*Larix kaempferi* (Lamb.) Carr.). Genome 42:837–848

Tuteja R, Tuteja N (1998) Nucleolin: a multifunctional major nucleolar protein. CRC Crit Rev Biochem Mol Biol (Moscow) 33:407–436

Tuteja N, Beven AF, Shaw PJ, Tuteja R (2001) A pea homologue of human helicase I is localized within the dense fibrillar component of the nucleolus and stimulated by phosphorylation with CK2 and cdc2 protein kinases. Plant J 25:9–17

Vakhitov VA, Chemeris AV, Akhmetzyanov AA (1989) Nucleotide sequence of rDNA intergenic and internal transcribed spacers in diploid wheat *Triticum urartu* Thum. ex Candil. Mol Biol 23:441–447

Van Campenhout S, Aert R, Volckaert G (1998) Orthologous DNA sequence variation among 5S ribosomal RNA genes pacer sequences on homoeologous chromosomes 1B, 1D, and 1R of wheat and rye. Genome 41:244–255

Vant'Hof J, Lamm S S (1992) Site of initiation of replication of the ribosomal genes of pea (*Pisum sativum*) detected by two-dimensional gel electrophoresis. Plant Mol Biol 20:377–382

Venema J, Tollervey D (1999) Ribosome synthesis in *Saccharomyces cerevisiae*. Annu Rev Genet 33:261–311

Viera A, Morais L, Barao A, Mello-Sampayo T, Viegas WS (1990a) 1R chromosome nucleolus organizer region activation by 5-azacytidine in wheat x rye hybrids. Genome 33:707–712

Viera A, Mello-Sampayo T, Viegas WS (1990b) Genetic control of 1R nucleolus organizer region in the presence of wheat genome. Genome 33:713–718

Volkov RA, Borisjuk NV, Kostishin SS, Panchuk II (1991) Variability of RNA genes in *Nicotiana* correlates with the chromosome reconstruction. Mol Biol (Moscow) 25:442–450

Volkov RA, Kostishin SS, Panchuk II (1993) rDNA organization in species from subfamily *Prunoideae*. Mol Biol (Moscow) 27:1356–1367

Volkov R, Kostishin S, Ehrendorfer F, Schweizer D (1996) Organization and molecular evolution of rDNA external transcribed spacer region in two diploid relatives of *Nicotiana tabacum*. Plant Syst Evol 201:117–129

Volkov R, Borisjuk N, Panchuk I, Komarova N, Hemleben V (1997) rDNA intergenic spacer region structure and taxonomic status of *Atropa belladonna* (Solanaceae). Biol Chem 378 (Supl.):115

Volkov RA, Borisjuk NV, Panchuk II, Schweizer D, Hemleben V (1999a): Elimination and rearrangement of parental rDNA in allotetraploid *Nicotiana tabacum*. Mol Biol Evol 16:311–320

Volkov RA, Bachmair A, Panchuk II, Kostyshyn SS, Schweizer D (1999b) 25S-18S rDNA intergenic spacer of *Nicotiana sylvestris* (Solanaceae): primary and secondary structure analysis. Plant Syst Evol 218:89–97

Volkov RA, Zanke C, Panchuk II, Hemleben V (2001) Molecular evolution of 5S rDNA of *Solanum* species (sect. *Petota*): application for molecular phylogeny and breeding. Theor Appl Genet 103:1273–1282

Volkov RA, Komarova NY, Panchuk II, Hemleben V (2003) Molecular evolution of rDNA external transcribed spacer and phylogeny of sect. *Petota* (genus *Solanum*) Mol Phylogenet Evol (in press and online, http: www.science direct. com)

Wachtler F, Hartung M, Devictor M, Wiegant J, Stahl A, Schwarzacher HG (1989) Ribosomal DNA is located and transcribed in the dense fibrillar component of human Sertoli cell nucleoli. Exp Cell Res 184:61–71

Wallace H, Langbridge WHR (1971) Differential amphiplasty and the control of ribosomal RNA synthesis. Heredity 27:1–13

Warrener P, Petryshyn R (1991) Phosphorylation and proteolytic degradation of nucleolin from 3T3-F442A cells. Biochem Biophys Res Comm 180:716–723

Warner JR (1989) Synthesis of ribosomes in *Saccharomyces cerevisiae*. Microbiol Rev 53:256–271

Warner JR (1999) The economics of the ribosome biosynthesis in yeast. Trends Biochem Sci 24:437–440

Warner JR (2001) Nascent ribosomes. Cell 107:133–136

Wegner M, Zastrow G, Klavinius A, Achwender S, Muller F, Luksza H, Hoppe J, Wienberg J, Grummt F (1989) Cis-acting sequences from mouse rDNA promote plasmid amplification and persistence in mouse cells: implication of HMG-1 in their function. Nucleic Acids Res 17:9909–9932

Weisenberger D, Scheer U (1995) A possible mechanism for the inhibition of ribosomal RNA gene transcription during mitosis. J Cell Biol 129:561–575

Wendel JF (2000) Genome evolution in polyploids. Plant Mol Biol 42:225–249

Wilcox LW, Lewis LA, Fuerst PA, Floyd GL (1992) Group 1 introns within the nuclear-encoded small-subunit rRNA gene of three green algae. Mol Biol Evol 9:1103–1118

Wyszko E, Barciszewska M (1997) Purification and characterization of transcription factor IIIA from higher plants. Eur J Biochem 249:107–112

Xie L, Wang BZ, Hu RG, Ji HB, Zhang L, Liu WY (2001) Structural and functional studies of cinnamomin, a new type II ribosome-inactivating proteins isolated from the seeds of the campor tree. Eur J Biochem 268:5723–5733

Yakura K, Kato A, Tanifuji S (1983) Structural organization of ribosomal DNA in four *Trillium* species and *Paris verticillata*. Plant Cell Physiol 24:1231–1240

Yano H, Sato S (2002) Combination of electron microscopic in situ hybridization and anti-DNA antibody labelling reveals a peculiar arrangement of ribosomal DNA in the fibrillar centres of the plant cell nucleolus. J Electron Microsc (Tokyo) 51:231–239

Zentgraf U, Ganal M, Hemleben V (1990) Length heterogeneity of the rRNA precursor in cucumber (*Cucumis sativus*). Plant Mol Biol 15:465–474

Zentgraf U, Velasco R, Hemleben V (1998) Different transcriptional activities in the nucleus. In: Progress in botany 59. Springer, Berlin Heidelberg New York, pp 131–168

Zhou Y, Santoro R, Grummt I (2002) The chromatin remodeling complex NoRC targets HDAC1 to the ribosomal gene promoter and represses RNA polymerase I transcription. EMBO J 21:4632–4640

Prof. Dr. Vera Hemleben
Dr. Roman A. Volkov
PD Dr. Ulrike Zentgraf
Lehrstuhl für Allgemeine Genetik
Center of Plant Molecular Biology (ZMBP)
University of Tübingen
Auf der Morgenstelle 28
72076 Tübingen, Germany
e-mail: vera.hemleben@uni-tuebingen.de

Dr. Francisco J. Medina
Centro de Investigaciones Biologicas (CSIC)
Velazquez 144
28006 Madrid, Spain
Tel.: +34-915-644562 #4261
Tel.: +34-915-644562 #4261
e-mail: fjmedina@cib.csic.es

Genetics of Phytopathology: Fungal Morphogenesis and Plant Infection

Stefan G.R. Wirsel, Sven Reimann, and Holger B. Deising

1 Introduction

Many fungi have evolved distinct cellular structures to facilitate the invasion of their host plants. These infection structures exhibit a range of morphological and physiological specialisations (Staples and Hoch 1987; Mendgen and Deising 1993; Deising et al. 1996; Perfect and Green 2001). The fact that fungi belonging to different taxa have developed a highly specialised cell type, called appressorium, to enter their host and to gain access to the nutrient sources inside plant cells or within the apoplast (Mendgen and Deising 1993) may indicate the existence of a "force" that drives evolution of fungal infection structures. In this respect it might be interesting to note that extant arbuscular mycorrhizas can also penetrate roots of their host by appressoria.

Parasitic relationships between fungi and plants had already developed in the Lower Devonian, as indicated by recent fossil findings in the Rhynie chert in Scotland. Taylor and coworkers (Taylor et al. 1999) discovered hyphae of a pathogenic ascomycete extending through the cuticle of the early land plant *Asteroxylon*. Successful invasion of and propagation on plants seen in these samples indicates that pathogenic relationships between fungi and the early land plants existed 400 million years ago (Taylor et al. 1999), and co-evolution of some plant pathogenic fungi and their host plants has thus taken place since (Bakkeren et al. 2000; Mugnier 2002). Recent molecular phylogenetic studies based on multiple protein analyses have indicated that interactions between plants and associated fungi might have been established much earlier (Heckman et al. 2001). These data suggested that land plants appeared around 700 million years ago and that the major lineages of fungi divided between 1.5 billion and 970 million years ago. It was hypothesised that colonisation of the land by plants was supported by symbiotic fungi, probably by glomalean type mycorrhizas (Pirozynski 1976; Selosse and Le Tacon 1998; Blackwell 2000).

Frank (1883) first coined the term 'appressorium' (= adhesion organ) to describe the apical swellings of germ tubes observed with *Fusicladium* sp. and *Gloeosporium* (*Colletotrichum*) *lindemuthianum* infecting poplar (*Populus tremula*) and bean (*Phaseolus vulgaris*), respectively. Emmett and Parbery (1975) extended this definition to "all structures adhering to host surfaces to achieve penetration, regardless of morphology".

Appressoria may be visible as inconspicuous swellings of the germ tube as in *Ustilago maydis* (Snetselaar and Mims 1993) or in basidiosporelings of

Progress in Botany, Vol. 65

rust fungi (Gold and Mendgen 1991), or discrete swollen, lobed or dome-shaped cells, separated from the germ tube by a septum as in rust uredinio- and aeciosporelings, in *Magnaporthe grisea* and *Colletotrichum* species, and in many other plant pathogens (Hahn et al. 1997a). Several reports indicate that infection structure differentiation is regulated by an ordered expression of a multitude of infection structure-related genes. In rust urediniosporelings, for example, single surface signals are capable of inducing a sequence of morphological responses leading to infection structure differentiation, and synthesis of various cell wall-degrading enzymes (Hoch et al. 1987; Mendgen et al. 1996; Read et al. 1997). Likewise, perception of single chemical signals present on or hydrophobicity or hardness of the plant surface have been described to trigger the expression of an array of genes required for appressorium formation (Podila et al. 1993; Lee and Dean 1994; Kolattukudy et al. 1995; Gilbert et al. 1996; Dean 1997; Kim et al. 2002, and references therein).

During the last decade, significant progress in understanding mechanisms and functions of fungal infection structures has been made. This has been possible mainly by modern cell and molecular biological approaches, leading to the identification of genes essential for infection-related morphogenesis and virulence or pathogenicity.

This review will describe the events beginning with the germination of fungal spores on the plant surface leading to its penetration and establishment of the pathogen within the host. We will draw attention to the link between the form and the function of fungal infection structures where appropriate, and will provide molecular data underlining more "classical" genetic and physiological results. Emphasis will be given to demonstrate that different fungi use different strategies to gain access to nutrients from the interior of plant cells or from apoplastic compartments, which is a requirement for successful colonisation and reproduction. We will also discuss that a detailed knowledge of morphogenic differentiation exhibited by fungal plant pathogens during various stages of infection might be helpful to understand the mode of action of fungicides currently in use. This might also be useful for rational drug design and screening procedures.

2 Infection Structures and Modes of Entry

The mode of entry depends on the infection structures differentiated. Fungi that form distinct and melanised infection structures are able to generate significant turgor pressure which is translated into force. Examples of fungi that differentiate melanised appressoria are *Magnaporthe grisea*, different *Colletotrichum* spp. and *Gaeumannomyces graminis*, the latter forming hyphopodia. In forceful invasion cell wall-degrading enzymes (CWDEs) are thought to play a minor role, although the exact impact of CWDEs needs to be analysed carefully in each fungal system. In contrast, fungi that do not form elaborate appressoria, e.g. *Botrytis cinerea*, *Claviceps purpurea*, or different *Cochliobolus* or *Fusarium* spp., are likely to depend on CWDEs for initial invasion. In order to mechanistically understand the different modes of infection, we will describe in detail force- and enzyme-dependent penetration of the host epidermis.

2.1 Turgor-Mediated Breaching of the Plant Cell Wall

Although the morphology of appressoria in different species varies significantly (Howard 1997), it has since long been accepted that these cells mediate penetration of plant cuticle and epidermal cell wall (Frank 1883; Miyoshi 1895). Irrespective of whether fungi use enzymes or force, or a combination of both to penetrate, they need to adhere tightly to the plant surface.

A clear indication of the strength of adhesion is provided by the fact that the appressorial base remained attached to the plant cuticle after sonication, which completely removed the upper part of the cell (Howard 1997). Adhesion is necessary to counteract the force exerted onto the underlying surface by the penetration peg arising from the appressorial base. A convincing illustration of appressorial adhesion is provided by studies employing optical waveguides (Bechinger et al. 1999; Bastmeyer et al. 2002).

Optical waveguides consist of a transparent high viscosity polydimethylsiloxane (PDMS) layer of 1-μm thickness, sandwiched between two thin films of aluminium. Light propagation in planar waveguides allows optical imaging and thus direct determination of vertical forces exerted by penetration pegs, and also forces needed to counteract these. While the penetration peg exerts force onto the waveguide, a ring-shaped area of the appressorial base surrounding the penetration peg lifts the elastic waveguide, indicating adhesion to counteract the invasive force. This technique allowed Bechinger and co-workers (1999) to directly measure forces of up to 25 μN (mean value 16.8 μN) exerted by single melanised appressoria of *C. graminicola*. For comparison, if a force of 17 μN μm^{-2} were exerted over the palm of a hand, a human could lift an 8,000-kg school bus (Money 1999). Howard and coworkers (1991b) demonstrated that *M. grisea* is able to forcefully penetrate hard synthetic surfaces such as Mylar membranes. In order to determine the turgor pressure of appressoria, these infection cells were incubated in solutions with increasing osmotic pressure, and the point of incipient cytorrhysis was taken as a measure of intracellular osmotic potential. These indirect turgor measurements showed that the rice blast fungus can generate pressures of up to 8 MPa (80 bar) within its appressorium (Howard et al. 1991b). Already in 1895, Miyoshi took a simple experimental approach, observing puncturing of plant cells by needles of known diameter and weight, and estimated that the pressure needed to breach the epidermis of *Allium cepae* and *Tradescantia procumbens* corresponds to 0.35 and 0.5 MPa, respectively (Miyoshi 1895). The pressure in melanised appressoria of *M. grisea* and *C. graminicola* corresponds to values between 5 and 8 MPa (Howard et al. 1991b; Bechinger et al. 1999), suggesting that these fungi could enter plant cells without enzymatic assistance.

The mechanism of turgor generation has been discussed for a long time, and in the 1980s it was recognised that melanin plays an important role in the penetration process. Kubo and coworkers have recently identified novel transcriptional activators of *Magnaporthe grisea* and *Colletotrichum lagenarium*, containing Cys2His2 zinc finger and Zn(II)2Cys6 binuclear cluster DNA-binding motifs, that regulate transcription of the melanin biosynthesis genes polyketide synthase, trihydroxynaphthalene reductase, and scytalone dehydratase in a development-specific manner (Kubo et al. 2000; Tsuji et al. 2000). In most fungal systems, melanins are thought to protect the mycelium against environmental stresses such as UV radiation, oxidants, antifungal drugs and macrophages formed by man and other animals in response to fungal infection (Henson et al. 1999; Nosanchuk et al. 1999). In addition, melanin acts as a scavenger of free radicals, a property essential to pathogenicity of the facultative human pathogen *Cryptococcus neoformans* (Nosanchuk et al. 1999). In the black yeast *Wangiella dermatitidis*, a polymorphic pathogen of man, melanin is incorporated into the outer cell wall and appears to mediate biomechanical properties and to support invasive hyphal growth into solid substrata (Brush and Money 1999).

Melanin-deficient mutants of *Colletotrichum* spp. and *M. grisea* as well as wild-type isolates treated with inhibitors of melanin biosynthesis were unable to penetrate their host (Kubo and Furusawa 1991; Henson et al. 1999, and references therein). In *M. grisea* and in *C. graminicola* melanin was shown to reduce the pore size in the appressorial wall to less than 1 nm (Howard et al. 1991b; J.A. Sugui and H.B. Deising, unpubl. data). As a consequence, water, but no larger molecules, can pass this layer, leading to water influx and turgor generation when osmotically active solutes are accumulated in the appressorium (Fig. 1). Talbot and coworkers showed that glycerol accumulates to concentrations of at least 3 M in wild-type appressoria of *M. grisea* (De Jong et al. 1997). Such high concentrations were neither present in appressoria of a melanin-deficient mutant nor in appressoria treated with the melanin biosynthesis inhibitor, tricyclazol. De Jong et al. (1997) also demonstrated that non-melanised appressorial walls are permeable to glycerol, whereas those of melanised appressoria are not. The turgor resulting from this glycerol concentration was calculated to be at least 5.8 MPa, which is close to that estimated from indirect turgor measurements (Howard et al. 1991b). Interestingly, glycerol accumulation in appressoria is regulated by a different signal transduction pathway than the

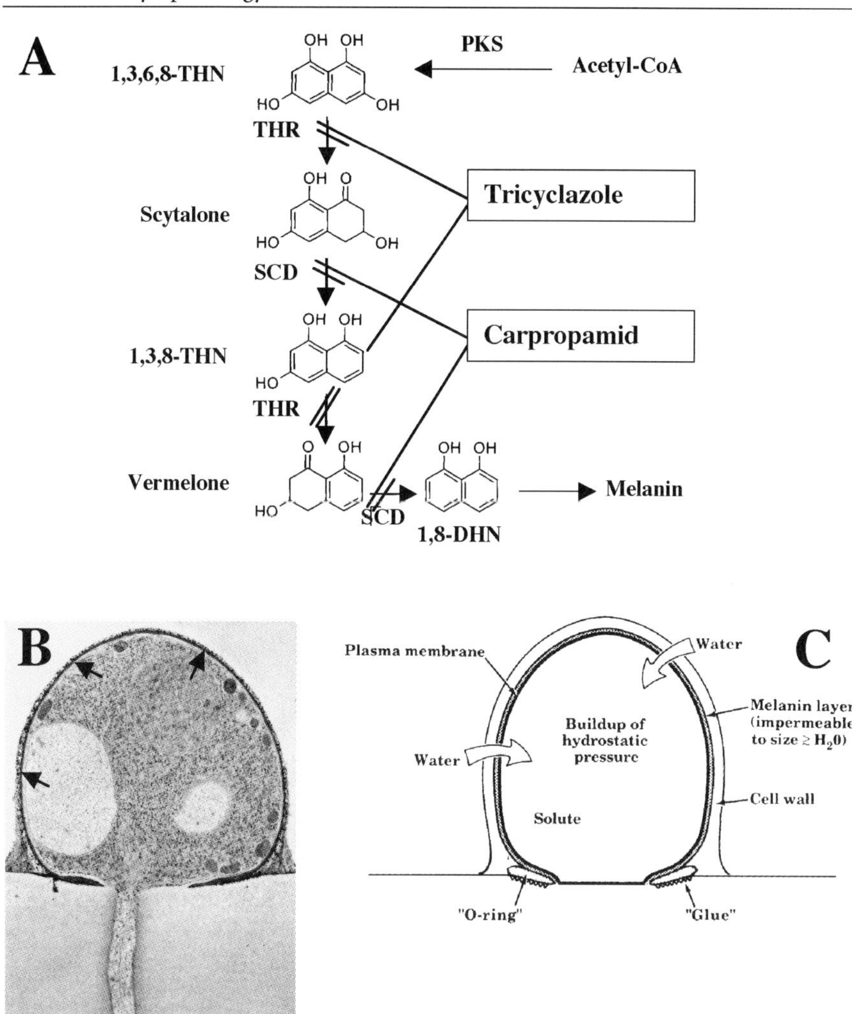

Fig. 1. Biosynthesis, localisation and function of appressorial melanin. **A** Melanin is synthesised from acetyl-CoA by polyketide synthase (PKS) and two subsequent reductase- (mediated by tetra- and trihydroxynaphthalene reductase, *THR*) and dehydratase reactions (mediated by scytalone dehydratase, *SCD*). THR is inhibited by tricyclazole fungicides, SCD by carpropamid. **B** The melanin layer (*arrows*) is localised in the appressorial cell wall, in close vicinity to the plasma membrane. **C** By incorporation of melanin the pore size of the appressorial wall is reduced to less than 1 nm, so that larger molecules such as osmotically active compounds, e.g. glycerol, cannot migrate through plasma membrane and cell wall. In contrast, water is taken up, resulting in generation of high turgor pressure, which is translated into force at the appressorial base. **B** and **C** are from Howard et al. (1991a)

cellular response to hyperosmotic stress, which is compensated by accumulation of arabitol (Dixon et al. 1999).

Appressoria of *C. graminicola* formed in the presence of the chitin synthase inhibitor Nikkomycin Z, incorporate melanin into their appressorial walls, but these walls are not rigid enough to withstand the developing turgor pressure of 5 to 8 MPa and burst (S. Werner and H.B. Deising, unpubl. results). Thus, chitin and probably also the other prominent cell wall polymer β-1,3 glucan mediate appressorium rigidity, whereas melanin serves as the selectively permeable wall layer. As indicated by dense labelling of appressorial walls of *M. grisea* by WGA gold conjugates (Howard et al. 1991a), chitin is a prominent structural component of these cells. Similar results have been obtained for *C. graminicola* (U. Rauchhaus, G. Hause and H.B. Deising, unpubl. data).

2.2 The Role of Cell Wall-Degrading Enzymes in Fungi with Elaborate Appressoria

In different *Colletotrichum* species, CWDEs have received considerable attention, with emphasis on pectic enzymes. Enzyme activities have been primarily been detected in necrotrophic growth phases, e.g. in *C. graminicola*-infected maize pith or in bean leaves infected with *C. lindemuthianum* (Nicholson et al. 1976; Wijesundera et al. 1989). Polygalacturonase activity was not detected in the latter interaction. Likewise, in the interaction between *C. gloeosporioides* f.sp. *malvae* and its host, *Malva pusilla* only pectic lyase activity was detected at the end of the biotrophic phase and it further increased in the necrotrophic phase of infection. A comparison of expression of *pel-1* and *pel-2* genes in culture and *in planta* with other pectinase genes of *C. gloeosporioides* f. sp. *malvae* and other plant pathogenic fungi suggests that during necrotrophic infection genes that are not under catabolite control are predominantly expressed (Shih et al. 2000). Lack of polygalacturonase activity may be due to the presence of proteinaceous inhibitors of these enzymes, which have been found in many plants (De Lorenzo et al. 2001; De Lorenzo and Ferrari 2002). Two *endo*-polygalacturonase genes (*clpg1, clpg2*) of *C. lindemuthianum* have been cloned (Centis et al. 1997). Experiments with the GFP reporter gene fused to the promoter of *clpg2* indicated that in spite of the high turgor generated by melanised appressoria, enzymes may assist the penetration process (Dumas et al. 1999). A similar conclusion can be drawn from a different line of evidence. Inactivation of the *Snf1* homologue (see below) of *C. graminicola* resulted in reduction of extracellular activities of CWDEs, reduced rates of penetration and reduced virulence towards maize leaves (M. Wernitz and

H.B. Deising, unpubl. data). Analyses of the *C. lindemuthianum clpg2* promoter identified a *cis*-acting sequence with similarity to a yeast filamentation and invasion response element, suggesting that invasive growth and depolymerising activity are co-ordinatedly regulated (Herbert et al. 2002). The *Kss1* MAPK cascade of *S. cerevisiae* controls both dimorphic development (formation of pseudohyphae and invasive growth) and a depolymerase (polygalacturonase) (Madhani et al. 1999). Also, in the mammalian pathogen *Candida albicans*, the same transcription factor (*CaTEC1*) controls hyphal development and expression of extracellular proteases (Schweizer et al. 2000).

However, in addition to hydrolytic enzymes, pectate or pectin lyases that cleave α-1,4-bonds by introducing a Δ-4,5-unsaturated bond, have received some attention. Antibodies raised against pectate lyase of *C. gloeosporioides* had no effect on spore germination, germ tube elongation or appressoria formation, but when conidia were mixed with the antibodies prior to inoculation, symptom development was inhibited on avocado, mango and banana fruits (Wattad et al. 1997). Targeted *C. gloeosporioides* mutants deficient in the pectate lyase (*pelB*) gene exhibited 25% lower pectate lyase (PL) and pectin lyase (PNL) activities and 15% higher polygalacturonase (PG) activity than the wild type. When *pelB* mutants were inoculated onto avocado fruits, a reduction in decay diameter by approximately 40% was observed compared with the wild type and undisrupted transformant controls, supporting the role of *pelB* as a virulence factor in *C. gloeosporioides* (Yakoby et al. 2001). Accordingly, when the *C. gloeosporioides pelB* gene was transformed into and expressed in *C. magna*, a pathogen of cucurbits that causes minor symptoms on watermelon seedlings and avocado fruits, the transformants showed increased pectate lyase activity and increased maceration capabilities on avocado pericarp, and more severe maceration and damping off developed on watermelon seedlings (Yakoby et al. 2000).

Wu et al. (1997) demonstrated that deletion of two *endo*-β-1,4-xylanase genes of *M. grisea* led to synthesis and secretion of additional hitherto unknown isozymes by the rice blast fungus. This again may be taken as evidence for the important role of these enzymes in host penetration and/or spread in the tissue. As this fungus is known to penetrate forcefully, little attention has so far been paid to the contribution of CWDEs for penetration in this system.

The wealth of data available indicates that it is very likely that both, enzymes and force are needed for initial penetration, the proportions being characteristic for each individual pathogen. This had been demonstrated for the infection process of the powdery mildew fungus of barley, *Blumeria graminis* f. sp. *hordei*. Although the appressoria of this economically important pathogen are not melanised, they develop a maximum turgor

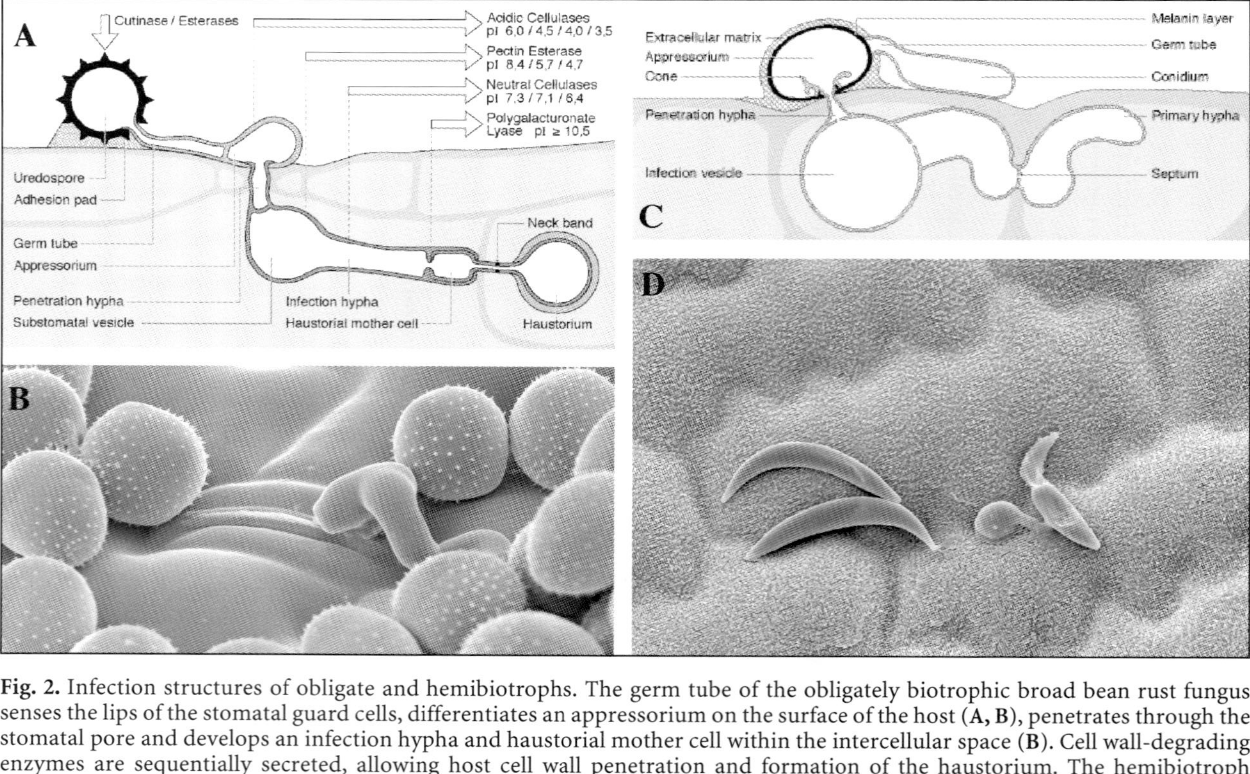

Fig. 2. Infection structures of obligate and hemibiotrophs. The germ tube of the obligately biotrophic broad bean rust fungus senses the lips of the stomatal guard cells, differentiates an appressorium on the surface of the host (**A, B**), penetrates through the stomatal pore and develops an infection hypha and haustorial mother cell within the intercellular space (**B**). Cell wall-degrading enzymes are sequentially secreted, allowing host cell wall penetration and formation of the haustorium. The hemibiotroph *Colletotrichum lindemuthianum* forms a melanised appressorium on the cuticle and invades the host cell directly (**C**). Within the host, infection vesicle and primary hyphae (biotrophic stage) develop. At later stages, necrotrophic secondary hyphae occur. Likewise, *C. graminicola* forms melanised appressoria after germination of falcate conidia (**D**). This fungus differentiates biotrophic hyphae only in the first host cell it infects. **A** and **C** are from Mendgen and Deising (1993), **B** and **D** courtesy of R. Guggenheim (REM laboratory of Basel University)

pressure of approx. 2–4 MPa, and the resulting force may act in concert with CWDEs such as cellulases, which have been shown to be present at the penetration site (Pryce-Jones et al. 1999). Irrespective of the mechanism of penetration, spread within the tissue will require the action of CWDEs in all invasively growing filamentous fungi. Also in other fungi with well-developed appressoria, for instance the rust fungus *Uromyces fabae*, CWDEs are thought to contribute to the infection process (Fig. 2) (Mendgen and Deising 1993; Mendgen et al. 1996).

2.3 The Role of Cell Wall-Degrading Enzymes in Fungi Without Elaborate Appressoria

CWDEs may not only decrease plant cell wall strength to facilitate inter- and intracellular growth and wall penetration, but also serve saprophytic functions, i.e. provide the fungus with nutrients. Targeted inactivation of genes encoding CWDEs would allow one to address the role(s) of these enzymes in pathogenesis directly, but their redundancy prevented to obtain clear results with this approach. In the polyphageous fungus *Botrytis cinerea*, causing disease in at least 235 plant species, 14 polygalacturonase isoforms (and 6 polygalacturonase-encoding genes) have been demonstrated in addition to other pectic enzymes (ten Have et al. 2002). Expression data have been presented for the polygalacturonase gene family of *B. cinerea*. The most prominent transcripts originate from the genes *Bcpg1* and *Bcpg2*. These genes do not appear to be under catabolite repression, whereas *Bcpg4* is currently the only one that is. The final degradation product, galacturonic acid, induces *Bcpg4* and *Bcpg6*, whereas *Bcpg3* is induced by low ambient pH and is not much affected by the carbon source present in the growth medium (Wubben et al. 2000; ten Have et al. 2002). As the genes differ in their expression patterns they may allow the fungus to quickly adapt to changes in the nutrient basis, and help to adapt to the different tissues of the various host plants. One of the most important findings was that *Bcpg1* knock-out mutants show reduced virulence on a number of hosts, including tomato leaves and fruits, apple fruits, and broad bean and *Arabidopsis thaliana* leaves (ten Have et al. 1998, 2002).

Endo-polygalacturonase genes of two closely related *Alternaria* species, i.e. *Alternaria citri* and *Alternaria alternata* rough lemon pathotype, exhibit 99% nucleotide sequence homology. Inactivation of the gene in *A. citri* led to significant reduction of virulence, whereas knock-out of the *A. alternata* gene had no effect (Isshiki et al. 2001). However, as *A. citri* seems to lack toxins, CWDEs may play a more important role in the infection process of this fungus, and this result may indicate that CWDEs represent basal virulence factors that act in addition to other known or hitherto unknown factors (ten Have et al. 2002). In other systems pectate lyases have been identified as virulence factors. The *Nectria haematococca* (*Fusarium solani* f. sp. *pisi*) genome contains a family of at least four pectate lyase genes, in addition to other genes encoding depolymerases such as polygalacturonases. Aggressive and non-aggressive isolates show minor differences in polygalacturonase activity in vitro, but aggressive isolates exhibit much higher pectate lyase activities, suggesting that these enzymes may determine virulence (Kolattukudy and Crawford 1987). While the pectate lyase gene *pelA* is induced when the fungus grows on pectate, *pelD* is only induced during *in planta* growth and not inducible in vitro. Inactivation of either gene alone does not interfere with virulence, but *pelA* / *pelD* double mutants show drastic virulence losses (Rogers et al. 2000).

In the ergot fungus *Claviceps purpurea*, not only polygalacturonase, but also xylanase is a virulence factor. This fungus attacks mainly rye, but several other grasses can also be infected. Infection by this fungus is organ-specific. It exclusively attacks young ovaries and replaces them with its own sclerotia. As only limited host cell death occurs, *C. purpurea* can be regarded as a biotroph (Tudzynski et al. 1995). Two *endo*-polygalacturonase genes, *cppg1* and *cppg2*, of this fungus have been cloned and characterised. They are closely linked in a head-to-tail arrangement and show a high degree of homology, suggesting that a recent gene duplication event has occurred (Tenberge et al. 1996). Mutants deleted for both genes were almost non-pathogenic on rye, and complementation of the mutants with wild-type copies of both genes fully restored virulence, proving that *endo*-polygalacturonase activity is essential to pathogenic development of the ergot fungus (Oeser et al. 2002). In addition to *endo*-polygalacturonase, xylanase has been shown to be essential for virulence of *C. purpurea*. In general, this can be anticipated for grass pathogens, as cell walls of the Poales contain high amounts of glucurono-arabino-xylans (Carpita and Gibeaut 1993). Two xylanase genes, *cpxyl1* and *cpxyl2*, were cloned from *C. purpurea*. While mutants deficient in *cpxyl1* showed wild-type virulence, *cpxyl2* mutants and *cpxyl1* / *cpxyl1* double mutants were drastically reduced in virulence (Giesbert et al. 1998; ten Have et al. 2002).

The maize pathogen *Cochliobolus carbonum* is currently the most intensively studied system with respect to CWDEs. This fungus does not differentiate distinct elaborate appressoria, but penetrates the host from hyphal swellings (Walton et al. 1995). Walton and coworkers have cloned and characterised several genes encoding CWDEs, including xylanase, cellulase, protease, polygalacturonase, β-1,3-glucanase and mixed-linked glucanases. Transformants with single to quadruple mutations do not show reduced virulence. However, as plant cell walls represent complex polymer networks (Carpita and Gibeaut 1993), simultaneous reduction in several different CWDE activities might be required to decrease the efficiency of invasion and spread within the plant tissue. In an elegant approach to overcome the problem of gene redundance, Tonukari et al. (2000) cloned and inactivated the ortholog of yeast *SNF1* (sucrose non-fermenting) gene. The SNF1 protein kinase phosphorylates the repressor of catabolite-repressed genes, MIG1p (the ortholog of which is CreAp in filamentous fungi), leading to de-repression of such genes. It is thought that many genes encoding CWDEs are subject to catabolite repression in filamentous fungi, and, indeed, growth of *C. carbonum ccsnf1* mutants was reduced by 50 to 95% on complex carbon sources such as xylan, pectin, or purified maize cell walls, correlating well with significantly reduced transcript levels and activities of CWDEs. In the *ccsnf1* deletion mutant, production of HC toxin (see paragraph 3.2) conidiation, conidial morphology and germination, in vitro appressorium formation and growth on glucose, fructose, or sucrose was normal. The mutant formed fewer spreading lesions and thus was less virulent on susceptible maize. The results suggest that ccSNF1 – and CWDEs – are required in pathogenesis by *C. carbonum*. However, as growth on galactose, galacturonic acid, maltose, xylose or arabinose was reduced, one may question the specificity of the approach, as basic metabolism is also affected by the mutation (Tonukari et al. 2000; ten Have et al. 2002). Several genes encoding CWDEs are not catabolite-controlled, so that this approach may not be useful in all plant pathogenic fungi.

Taken together, recent research has presented evidence that CWDEs contribute to the infection process in several plant pathogenic fungi. It can be anticipated that CWDEs are needed for initial penetration – particularly in those fungi that do not form elaborate appressoria – and for spreading within the host tissue.

3 Colonisational and Nutritional Strategies

Some taxa of plant-associated fungi (genus or higher levels) show the same type of colonisation and nutrition in all their species. Whereas the rusts and powdery mildews only include obligate biotrophs, genera like *Botrytis*, *Cochliobolus*, *Alternaria* and many others comprise necrotrophic species. Other fungal taxa exhibit rather a range of nutritional styles when comparing different species. Such groups offer the opportunity to identify the genes defining nutritional strategies and to study their evolution within a restricted phylogenetic context. An example of contrasting life styles within a single genus is *Cladosporium*, where saprotrophic, epiphytic, endophytic, facultative biotrophic or even mycoparasitic species have been described

(David 1997; Moricca et al. 1999; Wirsel et al. 2002). Also, the genus *Colletotrichum* exhibits diverse colonisation strategies ranging from asymptomatic endophytic or epiphytic non-pathogenic fungi to hemibiotrophic or necrotrophic plant pathogens (Perfect et al. 1999; Freeman et al. 2001; Latunde-Dada 2001). Making definitions even more complicated, this can vary for certain *Colletotrichum* species, depending on the colonised host species (Horowitz et al. 2002). Initial genetic evidence indicates that nutritional styles within this genus might be defined by just a few genes as shown for *Colletotrichum magna*, where mutations at single loci could change its mode from a hemibiotrophic pathogen to a non-pathogenic, mutualistic endophyte (Freeman and Rodriguez 1993; Redman et al. 1999). It remains to be shown which environmental factors might in addition influence this apparently finely tuned balance. In the following three paragraphs we will introduce one well-studied example for each of the most important nutritional types to demonstrate that plant pathogenic fungi have evolved quite different strategies to retrieve nutrients from their hosts after they have penetrated the plant surface (as outlined in Sect. 2).

3.1 Obligate Biotrophy (The Rusts)

This nutritional mode is typified by several remarkable features. First, in extant organisms it is only found in three phylogenetically very distant taxa, namely the rusts and powdery mildews, which are higher fungi, and in the downy mildews, which belong to the stramenopila that also include diatoms and some other groups of algae.

Second, one common specialised cell type, the haustorium, exists that evolved most likely independently in the ancestors of these organisms. After the early differentiation processes on the host surface culminating in appressorial development and the entering of the interior of the leave (Fig. 2B), dikaryotic hyphae of rusts grow in the apoplastic space. Unknown signals lead to the differentiation of haustorial mother cells that closely adhere to mesophyll cells and that breach the host wall with a penetration peg. Inside the mesophyll cell a terminal, globose haustorium is formed without disrupting the host plasma membrane (Fig. 2A). The extrahaustorial matrix separates fungal and host plasma membranes, and this interface allows exchange of metabolites and putatively also of molecular signals. Apparently, obligate biotrophs have recruited and modified existing morphogenic programs since haustorial mother cell differentiation shows some similarity to that of appressoria. In both cases terminal swellings of hyphae are used to initiate penetration of the rigid host cell wall. It is currently uncertain if haustorial mother cells are able to generate great osmotic pressure as demonstrated for appressoria of several fungi (see Sect. 2.1). In contrast, it has been indicated for *Uromyces fabae* that secreted hydrolases seem to locally weaken or even dissolve that area of the wall where the penetration peg grows through (Deising et al. 1995a,b).

Although it has been assumed for quite a while that haustoria serve as organs for absorption of nutrients from the host, it was only recently that molecular data have become available which indicated how this process operates. Several cDNAs encoding plasma membrane-localised transporters were recovered from a library enriched for haustorially expressed mRNAs (Hahn and Mendgen 1997).

Two amino acid permeases were analyzed in greater detail. AAT1p seems to have a rather broad substrate spectrum with highest activities for histidine and lysine as shown by heterologous expression in yeast and *Xenopus* oocytes (Struck et al. 2002). Its transcription does not appear to be under strict developmental control whereas that of the second putative amino acid permease, AAT2p, is (Hahn et al. 1997b). AAT2p was localized at haustorial plasma membranes by immunofluorescence microscopy, but its substrates are not yet known. The expression of a gene encoding a hexose transporter, HXT1p, is also developmentally regulated (Voegele et al. 2001). Heterologous expression experiments revealed that HXT1p specifically transports glucose and fructose. Immunofluorescence and electron microscopy have indicated that the localisation of HXT1p is restricted to the haustorial plasma membrane. A model for proton symport across the haustorial plasma membrane has been proposed for these transporters (Hahn and Mendgen 2001). A gene, *PMA1*, encoding a plasma membrane H^+ATPase that could energise this process has been cloned (Struck et al. 1998b).

Molecular evidence revealed that the regulation of several host genes for amino acid and carbon metabolism was also affected during biotrophic growth of rust (Wirsel et al. 2001). Alteration of host gene expression was not only observed in infected leaves, but also detectable in all vegetative organs, which might reflect changes in source-sink relationships in the whole plant. Apparently, through the differentiation of haustoria, obligate biotrophs have evolved an efficient way of retrieving nutrients from the host by preferentially localising at least some of their amino acid and carbohydrate transporters at the haustorial membrane.

A third remarkable feature of obligate biotrophy is the fact that this plant-fungal association is active for up to several months without visible host defense reactions occurring. Obviously, since obligate biotrophs depend on living cells, they must prevent the elicitation of the HR when the haustorial mother cell contacts a mesophyll cell and establishes the haustorium. Theoretically, obligate biotrophs can evade host defense reactions by at least two strategies for both of which only limited experimental evidence is available yet. First, they could prevent the expression of or modify those molecules that could be recognised by the surveillance systems of innate plant immunity. Work that had been carried out with dikaryotic *Uromyces fabae* shows that those cell types differentiating on the host leaf contain chitin as a prominent wall polymer (Deising and Siegrist 1995). Chitin, in contrast to citosan, is a target for plant-derived chitinases that generate highly elicitor-active chitin-oligomers (Ride and Barber 1990). Those infection structures that differentiate within the host leaf seem to have at least partially converted the surface-associated chitin into chitosan by means of chitin deacetylase. Chitosan is assumed to have reduced or no elicitor-activity (e.g. Vander et al. 1998). Labelling patterns of infection structures of several rust species suggest that infection-related modification of chitin is critical to the success of the pathogen (Deising et al. 1996; El Gueddari et al. 2002). A second strategy probably used by obligate biotrophs to prevent host defense reactions could involve active suppression, e.g. via affecting host signalling pathways somewhere between initial recognition of the pathogen and the final defense response. Genetic evidence comes from analysis of the maize lethal leaf spot 1 mutant (*lls1*) that forms spreading necrotic lesions, conferring enhanced resistance to fungal pathogens.

It was suggested that the wild-type gene *Lls1* suppresses cell death by preventing reactive oxidative species formation or removing a cell death mediator (Gray et al. 1997). Two fungal pathogens that exhibit compatible interactions with maize, one necrotroph, *Cochliobolus heterostrophus* race O (see below), and one obligate biotroph, *Puccinia sorghi*, were compared on *lls1* plants (Simmons et al. 1998). Both fungi were affected at developmental stages on the leaf surface, which was reflected by reduced rates of germination and appressorium formation. Whereas *C. heterostrophus* was severely impeded to spread through host tissues, *P. sorghi* still produced pustules at somewhat reduced frequency. Apparently, in cases where *P. sorghi* successfully penetrated, it spread without eliciting *lls1* lesions.

The ability of rust fungi to intersect plant defense seems to be host specific, since they elicit various defense responses in non-host plants (Heath 1997). Recent data suggest, that plasma membrane–cell wall adhesion is essential for the expression of cell wall–associated responses (Mellersh and Heath 2001).

In its monokaryotic form *Uromyces vignae* infects susceptible cultivars of its host cowpea by direct penetration of epidermal cells through appressoria. It has been shown for this interaction that disruption of the membrane-wall adhesion occurred exactly where the fungus penetrated the host cell and that this event suppressed further defense reactions (Mellersh and Heath 2001).

3.2 Necrotrophy (*Cochliobolus* spp.)

As the name implies, fungi exhibiting necrotrophy kill host tissues and use the contents of dead or dying cells along with the digestion products of polymeric cell wall materials for nutritional purposes. Two main features are known to characterise this life style, first, the secretion of large amounts of lytic enzymes degrading the constituents of plant cell walls, i.e. cellulases, pectinases, xylanases, proteases and in some cases lignin-degrading enzymes as well. Most of these enzymes are encoded by multigene families that confer functional redundancy to the system and make the analysis of their individual importance for pathogenicity by gene deletion approaches quite cumbersome as described in Section 2.

A second well-investigated feature of necrotrophic fungi is the production and secretion of phytotoxins that kill host cells in advance of the growing hyphae, thereby reducing the impact of host defense on the pathogen. Since most molecular investigations on fungal toxins have been carried out with *Cochliobolus* species, we focus the following discussion on this genus.

Over the years, *C. heterostrophus* (anamorph names: *Helminthosporium maydis*, *Drechslera maydis* and *Bipolaris maydis*) has become one of the best-studied systems for the evolution of fungal virulence factors. This fungus causes a disease called southern corn leaf blight and two races, race 0 and race T, were isolated from infected corn fields. Race T had been detected in the USA first in 1969 on corn cultivars with the Texas male sterile cytoplasm, which had been widely planted by then (Hooker 1974). It caused severe epidemics until it was recognized that the specific genetic constitution of the plant was apparently responsible for the spreading of this new race of the pathogen. By changing to host cultivars that did not rely on the Texas cytoplasm for producing hybrid male sterile seeds, *C. heterostrophus* was not a threat to corn crops any longer. Only race T produces T-toxin, which is a blend of linear polyketoles with a length of C35 to C41 (Kono and Daly 1979). All of them produced – when applied in pure form – large chlorotic lesions on Texas cytoplasm corn as did the fungus. T-toxin is a virulence factor, since race 0 which does not produce T-toxin is a milder pathogen causing only small necrotic lesions.

The molecular target identified to bind T-Toxin is a 13-kDa protein (urf13p) that is only found in the inner mitochondrial membrane of Texas cytoplasm corn but not in cultivars with the N-cytoplasm (normal cytoplasm) (Levings 1990). A mitochondrial locus, *T-urf13*, that had been selected for during the breeding for male sterility, encodes urf13p (Dewey et al. 1986; Wise et al. 1987). Therefore, plant breeders inadvertently introduced a strong selective advantage to a fungal race that was less competitive on the N-cytoplasm cultivars planted before. urf13p forms oligomers, probably tetramers, within the inner mitochondrial membrane (Rhoads et al. 1995). After binding of T-toxin, the complex changes its conformation, resulting in the creation of pores through which small molecules can diffuse thereby leading to the collapse of respiratory functions and cell death.

Classical genetic analyses have demonstrated a 1:1 segregation of parental phenotypes in a cross between race 0 and race T, suggesting that a single locus (*Tox1*) is responsible for producing T-toxin (Tegtmeier et al. 1982). It was also shown that *Tox1* is closely linked to a reciprocal translocation breakpoint (Bronson 1988; Chang and Bronson 1996). Sophisticated genetic analysis and electrophoretic karyotype comparisons of race 0 versus race T revealed that in the latter *Tox1* is actually spread over both of those breakpoints that led to their designation as *Tox1A* and *Tox1B* (Kodama et al. 1999). Cloning of the first of the genes involved in T-toxin production took advantage of the REMI (restriction enzyme-mediated integration of DNA) procedure (Lu et al. 1994). *ChPKS1* encodes a multifunctional polyketide synthase that is composed of six domains theoretically permitting to carry out most of the synthetic steps necessary for T-toxin production. While *ChPKS1* is located at *Tox1A*, two linked genes for T-toxin production, *ChDEC1* (acetoacetate decarboxylase) and *ChRED1* (medium-chain reductase), could be cloned from *Tox1B* (Rose et al. 2002). At both, *Tox1A* and *Tox1B*, probably as yet undiscovered genes are located that are involved in T-toxin production. Several unusual features of the three cloned genes suggested a horizontal gene transfer event in a race 0 progenitor to explain the appearance of race T. These genes exhibit a lower GC content than the average *Cochliobolus* gene, a codon usage deviating from that typically found in filamentous fungi, and they are lacking in race 0 and other species of *Cochliobolus* or related genera analyzed. However, the origin and the timing of the horizontal DNA transfer remain unknown.

A second *Cochliobolus* species, *C. carbonum* (anamorph names: *Helminthosporium carbonum* and *Bipolaris zeicola*) has been particularly well analysed with respect to toxin production. *C. carbonum* comes also in two races; race 1 is highly virulent on certain susceptible corn cultivars causing a disease called northern corn leaf spot, whereas race 2 has only limited potential as a pathogen. Only race 1 produces a phytotoxin, HC-toxin, which genetically cosegregates with a single locus, *Tox2* (Yoder 1980; Bronson 1991). HC-toxin is a cyclic tetrapeptide (cyclo D-Pro, L-Ala, D-Ala, L-Aeo; Aeo abbreviates 2-amino-9,10-epoxy-8-oxodecanoic acid) that is non-ribosomally synthesised (Walton et al. 1982). HC-toxin is a pathogenicity factor because wild-type strains or mutants at *Tox2* are practically non-pathogenic. Susceptibility to race 1 occurs when maize carries homozygous mutations at two unlinked loci, *Hm1* and *Hm2*, conferring the same function. It was suggested that this genotype (*hm1/hm1*; *hm2/hm2*) is a result of inbreeding and arose inadvertently in corn breeding programs in Kansas. *C. carbonum* was supposedly not prevalent or not a competitive pathogen under the dry conditions of that area which prevented disease outbreak (Multani et al. 1998). Transfer of this genotype to cultivars planted in the warm and humid corn belt of the USA, where *C. carbonum* had been present before, resulted in disease in the late 1930s.

Only the extracts from resistant but not from susceptible cultivars contained enzymatic activity, HC-toxin reductase (HCTR) that could eliminate the toxic potential in vitro (Meeley et al. 1992). The gene at the *HM1* locus has been cloned by a transposon tagging procedure. Its sequence showed similarity to dihydroflavonol-4-reductase genes from several plants (Johal and Briggs 1992). Some susceptible cultivars have been shown to carry transposon insertions at *hm1* and deletions at *hm2* (Multani et al. 1998). Interestingly, HCTR activity has been detected in other monocots. Homologues of *HM1* and *HM2* were discovered at syntenic positions in several other grasses but not in dicots (Multani et al. 1998). One HCTR homologue was cloned from barley and found to be highly similar to that of corn (Han et al. 1997). Apparently, monocots use an ancient defense mechanism for detoxifying fungal cyclic peptides and resistance provided by HCTR collapsed in the case of maize when new genotypes of the pathogen and the host interacted.

HC-toxin shows cytostatic activity towards susceptible host cells in contrast to the cytotoxic activity exhibited by T-toxin and other fungal phytotoxins. Pharmacological studies revealed that HC-toxin reversibly inhibits not only corn, but also mammal and oomycete histone deacetylase (Brosch et al. 1995). It was discussed that instead of killing host cells, HC-toxin rather prevents the modification of chromatin, which seems to be necessary to coordinate the transcription of plant defense genes. Again, it should be noted that in maize and probably other monocots the wild-type condition is resistance to HC-toxin, meaning that the plant counteracts a putative suppression of defense reactions by modifying the fungal toxin.

Electrophoretic karyotype analyses and physical mapping demonstrated that the genetic locus in *C. carbonum* race 1 responsible for HC-toxin production, *Tox2*, comprises more than 540 kb and indicated that three genes known to be involved in toxin production were present in multiple copies. The lack of *Tox2* in the genome of race 2 and suppression of recombination were discussed as reasons for the segregation of *Tox2* as a single unit in crosses between race 1 and race 2 (Ahn and Walton 1996). A refined map of *Tox2* showed that all copies of the genes known by then, with the exception of one, are combined in a huge gene cluster at that locus (Ahn et al. 2002). Currently, seven genes at *Tox2* have been cloned and analysed.

Race 1 carries two copies of the first cloned gene at *Tox2*, *HTS1*, and it was shown that loss of toxin production and pathogenicity occurred only after deleting both (Panaccione et al. 1992). Sequence analysis of *HTS1* revealed an intronless 15.7-kb open reading frame encoding a large cyclic peptide synthetase containing four domains. Each domain seems to catalyse aminoacylation and thioesterification steps for one of the four amino acids present in HC-toxin. In addition, an epimerisation activity was also presumed within *HTS1* (Scott-Craig et al. 1992). Interestingly, the *HTS1* regions are flanked by repetitive elements that exhibit similarity to fungal transposons of the Fot1 family (Panaccione et al. 1996). Besides *HTS1* additional genes appeared to be present in multiple functional copies at *Tox2*. *TOXA* is found in two copies, each closely linked to one of the *HTS1* copies. It encodes a putative efflux pump for HC-toxin, which is thought be responsible for secretion of the toxin and self-protection of the fungal cell, since deletions were only obtained when targeting one but not both copies (Pitkin et al. 1996). A mutant with deletions in all copies of *TOXC* was not able to produce HC-Toxin and was non-pathogenic. *TOXC* showed similarity to fatty acid synthases and it was proposed to be necessary for the synthesis of L-Aeo (Ahn and Walton 1997). Currently, the function of *TOXD* remains unknown. *TOXE* encodes a transcription factor whose proposed function is to coordinate the regulation of the *TOX* genes. *TOXE* has a unique modular structure, i.e. a bZIP basic DNA binding domain and four ankyrin repeats but no leucine zipper or helix-loop-helix domains. Deletion of *TOXE* resulted in a mutant lacking the capability to produce HC-toxin and to infect the host, indicating that it is pathway-specific (Ahn and Walton 1998). A ten-base motif (tox-box) had been located in the promoters of all *TOX* genes except *TOXE* itself and it was demonstrated that TOXEp binds in vitro and in vivo to the tox-box (Pedley and Walton 2001). The *TOXF* derived product shares similarity with branched-chain amino-acid transaminases and it was suggested to aminate a precursor of L-Aeo. A mutant with disruptions in all *TOXF* copies did not produce toxin and was non-pathogenic (Cheng et al. 1999). The last of the currently known *TOX* genes, *TOXG*, encodes an alanine racemase that provides the D-Ala found in HC-toxin. Disruptions of both *TOXG* copies resulted in a mutant still making a side product of the pathway that retained some toxic activity. This mutant exhibited only slightly reduced virulence in infection assays (Cheng and Walton 2000).

In conclusion, the two examples from the genus *Cochliobolus* detailed above display some resemblances. Genes responsible for producing host-selective fungal toxins have probably been acquired by horizontal gene transfer from presently unknown sources. These genes are organised into large gene clusters that are probably under a common regulatory control, in analogy to other fungal secondary metabolite pathways. In both cases a relatively benign pathogen turned highly virulent when meeting susceptible host plants. The latter became only available in the field after traditional breeding programs introduced new host cultivars carrying an altered trait. These cases also highlight the intricate interaction between the genomes and the physiology of fungal pathogens and their hosts.

3.3 Hemibiotrophy (*Colletotrichum* spp.)

As mentioned above, *Colletotrichum* harbours species with various nutritional modes. However, the majority of species seem to follow the hemibiotrophic mode. This mode involves two phases, a biotrophic phase which is followed by a necrotrophic phase that finally leads to the destruction of colonised host tissues. Biotrophy lasts from about 1 to 3 days, depending on the species, and is therefore much shorter than that of obligate biotrophs. *Colletotrichum* species provide an easier access than obligate biotrophs to study biotrophic interactions by molecular approaches since they can easily be cultivated in vitro and transformed. Once the total genome sequence of such an organism will be available, it will be a good system to directly compare by microarray technology biotrophic and necrotrophic nutritional modes in the same genetic background. This information might allow the identification of those genes defining these nutritional styles.

As outlined in Section 2, hemibiotrophic *Colletotrichum* spp. directly penetrate into epidermal cells by appressoria and infection hyphae (Fig. 1). The latter differentiate a specialised cell type, the infection vesicle, whose shape varies from spherical to branched, depending on the species

(Fig. 2C). This structure is similar to the haustorium of obligate biotrophs, and again the invaginated host cell stays alive and a special interface is created between the plasma membranes of the host and the pathogen. In some species like *C. truncatum* and *C. graminicola*, the biotrophic phase appears to be restricted to the first infected epidermal cell and is terminated when thin secondary hyphae emerge from it to invade surrounding cells that are killed upon (localised biotrophy) (O'Connell et al. 2000; Latunde-Dada 2001). If biotrophic primary hyphae arise from infection vesicles in these species they do not proliferate into neighbouring cells but rather remain in the infected epidermal cell. In other species, like *C. lindemuthianum* and *C. sublineolum*, the biotrophic phase expands with thick primary hyphae invading neighbouring cells. The biotrophic phase is therefore slowly spreading through host tissue and since biotrophy is also here transient in nature, cells that were infected earlier successively become necrotic (sequential biotrophy) (O'Connell et al. 2000; Latunde-Dada 2001).

Hemibiotrophic *Colletotrichum* spp. in their biotrophic phase do not appear to be recognised and rejected by their hosts defense systems, similar to the situation with powdery mildew or rust fungi. Also here, evidence is emerging that modification of fungal surfaces and active suppression of host pathways might be involved to escape defense. As is the case for *Uromyces fabae* and *Puccinia graminis* f. sp. *tritici*, *C. graminicola* seems to modify its chitin layer to chitosan as soon as it penetrates the epidermal cell as shown by fluorescently labelled wheat germ agglutinin detecting chitin and antibodies recognising chitosan (El Gueddari et al. 2002). A cDNA encoding a protein with similarity to cell-wall-associated receptor kinases was cloned from *C. gloeosporioides* cultures starved for nitrogen (Stephenson et al. 2000). Expression of CgDN3p *in planta* was highest in biotrophic infection vesicles as demonstrated by fusions to GFP. Interestingly, deletion mutants in that gene elicited an HR-like phenotype on a susceptible host where this is normally not occurring. In other words, wild-type *C. gloeosporioides* seems to suppress HR in a compatible interaction.

In another investigation, expression of PR-10 and chalcone synthase of sorghum was compared by Northern analyses in two incompatible interactions. The particular cultivar used displayed host-specific resistance to the tested strain of hemibiotrophic *C. sublineolum*, a species that is pathogenic to sorghum, and it displayed non-host resistance towards necrotrophic *Cochliobolus heterostrophus* (Lo et al. 1999). When inoculated with *C. heterostrophus*, transcripts of both genes became detectable after 4 to 6 h, whereas this took 36 to 48 h after inoculation with *C. sublineolum*. Maximum transcript levels were similar in both interactions. It was speculated that the delayed induction of these defense related genes in the interaction with a hemibiotrophic fungus might be based on the (partial) suppression of host defense mechanisms. Unfortunately, there was no host cultivar included that was fully susceptible to the *C. sublineolum* strain tested and therefore it is uncertain if in this case the induction of defense genes would be weaker and/or more delayed. Cytological comparison of the two interactions indicated that development of *C. heterostrophus* was blocked at the appressorial stage whereas *C. sublineolum* penetrated normally with appressoria and also developed infection vesicles. This fungus was arrested when it started to proliferate into neighbouring cells (Lo et al. 1999).

The switch from biotrophic to necrotrophic growth is under genetic regulation as was demonstrated for *C. lindemuthianum*. A tagged mutant in the gene *CLTA1* was recovered in a screen for reduced pathogenicity. The mutant elicited an HR-like phenotype on a susceptible cultivar of common bean (Dufresne et al. 2000). This phenotype also occurred when inoculating the wild type onto a non-susceptible host cultivar. Cytology showed that the mutant was arrested in the biotrophic stage since it appeared to differentiate normally from conidial germination to infection vesicles and to biotrophic primary hyphae. However, secondary hyphae never developed. CLTA1p has sequence similarity to transcription factors of the zinc cluster (Zn[II](2)Cys(6)) family, which typically regulate distinct metabolic pathways in fungi. It was speculated that CLTA1p might switch on genes necessary for necrotrophic nutrition (Dufresne et al. 2000). In any case, this gene appears to be important for regulating the transition in life style and it will be interesting to identify those genes, which are under its control. Recently, another mutant, which remained arrested in the biotrophic phase had been identified in *C. graminicola* by REMI mutagenesis (Thon et al. 2002). The tagged gene, *CPR1*, putatively encodes a subunit of the eukaryotic signal peptidase complex. A deletion of the gene was not achieved, indicating that it might be essential for growth and it appeared that the original mutant had only been recovered since the insertion occurred in the 3'-untranslated region close to the stop codon. The mutant exhibited reduced transcript levels, which was taken as evidence for an inability to secrete sufficient amounts of hydrolytic enzymes during necrotrophic growth, which would be not essential for biotrophic growth.

Thirteen infection specifically expressed sequence tags (ESTs) corresponding to genes activated during early stages of the interaction between *C. graminicola*, and its host, *Zea mays*, have been isolated by sequential subtractive hybridisation. Identification of the deduced proteins and RT-PCR studies suggest that considerable reprograming of the protein expression patterns occurs during the shift from biotrophic to necrotrophic infection phases. In addition, ESTs that share no similarity with known sequences in the database, thus representing hitherto unknown genes, have also been isolated (Sugui and Deising 2002). Functional analysis of these will show whether or not they play a role in the *C. graminicola* – maize interaction.

Necrotic lesions are typically characterised by anthracnose and blight symptoms. Cell wall-degrading enzymes – as discussed in Section 2 – and in at least some species also phytotoxins are produced, which leads to the killing of host cells in front of the spreading hyphae. Toxin production has been reported for *C. falcatum* (Naik and Vedamurthy 1997), *C. linde-*

muthianum (Fernandez et al. 2000), *C. dematium* (Yoshida et al. 2000), *C. gloeosporioides* (Barjau et al. 1995; Jayasankar et al. 1999), *C. fuscum* (Goodman 1960) and *C. acutatum* (Jayasinghe and Fernando 2001). To our knowledge, the molecular basis of toxin production by *Colletotrichum* is not known in any of these cases.

4 Infection Structures as Fungicide Targets

Depending on the crop, fungal plant pathogens cause significant pre- and post-harvest losses worldwide, which may amount to 20% or more. Accordingly, fungicides contribute approximately 20% to the world market of agrochemicals, corresponding to 6 billion $US spent on plant protection each year (Hewitt 1998). Although several fungicide classes differing in their mode of action have been established, increasingly stringent biosafety and specificity requirements, together with occurrence of fungicide resistance in field populations of fungal plant pathogens, make the development of new fungicides with novel modes of action necessary (FRAC, Fungicide Resistance Action Committee 2002, http://www.frac.info/frac.html; Deising et al. 2002, and references therein). Technical progress and increased knowledge of fungal biology allow one to search for specific targets in fungal cells, thus broadening classical fungicide development consisting of random synthesis of chemicals in combination with screening for antifungal activity (Knight et al. 1997). A promising alternative that may circumvent time consumption and reduce financial efforts is to identify fungicides produced by microbes. A well-known example is strobilurin A of *Strobilurus tenacellus* (Anke et al. 1977; Kraiczy et al. 1996). Although the microbial fungicide has been shown to be unstable in the light, chemical modification has led to several very successful fungicides that are well established on the market. Not only in fungal mycelia, but also in fruiting bodies substances with antifungal properties have been identified (Lorenzen and Anke 1998; Stadler and Sterner 1998), and it will be challenging to develop screening methods that allow one to identify not only activities directed against basic metabolism occurring in both saprophytic and pathogenic hyphae, but also against targets specifically needed for infection structure differentiation. Thus, infection structures (see above) may be regarded as excellent fungicide targets (Struck et al. 1998a), and some examples for this disease control strategy already exist.

Following landing on the plant surface, fungal spores adhere (Nicholson 1996) and germinate under permissive conditions. Germ tube growth and – in many cases – appressorium formation occurs, preceding invasion of the plant tissue. These developmental stages are targets of protective (or pre-invasion) fungicides. Fungicides that interfere with germ tube growth are multi-site inhibitors structurally classified as dithiocarbamates (e.g. mancozeb, maneb or thiram) or phthalimides (e.g. captan and folpet). The dialkyldithiocarbamate anion either interferes with enzymes containing copper ions such as laccases or ascorbate oxidases, or binds to SH-groups and thus inhibits protein function. Binding of SH-groups has also been reported for phthalimide derivatives.

Germination and germ tube growth are also inhibited by strobilurins, which are present in several modern fungicide mixtures. These inhibitors block mitochondrial electron transport via binding to ubiquinol oxidase of the cytochrome bc_1 complex. Like the strobilurins, the antifungal compound cyazofamide inhibits the mitochondrial electron transport. Cyazofamide is a specific inhibitor blocking ubiquinol reductase. Both fungicide groups thus efficiently interfere with mitochondrial ATP synthesis, which is essential for invasion of the plant and pathogenic development. An interesting group of fungicides are the anilino pyrimidins, as they not only inhibit methionine biosynthesis. Some members of this group, e.g. andoprim, mepanipyrim and pyrimethanil reduce secretion of depolymerising enzymes such as pectinases, cutinases, lipases or proteases, some of which have been shown to be virulence factors (see above). This mode of action is particularly valuable to control pathogens that cause extended tissue maceration, e.g. *B. cinerea* infecting strawberries or *C. gloeosporioides*, *Monilia fructigena*, *Penicillium expansum* and *P. italicum* causing apple rot.

Two groups of fungicides specifically interfere with two different targets of the melanin biosynthesis pathway and thus specifically interfere with appressorium function (Fig. 1). The rice blast fungus *M. grisea* and different *Colletotrichum* species synthesise melanin via 1,8-dihydroxynaphthalene (1,8-DHN) (Fig. 1; Kubo and Furusawa 1991), and reductase and dehydratase inhibitors are used as fungicides. The NADPH-dependent reductase that converts 1,3,6,8-tetrahydroxynaphthalene (1,3,6,8-THN) to scytalone and 1,3,8-trihydroxynaphthalene (1,3,8-THN) to vermelone is inhibited by isobenzofuranone, pyrrolquinoline and triazolbenzothiazole compounds, with tricyclazole being the most prominent fungicidal substance. The dehydratase converting scytalone to 1,3,8-THN and vermelone to 1,8-DNH, respectively, are inhibited by the cyclopropanecarbroxamide fungicide carpropamid (Kurahashi and Pontzen 1998; Thieron et al. 1998). Inhibition of melanin synthesis results in reduced turgor pressure and reduced generation of force not sufficient for mechanical breaching of the host cell wall. Interestingly, as shown with melanin-deficient *M. grisea* isolates the dehydratase inhibitor and rice blast fungicide carpropamid not only blocks melanin biosynthesis, but also stimulates resistance responses of rice plants

(Thieron et al. 1998). The dual mode of action may help to avoid the development of fungicide resistance (Deising et al. 2002).

Infection structure differentiation is initiated, as soon as the germ tube growing on the plant surface perceive signals typical of its host (Hoch et al. 1987; Jelitto et al. 1994; Kolattukudy et al. 1995). Compounds involved in fungal signal transduction (Dean 1997; Deising et al. 2000) would be excellent fungicide targets, as inhibition of signal cascades would specifically inhibit pathogenic fungi that rely on infection structure differentiation. Interestingly, phenyl pyrrol fungicides such as fenpiclonil and fludioxonil inhibit MAP kinases, and the quinoline quinoxyfen interferes with the G-protein-mediated signal transfer (FRAC 2002, http://www.frac.info/frac.html; Wheeler et al. 2000). As a consequence of application of the mildew-specific fungicide quinoxyfen, the fungus does not recognise its host plant surface, germ tube growth continues and appressorium differentiation and plant infection do not occur (Wheeler et al. 2000; Deising et al. 2002).

Another inhibitor, called glisoprenin A, interfering with signal transduction and appressorium formation has been isolated from submerged cultures of *Gliocladium roseum* (Thines et al. 1997). This compound inhibited appressorium formation of *M. grisea* on hydrophobic surfaces, resembling natural conditions, but infection structure induction by chemicals such as cAMP, 8,4-chlorophenylthio-adenosine-3',5'monophosphate, 3-isobutyl-1-methylxanthine or the plant wax compound 1,16-hexadecanediol, was not affected. These data showed that two different signal transduction pathways for appressorium formation exist in *M. grisea*. Again, inhibitors isolated from micro-organisms offer a rich source of lead compounds for fungicides.

The most prominent structural compounds of fungal hyphae are chitin and β-1,3-glucans, which interconnect to form a rigid network determining hyphal shape (Wessels 1993). As most fungi contain significant amounts of these polymers, and as even cellulosic oomycetes synthesise chitin (Werner et al. 2002) and β-1,3-glucans, chemicals that inhibit the synthesising enzymes may be excellent fungicides (Debono and Gordee 1994; Current et al. 1995; Knight et al. 1997). These fungicides inhibit growth of vegetative and pathogenic hyphae. The polyoxin chitin synthase inhibitors have been used since 1967 to control *Rhizoctonia solani*, *Cochliobolus miyabeanus* and *Alternaria alternata*, and glucan synthase inhibitors are important tools to control human invasive mycoses caused by *Cryptococcus neoformans*, *Aspergillus fumigatus* or *Pneumocystis carinii* (Debono and Gordee 1994; Urbina et al. 2000).

Like cell wall biosynthesis inhibitors, sterol biosynthesis inhibitors (SBIs) are general fungicides in the sense that they do not discriminate between pathogenic and saprophytic hyphae (Deising et al. 2002). Both groups of SBIs, C14-demethylase inhibitors and morpholines which block $\Delta^7 \rightarrow \Delta^8$-isomerase, are easily taken up by plants and are thus systemic fungicides. SBI fungicides interfere with plasma membrane function. Importantly, these compounds are ideally suited to access *in planta* growing pathogenic hyphae as well as infection structures growing on the plant surface (Deising et al. 2002, and literature therein).

5 Concluding Remarks

In this chapter, we have discussed several aspects of fungal morphogenesis, function of infection structures and different life styles of diverse plant pathogens. During the last years, several genes essential for efficient host invasion and colonisation have been identified. However, although many details are known, there still is a considerable lack of information, e.g. in regulatory control of morphogenesis, cell wall biogenesis and modification, and different metabolic pathways involved in synthesis of virulence or pathogenicity factors. Knowledge of factors required to induce infection structure differentiation may help to find highly specific antifungal agents that could be used as fungicides. Such compounds would be particularly interesting, as they hinder infection at an early stage and disease symptoms would not occur. As resistance against several of the most frequently used fungicides (strobilurins and SBIs) has been observed in different pathogens, discovery of new compounds is important not only in agriculture, but also in medicine.

References

Ahn JH, Walton JD (1996) Chromosomal organization of *TOX2*, a complex locus controlling host-selective toxin biosynthesis in *Cochliobolus carbonum*. Plant Cell 8:887–897

Ahn JH, Walton JD (1997) A fatty acid synthase gene in *Cochliobolus carbonum* required for production of HC-toxin, cyclo(D-prolyl-L-alanyl-D-alanyl-L-2-amino-9, 10-epoxi-8-oxodecanoyl). Mol Plant Microbe Interact 10:207–214

Ahn JH, Walton JD (1998) Regulation of cyclic peptide biosynthesis and pathogenicity in *Cochliobolus carbonum* by TOXEp, a novel protein with a bZIP basic DNA-binding motif and four ankyrin repeats. Mol Gen Genet 260:462–469

Ahn JH, Cheng YQ, Walton JD (2002) An extended physical map of the *TOX2* locus of *Cochliobolus carbonum* required for biosynthesis of HC-toxin. Fungal Genet Biol 35:31–38

Anke T, Oberwinkler F, Steglich W, Schramm G (1977) The strobilurins – new antifungal antibiotics from the basidiomycete *Strobilurus tenacellus*. J Antibiotics 30:806–810

Bakkeren G, Kronstad JW, Lévesque CA (2000) Comparison of AFLP fingerprints and ITS sequences as phylogenetic markers in Ustilaginomycetes. Mycologia 92:510–521

Barjau M, Druet D, Comeau LC (1995) Purification and partial characterization of phytotoxic glycopeptides secreted by *Colletotrichum gloeosporioides*. Appl Environ Microbiol 43:217–221

Bastmeyer M, Deising HB, Bechinger C (2002) Force exertion in fungal infection. Annu Rev Biophys Biomol Struct 31:321–341

Bechinger C, Giebel K-F, Schnell M, Leiderer P, Deising HB, Bastmeyer M (1999) Optical measurements of invasive forces exerted by appressoria of a plant pathogenic fungus. Science 285:1896–1899

Blackwell M (2000) Evolution. Terrestrial life-fungal from the start? Science 289:1884–1885

Bronson CR (1988) Ascospore abortion in crosses of *Cochliobolus heterostrophus* heterozygous for the virulence locus *Toxl*. Genome 30:12–18

Bronson CR (1991) The genetics of phytotoxin production by plant pathogenic fungi. Experientia 47:771–776

Brosch G, Ransom R, Lechner T, Walton JD, Loidl P (1995) Inhibition of maize histone deacetylases by HC toxin, the host-selective toxin of *Cochliobolus carbonum*. Plant Cell 7:1941–1950

Brush L, Money NP (1999) Invasive hyphal growth in *Wangiella dermatitidis* is induced by stab inoculation and shows dependence upon melanin biosynthesis. Fungal Genet Biol 28:190–200

Carpita NC, Gibeaut DM (1993) Structural models of primary cell walls in flowering plants: consistency of molecular structure with the physical properties of the walls during growth. Plant J 3:1–30

Centis S, Guillas I, Séjalon N, Esquerré-Tugayé M-T (1997) Endopolygalacturonase genes from *Colletotrichum lindemuthianum*: cloning of *CLPG2* and comparison of its expression to that of *CLPG1* during saprophytic and parasitic growth of the fungus. Mol Plant Microbe Interact 10:769–775

Chang HR, Bronson CR (1996) A reciprocal translocation and possible insertion(s) tightly associated with host-specific virulence in *Cochliobolus heterostrophus*. Genome 39:549–557

Cheng YQ, Walton JD (2000) A eukaryotic alanine racemase gene involved in cyclic peptide biosynthesis. J Biol Chem 275:4906–4911

Cheng YQ, Ahn JH, Walton JD (1999) A putative branched-chain-amino-acid transaminase gene required for HC-toxin biosynthesis and pathogenicity in *Cochliobolus carbonum*. Microbiology 145:3539–3546

Current WL, Tang J, Boylan C, Watson P, Zeckner D, Turner W, Rodriguez M, Dixon C, Ma D, Radding JA (1995) Glucan biosynthesis as a target for antifungals: the echinocandin class of antifungal agents. In: Dixon GK, Copping LG, Hollomon DW (eds) Antifungal agents: discovery and mode of action. Bios Scientific Publishers, Oxford, pp 143–160

David JC (1997) A contribution to the systematics of *Cladosporium*: Revision of the fungi previously referred to *Heterosporium*. CAB International, Oxon

Dean RA (1997) Signal pathways and appressorium morphogenesis. Annu Rev Phytopathol 35:211–234

Debono M, Gordee RS (1994) Antibiotics that inhibit fungal cell wall development. Annu Rev Microbiol 48:471–497

Deising H, Siegrist J (1995) Chitin deacetylase activity of the rust *Uromyces viciae-fabae* is controlled by fungal morphogenesis. FEMS Microbiol Lett 127:207–211

Deising H, Frittrang AK, Kunz S, Mendgen K (1995a) Regulation of pectin methylesterase and polygalacturonate lyase activity during differentiation of infection structures in *Uromyces viciae-fabae*. Microbiology 141:561–571

Deising H, Rauscher M, Haug M, Heiler S (1995b) Differentiation and cell-wall degrading enzymes in the obligately biotrophic rust fungus *Uromyces viciae-fabae*. Can J Bot 73:S624-S631

Deising H, Heiler S, Rauscher M, Xu H, Mendgen K (1996) Cellular aspects of rust infection structure differentiation: spore adhesion and fungal morphogenesis. In: Nicole M,

Gianinazzi-Pearson V (eds) Histology, ultrastructure and molecular cytology of plant-microorganism interactions. Kluwer, Dordrecht, pp 135–156

Deising HB, Werner S, Wernitz M (2000) The role of fungal appressoria in plant infection. Microb Infect 2:1631–1641

Deising HB, Reimann S, Peil A, Weber WE (2002) Disease management of rusts and powdery mildews. In: Kempken F (ed) The Mycota XI. Application in agriculture. Springer, Berlin Heidelberg New York, pp 243–269

De Jong JC, McCormack BJ, Smirnoff N, Talbot NJ (1997) Glycerol generates turgor in rice blast. Nature 389:244–245

De Lorenzo G, Ferrari S (2002) Polygalacturonase-inhibiting proteins in defense against phytopathogenic fungi. Curr Opin Plant Biol 5:295–299

De Lorenzo G, D'Ovidio R, Cervone F (2001) The role of polygalacturonase-inhibiting proteins (PGIPs) in defense against pathogenic fungi. Annu Rev Phytopathol 39:313–335

Dewey RE, Levings CS 3rd, Timothy DH (1986) Novel recombinations in the maize mitochondrial genome produce a unique transcriptional unit in the Texas male-sterile cytoplasm. Cell 44:439–449

Dixon KP, Xu J-R, Smirnoff N, Talbot NJ (1999) Independent signaling pathways regulate cellular turgor during hyperosmotic stress in appressorium-mediated plant infection by *Magnaporthe grisea*. Plant Cell 11:2045–2058

Dufresne M, Perfect S, Pellier AL, Bailey JA, Langin T (2000) A GAL4-like protein is involved in the switch between biotrophic and necrotrophic phases of the infection process of *Colletotrichum lindemuthianum* on common bean. Plant Cell 12:1579–1590

Dumas B, Centis S, Sarrazin N, Esquerré-Tugayé M-T (1999) Use of green fluorescent protein to detect expression of an endopolygalacturonase gene of *Colletotrichum lindemuthianum* during bean infection. Appl Environ Microbiol 65:1769–1771

El Gueddari NE, Rauchhaus U, Moerschbacher BM, Deising HB (2002) Developmentally regulated conversion of surface-exposed chitin to chitosan in cell walls of plant pathogenic fungi. New Phytol 156:103–112

Emmett RW, Parbery DG (1975) Appressoria. Annu Rev Phytopathol 13:147–167

Fernandez MT, Fernandez M, Centeno ML, Canal MJ, Rodriguez R (2000) Reaction of common bean callus to culture filtrate of *Colletotrichum lindemuthianum*: differences in the composition and toxic activity of fungal culture filtrates. Plant Cell Tissue Organ Cult 61:41–49

Frank B (1883) Ueber einige neue und weniger bekannte Pflanzenkrankheiten. Ber Dtsch Bot Ges 1:29–34

Freeman S, Rodriguez RJ (1993) Genetic conversion of a fungal plant pathogen to a nonpathogenic, endophytic mutualist. Science 260:75–78

Freeman S, Horowitz S, Sharon A (2001) Pathogenic and nonpathogenic lifestyles in *Colletotrichum acutatum* from strawberry and other plants. Phytopathology 91:986–992

Giesbert S, Lepping H-B, Tenberge KB, Tudzynski P (1998) The xylanolytic system of *Claviceps purpurea*: Cytological evidence for secretion of xylanases in infected rye tissue and molecular characterization of two xylanase genes. Phytopathology 88:1020–1030

Gilbert RD, Johnson AM, Dean RA (1996) Chemical signals responsible for appressorium formation in the rice blast fungus *Magnaporthe grisea*. Physiol Mol Plant Pathol 48:335–346

Gold RE, Mendgen K (1991) Rust basidiospore germlings and disease initiation. In: Cole GT, Hoch HC (eds) The fungal spore and disease initiation in plants and animals. Plenum Press, New York, pp 67–99

Goodman RN (1960) Colletotin, a toxin produced by *Colletotrichum fuscum*. Phytopathology 50:325–327

Gray J, Close PS, Briggs SP, Johal GS (1997) A novel suppressor of cell death in plants encoded by the *Lls1* gene of maize. Cell 89:25–31

Hahn M, Mendgen K (1997) Characterization of *in planta*-induced rust genes isolated from a haustorium-specific cDNA library. Mol Plant Microbe Interact 10:427–437

Hahn M, Mendgen K (2001) Signal and nutrient exchange at biotrophic plant-fungus interfaces. Curr Opin Plant Biol 4:322–327

Hahn M, Deising H, Struck C, Mendgen K (1997a) Specificity and recognition phenomena during infection. In: Hartleb H, Heitefuss R, Hoppe HH (eds) Resistance of crop plants against fungi. Gustav Fischer, Jena, pp 33–57

Hahn M, Neef U, Struck C, Gottfert M, Mendgen K (1997b) A putative amino acid transporter is specifically expressed in haustoria of the rust fungus *Uromyces fabae*. Mol Plant Microbe Interact 10:438–445

Han F, Kleinhofs A, Kilian A, Ullrich SE (1997) Cloning and mapping of a putative barley NADPH-dependent HC-toxin reductase. Mol Plant Microbe Interact 10:234–239

Heath MC (1997) Signalling between pathogenic rust fungi and resistant or susceptible host plants. Ann Bot 80:713–720

Heckman DS, Geiser DM, Eidell BR, Stauffer RL, Kardos NL, Hedges SB (2001) Molecular evidence for the early colonization of land by fungi and plants. Science 293:1129–1133

Henson JM, Butler MJ, Day AW (1999) The dark side of the mycelium: melanins of phytopathogenic fungi. Annu Rev Phytopathol 37:447–471

Herbert C, Jacquet C, Borel C, Esquerré-Tugayé M-T, Dumas B (2002) A cis-acting sequence homologous to the yeast filamentation and invasion response element regulates expression of a pectinase gene from the bean pathogen *Colletotrichum lindemuthianum*. J Biol Chem 277:29125–29131

Hewitt HG (1998) Fungicides in crop protection. CAB International, Wallingford

Hoch HC, Staples RC, Whitehead B, Comeau J, Wolf ED (1987) Signaling for growth orientation and cell differentiation by surface topography in *Uromyces*. Science 235:1659–1662

Hooker AL (1974) Cytoplasmic susceptibility in plant disease. Annu Rev Phytopathol 12:167–179

Horowitz S, Freeman S, Sharon A (2002) Use of green fluorescent protein-transgenic strains to study pathogenic and nonpathogenic lifestyles in *Colletotrichum acutatum*. Phytopathology 92:743–749

Howard RJ (1997) Breaching the outer barriers – cuticle and cell wall penetration. In: Carrol GC, Tudzynski P (eds) The mycota V. Plant relationships part A. Springer, Berlin Heidelberg New York, pp 43–60

Howard RJ, Bourette TM, Ferrari MA (1991a) Infection by *Magnaporthe*: an in vitro analysis. In: Mendgen K, Lesemann D-E (eds) Electron microscopy of plant pathogens. Springer, Berlin Heidelberg New York, pp 251–264

Howard RJ, Ferrari MA, Roach DH, Money NP (1991b) Penetration of hard substances by a fungus employing enormous turgor pressures. Proc Natl Acad Sci USA 88:11281–11284

Isshiki A, Akimitsu K, Yamamoto M, Yamamoto H (2001) Endopolygalacturonase is essential for citrus black rot caused by *Alternaria citri* but not brown spot caused by *Alternaria alternata*. Mol Plant Microbe Interact 14:749–757

Jayasankar S, Litz RE, Gray DJ, Moon PA (1999) Responses of embryogenic mango cultures and seedling bioassays to a partially purified phytotoxin produced by a mango leaf isolate of *Colletotrichum gloeosporioides* Penz. In Vitro Cell Dev Biol Plant 35:475–479

Jayasinghe CK, Fernando TH (2001) Toxic activity from liquid culture of *Colletotrichum acutatum*. Mycopathology 152:97–101

Jelitto TC, Page HA, Read ND (1994) Role of external signals in regulating the pre-penetration phase of infection by the rice blast fungus, *Magnaporthe grisea*. Planta 194:471–477

Johal GS, Briggs SP (1992) Reductase activity encoded by the HM1 disease resistance gene in maize. Science 258:985–987

Kim YK, Wang Y, Liu ZM, Kolattukudy PE (2002) Identification of a hard surface contact-induced gene in *Colletotrichum gloeosporioides* conidia as a sterol glycosyl transferase, a novel fungal virulence factor. Plant J 30:177–187

Knight SC, Anthony VM, Brady AM, Greenland AJ, Heaney SP, Murray DC, Powell KA, Schulz MA, Spinks CA, Worthington PA, Youle D (1997) Rationale and perspectives on the development of fungicides. Annu Rev Phytopathol 35:349–372

Kodama M, Rose MS, Yang G, Yun SH, Yoder OC, Turgeon BG (1999) The translocation-associated *Tox1* locus of *Cochliobolus heterostrophus* is two genetic elements on two different chromosomes. Genetics 151:585–596

Kolattukudy PE, Crawford MS (1987) The role of polymer degrading enzymes in fungal pathogenesis. In: Nishimura S, Vance VP, Doke N (eds) Molecular determinants of plant diseases. Springer, Berlin, Heidelberg New York, pp 75–95

Kolattukudy PE, Rogers LM, Li D, Hwang C-S, Flaishman MA (1995) Surface signaling in pathogenesis. Proc Natl Acad Sci USA 92:4080–4087

Kono Y, Daly JM (1979) Characterization of the host-specific pathotoxin produced by *Helminthosporium maydis*, race-T, affecting corn with Texas male sterile cytoplasm. Bioorg Chem 8:391–397

Kraiczy P, Haase U, Gencic S, Flindt S, Anke T, Brandt U, von Jagow G (1996) The molecular basis for the natural resistance of the cytochrome bc_1 complex from strobilurin-producing basidiomycetes to center Q_P inhibitors. Eur J Biochem 235:54–63

Kubo Y, Furusawa I (1991) Melanin biosynthesis. Prerequisite for successful invasion of the host by appressoria of *Colletotrichum* and *Pyricularia*. In: Cole GT, Hoch HC (eds) The fungal spore and disease initiation in plants and animals. Plenum, New York, pp 205–218

Kubo Y, Takano Y, Tsuji G, Horino O, Furusawa I (2000) Regulation of melanin biosynthesis genes during appressorium formation by *Colletotrichum lagenarium*. In: Prusky D, Freeman S, Dickman MB (eds) *Colletotrichum*. Host specificity, pathology, and host-pathogen interaction. APS Press, St. Paul, Minnesota, pp 99–113

Kurahashi Y, Pontzen R (1998) Carpropamid: a new melanin biosynthesis inhibitor. Pflanzenschutz Nachr Bayer 51:247–258

Latunde-Dada AO (2001) *Colletotrichum*: tales of forcible entry, stealth, transient confinement and breakout. Mol Plant Pathol 2:187–198

Lee Y-H, Dean RA (1994) Hydrophobicity of contact surface induces appressorium formation of *Magnaporthe grisea*. FEMS Microbiol Lett 115:71–76

Levings CS (1990) The Texas cytoplasm of maize: cytoplasmic male sterility and disease susceptibility. Science 250:942–947

Lo SC, Hipskind JD, Nicholson RL (1999) cDNA cloning of a sorghum pathogenesis-related protein (PR-10) and differential expression of defense-related genes following inoculation with *Cochliobolus heterostrophus* or *Colletotrichum sublineolum*. Mol Plant Microbe Interact 12:479–489

Lorenzen K, Anke T (1998) Basidiomycetes as a source for new bioactive natural products. Curr Org Chem 2:329–364

Lu S, Lyngholm L, Yang G, Bronson C, Yoder OC, Turgeon BG (1994) Tagged mutations at the *Tox1* locus of *Cochliobolus heterostrophus* by restriction enzyme-mediated integration. Proc Natl Acad Sci USA 91:12649–12653

Madhani HD, Galitski T, Lander ES, Fink GR (1999) Effectors of a developmental mitogen-activated protein kinase cascade revealed by expression signatures of signaling mutants. Proc Natl Acad Sci USA 96:12530–12535

Meeley RB, Johal GS, Briggs SP, Walton JD (1992) A biochemical phenotype for a disease resistance gene of maize. Plant Cell 4:71–77

Mellersh DG, Heath MC (2001) Plasma membrane-cell wall adhesion is required for expression of plant defense responses during fungal penetration. Plant Cell 13:413–424

Mendgen K, Deising H (1993) Infection structures of fungal plant pathogens – a cytological and physiological evaluation. New Phytol 124:193–213

Mendgen K, Hahn M, Deising H (1996) Mechanisms and morphogenesis of penetration by plant pathogenic fungi. Annu Rev Phytopathol 34:367–386

Miyoshi M (1895) Die Durchbohrung von Membranen durch Pilzfäden. Jahrb Wiss Bot 28:269–289

Money NP (1999) Fungus punches its way in. Nature 401:332–333
Moricca S, Ragazzi A, Mitchelson KR (1999) Molecular and conventional detection and identification of *Cladosporium tenuissimum* on two-needle pine rust aeciospores. Can J Bot 77:339–347
Mugnier J (2002) Coevolution of pathogenic fungi and grass hosts. In: Kempken F (ed) The mycota XI. Application in agriculture. Springer, Berlin Heidelberg New York, pp 359–373
Multani DS, Meeley RB, Paterson AH, Gray J, Briggs SP, Johal GS (1998) Plant-pathogen microevolution: molecular basis for the origin of a fungal disease in maize. Proc Natl Acad Sci USA 95:1686–1691
Naik GR, Vedamurthy AB (1997) In vitro evaluation of red rot toxin influence on sugarcane (*Saccharum officinarum* L.) var. CoC 671. Curr Sci 73:367–369
Nicholson RL (1996) Adhesion of fungal propagules: significance to the success of the fungal infection process. In: Nicole M, Gianinazzi-Pearson V (eds) Histology, ultrastructure and molecular cytology of plant-microorganism interactions. Kluwer, Dordrecht, The Netherlands, pp 117–134
Nicholson RL, Turpin CA, Warren HL (1976) Role of pectic enzymes in susceptibility of living maize pith to *Colletotrichum graminicola*. Phytopathol Z 87:324–336
Nosanchuk JD, Valadon P, Feldmesser M, Casadevall A (1999) Melanization of *Cryptococcus neoformans* in murine infection. Mol Cell Biol 19:745–750
O'Connell R, Perfect S, Hughes B, Carzaniga R, Bailey J, Green J (2000) Dissecting the cell biology of *Colletotrichum* infection processes. In: Prusky D, Freeman S, Dickamn MB (eds) *Colletotrichum*. Am Phytopathological Soc, St Paul, pp 57–77
Oeser B, Heidrich PM, Müller U, Tudzynski P, Tenberge KB (2002) Polygalacturonase is a pathogenicity factor in the *Claviceps purpurea*/rye interaction. Fungal Genet Biol 36:176–186
Panaccione DG, Scott-Craig JS, Pocard JA, Walton JD (1992) A cyclic peptide synthetase gene required for pathogenicity of the fungus *Cochliobolus carbonum* on maize. Proc Natl Acad Sci USA 89:6590–6594
Panaccione DG, Pitkin JW, Walton JD, Annis SL (1996) Transposon-like sequences at the *TOX2* locus of the plant-pathogenic fungus *Cochliobolus carbonum*. Gene 176:103–109
Pedley KF, Walton JD (2001) Regulation of cyclic peptide biosynthesis in a plant pathogenic fungus by a novel transcription factor. Proc Natl Acad Sci USA 98:14174–14179
Perfect SE, Green JR (2001) Infection structures of biotrophic and hemibiotrophic fungal plant pathogens. Mol Plant Pathol 2:101–108
Perfect SE, Hughes HB, O'Connell RJ, Green JR (1999) *Colletotrichum*: a model genus for studies on pathology and fungal-plant interactions. Fungal Genet Biol 27:186–198
Pirozynski KA (1976) Fossil fungi. Annu Rev Phytopathol 14:237–246
Pitkin JW, Panaccione DG, Walton JD (1996) A putative cyclic peptide efflux pump encoded by the TOXA gene of the plant-pathogenic fungus *Cochliobolus carbonum*. Microbiology 142:1557–1565
Podila GK, Rogers LM, Kolattukudy PE (1993) Chemical signals from avocado surface wax trigger germination and appressorium formation in *C. gloeosporioides*. Plant Physiol 103:267–272
Pryce-Jones E, Carver T, Gurr SJ (1999) The roles of cellulase enzymes and mechanical force in host penetration by *Erysiphe graminis* f. sp. *hordei*. Physiol Mol Plant Pathol 55:175–182
Read ND, Kellock LJ, Collins TJ, Gundlach AM (1997) Role of topography sensing for infection-structure differentiation in cereal rust fungi. Planta 202:163–170
Redman RS, Ranson JC, Rodriguez RJ (1999) Conversion of the pathogenic fungus *Colletotrichum magna* to a nonpathogenic, endophytic mutualist by gene disruption. Mol Plant Microbe Interact 12:969–975
Rhoads DM, Levings CS 3rd, Siedow JN (1995) URF13, a ligand-gated, pore-forming receptor for T-toxin in the inner membrane of cms-T mitochondria. J Bioenerg Biomembr 27:437–445

Ride JP, Barber MS (1990) Purification and characterization of multiple forms of endochitinase from wheat leaves. Plant Sci 71:185–197

Rogers LM, Kim Y-K, Guo W, González-Candelas L, Li D, Kolattukudy P (2000) Requirement for either a host- or a pectin-induced pectate lyase for infection of *Pisum sativum* by *Nectria haematococca*. Proc Natl Acad Sci USA 97:9813–9818

Rose MS, Yun SH, Asvarak T, Lu SW, Yoder OC, Turgeon BG (2002) A decarboxylase encoded at the *Cochliobolus heterostrophus* translocation-associated *Tox1B* locus is required for polyketide (T-toxin) biosynthesis and high virulence on T-cytoplasm maize. Mol Plant Microbe Interact 15:883–893

Schweizer A, Rupp S, Taylor BN, Rollinghoff M, Schroppel K (2000) The TEA/ATTS transcription factor CaTec1p regulates hyphal development and virulence in *Candida albicans*. Mol Microbiol 38:435–445

Scott-Craig JS, Panaccione DG, Pocard JA, Walton JD (1992) The cyclic peptide synthetase catalyzing HC-toxin production in the filamentous fungus *Cochliobolus carbonum* is encoded by a 15.7-kilobase open reading frame. J Biol Chem 267:26044–26049

Selosse MA, Le Tacon F (1998) The land flora: a phototroph-fungus partnership? Trends Ecol Evol 13:15–20

Shih J, Wei Y, Goodwin PH (2000) A comparison of the pectate lyase genes, *pel-1* and *pel-2*, of *Colletotrichum gloeosporioides* f.sp. *malvae* and the relationship between their expression in culture and during necrotrophic infection. Gene 243:139–150

Simmons C, Hantke S, Grant S, Johal GS, Briggs SP (1998) The maize lethal leaf spot 1 mutant has elevated resistance to fungal infection at the leaf epidermis. Mol Plant Microbe Interact 11:1110–1118

Snetselaar KM, Mims CW (1993) Infection of maize by *Ustilago maydis*: light and electron microscopy. Phytopatholol 83:843–850

Stadler M, Sterner O (1998) Production of bioactive secondary metabolites in the fruit bodies of macrofungi as a response to injury. Phytochemistry 49:1013–1019

Staples RC, Hoch HC (1987) Infection structures – form and function. Exp Mycol 11:163–169

Stephenson SA, Hatfield J, Rusu AG, Maclean DJ, Manners JM (2000) *CgDN3*: an essential pathogenicity gene of *Colletotrichum gloeosporioides* necessary to avert a hypersensitive-like response in the host *Stylosanthes guianensis*. Mol Plant Microbe Interact 13:929–941

Struck C, Hahn M, Mendgen K (1998a) Infection structures of plant pathogenic fungi – potential targets for plant disease control. J Plant Dis Protect 105:581–589

Struck C, Siebels C, Rommel O, Wernitz M, Hahn M (1998b) The plasma membrane H+-ATPase from the biotrophic rust fungus *Uromyces fabae*: molecular characterization of the gene (*PMA1*) and functional expression of the enzyme in yeast. Mol Plant Microbe Interact 11:458–465

Struck C, Ernst M, Hahn M (2002) Characterization of a developmentally regulated amino acid transporter (AAT1p) of the rust fungus *Uromyces fabae*. Mol Plant Pathol 3:23–30

Sugui JA, Deising HB (2002) Isolation of infection-specific sequence tags expressed during early stages of maize anthracnose disease development. Mol Plant Pathol 3:197–203

Taylor TN, Hass H, Kerp H (1999) The oldest fossil ascomycetes. Nature 399:648

Tegtmeier KJ, Daly JM, Yoder OC (1982) T-toxin production by near-isogenic isolates of *Cochliobolus heterostrophus* race-T and race-O. Phytopathology 72:1492–1495

Tenberge KB, Homann V, Oeser B, Tudzynski P (1996) Structure and expression of two polygalacturonase genes of *Claviceps purpurea* oriented in tandem and cytological evidence for pectic enzyme activity during infection of rye. Phytopathology 86:1084–1097

ten Have A, Mulder W, Visser J, van Kan JAL (1998) The endopolygalacturonase gene *Bcpg1* is required for full virulence of *Botrytis cinerea*. Mol Plant Microbe Interact 11:1009–1016

ten Have A, Tenberge K, Benen JAE, Tudzynski P, Visser J, van Kan JAL (2002) The contribution of cell wall degrading enzymes to pathogenesis of fungal plant pathogens.

In: Kempken F (ed) The Mycota XI. Application in agriculture. Springer, Berlin Heidelberg New York, pp 341–358

Thieron M, Pontzen R, Kurahashi Y (1998) Carpropamid: a rice fungicide with two modes of action. Pflanzenschutz Nachr Bayer 51:257–278

Thines E, Eilbert F, Sterner O, Anke H (1997) Glisoprenin A, an inhibitor of the signal transduction pathway leading to appressorium formation in germinating conidia of *Magnaporthe grisea* on hydrophobic surfaces. FEMS Microbiol Lett 151:219–224

Thon MR, Nuckles EM, Takach JE, Vaillancourt LJ (2002) *CPR1*: a gene encoding a putative signal peptidase that functions in pathogenicity of *Colletotrichum graminicola* to maize. Mol Plant Microbe Interact 15:120–128

Tonukari NJ, Scott-Craig JS, Walton JD (2000) The *Cochliobolus carbonum SNF1* gene is required for cell wall-degrading enzyme expression and virulence on maize. Plant Cell 12:237–248

Tsuji G, Kenmochi Y, Takano Y, Sweigard J, Farrall L, Furusawa I, Horino O, Kubo Y (2000) Novel fungal transcriptional activators, Cmr1p of *Colletotrichum lagenarium* and pig1p of *Magnaporthe grisea*, contain Cys2His2 zinc finger and Zn(II)2Cys6 binuclear cluster DNA-binding motifs and regulate transcription of melanin biosynthesis genes in a developmentally specific manner. Mol Microbiol 38:940–954

Tudzynski P, Tenberge KB, Oeser B (1995) *Claviceps purpurea*. In: Kohmoto K, Singh US, Singh RP (eds) Pathogenesis and host specificity in plant diseases: histological, biochemical, genetic and molecular bases. Pergamon, New York, pp 161–187

Urbina JM, Cortes JC, Palma A, Lopez SN, Zacchino SA, Enriz RD, Ribas JC, Kouznetzov VV (2000) Inhibitors of the fungal cell wall. Synthesis of 4-aryl-4-*N*-arylamine-1-butenes and related compounds with inhibitory activities on beta(1–3) glucan and chitin synthases. Bioorg Med Chem 8:691–698

Vander P, Varum KM, Domard A, El Gueddari NE, Moerschbacher BM (1998) Comparison of the ability of partially *N*-acetylated chitosans and chitoologosaccharides to elicit resistance reactions in wheat leaves. Plant Physiol 118:1353–1359

Voegele RT, Struck C, Hahn M, Mendgen K (2001) The role of haustoria in sugar supply during infection of broad bean by the rust fungus *Uromyces fabae*. Proc Natl Acad Sci USA 98:8133–8138

Walton JD, Earle ED, Gibson BW (1982) Purification and structure of the host-specific toxin from *Helminthosporium carbonum* race 1. Biochem Biophys Res Commun 107:785–794

Walton JD, Bronson CR, Panaccione DG, Braun EJ, Akimitsu K (1995) *Cochliobolus*. In: Kohmoto K, Singh US, Singh RP (eds) Pathogenesis and host specificity in plant diseases. Histopathological, biochemical, genetic and molecular bases. Elsevier Science, Oxford, pp 65–81

Wattad G, Kobiler D, Dinoor A, Prusky D (1997) Pectate lyase of *Colletotrichum gloeosporioides* attacking avocado fruits: cDNA cloning and involvement in pathogenicity. Physiol Mol Plant Pathol 50:197–212

Werner S, Steiner U, Becher R, Kortekamp A, Zyprian E, Deising HB (2002) Chitin synthesis during in planta growth and asexual propagation of the cellulosic oomycete and obligate biotrophic grapevine pathogen *Plasmopara viticola*. FEMS Microbiol Lett 208:169–173

Wessels JGH (1993) Wall growth, protein excretion and morphogenesis in fungi. New Phytol 123:397–413

Wheeler J, Hollomon DW, Longhurst C, Green E (2000) Quinoxyfen signals to stop infection by powdery mildews. Proc Br Crop Protection Conf Pests Dis 2000:841–846

Wijesundera RLC, Bailey JA, Byrde RJW, Fielding AH (1989) Cell wall degrading enzymes of *Colletotrichum lindemuthianum*: their role in the development of bean anthracnose. Physiol Mol Plant Pathol 34:403–413

Wirsel SG, Voegele RT, Mendgen KW (2001) Differential regulation of gene expression in the obligate biotrophic interaction of *Uromyces fabae* with its host *Vicia faba*. Mol Plant Microbe Interact 14:1319–1326

Wirsel SGR, Runge-Froböse C, Ahren DG, Kemen E, Oliver RP, Mendgen KW (2002) Four or more species of *Cladosporium* sympatrically colonize *Phragmites australis*. Fungal Genet Biol 35:99–113

Wise RP, Pring DR, Gengenbach BG (1987) Mutation to male-fertility and toxin insensitivity in Texas (T)-cytoplasm maize is associated with a frameshift in a mitochondrial open reading frame. Proc Natl Acad Sci USA 84:2858–2862

Wu SC, Ham KS, Darvill AG, Albersheim P (1997) Deletion of two *Endo*?β-1,4-xylanase genes reveals additional isozymes secreted by the rice blast fungus. Mol Plant Microbe Interact 10:700–708

Wubben JP, ten Have A, van Kan JAL, Visser J (2000) Regulation of endopolygalacturonase gene expression in *Botrytis cinerea* by galacturonic acid, ambient pH and carbon catabolite repression. Curr Genet 37:152–157

Yakoby N, Freeman S, Dinoor A, Keen NT, Prusky D (2000) Expression of pectate lyase from *Colletotrichum gloesosporioides* in *C. magna* promotes pathogenicity. Mol Plant Microbe Interact 13:887–891

Yakoby N, Beno-Moualem D, Keen NT, Dinoor A, Pines O, Prusky D (2001) *Colletotrichum gloeosporioides pelB* is an important virulence factor in avocado fruit-fungus interaction. Mol Plant Microbe Interact 14:988–995

Yoder OC (1980) Toxins in pathogenesis. Annu Rev Phytopathol 18:103–129

Yoshida S, Hiradate S, Fujii Y, Shirata A (2000) *Colletotrichum dematium* produces phytotoxins in anthracnose lesions of mulberry leaves. Phytopathology 90:285–291

Stefan G.R. Wirsel
Sven Reimann
Holger B. Deising
Phytopathologie und Pflanzenschutz
Martin-Luther-Universität Halle
Ludwig Wucherer Straße 2
06099 Halle, Germany
e-mail: deising@landw.uni-halle.de

Biotechnology: Production of Proteins for Biopharmaceutical and Industrial Uses in Transgenic Plants

Kerstin Stockmeyer and Frank Kempken

1 Introduction

Recent advances in plant molecular biology have opened new strategies for the production of genetically engineered plants and for the precise transfer of new genes into crop plants from diverse sources. Many genes have been transferred through various transformation techniques including genes for several agronomically important traits such as herbicide resistance, disease and insect resistance.

Over the last 8 to 10 years, it has even become apparent that plant systems may be particularly valuable for the expression and production of recombinant proteins. Foreign genes are integrated into the plant genome and therefore traditional breeding techniques can be used to generate transgenic seed stocks for easy and stable storage or distribution.

Plant cells are able to synthesize, target and process complex mammalian proteins in a very similar manner to the natural hosts. Therefore, they are an attractive alternative to bacterial or mammalian systems for the production of recombinant proteins.

Much experience has been accumulated in the manufacture of biopharmaceuticals in animal and bacterial cells, as well as in the biology and genetics of crop plants. Elements from these different disciplines can be brought together for the design of optimal production strategies.

2 Transgenic Plants as Bioreactors

Transgenic plants are attracting interest as efficient bioreactors for the safe and inexpensive production of large amounts of functional, recombinant therapeutically valuable proteins, such as blood substitutes, vaccines and antibodies. The production of recombinant blood factors for therapeutic use, such as antibodies or recombinant human hemoglobin, in transgenic plants has become an alternative to the isolation of these molecules from natural or other recombinant sources (Theisen 1999). Plant bioreactors

Progress in Botany, Vol. 65

offer the possibility of an inexpensive large-scale production of high volumes of recombinant proteins with increased safety concerning contaminations with human pathogens.

Bacteria have often been the expression system used for the production of recombinant proteins, with low production costs, their straightforward molecular biology and easy handling (Olins and Lee 1993). Complex mammalian proteins can also be produced in transformed plants or transformed plant suspension cells (Fischer et al. 1999a). Plants are attractive for the production of pharmaceutical proteins on a field scale because the expressed proteins are functional and almost identical to their mammalian counterparts. Regarding protein folding and structure, small peptides, polypeptides and even complex proteins can be expressed in plants that are fully assembled (Fischer and Emans 2000).

The first transgenic plants were reported in 1983 (Fraley et al. 1983; Zambryski et al. 1983). Many recombinant proteins have been expressed in several important agronomic plant species including tomato, potato, tobacco and corn since then (Austin et al. 1993). Examples are given in Table 1.

Bacterial or mammalian expression systems have significant limitations. Bacteria cannot perform the complex posttranslational modifications that are required for bioactivity of many human proteins and high-level expression often leads to accumulation of insoluble protein aggregates. While mammalian cell cultures perform the required protein modifications, instability of selected cell lines, low transgene expression levels, and the difficulties and high expense of scale-up are often limiting or severely impact cost. Furthermore, animal tissues bare potential viral or infectious agent contamination risks and allergic reaction problems.

Genetically engineered transgenic plants have several advantages as protein sources compared with human or animal fluids/tissues, recombinant microbes, transfected animal cell lines or transgenic animals:

- Plants are easy to grow and unlike bacteria or animal cells their cultivation is straightforward and does not require special equipment or media or involve toxic chemicals.
- They produce a large amount of biomass and protein production can be increased using plant suspension cell culture in fermenters (Fischer et al. 1999a), or by the propagation of stably transformed plant lines in the field.
- The storage of genes and gene products in plants can be very stable. Transgenic plants can be self-fertilized to produce stable breeding lines, propagated by conventional horticultural techniques and distributed and stored as seeds.

Table 1. Expression of recombinant biopharmaceuticals in transgenic plants

Protein	Plant host	Potential application or indication	Expression levels (% of total soluble protein)	Reference
Human interferon α	Rice	Hepatitis C and B treatment	Not reported	Zhu et al. (1994)
Human epidermal growth factor	Tobacco	Wound repair, control of cell proliferation	0.001	Higo et al. (1993)
Glucocerebrosidase	Tobacco	Gaucher's disease	1–10	Cramer et al. (1996)
Human hemoglobin α and β	Tobacco	Blood substitute	0.05	Dieryck et al. (1997)
Human serum albumin	Potato	Liver cirrhosis, burns, surgery	0.02	Sijmons et al. (1990)
E.coli enterotoxin B	Potato	Immunogenicity	0.003	Haq et al. (1995)
Hepatitis B surface antigen	Tobacco	Immunogenicity	0.007	Mason et al. (1992)
Cholera toxin B	Tobacco	Immunogenicity	Not reported	Hein et al. (1996)
α-Tricosanthin from TMW-U1	Tobacco	HIV therapies	4–5	Kumagai et al. (1993)
Human hirudin	Canola	Thrombin inhibitor	0.3	Parmenter et al. (1995)

- Plant cells carry out all of the posttranslational modifications required for optimal biological activity of proteins.

3 Plant Expression Strategies

Genetic transformation of the plant genome can be performed using *Agrobacterium*-mediated transformation (Horsch et al. 1985; Koncz and Schell 1986), micro-projectile bombardment (Sanford et al. 1987; Klein et al. 1992; Christou 1993), viral vectors (Grill 1993; Porta and Lomonossoff 1996) or transformation of protoplasts. The transfected plants can express the recombinant protein either constitutively or in a tissue-specific manner. Alternatively, genetically modified plant viruses can be used. Plants are infected with the modified virus and act as viral culture vessels. With this method large quantities of recombinant proteins can be generated quite rapidly, using relatively small numbers of plants. The expressed recombinant protein can be targeted to stable environments within the plant. Alternatively, tissue-specific promoters can be used to direct expression in storage organs such as seeds or tubers. Extraction and purification from these sites are simple.

Deposition into the extracellular apoplast may contribute to the stability of recombinant proteins by removing them from the more hydrolytic intracellular environment (Firek et al. 1993).

Two strategies have been used to produce recombinant protein in plant hosts. One is based upon the generation of transgenic plants by stable integration of a transgene in the plant genome (Cheng et al. 1997; Muller et al. 2001). The other uses plants as hosts for transient expression by using plant viruses as vectors (Brooks and Bruening 1995; Scholthof et al. 1996; Chowrira et al. 1998). Generating stable transgenic plants is more labor-intensive than generating recombinant virus vectors that can infect wild-type plants. On the other hand, transgenic plants do not require inoculation and can be obtained in very large amounts once a stable strain has been generated.

Stable transgenic plants can be used to produce leaves or seeds rich in the recombinant protein for long-term storage or direct processing (Hiatt et al. 1989; Pen et al. 1993; Fiedler and Conrad 1995; Stoger et al. 2000).

The choice of promoter, which determines the timing, tissue specificity, and level of transgene expression, is important for product yields of the transgene and recovery (Chinn and Comai 1996). The most widely used promoter in plant biology for overexpression of plant or recombinant proteins is the constitutive 35S promoter derived from the cauliflower mosaic virus (Benfey et al. 1989; Fang et al. 1989). However, constitutive expression of proteins has significant limitations. High constitutive expression is sometimes associated with cosuppression or gene silencing (Taylor 1997), resulting in little or no transgene product accumulation. For proteins that are not highly stable, constitutive expression can lead to synthesis-degradation cycles and contamination of the final product with degradation products. Use of inducible or tissue-specific promoters can avoid some of these limitations (Cramer and Weissenborn 1997).

3.1 Selection of Crop Species

In order to select a certain plant species as expression system it is important to take into account how readily it can be manipulated to produce a stable transgenic line, which tissue and subcellular compartment is suited for a stable expression, processing and stability of the recombinant protein. Some factors, like the amenability to transformation, the ability to regenerate whole plants and generation time, have a significant impact upon the time and resources required for product development (Lindsey 1996).

Tobacco, for example, is often used to test plant-based systems for production of recombinant proteins, because it can easily be genetically engineered, produces much biomass and seed (1 million seeds per plant). Other plants tested are potato, tomato, alfalfa, soybeans, canola, rice or barley (Zhu et al. 1994; Haq et al. 1995; Dieryck et al. 1997).

3.2 Choice of Tissue

For obtaining a maximum yield, it is important that the selected plant species is able to concentrate biomass in the tissue or organ where the foreign protein is expressed. Among higher plants a variety of options are available including vegetative storage organs (e.g., tubers), leaves and seeds, which contain large quantities of storage proteins and have a low hydrolytic intracellular environment. Seeds have the advantage over most other organs that they can be stored for several years without any appreciable degradation of proteins or loss of activity (Pen et al. 1993). Therefore, immediate processing is not required.

The chosen tissue should enable correct processing, stable accumulation, and efficient recovery. Mammalian antibodies, for example, have been produced in tobacco leaves following trafficking through the endoplasmic reticulum and Golgi complex (Ma et al. 1995). In tobacco leaves and some tissues of potato human serum albumin, a blood protein, has also been stably expressed (Sijmons et al. 1990), although a precise folding and functionality of the protein could not be established. Hirudin, an anticoagulant protein, has been produced in seeds of *Brassica napus* followed by targeting to protein and oil bodies (Parmenter et al. 1995).

4 Localization and Processing of Expressed Proteins

Many human therapeutic proteins of interest require complex posttranslational processing for their biological activity or appropriate targeting following administration to patients. The protein processing steps are very similar between plants and animals. Most human proteins that have been expressed in plants show remarkable biochemical, structural and functional identities to proteins from humans or animal cell cultures.

4.1 Posttranslational Processing

Posttranslational processing of proteins includes endomembrane targeting, cleavage of signal peptides, protein folding and oligomerization, glycosylation, disulfide bond formation (Fischer et al. 2000).

A remarkable fidelity with which plants recognize and correctly act upon most of the processing signals within mammalian polypeptides exists, which indicates a high degree of conservation in the protein processing machinery between plants and animals.

However, clear variations in protein processing, most obvious in glycoprotein processing, exist between plants and animals. This leads to the generation of glycans that are heterogenous or differ significantly from the native conformation of particular human proteins (Jenkins et al. 1996). The differences in glycosylation patterns of plant antibodies have no effect on antigen binding or specificity (Ma and Vine 1999).

Modifying the glycoprotein processing of plants through genetic engineering or in vitro enzymatic modification of the purified recombinant protein should enable commercialization of recombinant plant-derived glycoproteins for pharmaceutical uses (Chrispeels and Faye 1996).

4.2 Plant Purification Techniques

Plant-derived recombinant proteins can be purified by standard biochemical techniques (Gailliot et al. 1990; Fischer et al. 1999b). Impurities in plant-derived biopharmaceuticals may include biologically active plant metabolites or alkaloids, plant macro-molecules (like DNA, polysaccharides or lipids), herbicides and pesticides. Such impurities could affect the safety and efficacy of the product, including direct toxic effects on the recipient or effects on product stability. The relatively mild aqueous extraction procedures often used to solubilize recombinant proteins might not be very effective to extract many potential harmful substances, compared with the organic solvent treatments often employed in the extraction of other pharmaceuticals.

One method for the purification of recombinant proteins is the use of affinity tags. After the creation of a fusion between the interesting protein and another protein that exhibits affinity for a specific ligand, the fusion protein can be recovered by binding to the ligand immobilized onto a support matrix (Cramer et al. 1996).

For some proteins the affinity tag can be proteolytically removed from the fusion protein after purification. However, it is possible that the tag could alter folding or processing of the recombinant protein. In addition this strategy is not suitable for any large-scale production.

Another method used to simplify the purification of recombinant proteins is through compartmentalization. Using a signal peptide, the interesting protein can be targeted to a specific cellular location (Herbers et al. 1995). Purification of the desired recombinant protein is then facilitated through the possibility of its physical separation from other proteins in the cell. Examples for compartmentalization are expression on viral particles, extracellular secretion, and targeting to intracellular storage organelles (Parmenter et al. 1995; Boothe et al. 1997).

An attractive feature of secretion is that the signal peptide is removed from the recombinant protein during the normal processing without introducing additional proteolytic digestion steps.

5 Production of Immunogenic Proteins in Transgenic Plants

A considerable effort has focused on genetic engineering for the improvement of plant characteristics and agricultural properties, but plant biotechnology has also been applied to the production of "high-value" products, such as pharmaceutical compounds and vaccines. This approach is attractive, because it has the potential for growing immunotherapeutic reagents on an agricultural scale, thereby reducing the costs of production significantly. An important advantage of plants as a recombinant expression system is the ability to assemble full-length heavy chains with light chains to form full-length antibodies efficiently (Hiatt et al. 1989). In bacterial expression systems full-length antibodies are not readily assembled.

Originally, foreign antibody genes were introduced into plant cells by non-pathogenic strains of the natural plant pathogen *Agrobacterium tumefaciens* (Horsch et al. 1985) and regeneration in tissue culture resulted in the recovery of stable transgenic plants. Although the initial work to generate multichain proteins required crossing of plants expressing each chain, more recent studies have shown that multiple chains can be introduced via a single biolistic transformation event (Sanford 1988), greatly reducing the time to final assembly antibody.

The use of edible plants for the production and delivery of vaccines could provide an economical alternative to fermentation systems. Genes encoding bacterial and viral antigens are expressed in edible tissues to form immunogenic proteins. Studies in humans and animals demonstrated the production of antigen-specific antibodies in serum and mucosal secretions after ingestion of transgenic plants containing vaccine proteins.

5.1 Antibody Targeting

In general, the technology is limited by low expression levels for nuclear-integrated transgenes, but recent progress in plant organelle transformation shows promise for enhanced expression. Transgenes can be fusioned with signal peptide sequences to enable entry of proteins into the secretory pathway (Hiatt et al. 1989; Bednarek and Raikhel 1992). In the secretory system complete antibodies and Fab fragments can be assembled. Single-chain (sc) Fv constructs can be functionally expressed in the cytoplasm without any signal sequence (Owen et al. 1992; Tavladoraki et al. 1993). It is possible to target them to different compartments through fusion of targeting signals to either terminus of the scFv construct (Chrispeels 1991).

5.2 Optimization of Antibody Production

In 1989, the production of antibody in transgenic plants was first described. It was demonstrated, that co-expression of two recombinant gene products could lead to a correctly folded and assembled multimeric molecule in plants that was functionally identical to its mammalian counterpart (Hiatt et al. 1989). Originally, only low expression levels were achieved in plants. Various strategies have been devised to improve yields, e.g. the expression of genes for antibody fragments in fusion with the endoplasmic-reticulum retention signal KDEL. This strategy can increase scFv yield tenfold (Schouten et al. 1996) or up to 4–6.8% of total soluble protein (Fiedler et al. 1997).

Another successful approach is the targeting of scFv for expression in seeds, where the antibody fragment can accumulate to 3–4% of the total soluble seed protein (Fiedler et al. 1997).

Recombinant antibodies produced in tobacco plants have the same sensitivity, specificity and, importantly, the same affinity as monoclonal antibodies produced by the original hybridoma cell line (Hein 1996).

5.3 Developments in Plant-Derived Vaccines

Vaccines are aimed to prime the immune system to destroy specific disease-causing pathogens or agents, before they can multiply enough to cause symptoms. This has classically been achieved by presenting whole bacteria or viruses to the immune system that have been attenuated or killed to prevent their multiplication in the vaccinated person.

The first reported vaccine candidate to be expressed in transgenic plants was the cell surface-adhesion protein Spa A or streptococcal antigen I/II from *Streptococcus mutans* (Curtiss and Cardineau 1990). This is the major bacterial cause of tooth decay, and there is a considerable evidence that demonstrates protective immunity against dental caries following systemic or oral immunization with this antigen. The *spa A* gene was introduced into tobacco by agrobacterium-mediated transformation, and the protein was expressed at levels of up to 0.02% total leaf protein.

Hepatitis B surface antigen (HbsAg) has been expressed in transgenic potatoes, in addition to a chimeric gene encoding the M protein of hepatitis B virus. For oral vaccination with edible, transgenic plant material, potato may not be ideal, because the cooking process might denature the recombinant protein. Alternatively tomato was used for expression of the rabies virus glycoprotein (McGarvey et al. 1995). Using the complete unmodified gene under the control of the 35S promoter of cauliflower mosaic virus, expression of glycoprotein was found in leaf and fruit tissue. The tomato-expressed glycoprotein was recognized by specific antisera and a monoclonal antibody, suggesting preservation of important epitopes.

However, a persistent problem in the expression of vaccines in transgenic plants has been the low levels of expression, although several approaches are being adopted to overcome this technical problem. The exception is antibodies for which levels of 1–5% total protein are achieved consistently (Mason and Arntzen 1995; Daniell et al. 2001).

6 Regulatory and Safety Issues

6.1 General Concerns

The use of plants avoids many of the potential safety issues associated with contaminating mammalian viruses, as well as ethical concerns involving the use of animals. An important benefit of plants involves product safety. Plants are free from known human pathogens like HIV and also from unknown or poorly characterized agents as the prions that are responsible for bovine spongiform encephalopathy and the related Creutzfeld-Jakob disease (Roberts and James 1996; Vaughan 1996). Plant viruses, used in virus-based transient expression systems, are not known to infect humans or animals.

However, there are also some unresolved issues:

- How to easily identify seed or fruits from transgenic plants expressing therapeutic proteins? This is important as these are not acceptable in the food chain.
- Another issue is regarding the amount of antibodies necessary for an oral vaccination, as it is difficult to guarantee a certain level of antigen protein in a specific plant tissue.

6.2 Excision of Selectable Marker Genes from Transgenic Plants

Selectable marker genes are required to ensure the efficient genetic modification of crops. Safety considerations and economic incentives have prompted the development of several strategies (homologous recombination, site-specific recombination, and transposition) to eliminate these genes from the genome after they have fulfilled their purpose (Dale and Ow 1991; Fig. 1). Recently, chemically inducible site-specific recombinase systems have emerged as valuable tools for an efficient regulation of the

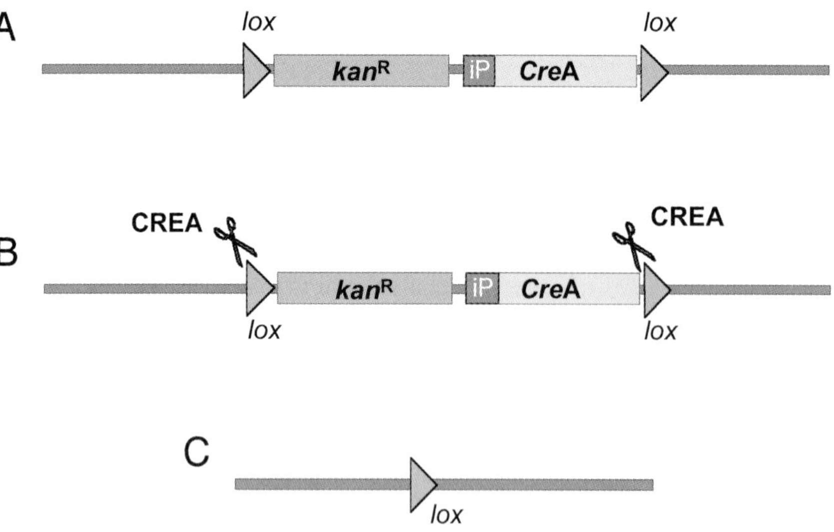

Fig. 1A–C. Excision of marker genes. Use of the Cre/lox system to remove marker genes from transgenic plants. **A** The marker gene (kan^R) and the *CreA* gene (shown as *rectangles*) are localized between two lox sequences (shown as *triangles*). The *CreA* gene is driven by an inducible promoter (iP). **B** Upon induction of the promoter, the CREA protein specifically recognizes the lox DNA sequences and the sequences between will be excised; **C** a single lox sequence remains

excision of transgenes when their expression is no longer required (Odell et al. 1994; Hajdukiewicz et al. 2001; Zuo et al. 2001). The implementation of these strategies in crops and the further improvement will help to expedite widespread public acceptance of agricultural biotechnology.

7 Outlook

Plant biotechnology is a rapidly expanding area. Its further development will bring many therapies that are now too expensive for wide use into the hands of most medical practitioners. One of the most obvious advantages of plants is the potential for scale-up production, in which large amounts of recombinant protein could be grown at minimal cost. However, some issues have not yet been properly addressed, i.e. how to comply with the requirements of "good manufacturing practice" (GMP), which is required for all human therapeutic proteins.

Advancing technology, such as oral DNA vaccines or edible plant-derived vaccines, may lead to a future of safer and more effective immunization. It seems likely that sufficient antibody could be "grown" in plants, for most medical applications, on only a few acres of land and at minimal cost relative to alternative production methods.

It is anticipated that the advances in the genetic engineering of plants and the demand for large quantities of new or improved therapeutics and diagnostics in the pharmaceutical markets will lead to a rapid expansion in this field.

Acknowledgement. K.S. acknowledges a grant received from the "Max-Buchner-Forschungsstiftung" (Frankfurt).

References

Austin S, Bingham ET, Koegel RG, Mathews DE, Shahan MN, Straub RJ, Burgess RR (1993) An overview of a feasibility study for the production of industrial enzymes in transgenic alfalfa. Ann N Y Acad Sci 721:234–244

Bednarek SY, Raikhel NV (1992) Intracellular trafficking of secretory proteins. Plant Mol Biol 20:133–150

Benfey P, Ling R, Chua NH (1989) The CaMV 35/S enhancer contains at least two domains which can confer different developmental and tissue-specific expression patterns. EMBO J 8:2195–2202

Boothe J, Saponja JA, Parmenter DL (1997) Molecular farming in plants: oilseeds as vehicles for the production of pharmaceutical proteins. Drug Dev Res 42:172–181

Brooks M, Bruening G (1995) Transient gene expression of antisense RNA and coat protein-encoding sequences reduced accumulation of cherry leafroll virus in tobacco protoplasts. Virology 208:132–141

Cheng M, Fry JE, Pang S, Zhou H, Hironaka CM, Duncan DR, Conner TW, Wan Y (1997) Genetic transformation of wheat mediated by *Agrobacterium tumefaciens*. Plant Physiol 115:971–980

Chinn A, Comai L (1996) Gene transcription. In: Owen M, Pen J (eds) Transgenic plants: a production system for industrial and pharmaceutical proteins. Wiley, Chichester, pp 27–48

Chowrira G, Cavileer TD, Gupta SK, Lurquin PF, Berger PH (1998) Coat protein-mediated resistance to pea enation mosaic virus in transgenic *Pisum sativum* L. Transgenic Res 7:265–271

Chrispeels MJ (1991) Sorting of proteins in the secretory system. Annu Rev Plant Physiol Plant Mol Biol 42:35–49

Chrispeels M, Faye L (1996) The production of recombinant glycoproteins with defined non-immunogenic glycans. In: Owen M, Pen J (eds) Transgenic plants: a production system for industrial and pharmaceutical proteins. Wiley, Chichester, pp 99–113

Christou P (1993) Particle gun-mediated transformation. Curr Opin Biotechnol 4:135–141

Cramer C, Weissenborn DL (1997) HMG2 promoter expression system and post-harvest production of gene products in plants and plant cell cultures. US Patent 5:349

Cramer C, Weissenborn DL, Oishi KK, Radin DN (1996) High-level production of enzymatically active human lysosomal proteins in transgenic plants. In: Owen M, Pen J (ed) Transgenic plants: a production system for industrial and pharmaceutical proteins. Wiley, Chichester, pp 299–309

Curtiss R, Cardineau GA (1990) Oral immunisation by transgenic plants. World Patent Application, WO 90/02484

Dale E, Ow DW (1991) Gene transfer with subsequent removal of the selection gene from the host genome. Proc Natl Acad Sci USA 23:10558–10562

Daniell H, Streatfield SJ, Wycoff K (2001) Medical molecular farming: production of antibodies, biopharmaceuticals and edible vaccines in plants. Trends Plant Sci 6:219–226

Dieryck W, Pagnier J, Poyart C, Marden MC, Gruber V, Bournat P, Baudino S, Merot B (1997) Human haemoglobin from transgenic tobacco. Nature 386:29–30

Fang R, Nagy F, Sivasubramanian S, Chua NH (1989) Multiple *cis*-regulatory elements for maximal expression of the cauliflower mosaic virus 35S promoter in transgenic plants. Plant Cell 1:141–150

Fiedler U, Conrad U (1995) High-level production and long-term storage of engineered antibodies in transgenic tobacco seeds. Biotechnology 13:1090–1093

Fiedler U, Phillips J, Atsaenko O, Conrad U (1997) Optimization of scFv antibody production in transgenic plants. Immunotechnology 3:205–216

Firek S, Draper J, Owen MRL, Gandecha A, Cockburn B, Whitelam GC (1993) Secretion of a functional single-chain Fv protein in transgenic tobacco plants and cell suspension cultures. Plant Mol Biol 23:861–870

Fischer R, Emans N (2000) Molecular farming of pharmaceutical proteins. Transgenic Res 9:279–299

Fischer R, Emans N, Schuster F, Hellwig S, Drossard J (1999a) Towards molecular farming in the future: using plant-cell-suspension cultures as bioreactors. Biotechnol Appl Biochem 30:109–112

Fischer R, Liao YC, Drossard J (1999b) Affinity-purification of a TMV-specific recombinant full-size antibody from a transgenic tobacco suspension culture. J Immunol Methods 24:1–10

Fischer R, Hoffmann K, Schilberg S, Emans N (2000) Antibody production by molecular farming in plants. J Biol Regul Homeost Agents 14:83–92

Fraley R, Rogers SG, Horsch RB, Sanders PR, Flick JS, Adams SP, Bittner ML, Brand LA, Fink CL, Fry JS, Galluppi GR, Goldberg SB, Hoffmann NL, Woo SC (1983) Expression of bacterial genes in plant cells. Proc Natl Acad Sci USA 80:4803–4807

Gailliot F, Gleason C, Wilson JJ, Zwarick J (1990) Fluidized bed adsorption for whole broth extraction. Biotechnol Prog 6:370–375

Grill L (1993) Tobacco mosaic virus as a gene expression vector. Agro Food Industry Hi Tech November/December:20–23

Hajdukiewicz PT, Gilbertson L, Staub JM (2001) Multiple pathways for Cre/lox-mediated recombination in plastids. Plant J 27:161–170

Haq T, Mason HS, Clements JD, Arntzen CJ (1995) Oral immunization with a recombinant bacterial antigen produced in transgenic plants. Science 268:714–716

Hein M, Yeo TC, Wang, F, Sturtevant A (1996) Expression of cholera toxin subunits in plants. In: Collins G, Shepherd RJ (eds) Engineering plants for commercial products and applications. New York Academy of Sciences, New York, pp 50–56

Herbers K, Wilke I, Sonnewald U (1995) A thermostable xylanase from Clostridium thermocellum expressed at high levels in the apoplast of transgenic tobacco has no detrimental effects and is easily purified. BioTech 13:63–66

Hiatt AC, Cafferkey R, Bowdish K (1989) Production of antibodies in transgenic plants. Nature 342:76–78

Higo K, Saito Y, Higo H (1993) Expression of a chemically synthesized gene for human epidermal growth factor under the control of cauliflower mosaic virus 35S promoter in transgenic tobacco. Biosci Biotechnol Biochem 57:1477–1481

Horsch R, Fry J, Hoffmann N, Eichholtz D, Rogers S, Fraley R (1985) A simple and general method for transferring genes into plants. Science 227:1229–1231

Jenkins N, Parekh RB, James DC (1996) Getting the glycosylation right: implications for the biotechnology industry. Nat Biotechnol 14:975–981

Klein TM, Arentzen R, Lewis PA, Fitzpatrick-McElligott S (1992) Transformation of microbes, plants and animals by particle bombardment. Biotech 10:286–291

Koncz C, Schell J (1986) The promotor of T_L-DNA gene 5 controls the tissue-specific expression of chimaeric genes carried by a novel type of *Agrobacterium* binary vector. Mol Gen Genet 204:383–396

Kumagai M, Turpen TH, Weinzettl N, della-Cioppa G, Turpen AM, Donson J, Hilf ME, Grantham GL, Dawson WO, Chow TP, Piatak MJ, Grill LK (1993) Rapid, high-level expression of biologically active alpha-trichosanthin in transfected plants by an RNA viral vector. Proc Natl Acad Sci USA 15:427–430

Lindsey K (1996) Plant transformation systems. In: Owen M, Pen J (eds) Transgenic plants: a production system for industrial and pharmaceutical proteins. Wiley, Chichester, pp 5–25

Ma JK, Vine ND (1999) Plant expression systems for the production of vaccines. Curr Top Microbiol Immunol 236:275–292

Ma JK-C, Hiatt A, Hein M, Vine N, Wang F, Stabila P, van Dolleweerd C, Mostov K, Lehner T (1995) Generation and assembly of secretory antibodies in plants. Science 268:716–719

Mason H, Arntzen CJ (1995) Transgenic plants as vaccine production systems. Trends Biotechnol 13:388–392

Mason H, Lam D, Arntzen CJ (1992) Expression of hepatitis B surface antigen in transgenic plants. Proc Natl Acad Sci USA 89:11745–11749

McGarvey P, Hammond J, Dienelt MM, Hooper DC, Fu ZF, Dietzschold B, Koprowski H, Michaels FH (1995) Expression of rabies virus glycoprotein in transgenic tomatoes. Biotechnology 13:1484–1487

Muller A, Iser M, Hess D (2001) Stable transformation of sunflower (*Helianthus annuus* L.) using a non-meristematic regeneration protocol and green fluorescent protein as a vital marker. Transgenic Res 10:435–444

Odell JT, Hoopes JL, Vermerris W (1994) Seed-specific gene activation mediated by the Cre/lox site-specific recombination system. Plant Physiol 106:447–458

Olins P, Lee SC (1993) Recent advances in heterologous gene expression in *Escherichia coli.* Curr Opin Biotechnol 4:520–525

Owen MR, Gandecha A, Cockburn W, Whitelam GC (1992) Synthesis of a functional anti-phytochrome single chain Fv protein in transgenic tobacco. Biotechnology 10:790–794

Parmenter D, Boothe JG, van Rooijen GJ, Yeung EC, Moloney MM (1995) Production of biologically active hirudin in plant seeds using oleosin partitioning. Plant Mol Biol 29:1167–1180

Pen J, Verwoerd TC, van Paridon PA, Buedeker RF, van den Elzen PJM, Geerse K, van der Klis JD, Versteegh JAJ, van Ooyen AJJ, Hoekema A (1993) Phytase-containing transgenic seed as a novel feed additive for improved phosphorus utilization. Biotechnology 11:811–814

Porta C, Lomonossoff GP (1996) Use of viral replicons for the expression of genes in plants. Mol Biotechnol 5:209–221

Roberts G, James S (1996) Prion diseases: transmission from mad cows? Curr Biol 6:1247–1249

Sanford JC (1988) The biolistic process. Trends Biotechnol 6:299–302

Sanford JC, Klein TM, Wolf ED, Allen N (1987) Delivery of substance into cells and tissues using a particle bombardment process. J Part Sci Technol 5:27–37

Scholthof H, Scholthof K, Jackson A (1996) Plant virus gene vectors for transient expression of foreign proteins in plants. Annu Rev Phytopathol 34:299–323

Schouten A, Roosien J, van Engelen FA, de Jong GA, Borst-Vrenssen AW, Zilverentant JF, Bosch D, Stiekema WJ, Gommers FJ, Schots A, Bakker J (1996) The C-terminal KDEL sequence increases the expression level of a single-chain antibody designed to be targeted to both the cytosol and the secretory pathway in transgenic tobacco. Plant Mol Biol 30:781–793

Sijmons P, Dekker BM, Schrammeijer B, Verwoerd TC, van den Elzen PJ, Hoekema A (1990) Production of correctly processed human serum albumin in transgenic plants. Biotechnology 8:217–221

Stoger E, Vaquero C, Torres E et al. (2000) Cereal crops as viable production and storage systems for pharmaceutical scFv antibodies. Plant Mol Biol 42:583–590

Tavladoraki P, Benvenuto E, Trinca S, de Martinis D, Cattaneo A, Galeffi P (1993) Transgenic plants expressing a functional single-chain Fv antibody are specifically protected from virus attack. Nature 366:469–472

Taylor CB (1997) Comprehending cosuppression. Plant Cell 9:1245–1249

Theisen M (1999) Production of recombinant blood factors in transgenic plants. Adv Exp Med Biol 464:211–220

Vaughan P (1996) Creutzfeld-Jakob disease – latest unknown in struggle to restore faith in the blood supply. Can Med Assoc J 155:565–568

Zambryski P, Joos H, Genetello C, Leemans J, Van Montagu J, Schell J (1983) Ti plasmid vector for the introduction of DNA into plant cell without alternation of their normal regeneration capacity. EMBO J. 2:2143–2150

Zhu Z, Hughes KW, Huang L, Sun B, Lui C, Li Y, Hou Y, Li X (1994) Expression of human α-interferon cDNA in transgenic rice plants. Plant Cell Tissue Organ Cult 36:197–204

Zuo J, Niu QW, Moller SG, Chua NH (2001) Chemical-regulated, site-specific DNA excision in transgenic plants. Nat Biotechnol 19:157–161

Kerstin Stockmeyer
Frank Kempken
Abteilung für Botanik mit Schwerpunkt
Genetik und Molekularbiologie
Botanisches Institut und Botanischer Garten
Christian-Albrechts-Universität zu Kiel
Olshausenstrasse 40
24098 Kiel, Germany
e-mail: fkempken@bot.uni-kiel.de

Plant Breeding: The ABCs of Flower Development in *Arabidopsis* and Rice

Günter Theißen and Annette Becker

1 Introduction

1.1 Functional Phylogenomics of Flower Development

Understanding the genetic basis of flower formation has been pioneered in higher eudicotyledonous flowering plants such as thale cress (*Arabidopsis thaliana*), snapdragon (*Antirrhinum majus*) and petunia (*Petunia hybrida*) (for a review, see Theißen and Saedler 1999). A deeper understanding of flower origin and diversification, however, requires studies in phylogenetic informative taxa outside the higher eudicots, including basal eudicots, monocots and basal angiosperms (Soltis et al. 2002). Such studies will comprise 'phylogenomic' approaches on gene families of interest (such as the MADS-box gene family), comparative EST sequencing and expression studies in a huge number of taxa (Soltis et al. 2002). These efforts, however, will allow only quite limited conclusions on the phylogeny of gene functions and thus should be complemented by comparative developmental genetics studies (Baum et al. 2002), implying that suitable model systems have to be established for different key non-eudicotyledonous flowering plant taxa. This will be a laborious effort, since very few plant species are amenable to the required technology (such as transformation).

Fortunately, nature and a strong human interest in agronomical important plants have provided us already with an almost ideal model system for the monocots: rice (*Oryza sativa*). Studies on flower development in rice are strongly catching up with those in *Arabidopsis*, so that comprehensive comparisons between both systems can start now, as exemplified below for the floral organ identity genes.

1.2 *Arabidopsis* and Rice: The Two Most Important Model Plants

For quite some time the tiny weed *Arabidopsis thaliana* (in brief, *Arabidopsis*) is serving now as the major model system for understanding gene

Progress in Botany, Vol. 65

functions in flowering plants. This is so due to a number of unrivaled advantages such as small genome size, short life cycle, low requirements for growth space, and easy transformability. Based on these features, a number of useful tools have been developed, such as transposon and T-DNA tagged populations for reverse genetics screens, and the genome has been almost completely sequenced (Arabidopsis Genome Initiative 2000). Therefore, *Arabidopsis* will certainly play its role as the primary plant model system also for quite some years to come.

Meanwhile, however, the genome of a second flowering plant, rice (*Oryza sativa*) has also been almost completely sequenced (Feng et al. 2002; Goff et al. 2002; Sasaki et al. 2002; Yu et al. 2002). Although rice has a genome size which is roughly about four times larger than that of *Arabidopsis* and has a longer life cycle, it has features which still qualify it as a reasonably good model system, and powerful tools for functional genomics are meanwhile also available for that species (Shimamoto and Kyozuka 2002).

Employing rice as a second major plant model system for genetic and functional genomics studies appears very much appropriate for a number of reasons.

- In sharp contrast to *Arabidopsis*, rice is of enormous agronomic importance. In fact, rice, with its two subspecies *Oryza sativa* ssp. *indica* and *Oryza sativa* ssp. *japonica*, is the major staple food for half of the world's population and belongs to the three most important crop plants on a global scale (besides wheat and maize). Moreover, rice is relatively closely related to all the other agronomically important cereal grasses, including not only wheat (*Triticum aestivum*) and maize (*Zea mays* ssp. *mays*), but also barley (*Hordeum vulgare*), rye (*Secale cereale*), oat (*Avena sativa*), and sorghum (*Sorghum bicolor*). The rice genome is largely colinear with those of all of these grasses, so that it is considered as the reference genome for the world's most important crop plants (Shimamoto and Kyozuka 2002). A transfer of knowledge from *Arabidopsis* to rice, and a detailed study of rice genetics and physiology may thus be very helpful for future plant breeding, including a future design of cereal crops by transgenic technology, or mutagenesis and marker assisted breeding (Meyerowitz 1994; Theißen 2000, 2001c, 2002a).
- The lineages that led to *Arabidopsis* (a eudicot) and rice (a monocot) separated quite early during flowering plant evolution, i.e., according to molecular data, about 200 million years ago (Theißen et al. 2000, and references cited therein). Thus a comparison between the functions of orthologous genes in rice and *Arabidopsis* promises a fairly good impression about both the conservation and divergence of gene functions throughout most of flowering plants.

- Rice has some physiological and morphological features which are quite different from those of *Arabidopsis*. For example, rice is a short-day plant rather than a facultative long-day plant like *Arabidopsis*, and it has a highly derived floral morphology. It is difficult, therefore, to generalize findings from *Arabidopsis* for the flowering plants in general (although this is often done, explicitly or implicitly) without looking at lineages that are so deviant like rice, representing the grasses (Poaceae).

1.3 The Flowers of *Arabidopsis* and Rice

The flowers of *Arabidopsis* and rice appear quite different. While for the reproductive organs, i.e., stamens and carpels, homology between *Arabidopsis* and rice is unquestionable, it is still quite controversial for other organs. An *Arabidopsis* flower is composed of (from inner to outer whorls) two fused carpels, being the female reproductive organs that constitute the gynoecium; six stamens, being the male reproductive organs; four white, showy petals; and four green, leaf-like sepals.

The tiny, wind-pollinated flowers of the grasses (Poaceae) are quite different from the flowers within other flowering plant families. They have carpels and stamens like their eudicot relatives, but they lack typical petals and sepals. Instead, lodicules, palea and lemma surround the carpels and stamens, thus constituting structures called florets. One or several florets are enclosed by typically two glumes, thus constituting structures called spikelets. Lodicules are small, glandular-like organs that swell at anthesis to spread the lemma and palea apart so that the wind can disperse the pollen produced by the stamens. According to their position in the flower, lodicules might be petal homologues. Palea, lemma and glumes are leaf-like organs. Palea and lemma could be homologues of eudioct sepals; alternatively, the palea may represent a prophyll, and the lemma a true bract (Clifford 1987).

In case of rice there is just a single carpel in the center of the flower; it is surrounded by six stamens, two lodicules, a palea and a lemma, and an upper and a lower sterile glume (Shimamoto and Kyozuka 2002).

1.4 The Developmental Genetic Context of Floral Organ Formation

Right from the beginning, flower development is under strict genetic control. This guarantees that flower development is initiated only under conditions which are favorable for reproduction, but that it, once it started, proceeds in a highly standardized way. Flower development can be subdivided into several major steps, such as floral induction, floral meristem formation, and floral organ development. Genetic control of the different steps of flower development is achieved by a hierarchy of interacting regulatory genes, most of which encode transcription factors (for a brief review, see Theißen 2001a). Near the top of that hierarchy are 'flowering time genes' which are triggered by developmental cues and environmental factors such as plant age, day length and temperature. 'Flowering time genes' mediate the switch from vegetative to reproductive development by activating 'meristem identity genes'. 'Meristem identity genes' control the transition from vegetative to inflorescence and floral meristems. When the transition from inflorescence to floral meristems has taken place, floral organs arise at defined positions from within these meristems under the control of different types of genes. 'Floral meristem size genes' regulate the size of the floral meristem and also influence floral organ number. 'Cadastral genes' are involved in setting the boundaries of floral organ identity gene functions. 'Floral organ pattern genes' act to establish floral organ primordia in specific numbers and positions. These primordia develop into the different types of floral organs under the control of specific homeotic selector genes, termed 'floral organ identity genes'.

The function of 'floral organ identity genes' was recognized during the study of homeotic mutants in which the identity of floral organs is changed. In *Arabidopsis* such mutants come in three classes, A, B and C. Ideal class A mutants have carpels in the first whorl instead of sepals, and stamens in the second whorl instead of petals. Class B mutants have sepals rather than petals in the second whorl, and carpels rather than stamens in the third whorl. Class C mutants have petals instead of stamens in the third whorl, and replacement of the carpels in the fourth whorl by sepals. In addition, these mutants are indeterminate, i.e. there is continued production of mutant floral organs inside the fourth whorl.

Based on these classes of mutants the 'ABC model' proposes three classes of combinatorially acting 'floral organ identity genes', called A, B and C, with A specifying sepals in the first floral whorl, A+B petals in the second whorl, B+C stamens in the third whorl, and C carpels in the fourth whorl. The model also maintains that the class A and class C genes negatively regulate each other. Based on studies in petunia (*Petunia hybrida*), the 'ABC model' was later extended by class D genes, specifying ovules. Meanwhile, it has been demonstrated by a reverse genetic approach that yet another class of 'floral organ identity genes', termed class E genes, is involved in specifying petals, stamens and carpels. The 'floral organ identity genes' can be interpreted as acting as major developmental switches that activate the entire genetic program for a particular organ (for a description of the ABCDE model, see Theißen 2001b).

In *Arabidopsis*, class A genes comprise *APETALA1* (*AP1*) and *APETALA2* (*AP2*). The class B genes are represented by *APETALA3* (*AP3*) and *PISTILLATA* (*PI*), and the class C gene is *AGAMOUS* (*AG*). Class D genes have been recognized only in petunia so far, where they have been termed *FLORAL BINDING PROTEIN7* (*FBP7*) and *FBP11*, but there is a putative ortholog in *Arabidopsis* termed *AGL11*. The class E genes in *Arabidopsis* comprise *SEPALLATA1* (*SEP1*), *SEP2* and *SEP3*, which have highly redundant functions. All these genes encode putative transcription factors. Thus the products of the 'floral organ identity genes' probably all control the transcription of other genes (largely unknown 'target genes') whose products are involved in the formation or function of the different floral organs. Except for *AP2*, all 'floral organ identity genes' are MADS-box genes, encoding MADS-domain transcription factors (Theißen 2001b).

Phylogeny reconstructions revealed that the MADS-box gene family is composed of several defined gene clades (Doyle 1994; Purugganan et al. 1995; Theißen et al. 1996, 2000). All class A, B, C, D and E genes identified from eudicots fall into separate clades, namely *SQUAMOSA*- (class A), *DEFICIENS*- or *GLOBOSA*- (class B), *AGAMOUS*- (class C and D) and *AGL2*-like genes (class E) (Doyle 1994; Purugganan et al. 1995; Theißen and Saedler 1995; Angenent and Colombo 1996; Theißen et al. 1996, 2000; Münster et al. 1997). If the establishment of the mentioned gene clades by gene duplication and the structural and functional diversification that established the different gene functions predated the radiation of crown angiosperms, class A, B, C, D and E genes from non-eudicots should also be found in these gene clades. This can be tested now in non-eudicot model systems such as rice. Such tests are not always straight forward, however, because of the unclear homology between some organs in the flowers of *Arabidopsis* and rice.

2 Floral Organ Identity Genes in *Arabidopsis* and Rice

2.1 *SQUA*-Like Genes (Class A Candidate Genes)

There are three *SQUA*-like genes in the *Arabidopsis* genome, termed *APETALA1* (*AP1*), *CAULIFLOWER* (*CAL*) and *AGL8* (Table 1; Fig. 1) (Mandel et al. 1992; Kempin et al. 1995; Gu et al. 1998). Only *AP1* is involved in specifying sepal and petal identity and thus, by definition, works as a class A floral organ identity gene (Mandel et al. 1992). In fact, *AP1* from *Arabidopsis* is the only MADS-box gene so far where a class A gene function can be genetically separated from other functions, such as specifying floral meristem identity.

Plants homozygous for mutations in *CAL* are phenotypically wild type, but together with mutations in *AP1*, the typical 'cauliflower' phenotype occurs, which is characterized by an extensive proliferation of meristems at each position that in wild type would give rise to a single flower (Kempin et al. 1995). These data indicate that *CAL* and *AP1* encode partially redundant functions in the specification of floral meristem identity. According to its single mutant phenotype *AGL8* is required for the normal pattern of cell division, expansion, and differentiation during morphogenesis of the silique, and was therefore renamed *FRUITFULL* (*FUL*) (Gu et al. 1998). The major part of the silique is provided by the carpel valves, and it appears that *FUL* is a valve identity gene (Gu et al. 1998). Although *ful* single mutants do not have an inflorescence-specific phenotype, it is also one of the first genes expressed after the transition to flowering at the inflorescence apex. In line with this, *ap1cal ful* triple mutants show an extreme enhancement of the *ap1 cal* phenotype, indicating partial functional redundancy between the *AP1*, *CAL* and *FUL* genes during flower initiation (Ferrándiz et al. 2000).

Phylogeny reconstructions suggest that the ancestral function of *SQUA*-like genes was in specifying inflorescence or floral meristem identity, which is maintained in many extant genes, and that functions in specifying sepal and petal, or fruit valve identity are probably derived (Theißen 2000; Theißen et al. 2000). Therefore, one may expect *SQUA*-type floral meristem identity genes in rice, but not necessarily class A gene functional equivalents.

Three different *SQUA*-like genes have been identified in the rice genome so far, termed *OsMADS14*, *OsMADS15* and *OsMADS18* (Table 1, Fig. 1). A number of other sequences published under other names probably represent the same genes; a few differences in sequences, if any, may reflect sequencing errors or allelic diversity, especially since sequences have often been obtained from different rice subspecies (*japonica* and *indica*).

OsMADS14 (Moon et al. 1999b) represents very likely the same genes as *FDRMADS6* (Jia et al. 2000) and *RAP1B* (Kyozuka et al. 2000). *OsMADS14* expression is confined to inflorescences. Expression occurs throughout the spikelet meristem. After organ primordia have been generated, *OsMADS14* expression becomes restricted to the sterile glumes, palea and lemma. Surprisingly, in the mature spikelet, *OsMADS14* expression is shut off in the sterile organs of the spikelet, and switches to the reproductive organs, i.e., stamens and carpels (Jia et al. 2000; Pelucchi et al. 2002).

OsMADS15 (Lim et al. 2000) is very likely the same gene as *RAP1A* (Kyozuka et al. 2000). *OsMADS15* is expressed throughout the spikelet meristem. After initiation of the different organs of the spikelet transcript accumulation becomes restricted to the non-reproductive organs, i.e., lodicules, palea, lemma, and empty glumes, with only a weak expression in the latter organs (Kyozuka et al. 2000). *SQUA*-like genes with a similar expression pattern have also been found in maize (Münster et al. 2002).

OsMADS18 (Moon et al. 1999b) is probably allelic with *FDRMADS7* (Jia et al. 2000) and *OsMADS28*. In contrast to the other rice *SQUA*-like genes, it has a quite ubiquitous expression in many rice organs and tissues. Within spikelets, *OsMADS18* is expressed in all organs except for very late stages, where transcript accumulation is not detectable in lodicules and sterile glumes (Fornara et al. 2003).

The question arises as to whether rice has a class A floral homeotic gene as defined by function (i.e., independent of gene relationships to class A genes from eudicots). Since grasses like rice have no sepals and petals, they should not be expected to have class A floral organ identity genes in a strict sense. And even in eudicots a function in the specification of perianth organ identity is hard to separate from evolutionary probably more ancient functions in providing meristem identity (Theißen et al. 2000). Nevertheless, if we assume that palea and lemma are homologues of eudicot sepals, and lodicules are petal homologues, some *SQUA*-like gene might be involved in the specification of these organs in rice. Unfortunately, loss-of-function mutants for any of the rice *SQUA*-like genes have not been reported yet. If we use the mRNA expression patterns as criteria, *OsMADS15* might be the best candidate for a class A-like gene involved in specifying the identity of lodicules, palea and lemma.

2.2 *DEF*-Like and *GLO*-Like Genes (Class B Candidate Genes)

In higher eudicots, including *Arabidopsis*, class B floral organ identity genes are involved in specifying the identity of petals and stamens. Class B gene loss-of-function mutants have petals replaced by sepals, and stamens by carpels. Studies on extant gymnosperms revealed that class B genes may have an ancestral function in specifying male reproductive organs (where B gene expression is on) and distinguishing them from female reproductive organs (B gene expression off) that was probably established already in the last common ancestor of extant seed plants about 300 million years ago (Albert et al. 1998; Mouradov et al. 1999; Sundström et al. 1999; Winter et al. 1999, 2002a; Fukui et al. 2001). After the establishment of distinct *DEF*- and *GLO*-like genes by gene duplication during angiosperm evolution, they were recruited also for specifying petal identity, and their gene products evolved from homodimerizing into obligatory heterodimerizing proteins (Winter et al. 2002b).

There is just one *DEF*-like and one *GLO*-like gene in the *Arabidopsis* genome, termed *APETALA3* (*AP3*) and *PISTILLATA* (*PI*), respectively (Table 1, Fig. 1; Jack et al. 1992; Goto and Meyerowitz 1994). Two *GLO*-like genes, termed *OsMADS2* and *OsMADS4*, and one *DEF*-like gene, *OsMADS16*, have been reported so far from rice (Chung et al. 1995; Moon et al. 1999a).

The presence of a paralogous pair of *GLO*-like genes is not a speciality of rice, but traces back to a gene duplication that predates the separation of the lineages that led to extant maize and rice 50–70 million years ago (Münster et al. 2002). *OsMADS2*, *OsMADS4* and *OsMADS16* are expressed in lodicules and stamens. Whether *OsMADS2* and *OsMADS4* are also expressed in carpels is controversial (compare results from Chung et al. 1995, with

Kyozuka et al. 2000), but mRNA accumulation in carpels has also been observed for the *GLO*-like genes of maize (Münster et al. 2001).

Transgenic plants expressing antisense RNA of *OSMADS4* exhibited homeotic transformations of lodicules into palea/lemma-like structures and stamens into carpel-like organs (Kang et al. 1998). A very similar phenotype was observed in mutants termed *superwoman1* (*spw1*), which are affected in *OsMADS16* (Nagasawa et al. 2003), and in the *silky1* mutant, in which the *DEF*-like gene from maize has lost its function (Ambrose et al. 2000). These data indicate that the *DEF*- and *GLO*-like genes from grasses are involved in specifying stamen and lodicule identity and thus suggest that considerable aspects of the *DEF*- and *GLO*-like gene function have been conserved throughout monocot and eudicot evolution. This is especially obvious for the function in the specification of male reproductive organs (stamens), which was very likely established in the most recent common ancestor of angiosperms and gymnosperms already (Theißen et al. 2000; Winter et al. 2002a); there is hardly any doubt that stamens are homologous organs throughout angiosperms.

Since in eudicot class B gene mutants petals are replaced by sepals, the loss-of-function phenotypes of *DEF*- and *GLO*-like genes in rice and maize have been used as evidence that lodicules are homologous to petals, and palea/lemma are homologous to sepals (Kang et al. 1998; Ambrose et al. 2000). While this is certainly a likely possibility, things are unfortunately not necessarily so straightforward. It could also be that orthologous *DEF*- and *GLO*-like genes have been recruited more than once for the specification of organs that originated independently in evolution and are thus not homologous. There is evidence that eudicot-type petals originated from stamens at the base of extant higher eudiots (and are hence termed 'andropetals') (Kramer and Irish 2000); thus, lodicules could have been derived from stamens independently in the lineage that led to extant monocots. This would also nicely explain why *DEF*- and *GLO*-like genes are involved in the specification of both kinds of organs.

Taken together, it is quite clear now that grasses have orthologs of both types of eudicot class B genes (*DEF*- and *GLO*-like genes), which share highly conserved functions in the specification of male reproductive organs with eudicot *DEF*- and *GLO*-like genes. Like their eudicot counterparts, being involved in the specification of petals, these genes are also involved in specifying the identity of the organs immediately outside of the stamens, i.e., the lodicules. Whether lodicules are homologous to petals, however, is not finally clear. The *DEF*- and *GLO*-like genes of grasses, including rice, have thus certainly a class B-like function, but the extent to which this function is related to that of eudicot class B genes is not fully understood.

2.3 *AG*-Like Genes (Class C and Class D Candidate Genes)

There are four different *AGAMOUS*- (*AG*-) like genes in the *Arabidopsis* genome, *AG*, *AGL11*, *SHATTERPROOF1* (*SHP1*) and *SHP2* (Table 1, Fig. 1)

(Yanofsky et al. 1990; Ma et al. 1991; Rounsley et al., 1995). *AG* is a typical class C floral homeotic gene; it is involved in specifying stamen and carpel identity, and in providing floral determinacy. *SHP1* and *SHP2* (formerly known as *AGL1* and *AGL5*, respectively) encode functionally redundant proteins that are required for the proper development of the fruit dehiscence zone of *Arabidopsis* (Liljegren et al. 2000). *AGL11* appears to be orthologous to the class D floral homeotic genes *FBP7* and *FBP11* from petunia (*Petunia hybrida*), which are master control genes of ovule identity (Angenent and Colombo 1996). In line with this, these genes are mainly expressed in developing ovules, which is also the case for *AGL11*. Thus *AGL11* may also be involved in the specification of ovule identity.

Two *AG*-like genes have been found in the rice genome, termed *OsMADS3* (Kang et al. 1995) and *OsMADS13* (Lopez-Dee et al. 1999). *OsMADS3* is very likely allelic to *RAG1* (Kyozuka et al. 2000). *OsMADS3* is expressed in both the stamens and carpels of rice (Kang et al. 1995; Kyozuka et al. 2000), while *OsMADS13* transcription is confined to developing ovules and the inner layer of the carpel wall (Lopez-Dee et al. 1999). These data suggest that *OsMADS3* represents a class C floral organ identity gene, while *OsMADS13* is a class D gene. Further studies enforce a more differentiated view, however.

In rice plants in which *OsMADS3* expression was silenced by an antisense approach, incomplete transformations of stamens into lodicules were observed, while carpels were replaced by abnormal flowers with undifferentiated stamens and carpels (Kang et al. 1998). These observations are compatible with a class C function of *OsMADS3*, assuming that there is functional redundancy in the rice genome, or, more likely, that *OsMADS3* gene function was not completely abolished in the antisense lines. However, Kang et al. (1998) observed also that lodicules were transformed into several types of unusual organs, including stamen-like structures, in the *OsMADS3* antisense lines. This may suggest that *OsMADS3* is also required for proper lodicule development, which would go beyond the classical class C floral homeotic function in stamens and carpels. Ectopic expression of *OsMADS3* in rice, but also in tobacco leads to the transformation of lodicules or petals, respectively, into stamen-like organs (Kang et al. 1995; Kyozuka and Shimamoto 2002). Thus, gain-of-function studies corroborate the view based on loss-of-function analyses that *OsMADS3* specifies reproductive organ identity.

Despite its expression in rice wild-type ovules, ectopic expression of *OsMADS13* in rice and *Arabidopsis* did not generate ectopic ovules; however, the same is also true for *FBP7* and *FBP11* in *Arabidopsis* (Favaro et al. 2002). Like the FBP7 and FBP11 proteins, however, OsMADS13 also interacts specifically with AGL2-like MADS-domain proteins, as revealed by yeast two-hybrid analyses, suggesting that FBP7/FBP11 and OsMADS13 share a conserved function. In line with its expression pattern, OsMADS13 may thus also have a class D function in the specification of ovules.

Phylogeny reconstructions (G. Theißen and A. Becker, unpubl.) suggest, however, that this class D function in monocots was not directly derived from the class D gene function represented in extant eudicots. Members of the clade of *AG*-like genes have been found in all major groups of seed plants (angiosperms, gnetophytes, conifers, cycads and *Ginkgo*), but not in nonseed plants so far (Theißen et al. 2000 and unpubl. data), suggesting that this clade was established 300–400 million years ago (Theißen et al. 2000). In diverse groups of gymnosperms *AG*-like genes are expressed in male and female reproductive organs (Tandre et al. 1995, 1998; Rutledge et al. 1998; Winter et al. 1999), which may thus represent the ancestral state of gene expression. It has been suggested that the ancestral function of these genes may be to specify reproductive organs (both male and female), and to distinguish them from nonreproductive organs (Haughn and Somerville 1988; Winter et al. 1999). A gene duplication close to the base of angiosperms may have led to the establishment of two clades: (1) a clade of ovule-specific genes, represented by the class D genes of extant eudicots, probably including *AGL11* from *Arabidopsis*; (2) and a clade of genes which remained expressed in both male and female reproductive organs. Members of this latter clade were recruited for specification of the carpel, and thus gave rise to the class C floral homeotic genes. According to phylogeny reconstructions, it appears likely that the ancestral ovule-specific genes have been lost in the lineage that led to monocots after the separation from the lineage that led to extant eudicots about 160–200 million years ago. Ovule- or carpel-specific angiosperm genes (monocot class D genes) may have originated a second time by splitting from the class C gene lineage close to the base of monocots. Thus, although there are ovule- or carpel-specific *AG*-like genes in both eudicots (e.g., *FBP7*) and monocots (*OsMADS13*, *ZAG2*, *ZMM1*) (Schmidt et al. 1993; Theißen et al. 1995; Lopez-Dee et al. 1999), phylogeny reconstructions suggest that they are not monophyletic, but that such genes have been lost and reestablished in the lineage that led to extant monocots. However, because of uncertainties of previous phylogeny reconstructions, monophyly of class D genes cannot be completely ruled out.

Taking together, some *AG*-like genes in angiosperms, such as *AG*, have probably an ancestral function in specifying both male and female reproductive organs, and some others, including *AGL11*, *SHP1* and *SHP2*, have derived or specialized functions restricted to female reproductive organs (including ovules and fruits) (Theißen 2000). Similarly, *OsMADS3* function may resemble more closely the broad ancestral function in the specification of both male and female reproductive organs; after a gene duplication subfunctionalization may have led to a ‘reinvention’ of an ovule-specific gene (*OsMADS13*), while the gene lineage that led to *OsMADS3* maintained its functions in carpels and stamens. The function of class D genes in ovules could also well be quite ancient (since gymnosperms have ovules, but not carpels), but at gymnosperm level it was not and is not separated from an additional function of *AG*-like genes in male reproductive organs. Thus the grasses, including rice, may have class C genes which are orthologous to those of the eudicots, while they may have genes with class D functions in ovule development that are members of the same gene subfamily, but which are possibly not orthologous to their putative eudicot functional equivalents (Table 1; Fig. 1).

2.4 *AGL2*-Like Genes (Class E Candidate Genes)

There are four *AGL2*-like genes in the *Arabidopsis* genome, termed *AGL2*, *AGL3*, *AGL4* and *AGL9* (Table 1, Fig. 1; Ma et al. 1991; Mandel et al. 1998). While *AGL3* is expressed in all major plant organs above ground (Huang et al. 1995), expression of the other genes is restricted to second, third and fourth whorl floral organ primordia (*AGL9*), or to floral organs of all four whorls (*AGL2* and *AGL4*) (Flanagan and Ma 1994; Savidge et al. 1995; Mandel and Yanofsky 1998). The similarity in both sequences and expression patterns between the *AGL2*-like genes from *Arabidopsis* (except *AGL3*) suggested that they encode proteins of redundant functions, and indeed, single gene mutations generated by reverse genetics produced only very weak phenotypes (Pelaz et al. 2000). However, in triple mutants where *AGL2*, *AGL4* as well as *AGL9* had lost their function, all flower organs resembled sepals, and the flowers became indeterminate (Pelaz et al. 2000). These findings indicate that *AGL2/4/9* have redundant functions required for petal, stamen and carpel development, and to prevent the indeterminate growth of the flower meristem. The triple mutant phenotype prompted Pelaz et al. (2000) to rename the genes *SEPALLATA1* (*SEP1*), *SEP2* and *SEP3*, respectively, meaning 'lots of sepals'. The *sep1/2/3* triple mutant phenotype is very similar to that of double mutants in a class B and a C class floral homeotic gene. However, *SEP1/2/3* are still expressed in B and C loss-of-function mutants, and the initial expression patterns of B and C class genes are not significantly altered in the *sep1/2/3* triple mutant. This indicates that *SEP1/2/3* do not act downstream of the class B and class C genes, and that they are not required for the initial activation of these genes (Pelaz et al. 2000).

These data suggest that the *SEP* genes constitute an additional class of floral homeotic genes, termed class E genes (Theißen 2001b). The SEP proteins constitute higher order complexes together with class A, B or C proteins, and certain versions of these complexes are obviously sufficient to transform leaves into floral organs; specifically, ectopic expression of class A or class E (*SQUA*- or *AGL2*-like) and class B (*DEF*- and *GLO*-like) genes results in the transformation of vegetative leaves into petals, while the expression of class B, class C (*AG*-like) and class E genes results in the transformation of leaves into stamen-like organs (Honma and Goto 2001). Therefore, the gene-based combinatorial ABC model of floral organ identity was refined to a protein-based combinatorial 'floral quartet model' (Theißen 2001b; Theißen and Saedler 2001). The 'floral quartet model' explains how the different floral organ identity genes interact at the molecular level, namely via their gene products, which form quaternary, DNA-binding transcription factors that may bind to two *cis*-regulatory elements termed CArG-boxes (consensus: 5'-CC(A/T)$_6$GG-3').

Table 1. *AG*-like, *AGL2*-like, *DEF*-like, *GLO*-like and *SQUA*-like MADS-box genes from *Arabidopsis* and rice. Genes are ordered alphabetically by gene subfamilies followed by gene names

Gene name[a]	Species[b]	Gene sub-family[c]	Synonymous gene name[d]	Acc. No. or MIPS code[e]	Gene function[f]
AG (Yanofsky et al. 1990)	*Arabidopsis*	AG		At4g18960	Class C floral homeotic gene
AGL1 (Ma et al. 1991)	*Arabidopsis*	AG	*SHP1* (Liljegren et al. 2000)	At3g58780	Fruit dehiscence zone development
AGL5 (Ma et al. 1991)	*Arabidopsis*	AG	*SHP2* (Liljegren et al. 2000)	At2g42830	Fruit dehiscence zone development
AGL11 (Rounsley et al. 1995)	*Arabidopsis*	AG		At4g09960	Candidate class D floral homeotic gene
OsMADS3 (Kang et al. 1995)	Rice	AG	*RAG* (Kyozuka et al. 2000)	AAA99964	Class C floral homeotic gene
OsMADS13 (Lopez-Dee et al 1999)	Rice	AG		AAF13594	Candidate class D floral homeotic gene
AGL2 (Ma et al. 1991)	*Arabidopsis*	AGL2	*SEP1* (Pelaz et al. 2000)	At5g15800	Class E floral homeotic gene
AGL3 (Ma et al. 1991)	*Arabidopsis*	AGL2		At2g03710	
AGL4 (Ma et al. 1991)	*Arabidopsis*	AGL2	*SEP2* (Pelaz et al. 2000)	At3g02310	Class E floral homeotic gene
AGL9 (Mandel and Yanofsky 1998)	*Arabidopsis*	AGL2	*SEP3* (Pelaz et al. 2000)	At1g24260	Class E floral homeotic gene
OsMADS1 (Chung et al 1994)	Rice	AGL2	*LHS1* (Jeon et al. 2000)	AAA66187	Floral meristem determinacy; identity of palea, lemma, lodicules; number of stamens and carpels
OsMADS5 (Kang and An 1997)	Rice	AGL2	*FDRMADS2*	AAB71434	
OsMADS24 (Greco et al. 1997)	Rice	AGL2	*OsMADS8* (Kang et al. 1997)	AAB53193	Candidate class E-like floral homeotic gene

Table 1. *Continued*

Gene name[a]	Species[b]	Gene sub-family[c]	Synonymous gene name[d]	Acc. No. or MIPS code[e]	Gene function[f]
OsMADS34 (Shinozuka et al. 1999; Pelucchi et al. 2002)	Rice	AGL2		BAA81882	
OsMADS45 (Greco et al. 1997)	Rice	AGL2	*OsMADS7* (Kang et al. 1997)	AAB50180	Candidate class E-like floral homeotic gene
AP3 (Jack et al. 1992)	*Arabidopsis*	DEF		At3g54340	Class B floral homeotic gene
OsMADS16 (Moon et al. 1999a)	Rice	DEF	*SPW1* (Nagasawa et al. 2003)	AAD19872	Class B-like floral homeotic gene
OsMADS2 (Chung et al 1995)	Rice	GLO	*NMADS1* (Yuan et al. 2000)	AAB52709	Candidate class B-like floral homeotic gene
OsMADS4 (Chung et al 1995)	Rice	GLO		AAC05723	Class B-like floral homeotic gene
PI (Goto and Meyerowitz 1994)	*Arabidopsis*	GLO		At5g20240	Class B floral homeotic gene
AGL8 (Rounsley et al. 1995)	*Arabidopsis*	SQUA	*FUL* (Gu et al. 1998)	At5g60910	Floral meristem and fruit valve identity
AP1 (Mandel et al. 1992)	*Arabidopsis*	SQUA	*AGL7*	At1g69120	Floral meristem identity and class A floral homeotic gene
CAL (Kempin et al. 1995)	*Arabidopsis*	SQUA	*AGL10*	At1g26310	Floral meristem identity gene
OsMADS14 (Moon et al. 1999b)	Rice	SQUA	*RAP1B* (Kyozuka et al. 2000), *FDRMADS6* (Jia et al. 2000)	AAF19047	
OsMADS15 (Moon et al. 1999b)	Rice	SQUA	*RAP1A* (Kyozuka et al. 2000)	AAF19048	Candidate class A-like floral homeotic gene

Table 1. *Continued*

Gene name[a]	Species[b]	Gene subfamily[c]	Synonymous gene name[d]	Acc. No. or MIPS code[e]	Gene function[f]
OsMADS18 (Moon et al. 1999b)	Rice	SQUA	*OSMADS28*, *FDRMADS7* (Jia et al. 2000)	AAF04972	

[a]Gene name: in case of synonymous gene names, the first published gene name is taken as the primary name, even when genes have been intentionally renamed.
[b]Species: *Arabidopsis*, *Arabidopsis thaliana*; rice, *Oryza sativa* (note that *FDRMADS* sequences have been isolated from subspecies *indica* rather than *japonica*).
[c]Gene subfamily: subfamilies represent special, defined gene clades and are named after the first identified member, whether or not from *Arabidopsis* (Theißen et al. 1996).
[d]Synonymous gene name: see gene name.
[e]Acc. No. or MIPS code: in case of *Arabidopsis* genes, the MIPS code rather than accession numbers are given. Information about the MIPS code is available at the following URL: http://www.mips.biochem.mpg.de/proj/thal/db/about/agicodes.html. In case of rice, the accession numbers of protein sequences are given.
[f]Gene function: gene function is classified or the functional context of gene action is indicated. A function is only given when defined by mutant phenotype. In case of *AGL1*, *AGL5*; *AP1*, *CAL* (for *CAL*); and *AGL2*, *AGL4* and *AGL9*, the informative phenotypes are double or triple mutant phenotypes, respectively, due to the redundancy of the genes. If the function of a gene has not been revealed by mutant analysis, but is hypothesized based on gene relationships and expression patterns only, the gene is called a 'candidate' for a respective function. Class A, B, C, D and E floral homeotic genes are involved in specifying the identity of sepals and petals (A), petals and stamens (B), stamens and carpels (C), ovules (D), or petals, stamens and carpels (E), respectively. If a function appears similar to the ABCDE functions, but involves (also) organs that are different from eudicot flower organs, the function is called class A-, B- or E-like, irrespective of hypothetical organ homology.

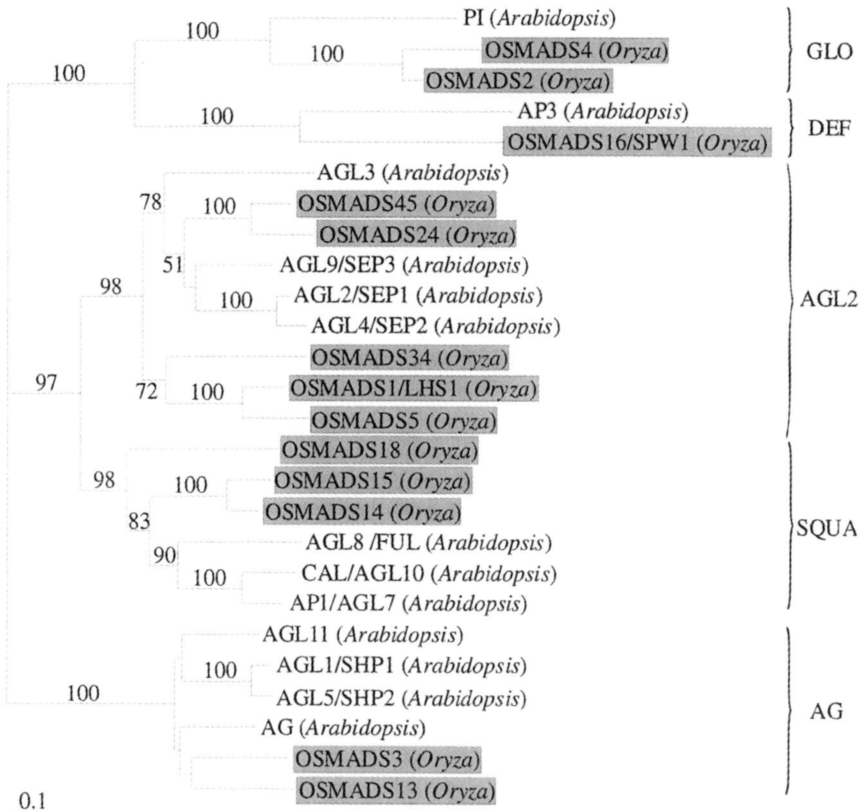

Fig. 1. Phylogenetic tree showing the relationships between AG-like, AGL2-like, DEF-like, GLO-like and SQUA-like MADS-domain proteins from *Arabidopsis* and rice. The tree was constructed using conceptual MIK-domain sequences aligned by CLUSTALX and the neighbor-joining algorithm. OSMADS sequences are from rice (*Oryza*), all others from *Arabidopsis*. The *numbers* next to some nodes give bootstrap percentages, shown only for relevant nodes and those defining gene subfamilies. Subfamilies, which generally represent monophyletic gene clades (Theißen et al. 1996; Münster et al. 1997), are labeled by *brackets* at the right margin

AGL2-like genes in grasses show relatively heterogeneous expression patterns, strongly suggesting that they are also functionally not a homogeneous class of genes. It seems that this subfamily underwent a relatively rapid functional diversification within the grasses, which possibly reflects the evolution of complex inflorescence structures (Theißen et al. 1996; Cacharrón et al. 1999).

Five different *AGL2*-like gene have been identified in the rice genome, termed *OsMADS1*, *OsMADS5*, *OsMADS24*, *OsMADS34* and *OsMADS45* (Chung et al. 1994; Greco et al. 1997; Kang and An 1997; Pelucchi et al. 2002).

OsMADS7 and *OsMADS8* (Kang et al. 1997) are allelic to *OsMADS45* and *OsMADS24*, respectively (Table 1); *OsMADS1* corresponds to the *leafy hull sterile1* (*lhs1*) mutant of rice (Jeon et al. 2000). For each of the five different genes mentioned, one or two putative orthologs from maize have been identified, indicating that the most recent common ancestor of maize and rice about 50–70 million years ago had already at least five different *AGL2*-like genes (Münster et al. 2002).

Only for one *AGL2*-like gene from rice a mutant analysis has been carried out so far. *lhs1* rice plants in which *OSMADS1* is mutated often develop several florets within the spikelets, while wild-type rice spikelets have only one floret (Jeon et al. 2000). This suggests that *OSMADS1/LHS1* plays an important role in floral meristem determination during the early development of rice florets (Jeon et al. 2000). Other phenotypic characteristics of plants with missense mutations in *OSMADS1* are elongated papery leafy paleas and lemmas, palea- and lemma-like lodicules, fewer stamens, and an increase in carpel number. The changes in lodicules and reproductive organs resemble class B loss-of-function mutants in grasses (Kang et al. 1998; Ambrose et al. 2000), suggesting that *OsMADS1* is required for this function. In addition to *OsMADS1* loss-of-function plants, putative gain-of-function plants have been studied. When *OsMADS1* is ectopically expressed, stunted panicles with irregular positioned branches and spikelets develop, and within spikelets, the glumes are transformed into palea/lemma-like organs (Prasad et al. 2001). Thus *OsMADS1* may be sufficient to confer palea/lemma identity to glumes, while the *lhs1* mutant suggests that it is not generally required for the specification of these organs. During spikelet development, *OsMADS1* expression is confined to the lemma, palea and carpel, but expression in the carpel is only weak (Chung et al. 1994; Prasad et al. 2001). This points again to the importance of *OsMADS1* for the development of palea and lemma. So even if we homologize these organs to the eudicot sepals, *OsMADS1* function would remarkably deviate from the class E gene function, which is confined to petal, stamen and carpel development. This is in line with phylogeny reconstructions, which indicate that within *AGL2*-like genes, *OsMADS24* and *OsMADS45* are more closely related to the *Arabidopsis* class E (*SEP*) genes than *OsMADS1*, *OsMADS5* and *OsMADS34* (Fig. 1; Münster et al. 2002). This again fits to the expression patterns known.

OsMADS5 and *OsMADS34* are transcribed in stamens and (weakly) in carpels, or in all plant tissues, respectively (Kang and An 1997; Pelucchi et al. 2002), which also makes them no obvious candidates for class E genes in a strict sense. *OsMADS24* and *OsMADS45* are expressed in lodicules, stamens and carpels (Greco et al. 1997). These genes thus show expression patterns very similar to that of *AGL9* (especially if we homologize lodicules to petals); they are thus better class E candidate genes than the other *AGL2*-like genes from rice.

Unfortunately, mutant phenotypes are not known for any *AGL2*-like gene from rice except *OsMADS1/LHS1*. Moreover, conservation of class E genes is still unclear even in eudicots, and both redundancy (*Arabidopsis*) and subfunctionalization into whorl-specific genes (*Gerbera*) complicate the definition of the class E gene function (Pelaz et al. 2000; Kotilainen et al. 2000; Theißen 2001b). Obviously, more work is needed to determine to which extent the functions of *AGL2*-like genes in rice resemble the class E gene function in *Arabidopsis*. A general key to the function of the AGL2-like proteins could be their ability to interact with members of other subfamilies in heterodimeric and higher order protein complexes ('floral quartets') (Fan et al. 1997; Honma and Goto 2001; Theißen 2001b; Theißen and Saedler 2001).

3 Concluding Remarks

Assuming conservation of the basic molecular mechanisms of flower development in angiosperms candidates for floral homeotic genes from rice are easily found by identifying putative rice orthologs of the floral homeotic genes from *Arabidopsis* (and other eudicots, such as *Antirrhinum* and *Petunia*). In first approximation these are identified in phylogenetic trees as members of the same MADS-box gene subfamilies (Theißen et al. 1996, 2000; Münster et al. 2002). This does not imply, however, that the functions of the rice genes thus identified are equivalent or even identical to their eudicot orthologs. There are several reasons for this:

1. Gene functions evolve, and even orthologous genes can functionally diverge during evolution (Theißen 2002b). Independent gene duplications in the lineages that led to extant eudicots and monocots have established orthologous relationships between clades of paralogous genes rather than individual genes (Münster et al. 2002), suggesting that subfunctionalization may have led to functional divergence between rice and *Arabidopsis* MADS-box genes.
2. The evolution of genes and morphological structures are not necessarily congruent, and homology between the organs of *Arabidopsis* flowers and of rice spikelets is unclear in some cases. For example, the expression of *DEF*- and *GLO*-like genes in the lodicules of rice and other grasses and in eudicot petals could well reflect homology between lodicules and petals, but would also be compatible with an independent origin of lodicules from stamens in the lineage that led to extant grasses. The class B-like function of *DEF*- and *GLO*-like genes in grasses and eudicots may thus involve exclusively the specification of homologous organs, i.e. stamens, and lodicules or petals (class B-like function in a strict sense), or it may include the specification of nonhomologous organs (lodicules and petals) (class B-like function in a less strict sense).
3. One trivial obstacle for a comprehensive comparison of floral homeotic gene functions in *Arabidopsis* and rice is, despite the 'model' character of both species, the fact that the extent to which candidate genes have been studied is still quite different in both species. For example, among the 13 *AG*-, *AGL2*-, *DEF*-, *GLO*- and *SQUA*-like genes in the *Arabidopsis* genome (candidate or proven class C+D, E, B, and A genes, respectively),

a mutant phenotype has been published for all except two genes (*AGL11*, *AGL3*). In contrast, for rice, the same total number of members of the mentioned gene subfamilies has been reported so far, but loss-of-function phenotypes have been reported for just four genes (*OsMADS1*, *OsMADS3*, *OsMADS4*, *OsMADS16*) (see Table 1). Thus while the functions of most *Arabidopsis* genes could be determined relatively precisely by the phenotypes of single or multiple null mutants, often only expression patterns, or gain-of-function phenotypes (mostly even only in heterologous plants such as tobacco) give clues to the function of the corresponding rice genes (e.g., Kang and An 1997). So quite a lot of work lies ahead before a detailed and comprehensive comparison of floral homeotic gene functions between rice and *Arabidopsis* can be done.

There is hardly any doubt, however, that intensive comparative studies of the floral homeotic genes (or their orthologs) from *Arabidopsis* and rice (Table 1) will tell us a great deal about the conservation and divergence of flower development throughout the angiosperms, especially if these are combined with comparative investigations on organ development and structure. One of the most relevant questions to answer is as to how gene evolution and morphological evolution are interlinked. The fact that gene evolution is often almost 'clock-like', while morphological evolution can be erratic, is one of the most intriguing observations to be explained. Studies on the small number of floral homeotic genes, and their huge impact on organogenesis, promise a better understanding of this enigmatic topic (Theißen et al. 2000). The 160–200 million years of evolutionary separation, the apparent differences between their flowers, and the obvious similarities between their (putative or proven) floral homeotic genes make rice and *Arabidopsis* an obvious and attractive choice for studies on the relationship between the evolution of genes and morphological structures. A better understanding of this relationship could not only help us to better understand the molecular genetic basis of biodiversity on earth; it might also help us to eventually develop homeotic genes into tools for the improvement of crop plants by transgenic technology, or mutagenesis and marker assisted breeding.

References

Albert VA, Gustafsson MHG, Di Laurenzio L (1998) Ontogenetic systematics, molecular developmental genetics, and the angiosperm petal. In: Soltis DE, Soltis PS, Doyle JJ (eds) Molecular systematics of plants II. Kluwer Academic, Boston, pp 349–374

Ambrose BA, Lerner DR, Ciceri P, Padilla CM, Yanofsky MF, Schmidt RF (2000) Molecular and genetic analyses of the *Silky1* gene reveal conservation in floral organ specification between eudicots and monocots. Mol Cell 5:569–579

Angenent GC, Colombo L (1996) Molecular control of ovule development. Trends Plant Sci 1:228–232

Arabidopsis Genome Initiative (2000) Analysis of the genome sequence of the flowering plant *Arabidopsis thaliana*. Nature 408:796–815

Baum DA, Doebley J, Irish VF, Kramer EM (2002) Response: missing links: the genetic architecture of flower and floral diversification. Trends Plant Sci 7:31–34

Cacharrón J, Saedler H, Theißen G (1999) Expression of the MADS-box genes *ZMM8* and *ZMM14* during inflorescence development of *Zea mays* discriminates between the upper and the lower floret of each spikelet. Dev Genes Evol 209:411–420

Chung Y-Y, Kim S-R, Finkel D, Yanofsky MF, An G (1994) Early flowering and reduced apical dominance result from ectopic expression of a rice MADS box gene. Plant Mol Biol 26:657–665

Chung Y-Y, Kim S-R, Kang H-G, Noh Y-S, Park MC, Finkel D, An G (1995) Characterization of two rice MADS box genes homologous to *GLOBOSA*. Plant Sci 109:45–56

Clifford HT (1987) Spikelet and floral morphology. In: Soderstrom TR, Hilu KW, Campbell CS, Barkworth ME (eds) Grass systematics and evolution. Smithsonian Institution Press, Washington, DC, pp 21–30.

Doyle JJ (1994) Evolution of a plant homeotic multigene family: towards connecting molecular systematics and molecular developmental genetics. Syst Biol 43:307–328

Fan HY, Hu Y, Tudor M, Ma H (1997) Specific interactions between the K domains of AG and AGLs, members of the MADS domain family of DNA binding proteins. Plant J 12:999–1010

Favaro R, Immink RGH, Ferioli V, Bernasconi B, Byzova M, Angenent GC, Kater M, Colombo L (2002) Ovule-specific MADS-box proteins have conserved protein-protein interactions in monocot and dicot plants. Mol Genet Genom 268:152–159

Feng Q, Zhang Y, Hao P et al. (2002) Sequence and analysis of rice chromosome 4. Nature 420:316–320

Ferrándiz C, Gu Q, Martienssen R, Yanofsky MF (2000) Redundant regulation of meristem identity and plant architecture by *FRUITFULL, APETALA1* and *CAULIFLOWER*. Development 127:725–734

Flanagan CA, Ma H (1994) Spatially and temporally regulated expression of the MADS-box gene *AGL2* in wild-type and mutant *Arabidopsis* flowers. Plant Mol Biol 26:581–595

Fornara F, Marziani G, Mizzi L, Kater M, Colombo L (2003) MADS-box genes controlling flower development in rice. Plant Biol 5:16-22

Fukui M, Futamura N, Mukai Y, Wang Y, Nagao A, Shinohara K (2001) Ancestral MADS box genes in Sugi, *Cryptomeria japonica*, D. Don (Taxodiaceae), homologous to the B function genes in angiosperms. Plant Cell Physiol 42:566–575

Goff SA, Ricke D, Lan TH et al. (2002). A draft sequence of the rice genome (*Oryza sativa* L. ssp. *japonica*). Science 296:92–100

Goto K, Meyerowitz EM (1994) Function and regulation of the Arabidopsis floral homeotic gene *PISTILLATA*. Genes Dev 8:1548–1560

Greco R, Stagi L, Colombo L, Angenent GC, Sari-Gorla M, Pe ME (1997) MADS box genes expressed in developing inflorescences of rice and sorghum. Mol Gen Genet 253:615–623

Gu Q, Ferrándiz C, Yanofsky MF, Martienssen R (1998) The *FRUITFULL* MADS-box gene mediates cell differentiation during *Arabidopsis* fruit development. Development 125:1509–1517

Haughn GW, Somerville CR (1988) Genetic control of morphogenesis in *Arabidopsis*. Dev Genet 9:73–89

Honma T, Goto K (2001) Complexes of MADS-box proteins are sufficient to convert leaves into floral organs. Nature 409:525–529

Huang H, Tudor M, Weiss CA, Hu Y, Ma H (1995) The *Arabidopsis* MADS-box gene *AGL3* is widely expressed and encodes a sequence-specific DNA-binding protein. Plant Mol Biol 28:549–567

Jack T, Brockmann LL, Meyerowitz EM (1992) The homeotic gene *APETALA3* of *Arabidopsis thaliana* encodes a MADS box and is expressed in petals and stamens. Cell 68:683–697

Jeon J-S, Jang S, Lee S, Nam J, Kim C, Lee S-H, Chung Y-Y, Kim S-R, Lee YH, Cho Y-G, An G (2000) *leafy hull sterile1* is a homeotic mutation in a rice MADS box gene affecting rice flower development. Plant Cell 12:871–884

Jia H-W, Chen R, Cong B, Cao K-M, Sun C-R, Luo D (2000) Characterization and transcriptional profiles of two rice MADS-box genes. Plant Sci 155:115–122

Kang H-G, An G (1997) Isolation and characterization of a rice MADS box gene belonging to the *AGL2* gene family. Mol Cells 7:45–51

Kang H-G, Noh Y-S, Chung Y-Y, Costa MA, An K, An G (1995) Phenotypic alterations of petal and sepal by extopic expression of a rice MADS box gene in tobacco. Plant Mol Biol 29:1–10

Kang H-G, Jang S, Chung JE, Cho YG, An G (1997) Characterization of two rice MADS box genes that control flowering time. Mol Cells 7:559–566

Kang H-G, Jeon J-S, Lee S, An G (1998) Identification of class B and class C floral organ identity genes from rice plants. Plant Mol Biol 38:1021–1029

Kempin SA, Savidge B, Yanofsky MF (1995) Molecular basis of the *cauliflower* phenotype in *Arabidopsis.* Science 267:522–525

Kotilainen M, Elomaa P, Uimari A, Albert VA, Yu D, Teeri TH (2000) *GRCD1*, an *AGL2*-like MADS-box gene, participates in the C function during stamen development in *Gerbera hybrida.* Plant Cell 12:1893–1902

Kramer EM, Irish VF (2000) Evolution of the petal and stamen developmental programs: evidence from comparative studies of the lower eudicots and basal angiosperms. Int J Plant Sci 161 (Suppl 6):S29-S40

Kyozuka J, Shimamoto K (2002) Ectopic expression of *OsMADS3*, a rice ortholog of *AGAMOUS*, caused a homeotic transformation of lodicules to stamens in transgenic rice plants. Plant Cell Physiol 43:130–135

Kyozuka J, Kobayashi T, Morita M, Shimamoto K (2000) Spatially and temporally regulated expression of rice MADS box genes with similarity to Arabidopsis class A, B and C genes. Plant Cell Physiol 41:710–718

Liljegren SJ, Ditta GS, Eshed Y, Savidge B, Bowman JL Yanofsky MF (2000) *SHATTERPROOF* MADS-box genes control seed dispersal in *Arabidopsis.* Nature 404:766–770

Lim J, Moon Y-H, An G, Jang SK (2000) Two rice MADS domain proteins interact with OsMADS1. Plant Mol Biol 44:513–527

Lopez-Dee ZP, Wittich P, Pè ME, Rigola D, Buono ID, Sari Gorla M, Kater MM, Colombo L (1999) *OsMADS13*, a novel rice MADS-box gene expressed during ovule development. Dev Genet 25:237–244

Ma H, Yanofsky MF, Meyerowitz EM (1991) *AGL1-AGL6*, an *Arabidopsis* gene family with similarity to floral homeotic and transcription factor genes. Genes Dev 5:484–495

Mandel MA, Yanofsky MF (1998) The *Arabidopsis AGL9* MADS box gene is expressed in young flower primordia. Sex Plant Reprod 11:22–28

Mandel MA, Gustafson-Brown C, Savidge B, Yanofsky MF (1992) Molecular characterization of the *Arabidopsis* floral homeotic gene *APETALA1.* Nature 360:273–277

Meyerowitz EM (1994) Flower development and evolution: new answers and new questions. Proc Natl Acad Sci USA 91:5735–5737

Moon Y-H, Jung J-Y, Kang H-G, An G (1999a) Identification of a rice APETALA3 homologue by yeast two-hybrid screening. Plant Mol Biol 40:167–177

Moon Y-H, Kang H-G, Jung J-Y, Jeon J-S, Sung S-K, An G (1999b) Determination of the motif responsible for interaction between the rice APETALA1/AGAMOUS-LIKE9 family proteins using a yeast two-hybrid system. Plant Physiol 120:1193–1203

Mouradov A, Hamdorf B, Teasdale RD, Kim JT, Winter K-U, Theißen G (1999) A *DEF/GLO*-like MADS-box gene from a gymnosperm: *Pinus radiata* contains an ortholog of angiosperm B class floral homeotic genes. Dev Genet 25:245–252

Münster T, Pahnke J, Di Rosa A, Kim JT, Martin W, Saedler H, Theißen G (1997) Floral homeotic genes were recruited from homologous MADS-box genes preexisting in the common ancestor of ferns and seed plants. Proc Natl Acad Sci USA 94:2415–2420

Münster T, Wingen LU, Faigl W, Werth S, Saedler H, Theißen G (2001) Characterization of three *GLOBOSA*-like MADS-box genes from maize: evidence for ancient paralogy in one class of floral homeotic B-function genes of grasses. Gene 262:1–13

Münster T, Deleu W, Wingen LU, Ouzunova M, Cacharrón J, Faigl W, Werth S, Kim JTT, Saedler H, Theißen G (2002) Maize MADS-box genes galore. Maydica 47:287–301

Nagasawa N, Miyoshi M, Sano Y, Satoh H, Hirano H, Sakai H, Nagato Y (2003) *SUPERWOMAN1* and *DROOPING LEAF* genes control floral organ identity in rice. Development 130:705–718

Pelaz S, Ditta GS, Baumann E, Wisman E, Yanofsky MF (2000) B and C organ identity functions require *SEPALLATA* MADS-box genes. Nature 405:200–203

Pelucchi N, Fornara F, Favalli C, Masiero S, Lago C, Pè E, Colombo L, Kater MM (2002) Comparative analysis of rice MADS-box genes expressed during flower development. Sex Plant Reprod 15:113–122

Prasad K, Sriram P, Kumar CS, Kushalappa K, Vijayraghavan U (2001) Ectopic expression of rice *OsMADS1* reveals a role in specifying the lemma and palea, grass floral organs analogous to sepals. Dev Genes Evol 211:281–290

Purugganan MD, Rounsley SD, Schmidt RJ, Yanofsky MF (1995) Molecular evolution of flower development: diversification of the plant MADS-box regulatory gene family. Genetics 140:345–356

Rounsley SD, Ditta GS, Yanofsky MF (1995) Diverse roles for MADS box genes in Arabidopsis development. Plant Cell 7:1259–1269

Rutledge R, Regan S, Nicolas O, Fobert P, Cote C, Bosnich W, Kauffeldt C, Sunohara G, Seguin A, Stewart D (1998) Characterization of an *AGAMOUS* homologue from the conifer black spruce (*Picea mariana*) that produces floral homeotic conversions when expressed in *Arabidopsis*. Plant J 15:625–634

Sasaki T, Matsumoto T, Yamamoto K et al. (2002) The genome sequence and structure of rice chromosome 1. Nature 420:312–316

Savidge B, Rounsley SD, Yanofsky MF (1995) Temporal relationships between the transcription of two *Arabidopsis* MADS box genes and the floral organ identity genes. Plant Cell 7:721–733

Schmidt RJ, Veit B, Mandel MA, Mena M, Hake S, Yanofsky MF (1993) Identification and molecular characterization of *ZAG1*, the maize homolog of the Arabidopsis floral homeotic gene *AGAMOUS*. Plant Cell 5:729–737

Shimamoto K, Kyozuka J (2002) Rice as a model for comparative genomics of plants. Annu Rev Plant Biol 53:399–419

Shinozuka Y, Kojima S, Shomura A, Ichimura H, Yano M, Yamamoto K, Sasaki T (1999) Isolation and characterization of rice MADS box gene homologues and their RFLP mapping. DNA Res 6:123–129

Soltis DE, Soltis PS, Albert VA, Oppenheimer DG, dePamphilis CW, Ma H, Frohlich MW, Theißen G (2002) Missing links: the genetic architecture of flower and floral diversification. Trends Plant Sci 7:22–31

Sundström J, Carlsbecker A, Svensson ME, Svenson M, Johanson U, Theißen G, Engström P (1999) MADS-box genes active in developing pollen cones of Norway spruce (*Picea abies*) are homologous to the B-class floral homeotic genes in angiosperms. Dev Genet 25:253–266

Tandre K, Albert VA, Sundas A, Engström P (1995) Conifer homologues to genes that control floral development in angiosperms. Plant Mol Biol 27:69–78

Tandre K, Svenson M, Svensson ME, Engström P (1998) Conservation of gene structure and activity in the regulation of reproductive organ development of conifers and angiosperms. Plant J 15:615–623

Theißen G (2000) Shattering developments. Nature 404:711–713

Theißen G (2001a) Flower development, genetics of. In: Brenner S, Miller JH (eds) Encyclopedia of genetics. Academic Press, London, pp 713–717

Theißen G (2001b) Development of floral organ identity: stories from the MADS house. Curr Opin Plant Biol 4:75–85

Theißen G (2001c) *SHATTERPROOF* oil seed rape: a *FRUITFULL* business? MADS-box genes as tools for crop plant design. Biotech News Int 6:13–15

Theißen G (2002a) Key genes of crop domestication and breeding: molecular analyses. In: Esser K, Lüttge U, Beyschlag W, Hellwig F (eds) Progress in Botany 63. Springer Berlin Heidelberg New York, pp 189–203

Theißen G (2002b) Orthology: secret life of genes. Nature 415:741

Theißen G, Saedler H (1995) MADS-box genes in plant ontogeny and phylogeny: Haeckel's 'biogenetic law' revisited. Curr Opin Genet Dev 5:628–639

Theißen G, Saedler H (1999) The golden decade of molecular floral development (1990–1999): a cheerful obituary. Dev Genet 25:181–193

Theißen G, Saedler H (2001) Floral quartets. Nature 409:469–471

Theißen G, Strater T, Fischer A, Saedler H (1995) Structural characterization, chromosomal localization and phylogenetic evaluation of two pairs of *AGAMOUS*-like MADS-box genes from maize. Gene 156:155–166

Theißen G, Kim J, Saedler H (1996) Classification and phylogeny of the MADS-box multigene family suggest defined roles of MADS-box gene subfamilies in the morphological evolution of eukaryotes. J Mol Evol 43:484–516

Theißen G, Becker A, Di Rosa A, Kanno A, Kim JT, Münster T, Winter K-U, Saedler H (2000) A short history of MADS-box genes in plants. Plant Mol Biol 42:115–149

Winter K-U, Becker A, Münster T, Kim JT, Saedler, H, Theißen G (1999) MADS-box genes reveal that genetophytes are more closely related to conifers than to flowering plants. Proc Natl Acad Sci USA 96:7342–7347

Winter K-U, Saedler H, Theißen G (2002a) On the origin of class B floral homeotic genes: functional substitution and dominant inhibition in *Arabidopsis* by expression of an ortholog from the gymnosperm *Gnetum*. Plant J 31:457–475

Winter K-U, Weiser C, Kaufmann K, Bohne A, Kirchner C, Kanno A, Saedler H, Theißen G (2002b) Evolution of class B floral homeotic proteins: obligate heterodimerization originated from homodimerization. Mol Biol Evol 19:587–596

Yanofsky MF, Ma H, Bowman JL, Drews GN, Feldman KA, Meyerowitz EM (1990) The protein encoded by the *Arabidopsis* homeotic gene *AGAMOUS* resembles transcription factors. Nature 346:35–39

Yu J, Hu S, Wang J et al. (2002) A draft sequence of the rice genome (*Oryza sativa* L. ssp. *indica*). Science 296:79–91

Yuan Z, Qian X, Liu J, Liu J, Qian M, Yang J, Liu J (2000) Cloning and characterization of two protein cDNAs encoding rice MADS box. Prog Nat Sci 10:357–363

Prof. Dr. Günter Theißen
Dr. Annette Becker
Friedrich-Schiller-Universität Jena
Lehrstuhl für Genetik
Philosophenweg 12
07743 Jena, Germany
Tel.: +49-3641-949550
Fax: +49-3641-949552
e-mail: guenter.theissen@uni-jena.de

Physiology

Molecular Responses of Halophytes to High Salinity

Dortje Golldack

1 Effects of Salinity on Plants

Excess amounts of salt (NaCl) occur as an abiotic environmental factor in certain habitats, e.g. as salt deserts in arid and semiarid areas, in coastal salt marshes, and along inland saline lakes. Next to the natural occurrence of salinity, salinization of soils has increasingly become a crucial problem for agriculture during the last decades. Salinization affects soils in particular in arid and semi-arid regions as well as in irrigated agronomic areas. For irrigation, availability of high-quality freshwater is often limited and water-containing traces of salt is used instead. Besides, salts are dissolved by irrigation within soil profiles, leading to increased accumulation in the soil surface layers.

Today, about one third of irrigated agricultural land is salinized worldwide (Schachtman and Liu 1999). Soil salinization severely limits agricultural productivity because almost all crop plants are highly sensitive to elevated concentrations of NaCl. Concentrations as low as 25 mM NaCl are not tolerated by many plants and concentrations of 150 mM NaCl are highly toxic for most crop plants (Serrano and Gaxiola 1994; Rubio et al. 1995). A recent example for salt desertification caused by mismanagement of soil irrigation is the Aral Sea Basin where soil salinity affects millions of people depending on agricultural production in this region, resulting in enormous economical and social problems (Saiko and Zonn 2000). The ecological and economic impact of soil salinization has led to intensive research on the physiological and molecular mechanisms of salt adaptation in plants with particular focus on the tolerance mechanisms in naturally salt-tolerant plants (e.g. Hasegawa et al. 2000; Bohnert et al. 2001, Läuchli and Lüttge 2002).

Plants having the genetic potential to grow on saline soils are halophytes that may tolerate elevated salt concentrations up to 500 mM NaCl. The capability to tolerate excess salt concentrations is of polyphyletic origin and an estimated amount of 5,000 to 6,000 angiosperm species from diverse plant families belong to the group of halophytes (Glenn et al. 1999). In

Progress in Botany, Vol. 65

general, halophytes might be subdivided into salt avoiders and salt includers. Salt-avoiding species exclude either excess NaCl by so-called root filtration as in some mangrove species or excrete NaCl via salt glands or vesiculated trichomes, e.g. as known for some halophytes of the Chenopodiaceae. Salt-including halophytes are often succulents, e.g. species of the Aizoaceae that hyperaccumulate salts in the leaves. In succulents, increased water content in the central vacuoles enables the accumulation of large quantities of salt and minimizes NaCl toxicity in the cytoplasmic compartment.

2 Main Mechanisms of Salt Adaptation in Halophytes

Exposure to excess salt results in both hyperosmotic and hyperionic stress effects in plants. High salt concentrations in the soil as well as salt accumulation within the apoplastic compartment cause osmotic stress and water deficit, whereas uptake and accumulation of salt in the cytoplasmic compartment causes damage of the metabolism by increased Na^+ and Cl^- concentrations and altered K^+/Na^+ ratios. Although most halophytes use Na^+ (Fig. 1) and Cl^- as osmolytes by accumulating these ions in the central vacuole, enhanced concentrations of Na^+ and Cl^- in the cytoplasm are highly toxic. According to in vitro studies, concentrations of 100 mM Na^+ inhibit protein synthesis by competing with K^+ for binding sites at proteins (Blumwald et al. 2000).

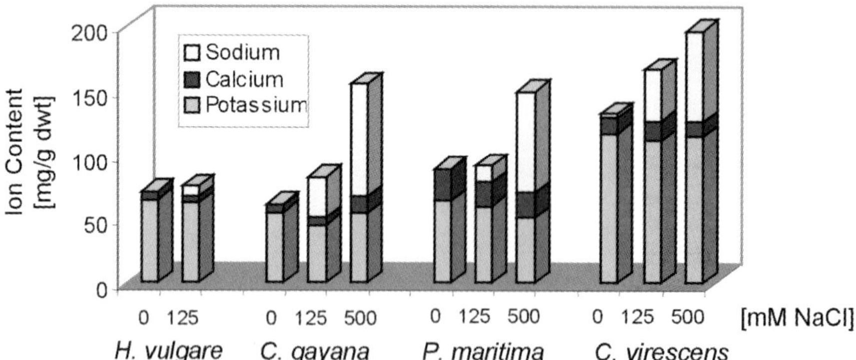

Fig. 1. Accumulation of sodium, calcium, and potassium in the crop plant barley (*Hordeum vulgare*) and in the halophytes *Chloris gayana* (Poaceae), *Plantago maritima* (Plantaginaceae), and *Carpobrotus virescens* (Aizoaceae). The plants were adapted either to control conditions (0 mM NaCl) or treated with 125 and 500 mM NaCl for 48 h

A main strategy of halophytes to cope with osmotic stress is the synthesis of osmolytes ('compatible solutes') and their cytoplasmic accumulation. Most halophytes accumulate the amino acid proline; besides, polyols are common osmolytes. The function of these hydrophilic, low molecular weight organic compounds consists in osmotic balancing as well as structural stabilization of membranes and proteins. The second main mechanism for salt adaptation is the maintenance of the intracellular ion homeostasis with particular importance of regulating the cytoplasmic K^+/Na^+ ratio.

2.1 Yeast Serves as a Model System for Plant Salt Tolerance Mechanisms

Yeast (*Saccharomyces cerevisiae*) has been serving as a model system for studies on the molecular responses of salt adaptation not only in fungi but also in plants due to similarities in the regulation of osmotic and ion homeostasis. Accordingly, molecular key mechanisms of yeast salt adaptation will be reviewed in the following.

Glycerol-3-phosphate dehydrogenase encoded by the genes *GPD1* and *GPD2* is the key enzyme for the synthesis of the compatible solute glycerol and is transcriptionally activated in salt-stressed yeast (Posas et al. 2000). Yeast regulates the cytoplasmic K^+/Na^+ ratio by excluding Na^+ from cells and by accumulation in pre-vacuolar compartments and vacuoles. The export of Na^+ from cells is mediated by the plasma membrane-localized Na^+/H^+-antiporter NHA1 and by the Na^+-ATPase ENA1, whereas the Na^+/H^+-antiporter NHX1 mediates the transport of Na^+ into the vacuolar compartments (e.g. Nass et al. 1997). In addition to Na^+ detoxification, NHX1 serves other functions in yeast as well. Hyperosmotic concentrations of sorbitol inhibited cell division in *nhx1* mutants, indicating involvement of Na^+ sequestered in the vacuoles in osmotically driven import of water via aquaporins (Nass and Rao 1998). Besides, increased Na^+ and H^+ concentrations are also the driving force for Cl^--uptake that is mediated by Cl^- channels of the Gef1 type in yeast (Flis et al. 2002).

Osmotic signaling in yeast is mediated by the HOG (high osmolarity glycerol response) pathway including the *HOG1* gene that codes for a mitogen-activated kinase and causes increased biosynthesis and accumulation of glycerol (Hohmann 2002). In addition, there is a calcineurin involving pathway responding specifically to ionic stress. This pathway regulates the homeostasis of K^+, Na^+, and Ca^{2+} in salt-stressed yeast and causes transcriptional activation of Na^+-ATPases and Ca^{2+}-ATPases (e.g. Matheos et al. 1997).

2.2 Salt-Dependent Regulation of Ion Homeostasis in Halotolerant Plants

Successful salt adaptation of halotolerant plants depends on maintenance of cytoplasmic K^+/Na^+ homeostasis with discrimination of K^+ over Na^+. Although no specific entry system for Na^+ is known on the molecular level yet, competition with the macronutrient K^+ for cellular influx is likely (Blumwald et al. 2000; Fig. 2). Candidates to mediate Na^+ influx are, e.g. voltage-independent monovalent cation channels (VIC; White 1999) that have not been identified on the molecular level yet, and LCT1-type low-affinity cation transporters that have been characterized in wheat (Clemens et al. 1998). HKT1-type transporters isolated from wheat, *Arabidopsis thaliana*, rice, and the halophyte *Mesembryanthemum crystallinum* mediate Na^+ influx when heterologously expressed in *Xenopus* oocytes (Schachtman and Schroeder 1994; Uozumi et al. 2000, Golldack et al. 2002; H. Su, E. Balderas, R. Vera-Estrella, D. Golldack, F. Quigley, C. Zhao, O. Pantoja, and H.J. Bohnert, unpubl. results). In salt-stressed *M. crystallinum* and in the moderately salt-tolerant rice line *Oryza sativa* var. Pokkali a transcriptional decline occurs for HKT1-type K^+-transporters (Golldack et al. 2002; H. Su, E. Balderas, R. Vera-Estrella, D. Golldack, F. Quigley, C. Zhao, O. Pantoja,

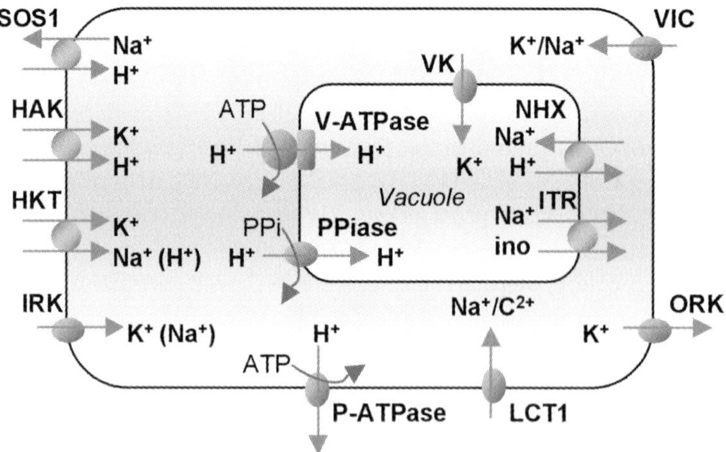

Fig. 2. Ion transport proteins involved in K^+ and Na^+ homeostasis in plant cells. *SOS1* Na^+/H^+ antiporter; *HAK* K^+ transporter; *HKT* K^+/Na^+ symporter or Na^+ transporter; *P-ATPase* P-type H^+-ATPase; *V-ATPase* V-type H^+-ATPase; *PPiase* vacuolar H^+-pyrophosphatase; *IRK* inward-rectifying K^+ channels; *ORK* outward-rectifying K^+ channels; *VIC* voltage-independent cation channels; *VK* vacuolar K^+ channels; *ITR* Na^+/*myo*-inositol symporter; *NHX* Na^+/H^+ antiporter; *LCT1* low-affinity cation transporter. See text for further details

and H.J. Bohnert, unpubl. results) as well as for AKT/KAT-type inward-rectifying K^+ channels (Su et al. 2001; Golldack et al. 2003) that are highly selective for K^+ but do not completely exclude Na^+ (Amtmann and Sanders 1999). In contrast, members of the HAK/KUP-type dual-affinity K^+-transporter family are transcriptionally activated in salt-stressed *M. crystallinum*. When expressed in a *trk1/trk2* yeast mutant deficient in K^+ uptake, McHAK1 and McHAK4 mediated K^+ but not Na^+ influx indicating that McHAK transporters become the major K^+ uptake system in salt-exposed *M. crystallinum* by efficiently excluding Na^+ (Su et al. 2002).

Whereas there is no evidence for Na^+-extruding Na^+-ATPases homologous to yeast *ENA1* in plants yet and the gene *AtSOS1* encoding a plasma membrane Na^+/H^+ antiporter homologous to bacterial and yeast Na^+/H^+ antiporters has been identified only in the glycophyte *A. thaliana* (Shi et al. 2000), but not in halophytes as yet, the vacuolar Na^+ sequestration is similar in yeast and in plants. Vacuolar uptake of Na^+ is mediated by a secondary active Na^+/H^+ antiport that is energized by the vacuolar H^+-ATPase (V-ATPase). The activity of this enzyme as well as its transcription and translation are stimulated by salt stress in halophytes, such as *M. crystallinum* and *Suaeda salsa* (Ratajczak et al. 1994; Golldack und Dietz 2001; Wang et al. 2001). Salt-induced expression of V-ATPase subunits has also been shown for salt-adapted suspension cells of tobacco and for salt-tolerant *Beta vulgaris* (Narasimhan et al. 1991; Kirsch et al. 1996; Lehr et al. 1999). In contrast, no salt-induced transcription of V-ATPase subunits has been observed for glycophytes, such as *A. thaliana* (Kluge et al. 1999) indicating salt-dependent activation of V-ATPase to be a specific response mechanism in halophytes that is missing in glycophytes. The vacuolar H^+-pyrophosphatase that is the second electrogenic H^+-pump at the tonoplast shows increased protein amounts in *S. salsa* (Wang et al. 2001) and increased transcript levels in the halotolerant *Lobularia maritima* (O.V. Popova and D. Golldack, unpubl. results) in response to salt stress.

NHX-type transporters have a key function in maintenance of cytoplasmic Na^+ homeostasis by mediating a secondary active Na^+/H^+ antiport at the tonoplast (Apse et al. 1999). Salt-induced Na^+/H^+ antiport has been shown at isolated tonoplast membrane vesicles of the halophytes *M. crystallinum* and *Salicornia bigelovii* (Barkla et al. 1995; Parks et al. 2002). In the salt-tolerant *Plantago maritima* a vacuolar Na^+/H^+ antiport has been identified that is missing in the closely related salt-sensitive species *P. media* (Staal et al. 1991). On the molecular level, the first NHX-type Na^+/H^+ antiporter from plants was identified and cloned with *AtNHX*1 from *A. thaliana* based on sequence information from the *Arabidopsis* genome sequencing initiative (Apse et al. 1999; Gaxiola et al. 1999; Quintero et al. 2000), whereas the first NHX-type transporter cloned from a halophyte was

AgNHX1 from *Atriplex gmelini* (Hamada et al. 2001). Exposure of *A. gmelini* to 100 and 400 mM NaCl, respectively, caused a three- to eightfold increase in the transcript amount of *AgNHX1* (Hamada et al. 2001). Reports on the transcriptional regulation of NHX-type transporters in glycophytes in response to salt are somewhat controversial. Whereas Gaxiola et al. (1999) in *A. thaliana* and Fukuda et al. (1999) in rice showed increased expression of this transporter type, Apse et al. (1999) and Chauhan et al. (2000) found no increase in transcript and protein amounts as well as Na^+/H^+-antiport activity of *AtNHX1* and *AtNHX2*, respectively, in salt-stressed *Arabidopsis*. Transgenic overexpression of *AtNHX1* caused increased vacuolar Na^+/H^+-transport activity and salt tolerance compared to wild-type *Arabidopsis*. Whereas a concentration of 200 mM NaCl was lethal to the wild type, this concentration was tolerated by the transgenic *Atnhx1* plants (Apse et al. 1999). Besides, also in tomato and *Brassica napus* overexpression of *AtNHX1* resulted in increased salt tolerance that was accompanied by increased Na^+ accumulation. The transgenic *B. napus* plants accumulated Na^+ up to 6% of its dry weight, whereas the concentration of K^+ declined to about 20% compared to non-stressed plants (Zhang und Blumwald 2001; Zhang et al. 2001).

Interestingly, *AtNHX1*-overexpressing *A. thaliana* displayed increased salt tolerance accompanied by stimulated Na^+/H^+-antiport activity but no significant increase in the *AtNHX1* transcripts and protein compared to control plants (Apse et al. 1999), indicating that overexpression of *AtNHX1* might activate additional mechanisms involved in salt tolerance, as e.g. salt-specific signaling, regulation of water uptake or homeostasis of other ions such as Cl^-. Mutation of *IpNHX1* in *Ipomoea nil* inhibited the development of blue color in the flowers that depends on the vacuolar pH, indicating additional function of NHX-type transporters for regulation of vacuolar pH and probably also involvement in the homeostasis of other cations in plant cells (Fukada-Tanaka et al. 2000).

2.3 Synthesis and Accumulation of Osmoprotective Compounds

Salt-tolerant plants accumulate the amino acid proline, quaternary ammonium compounds (e.g. glycine betaine), polyols, and sugars, respectively, as osmoprotective compounds that are synthesized via pathways that are connected to the basic plant metabolism.

Salt-induced synthesis and accumulation of proline has been reported from numerous halophytes (e.g. Maggio et al. 2000). In response to salinity, proline is synthesized de novo from glutamate (Nanjo et al. 1999). Accordingly, the requirement for nitrogen increases in salt-stressed plants. In *M.*

crystallinum, expression of the nitrate transporter homologue *McNRT1* increased in root epidermal cells and in the vasculature indicating stimulated uptake and *in planta* transport of nitrogen under salinity (O.V. Popova and D. Golldack, unpubl. results). Key enzymes of the biosynthesis of proline are Δ^1-pyrroline-5-carboxylate synthase (P5CS) and Δ^1-pyrroline-5-carboxylate reductase (P5CR) that catalyze the conversion of glutamate to proline. The NADP-specific isocitrate dehydrogenase (ICDH) has a key function in the biosynthesis of amino acids by generation of 2-oxoglutarate that is the primary acceptor for NH_3 assimilation. In leaves of *M. crystallinum*, enzyme activity as well as transcript and protein amounts of the NADP-specific ICDH *McICDH1* increased in response to salt stress. However, enzyme activity and transcript levels of ferredoxin-dependent glutamate synthase which catalyzes the synthesis of the proline precursor glutamate increased during short exposure of *M. crystallinum* to salinity but decreased in salt-adapted plants (Popova et al. 2002). These data indicate that de novo biosynthesis of proline is stimulated during the initial phase of salt stress but accumulation in salt-adapted plants is caused by inhibited degradation of the amino acid.

The second widely distributed compatible solute, the quaternary ammonium compound glycine betaine, is synthesized in chloroplasts in a two-step process by oxidation of choline to betaine aldehyde that is mediated by the salt-inducible choline monooxygenase (CMO) followed by a second oxidation step catalyzed by NAD-dependent betaine aldehyde dehydrogenase (Russel et al. 1998; McNeil et al. 1999). From halophytes of the Plumbaginaceae, four additional quaternary ammonium compounds are known that replace or supplement glycine betaine. β-Alanine betaine is synthesized by methylation of β-alanine. β-Alanine betaine is likely to be a favored compatible solute under the hypoxic conditions of salt marshes, because there is no direct requirement of O_2 for the biosynthesis of this compound (Hanson et al. 1994). In contrast, proline betaine and hydroxyproline betaine have been particularly found in Plumbaginaceae from dry or saline conditions (Hanson et al. 1994). Choline *O*-sulfate is likely to have not only a role in osmoprotection but also in detoxification of sulfate that may be found in excess amounts in saline environments as well. Choline is converted to choline *O*-sulfate by a 3'-phosphoadenosine-5'-phosphosulfate-dependent choline sulfotransferase. Increased activity of this enzyme is induced by NaCl in leaves and roots of e.g. *Limonium* species (McNeil et al. 1999).

Dimethylsulfonium propionate (DMSP) is a tertiary sulfonium compound that is an analogue of glycine betaine and has been found in halophytes of the Compositae, as e.g. the coastal plant *Wollastonia biflora* and in Poaceae, e.g., as the salt marsh grass *Spartina alternifolia* (Kocsis et al.

1998). The precursor of DMSP, methionine, is converted to *S*-methyl-L-methionine (SMM) in the cytosol. SMM is imported then into chloroplasts, deaminated and decarboxylated and converted to DMSP by a dehydrogenase (Trossat et al. 1998). A cDNA encoding the enzyme *S*-adenosyl Met:Met *S*-methyltransferase that catalyzes the SMM synthesis has been cloned from *W. biflora* (Bourgis et al. 1999).

Increased synthesis and cytosolic accumulation of compatible solutes of the group of polyols is known from salt-stressed *M. crystallinum*. *Myo*-inositol is synthesized from glucose-6-phosphate via *myo*-inositol 1-phosphate catalyzed by the enzymes *myo*-inositol 1-phosphate synthase (INPS) and *myo*-inositol 1-phosphate phosphatase. *Myo*-inositol is methylated to D-ononitol by the *myo*-inositol *O*-methyltransferase (IMT) and D-ononitol is epimerized to D-pinitol. In salt stressed *M. crystallinum*, the protein of IMT increased strongly in leaves and slightly in roots, whereas INPS was only upregulated in leaves and inhibited in roots but inositol and ononitol were transported to the roots via the phloem (Nelson et al. 1998, 1999). Tonoplast located Na^+/*myo*-inositol symporters (ITR) have been identified in *M. crystallinum* and have been suggested to function in the transport of Na^+ from root cell vacuoles to leaves (Chauhan et al. 2000; Fig. 2).

2.4 Protection Against ROS (Reactive Oxygen Species)

Next to osmotic and ionic stress under high salinity, plants are also affected by secondary stresses, e.g. the salt-induced excessive generation of reactive oxygen species (ROS), such as hydroxyl radicals, superoxide anions, and hydrogen peroxide by Fenton type reactions. Excess ROS formation in salt-stressed plants is caused by water deficits due to the osmotic component of salinity that leads to reduced CO_2 fixation and reduced regeneration of $NADP^+$ in the Calvin cycle (e.g. Vranová et al. 2002). ROS may damage e.g. cell walls, membranes, proteins, and DNA by non-specific oxidation (e.g. Schützendübel and Polle 2002).

In addition to osmotic balancing, another important function of osmolytes is the oxidative detoxification by scavenging of ROS. In transgenic tobacco, expression of a bacterial mannitol-1-phosphate dehydrogenase in chloroplasts caused accumulation of the polyol mannitol and increased tolerance to oxidative stress (Shen et al. 1997). Other antioxidant metabolites used by plants for ROS detoxification are, e.g., ascorbate, being a substrate for ascorbate peroxidase, thioredoxin, glutathione, and α-tocopherol (Asada 1999). Enzymatic detoxification of ROS is performed, e.g. by superoxide dismutase (SOD) that converts superoxide to H_2O_2, catalase that catalyzes the detoxification of H_2O_2 to water and oxygen, and peroxidases (Asada 1999). In calli of *Helianthus annuus* that are moderately tolerant to elevated concentrations of KCl, activities of SOD, catalase, and peroxidase increased (Santos et al. 2001). Overexpression of glutathione *S*-transferase and glutathione peroxidase resulted in increased growth of transgenic tobacco exposed to salt stress (Roxas et al. 1997). In *M. crystallinum*, salt stress causes increased enzyme activity of NADP-dependent isocitrate dehydrogenase, and up-regulated expression of *McICDH1* was particularly observed in leaf mesophyll cells (Popova et al. 2002). A role of NADP-dependent isocitrate dehydrogenase in supplying NADPH to the glutathione reductase/gluthatione peroxidase system for detoxification of H_2O_2 has been suggested (Popova et al. 2002).

2.5 Signal Transduction and Salt Adaptation

Regulatory signaling elements of salt-adaptation pathways have been rarely identified in halophytes yet but most knowledge on salinity-induced signal transduction is derived from the salt-sensitive model plant *A. thaliana*. Transcription factors of the MYC, MYB, and bZIP type interact with promoters of genes that are responsive to osmotic stress in *A. thaliana* (Shinozaki and Yamaguchi-Shinozaki 2000). Overexpression of a cDNA encoding the transcription factor DREB 1A that interacts with the *Arabidopsis cis*-acting 'dehydration response element' DRE caused increased salt and drought tolerance in *A. thaliana* (Kasuga et al. 1999). In *Arabidopsis*, a signaling pathway responsive to salt stress includes the Ca^{2+}-binding protein SOS3 that activates the serine/threonine protein kinase SOS2. Salt-induced expression of the plasma membrane Na^+/H^+ antiporter SOS1 that has been described from *A. thaliana* but not from halophytes yet, may be regulated by SOS3 and SOS2 (Halfter et al. 2000). A salt-induced mitogen-activated protein kinase (MAPK) has been identified from alfalfa with SIMK that is activated by the MAPK kinase SIMKK (Kiegerl et al. 2000).

In the salt-tolerant grass species *Lophopyrum elongatum* that is native to salt marshes of the Mediterranean region, expression of the serine/threonine kinase Esi47 is induced by salt stress of 250 mM NaCl and by abscisic acid (Shen et al. 2001). In *M. crystallinum*, transcription of the salt-inducible phosphoenolpyruvate carboxylase isoform *Ppc1* was stimulated by the Ca^{2+}-ionophore ionomycin and the Ca^{2+}-ATPase inhibitor thapsigargin indicating that Ca^{2+} has a role in signaling events that lead to salt-dependent *Ppc1*-induction (Taybi and Cushman 1999). Salt-inducible expression of the V-ATPase subunit E, of *Ppc1*, and of the key enzyme of polyol synthesis, myo-inositol *O*-methyltransferase (*Imt1*), could be inhibited by treatment with the activator of trimeric G-proteins mastoparan suggesting involvement of G-proteins in salt-induced signaling (Golldack and Dietz 2001).

3 Functional Genomics of Salt Adaptation

3.1 How Many Genes are Salt-Responsive in Plants?

Although a number of genes that are salt responsive in halophytes have been identified, the process of salt adaptation in its complexity is not understood yet. To understand the network of processes occurring under salt stress and adaptation to high salinity in salt-tolerant plants expression studies on the genomic scale are highly necessary.

The only halophyte that has been analyzed in depth for its molecular mechanisms involved in salt adaptation is the salt-tolerant succulent *Mesembryanthemum crystallinum* that has been becoming a model plant for salt stress research during the last decade. Based on EST-sequencings of *M. crystallinum* it is estimated that 10% of the total gene number of *M. crystallinum* is responsive to salt stress (Cushman et al. 1999). Different genome sizes in plants are mainly due to different amounts of interspersed repetitive sequences (Bennetzen et al. 1998). Accordingly, it is estimated that higher plants may have similar amounts of functional genes (Sommerville and Sommerville 1999). Based on a number of approximately 25,000 functional genes in the model plant *A. thaliana* (Kaiser 2000) the number of salt-responsive genes in *M. crystallinum* may be estimated to be up to 2,500.

3.2 Global Analyses of the Network of Salt-Responsive Genes in Halophytes

Taking into consideration that more than thousand genes may be involved in salt stress adaptation in halophytes it is obvious that the molecular network of salinity stress responses is not understood yet in the model halophyte *M. crystallinum* and nearly nothing is known from other salt-tolerant plant species. An innovative novel approach for systematic identification of salt-responsive genes in halotolerant plants on a genomic scale is the use of cDNA arrays that allow fast and comparable analyses of large transcript amounts in parallel. Using cDNA arrays it is possible to analyze expression profiles of unknown gene populations according to their function in salt adaptation and then select and identify genes that were shown to be regulated by stress exposure.

However, most halophyte species are not amenable for studies on the transcriptome level yet due to limited information on their gene sequences and accordingly the difficulty to identify salt-regulated genes and their products. In contrast, the model plants with complete sequencing of their genomes – *Arabidopsis* and rice – are typical glycophytes that are not very salt-tolerant. For comparative genomics studies, halophyte species would be most suitable that are closely related to the completely sequenced plant models to take advantage of the wealth of genomic data from these species.

One ideal halophyte model that meets these criteria is *Lobularia maritima* – a species of the Brassicaceae that is native to the Canary Islands and the Azores. *Lobularia* does not possess morphological specializations such as salt-accumulating bladder cells or salt-excreting glands but the salt tolerance of the species is due to adjustment in ion and osmotic homeostasis. Transcripts from *Lobularia* share in average 90% identity with homologous genes from *Arabidopsis* (D. Golldack, unpubl. results). According to the high level of molecular identity, cDNA arrays synthesized for *Lobularia* may be cross-hybridized with cDNA populations from *A. thaliana* and other Brassicaceae species and expression profiles can be directly compared in the closely related salt-tolerant and salt-sensitive species in response to salt stress.

In our studies, we used an array including 480 transcripts generated from a subtracted cDNA library of *Lobularia* that was enriched for salt-induced genes to study time and dose dependent responses of the species to salinity. Approximately 40% of the elements included in the array showed altered transcript abundance under salinity. An early, transient modification of gene expression could be distinguished from a second set of ESTs with modified transcript levels during long-term acclimation. The identified products with altered expression could be grouped into seven functional classes: (1) a large group of proteins of yet unknown function, (2) elements that control gene expression, (3) signal transduction elements, (4) metabolic enzymes, e.g. glutamine synthetase, (5) transport proteins, (6) photosynthetic proteins, and (7) the largest group with proteins involved in general stress response and defense (Fig. 3; O.V. Popova, S. Pabion, and D. Golldack, unpubl. results).

4 Perpectives

A complex network of salt-responsive genes involved in salt adaptation in halotolerant plants has been identified and novel salt stress tolerance genes are likely to be identified from large-scale transcriptome analyses in the near future. As a next step, the functions of these genes need to be investi-

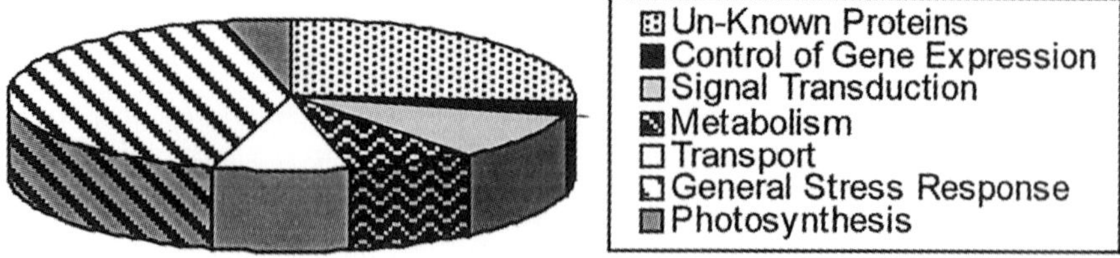

Fig. 3. Functional categories of salt-responsive genes identified from a cDNA array of the salt-tolerant *Lobularia maritima* (Brassicaceae; O.V. Popova, S. Pabion, and D. Golldack, unpubl.)

gated, e.g. by examining knock-out mutants or using specific antisense approaches as posttranscriptionally silenced genes by RNAi (Chuang and Meyerowitz 2000; Escobar 2001). Detailed functional studies of transcription factors, promoter elements, and signal transduction pathways that regulate stress responses will further enhance our knowledge of salt adaptation in naturally halotolerant plants.

Acknowledgement. Part of the work cited in this review was supported by the Deutsche Forschungsgemeinschaft, Bonn (GO 738/4). Performance of ICP AES measurements by Ms. Elfriede Reisberg (University of Würzburg) is gratefully acknowledged.

References

Amtmann A, Sanders D (1999) Mechanisms of Na^+ uptake by plant cells. Adv Bot Res 29:75–112

Apse MP, Aharon GS, Snedden WA, Blumwald E (1999) Salt tolerance conferred by overexpression of a vacuolar Na^+/H^+ antiport in *Arabidopsis*. Science 285:1256–1258

Asada K (1999) The water-water cycle in chloroplasts: scavenging of active oxygens and dissipation of excess photons. Annu Rev Plant Physiol Plant Mol Biol 50:601–639

Barkla BJ, Zingarelli L, Blumwald E, Smith JAC (1995) Tonoplast Na^+/H^+ antiport activity and its energization by the vacuolar H^+-ATPase in the halophytic plant *Mesembryanthemum crystallinum* L. Plant Physiol 109:549–556

Bennetzen JL, SanMiguel P, Chen MS, Tikhonov A, Francki M, Avramova Z (1998) Grass genomes. Proc Natl Acad Sci USA 95:1975–1978

Blumwald E, Aharon GS, Apse MP (2000) Sodium transport in plant cells. Biochim Biophys Acta 1465:140–151

Bohnert HJ, Ayoubi P, Borchert C, Bressan RA, Burnap RL, Cushman JC, Cushman MA, Deyholos M, Fischer R, Galbraith DW, Hasegawa PM, Jenks M, Kawasaki S, Koiwa H, Kore-eda S, Lee BH, Michalowski CB, Misawa E, Nomura M, Ozturk N, Postier B, Prade R, Song CP, Tanaka Y, Wang H, Zhu JK (2001) A genomics approach towards salt stress tolerance. Plant Physiol Biochem 39:295–311

Bourgis F, Roje S, Nuccio ML, Fisher DB, Tarczynski MC, Li CJ, Herschbach C, Rennenberg H, Pimenta MJ, Shen TL, Gage DA, Hanson AD (1999) S-methylmethionine plays a major role in phloem sulfur transport and is synthesized by a novel type of methyltransferase. Plant Cell 11:1485–1497

Chauhan S, Forsthoefel N, Ran Y, Quigley F, Nelson DE, Bohnert HJ (2000) Na^+/myo-inositol symporters and Na^+/H^+-antiport in *Mesembryanthemum crystallinum*. Plant J 24:511–522

Chuang CF, Meyerowitz EM (2000) Specific and heritable genetic interference by double-stranded RNA in *Arabidopsis thaliana*. Proc Natl Acad Sci USA 97:4985–4990

Clemens S, Antosiewicz DM, Ward JM, Schachtman DP, Schroeder JI (1998) The plant cDNA LCT1 mediates the uptake of calcium and cadmium in yeast. Proc Natl Acad Sci USA 95:12043–12048

Cushman MA, Bufford D, Fredrickson M, Ray A, Akselrod I, Landrith D, Stout L, Maroco J, Cushman JC (1999) An expressed sequencs tag (EST) database for the common ice plant, *Mesembryanthemum crystallinum*. ASPP 1999, Abstract 675

Escobar MA, Civerolo EL, Summerfelt KR, Dandekar AM (2001) RNAi-mediated oncogene silencing confers resistance to crown gall tumorigenesis. Proc Natl Acad Sci USA 98:13437–13442

Flis K, Bednarczyk P, Hordejuk R, Szewczyk A, Berest V, Dolowy K, Edelman A, Kurlandzka A (2002) The Gef1 protein of *Saccharomyces cerevisiae* is associated with chloride channel activity. Biochem Biophys Res Commun 294:1144–1150

Fukada-Tanaka S, Inagaki Y, Yamaguchi T, Saito N, Iida S (2000) Colour-enhancing protein in blue petals. Nature 407:581

Fukuda A, Nakamura A, Tanaka Y (1999) Molecular cloning and expression of the Na^+/H^+ exchanger gene in *Oryza sativa*. Biochim Biophys Acta 1446:149–155

Gaxiola RA, Rao R, Sherman A, Grisafi P, Alper SL, Fink GR (1999) The *Arabidopsis thaliana* proton transporters, AtNhx1 and Avp1, can function in cation detoxification in yeast. Proc Natl Acad Sci USA 96:1480–1485

Glenn EP, Brown JJ, Blumwald E (1999) Salt tolerance and crop potential of halophytes. Crit Rev Plant Sci 18:227–255

Golldack D, Dietz KJ (2001) Salt-induced expression of the vacuolar H^+-ATPase in the common ice plant is developmentally controlled and tissue specific. Plant Physiol 125:1643–1654

Golldack D, Su H, Quigley F, Kamasani UR, Munoz-Garay C, Balderas E, Popova OV, Bennett J, Bohnert HJ, Pantoja O (2002) Characterization of a HKT-type transporter in rice as a general alkali cation transporter. Plant J 31:529–542

Golldack D, Quigley F, Michalowski CB, Kamasani UR, Bohnert HJ (2003) Salinity stress-tolerant and -sensitive rice (*Oryza sativa* L.) regulate AKT1-type potassium channel transcripts differently. Plant Mol Biol 51:71–81

Halfter U, Ishitani M, Zhu JK (2000) The Arabidopsis SOS2 protein kinase physically interacts with and is activated by the calcium-binding protein SOS3. Proc Natl Acad Sci USA 97:3735–3740

Hamada A, Shono M, Xia T, Ohta M, Hayashi Y, Tanaka A, Hayakawa T (2001) Isolation and characterization of a Na^+/H^+ antiporter gene from the halophyte *Atriplex gmelini*. Plant Mol Biol 46:35–42

Hanson AD, Rathinasabapathi B, Rivoal J, Burnet M, Dillon MO, Gage DA (1994) Osmoprotective compounds in the Plumbaginaceae: a natural experiment in metabolic engineering of stress tolerance. Proc Natl Acad Sci USA 91:306–310

Hasegawa PM, Bressan RA, Zhu JK, Bohnert HJ (2000) Plant cellular and molecular responses to high salinity. Annu Rev Plant Phys 51:463–499

Hohmann S (2002) Osmotic stress signaling and osmoadaptation in yeasts. Microbiol Mol Biol Rev 66:300–372

Kaiser J (2000) From genome to functional genomics. Science 288:1715

Kasuga M, Liu Q, Miura S, Yamaguchi-Shinozaki K, Shinozaki K (1999) Improving plant drought, salt, and freezing tolerance by gene transfer of a single stress-inducible transcription factor. Nat Biotechnol 17:287–291

Kiegerl S, Cardinale F, Siligan C, Gross A, Baudouin E, Liwosz A, Eklof S, Till S, Bogre L, Hirt H, Meskiene I (2000) SIMKK, a mitogen-activated protein kinase (MAPK) kinase, is a specific activator of the salt stress-induced MAPK, SIMK. Plant Cell 12:2247–2258

Kirsch M, An Z, Viereck R, Löw R, Rausch T (1996) Salt stress induces an increased expression of V-type H^+-ATPase in mature sugar beet leaves. Plant Mol Biol 32:543–547

Kluge C, Golldack D, Dietz KJ (1999) Subunit D of the vacuolar H^+-ATPase of *Arabidopsis thaliana*. Biochim Biophys Acta 1419:105–110

Kocsis MG, Nolte KD, Rhodes D, Shen TL, Gage DA, Hanson AD (1998) Dimethylsulfoniopropionate biosynthesis in *Spartina alterniflora* – evidence that S-methylmethionine and dimethylsulfoniopropylamine are intermediates. Plant Physiol 117:273–281

Läuchli A, Lüttge U (ed) (2002) Salinity: environment – plants – molecules. Kluwer, Dordrecht

Lehr A, Kirsch M, Viereck R, Schiemann J, Rausch T (1999) cDNA and genomic cloning of sugar beet V-type H^+-ATPase subunit A and c isoforms: evidence for coordinate expression during plant development and coordinate induction in response to high salinity. Plant Mol Biol 39:463–475

Maggio A, Reddy MP, Joly RJ (2000) Leaf gas exchange and solute accumulation in the halophyte *Salvadora persica* grown at moderate salinity. Environ Exp Bot 44:31–38

Matheos DP, Kingsbury TJ, Ahsan US, Cunningham KW (1997) Tcn1p/Crz1p, a calcineurin-dependent transcription factor that differentially regulates gene expression in *Saccharomyces cerevisiae*. Genes Dev 11:3445–3458

McNeil SD, Nuccio ML, Hanson AD (1999) Betaines and related osmoprotectants. Targets for metabolic engineering of stress resistance. Plant Physiol 120:945–950

Nanjo T, Kobayashi M, Yoshiba Y, Sanada Y, Wada K, Tsukaya H, Kakubari Y, Yamaguchi-Shinozaki K, Shinozaki K (1999) Biological functions of proline in morphogenesis and osmotolerance revealed in antisense transgenic *Arabidopsis thaliana*. Plant J 18:185–193

Narasimhan ML, Binzel ML, Perez-Prat E, Chen Z, Nelson DE, Singh NK, Bressan RB, Hasegawa PM (1991) NaCl regulation of tonoplast ATPase 70-kilo-dalton subunit mRNA in tobacco cells. Plant Physiol 97:562–568

Nass R, Rao R (1998) Novel localization of a Na^+/H^+ exchanger in a late endosomal compartment of yeast. Implications for vacuole biogenesis. J Biol Chem 2732:1054–1060

Nass R, Cunningham KW, Rao R (1997) Intracellular sequestration of sodium by a novel Na^+/H^+ exchanger in yeast is enhanced by mutations in the plasma membrane H^+-ATPase. Insights into mechanisms of sodium tolerance. J Biol Chem 272:26145–26152

Nelson DE, Rammesmayer G, Bohnert HJ (1998) Regulation of cell-specific inositol metabolism and transport in plant salinity tolerance. Plant Cell 10:753–764

Nelson DE, Koukoumanos M, Bohnert HJ (1999) Myo-inositol-dependent sodium uptake in ice plant. Plant Physiol 119:165–172

Parks GE, Dietrich MA, Schumaker KS (2002) Increased vacuolar Na^+/H^+ exchange activity in *Salicornia bigelovii* Torr. in response to NaCl. J Exp Bot 53:1055–1065

Popova OV, Ismailov SF, Popova TN, Dietz KJ, Golldack D (2002) Salt-induced expression of NADP-dependent isocitrate dehydrogenase and ferredoxin-dependent glutamate synthase in *Mesembryanthemum crystallinum*. Planta 215:906–913

Posas F, Chambers JR, Heyman JA, Hoeffler JP, de Nadal E, Arino J (2000) The transcriptional response of yeast to saline stress. J Biol Chem 275:17249–17255

Quintero FJ, Blatt MR, Pardo JM (2000) Functional conservation between yeast and plant endosomal Na^+/H^+ antiporters. FEBS Lett 471:224–228

Ratajczak R, Richter J, Lüttge U (1994) Adaptation of the tonoplast V-type H^+-ATPase of *Mesembryanthemum crystallinum* to salt stress, C3-CAM transition and plant age. Plant Cell Environ 17:1101–1112

Roxas VP, Smith RK Jr, Allen ER, Allen RD (1997) Overexpression of glutathione S-transferase/glutathione peroxidase enhances the growth of transgenic tobacco seedlings during stress. Nat Biotechnol 15:988–991

Rubio F, Gassmann W, Schroeder JI (1995) Sodium-driven potassium uptake by the plant potassium transporter HKT1 and mutations conferring salt tolerance. Science 270:1660–1663

Russell BL, Rathinasabapathi B, Hanson AD (1998) Osmotic stress induces expression of choline monooxygenase in sugar beet and amaranth. Plant Physiol 116:859–865

Saiko TA, Zonn IS (2000) Irrigation expansion and dynamics of desertification in the circum-Aral region of Central Asia. Appl Geogr 20:349–367

Santos CLV, Campos A, Azevedo H, Caldeira G (2001) In situ and in vitro senescence induced by KCl stress: nutritional imbalance, lipid peroxidation and antioxidant metabolism. J Exp Bot 52:351–360

Schachtman D, Liu WH (1999). Molecular pieces to the puzzle of the interaction between potassium and sodium uptake in plants. Trends Plant Sci 4:281–287

Schachtman DP, Schroeder JI (1994) Structure and transport mechanism of a high-affinity potassium uptake transporter from higher plants. Nature 370:655–658

Shinozaki K, Yamaguchi-Shinozaki K (2000) Molecular responses to dehydration and low temperature: differences and cross-talk between two stress signaling pathways. Curr Opin Plant Biol 3:217–223

Serrano R, Gaxiola R (1994) Microbial models and salt stress tolerance in plants. Crit Rev Plant Sci 13:121–138

Schützendübel A, Polle A (2002) Plant responses to abiotic stresses: heavy metal-induced oxidative stress and protection by mycorrhization. J Exp Bot 53:1351–1365

Shen B, Jensen RG, Bohnert HJ (1997) Mannitol protects against oxidation by hydroxyl radicals. Plant Physiol 115:527–532

Shen W, Gomez-Cadenas A, Routly EL, Ho TH, Simmonds JA, Gulick PJ (2001) The salt stress-inducible protein kinase gene, Esi47, from the salt-tolerant wheatgrass *Lophopyrum elongatum* is involved in plant hormone signaling. Plant Physiol 125:1429–1441

Shi H, Ishitani M, Kim C, Zhu JK (2000) The *Arabidopsis thaliana* salt tolerance gene SOS1 encodes a putative Na^+/H^+ antiporter. Proc Natl Acad Sci USA 97:6896–6901

Sommerville C, Sommerville S (1999) Plant functional genomics. Science 285:380–383

Staal EM, Maathuis, FJM, Elzenga JTM, Overbeek JHM, Prins HBA (1991) Na^+/H^+ antiport activity in tonoplast vesicles from roots of the salt-tolerant *Plantago maritima* and the salt-sensitive *Plantago media*. Physiol Plant 82:179–184

Su H, Golldack D, Katsuhara M, Zhao C, Bohnert HJ (2001) Expression and stress-dependent induction of potassium channel transcripts in the common ice plant. Plant Physiol 125:604–614

Su H, Golldack D, Zhao C, Bohnert HJ (2002) The expression of HAK-type K^+ transporters is regulated in response to salinity stress in common ice plant. Plant Physiol 129:1482–1493

Taybi T, Cushman JC (1999) Signaling events leading to crassulacean acid metabolism induction in the common ice plant. Plant Physiol 121:545–556

Trossat C, Rathinasabapathi B, Weretilnyk EA, Shen TL, Huang ZH, Gage DA, Hanson AD (1998) Salinity promotes accumulation of 3-dimethylsulfoniopropionate and its precursor *S*-methylmethionine in chloroplasts. Plant Physiol 116:165–171

Uozumi N, Kim EJ, Rubio F, Yamaguchi T, Muto S, Tsuboi A, Bakker EP, Nakamura T, Schroeder JI (2000) The *Arabidopsis* HKT1 gene homolog mediates inward Na^+ currents in *Xenopus laevis* oocytes and Na^+ uptake in *Saccharomyces cerevisiae*. Plant Physiol 122:1249–1259

Vranová E, Inze D, Van Breusegem F (2002) Signal transduction during oxidative stress. J Exp Bot 53:1227–1236

Wang B, Lüttge U, Ratajczak R (2001) Effects of salt treatment and osmotic stress on V-ATPase and V-PPase in leaves of the halophyte *Suaeda salsa*. J Exp Bot 52:2355–2365

White PJ (1999) The molecular mechanism of sodium influx to root cells. Trends Plant Sci 4:245–246

Zhang HX, Blumwald E (2001) Transgenic salt-tolerant tomato plants accumulate salt in foliage but not in fruit. Nat Biotechnol 19:765–768

Zhang HX, Hodson JN, Williams JP, Blumwald E (2001) Engineering salt-tolerant *Brassica* plants: characterization of yield and seed oil quality in transgenic plants with increased vacuolar sodium accumulation. Proc Natl Acad Sci USA 98:12832–12836

Dr. Dortje Golldack
Department of Physiology and Biochemistry of Plants
Faculty of Biology
University of Bielefeld
33615 Bielefeld, Germany
Tel.: +49-521-1065594
Fax: +49-521-1066039
e-mail: dortje.golldack@uni-bielefeld.de

High Frequency or Ultradian Rhythms in Plants

Ulrich Lüttge and M.-Th. Hütt

1 The Quest for the Biological Clock or Circadian Oscillator and: "Can We Learn from Ultradian Rhythms?"

Biorhythm research is currently dominated by the quest for the biological clock or circadian oscillator on a genetic basis. One of us has reviewed the plant biological literature in a contribution to last year's volume of this series (Lüttge 2002). However, while spontaneous ultradian frequency rhythms remain a central topic in neurobiology with the electrical firing of neurons, and oscillations of cyclic adenosine-monophosphate (cAMP) secretion, attraction and pattern formation of amoebae of the slime mold *Dictyostelium* (e.g. Berridge and Rapp 1979) and the oscillatory behaviour of glycolysis (e.g., Berridge and Rapp 1979; Olsen and Degn 1985; Rapp 1986; see also below Sect. 2.2) are much studied, astonishingly little research effort is currently devoted to ultradian high frequency oscillations in plant biology. This is regrettable because much can be learnt from the study of high frequency rhythms. This is so not only because it is often much easier to keep systems studied under strict experimental control and obtain more readily sufficiently extended time series of measurements for theoretical analysis than with the much longer circadian oscillations. Although we must remember that in *Drosophila* both the circadian rhythm of eclosion and locomotor activity and the very short ultradian rhythm of the male courtship song seem to be governed by the same PER gene (Dunlap 1993), studying high frequency oscillations mostly implies that we set ourselves free from being paralyzed by the evolutionary and genetic biological clock enigma.

Among the ultradian cellular oscillators, Berridge and Rapp (1979) distinguished two types, (1) cytoplasmic oscillators, where the periodic phenomena are generated by an instability in metabolic pathways, and (2) membrane oscillators generating rhythms of membrane potential. There is a large variety of ultradian rhythms and there is no evidence for a common oscillatory mechanism at the cellular level (Berridge and Rapp 1979). Therefore, in contrast to circadian rhythms, we may be more open for

Progress in Botany, Vol. 65

system analysis in order to unravel governing principles of nonlinear dynamics. This is generic and it may be one of the reasons of the great success of studies of, e.g., *Dictyostelium* and glycolysis. It has much bearing also on our understanding of circadian rhythmicity (see Lüttge 2002).

Thus, it appears opportune to review here the respective information available in plant biology. This will be done in the following section describing phenomena of ultradian rhythms in membranes, metabolism and isolated enzymes, cell development, stomata and photosynthesis, movements of leaves and circumnutation, and root elongation. A theoretical section will then pick up key elements of nonlinear dynamics that became obvious while looking at the phenomena. This theoretical assessment will aim at giving an answer to the question "can we learn from ultradian rhythms?" by fathoming "*what* can we learn from ultradian rhythms?" Finally, we ask if ultradian rhythms may have functions and contribute to fitness achieved during evolution.

2 Phenomena

2.1 Membranes

The most conspicuous example on everybody's mind of high frequency ultradian rhythmicity involving membranes is the action potential firing of neurons. However, spontaneous rhythmic changes of electrical membrane potentials, i.e., series of spontaneous action potentials, membrane oscillators *sensu* Berridge and Rapp (1979), with period lengths in the range of seconds to minutes and several tens of minutes are also observed in fungi such as *Neurospora crassa* (Gradmann and Slayman 1975; Slayman et al. 1976), green algae such as *Nitella*, *Chara* and *Acetabularia* (Gradmann 1976; Ogata and Kishimoto 1976; Vucinic et al. 1978; Martens et al. 1979; Hayashi et al. 1982; Fisahn et al. 1986, 1989; Lucas and Fisahn 1989) and higher plants, e.g., *Rorippa nasturtium aquaticum* (Ping and Lou 1990). The literature up to the late 1980s is well reviewed in the introduction of Fisahn et al. (1989). Proton pumping at the plasma membrane evidently is involved (Gradmann and Slayman 1975; Fisahn et al. 1989; Tyerman et al. 2001). Sinusoidal electrical stimulation of giant internodal cells of *Nitella* with varied frequencies showed entrainment, quasiperiodic behaviour and deterministic chaos (Hayashi et al. 1982, 1983; Olsen and Degn 1985). Ultradian oscillations of H^+, K^+, Ca^{2+} and Cl^- fluxes with different period lengths of a few minutes and 50–80 min, respectively, at different regions of the root surface have been described by Shabala and Knowles (2002). They may be important in nutrient acquisition via the mature root zone. These observations also raise questions of synchronization/desynchronization between these different ion fluxes and also between different regions of the root surface.

Thus, the most conspicuous phenomena of nonlinear dynamics are expressed. Molecular entities involved are up to several different "electro-enzymes" such as channels, pumps and carriers (Gradmann 2001), e.g., in the plasma membrane of isolated root protoplasts, the H^+-ATPase, inward and outward K^+ rectifiers and gated proton channels (Tyerman et al. 2001). Ultradian oscillations of net flux of K^+ and external pH occur in osmo- and turgor-regulation after the onset of hyperosmotic shock in *Arabidopsis* epidermal root cells and appear to be related to voltage gated K^+ transporters in feedback loops of cell osmotic adjustment (Shabala and Lew 2002). To date the question appears to remain open if these oscillations of membrane system require cellular functions as suggested by the work with isolated protoplasts, where feedback control of cellular ion relations (Gradmann and Hoffstadt 1998) and metabolism (see below Sect. 2.2) may be involved (Tyerman et al. 2001), or if they may also be generated by isolated membranes. In fact, the electro-enzymes may be electrically coupled and interact with each other via voltage changes (Gradmann et al. 1993; Gradmann and Buschmann 1997; Gradmann 2001). The isolated wheat root protoplasts of Tyerman et al. (2001) exhibited oscillations in proton flux, and either free-running membrane potential oscillations under current clamp or free-running current oscillations under voltage clamp. Gradmann (2001) presents mathematical models, using two- and five-membrane transporters, respectively, in which oscillations are produced not only by local changes of ion concentrations, as for the apoplast of cells in parenchymas, but also by mere electrical coupling of the voltage-dependent electro-enzymes in the membrane.

Calcium ions are known to be important secondary messengers. Cytoplasmic Ca^{2+} concentrations control a vast variety of physiological processes (Trewavas 1999; Rudd and Franklin-Tong 2001). This includes regulation of channels, ion pumps and carriers by gating and Ca^{2+}-dependent phosphorylation. Ca^{2+} pumps (Liang and Sze 1998; Sze et al.2000) and Ca^{2+} channels (White 2000) participate in this regulation network. Calcium waves are involved in this signalling (Trewavas 1999). Cytoplasmic Ca^{2+} levels themselves show ultradian oscillations (Felle 1988; McAinsh et al. 1990, 1995; Gilroy et al. 1991; Read et al. 1993; Volotovski et al. 1993; Swillens et al. 1994; Bauer et al. 1998) and it has been suggested that the membrane oscillator may be a plasma membrane Ca^{2+} channel (McAinsh et al. 1995).

2.2 Metabolism and Enzymes

Ultradian oscillations in metabolic functions belong to the cytoplasmic oscillations *sensu lato* distinguished by Berridge and Rapp (1979; see also Rapp 1986). Various functions of isolated mitochondria have been shown to oscillate (Berridge and Rapp 1979; for photosynthesis see Sect. 2.5). Ultradian oscillations of glycolysis as a linear reaction chain of several enzymes with the now well-known allosteric control sites and feedback loops covered in all textbooks of biochemistry were discovered in the 1960s (Ghosh and Chance 1964). They have been extensively studied ever since (Berridge and Rapp 1979; Olsen and Degn 1985; Teusink et al. 1996a,b, 1998; Teusink 1999; Bier et al. 2000). It was shown that glycolysis can be driven into (deterministic) chaos under periodic supply of glucose (Markus et al. 1984). Ultradian oscillations as well as chaos occur in glycolysis of cell-free extracts of *Saccharomyces cerevisiae* (Chance et al. 1964; Markus et al. 1984; Olsen and Degn 1985). This demonstrates that oscillatory behaviour does occur in the absence of immediate genetic control. Theoretical models of glycolysis simulating glycolytic oscillations have been produced (Markus and Hess 1984; Teusink et al. 1996a), and it can be shown that oscillations are not dictated by a single key enzyme or "oscillophore" (Teusink et al. 1996a; Teusink 1999). The primary signalling molecule is glucose (Teusink et al. 1998).

The most simple conceivable oscillators consist of reactions which comprise only a single enzyme and its substrates (Hauser et al. 2002). Indeed, for the argument that oscillations can occur as a system-property totally independent of genetic oscillators it is even more relevant that oscillatory behaviour is produced by single enzymes. Rhythmic variations of ferredoxin-glutamate synthase activity in sunflower leaves with a period length of 4–5 h have been described as a phenomenon (Fernández-Conde et al. 1995). Ultradian oscillations of the catalytic capacity of malate dehydrogenase in extracts of leaves of the facultative CAM-plant *Kalanchoë blossfeldiana* cv. Tom Thumb have been explained by conformational changes due to changing water aggregates and entropy variations in the protein/solvent system, and a conceptual analogy to the totally inorganic Belusov-Zhabotinsky chemical oscillator was stated (Queiroz-Claret et al. 1988).

Detailed studies have been performed on the peroxidase-oxidase (PO) enzyme of roots of horse radish catalysing the reaction

$$O_2 + 2NADH + 2H^+ \rightarrow 2H_2O + 2NAD^+$$

which is involved in lignin biosynthesis. In a homogenous system with a stirred buffer medium, the enzyme shows nonlinear phenomena, such as bistability, oscillations and chaos due to the complicated branched chain nature of the reaction mechanism of the enzyme with its five distinct active forms (Olsen 1983; Olsen and Degn 1985; Hauser and Olsen 1996, 1999; Hauser et al. 1997; Bronnikova et al. 1998; Møller et al. 1998; Kirkor et al. 2000). The PO from other sources, e.g., soybean, a fungus and mammals, shows similar behaviour (see Kummer et al. 1997; Hauser and Olsen 1998, 1999; Hauser et al. 2000). The enzyme requires catalytic amounts of phenolics as electron mediators of electron flow in the reaction. Naturally occurring phenolics are effective in the reaction, and hence, it can be assumed that oscillations also occur in vivo (Kummer et al. 1997; Hauser and Olsen 1998).

The routes into chaos comprise a period-doubling-route (Feigenbaum route) and a period-adding-route (mixed-mode oscillations) depending on the pH of the medium (Hauser and Olsen 1996, 1999). Thus, the PO reaction may reproduce all the richness of the non-linear behaviour of the inorganic Belusov-Zhabotinsky oscillator (Larter et al. 1993; Hauser and Olsen 1999).

2.3 Cell Development and Pollen Tubes

In the marine microheterotroph *Traustochytrium* sp., closely related to heterokont algae, non-damping endogenous ultradian oscillations become evident during development. As cell development progresses from the zoospore stage towards maturity, oscillations in the period range of several minutes of fluxes of H^+, Na^+ and Ca^{2+} were expressed. They may be causally linked to cell development. H^+-fluctuations could be associated with changes in cytoskeleton organization required for cell cleavage and zoospore release (Shabala et al. 2001).

Another interesting example is rhythmic tip growth as it occurs in fungal hyphae (Kaminskyj et al. 1992; López-Franco et al. 1994; Bracker et al. 1995) and pollen tubes (Tang et al. 1992; Pierson et al. 1995; 1996; Geitmann et al. 1996; Geitmann and Cresti 1998; review Feijo et al. 2001). In the pulsating tip growth of pollen tubes two types of oscillations can be distinguished, i.e., sinusoidal oscillations with periods in the range of seconds and spike-shaped oscillations with periods of several minutes (Geitmann and Cresti 1998). Ca^{2+} influxes (Pierson et al. 1995, 1996) and a Ca^{2+}-channel oscillator may be involved in this pulsation growth. Ca^{2+}-channel blockers such as La^{3+} and Gd^{3+} reversibly suppress rhythmicity of growth (Geitmann and Cresti 1998). It is intriguing, however, that pollen tube growth and the intracellular calcium gradient oscillate in phase while extracellular calcium influx is delayed (Holdaway-Clark et al. 1997) and that periodic increases in elongation rate precede increases in cytosolic Ca^{2+} (Messerli et al. 2000). The cytoskeleton may play a role in these oscillations because there is a concerted action of Ca^{2+}, vesicle exocytosis and cytoskeleton, and cell wall in pollen tube growth (Derksen 1996). Fluctuations in the liberation of secretory vesicles (López-Franco et al. 1994; Bracker et al. 1995) possibly controlled by the cytoskeleton accompanied by changes in turgor pressure (Kaminskyj et al. 1992) have been discussed as causes of the oscillations. This also directed attention to mechanical stress or stretch-sensitive Ca^{2+} channels and gating by elastic tension in the membrane (Ramahaleo et al. 1996) or by the submembrane cytoskeleton (Sokabe et al. 1991) as possible mechanisms mediating the oscillations (Geitmann and Cresti 1998). A recent discovery of a new oscillatory component, namely chloride efflux at the pollen tube apex, also sheds new light on Ca^{2+} interactions. Oscillations of Cl^- efflux correlate with cell volume, turgor status and extension growth. They are targeted by the signalling molecule inositol 3,4,5,6-tetra*kis*phosphate known to regulate Ca^{2+}-activated Cl^- channels in mammalian cells and may interfere with vesicle movements during cell elongation (Zonia et al. 2002).

2.4 Stomata and Photosynthesis

Ever since the discovery of "*Stomata als Glieder eines schwingungsfähigen CO_2-Regelsystems*" (Raschke 1965), stomata have been a key example of cybernetics in plants. Ultradian oscillations of stomatal aperture and conductivity for water vapour have been reported over decades (Barrs 1971; Teoh and Palmer 1971; Cowan 1972a,b; Kaiser and Kappen 2001; Roelfsema and Hedrich 2002). Primary external signals of control parameters are hydraulic effects and leaf internal airspace CO_2 (Willmer 1988) affecting a highly complex internal regulation network. Cytoplasmic Ca^{2+} levels play a key role as a central node in this network (McAinsh et al. 1997), and a plasma membrane Ca^{2+} channel may be an important oscillator in the system (McAinsh et al. 1995).

However, just as for circadian rhythmicity of stomata and photosynthesis (see last year's review in this series: Lüttge 2002), due to the intimate interaction of photosynthetic metabolism and gas exchange it is difficult and often impossible to separate oscillations of stomata and photosynthesis. As Giersch (1994) reminds us, discovery of ultradian photosynthetic oscillations dates back to more than half a century ago (van der Veen 1949). They have been reviewed comprehensively (Walker 1992, Giersch 1994). Giersch presents an intriguing diagram (Giersch 1994: see Fig. 2), which poses the question "where is the oscillator"? just as in the enigma of and the quest for the central clock in circadian rhythmicity (see Lüttge 2002). Oscillations of about a period of 1 min are well sustained in an intact leaf, highly damped in isolated protoplasts and reduced to some irregularities in isolated chloroplasts. Notwithstanding the possibility of artefacts due to isolation, this strongly suggests that oscillations are consequences of the integration of functional subsystems. This encompasses oscillating metabolites, such as 3-phosphoglyceric acid, ribulosebisphosphate, NADPH/NADP, ATP/ADP and inorganic phosphate (Giersch 1994) and the regulatory metabolite fructose 2,6-bisphosphate (Stitt et al. 1988). The detailed metabolic mechanism of the oscillations to date is not clear (Giersch 1994). One explanation was temporary imbalance in the supply of NADPH and ATP to the photosynthetic reductive carbon cycle (Laisk et al. 1991), hence, imbalance between photochemical electron transport and CO_2-fixation capacity (Stitt and Schreiber 1988; Stitt et al. 1988). In this respect, however, Raghavendra et al. (1995) found interesting differences between the C_3 plant *Spinacia oleracea* and the C_4 plant *Amaranthus caudatus*. In C_4 plants oscillations are rarely observed under conditions readily producing oscillations in C_3 plants, i.e., dark/light-transitions at saturating CO_2, while oscillations occur after a sudden dark/light-transition at a lower light intensity, which would not allow oscillations in C_3 plants. It is suggested by these authors that in *Amaranthus* water oxidation does not occur as synchronously with CO_2 uptake as in *Spinacia* and that, by contrast to the C_3 plant, in the C_4-plant photosynthetic oscillations are due to co-operation of different cell types with the diffusional transport of photosynthetic intermediates between mesophyll and bundle-sheath cells.

Oscillations of stomata and photosynthetic output parameters begin after abrupt changes of external parameters such as light, CO_2, O_2, H_2O and temperature (Walker 1992; Giersch 1994). However, there are also spontaneous stomatal oscillations (Raschke 1975). In a seminal review, Giersch (1994) considers the various possible metabolic control points and discusses two different mathematical minimal or "skeleton" models (for rationale also see Hütt and Lüttge 2002) simulating oscillations. One model is based on a "two kinase-hypothesis" (see also Giersch et al. 1991), with phosphoglycerate and phosphoribulose kinases, and another one on a "coupling-hypothesis", with the interdependence of photosynthetic electron flow and photophosphorylation, and evidence is in favour of the first of these two hypotheses.

Oscillations can be a first step on the route to (deterministic) chaos (Giersch 1994). A theoretical hydraulic model predicted the possibility of Hopf bifurcation in stomatal responses (Rand et al. 1981). Chaos can be studied in entrainment experiments where the responses to externally imposed oscillations of different frequencies are analyzed. In this way chaotic time series where obtained (1) of electrical membrane activity by rhythmic stimulation of *Nitella* cells (Hayashi et al. 1982; see Sect. 2.1), (2) of glycolysis by rhythmic supply of glucose (Markus et al. 1984), and (3) of peroxidase-oxidase activity by rhythmic supply of oxygen (Olsen 1983; Olsen and Degn 1985).

For similar studies of ultradian oscillations of stomata and photosynthesis external oscillations of O_2 (Brogardh and Johnsson 1974; Cardon et al. 1994a) and especially light are used. Fast Fourier transform analysis (FFT) and resonant analysis approach (RAA) can then be applied to overt time series. RAA has been used to investigate kinetics of membrane-associated processes (Hansen 1978, Hansen et al. 1987, Vanselow et al. 1989). It unravels hidden oscillatory components and quantifies some parameters of feedback control systems. Thus, using extracellular bioelectric measurements of plant leaves under the influence of rhythmical illumination, Shabala (1997) was able to distinguish the light-induced bioelectric responses of the photosynthetically active mesophyll, i.e., a resonant mode of a 1.5-min period linked to thylakoid-related processes as well as a feedback interaction involving the membrane potential and the cytosolic pH in mesophyll cells (resonant mode at a period of 8 min), from stomatal movements based on ion exchange of the guard cells (resonant mode for a period of 40 min). Generally, stomata-related oscillations of CO_2 assimilation (Siebke and Weis 1995a) appear to have much lower frequencies than photosynthetic CO_2 fixation (Walker et al. 1983; Laisk et al. 1991; Walker 1992). The RAA-studies of Shabala et al. (1997a) revealed bifurcations of a period-doubling cascade, i.e., the Feigenbaum route to chaos, and other routes to chaos are also discussed.

Synchronization/desynchronization of individual oscillators or leaf patches also plays a role in whether or not overall gas exchange of a leaf is seen to oscillate. Cardon et al. (1994b) have demonstrated patchy distribution of stomatal oscillations in sunflower using chlorophyll fluorescence imaging and gas exchange techniques. Siebke and Weis (1995a,b) used chlorophyll fluorescence imaging. With the assimilation images obtained they observed that rapid changes in gas composition initiated oscillations in net gas exchange (H_2O and CO_2) which changed to non-synchronous oscillations due to slight local variations in the period. Obviously oscillations in net gas exchange died out, but assimilation persisted oscillating non-homogenously and non-synchronously over areas, patches or spots of the leaves (Siebke and Weis 1995a). Gas diffusion within the leaf may play a role (Cardon et al. 1994b; Siebke and Weis 1995a), but minor vein distribution and its mediation of transport processes (e.g., sugar export) also appears to be involved (Siebke and Weis 1995b).

2.5 Macroscopic Events of Turgor Movements: Leaves, Roots and Circumnutation

Among the macroscopic events of turgor movements the most spectacular ones appear to be the pulvinar movements of leaflets of *Desmodium motoricum* with a period length in the minute range related to oscillations of apoplastic K^+ and H^+ and electrical activities (Antkowiak and Engelmann 1989, 1995; Antkowiak et al. 1991; Antkowiak and Kirschfeld 1992; for circadian pulvinar movements see also last year's review by Lüttge 2002).

Another intriguing phenomenon is the rhythmic firing movement (ca. 7 min) of the stamens and style column of *Stylidium graminifolium* (Findlay and Findlay 1981).

Nutation of tendrils, shoots and roots is related to localized rhythmically changing turgor and perhaps also extension growth (see Sect. 2.6.). Nutations may occur as pendulum-type movements restricted to one plane or as elliptical or circular movements around the longitudinal axis of growth (circumnutation). Oscillatory nutation is known since Charles Darwin (Darwin and Darwin 1880) and has been documented and reviewed regularly (Gradmann 1926; Johnsson and Heathcote 1973; Johnsson 1979; Brown 1993; Barlow et al. 1994). Nutational oscillations with period lengths in the range of about one to a few hours can be overlain by so-called micronutations with a five to ten times higher frequency and a very small amplitude of maximally 1 mm (Heathcote 1966). Oscillations with very different frequencies exist simultaneously.

Rhythmic nutation was originally closely related to gravity responses in a gravity feedback model because with larger deviations from the vertical plumb-line the gravity vector becomes important (Israelsson and Johnsson 1967; Johnsson and Heathcote 1973; Andersen 1976; Johnsson 1979). However, studies under near weightlessness in space show that nutations persist independent of gravity although gravity may somewhat modulate them (Antonsen et al. 1995; Johnsson 1997), which is also supported by circumnutation observed in the absence of the gravity-sensing root cap (Shabala and Newman 1997). This suggests operation of an internal oscillator as already assumed in a model emanating from Darwin and Darwin (1880) in contrast to a gravitropic overshoot model (Israelsson and Johnsson 1967).

Hence, we must look for an endogenous oscillatory process to understand circumnutation. In roots this is clearly localized to the elongation zone (see Sect. 2.7) and correlated to H^+ flux oscillations, i.e., H^+ exchange between the cytosol and the cell wall (Shabala and Newman 1997). In twining shoots of *Phaseolus vulgaris* changes in the distribution of K^+ and putative related variations of water content and turgor in the cells of the bending zone were related to the movement (Badot et al. 1990; Caré et al.

1998). In tendrils of *Ipomoea* extracellular electrical potential variations had a period in accordance with the nutations (Claire et al. 1985).

In fact, for the nutational movements it is required that reactions travel around the plant organ and that there are waves of coupling/uncoupling of adjacent cells as for K^+ concentrations (Badot 1987; Badot et al. 1990), pH, Ca^{2+}, phosphate and chloride (Badot 1987). This has been described by nonlinear reaction–diffusion equations on an annulus corresponding to a ring of cells in a cross section of the circumnutating organ (Lubkin 1994; Johnsson 1997).

The role of Ca^{2+} is central for these oscillations. Based on studies of tendrils and noting the importance of compartmentation and the involvement of Ca^{2+} and calmodulin-regulated activities of the cellular cytoskeleton (Engelberth et al. 1995), Weiler (1997) developed the model of an autonomous calcium oscillator in plants.

2.6. Extension Growth

Extension growth is associated with H^+ extrusion into the apoplast in the elongation zone. Elongation shows ultradian rhythms in both shoots (Kristie and Jolliffe 1986) and roots.

Around roots there are oscillating electrical fields now known for almost fifty years (Scott 1957, 1962; Jenkinson and Scott 1961; Jenkinson 1962). They are strongest in the elongation zone (Toko et al. 1990, Shabala and Newman 1998). In bean roots they correlate in their period (6 to 12 min) with oscillations of elongation (Souda et al. 1990). There are oscillations of longer periods (4 and 7 h in roots of *Trifolium repens*, Macduff and Dhanoa 1996; 1.5 h in corn roots, Shabala et al. 1997b) and shorter periods (7 min in corn roots, Shabala et al. 1997b), i.e., different lengths of time in the same tissue. The ionic basis is given by H^+ and Ca^{2+} ions. H^+ and Ca^{2+} oscillations are closely correlated (Shabala et al. 1997b, Shabala and Newman 1998). Shabala et al. (1997b) argue that the fast 7-min oscillations in corn roots are due to oscillations of the activity of electrogenic H^+-ATPase, while the slow 1.5-h oscillations originate from physical coupling of passive H^+ transporters. The spatial periodic pattern along roots disappears under anoxia and depends on aerobic metabolism most likely because of respiration-dependent H^+-pumping (Yoshida et al. 1988; Hecks et al. 1992).

Not only transport and electrical activities at the root surface show ultradian oscillations but also transport across the roots. Xylem exudation displays 8-min pulses, where Gd^{3+}-inhibited Ca^{2+}-channels are involved (Schwenke and Wagner 1992).

3 Theory: "What Can We Learn from Ultradian Rhythms?"

In the sections above, describing phenomena of ultradian rhythms in plants, several keywords from the realm of nonlinear dynamics have come up, such as feedback, waves and nonlinear reaction-diffusion equations, synchronization, networks and many more. Routes to chaos with bistability, Hopf bifurcations and Feigenbaum series have been touched. A closer, more theoretical examination of these significant notations can tell us more about what we can generally learn from ultradian rhythms. We start from sorting the confusingly entangled topics related to these keywords.

Whenever one attempts to formulate a mathematical model of a biological system and relates the dynamic variables (e.g. metabolite pools, population sizes, concentrations) via linear interactions (fluxes, migration, diffusion processes, etc.), the theoretical description takes place within the realm of *linear systems theory*. Only when nonlinear interactions [feedback loops, (auto)catalytic effects, enzymatic regulations, cooperation, etc.] are included one enters the domain of *nonlinear dynamics*. This step leads to a far more complex repertoire of dynamical features possible in the model system, in particular stable (limit cycle) oscillations and, under certain conditions, deterministic chaos. When a next stage of complexity, namely noise and fluctuations, is taken into account one is confronted with additional dynamical effects. Noise can *destroy* order or, in certain nonlinear systems, it may *create* order. Noise-induced transitions and even noise-enhanced structure formation have been explored in theoretical model systems (Garcia-Ojalvo and Sancho 1999; Jung and Mayer-Kress 1995) as well as in nature (see, e.g., Beck et al. 2003 for an overview of recent activities in this field). Ever since the first theoretical characterization (Lorenz 1963) of deterministic chaos, attempts have been made to find such irregular, yet minimally generated (low-dimensional) dynamics in a natural system. While some of the early findings now have to be regarded as "random" (i.e. without a low-dimensional deterministic process behind it), in many cases we can now be certain that a given system, indeed, displays chaotic behaviour (see, e.g., Mosekilde and Mouritsen 1995 and Kendall et al. 1997, and the biological examples in Strogatz 1994 and Kaplan and Glass 1995). In spite of such success, it remains unclear to what extent biological systems functionally exploit this behaviour (Olsen and Degn 1985; Kuznetsov and Rinaldi 1996; Glass 2001; see also Fox and Hill 2001 for an interesting new point of view in this debate). One could also imagine these forms of dynamics to be only futile additions to the physiologically important regulational properties arising from the dynamical complexity of efficient regulation. In fact, similar questions have been posed with respect to ultradian oscillations themselves in the early 1980s (see the discussion in Hess and

Markus 1985; Olsen and Degn 1985). At that time it seemed possible that feedback regulation necessary for generating a stable steady state leads to stable oscillations in a certain range in parameter space as a by-product, which has no physiological significance. The remarkable property of frequency-coding in calcium oscillations has much contributed to appreciating ultradian oscillations as an important functional feature of biological systems (Woods et al. 1986; Meyer and Stryer 1991; Bock and Ackrill 1995). Indeed, it is well possible that even chaotic dynamics, if observed, serves a biological function. Adaptivity and plasticity may be important keywords, when discussing the value for a biological system to display chaos (Kuznetsov and Rinaldi 1996; Solé et al. 1996;).

Why are such questions important for biology? While noise and chaos often seem indistinguishable at a first glance, a major difference exists: For a chaotic system, one may have the hope to fully incorporate the dynamics in a (low-dimensional) deterministic mathematical model, whereas noise can only be accounted for by stochastic forces acting upon the deterministic part of the system. The discussion becomes even more difficult due to the complex interplay between noise and the dynamics of a natural system. Noise is a ubiquitous phenomenon and, in most cases, a disturbance to the measuring process. From the point of view of nonlinear dynamics, however, it may very well influence a system's dynamics in a more sophisticated way than just by overshadowing it. It is possible that some biological process (e.g., the detection of a signal) is only functioning in an optimal way at some intermediate noise level present in the system (Douglass et al. 1993; Bezrukov and Vodyanoy 1995; Gammaitoni et al. 1998; Anishchenko et al. 1999; Moss 2000). With lower or higher noise intensity the efficiency of the process will decrease. A schematic example of such a *stochastic resonance* is given in Fig. 1. This phenomenon is possible only in a nonlinear system (in our schematic example of Fig. 1, the threshold serves as nonlinearity). It is by now theoretically well understood and has been observed, together with the related effect of coherence resonance (Pikovsky and Kurths 1997), in a variety of natural and model systems (Longtin et al. 1991; Moss et al. 1994; Bezrukov and Vodyanoy 1995; Lee et al. 1998; Beck et al. 2001). Thus, in the last years it has been appreciated that unveiling the effect of noise in a system can substantially contribute to the understanding of the processes within the system. This development has taken the focus away from chaos. Nevertheless, it is still true that finding a chaotic behaviour in a natural system gives a lot of information on how to formulate a mathematical model of it.

Often one considers not only a single model system, but rather the collective behaviour of many identical (or almost identical) copies of such a system. Such investigations are situated in the vast domain of *self-organ-*

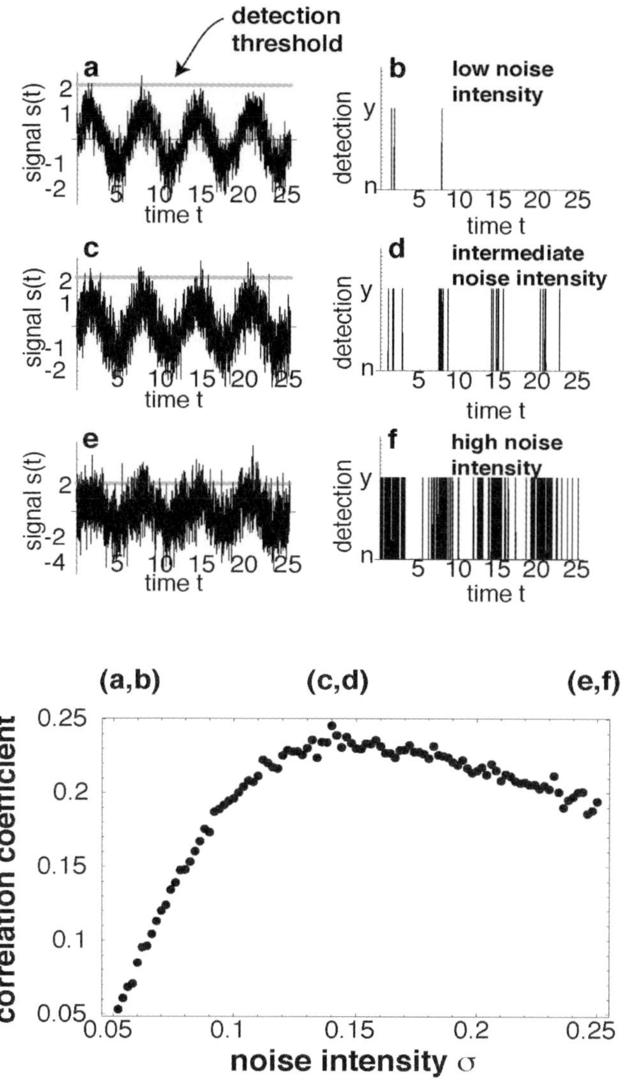

Fig. 1. Schematic example of a system displaying stochastic resonance. A signal *s(t)* consisting of a sine wave and additive noise enters some detection mechanism with a threshold; **a**, **c** and **e** denote different noise intensities σ. At each point in time, the signal is translated into "yes" (*y*) or "no" (*n*), depending upon whether the signal reaches above the threshold or not (**b**, **d** and **f**). One sees that the sequence of maxima present in the pure signal (i.e. in the sinusoidal oscillation without noise) is best reproduced by this device at an intermediate noise intensity. This visual impression is quantitatively supported by the (cross)correlation coefficient of the pure signal with the detected signal as a function of the noise intensity σ (*right-hand side*). In a natural system, the more frequent case is that a (almost) pure signal enters a system with a noisy threshold. Mathematically, these two cases are equivalent. (Adapted from Hütt and Lüttge 2002)

izing complex systems. One can then study the effect of noise on the dynamics of individual nonlinear elements and on the coupling between these elements in the formation of spatiotemporal patterns.

A yet more complicated setting emerges, when the concept of *networks* enters the description lined out so far. There are basically two distinct complications arising from the network aspect of interacting systems. First, the individual model system designed with the methods from nonlinear dynamics may, in a realistic biological situation, only be a part of a meta-system far more complicated than the isolated piece for which a mathematical description has been attempted. Second, the step from the dynamics of the individual model system to the behaviour of many identical (or similar) copies may require arrangement of the elements in a more complicated way than, e.g., on a regular grid. And this arrangement (the "architecture" at the system) may strongly influence the dynamical properties of the system (e.g., whether synchronization takes place, or how the system responds to external perturbations).

In light of this general background, one can now formulate the theoretical aims of investigating ultradian rhythms. From our point of view, three general questions are in the centre of interest in such a theoretical analysis. First: *time scales*. Which properties of the system determine the time scales present in the system? Can one find "universal time scales" (or ranges of time scales) shared by all or most systems capable of generating ultradian rhythms? Under which rules do time scales coexist? What observable properties are the result of interference of such coexisting time scales? Second: *responses*. How robust are ultradian rhythms with respect to external perturbations (e.g. external periodic driving or external noise)? Is such a "robustness", if it exists, a system property following immediately from the physiological realization of ultradian rhythms? Can a lack of robustness be seen as a key difference to circadian rhythms? Third: *relation to circadian oscillations*. Can one construct circadian rhythms (or certain forms of circadian oscillation) from ultradian rhythms (for example in the sense of producing a robust circadian oscillation by interference of many fast – ultradian – oscillations)? Some of these questions can be elucidated using simple mathematical model systems. This use of minimal models frequently applied to biological systems (see, e.g., Hütt and Lüttge 2002) has, however, a certain complication in the case of ultradian rhythms. As pointed out in sections 1 and 4, evolutionary pressure seems to act less strongly (or less coherently) on ultradian rhythms than, e.g. on circadian oscillations. One, therefore, expects a much wider range of physiological realizations. One of the key strengths of mathematical modelling, its unifying power, consequently seems not to apply. Indeed, comparatively small differences between the observed dynamics become much more important, as they may provide the key for distinguishing between the possible physiological mechanisms. Such small differences may be given by frequency ranges, curve forms, phase shifts between dynamic variables, possible additional forms of dynamics and their basins of attractions. Nevertheless, from a theoretical point of view the discussion of "minimality" is still an important issue in the study of ultradian rhythms. Using simple model systems one can see how a generic oscillator responds to a change of internal properties or external conditions.

How are ultradian rhythms related to the theoretical framework outlined above? In our opinion the key to putting this biological phenomenon into such a global theoretical perspective lies in the term "robustness".

When methods from nonlinear dynamics are applied to model or analyze ultradian rhythms an important topic is that of biological relevance of complex dynamical phenomena. In Section 2 we described a wide variety of empirical findings related to such complex dynamics. In the following, we will demonstrate that the notion of robustness is crucial for evaluating the relevance of such findings for biological systems in vivo. We have seen that a considerable amount of empirical investigations of ultradian rhythms focuses intensely on the notions and concepts of nonlinear dynamics. In particular the phenomenon of deterministic chaos and the different transitions from regular (e.g., oscillatory) to chaotic dynamics (the "routes to chaos") are frequently discussed. With a certain minimal model of a period-doubling route to chaos, the logistic finite-difference equation (or "logistic map") we will demonstrate what we understand by robustness. The most important future observation will be, how the bifurcation pattern changes when two such systems are interacting, and what properties of this route to chaos are robust with respect to such an insertion of additional degrees of freedom.

Biochemical reactions rarely occur as isolated phenomena. They rather are part of an entire network of interacting reaction channels and signalling pathways. This particular perspective, that the network structure of a biological system or biochemical process regulates system properties, has emerged in the last few years in the theoretical literature (Watts and Strogatz 1998; Albert and Barabasi 2002; Ravasz and Barabasi 2003) also because first full analyses of the topology of a real biological network has been performed (Jeong et al. 2000). One remarkable result of these recent investigations is that in contrast to theoretical expectations (see, e.g., the pioneering work of Kauffman 1974) only very few cases of deterministic chaos have been observed in real biological systems under in vivo conditions (i.e., when the entire network of interacting biological and biochemical processes is taken into account; see Fox and Hill 2001).

The concept of including additional degrees of freedom always comes into play, when the system (either in an experiment or represented in terms of a mathematical model) should not be considered as an isolated entity but rather as a subsystem within a complex network of systems permanently in interaction with each other. Such an additional layer of complexity arises in biological systems in two cases: First, when instead of the system at hand the spatial arrangement of many such systems becomes important. In this case, the synchronization and desynchronization of these systems have to be taken into account, when discussing the overt (integrated) dynamics. Instead of purely temporal dynamics, one then has to consider the spatiotemporal dynamics of the full system (Rascher et al. 2001; Beck et al. 2003). Second, when the system under consideration is part of a larger system interacting with the subsystem. In this case the network properties will influence the dynamics produced by the subsystem (Hartwell et al. 1999; Ravasz and Barabasi 2003).

Sometimes the influences of the system as a whole on a subsystem are acting on very different time scales than relevant for the discussion of the subsystem. In such cases the effect of the whole system on the subsystem can be approximated either by noise entering the subsystem (when the influence is restricted to fast dynamics) or as a drift in the subsystem's parameters (when the influence of the full system upon the subsystem consists of slow dynamics).

In all these cases the comparability of the model system (given either in terms of an experimentally isolated copy of a certain part of the biological system or as a mathematical model describing such a subsystem) with the real biological situation has to be studied carefully.

As already mentioned above, for estimating the biological relevance of a route to chaos it is instructive to look at one particular case, where robustness of a dynamical phenomenon with respect to increasing the complexity of the system is particularly obvious, namely the famous period-doubling route to deterministic chaos. To this end we extend the usual logistic finite-difference equation (or "logistic map")

$$x_{t+1} = f_x(x_t) = R_x x_t (1-x_t) \tag{1}$$

to a system consisting of two coupled equations

$$\begin{aligned} x_{t+1} &= f_x(x_t) + D\,[f_y(y_t) - f_x(x_t)] \\ y_{t+1} &= f_x(y_t) + D\,[f_x(x_t) - f_y(y_t)] \end{aligned} \tag{2}$$

or, more explicitly,

$$\begin{aligned} x_{t+1} &= (1-D)\,R_x x_t (1-x_t) + DR_y y_t (1-y_t) \\ y_{t+1} &= (1-D)\,R_y y_t (1-y_t) + DR_x x_t (1-x_t) \end{aligned} \tag{3}$$

The coupling terms (i.e. all terms proportional to the coupling constant D) could be named "migration terms" when keeping in mind the original biological motivation of the logistic map from population biology. The dynamics of the simple logistic map, Eq. (1), as well as some remarks of its role in plant biology can be found in (Hütt and Lüttge 2002). Due to the rich dynamics produced by Eq. (1) it is an excellent system for getting used to some of the concepts of nonlinear dynamics. Here, we will only very briefly mention its most important properties. For a more thorough review on this particular equation, see May (1976), Kaplan and Glass (1995) and Hütt (2001). Equation (1) contains a single control parameter, R_x, which regulates the general properties of the time evolution of the dynamical variable x_t. In order to fully determine the time course of x_t, the value of R_x has to be specified, as well as a starting value (the "initial condition") x_0. One can now study the dynamic behaviour of Eq. (1) under variation of R_x. The entire repertoire of possible time evolutions can be condensed into a single representation, the *bifurcation diagram*, which is shown in Fig. 2, together with three examples of time series. At low R_x the system has a stable fixed point, i.e. iteration transports all starting points to a value x_F, where it remains with continuing iteration. The value of x_F only depends slightly on R_x. At a certain value of R_x, however, the fixed point becomes unstable and a period-2 cycle takes over. Now the size (or amplitude) of the oscillation depends on the distance from this bifurcation point: As R_x increases, the amplitude becomes larger. At a further increase of R_x a series of period-doubling bifurcations appearing at ever smaller intervals in R_x takes the system via a period-4 cycle, a period-8 cycle, etc., to an irregular behaviour called deterministic chaos. In this regime small differences in the initial value lead after a few time steps to completely different (i.e. uncorrelated) time series.

Starting from this well-known model system of nonlinear dynamics, we can now formulate our thought experiment. Let us assume that Eq. (1) represents, to a certain extent a model of an ultradian oscillator: In a certain range of the control parameter R_x a stable oscillation is observed, followed, as R_x is increased, by a period-doubling route to deterministic chaos. Let us further assume that some upper and lower value for R_x exists, which delimit the biologically relevant parameter range. Values of R_x outside this range are assumed to be impossible or irrelevant for a biological system. Such a limitation is frequently encountered, when a description of a biological situation is attempted. In our case, this range serves as a reference point for visualizing the effect of coupling on the dynamics of the individual systems. Figure 3 shows some bifurcation diagrams for two coupled logistic equations given by Eqs. (2) and (3). Starting from the original (uncoupled) case (*upper left diagram*), coupling is introduced and the difference be-

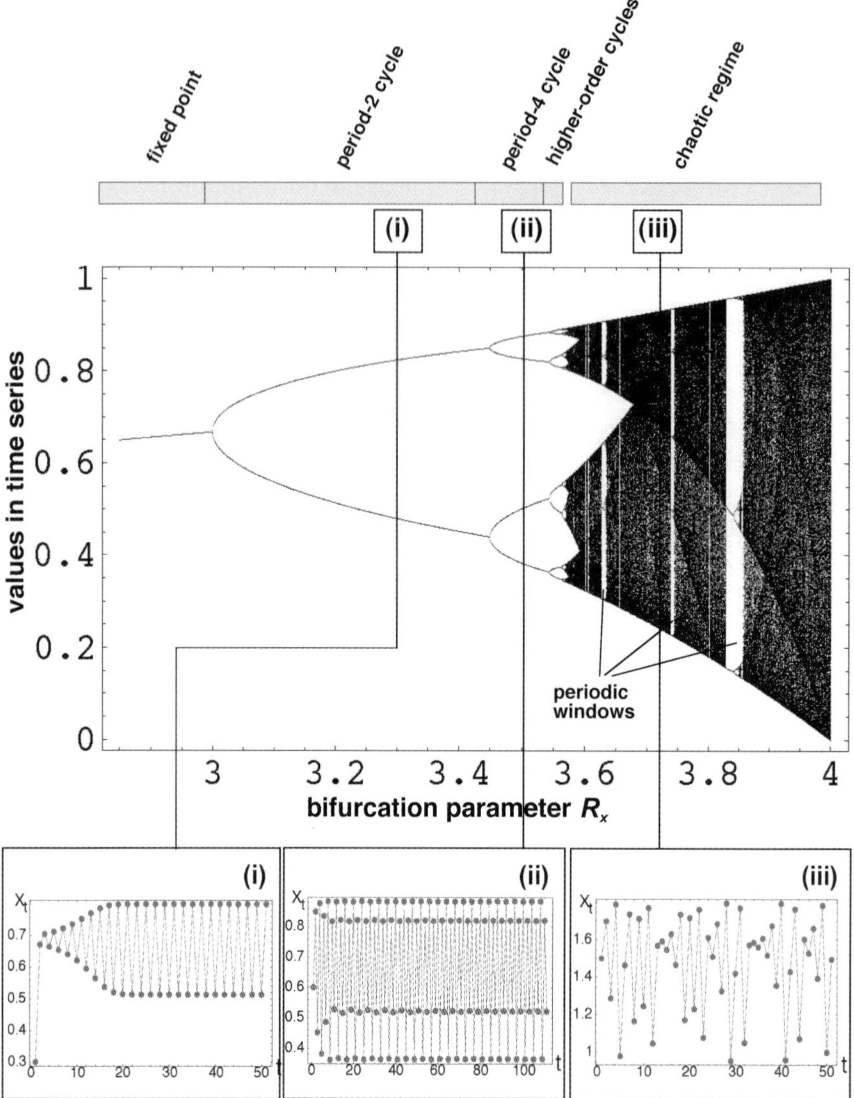

Fig. 2. Study of the dynamics produced by the logistic map, Eq. (1), for different values of the control parameter R_x. The most important information on the R_x dependence of the time development of x_t can be summarized in a bifurcation diagram, where the possible values of x_t are shown as a function of the control parameter R_x. In the bifurcation diagram of the logistic map, one can immediately see the various period-doubling bifurcations, which dominate the dynamics over a wide range of the control parameter R_x and ultimately lead to deterministic chaos, when R_x is further increased. Time series are shown for three values of R_x, namely R_x=3.3 (stable oscillation called period-2 cycle), R_x=3.5 (period-4 cycle) and R_x=3.72 (deterministic chaos)

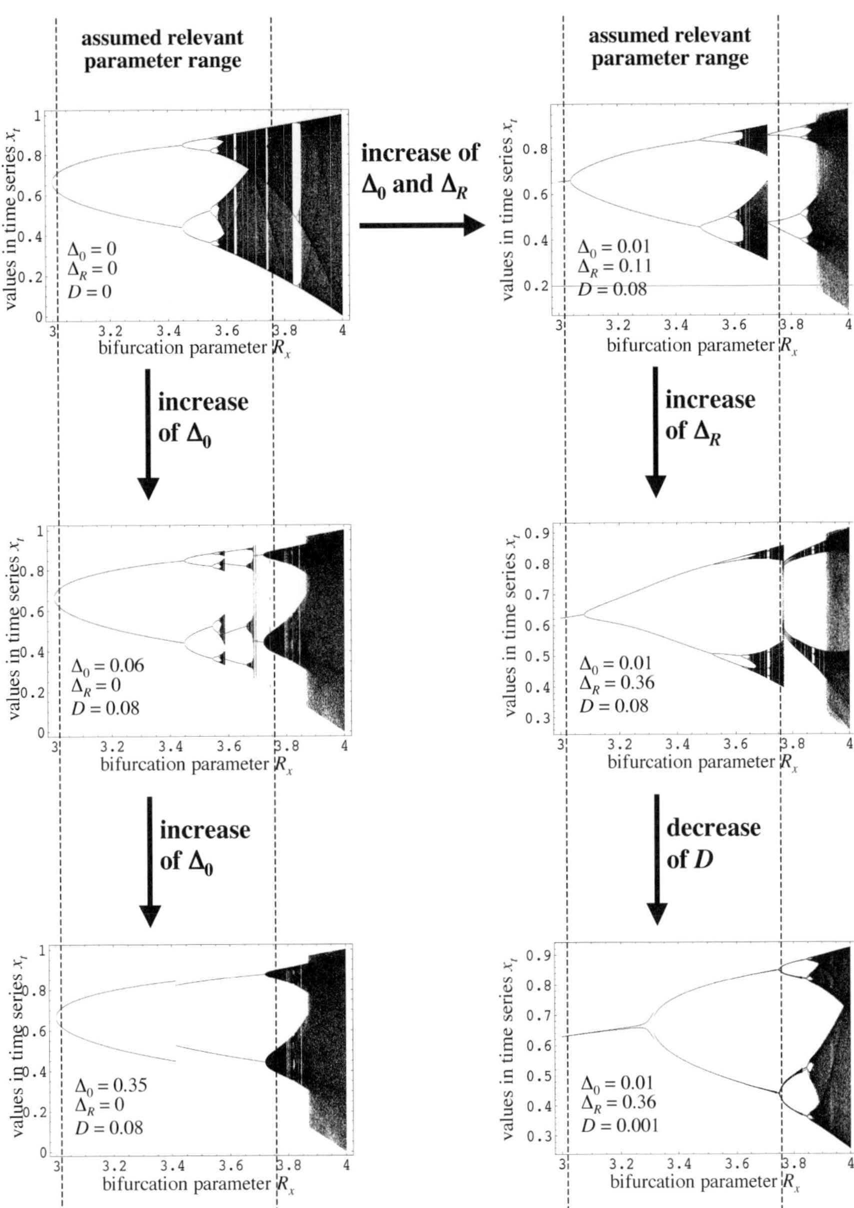

Fig. 3. Bifurcation diagrams of two coupled logistic maps, Eqs. (2) and (3), for different values of the bifurcation parameter difference Δ_R, coupling D and initial value difference Δ_0. Choices of parameter values are indicated as *insets* in each bifurcation diagram. The *upper left diagram* (uncoupled case; D=0) corresponds to the bifurcation diagram shown in Fig. 2. The *vertical dashed lines* show the assumed biologically motivated restrictions to the parameter range

tween initial values, $\Delta_0=y_0-x_0$, is increased (*left column*), as well as the difference between the bifurcation parameters, $\Delta_R=R_y-R_x$. It is seen that the bifurcation diagram undergoes drastic changes due to coupling. In particular, the ranges in R_x, where chaotic behaviour is observed, are in some cases highly suppressed. This is seen most strikingly, when one assumes biologically motivated restrictions to the parameter range, as outlined above. Figure 3 contains such a (fictitious) parameter range, which for $D=0$ contains a large portion of the chaotic regime. For small coupling and intermediate Δ_R, however, all remaining chaotic behaviour lies outside this range (*lower right diagram*).

The case of coupled logistic equations, Eqs. (2) and (3), together with some implications of coupling for our understanding of biological phenomena, have been discussed by Lloyd (1995), Kendall and Fox (1998) and Solé and Gamarra (1998). We consider them here as a very simple model representation of some of the striking phenomena of nonlinear dynamics easier to discuss and richer in dynamics than their continuous counterparts. Notwithstanding, the differential equations, finite-difference equations, in their own right, are an interesting starting point for both, mathematical simulation and analysis in biology (see Hütt et al. (2002) for an example from plant biology).

Both from the point of view of modelling as well as from data analysis the general concepts of nonlinear dynamics provide some guidelines (as well as some caveats!) for addressing the questions outlined above. Currently, one observes a focus on the study of isolated in vitro systems and relating these findings to the large catalogue of behaviours put forward by theoretical studies in nonlinear dynamics. In particular, the step, via simple interacting systems, towards complex biochemical and biophysical networks may eventually help to develop a clear and systematic understanding of the minimal realizations and the various biological implementations of ultradian rhythms.

4 Functions?

"In search for a biological hour" Koukkari et al. (1997) list a large number of ultradian oscillations of bacteria, amoebae, yeast, higher plants and mammals with periods ranging from 30 to 240 min and averaging 95 min. However, an analogy to the hour appears to be a curiosity rather than a functional trait. The hour is not a natural environmental period. Even in the discussion of circadian rhythmicity, we cannot be totally sure if oscillations are physiologically important and provide increased fitness or only a by-product of time constants of endogenous processes evolved under continuous entrainment by the natural diurnal cycle (Lüttge 2002). Thus, several authors conclude that ultradian oscillations are merely side effects of regulatory systems (Giersch 1994; Siebke and Weis 1995b).

Conversely, various cell physiological functions of ultradian oscillations have been suggested. They may be crucial for the synchronization and co-ordination of spatially separated physiological processes (Lloyd and Kippert 1993) as well as for the separation of incompatible processes within the same cellular compartment (Rapp 1987). In enzyme reactions involving potentially toxic reactive intermediates, oscillatory dynamics lower the average concentration of toxic compounds such as H_2O_2, superoxide and other reactive oxygen species in the peroxidase–oxidase reaction (see Sect. 2.2; Hauser et al. 2001). Oscillations increase the thermodynamic efficiency of metabolism lowering energy dissipation (Lazar and Ross 1990). Ca^{2+} oscillations may transmit stimulus-specific information to a downstream signalling machinery involving Ca^{2+}-dependent protein phosphorylation (McAinsh et al. 1995). In general, oscillations may encode information in signal transduction chains or networks (Berridge et al. 1998, Shabala et al. 2001), with, e.g., Ca^{2+}, or also the H_2O_2 signal emerging from the peroxidase–oxidase reaction, where frequency-encoded regulation is more accurate than amplitude-encoded regulation (Møller et al. 1998). That ultradian oscillations may represent a frequency-encoding mechanism for environmental information (Shabala et al. 2001) may be also supported by Fourier transform analysis of natural light fluctuations in Tasmania, which revealed frequency peaks of power spectra corresponding to periods of 3, 6 and 14 min, i.e., in the same range as in ultradian oscillations of photosynthesis (Shabala et al. 1997a). This is highly interesting in an evolutionary context.

References

Albert R, Barabási A-L (2002) Statistical mechanics of complex networks. Rev Mod Phys 74:47–97

Andersen H (1976) A mathematical model for circumnutations. Report 2/76, Department of Electrical Measurements, Lund Institute of Technology, Sweden

Anishchenko V, Moss F, Neiman A, Schimansky-Geier L (1999) Stochastic resonance: noise induced order. Sov Phys Usp 42:7–36

Antkowiak B, Engelmann W (1989) Ultradian rhythms in the pulvini of *Desmodium motorium* – an electrophysiological approach. J Interdiscip Cycle Res 20:164–165

Antkowiak B, Engelmann W (1995) Oscillations of apoplasmic K^+ and H^+ activities in *Desmodium motorium* (Houtt.) Merril. pulvini in relation to the membrane potential of motor cells and leaflet movements. Planta 196:350–356

Antkowiak B, Kirschfeld K (1992) Enflurane is a potent inhibitor of high conductance Ca^{2+}-activated K^+ channels of *Chara australis*. FEBS Lett 313:218–221

Antkowiak B, Mayer WE, Engelmann W (1991) Oscillations of the membrane potential of pulvinar motor cells in situ in relation to leaflet movements of *Desmodium motorium*. J Exp Bot 42:901–910

Antonsen F, Johnsson A, Perbal G, Driss-Ecole D (1995) Oscillatory growth movements of roots in weightlessness. Physiol Plant 95:596–603

Badot PM (1987) Approche cellulaire du mécanisme du movement révolutif des tiges volubile. Étude de quelques paramètres physico-chimiques. Ann Sci Univ Besançon 4ème Sèr 8:53–110

Badot PM, Melin D, Garree JP (1990) Circumnutation in *Phaseolus vulgaris*. II. Potassium content in the free-moving part of the shoot. Plant Physiol Biochem 28:123–130

Barlow PW, Parker JS, Brain P (1994) Oscillations of axial plant organs. Adv Space Res 14:149–158

Barrs HD (1971) Cyclic variations in stomatal aperture, transpiration, and leaf water potential under constant environmental conditions. Annu Rev Plant Physiol 22:223–236

Bauer CS, Plieth C, Bethmann B, Popescu O, Hansen UP, Simonis W, Schoenknecht G (1998) Strontium-induced repetitive calcium spikes in a unicellular green alga. Plant Physiol 117:545–557

Beck F, Blasius B, Lüttge U, Neff R, Rascher U (2001) Stochastic noise interferes coherently with a model biological clock and produces specific dynamic behaviour. Proc R Soc Lond B 268:1307–1313

Beck F, Hütt M-T, Lüttge U (eds) (2003) Nonlinear dynamics and the spatiotemporal principles of biology. Nova Acta Leopoldina, vol xx, Deutsche Akademie der Naturforscher Leopoldina, Halle

Berridge MJ, Rapp PE (1979) A comparative survey of the function, mechanism and control of cellular oscillators. J Exp Biol 81:217–279

Berridge MJ, Bootman MD, Lipp P (1998) Calcium – a life and death signal. Nature 395:645–648

Bezrukov SM, Vodyanoy I (1995) Noise-induced enhancement of signal transduction across voltage-dependent ion channels. Nature 378:362–364

Bier M, Bakker BM, Westerhoff HV (2000) How yeast cells synchronize their glycolytic oscillations: a perturbation analytic treatment. Biophys J 78:1087–1093

Bock GR, Ackrill K (eds) (1995) Calcium waves, gradients and oscillations. Ciba Foundation Symposia 188. Wiley, Chichester

Bracker CE, López-Franco R, Bartnicki-Garcia S, Murphy DM, Howard RJ (1995) Satellite Spitzenkörper, pulsed growth, and the determination of cell shape in growing hyphal tips of fungi. Proc R Microsc Soc 30:17

Brogardh T, Johnsson A (1974) Oscillatory transpiration and water uptake of *Avena* plants. 4. Transpiratory response to sine shaped light cycles. Physiol Plant 31:311–322

Bronnikova TV, Schaffer WM, Hauser MJB, Olsen LF (1998) Routes to chaos in the peroxidase – oxidase reaction. 2. The fat torus scenario. J Phys Chem B 102:632–640

Brown AH (1993) Circumnutations: from Darwin to space flights. Plant Physiol 101:345–348

Cardon ZG, Berry JA, Woodrow IE (1994a) Dependence of the extent and direction of average stomatal response in *Zea mays* L. and *Phaseolus vulgaris* L. on the frequency of fluctuations in environmental stimuli. Plant Physiol 105:1007–1013

Cardon ZG, Mott KA, Berry JA (1994b) Dynamics of patchy stomatal movements, and their contribution to steady-state and oscillating stomatal conductance calculated using gas-exchange techniques. Plant Cell Environ 17:995–1007

Caré AF, Nefed'ev L, Bonnet B, Millet B, Badot PM (1998) Cell elongation and revolving movement in *Phaseolus vulgaris* L. twining shoots. Plant Cell Physiol 39:914–921

Chance B, Hess B, Betz A (1964) DPNH oscillations in a cell-free extract of *S. carlsbergensis*. Biochem Biophys Res Commun 16:182–187

Claire A, Frachisse JM, Desbiez MO (1985) Oscillations électriques observées au cours du movement des tiges volubiles d'*Ipomoea purpurea*. C R Acad Sci 300:362–366

Cowan IR (1972a) Oscillations in stomatal conductance and plant functioning associated with stomatal conductance: observations and a model. Planta 106:185–219
Cowan IR (1972b) An electrical analogue of evaporation from, and flow of water in plants. Planta 106:221–226
Darwin CR, Darwin F (1880) The power of movement in plants. John Murray, London (Republished 1966, Da Capo Press, New York)
Derksen J (1996) Pollen tubes: a model system for plant cell growth. Bot Acta 109:341–345
Douglass JK, Wilkins L, Pantazelou E, Moss F (1993) Noise enhancement of information transfer in crayfish mechanoreceptors by stochastic resonance. Nature 365:337–340
Dunlap JC (1993) Genetic analysis of circadian clocks. Annu Rev Physiol 55:683–728
Engelberth J, Wanner G, Groth B, Weiler EW (1995) Functional anatomy of the mechanoreceptor cells in tendrils of *Bryonia dioica* Jacq. Planta 196:539–550
Feijo JA, Sainhas J, Holdaway-Clarke T, Cordeiro MS, Kunkel JG, Hepler PK (2001) Cellular oscillations and the regulation of growth: the pollen tube paradigm. BioEssays 23:86–94
Felle H (1988) Auxin causes oscillations of cytosolic free calcium and pH in *Zea mays* coleoptiles. Planta 174:495–499
Fernández-Conde ME, de la Haba P, Maldonado JM (1995) Rhythmic variations of ferredoxin-glutamate synthase activity in sunflower (*Helianthus annuus*) leaves. J Plant Physiol 145:682–685
Findlay GP, Findlay N (1981) Respiration-dependent movements of the column of *Stylidium graminifolium.* Aust J Plant Physiol 8:45–56
Fisahn J, Mikschl E, Hansen UP (1986) Separate oscillations of the electrogenic pump and of K^+-channel in *Nitella* are revealed by simultaneous measurement of membrane potential and of resistance. J Exp Bot 37:34–47
Fisahn J, McConnaughey T, Lucas WJ (1989) Oscillations in extracellular current, external pH and membrane potential and conductance in the alkaline bands of *Nitella* and *Chara.* J Exp Bot 40:1185–1193
Fox JJ, Hill CC (2001) From topology to dynamics in biochemical networks. Chaos 11:809–815
Gammaitoni L, Hänggi P, Jung P Marchesoni F (1998) Stochastic resonance. Rev Mod Phys 70:223–287
Garcia-Ojalvo J, Sancho JM (eds) (1999) Noise in spatially extended systems. Springer, Berlin Heidelberg New York
Geitmann A, Cresti M (1998) Ca^{2+} channels control the rapid expansions in pulsating growth of *Petunia hybrida* pollen tubes. J Plant Physiol 152:439–447
Geitmann A, Li YQ, Cresti M (1996) The role of the cytoskeleton and dictyosome activity in the pulsatory growth of *Nicotiana tabacum* and *Petunia hybrida.* Bot Acta 109:102–109
Ghosh A, Chance B (1964) Oscillations of glycolytic intermediates in yeast cells. Biochem Biophys Res Commun 16:174–181
Giersch C (1994) Photosynthetic oscillations: observations and models. Comments Theor Biol 3:339–364
Giersch C, Sivak MN, Walker DA (1991) A mathematical skeleton model of photosynthetic oscillations. Proc R Soc Lond B 245:77–83
Gilroy S, Fricker MD, Read ND, Trewavas AJ (1991) Role of calcium in signal transduction of *Commelina* guard cells. Plant Cell 3:333–344
Glass L (2001) Synchronization and rhythmic processes in physiology. Nature 410:277
Gradmann H (1926) Die Bewegungen der Ranken und die Ueberkrümmungstheorie. Jb Wiss Bot 65:224–278
Gradmann D (1976) "Metabolic" action potentials in *Acetabularia.* J Membr Biol 29:23–45
Gradmann D (2001) Models for oscillations in plants. Aust J Plant Physiol 28:577–590
Gradmann D, Buschmann P (1997) Oscillatory interactions between voltage gated electroenzymes. J Exp Bot 48:399–404
Gradmann D, Hoffstadt J (1998) Electrocoupling of ion transporters in plants: interaction with internal ion concentrations. J Membr Biol 166:51–59

Gradmann D, Slayman CL (1975) Oscillations of an electrogenic pump in the plasma membrane of *Neurospora*. J Membr Biol 23:181–212

Gradmann D, Blatt MR, Thiel G (1993) Electrocoupling of ion transporters in plants. J Membr Biol 136:327–332

Hansen UP (1978) Do light-induced changes in the membrane potential of *Nitella* reflect the feed-back regulation of a cytoplasmic parameter? J Membr Biol 41:197–224

Hansen UP, Kolbowski J, Dau H (1987) Relationship between photosynthesis and plasmalemma transport. J Exp Bot 38:1965–1981

Hartwell LH, Hopfield JJ, Leibler S, Murray AW (1999) From molecular to modular cell biology. Nature 402:C47-C52

Hauser MJB, Olsen LF (1996) Mixed-mode oscillations and homoclinic chaos in an enzyme reaction. J Chem Soc Faraday Trans 92:2857–2863

Hauser MJB, Olsen LF (1998) The role of naturally occurring phenols in inducing oscillations in the peroxidase – oxidase reaction. Biochemistry 37:2458–2469

Hauser MJB, Olsen LF (1999) Routes to chaos in the peroxidase-oxidase reaction. In: Müller SC, Parisi J, Zimmermann W (eds) Transport and structure. Their competitive roles in biophysics and chemistry. Springer, Berlin Heidelberg New York, pp 252–272

Hauser MJB, Olsen LF, Bronnikova TV, Schaffer WM (1997) Routes to chaos in the peroxidase – oxidase reaction: period-doubling and period-adding. J Phys Chem B 101:5075–5083

Hauser MJB, Lunding A, Olsen LF (2000) On the role of methylene blue in the oscillating peroxidase-oxidase reaction. Phys Chem Chem Phys 2:1685–1692

Hauser MJB, Kummer U, Larsen AZ, Olsen LF (2001) Oscillatory dynamics protect enzymes and possibly cells against toxic substances. Faraday Disc 120:215–227

Hauser MJB, Fricke N, Storb U, Müller SC (2002) Periodic and bursting pH oscillations in an enzyme model reaction. Z Phys Chem 216:375–390

Hayashi H, Nakao M, Hirakawa K (1982) Chaos in the self-sustained oscillation of an excitable membrane under sinusoidal stimulation. Phys Lett 88A:265–266

Hayashi H, Nakao M, Hirakawa K (1983) Entrained, harmonic, quasiperiodic and chaotic responses of the self-sustained oscillation of *Nitella* to sinusoidal stimulation. J Phys Soc Jpn 52:344–351

Heathcote DG (1966) A new type of rhythmic plant movement: micronutation. J Exp Bot 17:690–695

Hecks B, Hejnowicz Z, Sievers A (1992) Spontaneous oscillations of extracellular electrical potentials measured on *Lepidium sativum* L. roots. Plant Cell Environ 15:115–121

Hess B, Markus M (1985) The diversity of biochemical time patterns. Ber Bunsenges Phys Chem 89:642–651

Holdaway-Clark TL, Feijó JA, Hackett GR, Kunkel JG, Hepler PK (1997) Pollen tube growth and the intracellular calcium gradient oscillate in phase while extracellular calcium influx is delayed. Plant Cell 9:1999–2021

Hütt M-T (2001) Datenanalyse in der Biologie. Springer, Berlin Heidelberg New York

Hütt M-T, Lüttge U (2002) Nonlinear dynamics as a tool for data analysis and modeling in plant physiology. Plant Biol 4:281–297

Hütt M-T, Rascher U, Beck F, Lüttge U (2002) Period-2 cycles and 2:1 phase locking in a biological clock driven by temperature pulses. J Theor Biol 217:383–390

Israelsson D, Johnsson A (1967) A theory for circumnutations in *Helianthus annuus*. Physiol Plant 20:957–976

Jenkinson IS (1962) Bioelectric oscillations of bean roots: further evidence for a feedback oscillator. II. Intracellular plant root potentials. Aust J Biol Sci 15:101–114

Jenkinson IS, Scott BIH (1961) Bioelectric oscillations of bean roots: further evidence for a feedback oscillator. I. Extracellular response to oscillations in osmotic pressure and auxin. Aust J Biol Sci 14:231–247

Jeong H, Tombor B, Albert R, Oltavi ZN, Barabási A-L (2000) The large scale organization of metabolic networks. Nature 407:651–654

Johnsson A (1979) Growth movements not directed primarily by external stimuli. Circumnutations. In: Haupt W, Feinleib ME (eds) Encyclopedia of plant physiology new series, vol 7. Physiology of movements. Springer, Berlin Heidelberg New York, pp 627–646

Johnsson A (1997) Circumnutations: results from recent experiments on Earth and in space. Planta 203:S147-S158

Johnsson A, Heathcote D (1973) Experimental evidence and models on circumnutations. Z Pflanzenphysiol 70:371–405

Jung P, Mayer-Kress G (1995) Spatio-temporal stochastic resonance in excitable media. Phys Rev Lett 74:2130–2133

Kaiser H, Kappen L (2001) Stomatal oscillations at small apertures: indications for a fundamental insufficiency of stomatal feedback-control inherent in the stomatal turgor mechanism. J Exp Bot 52:1303–1313

Kaminskyj SGW, Garrill WA, Heath IB (1992) The relation between turgor and tip growth in *Saprolegnia ferax*: turgor is necessary, but not sufficient to explain apical extension rates. Exp Mycol 16:64–75

Kaplan D, Glass L (1995) Nonlinear dynamics and chaos. Springer, Berlin Heidelberg New York

Kauffman S (1974) The large scale structure and dynamics of gene control circuits. J Theor Biol 44:167–190

Kendall BE, Fox GA (1998) Spatial structure, environmental heterogeneity, and population dynamics: analysis of the coupled logistic map. Theor Population Biol 54:11–37

Kendall BE, Schaffer WM, Tidd CW, Olsen LF (1997) The impact of chaos on biology: promising directions for research. In: Grebogi C, Yorke JA (eds) The impact of chaos on science and society. United Nations University Press, Tokyo, pp 190–218

Kirkor ES, Scheeline A, Hauser MJB (2000) Principal component analysis of dynamical features in the peroxidase - oxidase reaction. Anal Chem 72:1381–1388

Koukkari WL, Bingham C, Hobbs JD, Duke SH (1997) In search of a biological hour. J Plant Physiol 151:352–357

Kristie DN, Jolliffe PA (1986) High-resolution studies of growth oscillations during stem elongation. Can J Bot 64:2399–2405

Kummer U, Hauser MJB, Wegmann K, Olsen LF, Baier G (1997) Oscillations and complex dynamics in the peroxidase - oxidase reaction induced by naturally occurring aromatic substrates. J Am Chem Soc 119:2084–2087

Kuznetsov YA, Rinaldi S (1996) Remarks on food chain dynamics. Math Biosci 134:1–33

Laisk A, Siebke K, Gerst U, Eichelmann H, Oja V, Heber U (1991) Oscillations in photosynthesis are initiated and supported by imbalances in the supply of ATP and NADPH to the Calvin cycle. Planta 185:554–562

Larter L, Olsen LF, Steinmetz CG, Geest T (1993) Chaos in biochemical systems: the peroxidase reaction as a case study. In: Field RJ, Györgyi L (eds) Chaos in chemistry and biochemistry. World Scientific, Singapore, pp 175–224

Lazar JG, Ross J (1990) Changes in mean concentration, phase shifts, and dissipation in a forced oscillatory reaction. Science 247:189–192

Lee S-G, Neiman A, Kim S (1998) Coherence resonance in a Hodgkin-Huxley neuron. Phys Rev E 57:3292–3297

Liang F, Sze H (1998) A high-affinity Ca^{2+} pump, ECA1, from the endoplasmic reticulum is inhibited by cyclopiazonic acid but not by thapsigargin. Plant Physiol 118:817–825

Lloyd AL (1995) The coupled logistic map; a simple model for the effects of spatial heterogeneity on population dynamics. J Theor Biol 173:217–230

Lloyd D, Kippert F (1993) Intracellular coordination by the ultradian clock. Cell Biol Int 17:1047–1052

Longtin A, Bulsara A, Moss F (1991) Time interval sequences in bistable systems and the noise induced transmission of information by sensory neurons. Phys Rev Lett 67:656–660

López-Franco R, Bartnicki-Garcia S, Bracker CE (1994) Pulsed growth of fungal hyphal tips. Proc Natl Acad Sci USA 91:12228–12232
Lorenz E (1963) Deterministic nonperiodic flow. J Atmos Sci 2:130–141
Lubkin S (1994) Unidirectional waves on rings: models for chiral preference of circumnutating plants. Bull Math Biol 56:795–810
Lucas WJ, Fisahn J (1989) Oscillations and inversions in the extracellular current profiles of *Chara* and *Nitella*. In: Dainty J, DeMichelis MI, Marre E, Rasi-Caldogno F (eds) Plant membrane transport: the current position. Elsevier Science, Amsterdam, pp 25–29
Lüttge U (2002) Circadian rhythmicity: is the "biological clock" hardware or software? Progress in botany 64. Springer, Berlin Heidelberg New York, pp 277–319
Macduff JH, Dhanoa MS (1996) Diurnal and ultradian rhythm in K^+ uptake by *Trifolium repens* under natural light patterns: evidence for segmentation at different root temperatures. Physiol Plant 98:298–308
Markus M, Hess B (1984) Transition between oscillatory modes in a glycolytic model system. Proc Natl Acad Sci USA 81:4394–4398
Markus M, Kuschmitz D, Hess B (1984) Chaotic dynamics in yeast glycolysis under periodic substrate input flux. FEBS Lett 172:235–238
Martens J, Hansen UP, Warncke J (1979). Further evidence for the parallel pathway model of the metabolic control of the electrogenic pump in *Nitella* as obtained from the high frequency slope of the action of light. J Membr Biol 48:115–139
May RM (1976) Simple mathematical models with very complicated dynamics. Nature 26:459–467
McAinsh MR, Brownlee C, Hetherington AM (1990) Abscisic acid-induced elevation of guard cell cytosolic Ca^{2+} precedes stomatal closure. Nature 343:186–188
McAinsh MR, Webb AAR, Taylor JE, Hetherington AM (1995) Stimulus-induced oscillations in guard cell cytosolic free calcium. Plant Cell 7:1207–1219
McAinsh MR, Brownlee C, Hetherington AM (1997) Calcium ions as second messengers in guard cell signal transduction. Physiol Plant 100:16–29
Messerli MA, Creton R, Jaffee LF, Robinson KR (2000) Periodic increases in elongation rate precede increases in cytosolic Ca^{2+} during pollen tube growth. Dev Biol 222, 84–98
Meyer T, Stryer L (1991) Calcium spiking. Annu Rev Biophys Biophys Chem 20:153–174
Møller AC, Hauser MJB, Olsen LF (1998) Oscillations in peroxidase-catalyzed reactions and their potential function in vivo. Biophys Chem 72:63–72
Mosekilde E, Mouritsen OG (eds) (1995) Modeling the dynamics of living systems: nonlinear phenomena and pattern formation. Springer, Berlin Heidelberg New York
Moss F (2000) Stochastic resonance: looking forward. In: Walleczek J (ed) Self-organized biological dynamics and nonlinear control. Cambridge Univ Press, Cambridge, pp 236–256
Moss F, Pierson D, O'Gorman D (1994) Stochastic resonance: tutorial and update. Int J Bifurc Chaos 4:1383–1397
Ogata K, Kishimoto U (1976) Rhythmic change of membrane potential and cyclosis of *Nitella* internode. Plant Cell Physiol 17:201–207
Olsen LF (1983) An enzyme reaction with a strange attractor. Phys Lett 94A:454–457
Olsen LF, Degn H (1985) Chaos in biological systems. Q Rev Biophys 18:165–225
Pierson ES, Li YQ, Zhang HQ, Willemse MTM, Linskens HF, Cresti M (1995) Pulsatory growth of pollen tubes: investigation of a possible relationship with the periodic distribution of cell wall components. Acta Bot Neerl 44:121–128
Pierson ES, Miller DD, Callaham DA, van Aken J, Hackett G, Hepler PK (1996) Tip-localized calcium entry fluctuates during pollen tube growth. Dev Biol 174:160–173
Pikovsky AS, Kurths J (1997) Coherence resonance in a noise driven excitable system. Phys Rev Lett 78:775–778
Ping Z, Lou CH (1990) Rhythmic excitation in *Rorippa nasturtium-aquaticum*. C R Acad Sci Paris 310 Sér III:545–549

Queiroz-Claret C, Lenk R, Queiroz O, Greppin H (1988) NMR studies of in vitro slow oscillations in enzyme properties and dissipative structures. Plant Physiol Biochem 26:333–338

Raghavendra AS, Gerst U, Heber U (1995) Oscillations in photosynthetic carbon assimilation and chlorophyll fluorescence are different in *Amaranthus caudatus*, a C_4 plant, and *Spinacia oleracea*, a C_3 plant. Planta 195:471–477

Ramahaleo T, Alexandre J, Lassalles J (1996) Stretch activated channels in plant cells. A new model for osmoelastic coupling. Plant Physiol Biochem 34:327–334

Rand RH, Upadhyana SK, Cooke JR, Stori DW (1981) Hopf bifurcation in a stomatal oscillator. J Math Biol 12:1–11

Rapp PE (1986) Oscillations and chaos in cellular metabolism and physiological systems. In: Holden A (ed) Chaos. Manchester University Press, pp 179–208

Rapp PE (1987) Why are so many biological systems periodic? Progr Neurobiol 29:261–273

Rascher U, Hütt M-T, Siebke K, Osmond CB, Beck F, Lüttge U (2001) Spatio-temporal variation of metabolism in a plant circadian rhythm: the biological clock as an assembly of coupled individual oscillators. Proc Natl Acad Sci USA 98:11801–11805

Raschke K (1965) Die Stomata als Gleider eines schwingungsfähigen CO_2-Regelsystems. Experimenteller Nachweis an *Zea mays* L. Z. Naturf 20b:1261–1270

Raschke K (1975) Stomatal action. Annu Rev Plant Physiol 26:309–340

Ravasz E, Barabási A-L (2003) Hierarchical organization in complex networks. Phys Rev E 67:026112

Read ND, Shacklock PS, Knight MR, Trewavas AJ (1993) Imaging calcium dynamics in living plant cells and tissues. Cell Biol Int 17:111–125

Roelfsema MRG, Hedrich R (2002) Studying guard cells in the intact plant: modulation of stomatal movement by apoplastic factors. New Phytol 153:425–431

Rudd JJ, Franklin-Tong VE (2001) Unravelling response-specificity in Ca^{2+} signalling pathways in plant cells. New Phytol 151:7–33

Schwenke H, Wagner E (1992) A new concept of root exudation. Plant Cell Environ 15:289–299

Scott BIH (1957) Electrical oscillations generated by plant roots and a possible feedback mechanism responsible for them. Aust J Bot 10:164–179

Scott BIH (1962) Feedback-induced oscillations of five-minute period in the electric field of the bean root. Annu N Y Acad Sci 98:890–900

Shabala SN (1997) Leaf Bioelectric responses to rhythmical light: identification of the contributions from stomatal and mesophyll cells. Aust J Plant Physiol 24:741–749

Shabala S, Knowles A (2002) Rhythmic patterns of nutrient acquisition by wheat roots. Funct Plant Biol 29:595–605

Shabala SN, Lew RR (2002) Turgor regulation in osmotically stressed *Arabidopsis* epidermal root cells. Direct support for the role of inorganic ion uptake as revealed by concurrent flux and cell turgor measurements. Plant Physiol 129:290–299

Shabala SN, Newman IA (1997) Proton and calcium flux oscillations in the elongation region correlate with root nutation. Physiol Plant 100:917–926

Shabala SN, Newmann IA (1998) Osmotic sensitivity of Ca^{2+} and H^+ transporters in corn roots: effect on fluxes and their oscillations in the elongation region. J Membr Biol 161:45–54

Shabala S, Delbourgo R, Newman I (1997a) Observations of bifurcation and chaos in plant physiological responses to light. Aust J Plant Physiol 24:91–96

Shabala SN, Newman IA, Morris J (1997b) Oscillations in H^+ and Ca^{2+} ion fluxes around the elongation region of corn roots and effects of external pH. Plant Physiol 113:111–118

Shabala L, Shabala S, Ross T, McMeekin T (2001) Membrane transport activity and ultraradian ion flux oscillations associated with cell cycle of *Thraustochytrium* sp. Aust J Plant Physiol 28:87–99

Siebke K, Weis E (1995a) Assimilation images of leaves of *Glechoma hederacea*: analysis of non-synchronous stomata related oscillations. Planta 196:155–165

Siebke K, Weis E (1995b) Imaging of chlorophyll-*a*-fluorescence in leaves: topography of photosynthetic oscillations in leaves of *Glechoma hederacea*. Photosyn Res 45:225–237

Slayman CL, Long WS, Gradmann D (1976) "Action potentials" in *Neurospora crassa*, a mycelial fungus. Biochim Biophys Acta 426:732–744

Sokabe M, Sachs F, Jing Z (1991) Quantitative video microscopy of patch clamped membranes: stress, strain, capacitance, and stretch channel activation. Biophys J 59:722–728

Solé RV, Gamarra JGP (1998) Chaos, dispersal and extinction in coupled ecosystems. J Theor Biol 193:539–541

Solé RV, Manrubia SC, Luque B, Delgado J, Bascompte J (1996) Phase transitions and complex systems. Complexity 2:13

Souda M, Toko K, Hayashi K, Fujiyoshi T, Ezaki S, Yamafuji K (1990) Relationship between growth and electric oscillations in bean roots. Plant Physiol 93:532–536

Stitt M, Gross H, Woo KC (1988) Interactions between sucrose synthesis and CO_2 fixation. II. Alterations of fructose 2,6-bis-phosphate during photosynthetic oscillations. J Plant Physiol 133:133–143

Stitt M, Schreiber U (1988) Interaction between sucrose synthesis and CO_2 fixation. II. Alterations of fructose 2,6-bisphosphate during photosynthetic oscillations. J Plant Physiol 133:263–271

Strogatz S (1994) Nonlinear dynamics and chaos with applications to physics. Addison-Wesley, Reading, Massachusetts

Swillens S, Combettes L, Campeil P (1994) Transient inositol 1,4,5-triphosphate-induced Ca^{2+} release: a model based on regulatory Ca^{2+} binding sites along the permeation pathway. Proc Natl Acad Sci USA 91:10074–10078

Sze H, Liang F, Hwang I, Curran AC, Harper JF (2000) Diversity and regulation of plant Ca^{2+} pumps: insights from expression in yeast. Annu Rev Plant Physiol Plant Mol Biol 51:433–462

Tang XW, Liu GQ, Yang Y, Zheng WI, Wu BC, Nie DT (1992) Quantitative measurement of pollen tube growth and particle movement. Acta Bot Sinica 34:893–898

Teoh CT, Palmer JH (1971) Nonsynchronized oscillations in stomatal resistance among sclerophylls of *Eucalyptus umbra*. Plant Physiol 47:409–411

Teusink B (1999) Exposing a complex metabolic system: glycolysis in *Saccharomyces cerevisae*. Academisch Proefschrift, Universiteit van Amsterdam, Print Partners Ipskamps BV, Enschede

Teusink B, Bakker BM, Westerhoff HV (1996a) Control of frequency and amplitudes is shared by all enzymes in three models for yeast glycolytic oscillations. Biochim Biophys Acta 1275:204–212

Teusink B, Larsson C, Diderich J, Richard P, van Dam K, Gustafsson L, Westerhoff HV (1996b) Synchronized heat flux oscillations in yeast cell populations. J Biol Chem 271:24442–24448

Teusink B, Diderich JA, Westerhoff HV, van Dam K, Walsh MC (1998) Intracellular glucose concentration in derepressed yeast cells consuming glucose is high enough to reduce the glucose transport rate by 50%. J Bacteriol 180:556–562

Toko K, Souda M, Matsuno T, Yamfuji K (1990) Oscillations of electrical potential along a root of a higher plant. Biophys J 57:269–279

Trewavas A (1999) Le calcium, c'est la vie: calcium makes waves. Plant Physiol 120:1–6

Tyerman SD, Beilby M, Whittington J, Juswono U, Newman I, Shabala S (2001) Oscillations in proton transport revealed from simultaneous measurements of net current and net proton fluxes from isolated root protoplasts: MIFE meets patch clamp. Aust J Plant Physiol 28:591–604

van der Veen R (1949) Induction phenomena in photosynthesis. II. Physiol Plant 2:287–296

Vanselow KH, Kolbowski J, Hansen UP (1989) Further evidence for the relationship between light-induced changes of plasmalemma transport and transthylakoid proton uptake. J Exp Bot 40:239–245

Volotovski LD, Sokolovsky SG, Nikiforov EL, Zinchenko VP (1993) Calcium oscillations in plant cell cytoplasm induced by red and far-red light irradiation. J Photochem Photobiol B Biol 20:95–100
Vucinic Z, Radenovic C, Damjanovic Z (1978) Oscillations of the vacuolar potential in *Nitella*. Physiol Plant 44:181–186
Walker DA (1992) Concerning oscillations. Photosynth Res 34:387–395
Walker DA, Sivak MN, Prinsley RT, Cheesbrough JK (1983) Simultaneous measurement of oscillations in oxygen evolution and chlorophyll *a* fluorescence in leaf pieces. Plant Physiol 73:542–549
Watts DJ, Strogatz SH (1998) Collective dynamics of 'small-world' networks, Nature 393:440–442
Weiler EW (1997) Sensorik der Pflanzen. In: Truscheit J (ed) Koordinaten der menschlichen Zukunft: Energie – Materie – Information – Zeit, S. Hirzel-Wiss Verlagsges, Stuttgart, pp 331–346
White PJ (2000) Calcium channels in higher plants. Biochim Biophys Acta 1465:171–189
Willmer CM (1988) Stomatal sensing of the environment. Biol J Linn Soc 34:205–217
Woods NM, Cuthbertson KSR, Cobbold PH (1986) Repetitive transient rises in cytoplasmic free calcium in hormone-stimulated hepatocytes. Nature 319:600–602
Yoshida T, Hayashi K, Toko K, Yamafuji K (1988) Effect of anoxia on the spatial pattern of electric potential formed along the root. Ann Bot 62:497–507
Zonia L, Cordeiro S, Tupý J, Feijó JA (2002) Oscillatory chloride efflux at the pollen tube apex has a role in growth and cell volume regulation and is targeted by inositol 3,4,5,6-tetra*kis*phosphate. Plant Cell 14:2233–2249

Note added in proof: Relevant to Section 2.3 a noteworthy comprehensive review on the roles of the major ions Ca^{2+}, H^+, K^+ and Cl^- involved and the phase behaviour of oscillations of their gradients and fluxes in relation to oscillatory pollen tube growth appeared at proof stage of this article: Holdaway-Clarke TL, Hepler PK (2003) Control of pollen tube growth: role of ion gradients and fluxes. New Phytol 159: 539–563.

Prof. Dr. Ulrich Lüttge
Prof. Dr. M.-Th. Hütt
Institut für Botanik
Technische Universität Darmstadt
Schnittspahnstrasse 3–5
64287 Darmstadt, Germany
e-mail: luettge@bio.tu-darmstadt.de

Nutritional Aspects of C_1 Unit Metabolism in Heterotrophic Tissues and Organisms (Excluding Prokaryotes) Under Particular Respect to Dissolved Organic Nitrogen (DON) from the Environment

Hartmut Gimmler and Wolfram Hartung

1 Introduction

One-carbon (C_1 unit) metabolism is essential to all eukaryotes and prokaryotes. In plants and fungi it supplies the C_1 units required for the synthesis of amino acids, proteins, nucleic acids (purines), some vitamins, and methylated compounds such as lignin, alkaloids or betaines (Fig. 1). Thus it is involved in such different physiological aspects such as regulation and differentiation via DNA and vitamin metabolism, cell wall formation and extension (lignin and pectin), osmoregulation (betaines), secondary plant metabolism (alkaloids), amino acid metabolism, internal and external detoxification (cyanides) or utilization of N- and C-containing solutes abundant in the external environment. Additionally it plays an important role in photorespiration of photosynthetic C_3 plants.

Serine is generally accepted as the major one-carbon donor in C_1 metabolism, but also glycine and formate; in some cases also cyanides can provide one-carbon units. In most cases the C_1 transfer is mediated by the co-factor tetrahydrofolate (THF). THF contains pteridine, *p*-aminobenzoate and glutamate entities and therefore its chemical name is tetrahydrofolate polyglutamate ($H_4PteGlu_n$). The synthesis of this compound starts with the generation of pteridine precursors from GTP and is described in detail e.g. by Cossins (1987), Cossins and Chen (1997) and Rébeillé and Douce (1999). Recently it was found that in few cases at least in bacteria also tetrahydromethanopterin (H_4MPT) can function as carrier for C_1 units (Grabarse et al. 1999; Pomper et al. 1999; Chistoserdova et al. 2000; Maden 2000).

In principle, in C_1 metabolism compounds such as glycine, serine or formate (in bacteria and fungi also methane or methanol), by catabolic reactions generate various specific derivatives of THF (Fig. 1). Subsequently these derivatives are interconverted enzymatically between different oxidation states. 10-Formyl-THF is the highest oxidized derivative, 5-methyl-THF the most reduced derivative, 5,10-methenyl-THF and 5,10 methylene-THF having intermediate redox states. In anabolic reactions the

Progress in Botany, Vol. 65

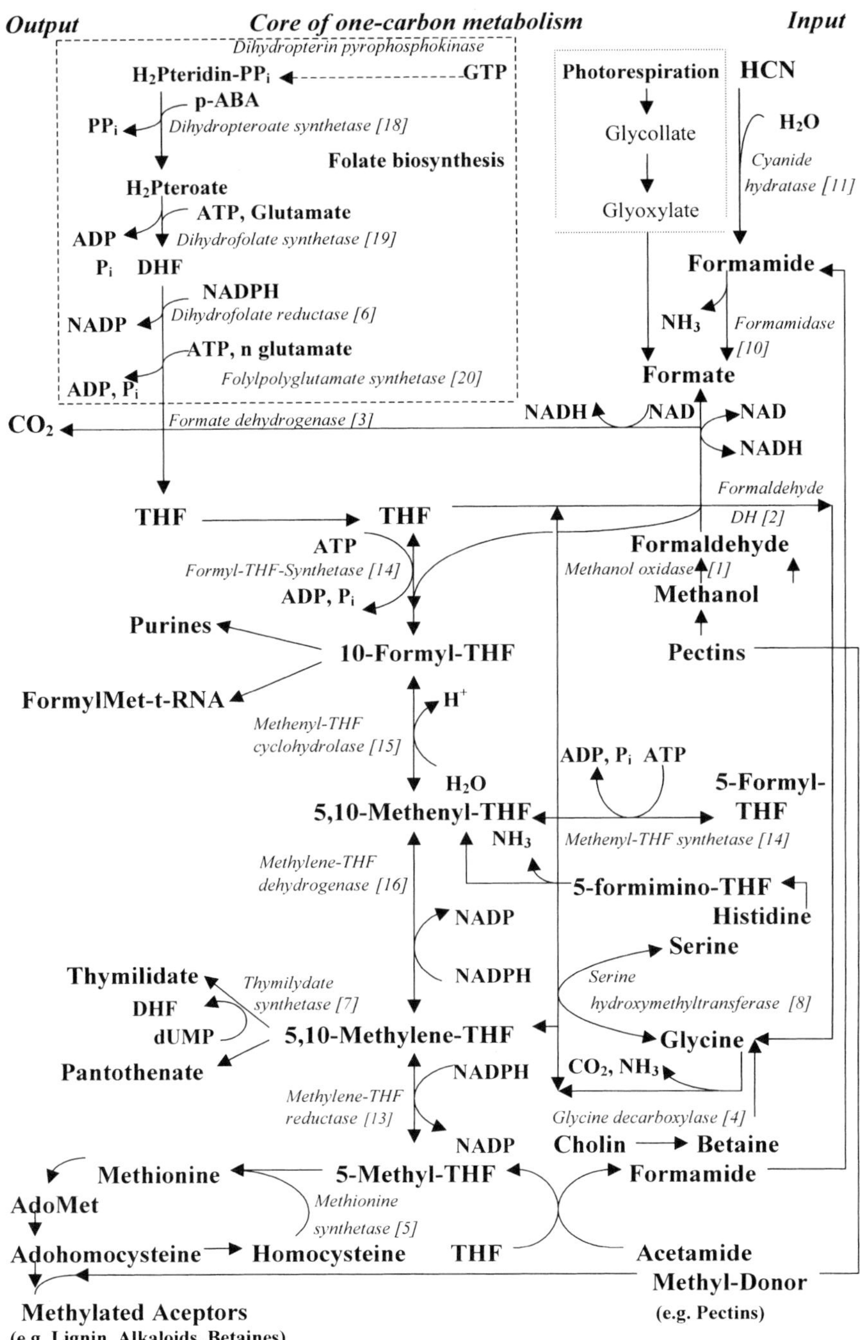

Fig. 1. Summarizing sketch of reactions of folate biosynthesis and C_1 unit metabolism in higher plants and fungi as covered in this review. *Numbers* refer to enzymes cited in Table 1 (column 1)

C_1 units are withdrawn from the various THF derivatives, yielding, e.g., purines, panthotenate, thymilidate or methionine. If C_1 units originate from formate, also serine can be synthesized (Igambierdiev et al. 1999; Wingler et al. 1999). The scheme in Fig. 1 neglects the intracellular compartmentation of C_1 metabolism in plastids, mitochondria, peroxisomes and the cytosol, which has been described in detail (Cossins 1987, McNeil et al. 1996, Neuburger et al. 1996, Kastanos et al. 1997, Chen et al. 1997, Atkins et al. 1997, Bourguignon et al. 1999, Rébeillé and Douce 1999 and Hanson et al. 2000).

Although C_1 metabolism is of crucial importance for all organisms, much of its details are unknown, because this metabolic pathway is difficult to investigate: The involved enzymes (compare Fig. 1 and Table 2) can be of low abundance or be inducible. They may exist in several iso-forms or multifunctional units. The key intermediates, the C_1-substituted THFs, occur only in µmolar concentrations in the cytosol and are labile, causing difficulties during quantitative determination. For this reason, recent progress made in the elucidation of C_1 metabolism came mainly through two new techniques: (1) the non-invasive ^{13}C-NMR-technique, which allows to determine in vivo fluxes of carbon through C_1 metabolism (for summary see Mouillon et al. 1999; Ratcliffe and Shachar-Hill 2001: higher plants; Pasternack et al. 1992; Pasternack et al. 1994a,b, 1996; Appling et al. 1997; Maaheimo et al. 2001: fungi) and (2) genomic-driven approaches, which means identification of genes by the use of mutants with specific deletions (Barlowe and Appling 1990a,b; Maier-Greiner et al. 1991; Wang and Vanetten. 1992; Cluness et al. 1993; Song and Rabinowitz, 1993; Wahls et al. 1993; West et al. 1993; Brown et al. 1995; Kirksey and Appling 1996; Bird and Bradshaw 1997; Todd et al. 1997; Parpinello et al. 1998; Toyama et al. 1998; Rover et al. 1999; Hanson et al. 2000; Ravanel et al. 2001; Hanson and Shahar-Hill 2002; Nikiforov et al. 2002; compare also Table 2).

The focus of this contribution is on physiological aspects of C_1 metabolism in heterotrophic tissues of plants and in fungi. It excludes, with few exceptions, prokaryotes. Special attention is directed to (1) nutritional aspects of C_1 metabolism and (2) its role in stress physiology. Beside a survey, it will show by two case studies that C_1 metabolism is of essential importance for plants and fungi suffering from nitrogen deficiency and that by proper modulation of C_1 metabolism plants and fungi can acclimatize to uptake and metabolisation of those organic N-containing compounds prevailing in their environment. Extreme stress conditions produce high selection pressure on organisms. This applies also to the use nutritional sources from the environment. Organisms which are able to use N-containing solutes from the soil solution, which cannot be exploited by other organisms competing for the same habitat, will finally dominate this ecological niche.

Because of this special focus, another interesting aspect of C_1 metabolism in plants, e.g. its function in photorespiration (Gardeström et al. 1985; Besson et al. 1993; Neuburger et al. 1996; Douce and Neuburger 1999; Wingler et al. 1999) had to be omitted. It is also not the aim of this paper to review the biochemistry and the underlying molecular biology of C_1 metabolism. Instead, we refer to the excellent reviews which have been written recently covering various biochemical aspects of C_1 metabolism (Appling 1991; Cossins and Chen 1997; Mouillon et al. 1999; Rébeillé and Douce 1999; Hanson et al. 2000).

To the best of our knowledge no corresponding articles on C_1 metabolism have been published so far in *Progress of Botany* and we neither found the terms "*carbon-one metabolism*" nor "*tetrahydrofolic acid*" in the subject indexes of the last twenty volumes of this series. Thus, we hope to close a "gap" by this review, even though the latter is not a general one, but prefers to put the emphasis on some selected physiological topics of C_1 metabolism and application to two case studies.

2 Higher Plants

2.1 Glycine and Serine as Potential Sources of Nitrogen in Higher Plants: General Aspects

Dissolved organic nitrogen (DON) in the soil provides an alternative to the usual inorganic forms in a wide range of plants and circumstances (Näsholm and Persson 2001). When nitrogen mineralisation is impaired, the concentration of inorganic nitrogen in the soil solution can be very low, and under such conditions organic nitrogen, often in the form of amino acids, may be the major nitrogenous compounds available to the roots. This process has been shown to be an important factor in the nitrogen nutrition of plants in alpine and arctic habitats (Chapin et al. 1993; Kielland 1994; Lipson and Monson 1998; Raab et al. 1999), as well as for plants in heathlands (Schmidt and Stewart 1997), boreal forests (Näsholm et al. 1998), coastal marsh (Henry and Jeffries 2002), grassland communities (Falkengren-Grerup et al. 2000; Streeter et al. 2000; Thornton 2001) and the aquatic resurrection plant *Chamaegigas intrepidus* (Heilmeier and Hartung 2001). Moreover utilization of glycine has been demonstrated in agriculturally important species grown under field conditions, suggesting that organic nitrogen may be a more important source of nitrogen for plants under cultivation than previously thought (Näsholm et al. 2000, 2001).

The observation that organic nitrogen compounds, such as amino acids, can be absorbed by roots under field conditions at rates that can make a substantial contribution to the nitrogen requirement of plants, is good evidence for the involvement of such compounds in plant nitrogen nutrition. The existence of a range of amino acid transporters in plant roots suggests a likely mechanism (Fischer et al. 1998), and indeed the use of dual-labelled amino acids, for example glycine labelled with both ^{13}C and ^{15}N (Näsholm et al. 1998, 2001), followed by the analysis of plant extracts using GC-MS provides a convenient method for investigating the contribution of direct and indirect uptake to the intracellular nitrogen pool. Demonstrating the quantitative significance of the uptake process has been the priority in most of these studies, with the result that the subsequent metabolism of the absorbed nitrogenous compounds has received much less attention, even though the pathways for their utilization may be unclear. Thus while the uptake of glycine has been shown to be a significant source of plant nitrogen in many of these studies, the extent to which the glycine decarboxylase complex (GDC; EC 2.1.2.10; compare Fig. 1) might complement the action of aminotransferases in the subsequent metabolism of the glycine has been investigated infrequently. In one such study, it was concluded that glycine was metabolized in the roots and cluster roots of *Hakea* seedlings via aminotransferase activity (Schmidt and Stewart 1999). This conclusion was consistent with earlier observations on the low GDC activity in pea root apices (Walton and Woolhouse 1986), but more recent data suggests that glycine metabolism via GDC in heterotrophic tissues may actually occur quite readily (Mouillon et al. 1999).

It has been argued on the basis of the results obtained with a heterotrophic *Acer pseudoplatanus* cell culture that metabolism through the GDC pathway must be an essential feature of heterotrophic plant metabolism (Mouillon et al. 1999), despite the generally low levels of extractable GDC activity in such tissues (Walton and Woolhouse 1986; Bourguignon et al. 1993).

2.2 Glycine and Serine as Potential Sources of Nitrogen for the Aquatic Resurrection Plant *Chamaegigas intrepidus*

Recently data of in vivo NMR studies with roots of *Zea mays* and *Chamaegigas intrepidus* that have been incubated with ^{15}N- and ^{13}C-labelled glycine have been published (Hartung and Ratcliffe 2002). The experiments with [^{15}N]glycine provided direct evidence for the release of [^{15}N] ammonium, and its subsequent incorporation into glutamine and glutamate via the GS/GOGAT (glutamine synthetase/glutamine 2-oxoglutarat-aminotransferase) path- way. The first explanation for the ammonium release is that glycine is metabolized by glycine decarboxylase (GDC), and indeed this explanation is supported by both experiments using inhibitors of GDC and [2-^{13}C]glycine. More complicated pathways for the release of ammonium can be envisaged, which however seem to be quite unlikely as discussed in detail by the authors. Secondly, the experiments with [2-^{13}C]glycine confirmed the involvement of GDC and showed that serine hydroxymethyltransferase (SHMT), both alone and in concert with GDC, was responsible for the conversion of glycine to serine. These conclusions can be deduced from the labelling patterns observed for the serine isotopomers in the ^{13}C-NMR spectrum: [2-^{13}C]serine is formed by the SHMT-mediated reaction of unlabelled CH_2-THF and [2-^{13}C]glycine (compare Fig. 1); [2,3-^{13}C]serine arises from the reaction of [^{13}C]–labelled CH_2-THF, generated by the action of GDC on [2-^{13}C]glycine, and [2-^{13}C]glycine; and [3-^{13}C]glycine arises from reaction of [^{13}C]-labelled CH_2-THF and unlabelled glycine. The characteristic signals of the three isotopomers are readily identifiable and they provide unequivocal evidence for the involvement of both GDC and SHMT in the metabolism of glycine.

Thus it can be seen that GDC and SHMT are directly involved in the metabolism of glycine by *C. intrepidus* and maize root tissues, and this provides further support for the emerging view that GDC plays an essential part in heterotrophic metabolism (Mouillon et al. 1999). This conclusion is also significant in relation to the nitrogen nutrition of a range of plants in their natural habitat, since glycine is often the most or second-most abundant nitrogen source that supports the plants. The metabolism of glycine can act as a direct source of ammonium for the GS/GOGAT pathway and thus define the pathway that permits plants such as *C. intrepidus* to utilize glycine as a nitrogen source. Labelling experiments with [2-^{13}C]- and [^{15}N]glycine point to the involvement of GDC and SHMT in the metabolism of glycine by the root tissues of *C. intrepidus* and maize. The increasing awareness of the contribution that organic nitrogen forms make to plant nitrogen nutrition in a wide range of habitats, including ones of agricultural importance, suggest that more detailed investigations of the pathways of both glycine and serine metabolism would be worthwhile.

3 Eukaryotic Micro organisms

3.1 General Physiological Aspects of C_1 Metabolism in Eukaryotic Micro-organisms

In fungi the three-carbon of serine is generally the major one-carbon donor of C_1 metabolism (Pasternack et al. 1994a). Occasionally glycine and formate can substitute for serine. If formate is the one-carbon donor, assimilation is primarily cytoplasmic (Pasternack et al. 1992), whereas when serine serves as C_1 donor, metabolism occurs preferentially via the mitochondria. The interconversion of serine and glycine in fungal C_1 metabolism is catalyzed by serine hydroxymethyltransferase (SHMT) (EC 2.1.2.1) following Eq. (1),

$$\text{Serine} + H_4PteGlu_n \xleftrightarrow{\textit{serine hydroxymethyltransferase}} 5,10\text{-}CH_2\text{-}H_4PteGlu_n + \text{glycine} \quad (1)$$

where $H_4PteGlu_n$ is 5,6,7,8 tetrahydropteropolyl-γ-glutamate (= THF) and n the number of glutamate moieties. SHMT is localized in the cytosol as well as in mitochondria of fungi (Cossins 1987; compare Table 1).

In yeast the formation of 10-HCO-$H_4PteGlu_n$ (10-formyl-THF) and its conversion to 5,10-CH_2-$PteGlu_n$ (5,10-methylene-THF) are catalyzed by a trifunctional peptide called C_1 tetrahydrofolate synthetase (Cossins and Chen 1997; compare Table 1). This protein has three distinct domains for

1. 10-HCO-H_4PteGlu synthetase (EC 6.3.4.3) (Eq. 2),

$$\text{Formate} + \text{H4PteGlun} \xleftrightarrow{\textit{10-formyl-THF synthetase}} 10\text{-HCO-}H_4PteGlu_n + \text{ADP} + P_i \quad (2)$$

2. 5,10-CH_2-H_4PteGlu cyclohydrolase (EC 3.5.4.9) (Eq. 3),

$$10\text{-CHO-}H_4PteGlu_n + H^+ \xleftrightarrow{\textit{5,10-methenyl-THF-cyclohydrolase}} 5,10\text{-}CH^+\text{-}H_4PteGlu_n + H + H_2O \quad (3)$$

and

3. 5,10-CH_2-H_4PteGlu dehydrogenase (EC 1.5.1.5) (Eq. 4).

$$5,10\text{-}CH^+\text{-}H_4PteGlu_n + \text{NADPH} \xleftrightarrow{\textit{5,10-methenyl-THF-dehydrogenase}} 5,10\text{-}CH_2\text{-}H_4PteGlu_n + \text{NADP} \quad (4)$$

The C_1-tetrahydrofolate synthetase is by no means the only multifunctional enzyme complex in fungal C_1 metabolism. Also the fungal glycine decarboxylase (GCV) and the thymilidate synthetase (TS) are catalyzed by multifunctional peptids (Table 1).

Table 1. Biochemical studies of microbial enzymes related to C_1 metabolism (compare Fig. 2; Cossins and Chen, 1997; Hanson and Shachar-Hill 2002). Information refers mainly to fungi, but some data on bacteria, protozoa and algae are also included. For molecular biological data see Table 2. Numbers in column 1 refer to Figs. 1 and 5

No.	Enzyme or function	E.C.	Substrates/products	Organism	Remarks	Km	Molecular weight (kDA) Native	Subunit	Reference
[1]	Methanol oxidase *(MOX)*	1.1.1.13	CH_3OH, O_2/ CH_2O, H_2O_2	*Pori, Candida, Hansenula polymorpha, Pichia, Phanerochaete, Saccharomyces*	Peroxisomal	0.79 mM	310	70–75	1, 22,25, 37, 38, 49, 51, 56
[2]	Formaldehyde dehydrogenase (FLD)	1.2.1.1	CH_2O, NAD/HCOOH, NADH; GSH as cofactor	*Candida boidinii, Saccharomyces, Hansenula polymorpha, Pichia pinus*	Glutathione (GSH) dependent	0.29 mM CH_2O, 0.025 mM NAD	82	40–43	3, 29, 38, 50, 56, 62, *68, 75, 79*
[3]	Formate *dehydrogenase (FDH)*	1.2.1.2	COOH, NAD/ *CO_2*, NADH	*Candida boidini, Pichia, Hansenula polymorpha, C. methylica, Saccharomyces*			81	41	2, 11, 23, 26, 30, 34, 35, 36, 42, 46, 52, 64, 66,79
[4]	Glycine decarboxylase (GDC), P-, H-, L-, T-protein	2.1.1.10	Glycine, NAD, THF/CO_2, CH_2-THF, NADH, NH_3	*Saccharomyces*	Inducible, mitochondrial, P protein causes	0.16–0.75 mM NAD, 0.6 mM glycine decarboxylation		94_P, 15_H, 60_L, 41_T	20, 44, 48, 54
[5]	Methionine synthetase (MS) THF, methionine	2.1.1.13	Methyl-THF, homocysteine/	*Schizosaccharomyces pombe*					45, 47
[6, 7]	Dihydrofolate reductase (DHFR) bifunctional with thymilidate synthetase (TS)	2.1.1.45	DHF, NADPH/THF, NADP dUMP, 5,10-CH_2-THF/dTMP, DHF	*Candida albicans, Aspergillus nidulans*	Homodimer, *mitochondrial,* x ray structure	1 µM DHF 22 µM CH2-THF		50–60	8, 12, 14, 47, 58, 81
[8]	Serine hydroxy-methyltransferase *(SHMT)*	2.1.2.1	Serine/THF, glycine, CH_2-THF	*Saccharomyces, Neurospora crassa, Aspergillus nidulans, Euglena gracilis*	Cytosolic and mitochondrial, pyridoxal as cofactor	1.0–1.5 mM serine	90	45–54	14, 28, 42, 44, 54, 65

Table1. *Continued*

No.	Enzyme or function	E.C.	Substrates/products	Organism	Remarks	Km	Molecular weight (kDA) Native	Subunit	Reference
[9]	Acetamidase	3.5.1.4	Acetamide, H_2O/ *acetate*, NH_3	*Aspergillus nidulans, Bispora, Fusarium venetatum, Candida famata, Phanero-chaetechrysosporium, Saccharomyces*	Thiolenzyme				6, 24, 39, 63, 71, 72, 74
[10]	Formamidase	3.5.1.9	Formamide, H_2O/ formate, NH_3	*Phanerochaete chrysosporium*	Inducible				74
[11]	Formamide hydrolyase (Cyanide hydratase) (FHL)	4.2.1.66	HCN, H_2O/formamide	*Pseudomonas fluorescens, Candida famata, Gloeocercospora sorghi, Stemphylium loti, Fusarium lateritium, F. solani, F. formo-sum, F. oxysporum*	Inducible metalloenzyme (Zn, Fe, Co)	1–12 mM (KCN)	200–300	43–45	4, 7, 13, 21, 33, 39, 57, 61, 78, 83
[12]	Nitrile hydratase		Nitriles/amides	Widespread in bacteria and fungi *Pseudomonas, Pichia pastoris*	Metalloenzyme				17, 32, 40, 60, 82
[13]	5,10-Methyl-THF reductase		CH2-THF, NADPH/ Methyl-THF, NADP						58
	Trifunctional C-1-THF-folate synthetase:								
[14]	10-Formyl-THF-synthetase	6.3.4.3	Formate, THF, ATP/ formyl-THF, ADP, Pi	*Saccharomyces, Schizosaccharomyces pombe*	Mitochondrial	8–20 mM format 40 μM THF 40–100 μM ATP	217	76	5, 18, 31, 44, 53, 58, 67, 69, 77, 80
[15]	Methenyl-THF-cyclohydrolase	3.5.4.9	Formyl-THF, H^+/ CH^+-THF, H_2O						
[16]	5,10 Methylene-THF-dehydrogenase (MTFDH)	1.5.1.5	CH^+-THF, NADPH/ CH_2-THF, NADP						

[17]	Cyanamide hydratase	4.2.1.69	Cyanamide/urea	*Mytothecium verrucaria*	Inducible, Zn-containing peptid			27.7	40
[18]	Dihydropterate synthetase (DHPS)		H_2pterinPP$_i$ p-ABA/ H_2pteroate, PP$_i$	*Saccharomyces, Plasmodium flaciparum*	Multifuntional with dihydropterin pyrophosphokinase (HPPK)	0.6 μM p-ABA	280–300	53–83	12, 14, 58, 73
[19]	Dihydrofolate *synthetase (DHFS)*	6.3.2.12	H_2pteroate, *glutamate, ATP/DHF,* ADP, P$_i$	*Neurospora crassa, Saccharomyces*	pH Optimum 8.0–10 monomeric	1 μM pteroate, 1.5 mM glutamate, 0.1 mM ATP		52	43
[20]	Monofunctional Folylpolyglutamate synthetase (FPGS)	6.3.2.15	THF-glutamate$_n$, glutamate, ATP/ADP, P$_i$, THF-glutamate$_{n+1}$	*Neurospora crassa*	Cytosolic, mitochondrial; alkaline pH optimum, monomeric	0	65–70		9, 12, 19, 43
[21]	DMSP-THF methyl transferase		DMSP/(DMS, acrylate) or MTPA	*Eubacterium limosum,* Acetobacterium woodii, Sporumusa ovata, S. spheroides				35	27
[22]	Chloromethane *dehalogenase*		CH_4Cl/var.	*Methylobacterium chloromethanicum*	Enzyme contains Zn and Vit. B12	0.27 μM CH_3Cl	45–67		16, 41, 70, 76

[a]References: 1) Aggelis et al. (1999), 2) Allen and Holbrook (1995), 3) Baerends et al. (2002), 4) Barclay et al. (1998a,b), 5) Barlowe and Appling (1990a,b), 6) Bird and Bradshaw (1997), 7) Brown et al. (1995), 8) Cella and Parisi (1993), 9) Chan et al. (1991), 11) Chow and Rajbhandary (1993), 12) Cherest et al. (2000), 13) Cluness et al. (1993), 14) Cossins and Chen (1997), 16) Coulter et al. (1999), 17) Cowan et al. (1998), 18) Suarez de Mata and Rabinowitz (1980), 19) DeSouza et al. (2000), 20) Douce el al. (2001), 21) Dumestre et al. (1997a,b), 22) Fall and Benson (1996), 23) Ferry (1990), 24) Gimmler et al. (2002), 25) Griffin (1994), 26) Illeova et al. (1993), 27) Jansen and Hansen (2000, 2001), 28) Kastanos et al. (1997), 29) Kato (1990a), 30) Kato (1990b), 31) Kirksey et al. (1996), 32) Kobayashi and Shimizu (1998), 33) Kunz et al. (1992, 1994), 34) Labrou and Clonis (1995), 35) Labrou (2000), 36) Labrou et al. (1995), 37) Leak (1992), 38) Lee et al. (2002), 39) Linardi et al. (1996), 40) Maier-Greiner et al. (1991), 41) McAnulla et al. (2001), 42) McClung et al. (1992), 43) McDonald et al. (1995), 44) McNeil et al. (1996), 45) Naula et al. (2002), 46) Mezentsev et al. (1997), 47) Nadolska et al. (1989), 48) Nagarajan and Storms (1997), 49) Nemecek-Marshall et al. (1995), 50) Neuhauser et al. (1998), 51) Nishida and Erikson (1987), 52) Overkamp et al. (2002), 53) Pasternack et al. (1992), 54) Pasternack et al. (1992, 1994a,b), 55) Pasternack et al. (1994b), 56) Pennincksx (2000), 57) Pereira et al. (1997), 58) Rébeillé and Douce (1999), 60) Rezende et al. (1999), 61) Rissler and Millar (1977), 62) Rosario et al. (1995), 63) Rover et al. (1999), 64) Sakai et al. (1997), 65) Sakamoto et al. (1996), 66) Serov et al. (2002), 67) Shannon and Rabinowitz (1986), 68) Shen et al. (1998), 69) Song and Rabinowitz (1993), 70) Studer et al. (1999), 71) Todd et al. (1997), 72) Tonon et al. (1990), 73) Triglia and Cowman (1994), 74) van der Walt et al. (1993), 75) van Ophem and Duine (1994), 76) Vannelli et al. (1998), 77) Wahls et al. (1993), 78) Wang and Vanetten (1992); Wang et al. (1992), 79) Wehner et al. (1993), 80) West et al. (1993, 1996), 81) Whitlow et al. (1997, 2001), 82) Wu et al. (1999), 83) Yanase et al. (2000)

Table 1 summarizes recent literature of biochemical studies of enzymes involved in C_1 metabolism and folate biosynthesis. Many genes of these enzymes haven been identified and sequenced lately (Table 2).

3.2 Amides and Nitriles as Nitrogen Sources During the Adaptation of Fungi to Extreme Environmental Conditions

Fe- or Co-containing nitrile-degrading enzymes (Kobayashi and Shimizu 1998) are widely dispersed in prokaryotes and lower orders of eukaryotes, but not in higher plants (Miller and Conn 1980). By means of hydrolyases (nitrile-hydratases, NHases) some bacteria and fungi are able to detoxify cyanides (nitriles). Cyanide hydratase, e.g. (EC 4.2.1.66 = formamide hydrolyase, FHL) converts cyanide to formamide according to

$$HCN+H_2O \xleftrightarrow{\textit{Formamide hydro-lyase}} HCON_2 \quad (5)$$

The resulting amides may be subsequently split by amidases into organic acids and ammonia. For instance, formamide is converted to formate and NH_3 according to

$$HCONH_2+H_2O \xleftrightarrow{\textit{Formamidase}} HCOOH_2+NH_3 \quad (6)$$

NH_3 is utilized as nitrogen source by conventional pathways via glutamate. A formate dehydrogenase (EC 1.2.1.2) may liberate CO_2 from formate

$$HCOOH+NAD \xleftrightarrow{\textit{Formate dehydrogenase}} CO_2+NADH \quad (7)$$

In such a case formamide can serve only as N-source but not as C-source. However, formate may be incorporated into N^{10}-formyl tetrahydrofolate (THF) (Rébeillé and Douce 1999; Rhodes et al. 1999) according to Eq. (2) and subsequently used e.g. for methionine synthesis from homocysteine. In such a case formamide is used both as C- and N-source. NADH of Eq. 7 can be used for the formation of ATP.

The two-step degradation of toxic cyanides to potentially useful substrates occurs in both bacteria (Miller and Gray 1982) and fungi (Rissler and Millar 1977; van der Walt et al. 1993; Brown et al.1995; Dumestre et al. 1997a,b). In yeast assimilation of amides can be used for systematical identification: Testing hydrolysis, e.g., of 23 different amides by 500 different yeasts and yeast-like strains demonstrated that 10 of these amides are sufficient to distinguish seven genera and 19 species (Mira et al. 1955). From the evolutionary view the first step may present an adaptation to the chemical environment in early history of the biosphere. However, also a correlation between FHL activity and the phytopathogenicity of some fungi to cyanogenic plants was observed (Fry and Evans 1977; Wang et al. 1992).

The occurrence of amidases may reflect a secondary adaptation, following the evolution of NHases. It allows the utilization of the end-products of NHase activities. Since nitrogen is often the limiting element in growth the use of any amides present in the environment is of evolutionary advantage. Thus, it is not surprising that some microbes are able to live with cyanides as the sole nitrogen source (Maier-Greiner et al. 1991; Kunz et al. 1994; Linardi et al. 1996; Dumestre et al. 1997a, b; Barclay et al. 1998a; Nauter et al. 1998; Gimmler et al. 2002) and additionally as the sole C-source (Miller and Gray 1982; Narayanasamy et al. 1990; Maier-Greiner et al. 1991; Dumestre et al. 1997a,b; Nauter et al. 1998).

Table 1 compiles progress made in previous years in enzymatic studies of fungal C_1 metabolism (see also Fig. 1), whereas the literature on identification and sequencing of genes of such enzymes is listed in Table 2. For information on corresponding enzymes of higher plants see Hanson et al. (2000). In Table 3 recent studies on microbial utilization of selected organic substrates related to C_1 metabolism are summarized with emphasis on (1) utilization of DON under N-limiting conditions, (2) detoxification of solutes of toxic potentials in organisms, (3) compounds of environmental concern such as climate gases, (4) mutualistic relations between different organisms.

3.3 Amides and Nitriles as Nitrogen Sources in the Extreme Acid-Tolerant Fungus *Bispora* sp.

Considering the fact that many polluted soils, industrial wastes and effluents contain considerable amount of various toxic cyanides, biodegradation of nitriles by microbes is of vast interest both from an environmental and health perspective. Therefore the potential of micro-organisms containing nitrile-hydratases and amidase for phytoremediation is studied with increasing intensity (Padmaja and Balagopal 1985; Narayanasamy et al. 1990; Shanker et al. 1990; Silva-Avalos et al. 1990; Shah et al. 1991; Barclay et al. 1998a,b; Dumestre et al. 1997a,b; Pereira et al. 1997; Cowan et al. 1998; Kobayashi and Shimizu 1998; Nauter et al. 1998). Recently it could be shown that the stress-resistant fungus genus *Scytalidium*, which comprises thermophilic and acidophilic species (Sigler and Carmichael 1974) is able to grow on metal-complexed cyanides, typically found in former gasworks sites (Barclay et al. 1998a). The genus *Scytalidium* is closely related to the recently isolated, extremely acid and heavy metal-resistant filamentous fungus *Bispora* sp., which has been investigated also in this study. Stress physiology of the latter fungus was recently analyzed in our department (Gimmler 2000, 2001; Carandang et al. 2001; Gimmler et al. 2000, 2001,

Table 2. Molecular biological studies of microbial enzymes related to C_1 metabolism and folate biosynthesis (compare Fig. 1; see also Cossins and Chen 1997). Information refers preferentially to fungi, but few data on bacteria, protozoa and algae are also included. For biochemical data see Table 1

Enzyme (E.C.)	Organism	Remarks	Gene	Sequence	Reference
Methanol oxidase (MOX) (1.1.1.13)		Peroxisomal	MUT3, MUT5		Vallini et al (2000)
Formaldehyde dehydrogenase (FLD) (1.2.1.1)	*Candida boidinii* *Saccharomyces*		FLD1 SFA, SFA1	+	Wehner et al. (1993), Grey et al. (1996), van den Berg and Steensma (1997), Shen et al. (1998), Lee et al. (2002), Baerends et al. (2002)
Formate dehydrogenase (FDH) *(1.2.1.2)*	*Saccharomyces, Candida methylica, C. boidinii, Hansenula polymorpha, Neurospora crassa*		fdh (FDH1, FDH2)	+	Ferry (1990), Chow et al. (1993), Illeova et al. (1993), Allen and Holbrook (1995), Sakai et al. (1997), Labrou and Ridgen (2001), Overkamp et al. (2002)
5,10-Methylene-THF dehydrogenase (1.5.1.5)	*Saccharomyces cerevisiae* *Aspergillus nidulans*	Cytosolic	MTD1	+	Nadolska et al. (1989), Barlowe and Appling (1990), West et al. 1993, 1996
Methionine synthetase (2.1.1.13)	*Aspergillus nidulans, Saccharomyces, Ascobulus immersus*		SAM1		Nadolska et al. (1989), Martinez and Benitez (1994), Mautino et al. (1996)
Serine hydroxymethyl-transferase (SHMT) (2.1.2.1)	*Saccharomyces, Neurospora, Aspergillus nidulans*	Mitochondrial Cytosolic	SHM1, SHM2 for		Nadolska et al. (1989), McClung et al. (1992), Kruschwitz et al. (1993, 1994a,b), McNeil et. (1994, 1996), Kastanos et al. (1997), Cossins and Chen, (1997)
Amidase (3.5.1.4)	*Aspergillus nidulans* *Fusarium venetatum*		fac B amdS		Todd et al. (1997), Bird and Bradshaw (1997), Rover et al. (1999)
Cyanide hydratase (Formamide hydrolyase) (FHL) (4.2.1.66)	*Gloecercospora sorghi* *Fusarium lateritium*		Cht Chy1	+	Wang and Vanetten (1992a,b), Cluness et al. (1993), Brown et al. (1995)

Cyanamide hydratase (4.2.1.69)	*Myrothecium verrucaria*			+	Maier et al. (1991)
Trifunctional C-1-THF synthetase complex 10-Formyl-THF-synthetase (6.3.4.3) Methenyl-THF-cyclohydrolase (3.5.4.9) 5,10 Methyl-THF-dehydrogenase (1.5.1.5)	*Saccharomyces cerevisiae* *Schizosaccharomyces pombe*	Cytosolic	ade3 Mitochondrial	+ MIS1	Appling and Rabinowitz (1983), Staben and Rabinowitz (1986), Shannon and Rabinowitz (1986, 1988), Barlowe and Appling (1990), Tanura et al. (1990), Appling (1991), Song and Rabinowitz (1993), Wahls et al.(1993), West et al. (1996), McNeil et al. (1996)
Glycine decarboxylase (GDC) (2.1.2.10) Glycine cleavage system (GCV)	*Thalassiosira weiisflogii* *Saccharomyces*	Mitochondrial	GCV1 (YAL044) GCV3	+	McNeil et al.(1996, 1997), Nagarajan and Storms (1997), Parker and Armbrust (2000)
Glycine synthesis	*Saccharomyces cerevisiae*		Gly1		McNeil et al. (1996)
Dihydropterate synthetase (DHPS)	*Plasmodium falciparum*			+	Triglia and Cowman (1994)
Dihydrofolate synthetase (DHFS) (6.3.2.12)	*Neurospora* *Saccharomyces cerevisiae*		folC FOL3 (YMR113w)		Bognar et al. (1985), Cherest et al. (2000)
Dihydrofolate reductase (DHFR)	*Neurospora,* *Aspergillus nidulans* *Candida albicans*	3-D structure of protein	YMR113 W (FOL3)	+	Nadolska et al. (1989), Blakley et al. (1993), Whitlow et al. (1997, 2001), Cherest et al. (2000)
Folylpolyglutamate synthetase (FPGS) (6.6.2.17)	*Saccharomyces,* *Neurospora crassa,* *Schizosaccharomyces pombe*		folC met-6, met7, met 11	+	Bognar et al. (1985), Atkinson et al. (1995, 1998); Chan et al. (1990, 1991), Cherest et al. (2000)
Chloromethane dehalogenase	*Methylobacterium chloro methanicum* *Hyphomicrobium chloromethanicum*		Cmu MA, CmuMB		Studer et al. (2000, 2001), McAnulla et al. (2001)
Dichloromethane dehalogenase	*Methylobacterium dichloromethanicum*		dcmA		Kayser et al. (2002)

Table 3. Microbial utilization of low-molecular substrates related to carbon one metabolism as source for nitrogen and/or carbon. Data in paratheses: Restricted to certain growth conditions.Most of the information refers to fungi, but some bactera are also included

Substrate	N source	C source	Remarks	Organism	Reference
Formamide	+			*Rhodococcus, Phanerochaete chrysosporium, Bispora, Emiliana huxleyi*	Miller and Gary (1982), Tonon et al. (1990), Gimmler et al. (200X), Palenik and Henson (1997)
Acetamide	+	+		*Bacillus sphaericum, B. gordonae, E. coli, Rhodocccus, Arthrobacter, Aspergillus nidulans, Bispora, Fusarium ve-nenatum, Phanerochaete chrysosporium, 500 yeast and yeast like strains, Emiliana huxleyi*	*Pinichoty (1988), Ramirez et al.* (1998), Nauter et al. (1998) Narayanasamy et al. (1990), Todd et al. (1997), Bird and Bradshow (1997), Mira et al (1995), Gimmler et al. (2002), Rover et al. (1999), Tonon et al. (1990), Palenik and Henson (1997)
Acrylamide	+	+		*Pseudomonas, Arthrobacter, Candida famata*	Shanker et al. (1990), Narayanasamy et al. (1990), Linardi et al. (1996)
Cyanamide	+			*Mycrothecium verrucaria*	Mayer et al. (1991)
Acetonitrile	+	+		*Rhodococcus, Arthobacter, E. coli, Candida, Debaryomy-ces, Aureobasidium, Geotrichum, Pichia Rhodotorula, Tre-mella, Hanseniaspora, Cryptococcus*	Narayanasamy et al. (1990), Miller and Gray (1982), Nauter et al. (1998), Linardi et al. (1996), Rezende et al. (1999)
Acrylnitrile	+	+		*Rhodococcus, Candiada famata*	Narayanasamy et al. (1990), Linardi et al. (1996)
Methacrylnitrile				*E. coli, Rhodotorula glutinis, Candida famata*	Nauter et al. (1998), Rezende et al. (1999), Linardi et al. (1996)
Proprionitrile	+	+		*Rhodococcus, E. coli, Chryptococcus, Candida famata*	Miller and Gry (1982), Nauter et al. (1998), Rezende et al. (1999), Linardi et al. (1996)

Isobytyronitrile	+	+		*Rhodococcus, E. coli, Chyptococcus flavus, Candida fa-mata*	*Miller and Gray (1982), Nauter et al. (1998),* Rezende et al. (1999), Linardi et al. (1996)
CN	+	+	1 of 6 fungi *pathogenic to* plants, 9 of 14 pathogens of non-cyanogenic plants, 11 (all) pathogens of cyano-genic lants, 21 of 31 fungi	*Pseudomomnas fluorescens, E. coli, Gloeocercosporon sorghi, Fusarium lateritium, F. oxysporium, Stemphylium loti, Phanerochaete chrysosporium, Rhizopus oryzae*	Dumestre et al. (1977a,b), Kunz et al. (1992, 1994), Nauter et al. (1998), Chen and Kunz (1997), Fry and Evans (1997), Cluness et al. (1995), Brown et al. (1995), Shah et al. (1991), Rissler and Millar (1977), Wang et al. (1992a), Pereira et al. (1997), Padmaya and Balaopal (1985)
Tetracyanonickelate	+			*Pseudomonas, Klebsiella, Fusarium solani, F.oxysporium, Trichoderma polysporum, Emiliana huxleyi*	Silva et al. (1990), Palenik and Henson (1997), Barclay et al. (1998a, b), Yanase et al. (2000)
Hexacyanoferrate	+	+		*Fusarium oxysporum, Scytalidium thermophilum,* ***Penicillium miczynski, Bispora***	Barlay et al. (1998a), Bird and Bradshaw (1997), Gimmler et al. (2002)
Hexacyanocobaltate	+	+		*Bispora*	Gimmler et al. (2002)
Thiocyanate				*Rhizopus oryzae*	Padmaya and Balaopal (1985)
Cyanogenic glycosides				*E. coli, Rhizopus oryzae*	Nauter et al. (1998), Padmaya and Balaopal (1985)
Dimethylsulfonio-proprionate (DMSP)	+		Higher plant and algal osmolyticum	*Eubacterium limosum, Acetobacterium woodii, Sporomusa ovata, S. sphaeroides*	Jansen and Hansen (2000, 2001)

Table 3. *Continued*

Substrate	N source	C source	Remarks	Organism	Reference
Chloromethane		+	+ Energy source + Energy source	*Methylobacterium chloromethanicum, Hyphobacterium chloromethanicum*	Coulter et al. (1999), Studer et al. (2001, 2002), McAnulla et al. (2001)
Dichloromethane		+	+ Energy source	*Methylobacterium dichloromethanicum*	Kayser et al. (2002)
Methanol		(+)	(+ Energy source)	*Candida boidinii, C. maltosa, Saccharomyses, Phanero-chaete chrysosporium, Hansenula polymorpha, Pichia, Lenzites, Polyporus, Kloeckera*	Erikson and Nishida (1988), Leak (1992), Maidan et al. *(1997)*, Shen et al. (1998), Parpinello et al. (1998), Aggelis et al. (1999, 2000), Penninckx (2000), Lee et al. (2002)
Formaldehyde		(+)	(Energy source)	*Hansenula polymorpha, Candida utilis*	Müller and Babel (1991)

2002). Because of its strong acid and metal resistance, *Bispora* sp. may be a good candidate for phytoremediation of polluted soils, e.g. former gas-works sites and industrial wastes.

Indeed, *Bispora* sp. is able to grow on metal-complexed cyanides as the sole N-source, even though growth rates were considerably lower (about 30%) than in the presence of NH_4^+ (Fig. 2A) or nitrate (not shown). Also putative products of NHase activity, namely short-chain amides, can serve as growth substrates in *Bispora* sp.: At both acid and neutral pH the fungus exhibited good growth with ammonium and acetamide as N-source (Fig. 2B), whereas the growth was much lower with formamide. *Bispora* sp. is unable to utilize urea as the nitrogen sources (Fig. 2C). Growth on acetamide was resumed only after a lag period of several days (Fig. 2C).

The nitrogen content of control samples (NH_4^+) and those grown on acetamide and formamide were relatively similar, whereas that of cultures deprived of nitrogen contained much less nitrogen (Fig. 3). The molar ratios carbon/nitrogen were close to 10 in control cells and in cultures grown with acetamide, but 2.5-fold higher in cultures deprived of nitrogen (Fig. 3B). C/N ratios in formamide cultures are significantly higher than those of control cells, but also significantly lower than in cells deprived of nitrogen. The data thus indicates that *Bispora* sp. exhibits good growth and a reasonable N-status with acetamide as sole N-source and glycerol as C-source. With formamide as sole N-source the N-status satisfactory, but the growth rate is reduced. This may indicate that ammonia deriving from a putative formamidase reaction is well used, but that C-supply is not adequate under these conditions.

Bispora sp. grown at pH 7.0 with NH_4^+ as N-source and glycerol as C-source neither expresses urease (EC 3.5.1.5) nor formamidase (EC 3.5.1.9) or acetamidase (EC 3.5.1.14) (Table 4). This agrees with the observation that in general NH_4^+ prevents the expression of these enzymes in filamentous fungi (Tonon et al. 1990). However, when *Bispora* is cultured in the presence of formamide as N-source and glycerol as C-source a formamidase is expressed (Table 4) even though growth is low under such conditions (Table 4, compare Fig. 1B). Surprisingly, when cultured with acetamide as N-source and glycerol as C-source, which permits good growth (Fig. 2B), neither acetamidase nor an urease, but a formamidase is expressed (Table 4).

Bispora sp. takes up ^{14}C-acetamide into control cells and cells pre-cultured with acetamide, but much less into cells pre-cultured with formamide (Fig. 4). This effect is observed both in cells pre-cultured with (Fig. 4A) and without glycerol (Fig. 4B) as additional carbon source. Part of the radioactivity from ^{14}C-acetamide is incorporated also into the water-insoluble, pelletable fraction of the fungus (not shown). TLC analysis reveals that

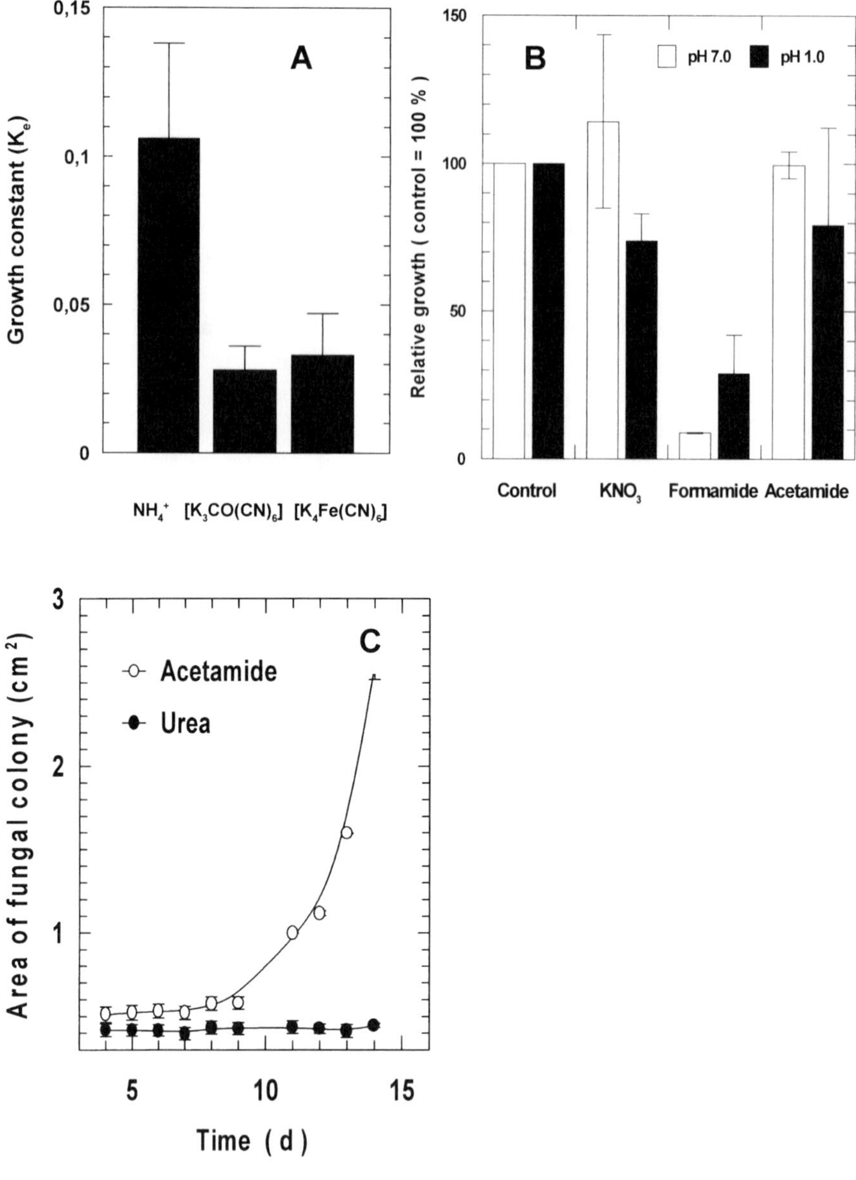

Fig. 2. Nitrogen sources of the filamentous fungus *Bispora* sp. (adapted from Gimmler et al. 2002). **A** Growth on $NH4^+$ and metal complexed cyanides as the sole nitrogen source (pH 7.0, glycerol as carbon source). **B** Growth with $NH4^+$ (control), nitrate, formamide and acetamide as N-source (glycerol as carbon source) at pH 1.0 and 7.0. **C** Kinetics of growth with acetamide or urea as the sole N- and C-source (pH 7.0)

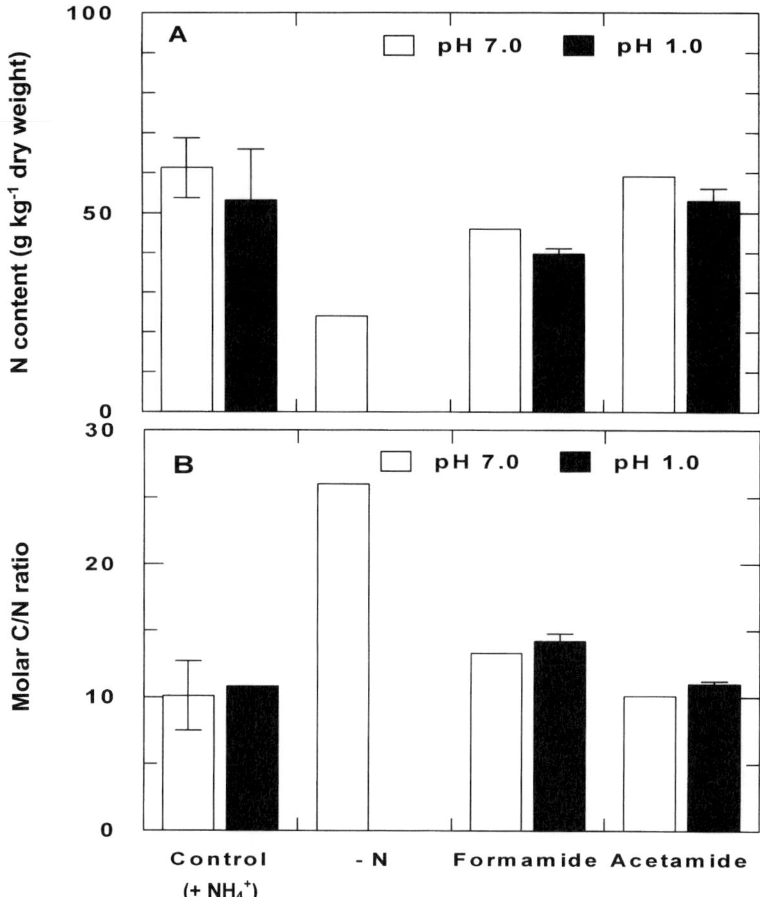

Fig. 3. A N-content and **B** molar C/N ratio in the filamentous fungus *Bispora* sp. as affected by the pH of the growth medium and various N-sources (carbon source: glycerol). (Adapted from Gimmler et al. 2002)

products labelled from ^{14}C-acetamide are (glycine+serine), malate, alanine and methyl-THF (Fig. 4C). *Bispora* sp. also incorporates ^{14}C-formate into the insoluble, pelletable fraction independent on the pH of the growth medium (Fig. 5).

In experiments crude enzyme extracts of *Bispora*, the subsequent TLC analysis demonstrated that the label of the substrates ^{14}C-formate and ^{14}C-methyl-THF are incorporated into THF and glycine/serine, respectively.

In summary, results of this case study indicate that the acid and heavy metal-resistant filamentous *Bispora* sp. can use acetamide as sole nitrogen

Table 4. In vitro amidase and urease activity of *Bispora* sp. as affected by the N-source during the pre-culture and the substrate during the assay

Experiment number	Pre-culture of the fungus in	Substrate during the measurement	Amidase activity measured as change in conductivity (Δ µS h^{-1}µg^{-1} protein)
I	NH_4^+ (control)	Formamide	0
		Acetamide	0
		Urea	0
	Formamide	Formamide	0.064±0.42
		Acetamide	0
	Acetamide	Formamide	0.121±0.008
		Acetamide	0
II	Formamide	Urea	0
	Acetamide	Urea	0
	Formamide	Formamide	0.046±0.006
		Acetamide	0
	Acetamide	Formamide	0.033±0.003
		Acetamide	0

and carbon source (Gimmler et al. 2002). To a lesser extent the fungus can also assimilate formamide and metal-complexed cyanides, whereas it is totally unable to assimilate urea.

It is unknown whether acetamide and formamide are taken up by a catalyzed transport across the plasma membrane or by simple diffusion. Both is thermodynamically possible. Some authors postulated that acetamide and formate are able to enter cells via aquaporines (Meinild et al. 1998). The reflection coefficient for acetamide was measured to be close to 0.91 and that for formamide close to 0.80.

After entry into the cells acetamide is immediately converted to other metabolites, indicated by a very low cytosolic pool of acetamide. The same is assumed to occur with formate. The supposed fate of both amides in *Bispora* (Fig. 6) is in principle agreement with data of the literature (Fig. 1). However, somehow unusual is the apparent absence of a classical acetamidase. Acetamide is supposed to be split into formamide and a C_1 unit. The latter is assumed to be fed into the C_1 metabolism by a methyl-tetrahydrofolate synthetase, which relates it to the metabolism of glycine, serine, and methionine (Pasternack et al. 1994a,b, 1996; Mouillon et al. 1999; Rébeillé and Douce 1999; Hanson et al. 2000). The internal level of folic acid in *Bispora* sp. is about 0.5 µg per gram of fresh weight (not shown), corresponding to about 1.1 nmol/g fresh weight. This level is comparable to that

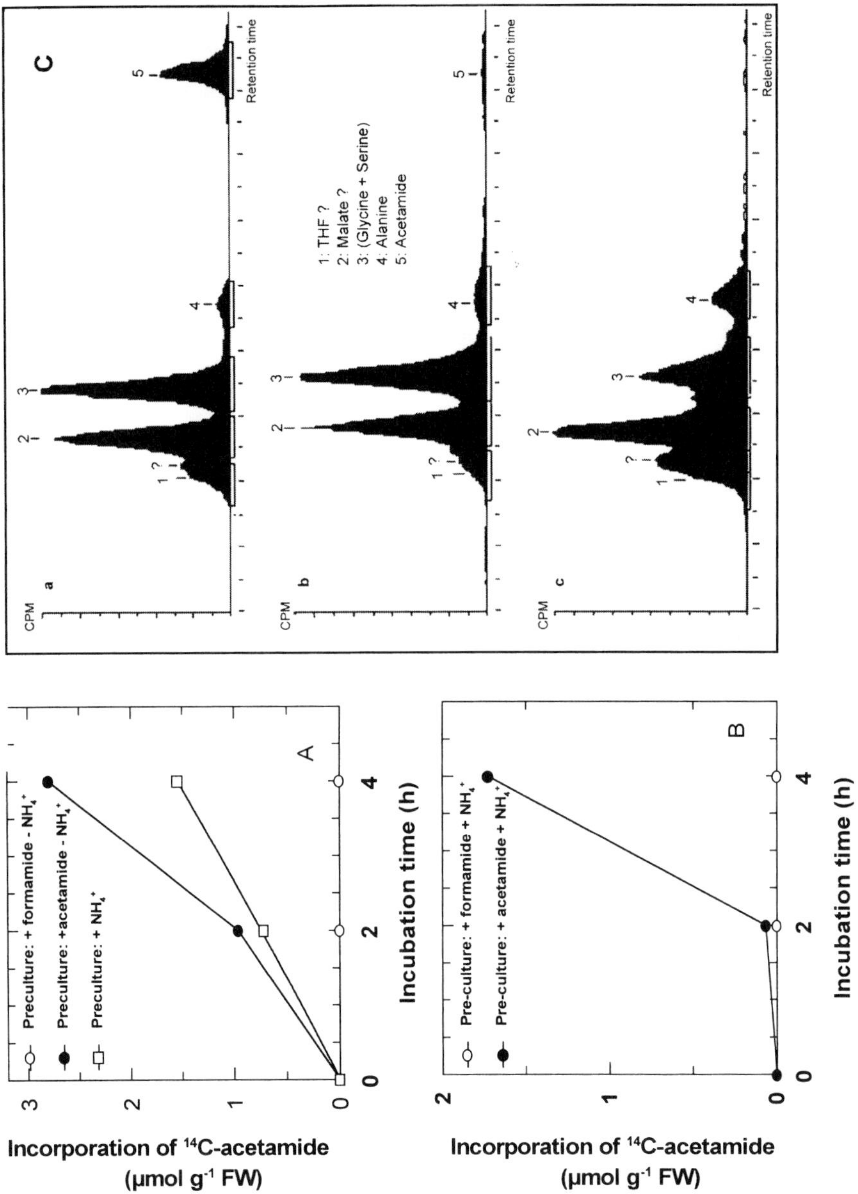

Fig. 4. Incorporation of ^{14}C-labelled solutes into intact cells of the filamentous fungus *Bispora* sp. (adapted from Gimmler et al. 2002). **A** Time-dependent uptake of ^{14}C-acetamide (final concentration 0.1 mM) into the ethanol-soluble fraction of *Bispora* sp. in the presence of different N-sources with glycerol as carbon source. **B** Time-dependent uptake of ^{14}C-acetamide (final concentration 0.1 mM) into the ethanol-soluble fraction of *Bispora* sp. in the presence of acetamide or formamide as N- and C-sources. **C** TLC analysis of the incorporation of ^{14}C-acetamide (AA) (4 h) into ethanol-soluble compounds of *Bispora* sp. (pre-culture: *a*) + glycerol + AA – NH_4^+; *b*) + glycerol + AA + NH_4^+; *c*) – glycerol + AA + NH_4^+. Compound 5 in AA from co-chromatography

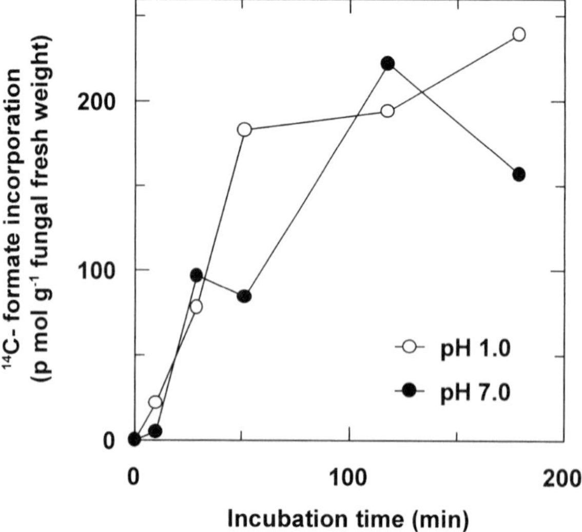

Fig. 5. Time-dependent incorporation of ^{14}C-formate (final concentration 0.01 mM) into the insoluble, pelletable fraction of *Bispora*. sp. at two different pH values of the growth medium. (Adapted from Gimmler et al. 2002)

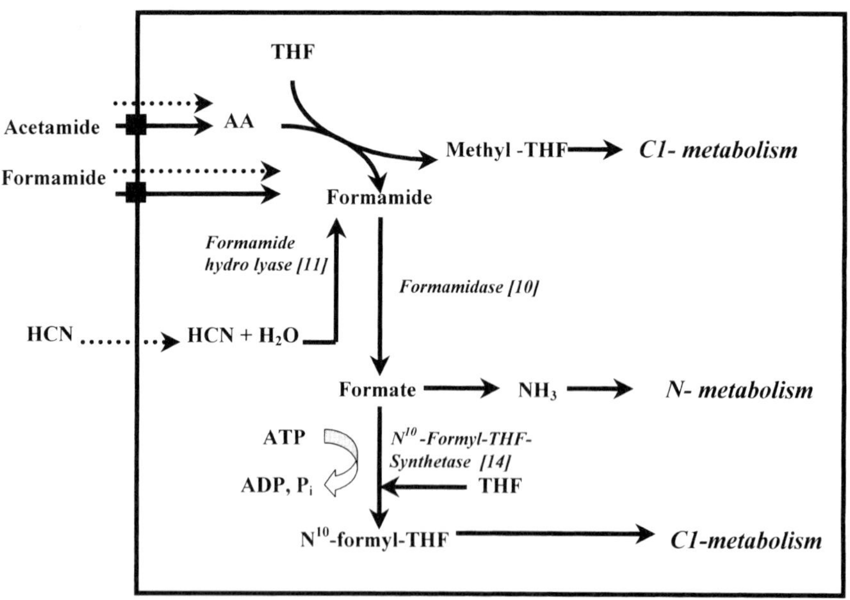

Fig. 6. Putative scheme of uptake and assimilation of acetamide, formamide and metal complexed cyanides in the acid and heavy-metal-resistant filamentous fungus *Bispora sp.* and its relation to N-metabolism and C1 metabolism as treated in this review. Dotted arrows indicate diffusional uptake, squares possible sites of catalyzed uptake (adapted from Gimmler et al. 2002). *Numbers* refer to enzymes cited in Table 1 (column 1)

of higher plants (Rébellé and Douce 1999) and should enable the fungus to carry out an efficient C_1 unit metabolism.

Formamide – independent of whether it is taken up as growth substrate or is a product of acetamide assimilation – is split into formate and NH_3 by classical formamidase, causing a second entry into C_1 metabolism by a formyl-tetrahydrofolate synthetase, which connects to the synthesis of purines, glycine, serine, and methionine (Pasternack et al. 1994a,b, 1996; Mouillon et al. 1999; Rébeillé and Douce 1999; Hanson et al. 2000). When *Bispora* sp. is grown on metal-complexed cyanides, a formamide hydrolyase is supposed to produce formamide, which is metabolized as described above. Formamide hydrolyases are high molecular weight proteins (native MW between 300 and 1210 kDa; subunits 41–45 kDa; Wang and Vanetten 1992; Wang et al. 1992; Brown et al. 1995; Dumestre et al. 1997a,b; Barclay et al. 1998a,b; compare Table 1). Enzyme activity is generally found only under aerobic conditions (which matches the fact that *Bispora* sp. is obligate aerobic) and takes HCN, but not CN^- as substrate. For this reason its activity is strongly dependent on the pH. It is reasonable to assume that the entry of the cyanides into the cells takes place in the form of the lipophilic, protonated form. The assimilation of metal-complexed cyanides by the acid and metal-resistant fungus makes *Bispora* sp. a possible candidate for the remediation of industrial polluted soils.

3.4 DMSP and Microbial C_1 Metabolism

Dimethylsulfonioproprionate (DMSP), a common osmolyticum of higher plants (Rhodes et al. 1997) and algae (Summers et al. 1998), does not fit exactly into the topic of this review, since it does not contain nitrogen. Nonetheless, DMSP accumulation is of large environmental and physiological interest. DMSP biosynthesis begins in the cytosol of higher plants and algae and finishes in chloroplasts (Trossat et al. 1996). Its environmental importance arises from the observation that DMSP is the major source of atmospheric dimethylsulfide (DMS), which is a crucial climate gas. Its formation is related to microbial C_1 metabolism. Some bacteria, which are able to grow on DMSP, metabolize DMSP to DMS and acrylate or via initial demethylation to methylthioproprionate (MTPA) (Jansen and Hansen 2000, 2001).The former reactions follows Eq. (8):

$$\text{DMSP} \xrightarrow{\textit{DMSP-THF Methyl transferase}} \text{DMS+acrylate} \quad (8)$$

This THF methyl transferase exhibits some specificity for DMSP, since glycine betaine, which is structurally an N-containing analogue of DMSP, does not function as methyl donor for the DMSP-THF methyl transferase.

The same applies to various S-containing derivates of DMSP. The physiological benefit of Eq. (8) is clear, since DMSP, preferentially of algal origin, supports growth of the bacteria containing the DMSP-THF methyl transferase. Obscure, however, is the physiological benefit of the stoichiometric demethylation of DMSP to MTPA. The (homo)acetogenic fermentation based on this demethylation did not support growth of five *Eubacterium limosum* strains, *Sporomosa ovata*, *Sporomosa spaheroides* and *Acetobacterium woodii*. This is in contrast to the analogous demethylation of glycine betaine, which strongly supports the growth of these bacteria.

3.5 Methanol Connects C_1 Metabolism of Higher Plants and Micro-organisms

Methanol has been identified beside methane and DMS as one of the major organic compounds in forest air and in the troposphere (Nemecek-Marshall et al. 1995; Fall and Benson 1996). Vast amounts are produced by the vegetation, mainly by C3 plants. Two main sources of methanol are discussed:

1. Demethylation of methylated pectins to polygalacturonans with methanol as by-product catalyzed in microsomal membranes by a pectin methyl transferase (PMT; EC 2.1.1) (Ishikawa et al. 2000; see Eq. 9, compare also bottom of Fig. 1):

   $$\text{Polygalacturonan methyl ester} \xrightarrow{\textit{Pectin methyl transferase}} \text{Polygalacturonan} + nCH_3OH \qquad (9)$$

   This demethylation of pectin serves to assist cell maturation (Fall and Benson 1996) by providing carboxyl groups in side chains of the pectin polymers in the primary layer of the cell walls (preferentially those of dicots and gymnosperms). These groups bind divalent cations such as calcium and magnesium and thereby support cross-linking of the polymer chains to a more rigid gel matrix of the primary layer.
2. Another potential source of biospheric methanol is lignin from the secondary layer of cell walls. Lignin is rich in methyl ether groups. Wood rotting fungi and a variety of bacteria are capable of enzymatic cleavage of these ether bonds leading to volatilization of methanol.

Independent on its origin, the liberated methanol relates to C_1 metabolism in completely different pathways: It can to a certain extent stimulate plant growth (Fall and Benson 1996). Leaf methanol is rapidly oxidized through the intermediate stages of formaldehyde and formate to CO_2 (Fig. 1). Formate may feed into the C_1 folate pool of the plant and thereby support growth, e.g. by the synthesis of serine, glycine, methionine, purines or thymilidates. CO_2 may be utilized by the photosynthetic machinery of the leaves. Similar as fungi and bacteria higher plants contain formate dehydrogenase (FDH) required for this process, preferentially in mitochondria and non-photosynthetic tissues (Cowan et al. 1998; Hourton-Cabassa et al. 1998; Suzuki et al. 1998; Igambierdiev et al. 1999; Wingler et al. 1999). Its activity is increased by environmental stress such as hypoxia of roots and Fe-deficiency (Hourton-Cabassa et al. 1998; Suzuki et al. 1998). It is not clear as yet, whether physicochemical and molecular properties of the FDH in higher plants are identical with those of the various FDHs in micro-organisms. The key enzyme for channelling formate into C_1 metabolism of higher plants is thought to be similar to that in micro-organisms, i.e. the serine hydroxymethyl transferase (SHMT). Additionally it is discussed that methanol may induce changes that cause an increase in the availability of photorespiratory carbon (Fall and Benson 1996).

Alternatively leaf methanol, emitted primarily through the stomata, or from other sources may serve as carbon source for methylotrophic bacteria or fungi, which leads again to C_1 metabolism. In more than 70 plant species the surface of that/those leaf side(s) which possess stomata is occupied by an extensive society of epiphyllic methylotrophic bacteria, that are bacteria which can utilize methanol as sole carbon source for growth.

The first step of the fungal methanol utilization is methanol oxidation by the methanol oxidase reaction (MOX, EC.1.1.1.13, Eq. 10) in fungal peroxisomes (Nishida and Erikson 1987; Erikson and Nishida 1988)

$$CH_3OH+O_2 \xrightarrow{\text{Methanol oxidase}} CH_2O+H_2O_2 \tag{10}$$

Formaldehyde is converted to formic acid by a glutathione-dependent formaldehyde dehydrogenase (EC 1.2.1.1, Eq. 11; compare also Eq. 7):

$$CH_3O+\text{glutathione}_{ox} \xrightarrow{\text{Formaldehyde dehydrogenase}} CHOOH+\text{glutathione}_{red} \tag{11}$$

The latter reaction is followed eventually by the formate dehydrogenase reaction, which converts formate into CO_2 (Eq. 7). It is under discussion, to which extent Eqs. (11) and (7) reflect a detoxification mechanism for formaldehyde or a pathway for carbon and/or energy utilization under conditions of methylotrophic growth of fungi or when co-cultured in the presence of other organic growth substrates (compare references cited in Table 4).

3.6 Halogenated Organic Compounds and Microbial C_1 Metabolism

Halogenated organic compounds (HOCs) are produced in large quantities world-wide both industrially (e.g. organochlorine pesticides) and naturally (Fetzner 1998). HOCs present an important class of environmental pollutants, since they are thought to be involved in stratospheric ozone depletion. A large proportion of total HOCs is chloromethane (CH_3Cl). Naturally it originates mainly from tropical and subtropical forests (Watling and Harper 1998). About 16,000 t of chloromethane are produced in these forests yearly by wood-rotting fungi (Hymnochaetaceae), among them genera such as *Phellinus* or *Inanotus* (Fetzner 1998; Saxena et al. 1998; Watling and Harper 1998). The fungal biosynsythesis of CH_3Cl is catalyzed by a methyl chloride transferase (Saxena et al. 1998). It is suggested that adenoysl-homocysteine (SAM) is the methyl donor (compare bottom of Fig. 1), which presents a link to C1 metabolism.

Some bacteria have evolved several strategies for enzyme-catalyzed dehalogenation and degradation of both haloaliphatic and haloaromatic compounds (Fetzner 1998). This is of two-fold environmental interest. On one side such systems present detoxification mechanisms for the environment. From the bacterial view it presents the utilization of unique organic nutrients: Some bacteria, e.g., are able to grow with CH_3Cl as the sole C and energy source (Vannelli et al. 1998; Coulter et al. 1999; McAnulla et al. 2001; Kayser et al. 2002; Studer et al. 2002). The substrate-inducible chloromethane dehalogenase (CMD) reaction involves a multi-step pathway from CH_3Cl via formaldehyde to formate (Fetzner 1998; Vannelli et al. 1998; Coulter et al. 1999; Studer et al. 1999; Studer et al. 2001) and thus clearly presents a link to core C_1 metabolism of bacteria. CMD has a molecular weight of about 67 kDa and contains cobalamin (vitamin B_{12} cofactor). Recently, the gene of this enzyme has been identified (Studer et al. 1999, 2001, 2002; McAnulla et al. 2001; Kayser et al. 2002).

In summary, CH_3Cl produced naturally by wood-rotting fungi in reactions related to C_1 metabolism is dehalogenated by certain bacteria again with the aid of reactions linked with C_1 metabolism.

Acknowledgement. We are grateful for financial support by the Deutsche Forschungsgemeinschaft (SFB 251, TP A2: H. Gimmler) and TP A3 (W. Hartung) and SFB 567, TP C_1 (W. Hartung) and the Bayerische Staatsministerium für Landesentwicklung und Umweltfragen (BayFORREST project F 172, H. Gimmler).

References

Aggelis G, Fakas S, Melissis S, Clonis YD (1999) Growth of *Candida boidiniii* in a methanol-limited continuous culture and the formation of methanol-degrading enzymes. J Biotechnol 72:127–139

Aggelis G, Margariti N, Kralli, C, Flouri F (2000) Growth of *Candida boidinii* on methanol and the activity of methanol-degrading enzymes as affected from formaldehyde and methylformate. J Biotechnol 80:119–125

Allen SJ, Holbrook JJ (1995) Isolation, sequence and overexpression of the gene encoding NAD-dependent formate dehydrogenase from the methylotrophic yeast *Candida methylica*. Gene 162:99–104

Appling DA (1991) Compartmentation of folate-mediated one-carbon metabolism in eukaryotes. FASEB J 5:2645–2651

Appling DR, Kastanos E, Pasternack LB, Woldmann YY (1997) The use of ^{13}C nuclear magnetic resonance to evaluate metabolic flux through folate one-carbon pools in *Saccharomyces cerevisiae*. Methods Enzymol 281:218–231

Atkins CA, Smith PMC, Storer PJ (1997) Reexamination of the intracellular localization of de novo purine synthesis in cowpea nodules. Plant Physiol 113:127–135

Atkinson IJ, Nargang FE, Cossins EA (1995) Folylglutamate synthesis in *Neurospora crassa*: transformation of polyglutamate-deficient mutants. Phytochemistry 38:603–608

Atkinson IJ, Nargang FE, Cossins EA (1998) Folylpolyglutamate synthesis in *Neurospora crassa*: primary structure of the folylpolyglutamate synthetase gene and elucidation of the MET-6 mutation. Phytochemistry 49:2221–2232

Baerends RJS, Sulter GJ, Jeffries TW, Cregg JM, Veenhuis M (2002) Molecular characterization of the *Hansenula polymorpha* FLD1 gene encoding formaldehyde dehydrogenase. Yeasts 19:37–42

Barclay M, Hart A, Knowles CJ, Meeussen JCL, Tett VA (1998a) Biodegradation of metal cyanides by mixed and pure cultures of fungi. Enzyme Microbial Technol 22:223–231

Barclay M, Tett VA, Knowles CJ (1998b) Metabolism and enzymology of cyanide/ metallocyanide biodegradation by *Fusarium solani* under neutral and acidic conditions. Enzyme Microbial Technol 23:321–330

Barlowe CK, Appling DR (1990a) Isolation and characterization of a novel eukaryotic monofunctional DAD-dependent 5,10 methylenetetrahydrofolate dehydrogenase. Biochemistry 29:7089–7094

Barlowe CK, Appling DR (1990b) Molecular genetic analysis of *Saccharomyces cerevisiae* C-1-tetrahydrofolate synthase mutants reveals a noncatalytic function of the ADE3 gene product and an additional folate-dependent enzyme. Mol Cell Biol 10:5679–5687

Besson V, Rébeillé F, Neuburger M, Douce R, Cossins EA (1993) Effect of tetrahydrofolate polyglutamates on the kinetic parameters of serine hydroxymethyltransferase and glycine decarboxylase from pea mitochondria. Biochem J 292:425–430

Bird D, Bradshaw R (1997) Gene targeting is locus dependent in the filamentous fungus *Aspergillus nidulans*. Mol Gen Genet 255:219–225

Bird IF, Cornelius MJ, Keys AJ, Whittingham CP (1972) Oxidation and phosphorylation associated with the conversion of glycine to serine. Phytochemistry 11:1587–1594

Bourguignon J, Vauclare P, Merand V, Forest E, Neuburger M, Douce R (1993) Glycine decarboxylase complex from higher plants. Mol cloning, tissue distribution and mass spectrometry analysis of the T-protein. Eur J Biochem 217:377–386

Bourguignon J, Rébeillé F, Douce R (1999) Serine and glycine metabolism in higher plants. In: Singh BK (ed) Plant amino acids. Marcel Dekker, New York, pp 111–146

Brown DT, Turner PD, O'Reilly C (1995) Expression of the cyanide hydratase enzyme from *Fusarium lateritium* in *Escherichia coli* and identification of an essential cysteine residue. FEMS Microbiol Lett 134:143–146

Carandang JS, Gimmler H, Hahn T, Greiser A (2001) Characterization of an acid tolerant filamentous fungus (*Deuteromycetes*) coexisting with the extreme acid-resistant green alga *Dunaliella acidophila* at pH 1.0. Acta Manilana 49:5-23

Cella R, Parisi B (1993) Dihydrofolate reductase and thymilidate synthase in plants: an open problem. Physiol Plant 88:509–521

Chan PY, Dale PL, Cossins EA (1991) Purification and properties of *Neurospora* folylpolyglutamate synthatase. Phytochemistry 30:3525–3532

Chapin FS, Moilanen L, Kielland K (1993) Preferential use of organic nitrogen for growth by a non-mycorrhizal arctic sedge. Nature 361:150–153

Chen JL, Kunz DA (1997) Cyanide utilization of Pseudomonas fluroescens NCIMB 11674 involves a putative siderophore. FEMS Microbiol Lett 156:61–67

Chen L, Chan SY, Cossins EA (1997) Distribution of folate derivates and enzymes for synthesis of 10-formyltetrahydrofolate in cytosolic and mitochondrial fractions of pea leaves. Plant Physiol 115:299–301

Cherest H, Thomas D, Surdin KY (2000) Polyglutamylation of folate enzymes is necessary for methionine biosynthesis and maintanance of intact mitochondrial genome in *Saccharomyces cerevisiae.* J Biol Chem 275:14056–14063

Chistoserdova L, Gomelsky L, Vorholt JA, Gomelsky M, Tsygankov YD, Lidstrom ME (2000) Analysis of two formaldehyde oxidation pathways in *Methlyobacillus flagellatus* KT, a ribulose monophosphate cycle methylotroph. Microbiology 146:233–238

Chow CM, Rajbhandary UL (1993) Developmental regulation of the gene for formate dehydrogenase in *Neurospora crassa.* J Bacteriol 175:3703–3709

Cluness MJ, Turner PD, Clements E, Brown DT, O'Reilly C (1993) Purification and properties of cyanide hydratase from *Fusarium lateritium* and analysis of the corresponding chy1 gene. J Gen Microbiol 139:1807–1815

Coulter C, Hamilton JTG, McRoberts WC, Kulakov L, Larkin MJ, Harper DB (1999) Halomethane:bisulfide/halide ion methyltransferase, an unusual corrinoid enzyme of environmental significance isolated from an aerobic methylotroph using chloromethane as the sole carbon source. Appl Environ Microbiol 65: 4301–43112

Cossins EA (1987) Folate biochemistry and the metabolism of one-carbon units. In: Conn EE, Stumpf PK (eds) The biochemistry of plants, vol 11. Academic Press, New York, pp 317–353

Cossins EA, Chen L (1997) Folates and one-carbon metabolism in plants and fungi. Phytochemistry 45:437–452

Cowan D, Cramp R, Pereira R, Graham D, Almatawah Q (1998) Biochemistry and biotechnology of mesophilic and thermophilic nitrile metabolizing enzyme. Extremophiles 2:207–216

DeSouza L, Shen Y, Bognar AL (2000) Disruption of cytoplasmic and mitochondrial folylpolyglutamate synthetase activity in *Saccharomyces cerevisiae.* Arch Biochem Biophys 376:299–312

Douce R, Neuburger M (1996) Biochemical dissection of photorespiration. Curr Opin Plant Biol 2:214–222

Douce R, Bourguignon J, Neuburger M, Rébeillé F (2001) The glycine decarboxylase system; a fascinating complex. Trends Plant Sci 6:167–176

Dumestre A, Bousserhin N, Berthelin J (1997a) Biodegradation of free cyanide by the fungi *Fusarium solani*: relation to pH and cyanide speciation in solution. Compt Rend Acad Sci Ser II A 325:133–138

Dumestre A, Chone T, Portal JM, Gerard M, Berthelin J (1997b) Cyanide degradation under alkaline conditions by a strain of *Fusarium solani* isolated from contaminated soils. Appl Environ Microbiol 63:2729–2734

Erikson KE, Nishida A (1988) Methanol oxidase of *Phanerochaete chrysosporium.* Methods in Enzymol 161:322–326

Falkengren-Grerup U, Mansson KF, Olsson MO (2000) Uptake capacity of amino acids by ten grasses and forbs in relation to soil acidity and nitrogen availability. Environ Exp Bot 44: 207–219

Fall R, Benson AA (1996) Leaf methanol – the simplest natural product from plants. Trends Plant Sci 1:296–301

Ferry JG (1990) Formate dehydrogenase. FEMS-Microbiol Rev 87:377–382

Fetzner S (1998) Bacterial dehalogenation. Appl Microbiol Biotechnol 50:633–657

Fischer WN, André B, Rentsch D, Krolkiewicz S, Tegeder M, Breitkreuz K, Frommer WB (1998) Amino acid transport in plants. Trends Plant Sci 3:188–195

Fry WE, Evans PH (1977) Association of formamide hydrolyase with fungal pathogenicity to cyanogenic plants. Phytopathology 67:1001–1006

Gardeström P, Bergman A, Ericson L (1981) Inhibition of the conversion of glycine to serine in spinach leaf mitochondria. Physiol Plant 53:439–444

Gardeström P, Edwards GE, Henricson D, Ericson I. (1985) The localization of serine hydromethyltransferase in leaves of C_3 and C_4 species. Physiol Plant 64:29–33

Gimmler H (2000) GABA and the acid resistance of the filamentous fungus *Bispora* sp. Manila J Sci 3:1–8

Gimmler H (2001) Mutualistic relationships between algae and fungi (excluding lichens). Prog Bot 62:194–214

Gimmler H, De Jesus J, Greiser A (2001) Heavy metal resistance of the extreme acidotolerant filamentous fungus *Bispora* sp. Microbial Ecol 42:87–98

Gimmler H, Boots A, De Jesus J (2002) Acetamide as potential C- and N-source of the extremely acid and heavy metal resistant filamentous fungus *Bispora* sp. Manila J Sci 3:1–8

Grabarse W, Vaupel M, Vorholt JA, Shima S, Thauer RK (1999) The crystal structure of methenyltetrahydromethanopterin cyclohydrolase from the hyperthermophilic archaeon *Methanopyrus kandleri.* Struct Lond 7:1257–1268

Grey M, Schmidt M, Brendel M (1996) Overexpression of ADH1 confers hyper-resistance to formaldehyde in *Saccharomyces cerevisiae.* Curr Genet 29: 437–440

Griffin DH (1994) Fungal physiology, 2nd edn. Wiley and Liss Publisher, New York, p 81

Hanson AD, Shachar-Hill Y (2002) Assigning gene function in the *Arabidopsis* one (C_1) metabolism network. Plant Physiol 129:409–410

Hanson AD, Gage DA, Shachar-Hill Y (2000) Plant one-carbon metabolism and its engineering. Trends Plant Sci 5:206–213

Hartung W, Ratcliffe RG (2002) Utilization of glycine and serine as nitrogene sources in the roots of *Zea mays* and *Chamaegigas intrepidus.* J Exp Bot 53:1–10

Heilmeier H, Hartung W (2001) Survival strategies under extreme and complex environmental conditions: the aquatic resurrection plant *Chamaegigas intrepidus.* Flora 196:245–260

Henry HAL, Jefferies RL (2002) Free amino acid, ammonium and nitrate concentrations in soil solutions of a grazed coastal marsh in relation to plant growth. Plant Cell Environ 25:665–676

Hourton-Cabasssa C, Ambard-Bretteville F, Moreau F, Davy de Virville J, Remy R, Colas des Francs-Small C (1998) Stress induction of mitochondrial formate dehydrogenase in potato leaves. Plant Physiol 116:627–635

Igambierdiev AU, Bykova NV, Kleczkowski LA (1999) Origins and metabolism of formate in higher plants. Plant Physiol Biochem 37:503–513

Illeova V, Certik M, Stefuca V, Bales V (1993) Formate dehydrogenases in methanol-utilizing yeasts. Microbios 76:29–33

Ishikawa M, Kuroyama H, Takeuchi Y, Tsumuraya Y (2000) Characterization of pectin methyl transferase from soybean hypocotyls. Plants 210:782–791

Jansen M, Hansen TA (2000) DMSP tetrahyrofolate methyltransferase from the marine sulfate-reducing bacterium strain WN. J Sea Res 43:225–231

Jansen M, Hansen TA (2001) Non-growth-associated demethylation of dimethylsulfonioproprionate by (homo)acetegenic bacteria. Appl Eviron Microbiol 67:300–306

Kastanos EK, WoldmanYY, Appling DR (1997) Role of mitochondrial and cytoplasmic serine hydroxymethyltransferase isoenzymes in *de novo* synthesis in *Saccharomyces cerevisiae*. Biochemistry 36:14956–14964

Kato N (1990a) Formaldehyde dehydrogenase from methylotrophic yeasts. Methods Enzymol 188:455–459

Kato N (1990b).Formate dehydrogenase. Methods in Enzymol 188:459–462

Kayser MF, Ucurum Z, Vuilleumier S (2002) Dichromomethane metabolism and C_1 utilization in *Methylobacterium* strains. Microbiology 148:1915–1922

Kielland K (1994) Amino acid absorption by arctic plants: implications for plant nutrition and nitrogen cycling. Ecology 75:2373–2383

Kirksey TJ, Appling DR (1996) Site-directed mutagenesis of a highly conserved aspartate in the putative 10-formyltetrahydrofolate binding site of yeast C_1-tetrahydrofolate synthase. Arch Biochem Biophys 333:251–259

Kobayashi M, Shimizu S (1998) Metalloenzyme nitrile hydratase: structure, regulation and application to biotechnology. Nat Biotechnol 16:733–736

Kruschwitz H, Schirch V (1994) Evidence that 5-formyltetrahydrofolate is a storage form of folate and one-carbon groups in *Neurospora crassa*. FASEB J 8:21–25

Kruschwitz HL, McDonald D, Cossins EA, Schirch V (1994) 5-Formyltetrahydroptero-polyglutamates are the major folate derivates in *Neurospora* conidiospores. J Biol Chem 269:28757–28763

Kunz DA, Nagappan O, Silva-Avalos J, Delong GT (1992) Utilization of cyanide as a nitrogenous substrate by *Pseudomonas fluorescens* NCIMB 11764 evidence for multiple pathways of metabolic conversion. Appl Environ Microbiol 58:2022–2029

Kunz DA, Wang CS, Chen JL (1994) Alternative routes of enzymic cyanide metabolism in *Pseudomonas fluorescens* NCIMB 11764. Microbiology 140:1705–1712

Labrou NE (2000) Improved purification of *Candida boidinii* formate dehydrogenase. Bioseparation 9:99–104

Labrou NE, Clonis YD (1995) The interaction of *Candida boidinii* formate dehydrogenase with a new family of chimeric biomimetic dye-ligands. Arch Biochem Biophys 316:169–178

Labrou NE, Karagouni A, Clonis YD (1995) Biomimetic-dye affinity adsorbents for enzyme purification: application to the one-step purification of *Candida boidinii* formate dehydrogenase. Biotech Bioeng 48:278-288

Labrou NE, Ridgen DJ (2001) Active site characterization of *Candida boidinii* formate dehydrogenase. Biochem J 354:455–463

Leak DJ (1992) Biotechnological and applied aspects of methane and methanol utilizers. In: Murrell JC, Dalton H (eds) Biotechnology handbooks, vol 5. Plenum Press, New York, pp 245–279

Lee B, Yurimoto H, Sakai Y, Kato N (2002) Physiological rol of the glutathione-dependent formaldehyde dehydrogenase in the methylotrophic yeast *Candida boidinii*. Microbiology 148:2697–2704

Linardi VR, Dias JCT, Rosa CA (1996) Utilization of acetonitrile and other aliphatic nitriles by a *Candida famata* strain. Microbiol Lett 144:67–71

Lipson DA, Monson RK (1998) Plant-microbe competition for soil amino acids in the alpine tundra: effects of freeze-thaw and dry-rewet events. Oecologia 113:406–414

Lovell CA, Przybyla A, Ljungdahl LG (1990) Primary structure of the thermostable formyltetrahydrofolate synthetase from *Clostridium thermoaceticum*. Biochemistry 29:5687–5694

Maaheimo H, Fiaux J, Cakar ZP, Bailey JE, Sauer U, Szyperski T (2001) Central carbon metabolism of *Saccharomyces cerevisiae* explored by biosynthetic fractional 13C labeling of common amino acids. Eur J Biochem 268:2464–2479

Maden BEH (2000) Tetrahydrofolate and tetrahydromethanopterin compared: functionally distinct carriers in C_1 metabolism. Biochem J 350:609–629

Maier-Greiner UH, Obermaier-Skrobanek BMM, Estermaier LM, Kammerloher W, Freund C, Wülfling C, Burkert UI, Matern DH, Breuer M (1991) Isolation and properties of a nitrile hydratase from the soil fungus *Myrothecium verrucaria* that is highly specific for the fertilizer cyanamide and cloning of its gene. Proc Natl Acad Sci USA 88:4260–4264

Martinez FE, Benizez T (1994) The SAM2 gene product catalyzes the formation of S-adenosyl-ethionine from ethionine in *Saccharomyces cerevisiae*. Curr Microbiol 28:339–343

Mautino MR, Goyon C, Rosa AL (1996) Cloning and sequence of the *Ascobulus immersus* S-adenosyl-L-methionine synthease encoding. Gene 170:155–156

McAnulla C, Woodall CA, McDonald IR, Studer A, Vuilleumier S, Leisinger T, Murrel JC (2001) Chloromethan utilization gene cluster from *Hyphomicrobium chloromethanicum* strain CM2T and development of functional gene probes to detect halomethane-degrading bacteria. Appl Environ Microbiol 67:307–316

McClung CR, Davis CR, Page KM, Denome SA (1992) Characterization of the formate FOR locus which encodes the cytosolic serine hydroxymethyltransferase of *Neurospora crassa*. Mol Cell Biol 12:1412–1421

McDonald D, Atkinson IJ, Cossins EA, Shane B (1995) Isolation of dihydrofolate and folylpolyglutamate synthetase activities from *Neurospora*. Phytochemistry 38:327–333

McNeil, JB, McIntosh EM, Taylor BV, Zhang FR, Tang S, Bognar AL (1994) Cloning and molecular characterization of three genes, including two genes encoding serine hydroxymethyl transferases, whose inactivation is required to render yeast auxotrophic for glycine. J Biol Chem 269:9155–9165

McNeil JB, Bognar AL, Pearlman RE (1996) In vivo analysis of folate coenzymes and their compartmentation in *Saccharomyces cerevisiae*. Genetics 142:371–381

McNeil JB, Zhang FR,Taylor BV, Sinclair D, Pearlman RE, Bognar AL (1997) Cloning, and molecular characterization of the GCV1 gene encoding the glycine cleavage T-protein from *Saccharomyces cerevisiae*. Gene 186:13–20

Meinhild AK, Klaerke DA, Zeuthen T (1998) Bidirectional water fluxes and specificity for small molecules in aquaporine 0–5. J Biol Chem 273:32446–32451

Mezentsev AV, Ustiniikova TB, Tikonova TV, Popov VO (1997) Isolation and kinectic mechnisms of NAD-dependent format dehydrogenase from the methylotrophic yeast *Hansenula polymorpha*. Prokladnaya Biokhim Mikrobiol 32:589–595

Miller JM, Conn EE (1980) Metabolism of hydrogen cyanide in higher plants. Plant Physiol 65:1199–1202

Miller JM, Gray DO (1982) The utilization of nitriles and amides by a *Rhodococcus* species. J Gen Microbiol 128:1803–1810

Miller LG, Kalin RM, McCauley SE, Hamilton JTG, Harper DB, Millet DB, Oremland RS, Goldstein AH (2001) Large carbon isotope fractionation associated with oxidation of methyl halides by methylotrophic bacteria. Proc Nat Acad Sci USA 98:5833–5837

Mira GJ, Garcia MP, Mira GAJ (1995) Identification of yeast by hydrolysis of amides. Mycoses 38:101–106

Mouillon JM, Aubert S, Bourguignon J, Gout E, Douce R, Rébeillé F (1999) Glycine and serine catabolism in non-photosynthetic higher plant cells: Their role in C_1 metabolism. Plant J 20:197–205

Müller RH, Babel W (1991) Formaldehyde as a carbon and energy source in yeast growth. Zentralblatt Mikrobiol 146:25–33

Nadolska-Lutyk J, Balinska M, Paszewski A (1989) Interrelated regulation of sulfur-containing amino acid biosynthesis enzymes and folate metabolizing enzymes in *Aspergillus nidulans*. Eur J Biochem 181:231–236

Nagarajan L, Storms RK (1997) Molecular characterization of GCV3, the *Saccharomyces cerevisiae* gene coding for the glycine cleavage system hydrogen carrier protein. J Biol Chem 272:4444–4450

Narayanasamyk K, Shukla, S, Parekh LJ (1990) Utilizaton of acrylonitrile by bacteria isolated from petrochemical waste waters. Indian J Exp Biol 28:968–971

Näsholm T, Persson J (2001) Plant acquisition of organic nitrogen in boreal forests. Physiol Plant 111:419–426

Näsholm T, Ekblad A, Nordi, A, Giesler R, Högberg M, Högberg P (1998) Boreal forest plants take up organic nitrogen. Nature 392:914–916

Näsholm T, Huss-Danell K, Hogberg P (2000) Uptake of organic nitrogen in the field by four agriculturally important plant species. Ecology 81:1155–1161

Näsholm T, Huss-Danell K, Hogberg P (2001) Uptake of glycine by field-grown wheat. New Phytol 150:59–63

Naula N, Walther C, Baumann D, Schweingruber ME (2002) Two non-complementing genes encoding enzymatically active methylenetetrahydrofolate reductases control methionine requirement in fission yeast *Schizosaccharomyces pombe*. Yeast 19:881–848

Nauter GCC, Figueira MM, Linardi VR (1998) Degradation of cyano-metal complexes and nitriles by an *Escherichia coli* strain. Rev Microbiol 29:12–16

Nemecek-Marshall M, McDonald RC, Franzen JJ, Wojciechwski Cl, Fall R (1995) Methanol emission from leaves. Enzymatic detection of gas-phase methanol and relation of methanol fluxes to stomatal conductance and leaf development. Plant Physiol 108:1359–1368

Neuburger M, Rébeillé F, Jourdain A, Nakamura S, Douce R (1996) Mitochondria are a major site for folate and thymilidate synthesis in plants. J Biol Chem 271:9466–9472

Neuhauser W, Steininger M, Haltrich D, Kulbe HD, Nidetzky B (1998) Rapid preparation in high yield of pure formate dehydrogenase from *Pichia angusta*. Biotechnol Techniques 12: 565–568

Nikiforov MA, Chandriani S, O'Connel B, Petrenko O, Kotenko J, Beavis A, Sedivy JM, Cole MD (2002) A functional screen for Myc-responsive genes reveals serine hydroxymethyltransferase, a major source of the one-carbone unit for cell metabolism. Mol Cell Biol 22:5793–5800

Nishida A, Erikson KE (1987) Formation, purification and partial characterization of methanol oxidase, a hydrogen peroxide producing enzyme in *Phanerochaete chrysosporium*. Biotechnol Appl Biochem 9:325–328

Overkamp KM, Koetter P, van der Hoek R, Schoondermark SS, Luttik MAH, van Dijken JP Pronk JT (2002) Functional analysis of structural genes for NAD^+-dependent formate dehydrogeanse in *Saccharomyces cerevisiae*. Yeast 19:509–520

Padmaja G, Balagopal C (1985) Cyanide degradation by *Rhizopus oryzae*. Can J Microbiol 31:663–669

Palenik B, Henson SE (1997) The use of amides and other nitrogen sources by the phytoplankton *Emiliana huxleyi*. Limnol Oceanogr 42:1544–1551

Parker MS, Armbrust EV (2000) Idendification of a glycine decarboxylase gene in *Thalliosira weissflogii*: a probe for photorespiration in diatoms. J Phycol 36 (Suppl):53–54

Parpinello G, Berardi E, Strabbioli R (1998) A regulatory mutant of *Hansenula polymorpha* exhibiting methanol utilization metabolism and peroxisome proliferation in glucose. J Bacteriol 180:2958–2967

Pasternack LB, Laude DA Jr, Appling DR (1992) Carbon-13 NMR detection of folate-mediated serine and glycine synthesis in vivo in *Saccharomyces cerevisiae*. Biochemistry 31: 8713–8719

Pasternack LB, Laude DA Jr, Appling DR (1994a) ^{13}C NMR analysis of intercompartmental flow of one-carbon units into choline and purines in *Saccharomyces cerevisiae*. Biochemistry 33:74–82

Pasternack LB, Laude DA Jr, Appling DR (1994b) Whole-cell detection by ^{13}C NMR of metabolic flux through the C_1-tetrahydrofolate synthase/serine hydroxymethyl transferase enzyme system and effect of antifolate exposure in *Saccharomyces cerevisiae*. Biochemistry 33:7166–7173

Pasternack LB, Littlepage LE, Laude DA Jr, Appling DR (1996) ^{13}C-NMR analysis of the use of alternative donors to the tetrahydrofolate-dependent one-carbon pools in *Saccharomyces cerevisiae*. Arch Biochem Biophys 326:158–165

Pennincksx M (2000) A short review on the role of glutathione in the response of yeasts to nutritional, environmental and oxidative stresses. Enzyme Microbial Technol 26:737–742

Pereira PT, Monteiro de Carvalho M, Arrabaca JD, Amaral CMT, Roseiro JC (1997) Alternative respiratory system and formamide hydrolyase activity as the key components of the cyanide-resistance mechanism in *Fusarium oxysporum*. Can J Microbiol 43:929–936

Pichinoty F (1988) Utilization of acetamide by *Bacillus gordonae* I. Isolation and characteristics of mutants able to use this compound as a source of carbon and energy. C R Seances Soc Biol Fil 182:177–180

Pomper BK, Vorholt JA, Chistoserdova L, Lidstrom ME, Thauer RK (1999) A methenyl tetrahydromethanopterin cyclohydrolase and a methenyl tetrahydrofolate cyclohydrolase in *Methylobacterium extorquens* AM1. Eur J Biochem 261:475–480

Raab TK, Lipson DA, Monson RK (1999) Soil amino acid utilization among species of the Cyperaceae: plant and soil processes. Ecology 80:2408–2419

Ramirez F, Monroy O, Favela E, Guyot JP, Cruz F (1998) Acetamide degradation by a continuous-fed batch culture of *Bacillus sphaericus*. Appl Biochem Biotechnol 70–72:215–223

Ratcliffe RG, Shachar-Hill Y (2001) Probing plant metabolism with NMR. Ann Rev Plant Physiol Plant Mol Biol 52:499–526

Ravanel S, Cherest H, Jabrin S, Grunwald D, Surdin KY, Douce R, Rébeillé F (2001) Tetrahydrofolate biosynthesis in plants: molecular and functional characterization of dihydrofolate synthestase and three isoformes of folylpolyglutamate synthetase in *Arabidopsis thaliana*. Proc Natl Acad Sci USA 98:15360–15365

Rébeillé F, Douce R (1999) Folate synthesis and compartmentation in higher plants. In: Kruger NJ (ed) Regulation of primary metabolic pathways in plants. Kluwer Academic Press, Amsterdam, pp 53–99

Rezende RP, Dias JCT, Rosa CA, Carazza F, Linardi VR. (1999) Utilization of nitriles by yeasts isolated from a Brazilian gold mine. J Gen Appl Microbiol 45:185–192

Rhodes D, Gage DA, Cooper AJL, Hanson AD (1997) S-Methylmethionine conversion to di-methylsulfonioproprionate: evidence for an unusual transamination reaction. Plant Physiol 115:1541–1548

Rhodes D, Verslues PE, Sharp RE (1999) In: Singh BJ (ed) Plant amino acids. Marcel Dekker, New York, pp 319–356

Rissler JF, Millar RL (1977) Contribution of a cyanide insensitive alternate respiratory system to increases in formamide hydrolyase activity and to growth in *Stemphylium loti* in vitro. Plant Physiol 60:857–861

Rosario FM, Biosca JA, Norin A, Jornvall H, Pares X (1995) Class III alcohol dehydrogenase from *Saccharomyces cerevisiae*: structural and enzymatic features differ toward the human/mammalian forms in a manner consistent with functional needs in formaldehyde detoxification. FEBS Lett 370:23–26

Rover JC, Christianos LM, Yoder WT, Gembetaa GA, Klotz AV, Morris CL, Brody H, Otani S (1999) Deletion of the trichodiene synthase gene of *Fusarium venenatum*: two systems for repeated gene deletions. Fungal Genet Biol 28:68–78

Sakai Y, Murdanoto AP, Konishi T, Iwamatsu A (1997) Regulation of the formate dehydrogenase gene, FDH1, in the methyltrophic yeast *Candida bidinii* and growth characetersitics of an FHD1-disrupted strain on methanol, methylamine and choline. J Bacteriol 179:4480–4485

Sakamoto M, Masuda T, Yanagimoto Y, Nakano Y, Kitaoka S (1991) Purification and characterization of cytosolic serine hydroxymethyltransferase from *Euglena gracilis* Z. Agr Biol Chem 55: 2243–2250

Sarojini G, Oliver DJ (1985) Inhibition of glycine oxidation by carboxymethoxylamine, methoxylamine and acethydrazide. Plant Physiol 77:786–789

Saxena D, Aouad S, Attieh J, Sain HS (1998) Biochemical charaterization of chloromethane emission from the wood-rotting fungus *Phellinus pomaceus*. Appl Environ Michrobiol 64: 2831–2835

Schmidt SGR (1999) Glycine metabolism by plant roots and its occurrence in Australian plant communities. Aust J Plant Physiol 26:253–264

Schmidt S, Stewart GR (1999) Waterlogging and fire impacts on nitrogen availability and utilization in a subtropical wet heathland (wallum). Plant Cell Environ 20:1231–1241

Serov AE, Popova AS, Fedorchuk VV, Tishov VI (2002) Engineering of coenzyme specificity of formate dehydrogenase from *Saccharomyces cerevisiae*. Biochem J 367:841-847

Shen S, Sulter G, Jeffries TW, Cregg JM (1998) A strong nitrogen source regulated promoter for controlled expression of foreign genes in the yeast *Pichia pastoris*. Gene 216:93-102

Shah MM, Grover TA, Aust SD (1991) Metabolism of cyanides by *Phanerochaete chrysosporium*. Arch Biochem Biophys 290:173–178

Shanker R, Ramakrishna C, Seth PK (1990) Microbial degradation of acrylamide monomer. Arch Microbiol 154:192–198

Sigler LJW, Carmichael JW (1974) Acidophilic *Scytalidium acidophilum* new species. Can J Microbiol 20:267–268

Silva-Avalos J, Richmond MG, Nagappan O, Kunz DD (1990) Degradation of the metal cyano complex tetracyanonickelat II by cyanide utilizing bacterial isolates. Appl Environ Microbiol 56:3664–3670

Song JM, Rabinowitz JC (1993) Function of yeast cytoplasmic C-1-tetrahydrofolate synthase. Proc Natl Acad Sci USA 90:2636–2640

Streeter T, Bol R, Bardgett RD (2000) Amino acids as a nitrogen source in temperate upland grasslands: the use of dual labelled (^{13}C, ^{15}C) glycine to test for direct uptake by dominant grasses. Rapid Comm Mass Spectrom 14:1351–1355

Studer A, Vuilleumier S, Leisinger T (1999) Properties of the methylcobalamin:H4folate methyltransferase involved in chloromethane utilization by *Methylbacterium* sp. strain. CM4. Eur J Biochem 264:242–249

Studer A, Stupperich E, Vuilleumier S, Leisinger T (2001) Chloromethane: tetrahydrofolate methyl transfer by two proteins from *Methylobacterium chloromethanicum* strain CM4. Eur J Biochem 268:2931–2938

Studer A, McAnulla C, Buchele R, Leisinger T, Vuilleumier S(2002) Chloromethane-induced genes define a third C_1 utilization pathway in *Methylbacterium chloromethnicum* CM4. J Bacteriol 184:3476–3484

Suarez de Mata Z, Rabinowitz JC (1980) Formylmethionyl methylene tetrahhydrofolate synthetase combined from yeast *Saccharomyces cerevisiae*. Biochemical characterization of the protein from an ADE-3 mutant lacking the formyl tetrahydrofolate synthetase function. J Biol Chem 253:2569–2577

Summers PS, Nolte KD, Cooper AJL, Borgeas H, Leustek T, Rhodes D, Hanson AD (1998) Identification and stereospecifity of the first three enzymes of 3-dimethylsulfonioproprionate biosynthesis in a chlorophytic alga. Plant Physiol 116:369–378

Suzuki K, Itai R, Suzuki K, Nakanishi H, Nishikawa NK, Yoshimura E, Mori S (1998) Formate dehydrogenase, an enzyme of anaerobic metabolism, is induced by iron deficiency in barley roots. Plant Physiol 116:725–732

Thornton B (2001) Uptake of glycine by non-mycorrhizal *Lolium perenne*. J Exp Bot 52: 1315–1322

Todd RB, Murphy RL, Martin HM, Sharp JA, Davis MA, Katz ME, Hynes MJ (1997) The acetate regulatory gene facB of *Aspergillus nidulants* encodes a Zn(II)2Cys6 transcriptional activator. Mol Gen Genet 254:495–504

Tonon F, Priorde CC, Odier E (1990) Nitrogen and carbon regulation of lignin peroxidase and enzymes of nitrogen metabolism in *Phanerochaete chrysosporium*. Exp Mycol 14:243–254

Toyama H, Anthony C, Lidstrom ME (1998) Construction of insertion and deletion mxa-mutants of *Methylobacterium extorquens* AM1 by electroporation. FEMS Microbiol 166:1–7

Trapp S, Larsen M, Christiansen H (2001) Experimente zum Verbleib von Cyanid nach Aufnahme in Pflanzen. Umweltwiss Schadstoff Forsch13:29–37

Triglia TL, Cowman AF (1994) Primary structure and expression of dihydropteroate synthase gene of *Plasmodium falciparum*. Proc Natl Acad USA 91:7149–7153

Trossat C, Nolte KD, Hanson AD (1996) Evidence that the pathway of dimethylsulfonioprorionate biosynthesis begins in the cytosol and ends in the chloroplast. Plant Physiol 111:985–973

Vallini V, Berardi E, Strabbioli R (2000) Mutations affecting the expr4ession of the MOX gene encoding peroxisomal methanol oxidase in *Hansenula* polymorpha. Curr Genet 38:163–170

Van den Berg M, Steensma HY (1997) Expression casettes for formaldehyde and fluoroacetate resistance, two dominant markers in *Saccharomyces cerevisiae*. Yeast 13:551–559

Van der Walt JP, Brewis EA, Prior B (1993) A note on the utilization of aliphatic nitriles by yeasts. Syst Appl Microbiol 16:330–332

Van Ophem PW, Duine JA (1994) NAD- and co-substrate (GSH or factor)-dependent formaldehyde dehydrogenase from methylotrophic microorganisms act as a class III alcohol dehydrogenase. FEMS Lett 116:87–93

Vannelli T, Studer A, Kertesz M, Leisinger T (1998) Chloromethane metabolism by *Methylobacterium* sp. strain CM4. Appl Environ Microbiol 64:1933–1936

Wahls WP, Song JM, Smith GR (1993) Single-standed DNA binding activity of C-1-tetrahydrofolate synthase enzymes. J Biol Chem 268:23792–23798

Walton NJ, Woolhouse HW (1986) Enzymes of serine and glycine metabolism in leaves and non-photosynthetic tissues of *Pisum sativum* L. Planta 167:119–128

Wang P, Vanetten HD (1992) Cloning and properties of a cyanide hydratase gene form the phytopathogenic fungus *Gloeocercospora sorghi.* Biochem Biophys Res Comm 187:1048–1054

Wang P, Matthews DE, Vanetten HD (1992) Purification and characterization of cyanide hydratase form the phytopathogenic fungus *Gloeocercospora sorghi*. Arch Biochem Biophys 298:569–575

Watling R, Harper DB (1998) Chloromethane production by wood-rotting fungi and an estimate of the global flux to the atmosphere. Mycol Res 102:769–787

Wehner EP, Rao E, Brendel M (1993) Molecular structure and genetic regulation of SFA, a gene responsible for resistance to formaldehyde in *Saccharomyces cerevisiae* and characterization of its protein product. Mol Gen Genet 237:351–358

West MG, Barlowe CK, Appling DR (1993) Cloning and characterization of the *Saccharomyces cerevisiae* gene encoding NAD-dependent 5,10-methylenetetrahydrofolate dehydrogenase. J Biol Chem 268:153–160

West MG, Horne DW, Appling DR (1996) Metabolic role of cytoplasmic isoenzymes of 5.10-methylenetetrahydrofolate dehydrogenase in *Saccharomyces cerevisiae*. Biochemistry 35:3122–3132

Whatley FR, Greenaway W, Dunstan RH (1986) Glycine metabolism in carrot suspension culture cells investigated by gas chromatography/mass spectrometry. Physiol Vegetale 24:3–13

Whitlow M, Howard AJ, Stewart D, Hardmann KD, Kuyper LF (1997) X-ray cristallographic studies of *Candida albicans* dihydrofolate reductase: high resolution structures of the holoenzymes and an inhibited ternary complex. J Biol Chem 272:30289–30298

Whitlow M, Howard AJ, Stewart D, Hardmann KD, Chan JH, Baccanari DP, Tansik RL, Hong JS, Kuyper LF (2001) X-ray crystal structure of *Candida albicans* dihydrofolate reductase: high resolution ternary complexes in which the dihydronicotinamide moiety of NADPS is displaced by an inhibitor. J Med Chem 44:2928–2932

Wingler A, Lea PJ, Leegood RC (1999) Photorespiratory metabolism of glycolate and formate in glycine-accumulation mutants of barley and *Amaranthus edulis.* Planta 207:518–526

Wu S, Fallon RD, Payne MS (1999) Engineering *Pichia pastoris* for stereo selective nitrile hydrolysis by co-producing heterologous proteins. Appl Microbiol Biotechnol 52:186–190

Yanase H, Sakamoto A, Okomoto K, Kit K, Sato Y (2000) Degradation of the metal-cyano complex tetracyanonickelate (II) by *Fusarium oxysporum* N-10. Appl Microbiol Biotechnol 53:328–334

Prof. Dr. Hartmut Gimmler
Prof. Dr. Wolfram Hartung
Julius-von-Sachs-Institut für Biowissenschaften
Lehrstuhl Botanik I der Universität Würzburg
Julius-von-Sachs-Platz 2
97082 Würzburg, Germany
e-mail: gimmler@botanik-uni-wuerzburg.de
e-mail: hartung@botanik-uni-wuerzburg.de

Communicated by U. Lüttge

Nuclear Magnetic Resonance Applications to Low-Molecular Metabolites in Plant Sciences

Bernd Schneider

1 Introduction and History

The basic feature of nuclear magnetic resonance (NMR) spectroscopy is the observation of magnetic properties of atomic nuclei and their changes under the influence of chemical bonds or adjacent atoms. Although restricted to atomic nuclei that possess a nuclear magnetic moment, NMR is universally applicable to analyze the occurrence of such nuclei in the steady state and in dynamic interactions with their chemical environment. Due to this general feasibility, after the discovery by Felix Bloch and Edward Purcel in 1946, who were awarded the 1952 Nobel Prize in physics, NMR was originally established in nuclear physics to accurately determine nuclear magnetic moments. After it had been demonstrated that the NMR frequency depends on the chemical environment (Knight 1949), this technique became an interesting tool in chemistry, e.g. for confirming structures of synthetic organic compounds. It rapidly expanded into different directions and additionally has been applied in various disciplines such as material science, medicine, and biology. Improvements in spectrometer technology (superconducting magnets, wide-bore magnets, ultra-high-field magnets), probe head design, Fourier-transform NMR, computer technology and progress in pulse sequences, especially 2D correlation, multiple resonance spectroscopy, and pulsed field gradients, further extended the possibilities to apply NMR techniques (Fig. 1). A second Nobel Prize was awarded to one of the pioneers of NMR spectroscopy, Richard Ernst, in 1991 for his contribution to NMR methodology. Biological applications allowed elucidation not only of metabolic processes but also of structures of biological macromolecules and their dynamic interaction with other molecules. The 2002 Nobel Prize in chemistry was awarded in part to Kurt Wüthrich for his contribution to determining the three-dimensional structures of biological macromolecules in solution by NMR. The present review covers various NMR applications, including high-resolution (HR-NMR) and magnetic resonance imaging (MRI), in natural product chemistry, biochemistry, and physiology with special emphasis on plants.

Progress in Botany, Vol. 65

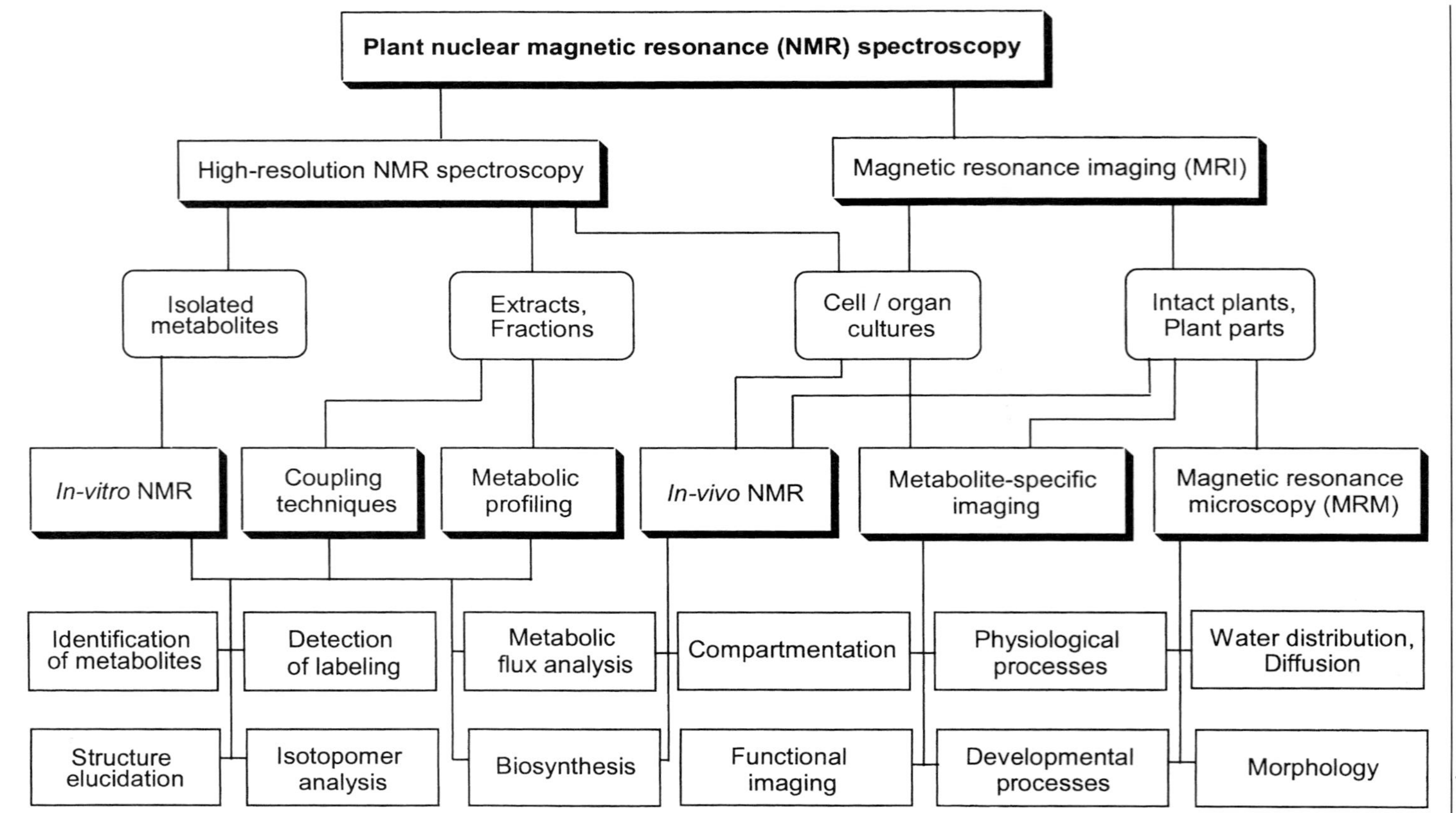

Fig. 1. Selected techniques and applications of NMR and MRI in plant sciences

2 Identification and Structure Elucidation

Structure elucidation of natural products has been an important task in drug discovery, the search for agricultural chemicals, and identification of compounds involved in ecological interspecies interactions. In addition, knowledge on the distribution of natural products in the plant kingdom is of considerable interest for chemotaxonomy. The majority of the more than 150,000 secondary metabolites (Steglich et al. 1997), among them a large number from plant sources, would not have been structurally elucidated without NMR spectroscopy. Natural product chemistry has been a domain of ^{1}H-, ^{13}C-, and ^{15}N-spectroscopy, proton homonuclear shift-correlation NMR spectroscopy (^{1}H-^{1}H-COSY, TOCSY) including detection of spatial proximities by NOESY and ROESY techniques. In addition, proton-detected heteronuclear shift-correlation experiments such as HMQC, HSQC, HMQC and a multitude of further variants have been developed. For details in structure elucidation, the reader is referred to detailed reviews on natural product NMR in general (Derome 1989; Sadler 1988; Duddeck et al. 1998; Croasmun and Carlson 1994) including specific classes of natural products such as alkaloids (Martin and Crouch 1994) and terpenoids (Bohlmann and Zeisberg 1975; Ferreira et al. 1998). Coupling of chromatographic separation techniques with NMR spectroscopy was first described by Watanabe and Niki in 1978 and has been established in recent years as an important supplementary methodology in natural product chemistry (Albert et al. 1999; Schneider 2000; Wolfender et al. 2001).

Online LC-NMR coupling represents one of the most powerful and time-saving methods for the identification and structure elucidation of natural products in mixtures. The continuous-flow mode is the most efficient technique but is restricted to major components of a mixture. It results in a pseudo-2D spectrum with a spectroscopic dimension (^{1}H chemical shift) and a time dimension (retention time). Stop-flow, loop-sampling, and solid phase extraction (SPE)-sampling (Nyberg et al. 2001) are more sensitive than the continuous-flow mode and, therefore, useful also to record ^{1}H-NMR spectra of less abundant compounds. Capillary LC-NMR coupled to a custom-built microcoil NMR probe (1.1 μl observed volume) has been used recently to identify 37 ng of α-pinene in a terpenoid mixture at 500 MHz (Lacey et al. 2001). In addition to ^{1}H-NMR spectra, various 2D spectra can be measured by stop-flow coupling techniques because the fraction can remain within the flow cell as long as necessary. UV or DAD detection, which usually are used for monitoring the separation of the mixture under study, provides absorption data, and coupling with mass spectrometry and additional analytical methods is possible as well. LC-NMR is especially suitable to solve the increasing dereplication problem in natural product isolation. Thus, several studies employed LC-NMR to screen extracts in order to identify new compounds, which subsequently have been isolated and subjected to off-line analytical techniques including conventional NMR.

The strategy of using HPLC-NMR coupling in addition to conventional NMR spectroscopy is exemplified by the structure elucidation of a phenylpropanoid-derived natural product, anigorootin (Fig. 2). Rhizomes of *Anigozanthos flavidus* (Haemodoraceae) were screened for new natural products by means of HPLC–NMR coupling (Hölscher and Schneider 1999). In addition to several known compounds, an unknown natural product was detected, the HPLC–^{1}H-NMR spectrum (Fig. 2A) of which showed signals indicating a compound of the phenylphenalenone type. Acetylation of the isolated substance yielded a product showing an acetyl signal at δ 2.2 in the ^{1}H-NMR spectrum due to acetylation of the equivalent aliphatic hydroxyls in the parent compound. In addition, short-term acetylation gave a second product, which showed a double set of ^{1}H signals (Fig. 2B) with only one acetyl resonance. From these data a homodimeric structure of the new compound was deduced. Heteronuclear correlation spectra (HMQC and HMBC) were used to elucidate the core structure and to assign all proton and carbon signals of anigorootin. The coupling constant between the methine protons H-7b and H-14b, which is essential to determine the stereochemistry of anigorootin, was obtained from satellite doublets in the parent compound and also from doublets of the nonequivalent methine protons of the monoacetyl derivative (Fig. 2B). The relative configuration of the four stereocenters were further established by NOESY experiments of the two acetyl derivatives and later confirmed by x-ray crystallography (Otálvaro et al. 2002).

The inherent low sensitivity of NMR spectroscopy is an obstacle, which has limited its application in natural product chemistry and plant sciences to a certain extent. There is an increasing demand to detect molecules and ions in biological samples and to elucidate structures of low-molecular compounds and biological macromolecules more sensitively and to enable measurement of reduced sample amounts, respectively. Consequently, there are several strategies to improve sensitivity of NMR and to overcome limitations. Inverse-detection methods in heteronuclear correlation techniques, the use of pulsed field gradients, and sophisticated pulse sequences have significantly contributed to improve performance of NMR spectroscopy. Hardware developments such as enhancement of magnetic field strength by construction superconducting magnets, ultra-high-field magnets making use of Joule-Thompson cooling units, miniaturization (Crouch and Martin 1992; Olson et al. 1995; Schlotterbeck et al. 2002; Wolters et al. 2002), and improvements in probe head design are further

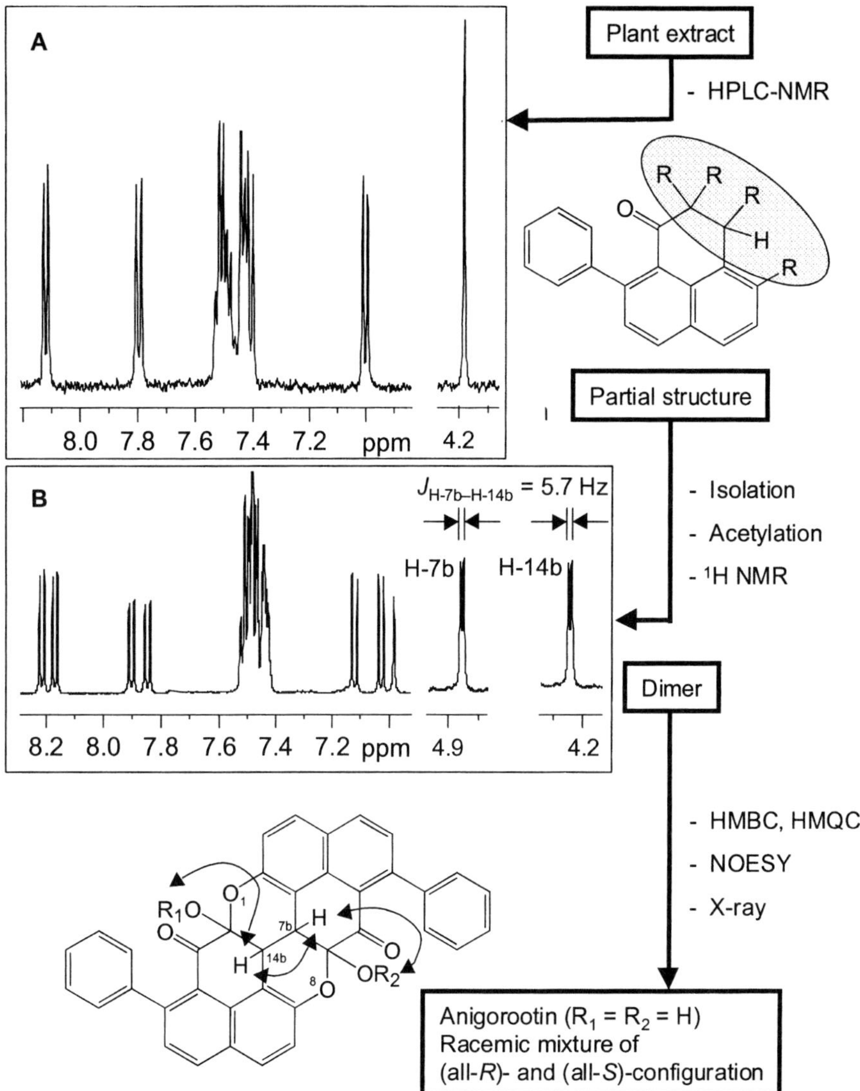

Fig. 2. Strategy of structure elucidation of anigorootin from *Anigozanthos preissii* (Haemodoraceae) using **A** HPLC–^{1}H-NMR coupling and **B** conventional ^{1}H NMR and 2D techniques. The *arrows* in the structure of anigorootin (R_1=R_2=H) and its acetyl derivatives (monoacetyl: R_1=Ac, R_2=H; diacetyl: R_1=R_2=Ac) indicate NOESY correlations between H-7b/H-14b and the acetyl protons, respectively. For further details, see text

important factors in this context. One of the most exciting developments in recent years has been the introduction of cryogenic probes (Styles et al. 1984, 1989). Since thermal noise is proportional to temperature and resistance of materials used, reduction of one or both of these parameters is useful to improve signal-to-noise ratio. Thus, probe heads have been developed in which the radio frequency coils and the preamplifiers are cooled to very low temperature around or below 20 K while the sample remains at room- or even elevated temperature. This leads to reduction of noise by factor 3 to 4, corresponding up to 16-fold shorter data acquisition time, to obtain comparable results for the same sample in a cryogenic probe compared to a conventional probe. This high sensitivity of cryogenic NMR technology is of special interest for the identification of small samples of natural products (Logan et al. 1999; Russell et al. 2000) but is extending to other fields of plant sciences as well. The recent availability of cryogenic flow probes provides the opportunity to combine the sensitivity of the cryoprobe technology with the advantages of LC-NMR hyphenation.

3 Metabolic Profiling

NMR-based metabolic profiling provides information which is complementary to genomics and proteomics, and in combination with those, can be used to develop models describing cellular functions. Standardized extraction procedures and NMR analyses of extracts generate datasets, which are evaluated by computational methods.

It has been demonstrated, for example, that a 1D-NMR approach is capable of accurately discriminating the mode of action of different herbicides. Using only a small amount of tissue, this method was able to detect minute differences in a plant's metabolic profile caused by treatment with various herbicides. Signals of the herbicide were not visible in the ^{1}H-NMR spectra since it was applied in low concentration and only the spectral manifestation of the variations in the pool of the plant metabolites was observed (Aranìbar et al. 2001).

However, in many cases, the occurrence of one or more individual metabolites in an extract may be of interest. To make sure that the metabolite of question is present, a fingerprint pattern of characteristic resonance lines in the expected proportions in the 1D spectrum may be sufficient. A set of characteristic lines often is visible even if the spectrum partially is crowded with overlapping signals. Homo- and heteronuclear 2D-NMR techniques, preferable in the ^{1}H-detected mode, are powerful tools to enhance resolution, to detect co-resonating components, and thereby identify desired components in a mixture. This approach has been successfully applied, for example, to an algal extract (Fan 1996). Using 2D-TOCSY spectra, a variety of amino acids and other metabolites have been identified unambiguously in *Chlorella* without separation of the sample. In addition to plant NMR, mass spectrometric techniques provide supplementary information in metabolic profiling. A number of components have been identified in crude root exudates of different gramineous plants by 2D NMR, GC-MS and HR-MS (Fan et al. 2001). Using this approach, the role of exudate metal ion ligands in the acquisition of Cd and transition metals has been examined.

Another recent example of metabolomic analysis has shown the consequences of Cd exposure to a heavy metal-resistant plant, *Silene cucubalus* (Bailey et al. 2003). These examples demonstrate that metabolic profiling is capable of detecting the metabolic consequences of altered environmental conditions. Further applications are expected in detection of changes of metabolite pattern in transgenic plants and mutants in comparison with the wild type and, in this context, metabolic profiling will frequently be used in future.

4 NMR in Biosynthesis and Metabolic Flux Analysis

The question of biosynthetic pathways of natural products has been an important area in plant research since radiolabeled precursors became available in the 1950s. The possibility to determine the labeling position in the target molecules by means of NMR was one of the most important reasons which resulted largely in substitution of radiolabel by stable isotopes such as ^{2}H, ^{13}C, and ^{15}N in tracer experiments. The use of labeled precursors in biosynthetic studies and aspects of labeling techniques have been reviewed several times (Vederas 1985; Simpson 1998; Schneider et al. 2003). Information on the number of isotopically enhanced atoms and their specific position in a molecule is mainly gained from enlarged resonance signals, coupling patterns and isotopic-induced changes in the chemical shifts. Data obtained from tracer experiments are useful to gain insight into precursor–product relationships. Modern developments in NMR spectroscopy such as LC-NMR hyphenation (Schmitt and Schneider 2001) and cryoprobe technology (Bringmann and Feineis 2001) have already been used in sensitive detection of stable isotope labeling in biosynthetic studies.

A recent ^{13}C-NMR study is shown as an example (Fig. 3). Incorporation of [1-^{13}C]phenylalanine into both phenylphenalenones, e.g. anigorufone (**1**) and phenylbenzoisochromenones (phenylnaphthalides) (e.g. compound **2**) in root cultures of *Wachendorfia thyrsiflora*, a South African member of the Haemodoraceae (Opitz and Schneider 2003) confirmed that

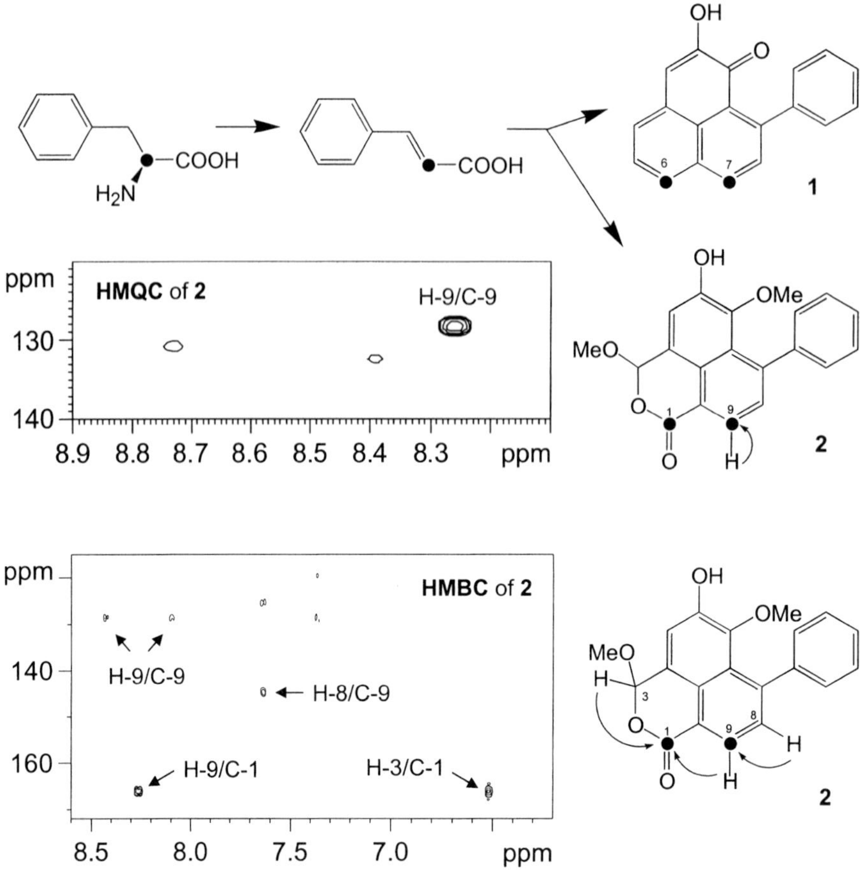

Fig. 3. Related biosynthesis of oxabenzochrysenones and phenylphenalenones in *Wachendorfia thyrsiflora* (Haemodoraceae). The HMQC spectrum of the oxabenzochrysenone **2** shows enhanced intensity of the cross-signal of H9/C-9 indicating ^{13}C enrichment of C-9. The HMBC spectrum of **2** exhibits enhanced cross-signals of H-9/C-1 and H-3/C-1 (which are due to three bond C–H coupling), indicating ^{13}C enrichment of C-1. The one bond C–H correlation of the enriched signal of C-9 and the correlation through two bonds between H-8 and C-9 are also visible in the HMBC spectrum and provide additional evidence of ^{13}C enrichment of C-9. Some weak natural abundance cross-signals are visible as well. The C-H-heterocorrelation spectra of anigorufone (**1**) (not shown) exhibited enrichment of analogous positions C-6 and C-7

both types of compounds are formed via a common pathway diverging only in late steps. Conclusively, oxidative formation of the oxalactone structure of the phenylbenzoisochromenones from the aromatic ring system is very likely. A dioxygenase-type mechanism, followed by decarboxylation, has been proposed for the key steps of this conversion.

In experiments using singly or doubly isotope-labeled precursors only a very limited number of isotopomers is observed. The complexity of isotopomer patterns increases with the number of enriched positions and the complexity of the pathway, e.g. whether or not bonds between stable isotope-labeled positions are formed or broken in the course of a biosynthetic process. More complex isotopomer mixtures are found after application of general metabolites, which are diverted to all branches of intermediary metabolism. An isotopomer pattern recognition approach has been developed to deduce the origin of secondary products from intermediary metabolism (Bacher et al. 1999). This retrobiosynthetic method is based on comparison between the labeling patterns of secondary metabolites and primary metabolites such as amino acids and ribonucleotides. It can be used to reconstruct the isotopomer pattern of central metabolic intermediates. The biosynthetic retroanalysis has been widely explored in a basic biosynthetic issue, namely to discriminate the origin of isoprenoids from the classical mevalonate (MVA) pathway and the more recently discovered 2C-methyl-D-erythritol 4-phosphate (MEP) alias 1-deoxy-D-xylulose-5-phosphate (DOXP) pathway (Rohmer et al. 1993).

In a recent example, the biosynthesis of the bitter compound amarogentin, produced by root cultures of *Swertia chirata* has been investigated (Wang et al. 2001). The results clearly showed that this compound originates from a mixed biosynthesis. The sweroside moiety was formed predominately via the DOXP route and the biphenylcarboxylic acid moiety surprisingly was not biosynthesized via benzoic acid or phenylpropanoid precursors but from an early shikimate precursor via a polyketide-type biosynthesis. As the DOXP pathway is operative in the plastids, conclusions about the compartmentation of the biosynthesis can be drawn indirectly from retrobiosynthetic NMR studies. However, the compartmental separation is not complete and crosstalk between the mevalonate pathway and the DOXP pathway occurs (Lichtenthaler 1999) and may play a role in stress situations, and in cell cultures where the architecture of the plant tissue is different from that of intact plants.

In this context, quantitative flux analysis is indispensable to determine the contribution of different pathways to a distinct metabolic pool. Since the intensity of ^{13}C-NMR signals largely depends on the relaxation properties, spectra of the isotope labeled compound under study and of natural abundance reference compounds have to be recorded under identical experimental conditions. NOE-based differences in signal intensities are usually eliminated by inverse-gated decoupling and, in addition, a relaxation reagent such as $Cr(acac)_3$ can be used to determine the absolute enrichment of ^{13}C in specific positions of the molecule.

The incorporation of [^{13}C]1-deoxy-D-xylulose into a variety of terpenes of the liverwort *Conocephalum conicum* has been studied by quantitative ^{13}C-NMR spectroscopy. The labeling patterns of the sesquiterpene cubebanol, for example, indicated a possible degradation of 1-deoxy-D-xylulose to acetate and subsequent incorporation via the mevalonic acid pathway (Thiel and Adam 2002). Moreover, in bornyl acetate, the labeling occurred only in the acetate moiety whereas the isoprene units remained unlabelled. Different levels of enrichment of the labeled positions of cubebanol and the acetate residue of bornyl acetate were found, which imply that different acetyl CoA pools are utilized for the acetylation of borneol and the IPP biosynthesis of cubebanol via the mevalonic acid pathway. From these data, it was assumed that the two reactions occur most probably in different cellular compartments or cell types of the plant.

Many investigations have shown that biosynthetic pathways frequently are not simple linear precursor-product relationships but in reality represent rather complex networks, which enable formation of products via different subroutes. Moreover, depending on various endogenous and exogenous factors, plants are able to use alternative pathways in a flexible way. Labeling patterns can be used to reveal quantitative aspects of carbon flux in the metabolic network of the organism under investigation. Consequently, in the postgenomic era, the quantitative analysis of complex metabolic flux patterns is becoming more and more important (Szyperski 1998; Roberts 2000; Roscher et al. 2000; Weichert 2001). This approach helps to understand how the genome determines plant growth, developmental and defense processes through the regulation of the metabolic network. The availability of molecular tools and the detailed knowledge of the genetic basis of metabolic processes offer possibilities for the manipulation and engineering of metabolic pathways such as the transfer of plant-specific biosynthetic genes to other plants or microbial producers and even the design of artificial natural products (Dixon 2001; Sato et al. 2001; Lee and Schmidt-Dannert 2002). NMR spectroscopy is an important tool in monitoring consequences and many other aspects of metabolic engineering of plants (Shachar-Hill 2002).

In biotechnology, the cellular metabolism of a particular organism can be rationally redesigned to produce desired products in expense of others, which are unwanted. The consequences of altering the metabolic flux through the benzophenanthridine alkaloid pathway, for example, were evaluated by two-dimensional ^{1}H-COSY-NMR spectroscopy (Park et al. 2003). In this study, changes in metabolic flux through benzylisoquinoline alkaloid biosynthesis were shown to affect the regulation of amino acid pools.

Isotopomer analysis, which represents an essential prerequisite for metabolic flux analysis, is often a challenge. A ^{1}H-NMR method has been developed which allows the determination of the complete isotopomer distribution in metabolites having a backbone consisting of up to at least four carbons (deGraaf et al. 2000). In a first application, the complete isotopomer distribution of aspartate isolated from an [1-^{13}C]ethanol-grown microorganism, *Ashbya gossypii*, was determined. The approach certainly will be useful in plant studies as well.

In ecology research it is particularly interesting how plants, attacked by predators and herbivores, defend themselves by the production of deterrent chemicals or enhancement of mechanical resistance. These costly defense strategies often require dramatic reorientation of metabolic fluxes. To understand the ecological function of low-molecular defense compounds and the biosynthetic networks by which they are formed, requires comparative quantitative flux analysis of the processes found in wild-type plants with those found in mutants, or which have been designed by recombinant DNA technology.

The web of metabolic processes involved in the balance between plant growth and reproduction, and defense strategies, on the other hand, is characterized by competition of divergent pathways for the same key intermediates or alternatively, biosynthetic pathways can contribute to one product. Quantitative analysis of isotope enrichment of metabolites by NMR spectroscopy enables determination of fluxes through a distinct pathway, i.e. the extent by which a certain metabolite is synthesized through alternative routes can be deduced quantitatively. This approach implies both spectroscopic skills and knowledge of mathematical data evaluation methods. The application of different strategies to determine metabolic fluxes quantitatively has been discussed in detail using an example from intermediary plant metabolism, the pathways interconverting hexose phosphates and triose phosphates (Roscher et al. 2000).

5 In Vivo NMR Spectroscopy

As a non-invasive technique, in-vivo plant NMR spectroscopy provides information on molecules or ions in plant tissue under physiological, or near physiological, conditions. In-vivo NMR spectroscopy has been applied to plant systems for many different purposes (Ratcliffe et al. 2001). Numerous publications aimed at the detection of ^{1}H, ^{13}C, ^{15}N, and ^{31}P in living plant tissue and several other nuclei have been employed in addition (Martin 1985; Ratcliffe 1994) in in-vivo-NMR experiments. Steady-state measurements of NMR-detectable inorganic ions (e.g. ammonium, nitrate, sodium, phosphate) and low-molecular plant metabolites (e.g. carbohydrates, amino acids, organic acids, triglycerides) can be performed to determine intracellular pool sizes, compartmentation, or pH (Lundberg et al. 1990; Ratcliffe 1996).

For example, the subcellular compartmentation of homoserine in *Acer pseudoplatanus* cells (Aubert et al. 1998) and proline in salt-stressed leaf pieces of *Pringlea antiscorbutica* (Aubert et al. 1999) has been studied by in vivo ^{13}C and ^{31}P NMR. The technique developed by these authors involved NMR analysis at pH 6.0 followed by alkalinization of the perfusion medium using NH_4^+. Under these conditions certain carbon resonances of cytosolic amino acids were shifted selectively to lower field and the signals were resolved in the spectra to two distinct peaks corresponding to the vacuolar and cytoplasmic pools of homoserine and proline, respectively. In *Pringlea antiscorbutica*, proline mainly accumulated in the cytoplasm with concentrations 2–3 times higher in comparison to that in the vacuole, underlining its role as cytoplasmic osmoregulator in this plant.

The possibility to gain information about dynamic processes is one of the most important advantages of in-vivo NMR spectroscopy. This possibility is due to the non-destructive character of this technique, i.e. it is possible to analyze the same sample many times under controlled constant or changing external conditions. This aspect is of special value to elaborate dynamic physiological processes in plants. Consequently, kinetic in-vivo NMR studies have been undertaken in energy metabolism, glycolysis, amino acid metabolism, Krebs cycle, and other areas of intermediary metabolism (Ratcliffe 1996).

Both in-vivo and in-vitro NMR techniques have been employed to study various functional aspects of ectomycorrhiza and arbuscular mycorrhiza, such as phosphorus nutrition, nitrogen and carbon metabolism, and xenobiotics and secondary metabolism (Pfeffer et al. 2001). Another area of application of in-vivo NMR is the adaptation of plant metabolism to specific environmental conditions. The pathways by which glycine and serine might be utilized as nitrogen sources for the aquatic resurrection plant *Chamaegigas intrepidus* were investigated by in-vivo NMR of roots incubated with [^{15}N]glycine, [^{15}N]serine and [2-^{13}C]glycine. The results were consistent with the involvement of the glycine decarboxylase complex and serine hydroxymethyltransferase in the utilization of glycine (Hartung and Ratcliffe 2002).

In-vivo NMR studies on the biosynthesis of secondary metabolites have been conducted rather rarely because in general their concentrations are lower in comparison with intermediary metabolites. Thus, suitable plant systems are required which do contain detectable amounts of the metabolites of interest or are able to form them from exogenously supplied precursors. In addition to abundance of the metabolite, obstacles such as susceptibility differences in the tissue due to air spaces and paramagnetic ions in the medium have to be taken into account. Examples of in-vivo NMR studies on secondary metabolites include [^{15}N]tropinone metabolism in transformed root cultures of *Datura stramonium* (Ford et al. 1996) and the conversion of monoterpenoid indole alkaloids in cell suspensions of *Rauvolfia serpentina* (Hinse et al. 2001). Remarkably, the latter investigation was carried out with unlabeled precursors. Vellosimine (final concentration approximately 1 mM) was fed to suspended cells and, in the first step, rapid reduction of the aldehyde group to give vellosimol was observed with a high-field spectrometer (800 MHz). In another step, hydroxylation in the aromatic ring of the indole moiety of vellosimol resulted in sarpagine. Assignment of the indole alkaloid metabolites was performed on the basis of cross signals in the aldehyde and aromatic region of gradient-assisted HSQC spectra, indicating the advantages of using 2D techniques in in-vivo NMR studies. In addition to assignment information and improved spectral resolution by the second dimension, heteronuclear 2D spectra acquired in the inverse-mode provide enhanced sensitivity of ^{1}H for the detection of heteronuclei such as ^{13}C or ^{15}N in comparison with direct detection techniques.

Figure 4 shows a gradient-assisted HMQC spectrum of root cultures of *Anigozanthos preissii* supplied with [2-^{13}C]cinnamic acid. Although the nutrient medium was substituted with sugar-free medium for the measurement, strong natural abundance signals of the carbohydrates are visible in the upper right part of the spectrum. The cross resonances in the lower left

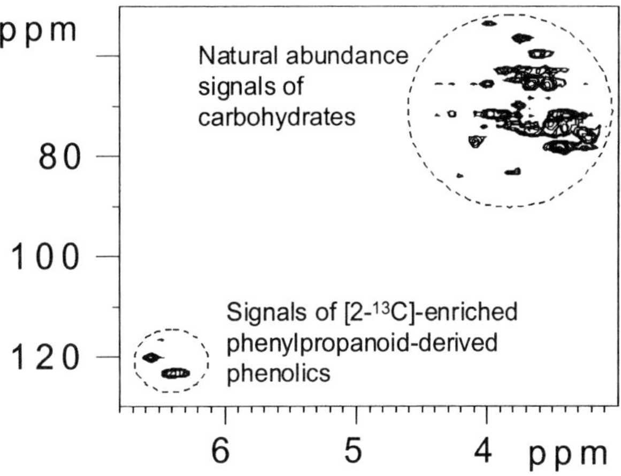

Fig. 4. In vivo NMR spectrum of root cultures of *Anigozanthos preissii* (Haemodoraceae) after feeding of [2-^{13}C]cinnamic acid

part are due to ^{13}C-enriched phenylphenalenones formed through the phenylpropanoid pathway from [2-^{13}C]cinnamic acid. Quantification of the metabolites formed from labeled precursors directly in the NMR spectrum is problematic because dilution with endogenous isotopically non-enriched metabolites is different at each biosynthetic step. Moreover, relaxation times in vivo may be different from those measured from isolated samples.

Frequently standard bore NMR magnets with 10 mm probe heads, which are designed for recording high-resolution NMR spectra, have been used for in-vivo plant studies. Since in high-resolution NMR only liquid samples give spectra of reasonable quality, it is necessary to submerge tissue under study in aqueous medium. Therefore, experimental conditions have to be chosen which take into account that oxygen concentration is much smaller in aqueous medium than in air. This is one of the reasons why most in-vivo NMR studies have been performed on cell suspensions, root cultures, and excised tissue. From these considerations it becomes obvious that the consequences of cellular oxygen deficiency, i.e. reduced respiration and induction of fermentation, are relevant to NMR studies of plants and must be involved in data interpretation. Aspects of low-oxygen stress thus have been extensively investigated (Roberts and Xia 1996; Ratcliffe 1997). Owing to improved tolerance of anoxia, hypoxically pretreated root tips (Saglio et al. 1980) have been frequently used. In order to ensure balanced oxygen supply, many experimental arrangements of in-vivo NMR studies make use of air-lift systems and/or circulation of oxygenated medium through the NMR tube (Ratcliffe 1994).

6 Magnetic Resonance Imaging

In addition to high-resolution NMR, magnetic resonance imaging (MRI) techniques are being used in plant studies (Chudek and Hunter 1997; Köckenberger 2001a,b). This method does not require submerged tissue, physiological conditions are easier to maintain and, therefore, intact tissues can be studied non-invasively. In MRI, magnetic field gradients are being used in three dimensions during the NMR pulse sequence. This technique provides information on the spatial origin of the resonance signals from heterogeneous systems, resulting in images depicting the distribution of the observed resonance signal in the sample. Special microimaging probeheads are available for standard-bore NMR magnets which make use of exchangeable inserts suitable for covering samples of a diameter from 1 to 10 mm in maximum. Homemade inserts and probe heads which are adapted to special requirements have been used successfully (Heidenreich et al. 1998).

Since the space in standard-bore magnets is very limited and does not suffice for accommodation of entire plants larger than Ø10 mm, wide-bore magnets of various bore diameters and vertical or horizontal orientation frequently are used for MRI studies. The largest magnet specifically dedicated to plant studies has been established at the Wageningen NMR Centre in The Netherlands (http://ntmf.mf.wau.nl/wnrmc/). The vertical 3 T magnet with a 50-cm bore contains a "green house", i.e. a climate chamber to control growth conditions. A portable magnetic resonance imaging system has been described, which is equipped with a permanent magnet and flat biplanar gradient coils. This instrument enables investigations of plant parts not exceeding a diameter of 12 mm in their natural environment (Rokitta et al. 2000). Electromagnets of low field below 1 T are useful for plant studies but are limited to measure content and flow of water. In a recent example, using a 0.47 T electromagnet and a custom-engineered 50-mm coil/gradient system, developmental changes and water status in tulip bulbs during storage have been studied (Van der Toorn 2000). Images reflecting water content, longitudinal (T_1) and transverse (T_2) relaxation times, and the apparent self-diffusion coefficient (ADC) were obtained for intact tulip bulbs stored for 12 weeks under chilled (4 °C) and non-chilled (20 °C) conditions, and after planting. The progressive development and elongation of the central flower bud was visualized and quantitatively measured by MRI. Chilled bulbs, in comparison with non-chilled ones, showed significantly faster elongation of the flower bud during storage. Redistribution of water was observed between different bulb organs for both storage protocols but the pattern of redistribution was different in chilled and non-chilled bulbs. The growing flower bud was a sink of water that is supplied by the basal plate and the scales.

In addition to imaging-based monitoring of water distribution and movement, imaging techniques are also feasible to monitor changes in metabolic pools and to provide information on spatial resolution of selective chemical shift data. Abundant metabolites such as sucrose, amino acids and lipids can be identified in living tissue by the chemical shift of characteristic resonances that appear well separated from other resonance lines of the same sample. Selective excitation of the desired resonance such as methyl protons, anomeric carbohydrate protons, or aromatic protons leads to images selectively reflecting the spatial distribution of the specified metabolite.

Chemical shift imaging (CSI) has been used to localize the site of essential oil accumulation in fruit of *Carum copticum* (Gersbach and Reddy 2002). A chemically selective fast-spin-echo sequence was employed to localize thymol, the major component of the essential oil in the secretory structures (canals) situated in the fruit walls. Selective imaging of the signal at δ 6.2, resulting from resonances of aromatic protons, indicated thymol. The obtained chemical shift image corresponded with the location of canals in the optical image. The non-invasive character of the technique was verified by normal germination and development of the seeds subsequent to the MRI experiment.

Since only an individual chemical shift is selectively excited in CSI, in practice it is difficult to discriminate between structurally closely related compounds, which often occur in the same tissue. In the example described above, thymol is accompanied by *p*-cymene (Gersbach and Reddy 2002) that shows similar chemical shift properties. Further discrimination and quantification is possible by means of independent analytical methods. Alternatively, more comprehensive magnetic resonance information can be gained by localized spectroscopy (spectroscopic imaging), which provides a three-dimensional data set with two spatial dimensions and a one-dimensional ^{1}H-NMR spectrum. The dynamic variation of the sucrose concentration within the phloem of castor bean seedlings has been observed, for example, by means of spectroscopic imaging with a time resolution of 70 min (Verscht et al. 1998). In sugarcane, distribution of sucrose has been monitored using ^{1}H-NMR imaging in combination with double-quantum filtering with magnetic field gradients (Wolf et al. 2000). NMR imaging and CSI-based detection of resins in special resin canals have been used as tools for noninvasive plant taxonomy (Glidewell et al. 2002). Morphological details of two species of the Podocarpaceae, which were difficult to investigate by conventional methods, were studied in this work.

Another strategy to overcome limited information of CSI was developed, namely the so-called correlation peak imaging (CPI) sequence. Similar to CSI, a characteristic chemical group of the metabolite under study is selectively monitored and its spatial distribution in plant tissue is mapped. However, a second spectroscopic dimension is being used additionally to enhance spectral resolution. The CPI experiment has been developed as a homonuclear correlation technique based on the 2D COSY sequence and, in addition to the two spectroscopic dimensions, covers two spatial dimensions, thus representing a 4D data set (Ziegler et al. 1996). Its feasibility was demonstrated with castor bean seedlings as a model plant. A number of carbohydrates and amino acids have been detected down to the order of 10 mM at 11.7 T with an experimental duration of approximately 4.5 h and a voxel size of 560 nl (Metzler et al. 1995). To reduce the extensive acquisition time required for recording correlation-peak images, carefully optimized conditions are necessary.

Radial spectroscopic ^{1}H-NMR images of *Ancistrocladus heynanthus* have been recorded, using both radial CSI and radial CPI, to demonstrate the spatial distribution of metabolites in concentric volume elements of this liana plant (Meininger et al. 1997). Using radial volume elements, considerable reduction of acquisition time compared to the conventional quadratic raster technique was achieved.

The imaging techniques and examples discussed so far in this article make use of the detection of protons because of the high sensitivity and natural abundance of this nucleus. However, other nuclei may be of interest for MRI in plants as well. The high relative sensitivity of ^{19}F, which is close to that of ^{1}H, makes it a highly suitable nucleus for MRI studies. It has been applied in a number of medical-oriented studies and a first plant CSI study has been published recently (Bringmann et al. 2001). Indirect detection of carbon by a cyclic polarization transfer technique has been utilized in monitoring biosynthesis and translocation of ^{13}C sucrose in the vascular bundle of the hypocotyl of castor bean seedlings (Heidenreich et al. 1998; Hudson et al. 2002). A few other nuclei such as ^{23}Na and ^{2}H have been used for MRI studies (Köckenberger 2001b). The distribution of sodium, for example, was mapped in seedlings of *Ricinus communis* by means of ^{23}Na-NMR microimaging at 11.75 T using a homebuilt probe head equipped with a climate chamber (Olt et al. 2000). ^{31}P imaging of a lepidopteran larvae, which was reared on a lima bean diet, was used to map both radial and longitudinal pH gradients within the digestive tract of the insect (Skibbe et al. 1996).

7 Conclusion and Perspectives

NMR and MRI are collective designations for a constantly increasing number of methods used in a wide variety of research areas. They are well established for the investigation of numerous questions arising in plant sciences. The diversity of techniques and applications of NMR and MRI requires close interdisciplinary cooperation of spectroscopists with plant physiologists, biochemists, geneticists, chemists, and IT specialists. It has been shown by several examples in this review that both steady-state and dynamic processes are accessible by HR-NMR techniques and imaging methods. Future prospects are expected by further improvement of both sensitivity and resolution. In HR-NMR, methods with higher sensitivity will allow the identification of minute amounts of samples in conjunction with computational data management. Automatic structure elucidation software will substantially reduce time required for spectra interpretation. Analysis of mixtures, either after separation using coupling to chromatographic methods, computational deconvolution, or fingerprint analysis, will certainly become more important in future. Metabolic profiling of plant extracts and NMR of living plant tissue are of increasing interest to examine the consequences of physiological perturbations by exogenously applied impact factors, such as elicitors, inhibitors, various types of stress, or altered gene expression. Quantitative analysis of metabolic flux of pathways in mutants and genetically transformed plants is another area in which NMR will contribute increasingly to unravel the function of individual plant genes and their role, for example, in physiological development and plant defence. MRI techniques certainly are among the most promising and rapidly developing approaches in instrumental plant analysis. Its non-invasive character is extremely useful to study physiological and metabolic processes in untransformed and genetically manipulated plants. In addition to pure plant studies, there is considerable potential of MRI to investigate dynamic aspects of pathogenic and symbiotic interactions between plants and microorganisms, fungi, or herbivores, in bi- or multitrophic systems.

Acknowledgements. Dr Daniel J Fowler, Jena/Cambridge, is gratefully acknowledged for linguistic support in the preparation of the manuscript. The Deutsche Forschungsgemeinschaft, Bonn, and the Fonds der Chemischen Industrie, Frankfurt, financially supported our work in this field.

References

Albert K, Dachtler M, Glaser T, Händel H, Lacker T, Schlotterbeck G, Strohschein S, Tseng L-H, Braumann U (1999) On-line coupling of separation techniques to NMR. J High Resolut Chromatogr 22:135–143

Aranìbar N, Bijay KS, Stockton GW, Ott K-H (2001) Automated mode-of-action detection by metabolic profiling. Biochem Biophys Res Commun 286:150–155

Aubert S, Curien G, Bligny R, Gout E, Douce R (1998) Transport, compartmentation and metabolism of homoserine in higher plant cells. Carbon-13 and phosphorus-31 nuclear magnetic resonance studies. Plant Physiol 116:547–557

Aubert S, Hennion F, Bouchereau A, Gout E, Bligny R, Dorne A-J (1999) Subcellular compartmentation of proline in the leaves of the subantarctic Kerguelen cabbage *Pringlea antiscorbutica* R. Br. In vivo ^{13}C-NMR study. Plant Cell Environ 22:255–259

Bacher A, Rieder C, Eichinger D, Arigoni D, Fuchs G, Eisenreich W (1999) Elucidation of novel biosynthetic pathways and metabolite flux patterns by retrobiosynthetic NMR analysis. FEMS Microbiol Rev 22:567–598

Bailey NJC, Oven M, Holmes E, Nicholson JK, Zenk MH (2003) Metabolomic analysis of the consequences of cadmium exposure in *Silene cucubalus* via ^{1}H NMR spectroscopy and chemometrics. Phytochemistry 62:851–858

Bohlmann F, Zeisberg R (1975) ^{13}C-NMR-Spektren von Monoterpenen. Org Magn Reson 7:426–432

Bringmann G, Feineis D (2001) Stress-related polyketide metabolism of Dioncophyllaceae and Ancistrocladaceae. J Exp Bot 52:2015–2022

Bringmann G, Wolf K, Meininger M, Rokitta M, Haase A (2001) In-vivo F-19 NMR chemical-shift imaging of *Ancistrocladus* species. Protoplasma 218:134–143

Chudek JA, Hunter G (1997) Magnetic resonance imaging of plants. Prog Nucl Magn Reson Spect 31:43–62

Crouch RC, Martin GE (1992) Micro inverse-detection – a powerful technique for natural product structure elucidation. J Nat Prod 55:1343–1347

Croasmun WR, Carlson RMK (eds) (1994) Two-dimensional NMR spectroscopy. VCH, New York

deGraaf AA, Mahle M, Mollney M, Wiechert W, Stahmann P, Sahm H (2000) Determination of full C-13 isotopomer distributions for metabolic flux analysis using heteronuclear spin echo difference NMR spectroscopy. J Biotech 77:25–35

Derome AE (1989) The use of NMR spectroscopy in the structure determination of natural products: two-dimensional methods. Nat Prod Rep 6:111–141

Dixon RA (2001) Natural products and plant disease resistance. Nature 411:843–847

Duddeck H, Dietrich W, Toth G (1998) Structure elucidation by modern NMR, 3rd edn. Steinkopf, Darmstadt

Fan TW-M (1996) Metabolic profiling by one- and two-dimensional NMR analysis of complex mixtures. Prog Nucl Magn Reson Spect 28:161–219

Fan TW-M, Lane AN, Shenker M, Bartley JP, Crowley D, Higashi RM (2001) Comprehensive chemical profiling of gramineous plant root exudates using high-resolution NMR and MS. Phytochemistry 57:209–221

Ferreira MJP, Emerenciano VP, Linia GAR, Romoff P, Macari PAT, Rodrigues GV (1998) ^{13}C NMR spectroscopy of monoterpenoids. Prog Nucl Magn Reson Spect 39:267–300

Ford Y-Y, Ratcliffe RG, Robins RJ (1996) In vivo NMR analysis of tropane alkaloid metabolism in transformed root and de-differentiated cultures of *Datura stramonium*. Phytochemistry 43:115–120

Gersbach PV, Reddy N (2002) Non-invasive localization of thymol accumulation in *Carum copticum* (Apiaceae) fruits by chemical shift selective magnetic resonance imaging. Ann Bot 90:253–257

Glidewell SM, Möller M, Duncan G, Mill RR, Masson D, Williamson B (2002) NMR imaging as a tool for noninvasive taxonomy: comparison of female cones of two Podocarpaceae. New Phytol 154:197–207

Hartung W, Ratcliffe RG (2002) Utilization of glycine and serine as nitrogen sources in the roots of *Zea mays* and *Chamaegigas intrepidus*. J Exp Bot 53:2305–2314

Heidenreich M, Köckenberger W, Kimmich R, Chandrakumar N, Bowtell R (1998) Investigation of carbohydrate metabolism and transport in castor bean seedlings by cyclic *J* cross polarization imaging and spectroscopy. J Magn Reson 132:109–124

Hinse C, Sheludko YV, Provenzani A, Stöckigt JHH (2001) In vivo NMR at 800 MHz to monitor alkaloid metabolism in plant cell cultures without tracer labeling. J Am Chem Soc 123:5118–5119

Hölscher D, Schneider B (1999) HPLC-NMR analysis of phenylphenalenones and a stilbene from *Anigozanthos flavidus*. Phytochemistry 50:155–161

Hudson AMJ, Köckenberger W, Heidenreich M, Chandrakumar N, Bowtell R (2002) H-1 detected C-13 planar imaging. J Magn Reson 155:64–71

Knight WD (1949) Nuclear magnetic resonance shift in metals. Phys Rev 76:1259–1260

Köckenberger W (2001a) Nuclear magnetic resonance micro-imaging in the investigation of plant cell metabolism. J Exp Bot 52:641–652

Köckenberger W (2001b) Functional imaging of plants by magnetic resonance experiments. Trends Plant Sci 6:286–292

Lacey ME, Tan ZJ, Webb AG, Sweedler JV (2001) Union of capillary high-performance liquid chromatography and microcoil nuclear magnetic resonance spectroscopy applied to the separation and identification of terpenoids. J Chromatogr A 922:139–149

Lichtenthaler HK (1999) The 1-deoxy-D-xylulose-5-phosphate pathway of isoprenoid biosynthesis in plants. Annu Rev Plant Physiol Plant Mol Biol 50:47–65

Lee PC, Schmidt-Dannert C (2002) Metabolic engineering towards biotechnological production of carotenoids in microorganisms. Appl Microbiol Biotech 60:1–11

Logan TM, Murali N, Wang G, Jolivet J (1999) Application of a high-resolution superconducting NMR probe in natural product structure determination. Magn Reson Chem 37:512–515
Lundberg P, Harmsen E, Ho C, Vogel HJ (1990) Nuclear magnetic resonance studies of cellular metabolism. Anal Biochem 191:193–222
Martin F (1985) Monitoring plant metabolism by ^{13}C, ^{15} N and ^{14} N nuclear magnetic resonance spectroscopy. A review of the applications to algae, fungi, and higher plants. Physiol Vég 23:463–490
Martin GE, Crouch RC (1994) Inverse-detected 2D-NMR applications in alkaloid chemistry. In: Linskens HF, Jackson JF (eds) Alkaloids. Modern methods of plant analysis, vol 15. Springer, Berlin Heidelberg New York, pp 25–89
Meininger M, Jakob PM, von Kienlin M, Koppler D, Bringmann G, Haase A (1997) Radial spectroscopic imaging. J Magn Reson 125:325–331
Metzler A, Izquierdo M, Ziegler A, Köckenberger A, Komor E, von Kienlin M, Haase A, Decorps M (1995) Plant histochemistry by correlation-peak imaging. Proc Natl Acad Sci USA 92:11912–11915
Nyberg NT, Baumann H, Kenne L (2001) Application of solid-phase extraction coupled to an NMR flow-probe in the analysis of HPLC fractions. Magn Reson Chem 39:236–240
Olson DL, Peck TL, Webb AG, Magin RL, Sweedler JV (1995) High-resolution microcoil H-1-NMR for mass-limited, nanoliter-volume samples. Science 270:1967–1970
Olt S, Krotz E, Komor E, Rokitta M, Haase A (2000) Na-23 and H-1 NMR microimaging of intact plants. J Magn Reson 144:297–304
Opitz S, Schneider B (2003) Oxidative biosynthesis of phenylbenzoisochromenones from phenylphenalenones. Phytochemistry. Phytochemistry 62:307–312
Otálvaro F, Görls H, Hölscher D, Schmitt B, Echeverri F, Quiñones W, Schneider B (2002) Dimeric phenylphenalenones from *Musa acuminata* and various Haemodoraceae species. Crystal structure of anigorootin. Phytochemistry 60:61–66
Park S-U, Yu M, Facchini PJ (2003) Modulation of berberine bridge enzyme levels in transgenic root cultures of California poppy alters the accumulation of benzophenanthridine alkaloids. Plant Mol Biol 51:153–164
Pfeffer PE, Bago B, Shachar-Hill Y (2001) Exploring mycorrhizal function with NMR spectroscopy. New Phytol 150:543–553
Ratcliffe RG (1994) In vivo NMR studies of higher plants and algae. In: Woolhouse HW (ed) Advances in botanical research, vol 20. Academic Press, London, pp 43–123
Ratcliffe RG (1996) In vivo NMR spectroscopy: biochemical and physiological applications to plants. In: Shachar-Hill Y, Pfeffer PE (eds) Nuclear magnetic resonance in plant biology. The American Society of Plant Physiologists, Rockville, Maryland, pp 1–32
Ratcliffe RG (1997) In vivo NMR studies of the metabolic response of plant tissues to anoxia. Ann Bot 7 (suppl A) 9:39–48
Ratcliffe RG, Roscher A, Shachar-Hill Y (2001) Plant NMR spectroscopy. Prog Nucl Magn Reson Spect 39:267–300
Roberts JKM (2000) NMR adventures in the metabolic labyrinth within plants. Trends Plant Sci 5:30–34
Roberts JKM, Xia J-H (1996) NMR contributions to understanding of plant responses to low oxygen stress. In: Shachar-Hill Y, Pfeffer PE (eds) Nuclear magnetic resonance in plant biology. The American Society of Plant Physiologists, Rockville, Maryland, pp 155–180
Rohmer M, Knani M, Simonin P, Sutter B, Sahm H (1993) Isoprenoid biosynthesis in bacteria: a novel pathway for the early steps leading to isopentenyl diphosphate. Biochem J 295:517–524
Rokitta M, Rommel E, Zimmermann U, Haase A (2000) Portable nuclear magnetic resonance imaging system. Rev Sci Instr 71:4257–4262
Roscher A, Kruger NJ, Ratcliffe RG (2000) Strategies for metabolic flux analysis in plants using isotope labelling. J Biotech 77:81–102

Russell DJ, Hadden CE, Martin, GE, Gibson AA, Zens AP, Carolan JL (2000) A comparison of inverse-detected heteronuclear NMR performance: conventional vs cryogenic microprobe performance. J Nat Prod 63:1047–1049

Sadler IH (1988) The use of NMR spectroscopy in the structure determination of natural products: One-dimensional methods. Nat Prod Rep 5:101–127

Saglio PH, Raymond P, Pradet A (1980) Metabolic activity and energy charge of excised maize root tips under anoxia. Plant Physiol 66:1053–1057

Sato F, Hashimoto T, Hachiya A, Tamura K-I, Choi K-B, Morishige T, Fujimoto H, Yamada Y (2001) Metabolic engineering of plant alkaloid biosynthesis. Proc Natl Acad Sci USA 98:367–372

Schlotterbeck G, Ross A, Hochstrasser R, Senn H, Kühn T, Marek D, Schett O (2002) High-resolution capillary tube NMR. A miniaturized 5-µL high-sensitivity TXI probe for mass-limited samples, off-line LC NMR, and HT NMR. Anal Chem 74:4464–4471

Schmitt B, Schneider B (2001) Phenylpropanoid interconversion in *Anigozanthos preissii* observed by high-performance liquid chromatography – nuclear magnetic resonance spectroscopy. Phytochem Anal 12:43–47

Schneider B (2000) Natural products: Liquid chromatography-nuclear magnetic resonance. In: Wilson ID, Adlard TR, Cooke M, Poole CF (eds) Encyclopedia of separation science, vol 7. Academic Press, London, pp 3434–3445

Schneider B, Gershenzon J, Graser G, Hölscher D, Schmitt B (2003) Carbon-13 and deuterium labeling in biosynthetic studies observed by one-dimensional ^{13}C NMR and HPLC-^{1}H NMR techniques. Phytochem Rev (in press)

Shachar-Hill Y (2002) Nuclear magnetic resonance and plant metabolic engineering. Metab Eng 4:90–97

Simpson TJ (1998) Application of isotopic methods to secondary metabolic pathways biosynthesis. Topics Curr Chem 195:1–48

Skibbe U, Cristeller JT, Callaghan PT, Eccles CD, Laing WA (1996) Visualization of pH gradients in the larval midgut of *Spodoptera litura* using ^{31}P-NMR microscopy. J Insect Physiol 42:777–790

Steglich W, Fugmann B, Lang-Fugmann S (eds) (1997) Römpp Lexikon Naturstoffe. Thieme, Stuttgart

Styles P, Soffe NF, Scott CA, Cragg DA, Row F, White DJ, White PCJ (1984) A high-resolution NMR probe in which the coil and preamplifier are cooled with liquid helium. J Magn Reson 60:397–404

Styles P, Soffe NF, Scott CA (1989) An improved cryogenically cooled probe for high-resolution NMR. J Magn Reson 84:376–378

Szyperski T (1998) ^{13}C-NMR, MS and metabolic flux balancing in biotechnology research. Q Rev Biophys 31:41–106

Thiel R, Adam KP (2002) Incorporation of [1-^{13}C]1-deoxy-D-xylulose into isoprenoids of the liverwort *Conocephalum conicum*. Phytochemistry 59:269–274

Van der Toorn A, Zemah H, Van As H, Bendel P, Kamenetsky R (2000) Developmental changes and water status in tulip bulbs during storage: visualization by NMR imaging. J Exp Bot 51:1277–1287

Vederas JC (1985) The use of stable isotopes in biosynthetic studies. Nat Prod Rep 4:77–337

Verscht J, Kalusche B, Köhler J, Köckenberger W, Metzler A, Haase A, Komor E (1998) The kinetics of sucrose concentration in the phloem of individual vascular bundles of the *Ricinus communis* seedling measured by magnetic resonance micro-imaging. Planta 205:132–139

Wang C-Z, Maier UH, Eisenreich W, Adam P, Obersteiner I, Keil M, Bacher A, Zenk MH (2001) Unexpected biosynthetic precursor of amarogentin – a retrobiosynthetic ^{13}C NMR study. Eur J Org Chem 1459–1465

Watanabe N, Niki E (1978) Direct coupling of FT-NMR to high-performance liquid-chromatography. Proc Jpn Acad Ser B Phys Biol Sci 54:194–199

Weichert W (2001) ^{13}C metabolic flux analysis. Metab Eng 3:195–206

Wolf C, van der Thoorn A, Hartmann K, Schreiber L, Schwab W, Haase A, Bringmann G (2000) Metabolite monitoring in plants with double-quantum filtered chemical shift imaging. J Exp Bot 51:2109–2117

Wolfender J-L, Ndjoko K, Hostettmann K (2001) The potential of LC-NMR in phytochemical analysis. Phytochem Anal 12:2–22

Wolters AM, Jayawickrama DA, Sweedler JV (2002) Microscale NMR. Curr Opin Chem Biol 6:711–716

Ziegler A, Metzler A, Köckenberger A, Izquierdo M, Komor E, Haase A, Decorps M, von Kienlin M (1996) Correlation-peak imaging. J Magn Reson B 112:141–150

Dr. Bernd Schneider
Max-Planck-Institute for Chemical Ecology
Beutenberg Campus
Winzerlaer Strasse 10
07745 Jena, Germany
e-mail: schneider@ice.mpg.de

The Old Arbuscular Mycorrhizal Symbiosis in the Light of the Molecular Era

Natalia Requena and Magdalene Breuninger

Introduction

Mycorrhiza, from the Greek terms myco (=fungus) and rhiza (=root), is the predominant root symbiosis. More than 90% of land plants form mycorrhiza with soil fungi belonging to three different phyla (Smith and Read 1997). A few plant families (e.g. Cruciferae, Chenopodiaceae, Proteaceae) as well as some genera (e.g. *Lupinus*) have been described as non-mycorrhizal. It is speculated that the mycorrhizal character has possibly arisen several times during evolution (Trappe 1987), but the mechanisms of mycorrhiza exclusion have not yet been identified. From the different types of mycorrhiza existing in nature arbuscular mycorrhizas are the oldest and most widespread symbiosis (Remy et al. 1994). Recent fossil studies and molecular data have tracked the presence of this symbiosis all the way to the Ordovician era, i.e. to be at least 460 million years old (Redecker et al. 2000b). The fungi forming this mutualistic symbiosis have been recently recognized as belonging to an independent phylum, the Glomeromycota (Schüßler et al. 2001b) with a monophyletic origin. The permanence of this mutualistic association during evolution reflects its importance in nature. The reciprocal benefit, achieved by the nutrient exchange between both partners in intimate contact, is possibly the reason for this durability. However, the benefits of the symbiosis to the maintenance of natural ecosystems cannot be only estimated in terms of improved plant nutrition. Thus, other important mycorrhizal contributions to the terrestrial ecosystem equilibrium might be achieved through phyto-protection effects often observed in arbuscular mycorrhiza (AM) colonized versus non-colonized plants (Garcia-Garrido and Ocampo 2002), maintenance of soil structure through the formation of soil aggregates with an active contribution of the extraradical fungal mycelium (Bethlenfalvay et al. 1998) or improvement of plant growth in degraded soils by restoring the microbial equilibrium in the rhizosphere (Requena et al. 1997). The overall impact of the mycorrhizal symbiosis is so large that mycorrhizal fungal diversity determines plant biodiversity, ecosystem variability and productivity (van der Heijden et al.

Progress in Botany, Vol. 65

1998). Not surprisingly, sustainable agriculture and ecosystem conservation have challenged the management of the symbiosis as an environment-friendly alternative to fertilizers and/or pesticides (Barea and Jeffries 1995), as well as in the conservation and regeneration of degraded areas (Requena et al. 1996, 2001).

Despite of its importance for the sustainability of the terrestrial ecosystems, many aspects of the AM symbiosis functioning are still unknown. The principal reasons have been the underground nature of the association with a great part of the fungal biomass hidden inside of the root and the extensive but unculturable mass of mycelium spreading into the soil. AM fungi are obligate biotrophs and their study under laboratory conditions has been limited by the need of co-cultivation with their hosts. However, significant progress has been made in the last decade with the outcome of new molecular techniques. They have helped first to position these fungi more accurately in the evolutionary scale and to describe many aspects of their particular idiosyncrasy. But additionally, many of the molecular and cellular aspects of the symbiosis development have been revised in the light of the disclosures achieved by these methodologies.

2 Molecular Characterization of Arbuscular Mycorrhizal (AM) Fungi

2.1 Molecular Identification of AM Fungi

The taxonomy of arbuscular mycorrhizal (AM) fungi have been traditionally based on the morphological characteristics of their spores, with the substructure of the spore walls being a decisive criterion for species identification. On the basis of the spore morphology approximately 150 species of AM fungi have been identified (Schenck and Pérez 1990; Walker 1992; Walker and Trappe 1993). Proper species identification is the basis for population dynamic studies of AM fungi, but it is a difficult task when based on spore morphology. On one hand, spore formation is highly variable depending on physiological and environmental conditions (Morton and Bentivenga 1994). On the other hand, morphological criteria are also difficult to assess in old or parasitized spores, as they are often found in the soil. Furthermore, spore diversity in soil often does not correlate with diversity of fungi in the root (Clapp et al. 1995). Morphology of mycorrhizal structures inside the root enables identification at best at the family level (Abbott 1983; Merryweather and Fitter 1998). In addition, several new species have been reported to be undetectable by classical root staining methods (Redecker et al. 2000a). Biochemical approaches to identify AM fungi in roots have been alternatively employed. Isoenzyme analysis (Hepper et al. 1986; Rosendahl et al. 1989; Tisserant et al. 1998) and immunological methods were developed to detect AM fungi in roots (Wright and Morton 1989; Sanders et al. 1992; Hahn et al. 1995; Cordier et al. 1996; Kozlova et al. 2001). However, such techniques are difficult to apply when more than one fungus colonizes the same root, the normal situation in field studies. With the outcome of PCR-based methods higher specificity in the identification has been achieved and problems related to the amount of material solved. Moreover, molecular methods have allowed establishing links between the genetic and the morphological diversity and in some cases also to the functional diversity.

The ribosomal genes have been a main target in molecular phylogenetic studies in many fungal groups, including mycorrhizal fungi (reviewed in Lanfranco et al. 1998). These multicopy genes are highly conserved and present in all organisms. In eukaryotes they consist of three coding regions of different sizes (18S, 5.8S and 25S), separated by two non-translated sequences (ITS, internal transcribed spacer, and IGS, intergenic spacer). The first published putatively *Glomales*-specific primer was VANS1, located in the small subunit (18S) (Simon et al. 1992), but other family-specific primers followed (Simon et al. 1993a). These primers were successfully used to detect fungal species colonizing plants in pot cultures (Simon et al. 1992, 1993a; Di Bonito et al. 1995), as well as in field plants (Chelius and Triplett et al. 1999). Using these group-specific primers in combination with a subtractive hybridization technique to selectively enrich in fungal sequences, Clapp et al. (1995) carried out the first molecular field population study of AM fungi. An important aspect they demonstrated was the discrepancy between root colonization rate by a particular fungus and the amount of spores formed by that fungus in the soil. Another interesting field study was performed by Helgason et al. (1998), who showed that AM species diversity in cultivated soils was much lower than in comparable woodland. Thus, 92% of AM fungal sequences analyzed from arable fields corresponded to *Glomus mosseae* or closely related taxa, while this species was absent in the woodland. This suggests that this species is more adapted to conditions of arable lands than of woodland, and it points out the importance of edaphic factors in controlling AM population structure.

PCR amplification of ribosomal sequences has also been used in combination with other techniques, such as SSCP (single-strand conformation polymorphism; Kjøller and Rosendahl 2000, 2001; Clapp et al. 2001) or RFLP (restriction fragment length polymorphism; Helgason et al. 1998) facilitating a pre-screening when multiple samples have to be analyzed. Interesting findings were obtained by van Tuinen et al. (1998) in studies on AM fungal population in defined microcosms. They employed a nested PCR procedure to circumvent the problem of PCR inhibitors present in many root samples and designed specific primers to four different AM fungal species based on polymorphisms in the large ribosomal subunit (25S rDNA). Their results allowed them to study the complex pattern of fungal species interaction with synergistic effects among species. A similar methodological approach was used to analyze the impact of sewage sludge on AM fungal populations (Jacquot et al. 2000) and to study AM fungi in heavy metal-polluted soils (Turnau et al. 2001). Other ribosomal primers were designed with the aim of identifying single species or isolates (Abbas et al. 1996; Millner et al. 1998, 2000; van Tuinen et al. 1998), therefore, their application to population studies is limited.

With the massive outcome of AM fungal ribosomal sequences in the databases, it became clear that the original primers designed by Simon et al. (1992, 1993a) did not universally recognize all AM fungi. Thus, most of the sites where the primers were designed were either not well conserved, as in *Scutellospora*, or even completely absent as in several newly defined ancestral lineages (Clapp et al. 1999; Redecker et al. 2000a). The analysis of a large number of sequences has proved that VANS1 and the family-specific primers are only specific for a small subgroup of AM fungi and consequently not suitable for most diversity studies (Schüßler et al. 2001a). To overcome this problem Redecker (2000) developed a set of primers to screen mycorrhizal roots for the presence of the major phylogenetic groups within the Glomales. These primers are even able to monitor the presence in the root of the recently described ancestral families Archaeosporaceae and Paraglomaceae (Redecker et al. 2000a).

All these studies have demonstrated the enormous potential of molecular identification techniques and have highlighted their importance for ecological studies on AM fungal communities. However, they also point out their limitations and the precautions to be considered. This challenges the design of reliable molecular markers able to identify individuals as well as populations.

2.2 Molecular Phylogeny

AM fungi have been until recently phylogenetically positioned in the order Glomales according to spore characteristics (Morton and Benny 1990; Walker and Trappe 1993). The three-family structure of the Glomales, comprising the families Glomaceae, Gigasporaceae and Acaulosporaceae, was initially supported by rDNA data (Morton and Benny 1990; Simon et al. 1993b). However, it was also clear that the genus *Glomus* had a much larger genetic diversity than previously envisaged according to morphological characters. This genus represented a conglomerate of fungi morphologically difficult to differentiate but exhibiting great genetic differences, as for instance those between the families Gigasporaceae and Acaulosporaceae (Walker 1992; Simon et al. 1993b; Simon 1996). The analysis of a large data set corresponding to the small subunit ribosomal RNA (SSU rRNA) gene sequences have shown that the genus *Glomus* is non-monophyletic, and that, at least, three distinct well-supported groups within *Glomus* can be distinguished (Schüßler et al. 2001a). The increasing amount of molecular sequence data are helping to clarify the AM fungal phylogeny. Recently, two deeply divergent lineages have been found and ascribed to two new genera *Achaeospora* and *Paraglomus* within the new families Archaeosporaceae and Paraglomeraceae, respectively. These fungi represent the ancestral lineages of AM fungi and exhibit a *Glomus*-like morphology (Redecker et al. 2000a; Morton and Redecker 2001).

Based on molecular, morphological and ecological characteristics AM fungi have been removed from the polyphyletic phylum Zygomycetes and placed into the new monophyletic phylum Glomeromycota, containing a single class Glomeromycetes (Schüßler et al. 2001b). Molecular data suggest that this new phylum probably diverged from the same common ancestor as the Ascomycetes and Basidiomycetes (Gehrig et al. 1996; Schüßler et al. 2001b). Three new orders have been added to the Glomerales: Archaeosporales, Paraglomerales and Diversisporales (Schüßler et al. 2001b). The order Diversisporales contains the families Gigasporaceae, Acaulosporaceae and the new family Diversiporaceae (Fig. 1). The order Archaeosporales includes two families: Archaeosporaceae and Geosiphonaceae. The order Paraglomerales contains a single family, the Paraglomeraceae. The order Glomerales still includes many of the "classical" taxa as described by Morton and Benny (1990).

A remaining open question is how to define genera and species within the AMF. For some groups, this seems to be possible by specific sequence patterns and specific PCR primers, as proposed for the families Archaeosporaceae and Paraglomeraceae (Morton and Redecker 2001). The studies of vegetative incompatibility between different isolates of AM fungi may represent a complementary tool for investigating the genetic relationships between isolates and the structure of the population (Giovanetti et al. 2003). A concerted search for morphological and molecular characters might help to solve the problem in the future, but it needs detailed investigations by AMF taxonomists and molecular biologists (Redecker 2001).

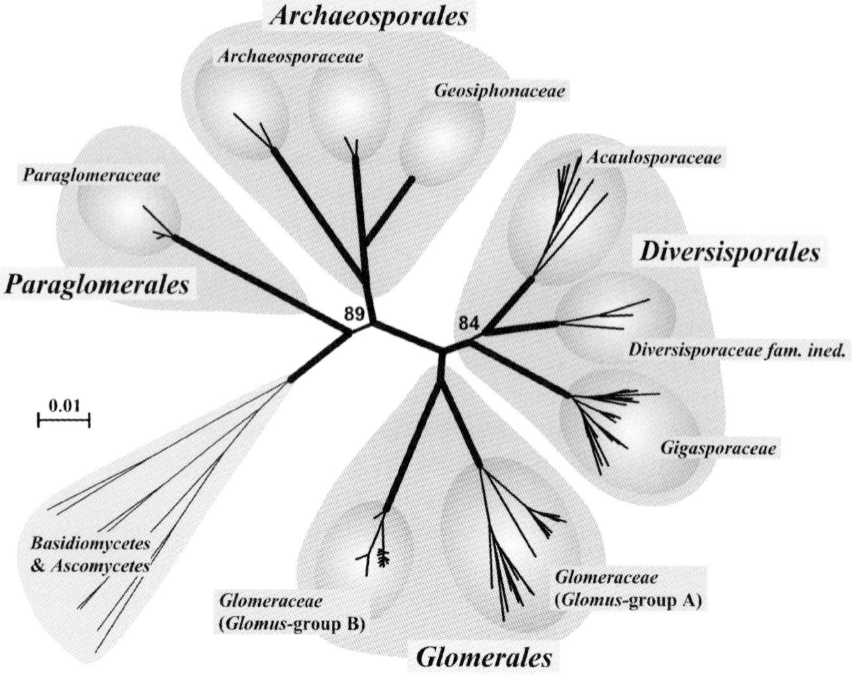

Fig. 1. Proposed taxonomic structure of the Glomeromycota and related fungi, based on SSU rRNA sequences. (Schüßler et al. 2001b, with permission)

2.3 *Geosiphon pyriformis* – A Cousin Helps to Solve the Puzzle

Fossil records (Remy et al. 1994; Taylor et al. 1995) and molecular data (Simon et al. 1993b) have suggested that AM fungi evolved together with land plants and may have been crucial in helping them to colonize land (Pirozynski and Malloch 1975). The occurrence of AM fungi has recently been dated back to at least 460 million years (Redecker et al. 2000b), further supporting this hypothesis. An important piece recently added to the puzzle has been the phylogenetic relation between *Geosiphon pyriformis* and AM fungi. This obligate symbiotic fungus forms a consortium with Cyanobacteria and in contrast to AM fungi the macrosymbiont is here the fungal partner (Schüßler et al. 1996; Schüßler and Kluge 2001). Morphological and ultrastructural investigations have shown that *Geosiphon* forms *Glomus*-like spores (Schüßler et al. 1994), but even more remarkable, analyzes of the SSU rRNA genes in *Geosiphon* have shown its close relationship to *Glomus* species (Gehrig et al. 1996). These data suggest that *Geosiphon* might represent a more ancestral type of AM-like association with photoautotrophic organisms (Gehrig et al. 1996; Schüßler and Kluge 2001). The proposed new phylogeny places *Geosiphon* within the Archaosporales with the oldest mycorrhizal fungi known being in accordance with this hypothesis.

2.4 Asexual and Ancient Organisms: A Contradiction to Evolutionary Theories?

The molecular studies based on ribosomal gene sequences have revealed that AM fungal diversity is much higher than previously envisaged according to morphological studies (Sanders et al. 1995; Lloyd Macgilp et al. 1996; Clapp et al. 1999; Redecker et al. 1997; Hijri et al. 1999; Hosny et al. 1999; Lanfranco et al. 1999; Schüßler 1999; Antonionelli et al. 2000; Pringle et al. 2000; Clapp et al. 2001). These results have been supported by studies of the whole genome using other techniques including RAPD (Wyss and Bonfante 1993, Abbas et al. 1996), PCR and DNA fingerprinting (Longato and Bonfante 1997; Zéze et al. 1997; Vandenkoornhuyse and Leyval 1998) and AFLP (Rosendahl and Taylor 1997). But it is unknown at the moment whether this high genetic diversity correlates to functional diversity (Kjøller and Rosendahl 2001).

In the life cycle of AM fungi no sexual stage has been observed (Smith and Read 1997). Thus, such high genetic diversity, observed as sequence heterogeneity even within single spores of one isolate, has been somehow a surprising finding (Sanders et al. 1996). Sequence heterogeneity has been correlated with the fact that AM fungal spores are multinucleated (Sanders et al. 1995; Lloyd-Macgilp et al. 1996). This raises the major question whether heterogeneity is the result of heterokaryotic spores or due to heterogeneity within single nuclei. The latter requires mutation and rearrangement events within a single nucleus. This is more likely in multicopy genes, where such mutations may be maintained even if they result in non-functional copies of the gene, and provided that sexual recombination processes do not occur (Sanders et al. 1996). This is supported by studies of Rosendahl and Taylor (1997) indicating that the genetic variation among the spores of AM fungi is significantly different from that expected in a recombinant population. Heterogeneity within single nuclei is therefore one of the possible reasons for the high genetic variability.

There are, however, experimental data showing that heterokaryotic status of AM could be an alternative explanation for the high sequence variability found. Experimental evidence that genetically different nuclei co-exist in individual AM fungi has been provided by the studies of Hijri et al. (1999), in which single nuclei from one spore were used as template for PCR reactions on the ribosomal genes. Data analysis revealed that different nuclei harbour different rDNA sequences. Further support of this hypothesis was given by DNA-DNA fluorescent in situ hybridization (FISH) experiments (Kuhn et al. 2001). These authors predicted that genetic variation is generated by accumulation of mutations in a predominantly clonal genome. However, recombination could eventually also occur among nuclei within individuals, homogenizing the nuclear population. This would not be surprising in a coenocytic organism where genetically different nuclei coexist. However, these recombination events seem to be rare and therefore insufficient to eliminate mutations accumulated in the genome (Kuhn et al. 2001). These results indicate that an individual of AM fungi is essentially a population of genetically different nuclei (Sanders 1999).

Besides the occurrence of mutation events, heterokaryosis can be achieved through exchange of nuclei following hyphal fusion (anastomosis). Anastomosis and exchange of nuclei has been recently observed in several *Glomus* species (Giovanetti et al. 1999, 2001, 2003). This suggests the interesting possibility that beyond a nutritional flow, an information and genetic exchange might also take place through the network of AM fungal communities. However, no anastomosis has been observed between different species or isolates from one species (Giovanetti et al. 2003), indicating that genetic exchange through anastomosis would be limited to very closely related isolates of the same species. More results on these aspects are required to elucidate the role of anastomosis in the AM fungal genetic heterogeneity.

3 Development of the AM Symbiosis

Briefly, the establishment of the AM symbiosis begins with the colonization of a compatible root by a small mycelium produced by the fungal soil propagules, either asexual spores or hyphae growing out of dead or alive AM colonized roots. After attachment to the root, the fungus penetrates into the cortex and forms distinct morphologically specialized structures (intercellular hyphae, coils and arbuscules), possible locations for the nutrient exchange between symbionts. The fungal mycelium then grows out of the root exploring the soil in search of mineral nutrients, and it can also colonize other susceptible roots. The fungal life cycle is completed after formation of asexual chlamydospores by the soil mycelium. Distinct morphological stages can be therefore identified during the development of AM (Smith and Read 1997) and they are schematically represented in Fig. 2.

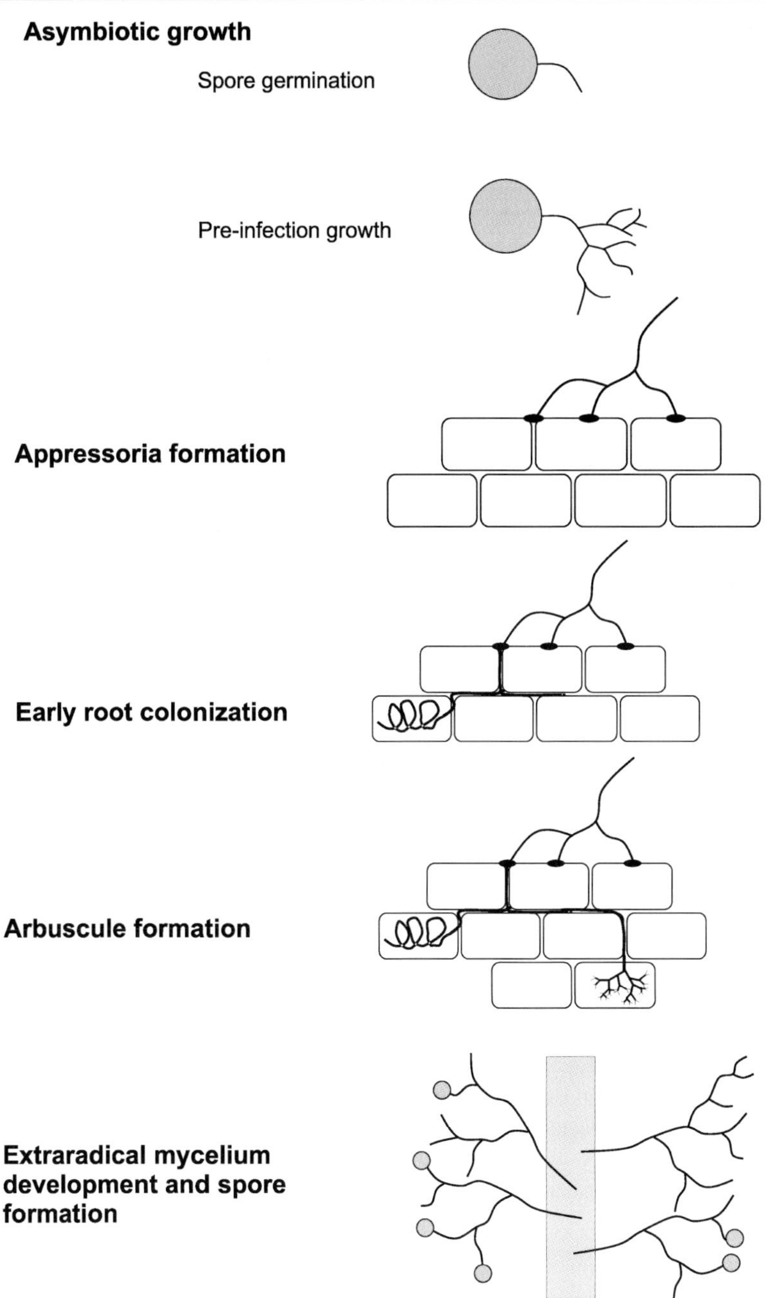

Fig. 2. Distinct developmental stages in the arbuscular mycorrhizal symbiosis

3.1 Asymbiotic Growth

Arbuscular mycorrhizal fungi are obligate biotrophs. In contrast to the plant partner they are unable of completing their life cycle during asymbiosis (Bonfante and Bianciotto 1995). The formation of the next spore generation able to germinate and establish a mycorrhiza has been, so far, only achieved from AM mycelium growing out of mycorrhizal roots. These spores, considered the resting propagule of the AM fungus, are the only plant-independent phase of the mycobiont. They are round-shape spores with a thick cell wall and average diameter between 50–100 μm. However, the most striking feature of these spores is their particular physiology. As a difference with other soil-borne fungi, these spores have the ability to germinate and arrest growth many times without the involvement of plant-derived signals (Mosse 1959; Koske 1981). AM spores germinate under appropriate water and temperature conditions and germlings extend their hyphal growth for about 2 to 3 weeks. In this time the fungal colonies extend no more than a couple of centimeters due to their inherent growth pattern with a marked apical dominance and infrequent branching. In the absence of a host root, the apical growth ceases and septation from the apex occurs (Mosse 1988). The apical septation starts with extensive vacuolization and extends towards the spore with parallel retraction of the protoplasm (Logi et al. 1998). Although the colonizing ability of these spores remains after several cycles of growth-arrest (Tommerup 1984), the infectivity decreases with time (Logi et al. 1998). During this asymbiotic phase, the fungus lives from its triacylglyceride reserves and the presence in the germination-growth medium of different carbon and nitrogen sources has little effect at increasing the length or extension of this development (Hepper 1979). Growth arrest occurs long before the spore reserves are depleted and therefore it is the absence of a host-derived signal what causes it. This phase of growth in the absence of signal from the plant is what is known as the asymbiotic stage. Cytological studies have shown that during asymbiotic growth, nuclear division occurs, although most nuclei remain arrested in the S phase or between G2-M phases (Bianciotto and Bonfante 1993; Bianciotto et al. 1995). A homologue gene of the cell cycle check-point *TOR2* from *Saccharomyces cerevisiae* was isolated from *Glomus mosseae*. The anti-inflammatory drug rapamycin, known to interfere with the role of TOR2 controlling the arrest of the cell cycle in G1 was found to decrease hyphal growth during asymbiosis although it did not affect spore germination (Requena et al. 2000). This result showed that DNA replication is not necessary for germination but for asymbiotic hyphal growth.

Attempts to cultivate AM fungi in vitro led to the interesting discovery that certain soil microorganisms could significantly increase the saprotrophic growth of the mycelia during this stage (Mosse 1959; Hepper 1979). Some of them improved spore germination but most of the organisms reported had a beneficial effect on hyphal growth, branching and the production of vegetative spores (reviewed in Requena 1998). Little is known about the mechanisms by which these microorganisms are able to improve asymbiotic growth. However, molecular studies have shown that the fungus is able to perceive these microorganisms and change its gene expression pattern in response to them (Requena et al. 1999). Besides soil microorganisms, plant exudates, including flavonoids and CO_2 have been found to exert a positive effect on hyphal growth during asymbiosis (Bécard and Piche 1989a; Gianinazzi-Pearson et al. 1989). However, mutant plants impaired in flavonoid production were found to be mycorrhizal, showing that flavonoids are not essential for host recognition (Bécard et al. 1995). It is likely that most of these factors, beneficial microorganisms and root exudates, play a more important role in the soil by stimulating AM fungal growth and facilitating host root encounter.

3.2 Host Recognition and Appressorium Development

The mycelium from germinated spores extends in the soil in search of a compatible host root. Although no directional growth has been observed towards the root, several experiments have shown that exudates from host root elicit growth stimulation in contrast to non-host exudates (Bécard and Piché 1989b; Gianinazzi-Pearson et al. 1989; Nair et al. 1991; Giovannetti et al. 1993a,b, 1996). This stimulation is hypothesized to turn on the fungus and make possible the colonization of the root. Buee et al. (2000) succeeded in partially purifying a lipophilic fraction only present in host root exudates able to promote growth and branching of the asymbiotic mycelium. However, despite this growth-promoting effect, only the physical presence of a root cell is able to trigger appressorium formation. In contrast to other plant-interacting fungi, appressorium in AM fungi is not induced by fake root surfaces such as nylon, polyamide, silk, cellulose or glass threads, even when additionally stimulated with host root exudates (Giovannetti et al. 1993a). Interestingly, an experiment performed with isolated cell walls of carrot roots showed that appressorium formation does not require a signal secreted from the host root or the presence of intact host cytoplasm (Nagahashi and Douds 1997). Further fungal penetration was not observed. Therefore, it appears that appressoria formation is specifically triggered by physical epitopes from the rhizodermis but other signals from living cells are necessary for further colonization. This agrees with the findings of

Bonfante et al. (2000) who showed that discrete steps in the root colonization are cell layer controlled (see below).

Whether a thigmotropic signal alone or a combination with chemical signals is required for eliciting appressoria formation, it is clear that the fungus must undergo a reorganization of its cellular program to accomplish this developmental change. Molecular non-targeted approaches such as suppressive subtractive hybridization (SSH) or EST (expressed sequence tag) sequencing have been employed to approach fungal gene expression at these early stages. Using SSH, Requena et al. (2002) have isolated a novel developmentally regulated gene mainly expressed during pre-symbiosis. This gene has homology to a new class of GTPases possibly post-transcriptionally spliced which might play a role in the signaling events prior to appressoria formation. The same approach has being used to study the genes controlling appressoria formation (Breuninger and Requena, unpubl. data).

Appressoria from arbuscular mycorrhizal fungi are swollen multinucleated structures forming a papilla, often hook-shaped, on the rhizodermis (Giovanetti et al. 1995; Bonfante et al. 2000; Requena and Breuninger, unpubl. results). The number, shape and size of appressoria are plant/fungal combination dependent. In *Lotus japonicus* appressoria formed by *Gigaspora margarita* have been described to be between 8 and 16 μm in diameter (Bonfante et al. 2000). The cell wall of the appressorium is thick, electron dense and rich in melanin granules (Grippiolo and Bonfante 1994; Bonfante and Perotto 1995). However, in contrast to other biotrophic fungi, there is no formation of septum behind the appressorium that could account for the dramatic increase in turgor pressure leading to cell wall penetration observed in several plant-colonizing fungi, as for instance *Magnaporthe grisea* (Talbot and Foster 2001). In contrast, it is likely that AM fungi use a combination of mechanical pressure and a moderate production of cell wall degrading enzymes to penetrate between two adjacent epidermal cells (Bonfante and Perotto 1995). In this sense, the production of endo- and exoglucanases, cellulases, xyloglucanases, polygalacturonases as well as pectinolytic enzymes have been shown in AM fungi (Garcia-Romera et al. 1991; Garcia-Garrido et al. 1992a,b, 1996; Peretto et al. 1995; Rejón-Palomares et al. 1996). The infection peg growing out of the appressoria produces a deep furrow when penetrating between the epidermal cells. In *Lotus japonicus*, these epidermal cells appear slightly deformed. Immediately thereafter, a short branch penetrates one of the radial walls and forms a coil in one of the adjacent epidermal cells before passing to colonize the cell of the cortex (Garriock et al. 1989; Bonfante et al. 2000).

3.3 Early Root Cortex Colonization

The symbiotic AM fungi only colonize the differentiated region of the root, more than 2 mm above the root tip. The fungus remains in the cortical parenchyma zone and never crosses the endodermis (Genre and Bonfante 1997). Meristems and vascular tissues are resistant to infection (Bonfante and Perotto 1992). It is noteworthy that during the whole colonization process, AM fungi, despite producing localized cortical cell wall perforation, never penetrate the plasma membrane of the host cell. In contrast, the host membrane invaginates and new membrane is formed surrounding the growing hypha, the perifungal membrane. New apoplastic material is laid down between the cell wall of both symbionts creating a new apoplastic space of ca. 80–100 nm. This new cell wall has characteristics common to primary cell walls β-1,4 glucans, non-sterified polygalacturonans, hemicellulose, hydroxyproline-rich proteins and arabinogalactan proteins. However, ultrastructural observations have shown that these components are not assembled into a fully structured cell wall (Balestrini et al. 1994; Bonfante and Perotto 1995).

During the colonization and penetration phase, plant cells harboring mycorrhiza undergo dramatic organelle reorganization. These changes include vacuole fragmentation and movement of nucleus and other organelles from a peripheral position towards the fungal branches (Bonfante and Peroto 1995; Balestrini et al. 1996, 1997). This reorganization has been shown to correlate with an intense cytoskeleton rearrangement not only in cortical cells harboring mycorrhiza (Genre and Bonfante 1997, 1998), but also in neighboring cells (Blancaflor et al. 2001). A similar phenomenon has been described in cells infected by rust fungi (Kobayashi et al. 1994, 1997). In AM, host microtubule (MT) structures undergo an increase in complexity, reorienting to accommodate the fungal cell (Genre and Bonfante 1997). New MT are synthesized and reoriented from an irregular helical disposition to three new orientations (i.e. along intercellular long hyphae; linking hyphae; and binding hyphae to the host nucleus). It is likely that these changes are achieved through the synthesis of a new tubulin on the basis of earlier evidence showing the activation of α-3 tubulin gene in response to mycorrhizal colonization (Bonfante et al. 1996). Actin microfilaments (MF) also undergo dramatic changes in their organization pattern in mycorrhizal-colonized cells. They change from a loose cortical and perinuclear localization network to a dense reticulum in close connection with the fungal branches, enveloping them in a dense coating network (Genre and Bonfante 1998). Cytoskeleton rearrangements are essential for the development of biotrophic plant-fungus interactions, which require not only the reorganization of the colonized plant cells for fungal physical accommodation, but also the establishment of an intimate contact area between partners for exchange of signals and nutrients (Genre and Bonfante 1998). This has been further demonstrated with the comparison of cytoskeleton organization in plant mutants affected in their ability to host mycorrhiza. There, a relationship between cytoskeleton integrity and cell viability during early stages of colonization has been proved (Genre and Bonfante 2002). Much less is known about the cytoskeletal changes of the fungus during colonization. Immunofluorenscence microscopy studies have shown that MT are found both in the cortical and central parts of the hyphae, while MF have been revealed only in the cortical parts (Aström et al. 1994). However, in contrast to the dramatic MT reorganization undergone by the host cell, it appears that the fungus shows little MT rearrangements during the symbiotic phase. It is likely that as a difference to the plant cell, which has to accommodate a growing fungus, the fungal cell organization is not altered since its apical dominance pattern of growth is maintained (Timonen et al. 2001). MF changes during symbiosis establishment have not been investigated yet.

A large amount of information concerning the morphogenetic changes undergone by the host during early mycorrhiza formation, has been achieved by the use of plant mutants defective in mycorrhiza formation. These can be analyzed in comparison to their parental wild-type strains to identify the genes and proteins responsible for the recognition and signaling events taking place during early stages of infection. Most of these mutants are legumes, originally investigated in their inability to form symbiosis with nitrogen-fixing bacteria. Interestingly, many of those mutants were also impaired in their ability to form symbiosis with AM fungi, displaying the so-called myc-phenotype (reviewed in Gianinazzi-Pearson et al. 1996; Gadkar et al. 2001; Marsh and Schultze 2001). The first *myc*-mutants studied were described in pea (Duc et al. 1989). From those, a mutant called P2 was found to have a genetic block leading to the failure in the epidermis penetration after appressoria formation. These appressoria were aberrant, consisting of an abnormally thick papilla with deposition of β-1,3 glucans (Gollote et al. 1993).

The study of nodulation/mycorrhization mutant plants has pointed out the many similarities between the *Rhizobium* and the mycorrhiza symbiosis with the recent discovery of shared genes involved in the formation of both mutualistic associations (Stracke et al. 2002). These authors have shown that a plant receptor-like kinase is required for the perception of symbiotic fungi and bacteria. The mutant plants for these genes are unable to develop the signal cascade, leading to the rapid activation of downstream genes required for intracellular microbial accommodation and symbiosis establishment. These genetic studies have corroborated earlier morphological data and pointed out the active role of the epidermis as a checkpoint for mycorrhizal colonization (Wegel 1998; Bonfante et al. 2000; Genre and Bonfante 2002; Novero et al. 2002). Phenotype comparison of mutant alleles of the plant locus responsible of symbiont signal perception have shown that plant responses to the fungus are cell layer dependent (Novero et al. 2002). This corroborates earlier findings regarding sequential nodulin induction by mycorrhizal colonization. Thus, in *Pisum sativum*, the nodulin PsENOD12 is transiently induced during appressoria formation and epidermis penetration. Later on, the nodulin PsENOD5 is accumulated during cortical colonization (Albrecht et al. 1998). Both genes are not induced by cytokinin, in contrast to nodulins MsENOD2 and MsENOD40, which are mycorrhiza- and cytokinin-induced in *Medicago sativa* (Van Rhijn et al. 1997). Interestingly, none of these nodulins are induced when the plant is elicited with fungal pathogens such as *Rhizoctonia solani* or *Fusarium oxysporum* (Scheres et al. 1990; Van Rhijn et al. 1997). This suggests their specific involvement in a plant recognition signal cascade specific to mutalistic symbioses.

Both targeted and non-targeted molecular approaches have been extensively used to analyze these mycorrhizal plant mutants and find downstream gene markers of the colonization process (Martin-Laurent et al. 1997; Lapopin et al. 1999; Franken and Requena 2001a,b; Roussel et al. 2001). The combined use of mutant plants unable to form mycorrhiza and reporter genes such as gusA has also enabled mycorrhizal researchers to more precisely determine the place of expression of these early genes. Candidate genes specifically expressed during different stages of nodulation have been tagged during mycorrhization. Thus, MtENOD11, a repetitive proline-rich protein possibly playing a role in the modification of cell wall plasticity, has been shown to be transiently induced in epidermal cells in contact with appressoria (Chabaud et al. 2002). The authors suggest that the involvement of this protein in the cell wall changes is necessary for hyphal penetration.

As seen above, legumes have led the research in mycorrhization by the use of mutant strains. However, those mutants were originally isolated by their inability to nodulate. Therefore, it is clear that there must exist other genes essential to the mycorrhizal symbiosis and not shared by the nodule symbiosis. Mutations unique to early stages of AM symbiosis can be expected at the developmental stages of pre-infection, hyphal branching and appressoria formation (Gadkar et al. 2001). Some of those mutations have been already envisaged in tomato or maize and are currently investigated (Barker et al. 1998; David-Schwartz et al. 2001; Paszkowski et al. 2001).

3.4 Intraradical Colonization: Coils and Arbuscules

Once the fungus has overcome the epidermis, it starts developing into the cortex and the way it does this depends mainly on the plant species but also on the fungal genome (Smith and Smith 1997). Two modes of colonization structures have been observed (Arum and Paris types) and were first described and named after the plants in which they developed: *Arum maculatum* and *Paris quadrifolia* (Gallaud 1905). In the Arum type, the fungus spreads intercellularly growing between the loose cells of the outer root cortex. Hyphal coils are also sometimes formed in the hypodermis and exodermis, but the fungus progresses quickly toward the inner cortex where it grows intracellularly. There it develops the so-called arbuscules by dichotomous profuse branching from a trunk hypha. Arbuscules are specialized haustoria where the nutrient exchange between symbiotic partners is hypothesized to occur. In contrast, in the Paris type the intercellular phase is absent. The fungus spreads from cell to cell crossing radial cell walls and forming intracellular coils. Arbuscules are formed as intercalary structures between coil hyphae. There are many reports of Paris-mycorrhiza where arbuscules are completely absent. This represents a current matter of debate because it opens the possibility that arbuscules may not be the only interface for nutrient exchange (Smith and Smith 1990).

Arbuscules are ephemeral structures with a short life span of 2 to 5 days, after which they collapse and leave the plant cell intact, able to harbor another arbuscule (Alexander et al. 1988, 1989). New arbuscules are formed in younger sections of the root. When arbuscules senesce the fibrillar material from the apoplastic interface encapsulates the fungal structures (Gianinazzi-Pearson 1996). The morphological characteristics of the arbuscule make it an excellent location for nutrient exchange. The profuse dichotomous fungal branching is accompanied by a parallel development of the plant plasma membrane, which invaginates surrounding the hypha and giving rise to the peri-arbuscular membrane (PAM). This increases dramatically (three- to seven-fold membrane increase) the surface of contact between symbionts (Alexander et al. 1989). The PAM, although not physically separated from the peripheral plasma membrane of the plant, appears to show significant differences which might be indicators of its new function. Thus, a higher ATPase activity was first demonstrated by immunocytochemistry (Gianinazzi-Pearson et al. 1991) and has been recently supported by molecular methods showing up-regulation of H^+-ATPase encoding genes (Murphy et al. 1997; Gianinazzi-Pearson et al. 2000). This, together with a developmental induction of a fungal plasma membrane H^+-ATPase isozyme during the *in planta* phase (Requena et al. 2003), might explain the acid nature of the apoplastic compartment inhabited by arbuscules (Guttenberger 2000). The consequences of the subsequent increase in the membrane electrochemical gradient for the nutrient exchange are discussed later. The PAM has been observed to present similarities to the peribacteroid membrane surrounding *Rhizobium* bacteria in the nodules. Thus, seven from eight monoclonal antibodies raised against specific epitopes of pea nodules also recognize the interfacial membrane surrounding arbuscules (Perotto et al. 1994). This again is pointing out the many similarities between both root symbioses.

Other major changes taking place at the arbuscule interface are the progressive thinness of the fungal cell wall with the subsequent hyphal branching and the unstructured nature of the plant cell wall. This one remains unorganized, although new cell wall material is constantly deposited at the newly created apoplastic space (Bonfante and Perotto 1995). This situation leaves the plasma membrane of both symbionts separated only by an amorphous matrix of plant origin possibly facilitating the exchange of phosphate and/or carbon. Molecular genetic evidence of the plant cell wall remodeling at the arbuscule interface has accumulated in the last years. Thus, a comparative genetic analysis between mycorrhizal and non-mycorrhizal *Medicago truncatula* plants, allowed van Buuren et al. (1999) to isolate an induced gene coding for putative arabinogalactan protein. The expression of this gene in non-mycorrhizal plants was almost not detectable and only transiently induced 15 days after inoculation with the AM fungus *Glomus versiforme*. In situ hybridization experiments showed that it localizes exclusively in arbuscule-containing cells. This location and the putative function of the encoded protein corroborates earlier immunological data showing the presence of arabinosylated β-1,6-galactan epitopes in the PAM (Balestrini et al. 1996). Another gene encoding a cell wall remodeling protein, xyloglucan endotransglycosylase-related protein, was also found up-regulated in the same study. The gene although expressed in control roots was highly induced upon mycorrhization. Xyloglucans are responsible for the cross-linking of cellulose microfibrils in the cell wall and therefore xyloglucan endotransglycosylase enzymes are likely to play an important role in cell wall remodeling (Fry et al. 1992). The induction upon mycorrhization suggests that it could play a role in the loosening of the plant cell wall to assist fungal penetration (van Buuren et al. 1999). Another example of up-regulation of cell wall remodeling enzymes during arbuscule formation is the nodulin ENOD11 from *M. truncatula*. This repetitive proline-rich protein was localized with the help of the GUS reporter gene in inner cortical cells containing recently formed arbuscules (Journet et al. 2001). As we have seen above, this cell wall remodeling enzyme also has a role during earlier colonization steps. The authors suggest that in this case it could help to accommodate the developing arbuscule or to elaborate the arbuscular matrix interface.

Besides changes in the PAM and the cell wall, other plant cell components suffer major rebuilding during arbuscule formation. The cytoskeleton reorganizes completely possibly to allow organelles to accommodate the growing fungus (see also above). The central vacuole fragments and gives rise to many small vacuoles. A tonoplast aquaporin-encoding gene has been found to be induced upon mycorrhiza formation, most likely in arbuscule-containing cells (Krajinski et al. 2000). Tonoplast aquaporins are proteins involved in the control of water transport from the vacuole to the cytosol. The authors discussed that the up-regulation could be a compensation mechanism to a decreased vacuolar volume in arbuscule-containing cells.

Much less is known about fungal genes or proteins specifically expressed during this stage. The small amount of fungal material and the inaccessibility of the arbuscule in the inner cortex hinder the isolation and localization of fungal components necessary for the morphological development into arbuscule. It is to be expected that in the next ten years large-scale sequencing approaches and more refined cytological techniques will shed more light on the fungal side.

3.5 Defense Reactions Elicited During AM Cortex Colonization

Plants have evolved to defend themselves from the colonization by other organisms by exhibiting a set of cell defense responses which might even involve the sacrifice and death of some cells to guard from pathogen entry (Dangl et al. 1996). This integrated response includes the reinforcement of cell walls, the production of low-molecular-weight antimicrobial phytoalexins, and accumulation of antimicrobial molecules such as the PR (pathogenesis related) proteins (Somssich and Hahlbrock 1998). This implies the recognition of signals generated by the pathogen (elicitors) and the translation into an ordered sequence of defense actions against the intruder. During the colonization of the root cortex by AM fungi several signs of this defense response have been observed. This response appears to be again cell layer dependent, and it can be distinguished between early and late defense responses. In any case, AM fungi fail to elicit a strong defense and it is often transient and uncoordinatedly expressed (see reviews by Gianinazzi-Pearson et al. 1996; García-Garrido and Ocampo 2002). Interestingly, while in non-host plant species the fungal attempts to penetrate the root produced a hypersensitive-like response (Allen et al. 1989), *myc*-mutants respond to penetration attempts with a stronger defense reaction, including callose deposition, PR1 protein and phenolics accumulation (Gollote et al. 1993). In these mutants, accumulation of defense transcripts was observed to be higher than in the corresponding wild type when challenged with AM fungi (Ruiz-Lozano et al. 1999). These reactions compared with the weak response observed in compatible hosts suggest that mycorrhization represses a full expression of the defense response. However, it is unknown whether this happens in response to a passive recognition of AM fungi as friendly colonizing guests, or by contrast AM fungi actively repress this defense response.

General defense reactions observed during early colonization include induction of a chalcone synthase expression, first enzyme of the phytoalexin biosynthesis (Bonamoni et al. 2001); an oxidative burst at the cortical cell penetration sites (Salzer et al. 1999); necrosis points at the colonization sites (Douds et al. 1998); and transient increases in catalase, chitinase and peroxidase activities (Spanu and Bonfante-Fasolo 1988; Blilou et al. 2000a). Blilou et al. (2000b) also showed, using GUS-promoter fusion, a transient induction of a specific plant lipid transferase gene (Ltp b1) in rice during appressoria formation and epidermis colonization. Concomitant to this induction they also observed a transient increase in the expression of the phenylalanine ammonia-lyase (Pal) gene and accumulation of salicylic acid giving further evidence of the activation of the signal cascade activating plant defense reaction. However, these responses are all subsequently repressed as the symbiosis progresses and possibly reflect that AM fungi only produce a partial elicitation of a general plant defense response (Gianinazzi-Pearson et al. 1996).

During later stages a general depression of the defense reaction takes place and this is more localized to cells harboring fungal structures. Arbuscule-containing cells have been shown to accumulate proteins related to plant defense such as HPRP (hydroxyproline-rich glycoproteins), phenylpropanoid metabolites, plant hydrolases or enzymes involved in the metabolism of reactive oxygen species (Balestrini et al. 1997; Harrison and Dixon 1993; Volpin et al. 1995; Blee and Anderson 1996; Lambais and Mehdy 1998; Salzer et al. 2000). The activation of the phenylpropanoid metabolism seems to be weak, localized and not universal. Hosts such as soybean or *Medicago* exhibited it (Harrison and Dixon 1993; Morandi et al. 1984; Volpin et al. 1995), but it has not been observed in others, such as bean, parsley or potato (Lambais and Mehdy 1993; Franken and Gnädinger 1994). Plant hydrolases such as chitinases, chitosanases and β-1,3-glucanases have been shown to be regulated in response to mycorrhization (Blee and Anderson 1996; Lambais and Mehdy 1998), and even the induction of new isoforms of chitinases, chitosanases and β-1,3-glucanases in response to mycorrhization has been documented at the protein level (Pozo et al. 1996, 1998, 1999).

Interestingly, factors such as phosphate levels and infectivity of the fungal isolate employed have been shown to regulate the expression level of some of these enzymes (Lambais and Mehdy 1996, 1998). However, elegant studies using transgenic tobacco plants constitutively expressing several hydrolases showed that mycorrhizal colonization did not stop mycorrhiza development while inhibiting growth of other pathogens (Vierheilig et al. 1993, 1995). It is known that mycorrhizal plants often offer a higher resistance to pathogen infection (Davis and Menge 1980; Cordier et al. 1998). The local and systemic defense mediated by mycorrhiza against *Phytophthora* (Pozo et al. 2002) has been attributed to the induction of new hydrolytic enzymes along with superoxide dismutase, a protector against oxidative stress. Therefore, it is likely that hydrolytic enzyme induction by mycorrhiza might have additional or different roles as a classic plant defense response. A hypothesis proposed by Salzer et al. (2000) suggests the involvement of these hydrolytic enzymes in the destruction of AM fungal elicitors (chitin or β-1,3-glucan from the cell wall) to avoid a full defense response.

3.6 The External Mycelium

The symbiosis is not completed without the formation of the external or extraradical mycelium. This hyphal network links the colonized areas of the root with the soil matrix from where mineral nutrients are taken up and translocated towards the plant. Not less important, this mycelium represents a major sink for carbon and other elements withdrawn from the plant what makes it an essential component of the soil microbiota (Dodd 1994). The extraradical mycelium is also able to colonize new roots as well as forming the resting propagules, the spores. Friese and Allen (1991) described two forms of hyphae growing out of colonized roots, the runner hyphae and the absorptive hyphae. Runner hyphae spread as long single strands with angular projections, that either re-colonize the same root or grow into the soil a couple of cm until encountering another root. The absorptive hyphal network consists of a series of dichotomously branching hyphae that develop into a fan-shaped network. These absorptive structures are short-lived and die after 5 to7 days. They have been also observed in in vitro mycorrhizal cultures and named "branched absorbing structures" (BAS; Bago et al. 1998).

Besides the main absorptive function, an important role in soil aggregation and erosion control has been attributed to the extraradical mycelium (Bethlenfalvay et al. 1998). The hyphal network connecting soil particles is possibly facilitated by the production of AM fungal specific glycoproteins. Glomalins contribute to increase soil particle hydrophobicity (Wright and

Upadhyaya 1996). In this way, water-stable soil aggregates can be formed helping to maintain soil structure and therefore plant re-establishment and ecosystem restoration (Requena et al. 2001).

4 The Bi-Directional Nutrient Exchange

The bi-directional nutrient transport between plant and fungus is a fundamental issue in the mycorrhizal mutualistic association. During symbiosis, the fungus receives from the plant a substantial amount of the fixed carbon allocated to the root, calculated to be up to 20% of the photoassimilated carbon (Jakobsen 1995). This carbon, delivered through the extraradical hyphae mainly in form of lipids (Bago et al. 2002), expands into the soil system where it can be later used by the rest of soil microbiota. In turn, the plant improves its mineral status (mainly in phosphate) due to the ability of the fungal external mycelium to extend beyond the nutrient depletion area surrounding the root (Jakobsen 1995). A fundamental question to the study of the mycorrhizal symbiosis is where and how the nutrient exchange between symbionts takes place. There is increasing evidence that phosphate is taken up at the extraradical mycelium by a specific high affinity phosphate transporter (Harrison and van Buuren 1995; Maldonado-Mendoza et al. 2001), and translocated from the soil through the fungus to be downloaded at the arbuscule interface, where it is taken up by plant transporters (Rausch et al. 2001; Harrison et al. 2002). However, little is known about the location where the exchange of carbon takes place. Besides the arbuscule, inter- or intracellular hyphae (i.e. coils) formed in upper cortical cells cannot be ruled out as locations for the nutrient exchange (Smith and Smith 1997).

A main issue is, of course, the form of carbon to be transported at these interfaces. The literature in this respect indicates that glucose or fructose could be the main form of carbohydrate that the fungus imports from the apoplastic space (Saito et al. 1995; Shachar-Hill et al. 1995; Pfeffer et al. 1999). It is known that both, phosphate as well as hexoses, are usually translocated by means of symporters which are supported by an electrochemical gradient in the plasma membrane created by H^+-ATPase enzymes. Therefore, these H^+-ATPases are likely to play an important role both at the plant and at the fungal symbiotic interfaces where either phosphate or carbon are translocated. In plants, H^+-ATPases form a large gene family whose members are either transcriptionally or/and posttranscriptionally regulated at different stages of plant development. A recent paper has shown that at least two of these plant H^+-ATPase isoforms are involved in the interaction with symbiotic mycorrhizal fungi (Gianinazzi-Pearson et al. 2000). Using GUS-fused promoters the authors showed gene induction for these two isoforms in arbuscule-containing cells, suggesting a possible involvement in the phosphate uptake. For the fungal membrane where carbon has to be taken up from the apoplastic space, much less is known. The nature of the carbon symporter is completely unknown and no published information exists about the proton pumps involved. In our laboratory, we have identified two fungal H^+-ATPases which are developmental and nutrient regulated at the transcriptional level. One of these H^+-ATPase genes is preferentially expressed during pre-symbiotic growth and in intraradical hyphae, while the other is specifically induced upon formation of the symbiosis and in response to phosphate (Requena et al. 2003). Our results indicate that, similarly to what happens in plants, AM fungi possibly recruit different proton pump isozymes at different developmental or nutritional stages. Our next goal is to localize these proton pumps at the cellular level during mycorrhization and to establish their role in the energization of the plasma membrane at the crucial symbiotic interfaces.

With regard to the uptake and translocation of nitrogen towards the plant much less is known. Although in a lower proportion than phosphorus, nitrogen is also taken up from the soil by the fungus and transferred to the host in substantial proportions. In contrast, the transfer of nitrogen from the host plant to the fungus and hence towards the soil is insignificant (Johansen et al. 1992, 1993). Therefore, in soils where the AM fungi are undisturbed they play an important role in the turnover of inorganic N by competing efficiently with other soil microorganisms (Johansen et al. 1996). The nitrogen forms taken up from the soil by the AM extraradical mycelium are variable. Ammonium seems to be the preferred nitrogen source but uptake of nitrate and amino acids has also been shown (Johansen et al. 1992, 1994; Tobar et al. 1994; Hawkins et al. 2000). In particular, nitrate uptake might be of great significance in dry soils (Tobar et al. 1994). In contrast to the knowledge about the forms of nitrogen used by AM fungi, there has not been any nitrogen transporter isolated from these fungi yet. Once inside the fungal cell, this nitrogen is possibly assimilated to satisfy internal demands prior to its transfer to the host plant. Asparagine, arginine and glutamine are the most abundant amino acids present in extraradical mycelium of AM fungi (Johansen et al. 1996; Bago et al. 1999). It is possible that the translocation of nitrogen towards the plant takes place in the form of these amino acids. Previous experiments using enzymatic methods have shown that glutamine synthetase (GS) is possibly the main enzyme responsible for N assimilation in these fungi (Smith et al. 1985). In our laboratory, we have isolated the first gene coding for GS from the AM fungus *Glomus mosseae* (Breuninger, Trujillo and Requena, unpubl. data) and we are currently characterizing it to ascertain its role in the nitrogen assimilation during symbiosis.

5 Conclusions and Perspectives

We have tried to summarize in this chapter some of the main achievements from the last years in the field of AM symbiosis in which the molecular methods have played an important role. Many new results are surely coming and the interaction between different disciplines such as biochemistry, cell biology, molecular biology and ecology will be absolutely essential to understand this complex but fascinating interaction.

References

Abbas JD, Hetrick, BAD, Jurgensen JE (1996) Isolate specific detection of mycorrhizal fungi using genome specific primer pairs. Mycologia 88:939–946

Abbott LK (1983) Comparative anatomy of vesicular-arbuscular mycorrhizae formed on subterranean clover. Aust J Bot 30:485–495

Albrecht C, Geurts R, Lapeyrie F, Bisseling T (1998) Endomycorrhizae and rhizobial Nod factors both require SYM8 to induce the expression of the early nodulin genes PsENOD5 and PsENOD12A. Plant J 15:605–614

Alexander T, Meier R, Toth R, Weber HC (1988) Dynamics of arbuscule development and degeneration in mycorrhizas of *Triticum aestivum* L. and *Avena sativa* L. with reference to *Zea mays* L. New Phytol 110:363–370

Alexander T, Meier R, Toth R, Weber HC (1989) Dynamics of arbuscule development and degeneration in onion, bean and tomato with reference to vesicular-arbuscular mycorrhizas with grasses. Can J Bot 67:2505-2513

Allen MF, Allen EB, Friese CF (1989) Responses of the non-mycotrophic plant *Salsola kali* to invasion by vesicular-arbuscular mycorrhizal fungi. New Phytol 111:45–49

Antonionelli ZI, Schachtman DP, Ophel-Keller K, Smith SE (2000) Variation in rDNA ITS sequences in *Glomus mosseae* and *Gigaspora margarita* spores from a permanent pasture. Mycol Res 104:708–715

Aström H, Giovannetti M, Raudaskoski M (1994) Cytoskeletal components in the arbuscular mycorrhizal fungus *Glomus mosseae*. Mol Plant Microbe Interact 7:309–312

Bago B, Bentivenga SP, Brenac V, Dodd JC, Piche Y, Simon L (1998) Molecular analysis of *Gigaspora* (Glomales, Gigasporaceae). New Phytol 139:581–588

Bago B, Pfeffer PE, Douds DD Jr, Brouillette J, Becard G, Shachar-Hill Y (1999) Carbon metabolism in spores of the arbuscular mycorrhizal fungus *Glomus intraradices* as revealed by nuclear magnetic resonance spectroscopy. Plant Physiol 121:263–272

Bago B, Zipfel W, Williams RM, Jun J, Arreola R, Lammers PJ, Pfeffer PE, Shachar-Hill Y (2002) Translocation and utilization of fungal storage lipid in the arbuscular mycorrhizal symbiosis. Plant Physiol 128:108–124

Balestrini R, Romero C, Puigdoménech P, Bonfante P (1994) Location of a cell-wall hydroxyproline-rich glycoprotein, cellulose and (β-1,3-glucans in apical and differentiated regions of maize mycorrhizal roots. Planta 195:2016209

Balestrini R, Hahn MG, Faccio A, Mendgen K, Bonfante P (1996) Differential localization of carbohydrate epitopes in plant cell walls in the presence and absence of arbuscular mycorrhizal fungi. Plant Physiol 111:203–213

Balestrini R, José-Estanyol M, Puigdoménech P, Bonfante P (1997) Hydroxyproline-rich glycoprotein mRNA accumulation in maize root cells colonized by an arbuscular mycorrhizal fungus as revealed by in situ hybridization. Protoplasma 198:36–42

Barea JM, Jeffries PJ (1995) Arbuscular mycorrhizas in sustainable soil-plant systems. In: Varma A, Hock B (eds) Mycorrhiza. Springer, Berlin Heidelberg New York, pp 521–560

Barker SJ, Tagu D, Delp G (1998) Regulation of root and fungal morphogenesis in mycorrhizal symbiosis. Plant Physiol 116:1201–1207

Bécard G, Piché Y (1989a) New aspects on the acquisition of biotrophic status by a vesicular-arbuscular mycorrhizal fungus, *Gigaspora margarita*. New Phytol 112:77–83

Bécard G, Piché Y (1989b) Fungal growth stimulation by CO_2 and root exudates in vesicular-arbuscular mycorrhizal symbiosis. Appl Environ Microb 55:2320–2325

Bécard G, Taylor LP, Douds DD, Pfeffer PE, Doner LW (1995) Flavonoids are not necessary plant signal compounds in arbuscular mycorrhizal symbioses. Mol Plant Microbe Interact 8:252–258

Bethlenfalvay GJ, Cantrell IC, Mihara KL, Schreiner RP (1998) Relationships between soil aggregation and mycorrhizae as influenced by soil biota and nitrogen nutrition. Biol Fertil Soils 28:356–363

Bianciotto V, Bonfante P (1993) Evidence of DNA replication in an arbuscular mycorrhizal fungus in the absence of the host plant. Protoplasma 176:100–105

Bianciotto V, Barbiero G, Bonfante P (1995) Analysis of the cell cycle in an arbuscular mycorrhizal fungus by flow cytometry and bromodeoxyuridine labeling. Protoplasma 188:161–169

Blancaflor EB, Zhao LM, Harrison MJ (2001) Microtubule organization in root cells of *Medicago truncatula* during development of an arbuscular mycorrhizal symbiosis with *Glomus versiforme*. Protoplama 217:154–165

Blee KA, Anderson AJ (1996) Defense related transcript accumulation in *Phaseolus vulgaris* L. colonized by the arbuscular mycorrhiizal fungus *Glomus intraradices*. Schenk and Smith. Plant Physiol 110:675–688

Blilou I, Bueno P, Ocampo JA, García-Garrido JM (2000a) Induction of catalase and ascorbate peroxidase activities in tobacco roots inoculated with the arbuscular mycorrhizal fungus *Glomus mosseae*. Mycol Res 104:722–725

Blilou I, Ocampo JA, García-Garrido JM (2000b) Induction of Ltp (Lipid tranfer protein) and Pal (phenylalanine ammonia-lyase) gene expression in rice roots colonized by the arbuscular mycorrhizal fungus *Glomus mosseae*. J Exp Bot 51:1969–1977

Bonamoni A, Oetiker JH, Guggenheim R, Boller T, Wiemken A, Vögeli-Lange R (2001) Arbuscular mycorrhizas in mini-mycorrhizotrons:first contact of *Medicago truncatula* roots with *Glomus intraradices* induces chalcone synthase. New Phytol 150:573–582

Bonfante P, Bianciotto V (1995) Presymbiontic versus symbiontic phase in arbuscular endomycorrhizal fungi:morphology and cytology. In: Varma A, Hock B (eds) Mycorrhiza. Springer, Berlin Heidelberg New York, pp 229–247

Bonfante P, Perotto S (1992) The cellular basis of plant-fungus interchanges in mycorrhizal associations. In: Allen MG (ed) Functioning of mycorrhizae. Chapman and Hall, New York, pp 65–101

Bonfante P, Perotto S (1995) Strategies of arbuscular mycorrhizal fungi when infecting host plants. New Phytol 130:3–21

Bonfante P, Bergero R, Uribe X, Romera C, Rigau J, Puigdomenech P (1996) Transcriptional activation of a maize alpha-tubulin gene in mycorrhizal maize and transgenic tobacco plants. Plant J 9:737–743

Bonfante P, Genre A, Faccio A, Martinin I, Schauser L, Stougaard J, Webb J, Parniske M (2000) The *Lotus japonicus LjSym4* gene is required for the successful symbiotic infection of root epidermal cells. Mol Plant Microbe Interact 13:1109–1120

Buee M, Rossignol M, Jauneau A, Ranjeva R, Becard G (2000) The pre-symbiotic growth of arbuscular mycorrhizal fungi is induced by a branching factor partially purified from plant root exudates. Mol Plant Microbe Interact 13:693–698

Chabaud M, Venard C, Defaux-Petras A, Bécard G, Barker DG (2002) Targeted inoculation of *Medicago truncatula* in vitro root cultures reveals *MtENOD11* expression during early stages of infection by arbuscular mycorrhizal fungi. New Phytol 156:265–273

Chelius M, Triplett E (1999) Rapid detection of arbuscular mycorrhizae in roots and soil of an intensively managed turfgrass system by PCR amplification of small subunit rDNA. Mycorrhiza 9:61–64

Clapp JP, Young JPW, Merryweather JW, Fitter AH (1995) Diversity of fungal symbionts in arbuscular mycorrhizas from a natural community. New Phytol 130:259–265

Clapp JP, Fitter AH, Young JPW (1999) Ribosomal small sunbunit sequence variation within spores of an arbuscular mycorrhizal fungus, *Scutellospora* sp. Mol Ecol 8:915–921

Clapp JP, Rodriguez A, Dodd JD (2001) Inter- and intra-isolate rRNA large subunit variation in *Glomus coronatum* spores. New Phytol 149:539–554

Cordier C, Gianinazzi-Pearson V, Gianinazzi S (1996) An immunological approach for the study of spatial relationships between mycorrhizal fungi in planta. In: Azcón-Aguilar C, Barrea JM (eds) Mycorrhizas in integrated systems:from genes to plant development. European Commision, EUR 16728, Luxembourg, pp 25–30

Cordier C, Pozo MJ, Barea JM, Gianinazzi S, Gianinazzi-Pearson V (1998) Cell defense responses associated with localized and systemic resistance to *Phytophthora* induced in tomato by an arbuscular mycorrhizal fungus. Mol Plant Microbe Interact 11:1017–1028

Dangl JL, Dietrich RA, Richberg MH (1996) Death don´t have no mercy:cell death programs in plant-microbe interactions. Plant Cell 8:1793–1807

David-Schwartz R, Badani H, Smadar W, Levy AA, Galili G, Kapulnik Y (2001) Identification of a novel genetically controlled step in mycorrhizal colonization:plant resistance to infection by fungal spores but not extra-radical hyphae. Plant J 27:561–569

Davis RM, Menge JA (1980) Influence of *Glomus fasciculatus* and soil phosphorus on *Phytophthora* root rot of citrus. Phytopathology 70:447–452

Di Bonito R, Elliott ML, Jardin EA (1995) Detection of an arbuscular mycorrhizal fungus in roots of different plant species with the PCR. Appl Environ Microbiol 61:2809–2810

Dodd JC (1994) Approaches to the study of the extraradical mycelium of arbuscular mycorrhizal fungi. In: Gianinazzi S, Schüepp H (eds) Impact of arbuscular mycorrhizas on sustainable agriculture and natural ecosystems. Birkhäuser, Basel, pp 147–166

Douds DD, Galvez L, Bécard G, Kapulnik K (1998) Regulation of arbuscular mycorrhizal development by plant host and fungus species in alfalfa. New Phytol 138:27–35

Duc G, Trouvelot A, Gianinazzi-Pearson V, Gianinazzi S (1989) First report of non-mycorrhizal plant mutants (Myc^-) obtained in pea (*Pisum sativum* L.) and fababean (*Vicia faba* L.). Plant Sci 60:215–222

Franken P, Gnädinger F (1994) Analysis of parsley arbuscular endomycorrhizal: infection development and mRNA levels of defense related genes. Mol Plant Microbe Interact 7:612–620

Franken P, Requena N (2001a) Mol ecular analyisis of arbuscular mycorrhiza. In: Hock B (ed) The mycota, vol IX. Springer, Berlin Heidelberg New York, pp 19–28

Franken P, Requena N (2001b) Analysis of gene expression in arbuscular mycorrhizas:new approaches and challenges. New Phytol 150:517–523

Friese CF, Allen MF (1991) The spread of VA mycorrhizal fungi in the soil:inoculum types and external hyphal architecture. Mycologia 83:409–418

Fry SC, Smith RC, Renwick KF, Martin DJ, Hodge SK, Matthews KJ (1992) Xyloglucan endotransglycosylase, a new wall-loosening enzyme activity from plants. Biochem J 282:821–828

Gadkar V, David-Schwartz R, Kunik T, Kapulnik Y (2001) Arbuscular mycorrhizal fungal colonization. Factors involved in host recognition. Plant Physiol 127:1493–1499

Gallaud I (1905) Études sur les mycorrhizes endotrophs. Rev Générale Bot 17:5–48, 66–83, 123–135, 223–239, 313–325, 425–433, 479–500

García-Garrido JM, Ocampo JA (2002) Regulation of the plant defence response in arbuscular mycorrhizal symbiosis. J Exp Bot 53:1377–1386

Garcia-Garrido JM, Cabello MN, Garcia-Romero I, Ocampo JA (1992a) Endonuclease activity in letucce plants colonized with the vesicular-arbuscular mycorrhizal fungus *Glomus fasciculatum*. Soil Biol Biochem 24:955–959

Garcia-Garrido JM, Garcia-Romero I, Ocampo JA (1992b) Cellulase production by the vesicular arbuscular mycorrhizal fungus *Glomus mosseae*. New Phytol 121:221–226

Garcia-Garrido JM, Garcia-Romero I, Parra-Garcia MD, Ocampo JA (1996) Purification of an arbuscular mycorrhizal endoglucanase from onion roots colonized by *Glomus mosseae*. Soil Biol Biochem 25:1443–1449

Garcia-Romera I, Garcia-Garrido JM, Ocampo JA (1991) Pectinolytic enzymes in vesicular-arbuscular mycorrhizal fungus *Glomus mosseae*. FEMS Microbiol Lett 78:343–346

Garriock ML, Peterson RL, Ackerley CA (1989) Early stages in colonization of *Allium porrum* (leek) roots by the vesicular-arbuscular mycorrhizal fungus *Glomus versiforme*. New Phytol 112:85–92

Gehrig H, Schüßler A, Kluge M (1996) *Geosiphon pyriformis*, a fungus forming endocytobiosis with *Nostoc* (Cyanobacteria), is an ancestral member of the Glomales:evidence by SSU rRNA analysis. J Mol Evol 43:71–81

Genre A, Bonfante P (1997) A mycorrhizal fungus changes microtubule orientation in tobacco root cells. Protoplasma 199:30–38

Genre A, Bonfante P (1998) Actin versus tubulin configuration in arbuscule-containing cells from mycorrhizal tobacco roots. New Phytol 140:745–752

Genre A, Bonfante P (2002) Epidermal cells of a symbiosis-defective mutant of *Lotus japonicus* show altered cytoskeleton organisation in the presence of a mycorrhizal fungus. Protoplasma 219:43–50

Gianinazzi-Pearson V (1996) Plant cell responses to arbuscular mycorrhizal fungi:getting to the roots of the symbiosis. Plant Cell 8:1871–1883

Gianinazzi-Pearson V, Branzanti B, Gianinazzi S (1989) In vitro enhancement of spore germination and early hyphal growth of a vesicular-arbuscular mycorrhizal fungus by host root exudares and plant flavonoids. Symbiosis 7:243–255

Gianinazzi-Pearson V, Smith SE, Gianinazzi S, Smith FA (1991) Enzymatic studies on the metabolism of vesicular-arbuscular mycorrhizas: V. H^+- ATPase a component of ATP-hydrolyzing enzyme activities in plant-fungus interfaces? New Phytol 117:61–74

Gianinazzi-Pearson V, Dumas-Gaudot E, Gollotte A, Tahiri-Alaoui A, Gianinazzi S (1996) Cellular and molecular defense-related root responses to invasion by arbuscular mycorrhizal fungi. New Phytol 133:45–57

Gianinazzi-Pearson V, Arnould C, Oufattole M, Arango M, Gianinazzi S (2000) Differential activation of H^+-ATPase genes by an arbuscular mycorrhizal fungus in root cells of transgenic tobacco. Planta 211:609–613

Giovannetti M, Avio L, Sbrana C (1993a) Factors affecting appressorium development in the vesicular-arbuscular mycorrhizal fungus *Glomus mosseae* (Nicol. & Gerd.) Gerd. & Trappe. New Phytol 123:114–122

Giovannetti M, Sbrana C, Avio L (1993b) Differential hyphal morphogenesis in arbuscular mycorrhizal fungi during preinfection stage. New Phytol 125:587–593

Giovannetti M, Sbrana C, Citernesi AS, Avio L, Gollote A, Gianinazzi-Pearson V, Gianinazzi S (1995) Recognition and infection process, basis for host specificity of AMF. In: Gianinazzi S, Schüepp H (eds) Impact of arbuscular mycorrhizas on sustainable agriculture and natural ecosystems. Birkhäuser, Basel, pp 61–72

Giovannetti M, Sbrana C, Citernesi AS, Avio L (1996) Analysis of factors involved in fungal recognition responses to host derived signals by arbuscular mycorrhizal fungi. New Phytol 133:65–71

Giovannetti M, Azzolini D, Citernesi AS (1999) Anastomosis formation and nuclear and protoplasmic exchange in arbuscular mycorrhizal fungi. Appl Environ Microbiol 65:5571–5575

Giovannetti M, Fortuna P, Citernesi AS, Morini S, Nuti MP (2001) The occurrence of anastomosis formation and nuclear exchange in intact arbuscular mycorrhizal networks. New Phytol 151:717–724

Giovannetti M, Sbrana C, Strani P, Agnolucci M, Rinaudo V, Avio L (2003) Genetic diversity of isolates of *Glomus mosseae* from different geographic areas detected by vegetative compatibility testing and biochemical and molecular analysis. Appl Environ Microbiol 69:616–624

Gollote A, Gianinazzi-Pearson V, Giovannetti M, Sbrana C, Avio L, Gianinazzi S (1993) Cellular localization and cytochemical probing of resistance reactions to arbuscular mycorrhizal fungi in a "locus a" mutant *Pisum sativum* L. Planta 191:112–122

Grippiolo R, Bonfante P (1994) Sporopollenin and melanin-like pigments in the wall of a *Glomus* spore. Giorn Bot Ital 118:88–90

Guttenberger M (2000) Arbuscules of vesicular-arbuscular mycorrhizal fungi inhabit an acid compartment within plant roots. Planta 211:299–304

Hahn A, Horn K, Hock B (1995) Serological properties of mycorrhizas. In: Varma A, Hock B (eds) Mycorrhiza. Springer, Berlin Heidelberg New York, pp 181–201

Harrison M, Dixon R (1993) Isoflavonoid accumulation and expression of defense gene transcripts during the establishment of vesicular arbuscular mycorrhizal associations in roots of *Medicago truncatula.* Mol Plant Microbe Interact 6:643–659

Harrison MJ, van Buuren ML (1995) A phosphate transporter from the mycorrhizal fungus *Glomus versiforme.* Nature 378:626–629

Harrison MJ, Dewbre GR, Liu J (2002) A phosphate transporter from *Medicago truncatula* involved in the acquisition of phosphate released by arbuscular mycorrhizal fungi. Plant Cell 14:2413–2429

Hawkins HJ, Johansen A, George E (2000) Uptake and transport of organic and inorganic nitrogen by arbuscular mycorrhizal fungi. Plant Soil 226:275–285

Helgason T, Daniell TJ, Husband R, Fitter AH, Young YPW (1998) Ploughing up the wood-wide web? Nature 394:431

Hepper CM (1979) Germination and growth of *Glomus caledonius* spores:the effect of inhibitors and nutrients. Soil Biol Biochem 2:269–277

Hepper CM, Sen R, Maskall CS (1986) Identification of vesicular-arbuscular mycorrhizal fungi in roots of leek (*Allium porrum* L.) and maize (*Zea mays* L.) on the basis of enzyme mobility during polyacrylamide gel electrophoresis. New Phytol 102:529–539

Hijri M, Hosny M, van Tuinen D, Dulieu H (1999) Intraspecific ITS polymorphism in *Scutellospora castanea* (Glomales, Zygomycota) is structured within multinucleate spores. Fungal Genet Biol 26:141–151

Hosny M, Hijri M, Passerieux E, Dulieu H (1999) rDNA units are highly polymorphic in *Scutellospora castanea* (Glomales, Zygomycetes). Gene 226:61–71

Jacquot D, van Tuinen D, Gianinazzi S, Gianinazzi-Pearson V (2000) Monitoring species of arbuscular mycorrhizal fungi *in planta* and in soil by nested PCR: application to the study of the impact of sewage sludge. Plant Soil 226:179–188

Jakobsen I (1995) Transport of phosphorus and carbon in VA mycorrhizas. In: Varma A, Hock B (eds) Mycorrhiza. Springer, Berlin Heidelberg New York, pp 297–324

Johansen A, Jakobsen I, Jensen ES (1992) Hyphal transport of ^{15}N-labelled nitrogen by a vesicular-arbuscular mycorrhizal fungus and its effect on depletion of inorganic soil N. New Phytol 122:61–68

Johansen A, Jakobsen I, Jensen ES (1993) External hyphae of vesicular-arbuscular mycorrhizal fungi associated with *Trifolium subterraneum* L. 3. Hyphal transport of ^{32}P and ^{15}N. New Phytol 124:61–68

Johansen A, Jakobsen I, Jensen ES (1994) Hyphal transport by vesicular-arbuscular mycorrhizal fungus on N applied to the soil as ammonium or nitrate. Biol Fertil Soils 16:66–70

Johansen A, Finlay RD, Olsson PA (1996) Nitrogen metabolism of external hyphae of the arbuscular mycorrhizal fungus *Glomus intraradices*. New Phytol 133:705–712

Journet E-P, El-Gachtouli N, Vernoud V, de Billy F, Pichon M, Dedieu A, Arnould C, Morandi D, Barker DG, Gianinazzi-Pearson V (2001) *Medicago truncatula* ENOD11:a novel RPRP-encoding early nodulin gene expressed during mycorrhization in arbuscule-containing cells. Mol Plant Microbe Interact 14:737–748

Kjøller R, Rosendahl S (2000) Detection of arbuscular mycorrhizal fungi (Glomales) in roots by nested PCR and SSCP (Single Stranded Conformation Polymorphism). Plant Soil 226:189–196

Kjøller R, Rosendahl S (2001) Mol ecular diversity of glomalean (arbuscular mycorrhizal) fungi dtermined as distinct *Glomus* -specific DNA sequences from roots of filed grown peas. Mycol Res 105:1027–1032

Kobayashi I, Kobayashi Y, Hardham AR (1994) Dynamic reorganization of microtubules and microfilaments in flax cells during the resistance response to flax rust infection. Planta 195:237–247

Kobayashi Y, Kobayashi I, Funaki Y, Fujimoto S, Takemoto T, Kunoh H (1997) Dynamic reorganization of microfilaments and microtubules is necessary for the expression of non-host resistance in barley coleoptile cells. Plant J 11:525–537

Koske RE (1981) Multiple germination by spores of *Gigaspora gigantea*. Trans Br Mycol Soc 76:320–330

Kozlova N, Strunnikova O, Labutova N, Muromtsev G (2001) Production and specificity of polyclonal antibodies against soluble proteins from the arbuscular mycorrhizal fungus *Glomus intraradices*. Mycorrhiza 10:301–305

Krajinski F, Alexander B, Schubert D, Gianinazzi-Pearson V, Kaldenhoff R, Franken P (2000) Arbuscular mycorrhiza development regulates the mRNA abundance of *Mtaqp1* encoding a mercury-intensitive aquaporin of *Medicago truncatula*. Planta 211:85–90

Kuhn G, Hijri M, Sanders IR (2001) Evidence for the evolution of multiple genomes in arbuscular mycorrhizal fungi. Nature 414:745–748

Lambais MR, Mehdy MC (1993) Suppression of endochitinase, (β-1,3-endoglucanase and chalcone isomerase expression in bean vesicular arbuscular mycorrhizal roots under different soil phosphate conditions. Mol Plant Microbe Interact 6:75–83

Lambais MR, Mehdy MC (1996) Soybean roots infected by *Glomus intraradices* strains differing in infectivity exhibit differential chitinase and (β-1,3-endoglucanase expression. New Phytol 134:531–538

Lambais MR, Mehdy MC (1998) Spatial distribution of chitinases and (β-1,3-glucanase transcripts in bean arbuscular mycorrhizal roots under low and high phosphate conditions. New Phytol 140:33–42

Lanfranco L, Perotto S, Bonfante P (1998) Application of PCR for studying the biodiversity of mycorrhizal fungi. In: Bridge PD, Arora DK, Reddy CA, Elander RP (eds) Application of PCR in mycology. CAB International, Wallingford, pp 107–124

Lanfranco L, Delpero M, Bonfante P (1999) Intrasporal variability of ribosomal sequences in the endomycorrhizal fungus *Gigaspora margarita*. Mol Ecol 8:37–45

Lapopin L, Gianinazzi-Pearson V, Franken P (1999) Comparative differential RNA display analysis of arbuscular mycorrhiza in *Pisum sativum* wild type and a mutant defective in late stage development. Plant Mol Biol 41:669–677

Lloyd-Macglip SA, Chambers SM, Dodd JC, Fitter AH, Walker C, Young YPW (1996) Diversity of the ribosomal internal spacers within and among isolates of *Glomus mosseae* and related mycorrhizal fungi. New Phytol 133:103–111

Logi C, Sbrana C, Giovanetti M (1998) Cellular events involved in survival of individual arbuscular mycorrhizal symbionts growing in the absence of the host. Appl Environ Microbiol 64:3473–3479

Longato S, Bonfante P (1997) Mol ecular identification of mycorrhizal fungi by direct amplification of microsatellite regions. Mycol Res 101:425–432

Maldonado-Mendoza IE, Dewbre GR, Harrison MJ (2001) A phosphate transporter gene from extra-radical mycelium of an arbuscular mycorrhizal fungus *Glomus intraradices* ie regulated in response to phosphate in the environment. Mol Plant Microbe Interact 10:1140–1148

Marsh JF, Schultze M (2001) Analysis of arbuscular mycorrhizas using symbiosis-defective plant mutants. New Phytol 150:525–532

Martin-Laurent F, van Tuinen D, Dumas-Gaudot E, Gianinazzi-Pearson V, Gianinazzi S, Franken P (1997) Differential display analysis of RNA accumulation in arbuscular mycorrhiza of pea and isolation of a novel symbiosis-regulated plant gene. Mol Gen Genet 256:37–44

Merryweather J, Fitter A (1998) The arbuscular mycorrhizal fungi of *Hyacinthoides non-scripta*:I. Diversity of fungal taxa. New Phytol 138:117–129

Millner PD, Mulbry WW, Reynolds SL, Patterson CA (1998) A taxon-specific oligonucleotide probe for temperate soil isolates of *Glomus mosseae*. Mycorrhiza 8:19–27

Millner PD, Mulbry WW, Reynolds SL (2001) Taxon-specific oligonucleotide primers for detection of *Glomus etunicatum*. Mycorrhiza 10:259–265

Morandi D, Balley JA, Gianinazzi-Pearson V (1994) Isoflavonoid accumulation in soybean roots infected with vesicular arbuscular mycorrhizal fungi. Physiol Plant Pathol 24:357–364

Morton JB, Benny GL (1990) Revised classification of arbuscular mycorrhizal fungi (Zygomycetes):a new order, Glomales, two new suborders Glomineae and Gigasporineae, and two new families, Acaulosporaceae and Gigasporaceae, with an endemedation of Glomaceae. Mycotaxon 37:471–491

Morton JB, Bentivenga SP (1994) Levels of diversity in endomycorrhizal fungi (Glomales, Zygomycetes) and their role in defining taxonomic and non taxonomic groups. Plant Soil 159:47–59

Morton JB, Redecker D (2001) Two new families of Glomales, Archaeosporaceae and Paraglomaceae, with two new genera, *Archaeospora* and *Paraglomus*, based on concordant molecular and morphological characters. Mycologia 93:185–195

Mosse B (1959) The regular germination of resting spores and some observations in the growth requirements of an *Endogone* sp. causing vesicular-arbuscular mycorrhiza. Trans Br Mycol Soc 42:273–286

Mosse B (1988) Some studies relating to "independent" growth of vesicular-arbuscular endophytes. Can J Bot 66:2533–2540

Murphy PJ, Langridge P, Smith SE (1997) Cloning plant genes differentially expressed during colonization of roots of *Hordeum vulgare* by the vesicular-arbuscular mycorrhizal fungus *Glomus intraradices*. New Phytol 135:291–301

Nagahashi G, Douds DD (1997) Appressorium formation by AM fungi on isolated cell walls of carrot roots. New Phytol 136:299–304

Nair MG, Safir GR, Siqueira JO (1991) Isolation and identification of vesicular-arbuscular mycorrhiza stimulatory compounds from clover (*Trifolium repens*) root. Appl Environ Microbiol 57:434–439

Novero M, Faccio A, Genre A, Stougaard J, Webb KJ, Mulder L, Parniske M, Bonfante P (2002) Dual requirement of the *LjSym4* gene for mycorrhizal development in epidermal and cortical cells of *Lotus japonicus* roots. New Phytol 154:741–749

Paszkowski U, Shi L, Bi-Yu Li, Xun Wang, Briggs S, Boller T (2001) A single gene mutation in the maize nopel mutant abolishes the recognition of arbuscular mycorrhizal fungi. In: 10th International Congress, Molecular Plant-Microbe Interactions, Madison, WI, July10–14, Poster abstract no 686

Perotto S, Brewin NJ, Bonfante P (1994) Colonization of pea roots by the mycorrhizal fungus *Glomus versiforme* and by *Rhizobium* bacteria:immunological comparison using monoclonal antibodies as probes for plant cell surface components. Mol Plant Microbe Interact 7:91–98

Peretto P, Bettini V, Favaron F, Alghisi P, Bonfante P (1995) Polygalacturonase activity and location in arbuscular mycorrrhizal roots of *Allium porrum* L. Mycorrhiza 5:157–163

Pfeffer PE, Douds DD Jr, Bécard G, Shachar-Hill Y (1999) Carbon uptake and the metabolism and transport of lipids in an arbuscular mycorrhiza. Plant Physiol 120:587–598

Pirozynski KA, Malloch DW (1975) The origin of land plants a matter of mycotrophism. Biosystems 6:153–164

Pozo MJ, Dumas-Gaudot E, Slezack S, Cordier C, Asselin A, Gianinazzi S, Gianinazzi-Pearson V (1996) Detection of new chitinase isoforms in arbuscular mycorrhizal tomato roots:possible implications in protection against *Phytophthora nicotianae* var. *parasitica*. Agronomie 16:689–697

Pozo MJ, Azcón-Aguilar C, Dumas-Gaudot E, Barea JM (1998) Chitosanase and chitinase activities in tomato roots during interactions with arbuscular mycorrhizal fungi or *Phytophthora parasitica*. J Exp Bot 49:1729–1739

Pozo MJ, Azcón-Aguilar C, Dumas-Gaudot E, Barea JM (1999) (β-1,3-Glucanase activities in tomato roots inoculated with arbuscular mycorrhizal fungi and/or *Phytophthora parasitica* and their possible involvement in bioprotection. Plant Sci 141:149–157

Pozo MJ, Cordier C, Dumas-Gaudot E, Gianinazzi S, Barea JM, Azcón-Aguilar C (2002) Localized versus systemic effect of arbuscular mycorrhizal fungi on defense responses to *Phytophthora* infection in tomato plants. J Exp Bot 53:525–534

Pringle A, Moncalvo JM, Vilgalys R (2000) High levels of variation in ribosomal DNA sequences within and among spores of natural population of the arbuscular mycorrhizal fungus *Acaulospora colossica*. Mycologia 92:259–168

Rausch C, Daram P, Brunner S, Jansa J, Laloi M, Leggewie N, Bucher M (2001) A phosphate transporter expressed in arbuscule-containing cells in potato. Nature 414:462–466

Redecker D (2000) Specific PCR primers to identify arbuscular mycorrhizal fungi within colonized roots. Mycorrhiza 10:73–80

Redecker D (2001) Mol ecular identification and phylogeny of arbuscular mycorrhizal fungi. Plant Soil 244:67–73

Redecker D, Thierfelder H, Walker C, Werner D (1997) Restriction analysis of PCR-amplified internal transcribed spacers of ribosomal DNA as a tool for species identification in different genera of the order Glomales. Appl Environ Microbiol 63:1756–1761

Redecker D, Morton JB, Bruns TD (2000a) Ancestral lineages of arbuscular mycorrhizal fungi (Glomales). Mol Phylogenet Evol 14:276–284

Redecker D, Kodner R, Graham LE (2000b) Glomalean fungi from the Ordovician. Science 289:1920–1921

Rejón-Palomares A, Garcia-Garrido JM, Ocampo JA, Garcia-Romera I (1996) Presence of xyloglucan-hydrolysing glucanases (xyloglucanases) in arbuscular mycorrhizsl symbiosis. Symbiosis 21:249–261

Remy W, Taylor TN, Hass H, Kerp H (1994) Four hundred-million-year-old mycorrhizal vesicular arbuscular mycorrhizae. Proc Natl Acad Sci USA 91:11841–11843

Requena N (1998) Mycorrhizal interactions in the rhizosphere. In: Varma A (ed) Microbes: for health, wealth and sustainable environment. MPH New Delhi, India, pp 726–735

Requena N, Jeffries P, Barea JM (1996) Assessment of natural mycorrhizal potential in a desertified semiarid ecosystem. Appl Environ Microbiol 62:842–847

Requena N, Jimenez I, Toro M, Barea JM (1997) Interactions between plant-growth-promoting rhizobacteria (PGPR), arbuscular mycorrhizal fungi and *Rhizobium* spp. in the rhizosphere of *Anthyllis cytisoides*, a model legume for revegetation in mediterranean semi-arid ecosystems. New Phytol 136:667–677

Requena N, Füller P, Franken P (1999) Mol ecular characterization of *GmFOX2* an evolutionarily highly conserved gene from the mycorrhizal fungus *Glomus mosseae*, down-regulated during interaction with rhizobacteria. Mol Plant Microb Interact 12:934–942

Requena N, Mann P, Franken P (2000) A homologue of the cell-cycle check-point TOR2 from yeast exists in the arbuscular mycorrhizal fungus *Glomus mosseae*. Protoplasma 211:89–98

Requena N, Pérez-Solis E, Azcón-Aguilar C, Jeffries P, Barea M (2001) Increased establishment of mycorrhizal plants inoculated with native endophytes in semi-arid degraded mediterranean ecosystems. Appl Environ Microbiol 67:495–498

Requena N, Mann P, Hampp R, Franken P (2002) Early developmentally regulated genes in the arbuscular mycorrhizal fungus *Glomus mosseae*:identification of *GmGIN1*, a novel gene with homology to the C-terminus of metazoan hedgehog proteins. Plant Soil 244:129–139

Requena N, Breuninger M, Franken P, Ocón A (2003) Symbiotic status, phosphate and sucrose regulate the expression of two plasma membrane H^+-ATPase genes from the arbuscular mycorrhizal fungus *Glomus mosseae*. Plant Physiol 132 (in press)

Rosendahl S, Taylor JW (1997) Development of multiple genetic markers for studies of genetic variation in arbuscular mycorrhizal fungi using AFLP. Mol Ecol 6:821–829

Rosendahl S, Sen R, Hepper CM, Azcon-Aguilar C (1989) Quantification of three vesicular-arbuscular mycorrhizal fungi (*Glomus* spp) in the roots of leek (*Allium porum)* on the basis of the activity of diagnostic enzymes after polyacrylamide gel electrophoresis. Soil Biol Biochem 21:519–522

Roussel H, van Tuinen D, Franken P, Gianinazzi S, Gianinazzi-Pearson V (2001) Signalling between arbuscular mycorrhizal fungi and plants: identification of a gene expressed during early interactions by differential RNA display analysis. Plant Soil 232:13–19

Ruiz-Lozano JM, Roussel H, Gianinazzi S, Gianinazzi-Pearson V (1999) Defense genes are differentially induced by a mycorrhizal fungus and *Rhizobium* sp. in wild-type and symbiosis-defective pea genotypes. Mol Plant Microbe Interact 12:976–984

Saito M (1995) Enzyme activities of the internal hyphae and germinated spores of an arbuscular mycorrhizal fungus *Gigaspora margarita* (Becker & Hall). New Phytol 129:425–431

Salzer P, Corbière H, Boller T (1999) Hydrogen peroxide accumulation in *Medicago truncatula* roots colonized by the arbuscular mycorrhiza-forming fungus *Glomus mosseae*. Planta 208:319–325

Salzer P, Bonamoni A, Beyer K, Vögeli-Lange R, Aeschbacher RA, Lang J, Wiemken A, Kim D, Cook DR, Boller T (2000) Differential expression of eight chitinase genes in *Medicago truncatula* roots during mycorrhiza formation, nodulation and pathogen infection. Mol Plant Microbe Interact 13:763–777

Sanders IR (1999) No sex please, we're fungi. Nature 399:737–739

Sanders IR, Ravolanirina F, Gianinazzi-Pearson V, Pearson S, Lemoine MC (1992) Detection of specific antigens in the vesicular-arbuscular mycorrhizal fungi *Gigaspora margarita* and *Acaulospora laevis* using polyclonal antibodies to soluble spore fractions. Mycol Res 96:477–480

Sanders IR, Alt M, Groppe K, Boller T, Wiemken A (1995) Identification of ribosomal DNA polymorphism among and within spores of the Glomales: application to studies on the genetic diversity of AMF communities. New Phytol 130:419–427

Sanders IR, Clapp JP, Wiemken A (1996) The genetic diversity of AMF in natural ecosystems–a key to understanding the ecology and functioning of the mycorrhizal symbiosis. New Phytol 133:123–134

Schenck NC, Pérez Y (1990) Manual for the identification of vesicular - arbuscular mycorrhizal fungi, 3rd edn. INVAM, Gainsville, Florida, USA.

Scheres B, van de Wiel C, Zalensky A, Horvath B, Spaink H, van Eck H, Zwartkruis F, Wolters AM, Gloudemans T, van Kammen A (1990) The ENOD12 gene product is involved in the infection process during the pea-*Rhizobium* interaction. Cell 60:281–294

Schüßler A (1999) Glomales SSU rRNA gene diversity. New Phytol 144:205–207

Schüßler A, Kluge M (2001) *Geosiphon pyriformis*, an endocytobiosis between fungus and cyanobacteria, and its meaning as a model system for arbuscular macorrhizal research. In: Hock B (ed) The Mycota, vol IX. Springer, Berlin Heidelberg New York, pp 151–161

Schüßler A, Mollenhauer D, Schnepf, E, Kluge M (1994) *Geosiphon pyriformis*, an endosymbiotic association of fungus and cyanobacteria:the spore structure resembles that of arbuscular mycorrhizal (AM) fungi. Bot Acta 107:36–45

Schüßler A, Bonfante P, Schnepf E, Mol lenhauer D, Kluge M (1996) Characterization of the *Geosiphon pyriformis* symbiosome by affinity techniques:confocal laser scanning microscopy (CLSM) and electron microscopy. Protoplasma 190:53–67

Schüßler A, Gehrig H, Schwarzott D, Walker C (2001a) Analysis of partial Glomales SSU rRNA gene sequences:implications for primer design and phylogeny. Mycol Res 105:5–15

Schüßler A, Schwarzott D, Walker C (2001b) A new fungal phylum, the Glomeromycota: phylogeny and evolution. Mycol Res 105:1413–1421

Schwarzott D, Walker C, Schüßler A (2001) *Glomus*, the largest genus of the arbuscular mycorrhizal fungi (Glomerales) is non–monophyletic. Mol Phylogenet Evol 21:190–197

Shachar-Hill Y, Pfeffer PE, Douds D, Osman SF, Doner LW, Ratcliffe RG (1995) Partitioning of intermediary carbon metabolism in vesicular-arbuscular mycorrhizal leek. Plant Physiol 108:7–15

Simon L (1996) Phylogeny of the Glomales:deciphering the past to understand the present. New Phytol 133:95–101

Simon L, Lalonde M, Bruns TD (1992) Specific amplification of 18S ribosomal genes from VAM – fungi colonizing roots. Appl Environ Microbiol 58:291–295

Simon L, Levesque RC, Lalonde M (1993a). Identification of endomycorrhizal fungi colonizing roots by fluorescent single-strand conformation polymorphism–polymerase chain reaction. Appl Environ Microbiol 59:4211–4215

Simon L, Bousquet J, Levesque RC, Lalonde M (1993b). Origin and diversification of endomycorrhizal fungi and coincidence with vascular plants. Nature 363:67–69

Smith FA, Smith SE (1997) Structural diversity in (vesicular)-arbuscular mycorrizal symbiosis. New Phytol 137:373–388

Smith SE, Read DJ (1997) Mycorrhizal symbiosis, 2nd edn. Academic Press, London

Smith SE, Smith FA (1990) Structure and function of the interfaces in biotrophic symbiosis as they relate to nutrient transport. New Phytol 114:1–38

Smith SE, St John BJ, Smith FA, Nicholas DJD (1985) Activity of glutamine synthetase and glutamate dehydrogenase in *Trifolium subterraneum* L. and *Allium cepa* L.:effects of mycorrhizal infection and phosphate nutrition. New Phytol 99:211–227

Somssich IE, Hahlbrock K (1998) Pathogen defence in plants:a paradigm of biological complexity. Trends Plant Sci 3:86–90

Spanu P, Bonfante-Fasolo P (1988) Cell wall-bound peroxidase activity in roots of mycorrhizal *Allium porrum*. New Phytol 109:119–124

Stracke S, Kistner C, Yoshida S, Mulder L, Sato S, Kaneko T, Tabata S, Sandal N, Stougaard J, Szczyglowski K, Parniske M (2002) A plant receptor-like kinase required for both bacterial and fungal symbiosis. Nature 417:959–962

Talbot NJ, Foster AJ (2001) Genetics and genomics of the rice blast fungus *Magnaporthe grisea*:developing an experimental model for understanding fungal deseases of cereals. Adv Bot Res 34:263–287

Taylor TN, Remy W, Hass H, Kerb H (1995) Fossil arbuscular mycorrhizae from the early Devonian. Mycologia 87:560–573

Timonen S, Smith FA, Smith SE (2001) Microtubules of the mycorrhizal fungus *Glomus intraradices* in symbiosis with tomato roots. Can J Bot 79:307–313

Tisserant B, Brenac V, Requena N, Jeffries P, Dodd JC (1998) The detection of *Glomus* spp. (arbuscular mycorrhizal fungi) forming mycorrhizas in three plants, at different stages of seedling development, using mycorrhiza-specifi isozymes. New Phytol 138:225–239

Tobar R, Azcon R, Barea JM (1994) Improved nitrogen uptake and transport from ^{15}N-labelled nitrate by external hyphae of arbuscular mycorrhiza under water-stressed conditions. New Phytol 126:119–122

Tommerup IC (1984) Persistence of infectivity by germinated spores of vesicular-arbuscular mycorrhizal fungi in soil. Trans Br Mycol Soc 82:275–282

Trappe JM (1987) Phylogenetic and ecological aspects of mycotrophy in the angiosperms from an evolutionary standpoint. In: Safir GR (ed) Ecophysiology of VA mycorrhizal plants.CRC Press, Boca Raton, FL, pp 2–25

Turnau K, Ryszka P, Gianinazzi-Pearson V, van Tuinen D (2001) Identification of arbuscular mycorrhizal fungi in soils and roots of plants colonizing zinc wastes in southern Poland. Mycorrhiza 10:169–174

van Buuren ML, Maldonado-Mendoza IE, Trieu AT, Blaylock LA, Harrison MJ (1999) Novel genes induced during an arbuscular mycorrhizal (AM) symbiosis formed between *Medicago truncatula* and *Glomus versiforme*. Mol Plant Microbe Interact 12:171–81

Vandenkoornhuyse P, Leyval C (1998) SSU rDNA sequencing and PCR–fingerprinting reveal genetic variation within *Glomus mosseae*. Mycologia 90:791–797

van der Heijden MGA, Klironomos JN, Ursic M, Moutoglis P, Streitwolf-Engel R, Boller T, Wiemken A, Sanders IR (1998) Mycorrhizal fungal diversity determines plant biodiversity, ecosystem variability and productivity. Nature 396:69–72

van Rhijn P, Fang Y, Galili S, Shaul O, Atzmon N, Wininger S, Eshed Y, Lum M, LiY, To V, Fujishige N, Kapulnik Y, Hirsch AM (1997) Expression of early nodulin genes in alfalfa mycorrhizae indicates that signal transduction pathways used in forming arbuscular mycorrhizae and *Rhizobium*-induced nodules may be conserved. Proc Natl Acad Sci USA 94:5467–5472

van Tuinen D, Jacquot E, Zhao B, Gollotte A, Gianinazzi-Pearson V (1998) Characterization of root colonization profiles by microcosm community of arbuscular mycorrhizal fungi using 25S rRNA-target nested PCR. Mol Ecol 7:879–887

Vierheilig H, Alt M, Neuhaus JM, Boller T, Wiemken A (1993) Colonization of transgenic *Nicotiana sylvestris* plants, expressing different forms of *Nicotiana tabacum* chitinase, by root pathogen *Rhizoctonia solani* and the mycorrhizal symbiont *Glomus mosseae*. Mol Plant Microbe Interact 6:261–264

Vierheilig H, Alt M, Lange J, Gut-Rella M, Wiemken A, Boller T (1995) Colonization of transgenic tobacco constitutively expressing pathogenesis-related proteins vesicular-arbuscular mycorrhizal fungus *Glomus mosseae*. Appl Environ Microbiol 61:3031–3034

Volpin H, Phillips D, Ocón Y, Kapulnik Y (1995) Suppression of an isoflavonoid phytoalexin defense response in mycorrhizal alfalfa roots. Plant Physiol 108:1449–1454

Walker C (1992) Systematics and taxonomy of the arbuscular endomycorrhizal fungi (Glomales)–a possible way forward. Agronomie 12:887–897

Walker C, Trappe JM (1993) Names and epithets in the Glomales and Endogonaceae. Mycol Res 97:339–344

Wegel E, Schauser L Sandal N Stougaard J Parniske M (1998) Mycorrhiza mutants of *Lotus japonicus* define genetically independent steps during symbiotic infection. Mol Plant Microbe Interact 11:933–936

Wright SF, Morton JB (1989) Detection of vesicular-arbuscular mycorrhizal fungus colonization of roots by using a dot-immunoblot assay. Appl Environ Microbiol 55:761–763

Wright SF, Upadhyaya A (1996) Extraction of an abundant and unusual protein from soil and comparison with hyphal protein of arbuscular mycorrhizal fungi. Soil Sci 161:575–586

Wyss P, Bonfante P (1993) Amplification of genomic DNA of arbuscular-mycorrhizal (AM) fungi by PCR using short arbitrary primers. Mycol Res 97:1351–1357

Zéze A, Sulistyowati E, Ophel-Keller K, Barker S, Smith SE (1997) Intrasporal genetic variation of *Gigaspora margarita*, a vesicular arbuscular mycorrhizal fungus, revealed by M13 minisatellite-primed PCR. Appl Environ Microbiol 63:676–678

Natalia Requena
Magdalene Breuninger
Botanical Institute, Department of
Physiological Ecology of Plants
University of Tübingen
Auf der Morgenstelle 1
72076 Tübingen, Germany
e-mail: natalia requena@uni-tuebingen.de

The Role of Nitrate Reduction in Plant Flooding Survival

M. Stoimenova and W.M. Kaiser

1 Introduction

Higher plants (for the most part of their life cycle strict aerobes) are often challenged by environmental conditions (flooding, ice crusts sealing soil surface etc.) that deprive them of oxygen. In water-filled soil pores the diffusional resistance for gases is several orders of magnitude higher than in air-filled pores (Armstrong 1979). As a result, the oxygen concentration at the root surface drops dramatically, which inhibits mitochondrial respiration in root cells, since the requirement for oxygen as terminal electron acceptor is absolute (Aldrich et al. 1985; Andreev et al. 1991). Even the so-called flood-tolerant species (*Oryza sativa, Erythrina caffra, Trapa natans, Echinochloa crus-galli* etc.) could tolerate anaerobiosis for only a short time (Kennedy et al. 1992). Consequently, hypoxic/anoxic conditions often cause severe losses in crop production (an important practical problem in agriculture). Tolerance by plant roots of phases of partial or complete oxygen deficiency indeed differs greatly with plant species, but also with other environmental factors. Among the latter, the type of nitrogen source (nitrate or ammonium) appears to affect plant tolerance to hypoxia or anoxia, as already noticed by Arnon (1937):

'The lack of forced aeration although strikingly limiting the growth of ammonium plants hardly affected the total growth of nitrate plants.'

Amazingly, today – 65 years since the first reports of nitrate to be beneficial for plant flooding survival appeared, and although it is an established agricultural practice to apply inorganic nitrogen (mainly nitrate) to cereal and grass crops to help their recovery from temporary water logging – we are still far from a complete understanding how nitrate and/or nitrate reduction help plants survive flooding. Over the years the different aspects of N assimilation involvement and importance in anaerobic survival of plant cells have been discussed in more than 40 papers covering the responses of a variety of flood-tolerant and -sensitive plants. Here, we con-

Progress in Botany, Vol. 65

sider different possibilities for the apparent beneficial effects of nitrate and nitrate reduction on anoxia tolerance of roots, and summarize some more recent experimental evidence supporting and elucidating the beneficial nitrate effects.

2 What Are the Major Problems Encountered by Anoxic Cells?

The ATP that is available at a given time in aerobic plant cells could support their metabolism for only few minutes (Roberts et al. 1984, 1989), if not continuously resynthesized. When cells are deprived of oxygen, nucleoside-tri-phosphate (NTP) levels drop rapidly and reversible structural changes (swelling and elongation) of mitochondria are observed almost immediately (Saglio and Pradet 1980; Aldrich et al. 1985; Saglio et al. 1988; Andreev et al. 1991).

Another common response of cells to anoxia is a decrease in cytosolic pH which by itself may directly affect many cellular reactions. There are a number of possible reasons for cytosolic acidosis:

1. Passive proton release from the (usually acidic) vacuole and/or proton influx from an acidic apoplast,
2. a switch from a proton-neutral aerobic to a proton-producing anaerobic metabolism and fermentation.

A major reason for (1) is probably the low ATP level as such, which may hinder proton pumping into the apoplast or into the vacuole (Gout et al. 2001), due to the low Km of H^+-ATPases for ATP. In (2), major proton-producing reactions are probably NTP hydrolysis, proton-producing glycolysis and accumulation of strong organic acids, especially of lactate. In addition to cellular acidosis, production of potentially toxic compounds like acetic aldehyde or ethanol (Crawford's metabolic theory for flooding tolerance, McManmon and Crawford 1971; Perata and Alpi 1991a,b; Pfister-Sieber and Brändle 1994) may contribute to limited cell survival under anoxia. Without going into the details of these aspects, which have been repeatedly considered in a number of excellent reviews (Kennedy et al. 1992; Perata and Alpi 1993; Drew 1997; Ratcliffe 1997; Vartapetian and Jackson 1997; De Sousa and Sodek 2002 and others), any positive effect of nitrate on survival of anoxia should somehow affect one or more of these processes.

3 How Could Nitrate and Its Reduction Contribute to Survival of Anoxia?

1. In Arnon's original paper, he compared ammonium- versus nitrate-grown plants. This system has been used frequently (Apostolova and Georgieva 1990; Botrel at al. 1996; Botrel and Kaiser 1997; Scheible et al. 1997; Walch-Liu et al. 2001 among others). It is well known that ammonium-grown plants acidify their external water phase (Raven and Smith 1976; Mengel et al. 1976), whereas nitrate plants generally alkalize it (Kirkby and Armstrong 1980; Marschner 1995). Avoidance of cytosolic acidification should be the more difficult, the steeper the pH gradient is (i.e. the more acidic the external medium is). In slightly oxygen-limited cell suspensions, ammonium instead of nitrate indeed caused a decrease in cytosolic pH by 0.2 units (Caroll et al. 1994). However, no difference in sensitivity to acidification or in pH regulation was observed in NMR experiments on leaves from pea plants grown on nitrate or ammonium as sole N source, respectively (Bligny et al. 1997). The reduced growth frequently observed in ammonium grown plants could be explained by interference between uptake of ammonium and other cations through the root system, rather than by increased cellular acidification.
2. Reduction of nitrate to ammonium potentially consumes H^+, according to

   $$NO_3^- + NADH + H^+ \rightarrow NO_2^- + NAD^+ + H_2O \quad (1)$$

   $$NO_2^- + 3NADH + 5H^+ \rightarrow NH_4^+ + 3NAD^+ + 2H_2O \quad (2)$$

   and might therefore help to stabilize cytosolic pH. However, under anoxia, nitrite usually accumulates and reduction of exogenous nitrite is very low or absent (Lee 1978, 1979; Glaab and Kaiser 1993; Botrel et al. 1996). Thus, nitrate is reduced mainly to the level of nitrite, which would not contribute to pH stabilization, and in the absence of respiration would rather cause acidification according to the overall balance depicted in Eq. (3) (Gerendás and Ratcliffe 2002).

   $$C_6H_{12}O_6 + 2NO_3^- \rightarrow 2C_3H_3O_3^- + 2NO_2^- + 2H^+ + 2H_2O \quad (3)$$

3. Reduction of nitrate to nitrite consumes NAD(P)H. Thereby, nitrate reduction might either improve total glycolytic flux and ATP production by adding to the NAD^+ recycling via alcohol and lactic acid formation, or it might replace ethanol and lactic acid production, thereby avoiding accumulation of toxic (acetaldehyde, ethanol) and acidifying (lactate) metabolites.

4. Nitrate seems to have a beneficial effect in anoxia via preserving membrane structure and stability. It postponed free fatty acid release in potato cells for about 6 h compared with ammonium within 24 h of anoxic treatment. The increased membrane lipid stability of nitrate-treated cells under anoxia was correlated with higher nitrate reduction capability and an improved energy status, which in turn (probably) delayed the activation of lipolytic acyl hydrolase and thus had a beneficial effect on anoxia survival (Oberson et al. 1999). Anaerobic incubation of detached rice cotyledons in KNO_3-free medium for 48 h resulted in complete destruction of mitochondrial and other cell membranes, whereas 10 mM KNO_3 completely prevented destruction, although the mitochondrial structure was modified (Vartapetian and Polyakova 1999).

Since the early attempts of Arnon, most investigations to unravel the role of nitrate reduction for anoxia tolerance have been based on a comparison of plants grown either on ammonium or on nitrate (Apostolova and Georgieva 1990; Botrel at al. 1996; Botrel and Kaiser 1997; Scheible et al. 1997; Walch-Liu et al. 2001; among others). The former usually do not express NR and do not reduce nitrate (Ullrich 1983) in contrast to nitrate-grown plants which frequently showed elevated NR activity in hypoxic or anoxic conditions (Ferrari and Varner 1970; Ferrari et al. 1973; Atkins and Canvin 1975; Mann et al. 1979; Kenis and Trippi 1986), and released nitrite to the root medium in the presence of nitrate (Nance 1950). In most plants nitrite reduction is the step in the overall nitrate assimilation process that is most impaired from a lack of oxygen (Lee 1979; Glaab and Kaiser 1993; Botrel et al. 1996). However, species that are able to germinate in anoxic conditions, such as rice, appear to be a special case since they exhibit higher activities of nitrite reductase, ferredoxin-dependent glutamate synthase, Fd-NADP oxidoreductase in the presence of exogenous nitrate (5 mM). Northern blot analysis revealed the presence of mRNAs for NR, NiR, cytosolic GS, Fd-GOGAT, and root FNR. These data indicate that in rice, the nitrate assimilation pathway is operative under anoxia (Mattana et al. 1994, 1997).

It has already been mentioned above that nitrate versus ammonium uptake by itself may affect cellular pH stabilization. In addition, ammonium- and nitrate-grown plants differ in many other aspects, like root/shoot ratios, root morphology, ion contents and sugar and amino acid composition of cellular solutes (Lang and Kaiser 1994). When nitrate was made available for reduction by rice roots exposed to anaerobic conditions, less accumulation of NADH took place, which may indicate additional NADH consumption for nitrate reduction (Reggiani et al. 1985a). In potato cells incubated in anoxic medium containing ammonium or nitrate as N source for 6 h, ATP levels were 20±2% of the aerobic control for ammonium grown plants and 52±3% of the aerobic control for nitrate grown plants, and the adenylate energy charge was 0.57 and 0.76, respectively (Oberson et al. 1999). Ethanol production was higher, but lactate acumulation was lower in rice roots on nitrate than on ammonium (Reggiani et al. 1985b,c). Both, lower lactate and lower ethanol production were observed in anoxic roots of *Carex pseudocyperus* L. and *Carex sylvatica* Huds. grown on nitrate (Müller et al. 1994). Nitrate reduced the amount of fermentation end products, helped maintain a higher free NTP concentration during hypoxia and increased the rate of overall recovery from hypoxia in mature maize roots (Fan et al. 1988).

Although the 'nitrate respiration' hypothesis (which considers nitrate as an alternative electron acceptor in the absence of oxygen) certainly gained some attention over the years, a number of observations do not support it. Drew and Pattiradjawene (1975) reported that beneficial effects of nitrate on flooded roots were only observed when part of the root system was aerobic and concluded that its positive role might be limited to being an N source for metabolism and growth. A higher concentration of NADH was observed in roots of soybean and sunflower grown on nitrate than in roots of plants grown on ammonium (Wiskish 1977). Recently it was reported that supplementation of soybean plants with nitrate improved their tolerance to flooding relative to those relying on N_2 fixation, probably via a lower O_2-requirement of nitrate uptake and assimilation compared to that of N_2 fixation (Bacanamwo and Purcell 1999).

Extensive work from the 1970s (Lee 1978, 1979) suggested that occurrence of dissimilatory reduction of inorganic N in roots could be detected by the following criteria: appreciable stimulation of one or more of the reactions of the assimilatory pathway in the absence of oxygen would indicate that the pathway was replacing oxygen as electron acceptor; a decrease in synthesis of a normal end product of fermentation (i.e. ethanol, lactate), accompanied by a proportional stimulation of nitrate assimilation would also indicate a diversion of reducing power, and least but not last the total quantity of inorganic nitrogen reduced would exceed the requirement of growth for nitrogen, leading to accumulation of the reduced N-compounds (ammonium or amino acids). This last criterion needs to be treated with caution, however, for such an increase in reduced nitrogen in anaerobic conditions might be caused at least in part by a smaller requirement of N for growth. Applying his criteria to aerobic and anaerobic barley roots, Lee failed to find evidence that any form of dissimulatory reduction of inorganic N occurred in anoxic conditions.

Altogether, data obtained from such comparisons of nitrate- and ammonium-grown plants are contradictory, and indeed, comparing the anoxic response of ammonium-grown plants with that of nitrate-grown plants, although providing a wealth of information, may appear like a comparison of "apples and pears" since the form of nitrogen available affects whole plant growth and metabolism as a substrate. In addition, nitrate itself (Crawford 1995), or products downstream of nitrate reduction may act as signalling compounds which not only control the expression of genes directly involved in nitrate transport and reduction, but also affect many other aspects of metabolism and plant morphology (Cooper and Clarkson 1989).

4 A New Tool: NR-Deficient Mutants or Transformants

We have recently chosen another approach to the nitrate/anoxia problem. The expression of the NR structural gene *nia2* under the control of a leaf-specific promoter in the NR-deficient tobacco mutant Nia30 (Hänsch et al. 2001) has resulted in transformant plants that have no nitrate reductase in the roots but almost normal NR activities in the shoots. These plants (called LNR-H, which stands for leaf nitrate reductase line H) offer an alternative way of investigating the contribution of nitrate reduction to anaerobic metabolism in roots when compared with the response of wild-type tobacco (*Nicotiana tabacum* cv. Gatersleben) expressing nitrate reductase.

Since our evaluation of LNR-H metabolism that we presented previously (Hänsch et al 2001) was not sufficient to validate a comparison between the two lines in anoxic stress conditions, we first concentrated our attention on analyzing the similarity of parameters in aerated conditions, that might be of vital importance for plants exposed to anaerobiosis. We compared the two lines with respect to their root morphology, root respiration and the root content of inorganic cations, anions, and metabolites. Leaf transpiration in relation to root morphology was also determined. Plants were grown in hydroponics where localized nutrient supply was avoided (Stoimenova et al. 2003a,b).

4.1 Basic Properties of Roots with or Without Nitrate Reduction

Briefly, growth of the NR-free LNR-H transformants was somewhat retarded compared to WT, and LNR-H transformants had shorter and slightly thicker roots with a lower root surface area per g leaf FW. Root nitrate concentrations were very similar to WT, whereas leaf nitrate concentrations were higher. LNR-H roots had usually somewhat higher sugar and a higher total amino acid content. The observed growth differences are surprising as total soluble protein levels and nitrate content in WT and LNR-H roots were not significantly different. Negative correlation between internal N status (amino acids) and root growth has been reported before (Scheible et al. 1997).

Certainly, the above-described differences in root morphology and root/shoot ratios of our LNR-H and WT lines were not due to N- or C-deficiency, nor to differences in the major anion and cation contents. The latter proved to be practically identical, quite in contrast to nitrate versus ammonium grown plants (Lang and Kaiser 1994).

Levels of abscisic acid in roots of both lines were also not significantly different (W. Hartung, pers. comm.). Other hormones have not been measured. Roots with high specific root length are often found in nutrient-deficient plants (Fitter 1985) and in barley and wheat abundance of nitrate provoked an increase in root diameter (Hackett 1972; Drew et al. 1973; Cruz et al. 1997). Taken together, root nitrate reduction apparently has a far more complicated role on plant development and metabolism than just supplying reduced nitrogen.

The reason for the different root size and morphology of LNR-H versus WT plants is still not fully understood, but the resulting differences in root/shoot ratio, and specifically the lower root surface area per leaf FW of LNR-H versus WT may be one reason for the observed differences in leaf transpiration: At a whole plant level, the LNR-H plants were more prone to wilting when the root system was deprived of oxygen, but this commonly observed anoxic response may as well have its origin in a reduction in root hydraulic conductivity (Vartapetian and Jackson 1997).

For further comparing the anoxic response of LNR-H roots over WT roots, data on root respiration were required. Respiration of LNR-H roots was higher than in WT roots, and that was paralleled by higher nucleotide levels in LNR-H (see below). The higher root respiration of LNR-H plants as compared to WT may indicate that nitrate reduction (present in WT only) can compete with respiration for reductant. However, respiration rates of WT roots (28.11 μmol/g fresh wt. h^{-1}) greatly exceeded extractable NR activity, (0.1 μmol/g fresh wt. h^{-1}, measured in the presence of Mg^{2+}, or 0.4 μmol/g fresh wt. h^{-1} measured in the presence of EDTA). Therefore, a simple competition for reductant cannot be the reason for the different respiration rates between roots with (WT) and without (LNR-H) nitrate reduction.

While the above analysis has elucidated some metabolic and morphological differences between the root systems of WT and LNR-H, the two lines seemed similar enough in other basic features, such as cation and anion contents, total soluble protein and sugar contents to be used for a comparison of their response to anoxia, in order to reexamine the role of nitrate reduction for survival of anoxia.

4.2 Absence or Presence of Nitrate Reduction Greatly Affects Fermentation Rates

Two questions are of special importance when evaluating the importance of nitrate and nitrate reduction in flooding survival

1. Does nitrate reduction increase metabolism under anoxia by increasing the recycling of NADH?
2. (How) does nitrate reduction under anoxia affect cytosolic pH?

Note that (as mentioned above) although nitrate reduction to either nitrite or ammonium consumes H^+ under aerobic conditions, the balance switches to H^+ production under anoxia where the only source of NADH is glycolysis (Gerendás and Ratcliffe 2002).

A number of enzymatic and metabolic measurements were used to address the first question, and ^{31}P-NMR spectroscopy was applied to examine cytosolic pH under anoxia (Stoimenova et al. 2003b).

The most important difference in the anoxic response of LNR-H roots versus WT roots was a strikingly greater production of fermentation end products in LNR-H (Table 1). This appeared a genuine difference between WT and LNR-H plants. It was not an effect of carbon starvation of WT roots, as might be suspected from their slightly lower initial sugar contents (see above), since sucrose feeding to WT plants hardly increased their accumulation of fermentation products (Stoimenova et al. 2003b). Further support for a decisive role of nitrate reduction for fermentation came from the observation that WT roots also produced more ethanol and lactate under anoxia when their NR activity was abolished by tungstate feeding (Table 1).

4.3 Can Nitrate Reduction Decrease Fermentation by Competition for Reductants?

It has been mentioned above that in WT roots, the capacity for nitrate reduction, derived from the maximum NR activity, was about 0.4 μmol/g FW h. In LNR-H-roots, it was zero. In Table 1, the rate of ethanol plus lactate production in WT roots (and the resulting NADH consumption) was about 0.9, whereas in LNR-H it was about 7 (μmol/g fresh wt. h^{-1}). The difference in the absence and presence of nitrate reduction was thus 6.2 (μmol/g fresh wt. h^{-1}), and thus more than 15 times higher than the nitrate reduction

Table 1. Ethanol and L-lactate production of hydroponically grown wild-type tobacco (*Nicotiana tabacum* cv. GAT, WT), LNR-H and WT tungstate-treated roots (μmol g^{-1} fresh wt. ± SE). Full-strength nutrient solution (pH 6.3) contained: 5 mM KNO_3, 1 mM $CaCl_2$, 1 mM $MgSO_4$, 0.025 mM NaFe-EDTA, 1 mM K_2HPO_4, 2 mM KH_2PO_4 and trace elements according to (Johnson at al. 1957). Roots of 6- to 10-week-old plants were used. Tungstate-treated WT plants were fed daily with 20 μM tungstate (to inhibit NR) and 2 mM NH_4Cl (to avoid N starvation) in addition to nutrient solution for 8 to 10 days. Ethanol was measured in the surrounding medium only and L-lactate in both roots plus the surrounding medium. Data are means of 4 to 12 plants each. For more details, see Stoimenova et al. (2003b)

Ethanol	WT	LNR-H	WT+Tungstate
Time 0	0	0	0
1 h, aerated	0.02±0.01	0.11±0.01	0.11±0.005
1 h, anoxic	0.05±0.02	1.16±0.27	1.66±0.55
4 h, aerated	0.003±0.003	0.04±0.02	0.05±0.035
4 h, anoxic	0.22±0.05	4.47±0.75	3.00±0.56
L-Lactate	**WT**	**LNR-H**	**WT+Tungstate**
Time 0	0.13±0.02	0.73±0.31	0.09±0.02
1 h, aerated	0.15±0.04	2.03±0.29	0.04±0.02
1 h, anoxic	0.88±0.17	5.78±0.33	2.91±0.27
4 h, aerated	0.17±0.02	0.73±0.10	nd
4 h, anoxic	0.75±0.17	7.46±0.49	2.34±0.03

capacity. Clearly, nitrate reduction could not decrease fermentation through competition for NADH.

4.4 Absence of Nitrate Reduction Under Anoxia Impairs Cytosolic pH

^{31}P-NMR measurements showed that in LNR-H roots, under anoxia the cytosolic pH dropped lower than in WT roots (Fig. 1). As shown previously (Stoimenova et al. 2003a), ATP-levels in LNR-H roots were generally higher than in WT roots. Under anoxia, ATP levels dropped drastically in both cases, but they were still higher in LNR-H (0.036 μmol/g fresh wt.) than in WT roots (0.009 μmol/g fresh wt.). However, ADP levels in LNR-H roots were also lower, and thus the ATP/ADP ratio (or the energy charge) which may determine the rate of proton pumping, was not different. Thus, pres-

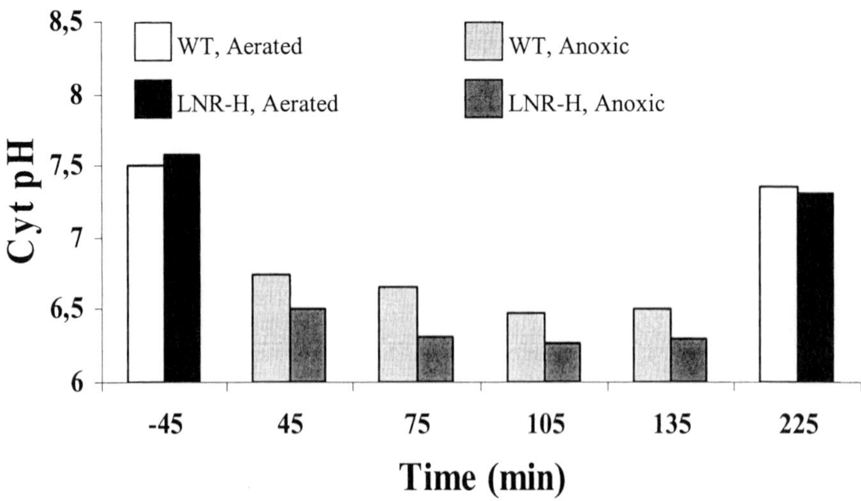

Fig. 1. Cytoplasmic pH values derived from the ^{31}P-NMR spectra of WT and LNR-H root segments. Growing conditions as in Table 1 except that roots were supplied with 50 mM sucrose during NMR measurements. A series of spectra were recorded before, during, and after a period of oxygen deprivation and the time refers to the midpoint of each 90-min spectrum. The switch to anoxia took place at time zero and the return to oxygenated conditions occurred after 180 min. Data are means of up to five plants. For more details, see Stoimenova et al. (2003b)

ence or absence of nitrate reduction should not affect cytosolic pH via differences in proton pumping. The conclusion is that it must be the higher metabolic rate (and especially the higher lactic acid formation) in the LNR-H roots that leads to the greater cytoplasmic acidification under anoxia despite the absence of a contribution from the metabolism of nitrate. For further details compare Stoimenova et al. (2003a,b).

Thus, in agreement with some other studies (e.g. Saglio et al. 1988), the frequently discussed proposal that NR should be able to promote carbohydrate metabolism under anoxia by providing an extra sink for NADH needs to be treated with caution.

4.5 How Does Nitrate Reduction Affect Carbohydrate Metabolism Under Anoxia?

There appear to be two speculative explanations for the reduced metabolic rate in the presence of nitrate reduction (WT): (1) down-regulation of metabolism mediated by an unknown metabolite downstream from NR; or (2) a difference in the physiology of the NR-free LNR-H roots, perhaps reflected in the higher respiration rate of the LNR-H plants under aerobic conditions (Stoimenova et al. 2003b). However, some support for the first explanation, and for the comparability of the WT and LNR-H roots, is provided by the observation that the tungstate-treated WT plants produced substantially more lactate and ethanol than untreated plants under anoxia (Table 1).

One other intriguing possibility is that the lower and less acidifying metabolism of WT roots containing NR may be achieved under anoxia through a side reaction of NR, the reduction of nitrite to nitric oxide (NO), that has recently become the focus of attention (Yamasaki et al. 1999; Rockel et al. 2002). Under anoxia, NR is usually activated, nitrite accumulates and NO emission increases (Rockel et al. 2002). NO is known to inhibit respiratory electron transport and to induce the alternative electron transport pathway (Leshem 2000). NO also inhibits many enzymes with prosthetic heme groups, and it may also interact with ethylene production (Leshem 2000). Thus it appears that NO may exert hitherto unknown regulatory functions in plant growth and metabolism, and future experiments on NO production and its function under anoxia could bring new insights into the old problem of the relationship between anoxia tolerance and nitrate reduction.

5 Conclusions

By comparing the anoxic response of tobacco roots with (WT) or without NR (LNR-H), we found that in the absence of nitrate reduction, roots have much higher fermentation rates, representing a more acidifying metabolism and resulting in a somewhat greater cytosolic acidification. The lower fermentation in the presence of nitrate reduction is not due to a competition for NADH between nitrate reduction and fermentation, because the capacity for nitrate reduction was too low compared to fermentation. The reason for the greater fermentation in the absence of nitrate reduction remains obscure, but is tentatively attributed to some regulatory metabolite downstream nitrate reduction, eventually nitric oxide (NO), which may affect anoxic metabolism.

Acknowledgement. The authors would like to thank Prof. Dr. R.G. Ratcliffe and I.G.L. Libourel for the many fruitful discussions and valuable ideas. The support of the DFG (Ka 456/13-1 to W.M. Kaiser) is gratefully acknowledged.

References

Aldrich HC, Fel RJ, Hils MH, Akin DE (1985) Ultra structural correlates of anaerobic tress in corn roots. Tissue Cell 17:341–46

Andreev VY, Generosova IP, Vertapetian BB (1991) Energy status and mitochondrial ultra structure of excised pea root at anoxia and post anoxia. Plant Physiol Biochem 29:171–76

Apostolova E, Georgieva M (1990) Effect of nitrate and ammonium on the accumulation of biomass nitrogen and ash elements in the overground parts and roots of *Virginia* cultivar tobacco. Pochvoznanie Agrochim 25: 47–54

Armstrong W (1979) Aeration in higher plants. Adv Bot Res 7:225–332

Arnon DI (1937) Ammonium and nitrate nitrogen nutrition of barley at different seasons in relation to hydrogen-ion concentration, manganese, copper, and oxygen supply. Soil Sci 44:91–121

Atkins CA, Canvin DT (1975) Nitrate, nitrite and ammonium assimilation by leaves: effect of inhibitors. Planta 123:41–51

Bacanamwo M, Purcell LC (1999) Soybean dry matter and N accumulation responses to flooding stress, N sources and hypoxia. J Exp Bot 50 (334):689–696

Bligny R, Gout E, Kaiser WM, Heber U, Walker D, Douce R (1997) pH regulation in acid-stressed leaves of pea plants grown in the presence of nitrate or ammonium salts: studies involving 31P-NMR spectroscopy and chlorophyll fluorescence. Biochim Biophys Acta 1320:142–152

Botrel A, Kaiser WM (1997) Nitrate reductase activation state in barley roots in relation to the energy and carbohydrate status. Planta 201:496–501

Botrel A, Magné C, Kaiser WM (1996) Nitrate reduction, nitrite reduction and ammonium assimilation in barley roots in response to anoxia. Plant Physiol Biochem 34:645–652

Carroll AD, Fox GG, Laurie S, Phillips R, Ratcliffe RG, Stewart GR (1994) Ammonium assimilation and the role of γ-aminobutiric acid in pH homeostasis in carrot cell suspensions. Plant Physiol 106:513–520

Cooper HD, Clarkson DT (1989) Cycling of amino nitrogen and other nutrients between shoots and roots in cereals: a possible mechanism integrating shoot and root in the regulation of nutrient uptake. J Exp Bot 40:753–762

Crawford NM (1995) Nitrate: nutrient and signal for plant growth. Plant Cell 7:859–868

Cruz C, Lips SH, Martins-Loucao MA (1997) Changes in the morphology of roots and leaves of carob seedlings induced by nitrogen source and atmospheric carbon dioxide. Ann Bot 80:817–823

De Sousa CAF, Sodek L (2002) The metabolic response of plants to oxygen deficiency. Braz J Plant Physiol 14(2):83–94

Drew MC (1997) Oxygen deficiency and root metabolism: injury and acclimation under hypoxia and anoxia. Annu Rev Plant Physiol Plant Mol Biol 48:223–250

Drew MC, Sistoro EJ (1977) Early effects of flooding on nitrogen deficiency and leaf chlorosis in barley. New Phytol 79(3):567–572

Drew MC, Saker LR, Ashley TW (1973) Nutrient supply and the growth of the seminal root system in barley. I. The effect of nitrate concentration on the growth of axes and laterals. J Exp Bot 24:1189–1202

Fan TW-M, Higashi RM, Lane AN (1988) An in vivo ^{1}H and ^{31}P NMR investigation of the effect of nitrate on hypoxic metabolism in maize roots. Arch Biochem Biophys 266:592–606

Ferrari TE, Varner JE (1970) Control of nitrate reductase activity in barley aleurone layers. Proc Natl Acad Sci USA 65:729–736

Ferrari TE, Yoder OC, Filner P (1973) Anaerobic nitrite production by plant cells and tissues: evidence for two nitrate pools. Plant Physiol 51:423–431

Fitter AH (1985) Functional significance of root morphology and root system architecture. In: Fitter AH et al. (eds) Ecological interactions in soil. Special publication of the British ecological survey, vol 4. Blackwell, Oxford, pp 87–106

Gerendas J, Ratcliffe RG (2002) Root pH regulation. In: Waisel Y, Eshel A, Kafkafi U (eds) Plant roots. The hidden half. Marcel Dekker, New York, pp 553-570

Glaab J, Kaiser WM (1993) Rapid modulation of nitrate reductase in pea roots. Planta 191(2):173–179

Gout E, Boisson AM, Aubert S, Douce R, Bligny R (2001) Origin of the Cytoplasmic pH changes during anaerobic stress in higher plant cells. Carbon-13 and Phosphorous-31 nuclear magnetic resonance studies. Plant Physiol 125:912–925

Hackett C (1972) A method of applying nutrients locally to roots under controlled conditions and some morphological effects of locally applied nitrate on the branching of wheat roots. Aust J Biol Sci 25:1169–1180

Hänsch R, Gómez Fessel D, Witt C, Hesberg C, Hoffmann G, Walch-Liu P, Engels C, Jörg Kruse, Rennenberg H, Kaiser WM, Mendel RR (2001) Tobacco plants that lack expression of functional nitrate reductase in roots show changes in growth rates and metabolite accumulation. J Exp Bot 52:1251–1258

Johnson C, Stout P, Broyer T, Carlton A (1957) Comparative chlorine requirements of different plant species. Plant Soil 8:337–353

Kenis JD, Trippi VS (1986) Regulation of nitrate reductase in detached oat leaves by light and oxygen. Physiol Plantarum 69:387–390

Kennedy RA, Rumpho ME, Fox TC (1992) Anaerobic metabolism in plants. Plant Physiol 100:1–6

Kirkby EA, Armstrong MJ (1980) Nitrate uptake by roots as regulated by nitrate assimilation in the shoot of castor oil plants. Plant Physiol 65:286–290

Lang B, Kaiser WM (1994) Solute content and energy status of roots of barley plants cultivated at different pH on nitrate- or ammonium-nitrogen. New Phytol 128:451–459

Lee RB (1978) Inorganic nitrogen metabolism in barley roots under poorly aerated conditions. J Exp Bot 29:693–708

Lee RB (1979) The release of nitrite from barley roots in response to metabolic inhibitors, uncoupling agents and anoxia. J Exp Bot 30:119–133

Leshem YY (2000) Nitric oxide in plants. Occurrence, function and use. Kluwer Academic Publishers, Dordrecht. 154 pp

Mann AF, Hucklesby DP, Hewitt EJ (1979) Effect of aerobic and anaerobic conditions on the in vivo nitrate reductase assay in spinach leaves. Planta 146:83–89

Marschner H (1995) Mineral nutrition of higher plants. Academic Press, London

Mattana M, Coragglio I, Bertani A, Reggiani R (1994) Expression of the enzymes of nitrate reduction durind the anaerobic germination of rice. Plant Physiol 106:1605–1608

Mattana M, Bestini F, Bertani A, Reggiani R (1997) Nitrate assimilation under anoxia in rice. Physiol Rastenii 44(4):547-551

McManmon M, Crawford RMM (1971) A metabolic theory of flooding tolerance: the significance of enzyme distribution and behaviour. New Phytol 70 (2):299–306

Mengel K, Viro M, Hehl G (1976) Effect of potassium on uptake and incorporation of ammonium-nitrogen of rice plants. Plant Soil 44:547–558

Müller E, Albers BP, Janiesch P (1994) Influence of nitrate and ammonium nutrition on fermentation, nitrate reductase activity and adenylate energy charge of roots of *Carex pseudocyperus* L. and *Carex sylvatica* Huds. exposed to anaerobic nutrient solutions. Plant Soil 166:221–230

Nance JF (1950) Inhibition of nitrate assimilation in excised wheat roots by various respiratory poisons. Plant Physiol 25:722–735

Oberson J, Pavelic D, Braendle R, Rawyler A (1999) Nitrate increases membrane stability of patato cells under anoxia. J Plant Physiol 55:792–794

Perata P, Alpi A (1991a) Ethanol induced injuries to carrot cells; the role of acetaldehyde. Plant Physiol 95 (3):748–752

Perata P, Alpi A (1991b) Ethanol metabolism in suspention cultured carrot cells. Physiol Plant 82 (1):103–108

Perata P, Alpi A (1993) Plant responses to anaerobiosis. Plant Sci 93(1–2):1–17

Pfister-Sieber M, Brandle R (1994) Aspects of plant behaviour under anoxia and post-anoxia. Proc R Soc Edinb Sec Biol Sci 102(0):313–324

Ratcliffe RG (1997) In vivo NMR studies of the metabolitic response of plant tissues to anoxia. Ann Bot 79 (Suppl A):39–48

Raven JA, Smith FA (1976) Nitrogen assimilation and transport in vascular land plants in relation to intracellular pH regulation. New Phytol 76:415–431

Reggiani R, Brambilla I, Bertani A (1985a) Effect of exogenous nitrate on anaerobic metabolism in excised rice roots I. Nitrate reduction and pyridine nucleotide pools. J Exp Bot 36:1193–1199

Reggiani R, Brambilla I, Bertani A (1985b) Effect of exogenous nitrate on anaerobic metabolism in excised rice roots II. Fermentative activity and adenylic energy charge. J Exp Bot 36:1698–1704

Reggiani R, Brambilla I, Bertani A (1985c) Effect of exogenous nitrate on anaerobic metabolism in excised rice roots III. Glycolitic intermediants and enzymatic activities. J Exp Bot 37(183):1472–1478

Roberts JKM, Callis J, Wemmer D, Walbot V, Jardetzky O (1984) Mechanims of cytoplasmic pH regulation in hypoxic maize Zea Mays root tips and its role in survival under hypoxia. Proc Natl Acad Sci USA 81(11):3379–3383

Roberts JKM, Chang K, Webster C, Callis J, Walbot V (1989) Dependance of ethanolic fermentation cytoplasmic pH regulation and viability of the activity of ADH in hypoxic maize root tips. Plant Physiol 89 (4):1275–1278

Rockel P, Strube F, Rockel A, Wildt J, Kaiser WM (2002) Regulation of nitric oxide (NO) production by plant nitrate reductase in vivo and in vitro. J Exp Bot 53:103–110

Saglio PH, Drew MC, Pradet A (1988) Metabolic acclimation to anoxia induced by low (2–4kPa partial pressure) oxygen pretreatment (hypoxia) in root tips of *Zea mays*. Plant Physiol 86:61–66

Saglio PH, Pradet A (1980) Soluble sugars, respiration, and energy charge during aging of excised maize root tips. Plant Physiol 66:516–519

Scheible WR, Lauerer M, Schulze ED, Caboche M, Stitt M (1997) Accumulation of nitrate in the shoot acts as a signal to regulate shoot-root allocation in tobacco. Plant J 11:671–691

Stoimenova M, Hänsch R, Mendel R-R, Gimmler H, Kaiser WM (2003a) The role of root nitrate reduction in the anoxic metabolism of roots I. Characterisation of root morphology and normoxic metabolism of a tobacco transformant lacking root nitrate reductase. Plant Soil 253:145–153

Stoimenova M, Libourel I, Ratcliffe RG, Kaiser WM (2003b) The role of nitrate reduction in the anoxic metabolism of roots. II. Anoxic metabolism of tobacco roots with or without nitrate reductase activity. Plant Soil 253:155–157

Ullrich WR (1983) Uptake and reduction of nitrate: algae and fungi. In: Epstein E, Läuchli A, Bieleski RL (eds) Encyclopaedia of plant physiology, new series, vol 15a. Springer, Berlin Heidelberg New York, pp 376–397

Vartapetian BB, Jackson MB (1997) Plant adaptations to anaerobic stress. Ann Bot 79 (Suppl A):3–20

Vartapetian BB, Polyakova LI (1999) protective effect of exogenous nitrate on the mitochondrial ultrastructure of oriza sativa coleoptiles under strict anoxia. Protoplazma 206(1–3):163–167

Walch-Liu P, Neumann G, Engels C (2001) Response to shoot and root growth to supply of different nitrogen forms is not related to carbohydrate and nitrogen status of tobacco plants. J Plant Nutr Soil Sci 164:97–103

Wiskich JT (1977) Mitochondrial metabolite transport. Annu Rev Plant Physiol 28:45–69

Yamasaki H, Sakihama Y, Takahashi S (1999) An alternative pathway for nitric oxide production in plants: new features of an old enzyme. Trends Plant Sci 4:128–129

M. Stoimenova
W.M. Kaiser
Lehrstuhl Botanik I
Julius-von-Sachs-Institut
für Biowissenschaften
Universität Würzburg
Julius-von-Sachs-Platz 2
97082 Würzburg, Germany
Tel.: 0049 931 888 6120
Fax: 0049 931 888 6158
e-mail: kaiser@botanik.uni-wuerzburg.de

Phloem-Mediated Remote Control by Long-Distance Signals

Aart J.E. van Bel, Frank Gaupels, Torsten Will, and Karl-Heinz Kogel

1 Pathways of Long-Distance Signalling in the Phloem

1.1 Introduction

Long-distance communication between organs by physical and chemical signals along the phloem is an established phenomenon. For instance, action potentials effecting on distant ion channels in *Mimosa pudica* are well-known long-distance messengers (Lüttge et al. 2002). Propagation of electric signals may be mainly associated with the phloem pathway (e.g. Sibaoka 1969; Fromm and Eschrich 1988; Rhodes et al. 1996). In response to an action potential propagated along the sieve tubes, ion channels and possibly aquaporins are gated selectively in distant pulvinus tissues. The resulting water flux from extensor to flexor parenchyma causes downward leaf movement in *Mimosa*.

First indication for chemical long-distance signal transmission was the presence of phytohormones in phloem sap (Ziegler 1975; Beveridge 2000). During the last decade, it has become apparent that mass flow through the sieve tubes (Sjolund 1997; Knoblauch and van Bel 1998; Oparka and Turgeon 1999; Thompson and Schulz 1999; van Bel et al. 2002) carries many other chemical signals. Responses known as systemic acquired resistance (SAR) are translocated from plant parts attacked by microbial or fungal phytopathogens, chewing insects or herbivores to unaffected organs (Enyedi et al. 1992; Uknes et al. 1992; Ryan 2000). These signals increase the resistance against pathogens and pests in distant organs by stimulating expression of genes coding for defense compounds (Farmer and Ryan 1992; Uknes et al. 1992; Pena-Cortes et al. 1995). To date, the nature of signals triggering defense reactions have not been elucidated conclusively. Other forms of remote control may be exerted on developmental programs in meristems. Proteins and RNA detected in the phloem sap may be involved in remote control on gene expression in sinks (Citovsky and Zambryski 2000; Lucas et al. 2001). Even photoassimilates themselves function as

Progress in Botany, Vol. 65

signals being engaged in remote control of gene expression (Rolland et al. 2002).

1.2 Modular Structure of Signalling Pathways in the Phloem of Angiosperms

Sieve tubes in angiosperms are arrays of sieve elements (SEs), each of which is associated with one or a few companion cells (CCs). SE and CC originate from the same meristematic mother cell and are strongly interdependent (van Bel et al. 2002). Given their phylogeny and ontogeny, it may be more appropriate to define sieve tubes as arrays of SE/CC-complexes. The advantage of the modular construction is that pathways can be extended virtually infinitely and that damaged sieve tubes can be locally repaired by insertion of new modules.

During ontogeny of the SE/CC-complex, SEs degenerate in a programmed fashion by which their cellular equipment is reduced to a plasma membrane, ER, a few mitochondria, parietal ER, SE plastids and to various forms of phloem-specific proteins (van Bel 2003).

Concurrently, three types of plasmodesmal connections develop (van Bel 2003). At the interface between CC and phloem parenchyma, conventional simple plasmodesmal connections persist. Between SEs, meristematic plasmodesmata transform into sieve pores. At the SE/CC-interface, special plasmodesmata (pore/plasmodesma units, PPUs) develop, which are branched at the CC-side and possess one corridor at the SE-side. These PPUs are held responsible for molecular trafficking from SEs to CCs (Oparka and Turgeon 1999).

Lack of studies pertinent to the phloem section (either collection, transport or release phloem) leaves us to speculate, where phloem-mobile signals are borne. Electrical signals may be generated mainly in the collection phloem zone and propagated along the transport phloem. With regard to macromolecular signalling, it appears plausible that capacity of signal generation is distributed along the entire phloem pathway. While many signals may be generated in the collection phloem zone, SE/CC-complexes in transport phloem may be able to multiply macromolecular signals or act as relay stations in signalling chains (Oparka and Turgeon 1999; Lucas et al. 2001). Given the correlation between phloem metabolism and translocation, involvement in sugar signalling may be equally distributed over the sieve tube modules along the pathway.

2 Long-Distance Signalling by Sugars

2.1 Sugar Concentration in the Sieve Tubes

Sugars function simultaneously as fuel and cargo (van Bel 2003). The pivotal role of sugars in their own uptake and retrieval is shown by genetic manipulation of glycolysis in CCs of tobacco plants. In transformants, in which pyrophosphate was hydrolyzed, phloem loading was impaired and massive photoassimilate loss was found along the phloem pathway due to a lack of ATP (Lerchl et al. 1995). Insertion of an invertase gene from *Saccharomyces* in the latter transformants led to a restoration of the ATP level. Further, phloem loading and phloem transport rates in the double-transformants are similar to those in non-transformed plants (Geigenberger et al. 1996). The CC-specific expression of sucrose synthase, an enzyme associated with a high respiratory demand of CCs, is probably related to the dramatic impact of sugar metabolism on phloem physiology (Nolte and Koch 1993; Koch and Nolte 1995).

Sophisticated sugar release/retrieval mechanisms along the pathway allow respiration and growth of heterotrophic cells along the pathway and temporary storage in axial sinks. Following metabolic processing, carbohydrates retrieved by the sieve tubes may strongly differ from those having escaped from the sieve tubes before. Obviously, carbohydrate metabolism along the pathway strongly impacts on the amount and nature of carbohydrates arriving in the terminal sinks. By acting so the entire phloem physiology exerts a remote control on processes in sinks. Carbohydrates arriving in sinks are monitored and become involved in regulation of gene expression (Lalonde et al. 1999; Hellmann et al. 2000; Smeekens 2000; Rolland et al. 2002) or function as osmotic equivalents in regulating phloem unloading (Patrick 1997).

2.2 Turgor Sensing in Sinks

The sugar concentration of phloem sap may exert an effect on turgor pressure sensing which, in turn, has a dramatic impact on phloem unloading. In many sink types, sieve element unloading occurs symplasmically (Patrick 1997; Patrick and Offler 2001; Patrick et al. 2001; Lalonde et al. 2003), most likely along a steep concentration gradient through an array of cells (Fisher and Wang 1975; Winch and Pritchard 1999). Gating of plasmodesmata, the key corridors in such a concept, may therefore regulate symplasmic phloem unloading (Lalonde et al. 2003). It is conceivable though that flux rates can be also modulated by regulating osmolarity of the apoplasmic fluid around sink cells. A high apoplasmic osmolarity would create a surge through the sink system resulting in an enhanced supply with nutrients (Dick and ap Rees 1975; Schulz 1994). Both mechanisms may operate in a concerted fashion; bathing pea root tips in high-osmolarity media was shown to result in a transient widening of plasmodesmata (Schulz 1995) and a concomitant enhanced supply with sugars (Schulz 1994). A key observation may be that phloem unloading in roots is stimulated by high concentrations of slowly permeant non-metabolizable mannitol and to a lesser extent by high concentrations of sucrose, the uptake of which reduces the diffusion gradient toward the sinks (Schulz 1994).

An entirely different situation occurs in sinks such as legume seeds where sink tissues and sieve tubes are separated by an apoplasmic interface (Patrick 1997; Patrick and Offler 2001; Patrick et al. 2001; Lalonde et al. 2003). In the seed coat, the cells are unloaded along a symplasmic route up to the apoplasmic gap between maternal and filial tissues (Patrick and Offler 2001; Lalonde et al. 2003). When the embryonic demand for nutrition exceeds supply by the seed coat phloem, osmolarity in the small apoplasmic space between seed coat and embryo readily decreases (Patrick and Offler 2001). This results in an immediate increase of turgor pressure in the seed coat parenchyma cells since the osmotic difference between apoplast and seed coat cells is relatively small (0.1–0.2 MPa). The increase of turgor above a certain turgor set point causes an error signal leading to an immediate compensatory increase in photoassimilate efflux (Patrick 1997). It infers that turgor-sensitive membrane-bound transporters are deemed to play a crucial role in sink nutrition. Obviously, the ability to maintain a turgor homeostasis ensures a continuous and sufficient photoassimilate supply (Patrick 1997). All sink systems may have in common that alterations in osmotic potential differences – which may be brought about by changes in phloem import – between symplast and apoplast are sensed as turgor pressure changes and adjusted by turgor-sensitive carriers and aquaporins (Crispeels et al. 1999).

2.3 Impact of Sugars and Sugar Levels on Organogenesis

Exemplary for the influence of sugars on the onset of developmental programs is the effect on flower (Corbesier et al. 1998) and seed development (Borisjuk et al. 1995; Wobus and Weber 1999). Flower induction seems to be associated with a sudden pulse of increased sugar export coincident with an enhanced phloem-loading capacity (Corbesier et al. 1998). However, it is difficult to imagine how phloem loading capacity is directly induced by sucrose concentration, as phloem loading appears to be independent of the amount of sucrose available in the leaf (Corbesier et al. 1998).

In seed development, causes and consequences are also hard to entangle in the apparent correlation between sugar levels and embryonic stages (Wobus and Weber 1999). In early stages of fava bean development, sucrose/glucose ratio in the seed coat exceeds by far those in endosperm cavity and embryo (Borisjuk et al. 1995; Weber et al. 1995). High hexose levels in the latter tissues are ascribed to activity of acid invertase (VfCWINV1) produced by the inner parenchyma layer of the seed coat (Weber et al. 1995). Glucose accumulated by the cotyledons display a steep inward gradient which correlates positively with cell division (Borisjuk et al. 1995, 1998). With degradation of the inner seed coat layer, sucrose/glucose ratio in the apoplasm around the embryo increases (Weber et al. 1995), possibly coincident with differentiation of the cotyledonary epidermis facing the seed coat into a transfer cell layer (Harrington et al. 1997; Weber et al. 1997).

There are several arguments in favor of sugar-controlled differentiation: (1) hexose and invertase levels are positively correlated with the seed size indicating a higher cell division rate in the presence of glucose (Weber et al. 1996). In turn, invertase production is linked with the thickness of the seed coat parenchyma (Weber et al. 1996). (2) Storage characteristics in dividing cells of isolated cotyledons is induced by a high sucrose concentration (Weber et al. 1996). (3) In cultured cotyledons, application of high hexose concentrations induces wall invaginations in cells located deeper inside the cotyledonary tissue (Offler et al. 1997; Farley et al. 2000). Their emergence suggests that these cells are not strictly pre-programmed and are under developmental control of apoplasmic sugar.

Sugar-induced control of gene expression may be deeply embedded into the genetic program of the cotyledonary cells, as deposition of cell wall material goes along with a number of events. The unique aggregation of microtubules in the outer periclinal cortical regions in adaxial epidermis cells suggests microtubular involvement in a polarized distribution of cellular organelles (Bulbert et al. 1998). Deposition of cell wall ingrowths also coincides with an increased expression of sucrose carriers and of plasma membrane H+/ATPase both in vivo (Weber et al. 1997) and in vitro (Farley et al. 2000). The relationship between apoplasmic sugar and cotyledonary differentiation seems to be complex. In-vivo experiments suggest that high hexose concentrations trigger transfer cell differentiation, high sucrose concentrations rather suppress it (Offler et al. 1997). This matches the observation that high sucrose concentrations suppress the production of VfSUT1 (Weber et al. 1997). Both observations render the idea plausible that transfer cell differentiation is meant to achieve accumulation of low concentrations of apoplasmic sucrose (Offler et al. 1997; Weber et al. 1997). However, cell wall invaginations may appear at a stage that the endosperm glucose concentration decreases and the sucrose/glucose ratio is increasing

(Wobus and Weber 1999). Therefore, signals produced by the maternal seed coat in response to contact with the growing cotyledon body are invoked as auxiliary tools to explain transfer cell development (Weber et al. 1997).

2.4 Sensing of Extracellular Sugar Concentrations

Sugar concentration may be directly sensed by the plant cells in analogy to sugar sensing by animal and yeast cells (Lalonde et al. 1999; Hellmann et al. 2000; Smeekens 2000; Rolland et al. 2002). Sucrose being the predominant transport sugar in many species may be sensed by sensors positioned on the plasma membrane. Strongly supportive of sucrose sensing by plasma membrane-bound proteins is the fact that cytosolic sucrose levels elevated by the supply of hexoses after metabolic processing are not sensed in *Arabidopsis* seedlings (Dijkwell et al. 1996). In contrast, exogenous supply of sucrose in physiological concentrations represses transcription and translation of a basic leucine zipper *AtB2* gene the expression of which is associated with newly established sink tissues (Rook et al. 1998a,b). In sugar beet leaves, high sucrose concentrations depressed transport activity of a H^+/sucrose symporter with correspondingly lower steady-state mRNA levels of Bv SUT1 (Chiou and Bush 1998). Decreased sucrose uptake capacity was caused by a reduction in the abundance of the symporter which is located exclusively on the plasma membrane of companion cells in leaf veins (Vaughn et al. 2002). The strikingly high turnover rate of message and symporter protein indicates a ready response to changes in apoplasmic sucrose concentrations (Vaughn et al. 2002). Membrane-bound proteins involved in sucrose sensing have not been identified yet with certainty, but their presence is unmistakable given the sensing of membrane-impermeant disaccharide analogs such as palatinose and turanose (Loreti et al. 2000; Fernie et al. 2001).

A putative candidate for being a sucrose sensor is the SUT2 sucrose transporter-like protein located in plasma membranes of tomato and *Arabidopsis* which contains extended intracellular domains, thus resembling the yeast sugar sensors SNF3 and RTG2 (Barker et al. 2000). SUT2 co-localizes with SUT1 and SUT4 on the plasma membrane of sieve elements in transport and release phloem and its expression is induced by high sucrose concentrations (Barker et al. 2000).

Extracellular glucose sensing may be carried out by membrane-bound transporter-like proteins analogous to the SNF3 and RTG2 sensors in yeast (Öczan et al. 1998). The ability of 3-O-methylglucose and 6-deoxyglucose to regulate gene expression points to the existence of glucose sensors located in the plasma membrane (Martin et al. 1997; Roitsch 1999).

2.5 Intracellular Signalling Pathways Associated with Sugar Sensing

Intracellular sensing of sugar concentrations in sinks may be executed mainly by monosaccharide sensors. In intracellular glucose sensing, hexokinase activity plays a key role (Jang et al. 1997, Hellmann et al. 2000, Rolland et al. 2002). Sugar-regulated gene expression is not restricted to the production of enzymes involved in sugar metabolism, but also extends to genes involved in phytohormone fabrication (Smeekens 2000, Rolland et al. 2002). A complete overview of sugar-induced regulation of gene expression goes far beyond the scope of this review. The reader is referred to the numerous reviews that provide an all-over complete picture of sugar impact on the genetic apparatus (e.g. Hellmann et al. 2000; Smeekens 2000; Rolland et al. 2002).

3 Long-Distance Signalling by Macromolecules

3.1 Trafficking Metabolites and Macromolecules Through Pore/Plasmodesma Units

In conjunction with long-distance signalling, it is important to realize that the enucleate SEs are largely dependent on their CCs (van Bel and Knoblauch 2000) and may be unable to produce most signalling compounds. Many of the long-distance signals translocated by the sieve tubes are therefore presumably manufactured in CCs and transferred to SEs (Thompson and Schulz 1999; Lucas et al. 2001).

Given the intimate relationship between SE and CC, a wide range of molecules was expected to be transferred through the PPUs (Raven 1991; Fisher et al. 1992). The wealth of proteins in sieve tube sap (e.g. Nakamura et al. 1993; Sakuth et al. 1993; Clark et al. 1997; Dannenhoffer et al. 1997; Schobert et al. 1998; Haebel and Kehr 2001) indicates a high degree of protein turnover. Turnover of phloem-specific proteins was demonstrated by incorporation of ^{35}S-methionine into phloem-specific proteins collected from phloem exudate (Fisher et al. 1992; Sakuth et al. 1993). Most, if not all, proteins present in the sieve tube sap are synthesized in CCs. For instance, mRNA of the 42 kDa PP2 lectin, present in the phloem sap of pumpkin, was localized exclusively in CCs and not in SEs (Bostwick et al. 1992).

Trafficking of macromolecules requires special properties of the PPUs. Molecular exclusion limits of PPUs seem to exceed by far those between parenchymatous cells. Using diverse methods, molecular exclusion limits close to 30 kDa were indicated for PPUs in tomato (Kempers and van Bel 1997) and *Arabidopsis* (Imlau et al. 1999). The enlarged exclusion diameter may be ascribed to certain phloem-specific proteins (sieve tube exudation proteins, STEPs) in the sieve tube sap. These proteins are able to increase the usual molecular exclusion diameter between mesophyll cells from 1 kDa up to about 30 kDa (Balachandran et al. 1997; Ishiwatari et al. 1998). Continuous presence of certain STEPs may result in a permanent "open-state" of PPUs.

The mode of macromolecular transfer through PPUs is still a matter of debate. Heat-shock proteins or heat-shock protein relatives may act as chaperones involved in the unfolding of the macromolecules as a prerequisite of PPU-trafficking (Schobert et al. 1995; Aoki et al. 2002). The chaperones may pick up and unfold macromolecules, dock onto receptor proteins on the ER-membrane and carry the macromolecules along the ER-rod traversing the PPU while widening the PPU diameter in a largely unknown fashion (Kragler et al. 1998, 2000; Oparka and Turgeon 1999; Oparka and Santa Cruz 2000). Specific proteins operating at the orifice of the PPU may be engaged in plasmodesmal dilation (Kragler et al. 1998, 2000; Lee et al. 2003). These findings suggest a sophisticated mode of PPU channelling, in which the proteinaceous complex (docking ER protein, chaperone and protein to be trafficked) docks onto two receptor proteins at the entry of PPUs before being channelled (Lucas et al. 2001). At the site of entry, the chaperone unfolds the protein and the complex crosses the PPU in the unfolded state (Lucas et al. 2001).

Trafficking of mRNA through PPUs from CC to SE may operate in a similar way (Lucas et al. 2001; Ruiz-Medrano et al. 2001). For example, the CmPP16 protein is able to specifically promote the transfer of its own RNA from CC to SE (Xoconostle-Cazares et al. 1999). Movement proteins of viruses are regarded as functional analogues to plant chaperones (Lucas et al. 2001). Potato leaf roll virus (PLRV) may imitate chaperone-mediated transfer of plant-borne mRNA in order to traffick its genome through PPUs (Prüfer et al. 1997; Sokolova et al. 1997). The acidic terminus of PLRV movement protein is probably phosphorylated by a membrane-bound kinase near the PPUs (Sokolova et al. 1997). Hence, phosphorylation may be involved in a conformational change of viral RNA precedent to passage of the PPU. The previous models render a function to ER as a sorting device for proteins and mRNA within the SE. This idea matches the observation that movement proteins of cucumber mosaic virus crept over the parietal ER in SEs (Blackman et al. 1998).

It has been a matter of debate whether macromolecules such as mRNA and proteins detected in phloem exudate are waste products of CCs (Oparka and Santa Cruz 2000). The rationale was that mRNA occurring in SEs could not possibly have a local effect given the absence of ribosomes. It was suspected that chaperones would only 'widen' the PPU-corridor to allow non-selective transport of macromolecules inclusive the escape of mRNA from CCs. Insertion of a *GFP* gene (for green fluorescent protein, GFP, 27 kDa) behind the promoter of the CC-specific *SUC2* gene resulted in GFP production in CCs and subsequent emergence of GFP in SEs of transgenic *Arabidopsis* plants (Imlau et al. 1999). The ability to traffick green fluorescent protein which is of animal origin speaks against a selective trafficking mechanism through PPUs.

However, the complex events at the orifice of the plasmodesma including the recognition by receptor proteins rather plea for the contrary (Kragler et al. 2000; Yoo et al. 2002; Lee et al. 2003). Another argument in favour of a selective mode of transfer is the asymmetric distribution of mRNA and its corresponding protein between SE and CC. Thioredoxin h is mainly found in the SEs, the corresponding RNA is localized in the CCs (Ishiwatari et al. 1998), while the opposite applies to CmNACP, a protein with an unknown function (Ruiz-Medrano et al. 1999). Given the sequence similarities with the NAC-family, a role in apical meristem development is possible. Supportive of selective protein trafficking through PPUs is the low number (five) of kinases in SEs of *Cucurbita* in comparison to that of several thousands in the parenchymatous cells (Yoo et al. 2002).

3.2 Long-Distance Signalling by RNA

Evidence is mounting that mRNA and other forms of RNA, alone or complexed with proteins, in the sieve tube sap exert remote control on gene expression in sinks (Jorgensen et al. 1998; Lucas 1999; Lucas et al. 2001). For instance, plant transcript delivery into SEs has been claimed for the sucrose carrier SUT1 in potato (Kühn et al. 1997). Collection of phloem sap from cut aphid stylets provides a method for identification of transcripts in the sieve tubes without transcript contamination from other cells (Sasaki et al.1998). Using stylectomy, mRNAs encoding the proton/sucrose cotransporter SUT1, a putative aquaporin and the H^+/ATPase PPA1 were detected in phloem exudate of barley (Doering-Saad et al. 2002). Ruiz-Medrano and coworkers (1999) identified ten representative transcripts within pumpkin phloem sap. Among them was *CmNACP* mRNA for which systemic transport through the phloem into the apical meristem has been proven (Ruiz-Medrano et al. 1999).

First evidence in favor of mRNA signalling is provided by the phenotypic impact of mRNA translocated via the phloem in heterografts of tomato plants (Kim et al. 2001). Xa-scions (yellowish, wild-type mutants with acute lobes) were grafted onto Me-stocks (octopinnate green leaves with unlobed leaflets). Grafting induced changes in the Xa-scions. In comparison to the Xa-plants, the leaves had a higher order of pinnation and the leaflet lobes were more rounded (Kim et al. 2001).

Further proof of remote control by RNA-species in the sieve tube sap is inferred from the transmission of signals imposing post-transcriptional gene silencing (PTGS) in sinks (Fagard and Vaucheret 2000; Lucas et al. 2001). PTGS is a phenomenon, by which aberrant RNA (transgene RNA, viral RNA) or excess RNA are recognized and deactivated/degraded (Ratcliff et al. 1997; Plasterk 2002). A major role in this protection mechanism is attributed to siRNAs (small interfering RNAs). Long-distance transport of PTGS signals may provide a remote control on developmental processes.

First indication for systemic movement of PTGS was provided by transgenic tobacco plants in which nitrite and nitrate reductase were overexpressed. Such plants showed PTGS of these genes beginning in small patches of the leaf. Later on, PTGS proliferated in the leaf and was apparently transmitted to other plant parts (Palauqui et al. 1996). A sequence-specific signal appeared to be translocated from a silenced stock to a non-silenced scion, in which exclusively expression of nitrate/nitrite reductase was suppressed (Palauqui et al. 1997). These phenomena suggest systemic movement of the signal (Voinnet and Baulcombe 1997). That the systemic signal conferring breakdown of GFP moved through three-way grafts indicated translocation via the phloem (Voinnet et al. 1998). Progression of silencing reminded of the phloem unloading pattern of GFP in sink leaves (Voinnet et al. 1998; Imlau et al. 1999).

Other indications in support of phloem-mediated PTGS transmission has been obtained with heterografts of pumpkin stock and cucumber scions (Xoconostle-Cazares et al. 1999; Lucas et al. 2001). Sieve tube sap of the scions contained pumpkin-specific transcripts (*CmPP16–1 and CmPP16–2*) and their respective proteins (CmPP16–1 and CmPP16–2), but were devoid of orthologous cucumber phloem sap proteins (Cs-PP16–1 or CsPP16–2). Absence of the cucumber protein CsPP16 in sieve tube sap of cucumber scion sap may be due to a blockage of the PPUs by CmPP-16 (Xoconostle-Cazares et al. 1999). More likely, however, expression of CsPP16 was suppressed in the presence of orthologous mRNA of pumpkin, although the substances responsible for PTGS of CsPP16 are undetected thus far. Among the candidates are siRNAs (Hamilton and Baulcombe 1999) which may operate in combination with proteins as systemic PTGS-signals (Waterhouse et al. 2001). Other candidates belong to a class of small RNAs – similar to microRNAs in animals – which have recently been discovered in plants (Llave et al. 2002; Jones 2002). Members of this group of more than 100 endogenous non-coding small RNAs (Jones 2002) are proposed to be also involved in PTGS (Llave et al. 2002) or in the regulation of transcription factors (Jones 2002).

As a final note, mRNA in the phloem sap has recently been related to systemic resistance induction (Ruiz-Medrano et al. 1999). mRNAs encoding WRKY transcription factors could be involved in triggering defense reactions (Rushton et al. 2002). Members of the WRKY family of transcription factors bind to W-box [(T)TGAC(C/T)] motives of promoters of genes potentially involved in pathogen defense pathways. There is growing evidence that W-boxes are a major class of *cis*-acting elements responsible for pathogen-induced expression of many plant genes. The importance of W-boxes has been illustrated recently by studies of the transcription during SAR expression in *Arabidopsis* (Maleck et al. 2000; Petersen et al. 2000).

3.3 Long-Distance Signalling by Proteins?

In general, phloem-specific proteins are transported over longer distances as is shown convincingly in grafts which allow to distinguish between endogenous and foreign phloem-specific proteins. In intergeneric grafts of Cucurbitaceae, stock-specific proteins were present in sieve tube exudates collected from the scion (Golecki et al. 1999). Several molecular techniques indicate that proteins rather than their transcripts are translocated (Golecki et al. 1998, 1999). The filamentous protein PP1 of *Cucurbita maxima* can undergo conformational changes and is translocated in a 88 kDa globular form which easily passes sieve pores (Leineweber et al. 2000).

That they can cross intergeneric graft borders (Golecki et al. 1998, 1999) raises the question whether phloem-specific proteins also impact on events in sink tissues. CmPP16, a plant paralog of a viral movement protein, that trafficks its own mRNA may effect plasmodesmal gating in distant organs. Several kinases (Nakamura et al. 1993; Yoo et al. 2002) and thioredoxin h and glutaredoxin (Schobert et al. 1998) could be engaged in targeting of unknown signal molecules. Dehydroascorbate reductase, superoxide dismutase and peroxidase protect SEs against oxidative stress (Walz et al. 2002) by catalyzing reactions that modulate reactive oxygen intermediates. The latter are known to induce defense reactions and to be involved in systemic acquired resistance against plant pathogens (Alvarez et al. 1998, Hückelhoven et al. 1999; Schultheiss et al. 2002; Hückelhoven and Kogel 2003). Another phloem sap protein potentially involved in defense and resistance signalling is lipoxygenase (Avdiushko et al. 1994). Essential components for the establishment of systemic acquired resistance probably are DIR1 lipid transfer proteins (Maldonado et al. 2002).

The phloem-translocated short peptide systemin (18 amino acids) increases the resistance against chewing insects in distant organs by inducing the production of proteinase inhibitors (Narvaez-Vasquez et al. 1995; Jacinto et al. 1997; Ryan 2000). Enhanced production of proteinase inhibitors interferes with events in the digestive tract of insects.

3.4 Pathway for Macromolecular Signals into the Sinks

In most unloading frames, photoassimilates are unloaded symplasmically into the sink cells (Lalonde et al. 2003). The logic that macromolecules follow the same plasmodesmal pathway highlights the role of the plasmodesmata in sinks. Experiments with *Arabidopsis* transformants demonstrate that the 27 kDa GFP is translocated through sieve tubes and can enter sink tissues (Imlau et al. 1999). Implicitly, plasmodesmata in sink tissues must have a molecular exclusion limit of at least approx. 30 kDa at some stage. Plasmodesmata of sink tobacco leaves indeed possess molecular exclusion limits close to 50 kDa as shown in epidermal cells bombarded with several plasmid constructs encoding GFP fusion proteins with a size up to 67 kDa (Oparka et al. 1999). During the transition to source cells, intercellular transport was very much restricted and the molecular exclusion limit drops radically, often down to 1 kDa (Oparka et al. 1999). Concurrently, the "large-sized" mode of molecular trafficking (but compare with Foster et al. 2002) is modified into a strongly selective mode. During transformation, plasmodesmata develop the ability of selective trafficking of macromolecules including viral genomes (Oparka et al. 1999). The decrease in molecular exclusion limit is probably correlated with a structural transformation of plasmodesmata during cell maturation (Ding et al. 1992). How plasmodesmal transformation relates timely with the temporary closure associated with cell differentiation (e.g. van Bel and van Rijen 1994; Ehlers et al. 1999; Ehlers and Kollmann 2000) remains to be investigated.

It appears that endogenous RNA molecules can be selectively trafficked into the shoot apex. In contrast, most viruses and PTGS signals are confronted with a selective barrier on their way to the sink cells (Lucas et al. 2001; Foster et al. 2002). A so-called surveillance system (Lucas et al. 2001) seemingly recognizes undesired RNA signals. Impairing the surveillance system by ectopic expression of a potexvirus movement protein led to various degrees of abnormal leaf polarity (Foster et al. 2002). These transgenic plants were compromised in their capacity to exclude viral RNA and specific PTGS signals. Given the properties of plasmodesmata in sink organs (Oparka et al. 1999; Roberts and Oparka 2003), one may speculate that the recognition sites for macromolecular signals reside at the plasmodesmal entry sites between protophloem SEs and sink cells.

What emerges is a regulatory complex of mRNA and protein messages imported via the phloem into the sinks along plasmodesmal checkpoints which may be selective depending on the nature of the molecules and the developmental stage of the plasmodesmata. The imported messages integrate the developing sink into the plant frame, whereas other sets of proteins are responsible for local fine-tuning of cell development in keeping with plasmodesmal permeability. In this context, it may be meaningful that transcription factors like KNOTTED 1 (Lucas et al. 1995), DEFICIENS and GLOBOSA (Perbal et al. 1996) and LEAFY (Sessions et al. 2000) which are transferred between sink cells, have not been detected in phloem sap. These proteins may be responsible for local implementation of the developmental program.

3.5 Long-Distance Signalling by Action Potentials

There is ample evidence for electric signal propagation in so-called excitable plants (e.g. Sibaoka 1962). That electric propagation would only occur in a few isolated species, such as e.g. *Mimosa* and *Dionea,* lacks evolutionary logic. Evidence is accumulating that electric propagation also takes place in plants that do not display readily visible reactions. Basic issues have been already covered in a textbook (Lüttge et al. 2002), but a few studies are added here.

There are astonishing similarities between action potentials in animals and plants. Triggering of action potentials in plants also follows the all-or-nothing principle (Shiina and Tazawa 1986; Fromm and Spanswick 1993; Malone 1996). The sensitivity to stimuli seems to depend on the individual plant (Fromm and Spanswick 1993) and on diurnal and seasonal rhythms (Zawadski et al. 1991).

Other similarities between animals and plants pertain to the existence of absolute and relative refractory periods. Refractory periods and propagation velocities in plants are much longer and much slower, respectively, than those in animals. Refractory periods were found to range between 2 min and 5 h! (Zawadski et al. 1991; Fromm and Spanswick 1993). Reported propagation speeds lie much closer and are between 2 and 10 cm/s (Fromm and Eschrich 1988; Fromm and Spanswick 1993; Mancuso 1999).

In higher plants, electric self-amplifying signals (Malone and Stankovic 1991) are likely propagated along sieve tubes (Fromm and Eschrich 1988; Wildon et al. 1992; Rhodes et al. 1996) by successive opening and closure of SE-membrane ion channels. Most likely, potassium and chloride channels are involved in the generation of action potentials in higher plants (Kourie 1994). Calcium channels seem to participate directly or indirectly in action potentials since the latter are inhibited in Ca^{2+}-free medium or in the presence of calcium channels blockers (Fromm and Spanswick 1993; Kourie 1994). Ion channels postulated to be involved in electrical propagation have been identified on SE/CC-membranes. The *ak2/ak3* gene (for the AK2/AK3 potassium channel) is expressed in the phloem of *Arabidopsis thaliana* (Marten et al. 1999; Deeken et al. 2000; Lacombe et al. 2000). A phloem-localized VFK1 K^+ channel in *Vicia faba* was postulated to be associated with phloem unloading (Ache et al. 2001). Phloem-specific (DHP-type) Ca^{2+} channels have been localized by immunocytochemistry and immunofluorescence in phloem of leaf veins of *Nicotiana tabacum* and *Pistia stratiotes* (Volk and Franceschi 2000).

Thus far, the function of electric signalling is a mystery for most plants. Apart from the impact of tactile stimuli in carnivorous plants and *Mimosa*, no apparent effects have been found. In one case, a concrete relationship between excitation (electrical stimulation) and response (increased production of proteinase inhibitors) was found (Stankovic and Davies 1997). Some authors value their existence as a means to achieve an integrated adaptive response to sudden changes in environmental conditions (Davies 1985; Zawadski et al. 1995). Action potentials appear to only occur in response to abrupt cold shocks or rigorous treatments such as burning (Sibaoka 1969; Fromm and Bauer 1994). To the best of our knowledge minor stimuli such as those induced by viral or fungal infection or penetration of aphid stylets could not be associated with action potentials. Sophisticated mechanisms may underlie the triggering of action potentials. In the upper parts of a plant, for instance, action potentials may even be subject to a biological clock (Zawadski et al. 1995). Expectedly, recording of causes and consequences in electrical signalling will strongly benefit from future improvement of methodical resolution.

3.6 Concluding Remarks

We are probably seeing glimpses of a multidimensional and sophisticated long-distance signalling system in the plant. Obviously, we have not yet an idea of the genetic and metabolic complexity and the degree of integration at various levels executed by a countless number of trade-off events. Undoubtedly, the system code will be extremely difficult to crack due to cross-talk between the signalling systems. In addition to that, the signalling network may respond to a variety of environmental challenges – including biotic and abiotic signals – and may be age-dependent and subject to day-time control.

Research in this field may also meet central questions regarding biotechnological improvement of agronomic plant quality with respect to abiotic stress, resistance to diseases and yield improvement. For instance, elucidation of systemic signalling processes may be a prerequisite for understanding the successful interaction between biotrophic parasites – such as powdery mildew and host plants – and provide information for disease resistance targets. Moreover, this information may indicate ways for yield improvement by modulating physiological source/sink interactions. Finally, exogenous modifiers of physiological source/sink balance – produced by biotrophic parasites – might provide examples for refined antimicrobial compounds in plant protection.

References

Ache P, Becker D, Deeken R, Dreyer I, Weber H, Fromm J, Hedrich R (2001) VFK1, a *Vicia faba* K^+ channel involved in phloem unloading. Plant J 27:571–580

Alvarez ME, Pennell RI, Meijer P-J, Ishikawa A, Dixon RA, Lamb C (1998) Reactive oxygen intermediates mediate a systemic signal network in the establishment of plant immunity. Cell 92:773–784

Aoki K, Kragler F, Xoconostle-Cazares B, Lucas W (2002) A subclass of plant heat shock cognate 70 chaperones carries a motif that facilitates trafficking through plasmodesmata. Proc Natl Acad Sci USA 99:16342–16347

Avdiushko SA, Ye XS, Kuc J, Hildebrand DF (1994) Lipoxygenase is an abundant protein in cucumber exudates. Planta 193:349–357

Balachandran S, Xiang Y, Schobert C, Thompson GA, Lucas WJ (1997) Phloem sap proteins from *Cucurbita maxima* and *Ricinus communis* have the capacity to traffic cell to cell through plasmodesmata. Proc Natl Acad Sci USA 94:14150–14155

Barker L, Kühn C, Weise A, Schulz A, Gebhardt C, Hirner B, Hellmann H, Schulze W, Ward JM, Frommer WB (2000) SUT2, a putative sucrose sensor in sieve elements. Plant Cell 12:1153–1164

Beveridge C (2000) The ups and downs of signalling between root and shoot. New Phytol 147:413–416

Blackman LM, Boevink P, Santa Cruz S, Palukaitis P, Oparka KJ (1998) The movement protein of cucumber mosaic virus trafficks into sieve elements in minor veins of *Nicotiana clevelandii*. Plant Cell 10:525–537

Borisjuk L, Weber H, Panitz R, Manteuffel R, Wobus U (1995) Embryogenesis in *Vicia faba* L.: histodifferentiation in relation to starch and storage protein synthesis. J Plant Physiol 147:203–218

Borisjuk L, Walenta S, Weber H, Müller-Klieser W, Wobus U (1998) High resolution histographical mapping of glucose concentrations in developing cotyledons of *Vicia faba* in relation to mitotic activity and storage processes: glucose as a possible developmental trigger. Plant J 15:583–591

Bostwick DE, Dannenhoffer JM, Skaggs MI, Lister RM, Larkins BA, Thompson GA (1992) Pumpkin phloem lectin genes are specifically expressed in companion cells. Plant Cell 4:1539–1548

Bulbert MW, Offler CE, McCurdy DW (1998) Polarized microtubule deposition coincides with wall ingrowth formation in transfer cells of *Vicia faba* L. cotyledons. Protoplasma 201:8–16

Chiou TJ, Bush DR (1998) Sucrose is a signal molecule in assimilate partitioning. Proc Natl Acad Sci USA 95:4784–4788

Citovsky V, Zambryski P (2000) Systemic transport of RNA in plants. Trends Plant Sci 5:52–54

Clark AM, Jacobsen, KR, Dannenhoffer JM, Skaggs MI, Thompson GA (1997) Molecular characterization of a phloem-specific gene encoding for the filament protein, phloem protein 1 (PP1) from *Cucurbita maxima*. Plant J 12:49–61

Corbesier L, Lejeune P, Bernier G (1998) The role of carbohydrates in the induction of flowering in *Arabidopsis thaliana*: comparison between the wild type and a starchless mutant. Planta 206:131–137

Crispeels MJ, Crawford NM, Schroeder JL (1999) Proteins for transport of water and mineral nutrients across the membranes of plant cells. Plant Cell 11:661–675

Dannenhoffer JM, Schulz A, Skaggs MI, Bostwick DE, Thompson GA (1997) Expression of the phloem lectin is developmentally linked to vascular differentiation in cucurbits. Planta 201:405–414

Davies E (1985) Action potentials as multifunctional signals in plants: a unifying hypothesis to explain apparently disparate wound responses. Plant Cell Environ 10:623–631

Deeken R, Sanders D, Ache P, Hedrich R (2000) Development and light-dependent regulation of a phloem-localised K^+-channel of *Arabidopsis thaliana*. Plant J 23:285–290

Dick PS, ap Rees T (1975) The pathway of sugar transport in roots of *Pisum sativum*. J Exp Bot 26:305–314

Dijkwell PP, Kock P, Bezemer R, Weisbeek P, Smeekens SCM (1996) Sucrose represses the deveopmentally controlled transient activation of the plastocyanin gene in *Arabidopsis thaliana* seedlings. Plant Phys 110:455-463

Ding B, Turgeon R, Parthasarathy MV (1992) Substructure of freeze substituted plasmodesmata. Protoplasma 169:28–41

Doering-Saad C, Newbury HJ, Bale JS, Prichard J (2002) Use of aphid stylectomy and RT-PCR for the detection of transporter mRNAs in sieve elements. J Exp Bot 53:631–637

Ehlers K, Kollmann R (2000) Primary and secondary plasmodesmata:structure, origin, and functioning. Protoplasma 216:1–30

Ehlers K, Binding H, Kollmann R (1999) The formation of symplasmic domains by plugging of plasmodesmata:a general event in plant morphogenesis? Protoplasma 209:181–192

Enyedi AJ, Yalpani N, Silverman P, Raskin I (1992) Signal molecules in systemic plant resistance to pathogens and pests. Cell 70:879–886

Fagard M, Vaucheret M (2000) (Trans)gene silencing in plants: how many mechanisms? Annu Rev Plant Physiol Plant Mol Biol 51:167–194

Farley SJ, Patrick JW, Offler CE (2000) Functional transfer cells differentiate in cultured cotyledons of *Vicia faba* L. seeds. Protoplasma 214:102–117

Farmer EE, Ryan CA (1992) Octadecanoid precursors of jasmonic acid activate the synthesis of wound-inducible proteinase inhibitors. Plant Cell 4:129–134

Fernie AR, Roessner U, Geigenberger P (2001) The sucrose analog palatinose leads to a stimulation of sucrose degradation and starch synthesis when supplied to discs of growing potato tubers. Plant Physiol 125:1967–1977

Fisher DB, Wang N (1975) Sucrose concentration gradients along the post-phloem transport pathway in the maternal tissues of developing wheat grains. Plant Physiol 109:587–592

Fisher DB, Wu Y, Ku MSB (1992) Turnover of soluble proteins in the wheat sieve tube. Plant Physiol 100:1433–1441

Foster TM, Lough TJ, Emerson SJ, Lee RH, Bowman JL, Forster RLS, Lucas WJ (2002) A surveillance system regulates selective entry of RNA into the shoot apex. Plant Cell 14:1497–1508

Fromm J, Bauer T (1994) Action potentials in maize sieve tubes change phloem translocation. J Exp Bot 45:463–469

Fromm J, Eschrich W (1988) Transport processes in stimulated and non-stimulated leaves of *Mimosa pudica*. I. The movement of ^{14}C-labelled photoassimilates. Trees 2:7–17

Fromm J, Spanswick R (1993) Characteristics of action potentials in willow (*Salix viminalis* L.). J Exp Bot 44:1119–1125

Geigenberger P, Lerchl J, Stitt M, Sonnewald U (1996) Phloem-specific expression of pyrophosphatase inhibits long-distance transport of carbohydrates and amino acids in tobacco plants. Plant Cell Environ 19:43–55

Golecki B, Schulz A, Carstens-Behrens U, Kollmann R (1998) Evidence for graft transmission of structural phloem proteins or their precursors in heterografts of Cucurbitaceae. Planta 206:630–640

Golecki B, Schulz A, Thompson GA (1999) Translocation of structural P proteins in the phloem. Plant Cell 11:127–140

Haebel S, Kehr J (2001) Matrix-assisted desorption/ionization time of flight mass spectrometry peptide mass fingerprints and post source decay: a tool for the identification and analysis of phloem proteins from *Cucurbita maxima* Duch. separated by two-dimensional polyacrylamide gel electrophoresis. Planta 213:586–593

Hamilton AJ, Baulcombe DC (1999) A species of small antisense RNA in posttranscriptional gene silencing in plants. Science 286:950–952

Harrington GN, Franceschi VR, Offler CE, Patrick JW, Tegeder M, Frommer WB, Harper JF, Hitz WD (1997) Cell specific expression of three genes involved in plasma membrane sucrose transport in developing *Vicia faba* seed. Protoplasma 197:160–173

Hellmann H, Barker L, Funck D, Frommer WB (2000) The regulation of assimilate allocation and transport. Aust J Plant Physiol 27:583–594

Hückelhoven R, Kogel K-H (2003) Who is who in powdery mildew resistance? Planta 216:891-902

Hückelhoven R, Fodor J, Preis C, Kogel K-H (199) Hypersensitive cell death and papilla formation in barley attacked by the powdery mildew fungus are associated with H_2O_2 accumulation but are not accompanied by enhanced concentrations of salicylic acid. Plant Physiol 119:1251–1260

Imlau A, Truernit E, Sauer N (1999) Cell-to-cell and long-distance trafficking of the green fluorescent protein in the phloem and symplastic unloading of the protein into sink tissues. Plant Cell 11:309–322

Ishiwatari Y, Fujiwara T, McFarland KC, Nemoto K, Hayashi H, Chino M, Lucas WJ (1998) Rice phloem thioredoxin h has the capacity to mediate its own cell-to-cell transport through plasmodesmata. Planta 205:12–22

Jacinto T, McGurl B, Franceschi V, Delano-Freier J, Ryan CA (1997) Tomato prosystemin promoter confers wound-inducible, vascular bundle-specific expression of the ß-glucuronidase gene in transgenic tomato plants. Planta 203:406–412

Jang J-C, Leon P, Zhou L, Sheen J (1997) Hexokinase as a sugar sensor in higher plants. Plant Cell 9:5–19

Jones L (2002) Revealing micro-RNAs in plants. Trends Plant Sci 7:473–475

Jorgensen RA, Atkinson RG, Forster RL, Lucas WJ (1998) An RNA-based information superhighway in plants. Science 279:1486–1487

Kempers R, van Bel AJE (1997) Symplasmic connections between sieve elements and companion cells in the stem phloem of *Vicia faba* L. have a size exclusion limit of at least 10 kDa. Planta 201:195–201

Kim M, Canio W, Kessler S, Sinha N (2001) Developmental changes due to long-distance movement of a homeobox fusion transcript in tomato. Science 293:287–289

Knoblauch M, van Bel AJE (1998) Sieve tubes in action. Plant Cell 10:35–50

Koch KE, Nolte KD (1995) Sugar-modelled expression of genes for sucrose metabolism and their relationship to transport pathways. In: Madore MA, Lucas WJ (eds) Carbon partitioning and source-sink interactions in plants. American Society of Plant Physiologists, Rockville, pp 145–155

Kouri JI (1994) Transient Cl^- and K^+ currents during the action potential in *Chara inflata*. Effect of external sorbitol, cations, and ion channel blockers. Plant Physiol 106:651–660

Kragler F, Monzer J, Shash K, Xoconostle-Cazares B, Lucas WJ (1998) Cell-to-cell transport of proteins:requirement for unfolding and characterization of binding to a putative plasmodesmal receptor. Plant J 15:367–381

Kragler F, Monzer J, Xoconostle-Cazares B, Lucas WJ (2000) Peptide antagonists of the plasmodesmal macromolecule trafficking pathway. EMBO J 19:2856–2868

Kühn C, Franceschi VR, Schulz A, Lemoine, Frommer WB (1997) Macromolecular trafficking indicated by localization and turnover of sucrose transporters in enucleate sieve elements. Science 275:1298–1300

Lacombe B, Pilot G, Michard E, Gaymard F, Sentenac H, Thibaud J-B (2000) A shaker-like K^+ channel with weak rectification is expressed in both source and sink phloem tissues of *Arabidopsis*. Plant Cell 12:837–851

Lalonde S, Boles E, Hellmann H, Barker L, Patrick JW, Frommer WB, Ward JM (1999) The dual function of sugar carriers:transport and sugar sensing. Plant Cell 11:707–726

Lalonde S, Tegeder M, Throne-Holst M, Frommer WB, Patrick JW (2003) Phloem loading and unloading of sugars and amino acids. Plant Cell Environ 26:37–56

Lee J-Y, Yoo B-C, Rojas M, Gomez-Ospina N, Staehelin LA, Lucas WJ (2003) Selective trafficking of non-cell-autonomous proteins mediated by NtNCAPP1. Science 299:392–396

Leineweber K, Schulz A, Thompson GA (2000) Dynamic transitions in the translocated phloem filament protein. Aust J Plant Physiol 27:733–741

Lerchl J, Geigenberger P, Stitt M, Sonnewald U (1995) Impaired photoassimilate partitioning caused by phloem-specific removal of pyrophosphate can be complemented by a phloem-specific cytosolic yeast-derived invertase in transgenic plants. Plant Cell 7:259–270

Llave C, Kasschau KD, Rector MA, Carrington JC (2002) Endogenous and silencing-associated small RNAs in plants. Plant Cell 14:1605–1619

Loreti E, Alpi A, Perata P (2000) Glucose and disaccharide-sensing mechanisms modulate the expression of a-amylase in barley embryos. Plant Physiol 123:939–948

Lucas WJ (1999) Plasmodesmata and the cell-to-cell transport of proteins and nucleoprotein complexes. J Exp Bot 50:979–987

Lucas WJ, Bouche-Pillon S, Jackson DP, Nguyen L, Baker L, Ding B, Hake S (1995) Selective trafficking of KNOTTED1 homeodomain protein and its mRNA through plasmodesmata. Science 270:1980–1983

Lucas WJ, Yoo B-C, Kragler F (2001) RNA as a long-distance information macromolecule in plants. Nat Rev Mol Cell Biol 2:849–857

Lüttge U, Kluge, Bauer G (2002) Botanik, 4th edn. VCH Verlag, Weinheim, pp 513–518.

Maldonado AM, Doerner P, Dixon RA, Lamb CJ, Cameron RK (2002) A putative lipid transfer protein involved in systemic resistance signalling in *Arabidopsis*. Nature 419:399–403

Maleck K, Levine A, Eulgem T, Morgen A, Schmid J, Lawton K, Dangl JL, Dietrich RA (2000) The transcriptome of *Arabidopsis thaliana* during systemic acquired resistance. Nat Genet 26:403–410

Malone M (1996) Rapid, long-distance signal transmission in higher plants. Adv Bot Res 22:163–228

Malone M, Stankovic B (1991) Surface potentials and hydraulic signals in wheat leaves following localized wounding by heat. Plant Cell Environ 14:431–436

Mancuso S (1999) Hydralic and electrical transmission of wound-induced signals in *Vitis vinifera*. Aust J Plant Physiol 26:55–61

Marten I, Hoth S, Deeken R, Ketchum KA, Hoshi T, Hedrich R (1999) *AKT3*, a phloem-localised K^+ channel is blocked by protons. Proc Natl Acad Sci USA 96:7581–7586

Martin T, Hellmann H, Schmidt R, Willmitzer L, Frommer WB (1997) Identification of mutants in metabolically regulated gene expression. Plant J 11:53–62

Nakamura S, Hayashi H, Mori S, Chino M (1993) Protein phosphorylation in the sieve tubes of rice plants. Plant Cell Physiol 34:927–933

Narvaez-Vasquez J, Orozco-Cardenas ML, Franceschi VR, Ryan CA (1995) Autoradiographic and biochemical evidence for the systemic translocation of systemin in tomato plants. Planta 195:593–600

Nolte KD, Koch KE (1993) Companion-cell specific localization of sucrose synthase in zones of phloem loading and unloading. Plant Physiol 101:899–905

Öczan S, Dover J, Johnston M (1998) Glucose sensing and signaling by two glucose receptors in the yeast *Saccharomyces cerevisiae*. EMBO J 17:2566–2673

Offler CE, Liet E, Sutton EG (1997) Transfer cell induction in cotyledons of *Vicia faba*. Protoplasma 200:51–64

Oparka KJ, Santa Cruz S (2000) The great escape: phloem transport and unloading of macromolecules. Annu Rev Plant Physiol Plant Mol Biol 51:323–347

Oparka KJ, Turgeon R (1999) Sieve elements and companion cells – traffic control centers of the phloem. Plant Cell 11:739–750

Oparka KJ, Roberts GA, Boevink P, Santa Cruz S, Roberts I., Pradel KS, Imlau, A, Kotlitzky G, Sauer N, Epel B (1999) Simple, but not branched, plasmodesmata allow the nonspecific trafficking of proteins in developing tobacco leaves. Cell 97:743–754

Palauqui J-C, Elmayan T, Dorlhac de Borne F, Crété P, Charles C, Vaucheret H (1996) Frequencies, timing, and spatial patterns of co-suppression of nitrate reductase and nitrite reductase in transgenic tobacco plants. Plant Physiol 112:1447–1456

Palauqui J-C, Elmayan T, Pollien JM, Vaucheret H. (1997) Systemic acquired silencing:transgene-specific post-transcriptional silencing is transmitted by grafting from silenced stocks to non-silenced scions. EMBO J 16:4738–4745

Patrick JW (1997) Phloem unloading: sieve element unloading and post-sieve element transport. Annu Rev Plant Physiol Plant Mol Biol 28:165–190

Patrick JW, Offler CE (2001) Compartmentation of transport and transfer events in developing seeds. J Exp Bot 52:551–564

Patrick JW, Zhang W, Tyerman SD, Offler CE, Walker NA (2001) Role of membrane transport in phloem translocation of assimilates and water. Aust J Plant Physiol 28:695–707

Pena-Cortes H, Fisahn J, Willmitzer L (1995) Signals involved in wound-induced proteinase inhibitor II gene expression in tomato and potato plants. Proc Natl Acad Sci USA 92:4106–4113

Perbal M-C, Haughn G, Saedler, H, Schwarz-Sommer Z (1996) Non-cell-autonomous function of the *Antirrhinum* floral homeotic proteins DEFICIENS and GLOBOSA is exerted by their polar cell-to-cell trafficking. Development 122:3433–3441

Petersen M, Brodersen P, Naested H et al. (2000) *Arabidopsis* MAP kinase 4 negatively regulates systemic acquired resistance. Cell 103:1111–1120

Plasterk RHA (2002) RNA silencing: the genome's immune system. Science 246:1263–1265

Prüfer D, Schmitz J, Tacke E, Kull B, Rohde W (1997) In vivo expression of full-length cDNA copy of potato leafroll virus (PLRV) in protoplasts and transgenic plants. Mol Gen Gen 253:609–614

Ratcliff F, Harrison BD, Baulcombe DC (1997) A similarity between viral defense and gene silencing in plants. Science 276:1558–1560

Raven JA (1991) Long-term functioning of enucleate sieve elements: possible mechanisms of damage avoidance and damage repair. Plant Cell Environ 14:139–146

Rhodes J, Thain JF, Wildon DC (1996) The pathway for systemic electrical signal transduction in the wounded tomato plant. Planta 200:50–57

Roberts AG, Oparka KJ (2003) Plasmodesmata and the control of symplastic transport. Plant Cell Environ 26:103–124

Roitsch T (1999) Source-sink regulation by sugar and stress. Curr Opin Plant Biol 2:198–206

Rolland F, Moore B, Sheen J (2002) Sugar sensing and signaling in plants. Plant Cell 14:S185-S205

Rook F, Gerrits N, Kortstee A, van Kampen M, Borrias M, Weisbeek P, Smeekens S (1998a) Sucrose-specific signalling represses translation of the *Arabidopsis* ATB2 bZIP transcription factor gene. Plant J 15:253–263

Rook F, Weisbeek PJ, Smeekens SCM (1998b) The light-controlled *Arabidopsis* bZIP transcription factor gene ATB2 encodes a protein with an unusually long leucine zipper domain. Plant Mol Biol 37:171–178

Ruiz-Medrano R, Xoconostle-Cazares B, Lucas WJ (1999) Phloem long-distance transport of CmNACP mRNA: implications for supracellular regulation in plants. Development 126:4405–4419

Ruiz-Medrano R, Xoconostle-Cazares B, Lucas WJ (2001) The phloem as a conduit for inter-organ communication. Curr Opin Plant Biol 4:202–209

Rushton PJ, Reinstädler A, Lipka V, Lippok B, Somssich IE (2002) Synthetic plant promoters containing defined regulatory elements provide novel insights into pathogen- and wound-induced signaling. Plant Cell 14:749–762

Ryan CA (2000) The systemin signaling pathway: differential activation of plant defensive genes. Biochim Biophys Acta 1477:112–121

Sakuth T, Schobert C, Pecsvaradi A, Eichholz A, Komor E, Orlich G (1993) Specific proteins in the sieve-tube exudate of *Ricinus communis* L. seedlings: separation, characterization and in-vivo labelling. Planta 191:207–213

Sasaki T, Chino M, Hayashi H, Fujiwara T (1998) Detection of several mRNA species in rice phloem sap. Plant Cell Physiol 39:895–897

Schobert C, Grossmann P, Gottschalk M, Komor E, Pecsvaradi A, zur Nieden U (1995) Sieve-tube exudate from *Ricinus communis* L. seedlings contains ubiquitin and chaperones. Planta 196:205–210

Schobert C, Baker L, Szederkenyi J, Grossmann P, Komor E, Hayashi H, Chino M, Lucas WJ (1998) Identification of immunologically related proteins in sieve-tube exudate collected from monocotyledonous and dicotyledonous plants. Planta 206:245–252

Schultheiss H, Dechert C, Kogel K-H, Hückelhoven R (2002) A small GTP-binding host protein is required for entry of powdery mildew fungus into barley cells. Plant Physiol 128:1447–1454

Schulz A (1994) Phloem transport and differential unloading in pea seedlings after source and sink manipulations. Planta 192:239–248

Schulz A (1995) Plasmodesmal widening accompanies the short-term increase in symplasmic phloem unloading in pea root tips under osmotic stress. Protoplasma 188:22–37

Sessions A, Yanovsky MF, Weigel D (2000) Cell-cell signaling and movement by the floral transcription factors LEAFY and APETALA 1. Science 289:779–781

Shiina T, Tazawa M (1986) Action potential in *Luffa cylindrica* and its effect on elongation growth. Plant Cell Physiol 27:1081–108

Siboaka T (1962) Excitable cells in *Mimosa*. Science 137:226–227

Sibaoka T (1969) Physiology of rapid movements in higher plants. Annu Rev Plant Physiol 20:165–184

Sjolund RD (1997) The phloem sieve element. A river flows through it. Plant Cell 9:1137–1146

Smeekens S (2000) Sugar-induced signal transduction in plants. Annu Rev Plant Physiol Plant Mol Biol 51:49–81

Sokolova M, Prüfer D, Tacke E, Rohde W (1997) The potato leaf roll virus 17 K movement protein is phosphorylated by a membrane-associated protein kinase from potato with biochemical features of protein kinase C. FEBS Lett 400:201–205

Stankovic B, Davies E (1997) Intercellular communication in plants: electrical stimulation of proteinase inhibitor gene expression in tomato. Planta 202:402–406

Thompson GA, Schulz A (1999) Long-distance transport of macromolecules. Trends Plant Sci 4:354–360

Uknes S, Mauch-Mani B, Moyer M, Potter S, Williams S, Dincher S, Chandler D, Slusarenko A, Ward E, Ryals J (1992) Acquired resistance in *Arabidopsis.* Plant Cell 4:645–656
van Bel AJE (2003) The phloem: a miracle of ingenuity. Plant Cell Environ 26:125–150
van Bel AJE, Knoblauch M (2000) Sieve element and companion cell:the story of the comatose patient and the hyperactive nurse. Aust J Plant Physiol 27:477–487
van Bel AJE, van Rijen HVM (1994) Microelectrode-recorded development of the symplasmic autonomy of the sieve element/companion cell complex in the stem phloem of *Lupinus luteus* L. Planta 192:165–175
van Bel AJE, Ehlers K, Knoblauch M (2002) Sieve elements caught in the act. Trends Plant Sci 7:126–132
Vaughn MW, Harrington GN, Bush DR (2002) Sucrose-mediated transcriptional regulation of sucrose symporter activity in the phloem. Proc Natl Acad Sci USA 99:10876–10880
Voinnet O, Baulcombe DC (1997) Systemic signaling in gene silencing. Nature 389:553
Voinnet O, Vain P, Angell S, Baulcombe DC (1998) Systemic spread of sequence-specific transgene RNA degradation in plants is initiated by localized introduction of ectopic promoterless DNA. Cell 95:177–187
Volk G, Franceschi VR (2000) Localization of a calcium-channel-like protein in the sieve element plasma membrane. Aust J Plant Physiol 27:779–786
Walz C, Juenger M, Schad M, Kehr J (2002) Evidence for the presence and activity of a complete antioxidant defence system in mature sieve tubes. Plant J 31:189–197
Waterhouse PM, Wang M-B, Finnegan EJ (2001) Role of short RNAs in gene silencing. Trends Plant Sci 6:297–301
Weber H, Borisjuk L, Heim U, Buchner P, Wobus U (1995) Seed coat-associated invertases of faba bean control both unloading and storage functions:cloning of cDNAs and cell-type specific expression. Plant Cell 7:1835–1846
Weber H, Borisjuk L, Wobus H (1996) Controlling seed development and seed size in *Vicia faba*: a role for seed coat-associated invertases and carbohydrate state. Plant J 10:823–830
Weber H, Borisjuk L, Heim U, Sauer N, Wobus U (1997) A role for sugar transporters during seed development: molecular characterization of a hexose and a sucrose carrier in faba bean seeds. Plant Cell 9:895–908
Wildon DC, Thain JF, Minchin PEH, Gubb IR, Reilly AJ, Skipper YD, Doherty HM, O'Donnell PJ, Bowles DJ (1992) Electrical signalling and systemic proteinase inhibitor induction in the wounded plant. Nature 360:62–65
Winch S, Pritchard J (1999) Acid-induced wall loosening is confined to the accelerating region of the root growing zone. J Exp Bot 50:1481–1488
Wobus U, Weber H (1999) Sugars as signal molecules in plant seed development. Biol Chem 380:937–944
Xoconostle-Cazares B, Xiang Y, Ruiz-Medrano R, Wang H-L, Monzer J, Yoo B-C, McFarland KC, Franceschi VR, Lucas WJ (1999) Plant paralog to viral movement protein that potentiates transport of mRNA into the phloem. Science 283:94–98
Yoo B-C, Lee J-Y, Lucas WJ (2002) Analysis of the complexity of protein kinases within the phloem sieve tube system. J Biol Chem 277:15325–15332
Zawadzki T, Davies E, Dziubinska H, Trebacz K (1991) Characteristics of action potentials in *Helianthus annuus.* Physiol Plant 83:601–604
Zawadzki T, Dziubinska H, Davies E (1995) Characteristics of action potentials generated spontaneously in *Helianthus annuus.* Physiol Plant 93:291–297
Ziegler H (1975) Nature of transported substances. In: Zimmermann MH, Milburn JA (eds) Encycopledia of plant physiology, vol I. Phloem transport. Springer, Berlin Heidelberg New York, pp 59–100

Aart J.E. van Bel
Frank Gaupels
Torsten Will
Plant Cell Biology
Institute of General Botany and Plant Physiology
Justus-Liebig University
Senckenbergstrasse 17
35390 Giessen, Germany
e-mail: Aart.v.Bel@bot1.bio.uni-giessen.de

Frank Gaupels
Karl-Heinz Kogel
Institute of Phytopathology
Justus-Liebig University
IFZ Heinrich-Buff-Ring 26–32
35392 Giessen, Germany

Systematics

Systematics of Himalayan Seed Plants

Shinobu Akiyama and Hideaki Ohba

1 Present Situation of Floristic Studies

Many national and international projects to prepare floristic treatments for Asian countries are in progress throughout the world. Amongst these are a number of projects focusing on the Himalayan region. The Flora of Bhutan, of which the first part was published in 1983, was completed when volume 3, part 2, dealing with the Orchidaceae, was published in 2002. The completion of the Flora of Bhutan marked the start of a new epoch in floristic research on the Himalayan region following the last major work for the area, Hooker's *The Flora of British India* (1872–1897). This major achievement was brought about by the late Andrew J. C. Grierson and his successor, David G. Long, of the Royal Botanic Garden Edinburgh, and other botanists. In Sikkim, now a part of India, a flora project was begun by the Botanical Survey of India after the establishment of the Sikkim Himalayan Circle of the Botanical Survey of India at Gangtok in 1979. In 1996, the first volume of that flora was published. For the western Himalaya, Karakorum and Kashmir have been treated as a part of the Flora of Pakistan compiled first by E. Nasir and S.I. Ali, and later by S.I. Ali and M. Qaiser since 1970.

In spite of its situation in the central part of the Himalayan region, there has been no comprehensive work on the flora of Nepal since the historical work by D. Don, *Prodromus Florae Nepalensis* (1825). *An Enumeration of the Flowering Plants of Nepal*, vols. 1–3, was published through a collaboration between British and Japanese workers by the British Museum (Natural History), now Natural History Museum, London, in 1978, 1979 and 1982, and represents an invaluable checklist for the area. The late Hiroshi Hara, University of Tokyo, was the central figure on the editorial board and one of the main authors of the *Enumeration*. He presided over the University of Tokyo's botanical research in the Himalayan region from 1960 and successfully established a center of Himalayan botany in Tokyo. After the Department of Medicinal Plants (reorganized as the Department of Forestry and Plant Research in 1989, and now the Department of Plant Resources) was established in Nepal in 1961, Hara cooperated with the De-

Progress in Botany, Vol. 65

partment, and sent Hiroo Kanai to Nepal as a Colombo Plan advisor to establish a herbarium and to initiate field surveys of the flora of Nepal in 1969. After the publication of the first local flora, *Flora of Phulchoki and Godavari*, for Nepal by the Department of Medicinal Plants in 1969, a number of local floras for central and western Nepal were published. The Department of Medicinal Plants played an important role in promoting international botanical studies as an offshoot and continues to conduct field surveys in collaboration with many foreign institutes and researchers. The latest publications through these efforts are *Flowering Plants of Nepal (Phanerogams)* (Singh 2001) and *Pteridophytes of Nepal* (Thapa 2002).

Upon Hara's retirement, Hideaki Ohba succeeded him and further developed the research program in Himalayan botany. As a first step to a comprehensive *Flora of Nepal*, a number of regional floras were published based on expeditions to central and eastern Nepal: *The Alpine Flora of the Jaljale Himal, East Nepal* (Ohba and Akiyama 1992), *A Contribution to the Flora of Ganesh Himal, Central Nepal* (Ohba and Ikeda 1999), and *The Flora of Hinku and Hunku Valleys, East Nepal* (Ohba and Ikeda 2000). The *Name List of the Flowering Plants and Gymnosperms of Nepal* (compiled by Koba et al. 1994) is an alphabetical list of all the seed plants recorded in *An Enumeration of the Flowering Plants of Nepal*. Based on that list, a database of specimens collected in Nepal (Flora of Nepal Database) was started by the Society of Himalayan Botany. The Society of Himalayan Botany has also published a newsletter, *Newsletter of Himalayan Botany*, since 1986 to promote both floristic and new research and to facilitate cooperation and communication among botanists interested in Himalayan botany.

In May 2001 a commemorative conference celebrating 40 years of joint research by Nepalese and Japanese botanists was held in the National Herbarium and Plant Laboratories at Godawari, Nepal. The aim of the conference was to review past and present developments in the study of Himalayan plants and to provide a prospective for the future of plant sciences in the region (Bista and Ohba 2002). The proceedings, with some additional articles, were issued as *Himalayan Botany in the Twentieth and Twenty-First Centuries* by the Society of Himalayan Botany (Noshiro and Rajbhandari 2002). The work consists of three parts: part 1, *History of Himalayan Botany*, part 2, *Achievements in Himalayan Botany* (recent research on Himalayan plants), and part 3, *Sharing of Field Experience*.

2 The Flora of Nepal Project

The first meeting to discuss the preparation of a *Flora of Nepal* was held on 2 September 1993 during the 15th International Botanical Congress in

Tokyo. It was agreed that *The Flora of Nepal* should be published in the very near future through an international cooperative project (see *Newsletter of Himalayan Botany*, no. 15 in 1994).

The Natural History Museum, London, decided to proceed with the preparation of the Nepal flora through the Darwin Initiative Programme between 1997 and 1999. That program resulted in the *Annotated Checklist of the Flowering Plants of Nepal* published by Press et al. (2000). The checklist lists 216 families, 1534 genera, 5345 species, 163 subspecies, 517 varieties, and 51 forms of flowering plants of Nepal, but according to Rajbhandari (2002) many species previously recorded from Nepal were omitted. The *Catalogue of Type Specimens from Nepal* (Shrestha and Press 2000) was also published as a result of that program. Those publications form important bases for the preparation of a *Flora of Nepal*.

In May 2002 the first editorial meeting for the Flora of Nepal was held at the Royal Botanic Garden Edinburgh, marking the actual start of the project (Watson and Blackmore 2003). The first volume of the Flora will appear in 2004 after 11 years of national and international preparation.

3 Noteworthy Taxonomic Works

Several taxonomic works on Himalayan plants and some long-term revisions or monographic studies related to the floristic projects mentioned above have also been published during this period. Taxonomic works are those by Dickoré (2001), Mill (2001b), Pearce et al. (2001), and Abid and Qaiser (2002). Revision and monographic studies include those by Robson (2001, 2002) on *Hypericum*, Al-Shehbaz (2001, 2002a,b,c) for Brassicaceae, Miyamoto and Ohba (2002) and Snogerup et al. (2002) on *Juncus*, Yonekura and Ohashi (2001, 2002) on *Bistorta* and Fujikawa and Ohba (2002) on *Saussurea*.

In some large families like the Asteraceae the full range of species-level diversity is still not well understood in the Himalayan region. The floristic projects have helped to improve the species-level taxonomy in such large families. We therefore wish to remark on papers in the following families.

3.1 Asteraceae

Dickoré (2001), who is preparing the flora of the Karakorum area, studied the taxonomy of three small, alpine species of *Saussurea*; *S. atkinsonii*, *S. ovata*, *S. elliptica*, and their distributions. He concluded that the major ranges of the W Kunlun-Karakorum-W Himalaya system have been colonized to about the same elevation and in similar habits by the three species, but that they apparently represent independent, isolated, taxonomic units. To consider the geo-ecological history of these high mountain habitats the phylogenetic relationships of species such as these will require detailed analysis.

Abid and Qaiser (2002) recognized 11 species including a new species, *I. stewartii*, in their regional revision of the species of *Inula* (in the strict sense) in Pakistan and Kashmir.

Saussurea subgenus *Eriocoryne* occurs in the Sino-Himalayan and central Asian alpine and subnival zones. Species of *Eriocoryne* exhibit a unique snow-ball like habit characterized by thick, unbranched, hollow, flowering stems with a terminal synflorescence usually covered by long, white or pinkish hairs. Fujikawa and Ohba (2002) discovered two new species in the Nepal Himalaya during their monographic work, which they described as *S. bhutkesh* and *S. kanaii*, and published with detailed illustrations and a key to the species.

3.2 Scrophulariaceae

Mill (2001a) discussed problems with the infrageneric classification of *Pedicularis* and its phytogeography in the Flora of Bhutan area. *Pedicularis* is the largest genus in the family with perhaps as many as 800 species. The main center of diversity, by far, is in China where 352 species have been recorded. The genus is represented by only 76 species in Bhutan.

The infrageneric classifications of *Pedicularis* have been numerous, but none has covered both the entire distributional range and all currently known species. Mill believed it appropriate to use the same classification systems employed recently in Nepal and China, since the species of *Pedicularis* in Bhutan are strongly related to those in both of those countries. Categories grex and subgrex, which are not acceptable under the present International Code of Botanical Nomenclature, were used as ranks in the classification system in the Chinese language flora of China, *Flora Reipublicae Popularis Sinicae*. Yamazaki's sections in his treatment of *Pedicularis* in Nepal (Yamazaki 1988) were based on those invalid categories in *Flora Reipublicae Popularis Sinicae*. Mill validated the sectional names adopted by Yamazaki. Mill's (2001a) paper, in which six new species and one variety were described, is useful for understanding the taxonomy of the Sino-Himalayan species of *Pedicularis*.

3.3 Apiaceae

In the eastern Himalaya field research in the alpine and subnival zones has been concentrated in the spring and summer and less often in autumn, since dangerous snows and snow deposits are possible at that time of year. Studies of the Apiaceae are therefore not easy particularly in the Himalaya eastward from Nepal, since the fruits, on which the taxonomy in the family is based, are produced late in the season. Mark Watson's work on the Apiaceae in relation to the Flora of Bhutan has been a remarkable contribution to our understanding of Himalayan Apiaceae.

Pimenov and Kljuykov and their collaborators at the Botanical Gardens, Moscow State University, have worked on the family mainly in Nepal and in the western Himalaya (Pimenov et al. 2001, Pimenov and Kljuykov 2002a,b). Although Farille et al. (1985) maintained the generic status of *Oreocome*, the genus has not been accepted in the majority of the regional floristic treatments and has been regarded as a part of the genus *Selinum*, s. l. Based on carpological and petiolar anatomy, Pimenov et al. (2001), in a revisionary work on *Oreocome* Edgew., redefined it as a distinct genus consisting of 6 species distributed from Pakistan to SW China, with a center of diversity in the Himalayas. They provided keys to the six species they recognize and to the Himalayan genus *Ligusticopsis*, which resembles *Oreocome* and *Selinum*. Two species, *O. involucellata* and *O. depauperata*, from Nepal and one, *O. hindukushensis*, from Pakistan, are described as new species. They also provided anatomical data on the accepted taxa and detailed descriptions. It is difficult to evaluate their work critically, however, since they cite only the specimens examined for the new species.

Pimenov and Kljuykov (2002a) reported on the carpel anatomy of all the species of *Lalldhowjia*, a small genus of the eastern Himalaya and related to *Peucedanum*. They described as new *L. pastinacifolia*, based on one of their collections from central Nepal near Gosain Kund.

Pimenov and Kljuykov (2002b) examined the type of *Archangelica roylei* Lindl. in trying to determine the taxonomic status of *Ligusticum alboalatum* Haines. They concluded that the type of *Archangelica roylei* agrees exactly with plants now known in numerous taxonomic and floristic works as *Ligusticum elatum* Edgew. The plant belongs neither to *Angelica* nor *Ligusticum*, but is closer to *Seseli mucronatum* (Schrenk) Pimenov & Sdobnina; they made the new combination, *Seseli roylei*, to reflect their findings.

3.4 Araliaceae

Jun Wen and her collaborators studied chloroplast DNA variation of *Panax* in Nepal (Yoo et al. 2001). Since 1991 Wen has studied and published numerous papers on Araliaceae. Wen also studied *Panax*, which includes the famous oriental drug plant, ginseng, more widely, initially starting from an interest in the eastern Asian-eastern North American disjunct pattern of distribution and then expanding to include both molecular approaches (Wen and Zimmer 1996; Wen 2001; Wen et al. 2001) and the study of the ultrastructure of the pollen grains (Wen and Nowicke 1999).

Yoo et al. (2001) examined restriction sites and size variation of five PCR-amplified fragments of non-coding chloroplast DNA (cpDNA) in materials from 13 populations from Nepal and China. Fourteen restriction endonucleases produced 81 restriction site and length variations from the large single-copy region of cpDNA, 27 of which are polymorphic. The dataset indicated that two distinct groups of *Panax* are in Nepal: clade I is referable to *P. pseudoginseng* subsp. *pseudoginseng*, and clade II comprises *P. pseudoginseng* subsp. *himalaicus* (with three varieties). *Panax pseudoginseng* subsp. *japonicus* and *P. ginseng* differed from the Himalayan *Panax* in cpDNA characters. The status of *Panax pseudoginseng* subsp. *pseudoginseng* sensu Hara (1970) as a distinct taxon from central Nepal was supported from its extremely restricted distribution and its morphology. This paper presents a new means for analyzing speciation in Himalayan plants at the molecular level.

3.5 Balsaminaceae

Grey-Wilson, and Akiyama and Ohba have made critical studies on the Indo-Himalayan and Sino-Himalayan species of *Impatiens*, but the taxonomy of the genus is still controversial. Akiyama and her collaborators have used morphological and cytological methods to study Sino-Himalayan *Impatiens* (Akiyama et al. 1991, 1992; Akiyama and Ohba 2000).

Fujihashi et al. (2002) used a molecular approach to examine 26 accessions representing 25 species mainly collected from the Sino-Himalayan region, plus *Tropaeolum major* as the outgroup. They used a combined sequence of *rbcL* (1291 bp of homologous region) and the spacer between *trnL* and *trnF* (344–412 bp). This is probably the first molecular approach to a phylogenetic study of *Impatiens*. The data set revealed the presence of two large clades: clade A consisting of only the Himalayan species, and clade B composed of species distributed throughout various parts of the world. The two clades are strongly correlated with basic chromosome numbers:

clade A has x=7 or 9, clade B has x=8 or 10. The phylogenetic tree obtained also supported the evolutionary trend of inflorescences obtained from morphology, but did not agree with previous thoughts on the evolution of the lower sepals, which are significant in determining the form and shape of the flowers.

Their work is interesting in that it suggests the heterogeneous nature of the Sino-Himalayan flora, even though it is preliminary and was based on only 25 species out of more than 100 species estimated to occur in the Sino-Himalayan region.

3.6 Brassicaceae

Al-Shehbaz has continuously worked on the taxonomy of the Brassicaceae and has published four special papers during this period. His greatest contribution is the treatment of the entire family with O. Appel (Appel and Al-Shehbaz 2002) for Kubitzki and Bayer's *The Families and Genera of Vascular Plants*. Through their work the generic diversification of the family in the Himalayan region has been made clear.

One species of *Cardamine* collected in Bhutan in 1938 has been described as new (Al-Shehbaz 2001). It is known only from the type collection and is most closely related to *C. violacea* (D. Don) Hook. f. & Thomson and *C. nepalensis* Kurosaki & H. Ohba. *Cardamine nepalensis* which has been known previously from Nepal is reported for the first time from Sikkim.

Al-Shehbaz (2002a) described a considerable number of new species in several genera from the Himalayas, including *Alyssum klimesii* and *Aphragmus ladakiana* both from Ladak. *Aphragmus*, a small genus ranging from northern and central Asia, the Himalayas and disjunctly in North America, now includes six species. He treated five species in a previous paper (Al-Shehbaz 2000). Al-Shehbaz (2002b) also described six new species of *Draba* from the Himalayas, including Tibet, and published another paper (Al-Shehbaz 2002c) on *Draba* and the allied genera *Hemilophia* and *Lepidostemon*, which included a key to the three genera and to the species of *Hemilophia* and *Lepidostemon*. *Hemilophia*, with five species, is endemic to China. A new combination, *H. serpens* (O.E. Schulz) Al-Shehbaz was made. *Lepidostemon* was previously recognized as comprising five species from Tibet and the eastern Himalaya (Nepal, Sikkim, Bhutan), but a sixth species, *L. williamsii* (H. Hara) Al-Shehbaz, has been added.

3.7 Fabaceae

The 78 previously recognized species and 12 infraspecific taxa of *Campylotropis* were revised into 37 species and 12 infraspecific taxa by Iokawa and Ohashi (2002). *Campylotropis* ranges from Pakistan to Korea through the Himalayas, China, Java, and Taiwan. The following species occur in the Himalayan region: *C. falconeri* (Prain) Schindl. (West Himalaya and Kashmir), *C. griffithii* Schindl. (Bhutan), *C. hirtella* (Franch.) Schindl. (Assam), *C. macrostyla* (D. Don) Miq. (Kashmir to Nepal), *C. speciosa* (Royle ex Schindl.) Schindl. (Himachal Pradesh to Assam), *C. stenocarpa* (Klotzsch) Schindl. (Kashmir to Nepal), *C. thomsonii* (Benth. ex Baker) Schindl. (Assam). In their paper, Iokawa and Ohashi (2002) provided a key to the taxa, synonyms with references, morphological descriptions, distribution maps, and a list of specimens examined. They also studied chromosomes and pollen grains, and conducted SEM examinations of some vegetative and reproductive features.

The genus *Vigna* has been studied by several research groups. One new species, *V. nepalensis* Tateishi & Maxted from E Nepal, Bhutan and India (Sikkim, Darjeeling and Assam) (Tateishi and Maxted 2002), was described. The species resembles *V. angularis* var. *nipponensis*, which occurs from Japan to the Himalayas through Korea, China and northern Myanmar, but differs in having a short hilum, glabrous bracts and flattened (not boat-shaped) bracteoles.

3.8 Hypericaceae

Norman Robson has been studying the genus *Hypericum* since the 1960s. Since his infrageneric classification appeared in 1977, he has used it as a basis for several monographic treatments. Part 4 of his monograph (Robson 2001, 2002) includes treatments of sections 7. *Roscyna*, 8. *Bupleuroides* and 9. *Hypericum*, all of which are considered to be derived directly or indirectly from section 3, *Ascyreia*. Parts 4(1) and 4(2) provide an introduction to the group and systematic treatments of section 7 and part of section 9, and treat *H. bupleuroides* in the monotypic section 8. *Bupleuroides*. Robson considered *H. elatoides*, which he erroneously included in section 7 in 1977, to be classified into section 3, *Ascyreia*. He also published a new genus, *Lianthus*, based on *H. ellipticifolium*, which he previously included in section 9.

Hui-Lin Li described *Hypericum ellipticifolium* in 1944 from the Yunnan-Myanmar border, but it differs from *Hypericum* in having white petals and fasciclodes, and most notably, two systems of foliar glands. It does not appear to be derived from any species of *Hypericum*, but appears similar to

Triadenum and *Thornea*, two genera placed in the Hypericoideae–Hypericeae by Robson. He discussed their relationship and evolutionary history. Section *Hypericum*, consisting of 66 species, was revised and divided into 6 sections: *Hypericum*, *Concinna*, *Graveolentia*, *Sampsonia*, *Elodeoida*, and *Monanthema*. Except for section *Hypericum*, the others were newly described. Three of the five species of section *Elodeoida*, *H. petiolatum*, *H. kingdonii*, and *H. elodeoides*, range into the Indo-Himalayan region.

3.9 Polygonaceae

Polygonum has recently been separated into several genera by specialists of the family. *Bistorta*, which was delimited as a small group of *Persicaria* by some, is a diverse genus, especially in the Sino-Himalayan region, and comprises ca. 30 species. Two groups, Yonekura and Ohashi (Sendai) and Miyamoto and Ohba (Tokyo) (Miyamoto et al. 2002) have studied the Sino-Himalayan species. Yonekura and Ohashi (2001, 2002) determined the taxonomic delimitation of several previously poorly known species and also described *B. griersonii*, *B. longispicata*, *B. ludlowii*, *B. burmanica* from the Sino-Himalayan region. They also revised the *B. amplexicalis* complex and proposed a new system of classification based on a comparative study of the leaf sheaths of the cauline leaves. They also discussed the lack of leaf sheaths, which they consider to be apomorphic. In the near future we expect to have the first regional monographic treatment of *Bistorta* in the Sino-Himalayan region.

3.10 Ranunculaceae

Hepatica falconeri which occurs in Kashmir and has been known as *Anemone falconeri* was studied morphologically, phenologically and cytologically by Ogisu et al. (2002). They made clear that the species is more similar to *H. nobilis* than to *Anemone flaccida*. The chromosome number of 2n=14, consisting of six pairs of median- and one pair of subterminal-centromeric chromosomes bearing satellites, indicates that *H. falconeri* is better placed in *Hepatica* than in *Anemone*.

3.11 Juncaceae

A monographic account of Juncaceae by Kirschner et al. has recently been published in the *Species Plantarum–Flora of the World* series. Since nomenclatural changes are not published within that series, Snogerup et al. (2002) published a related paper in which *Juncus leschenaultii* Laharpe was reduced to a subspecies of *J. prismatocarpus* R. Br. *Juncus leschenaultii*, distributed widely from the eastern Himalaya and Malesia to Kamchatka through Indo-China, China, and Japan, has been treated as a distinct species or conspecific with *J. prismatocarpus*.

Snogerup distinguished it from subsp. *prismatocarpus* by differences in the shape of the capsules, the length of the rostrum, the width of the sheath margins and the width of the stems. About 12 taxa described as species, including *J. unibracteatus* Griff., *J. hizenensis* Satake and *J. latior* Satake, and varieties are now treated as synonyms of subsp. *leschenaultii*.

Miyamoto and Ohba (2002) examined the Sino-Himalayan Junci and have published their studies on the *Juncus duthiei* group. This group, according to them, is characterized by cylindrical leaves, the absence of a sheathing bract on the peduncle, the lowest bract longer than or as long as the flowers, the clustered flowers in heads, and stamens as long as or shorter than the perianth. The group differs from the *J. himalensis* group, which it resembles, in lacking a sheathing bract on the peduncle. All the species of the *J. duthiei* group have similar morphology and the taxonomy has often been confused. The seven species within the group are distinguished by a combination of several characters. Miyamoto and Ohba (2002) illustrated all of the species in detail, with the exception of *J. longiflorus* (A. Camus) Noltie.

The Sino-Himalayan species of *Juncus* also have been studied by Noltie, especially as they relate to the Flora of Bhutan. Miyamoto and Ohba differ from Noltie in their treatment of *J. rohtangensis* Goel & Aswal and *J. harae* Miyam. & H. Ohba. Miyamoto and Ohba emphasized the length and ratio of the anthers and filaments, but Noltie regarded those as insignificant characters. In *J. duthiei* Noltie considered the ratio to be variable and did not distinguish *J. rohtangensis* and *J. harae* from *J. duthiei*. Miyamoto and Ohba's illustrations and dissected figures are helpful in showing the diversity and variation in features of the Sino-Himalayan Junci.

3.12 Orchidaceae

Pearce et al. (2001) treated the Orchidaceae for the Flora of Bhutan. One of the authors, the late Jany Renz, established the genus *Bhutanthera*, of which the type species, *B. albomarginata*, is a dwarf terrestrial orchid similar to both *Habenaria* and *Platanthera*. The new genus is characterized by having globose tubers, a three-lobed lip, and conjoined cushion-like stigmas. Renz overlooked the work of Ohba and Tsukaya (2000), who had described the morphology of the type species. Of the five species recognized by Renz, two of them, *B. albosanguinea* and *B. albovirens*, appear to be conspecific with the type of *B. albomarginata*.

4 Conclusion

In spite of the wealth of excellent research being conducted in the region, our knowledge of the Himalayan flora is still insufficient. In particular, field and taxonomic studies are much in need in the flora of the tropical regions extending from the foot of the Himalayan range to the Indian Plain. No contributions to the tropical floras of this area have been published during the recent period. Recent extensive research on the flora of the interior regions of China has revealed the close floristic similarity of the Himalayan region and SW China (Yunnan, Sichuan and Tibet). As a result, the term Sino-Himalaya has appeared more frequently in botanical papers to indicate the close affinities of the two areas. Further taxonomic comparisons of corresponding taxa in the Himalayan region and in SW China are still urgent, however. For political reasons, frequent visits and exchanges of materials between the Himalayan countries and China have been limited. Taxonomic work on Himalayan plants and detailed revisions and monographic studies related to the floristic projects are also needed. Species-level diversity is still not well understood in the Himalayan region and it is inevitable that more new taxa will be reported from this region.

The Himalayas are the world's highest mountains. Alpine and subnival plants have many unique representatives in the Himalayas and the mountain range is certainly one of the best areas on Earth to study speciation in relation to elevation.

Research on Himalayan plants using new methods has barely begun: molecular level studies on the phylogeny of Himalayan plants started only in 2001 and 2002 (Yoo et al. 2001; Fujihashi et al. 2002).

References

(Those appearing in the list of taxonomic and floristic studies of Himalayan plants in 2001 and 2002, which follows this reference list, are omitted.)

Akiyama S, Ohba H (2000) Inflorescences of the Himalayan species of *Impatiens* (Balsaminaceae). J Jpn Bot 75:226–240

Akiyama S, Ohba H, Wakabayashi M (1991) Taxonomic notes of the east Himalayan species of *Impatiens* (Balsaminaceae). In: Ohba H, Malla SB (eds) The Himalayan plants, vol 2. University of Tokyo Press, Tokyo, pp 67–94

Akiyama S, Wakabayashi M, Ohba H (1992) Chromosome evolution in Himalayan *Impatiens* (Balsaminaceae). Bot J Linn Soc 109:247–257
Al-Shehbaz IA (2000) *Staintoniella* is reduced to synonymy of *Aphragmus* (Brassicaceae). Harvard Pap Bot 5:109–112
Appel O, Al-Shehbaz I (2002) Cruciferae. In: Kubitzki K, Bayer C (eds) The Families and genera of vascular plants, vol 5. Springer, Berlin Heidelberg New York, pp 75–174
Bista M, Ohba H (2002) Foreword. In: Noshiro S, Rajbhandari KR (eds) Himalayan botany in the twentieth and twenty-first centuries. The Society of Himalayan Botany, Tokyo, p vii
Farille MA, Cauwet-Marc A-M, Malla SB (1985) Apiaceae himalayenses III. Candollea 40:509–562
Hara H (1970) On the Asiatic species of the genus *Panax*. J Jpn Bot 45:197–212, pls 10–12
Koba H, Akiyama S, Endo Y, Ohba H (1994) Name list of the flowering plants and gymnosperms of Nepal. The University Museum, The University of Tokyo, Material Reports No 32, 569 pp
Ohba H, Akiyama S (1992) The alpine flora of the Jaljale Himal, East Nepal. The University Museum, The University of Tokyo, Nature and Culture No. 4, 83 pp
Ohba H, Ikeda H (1999) A contribution to the flora of Ganesh Himal, Central Nepal. The University Museum, The University of Tokyo, Nature and Culture No 5, 84 pp
Ohba H, Ikeda H (2000) The flora of Hinku and Hunku Valleys, East Nepal. The University Museum, The University of Tokyo, Nature and Culture No. 6, 272 pp
Ohba H, Tsukaya H (2000) A taxonomic note on *Habenaria albomarginata* King and Prantl (Orchidaceae). J Jpn Bot 75:311–313
Press JR, Shrestha KK, Sutton DA (2000) Annotated checklist of the flowering plants of Nepal. The Natural History Museum, London, 430 pp
Rajbhandari KR (2002) Flora of Nepal: 200 years' march. In: Noshiro S, Rajbhandari KR (eds) Himalayan botany in the twentieth and twenty-first centuries. The Society of Himalayan Botany, Tokyo, pp 76–93
Shrestha KK, Press JR (2000) Catalogue of type specimens from Nepal. The Natural History Museum, London, 123 pp
Watson MF, Blackmore S (2003) The first editorial meeting for the flora of Nepal. Newsl Himalayan Bot 31:20–22
Wen J (2001) Evolution of the *Aralia-Panax* complex (Araliaceae) as inferred from nuclear ribosomal its sequences. Edinb J Bot 58:243–257
Wen J, Nowicke JW (1999) Pollen ultrastructure of *Panax* (the ginseng genus, Araliaceae). An eastern Asian and eastern North American disjunct genus. Am J Bot 86:1624–1636
Wen J, Zimmer EA (1996) Phylogeny and biogeography of *Panax* L. (the ginseng genus, Araliaceae): inferences from ITS sequences of nuclear ribosomal RNA. Mol Phylogenet Evol 6:167–177
Wen J, Plunkett GM, Mitchell AD, Wagstaff SJ (2001) The evolution of Araliaceae: a phylogenetic analysis based on ITS sequences of nuclear ribosomal DNA. Syst Bot 26:144–167
Yamazaki T (1988) A revision of the genus *Pedicularis* in Nepal. In: Ohba H, Malla SB (eds) The Himalayan plants, vol 1. University of Tokyo Press, Tokyo, pp 91–161

Lists of taxonomic and floristic studies of Himalayan Plants in 2001 and 2002

Abid R, Qaiser M (2002) Genus *Inula* L. (s. str.) (Compositae-Inuleae) in Pakistan and Kashmir. Candollea 56:315–325

Acharya N, Yonekura K, Suzuki M (2002) A new species and a new variety of *Boehmeria* (Urticaceae) from the Himalaya with special reference to the status of *B. pedunliflora* Wedd. ex D.G. Long. Acta Phytotax Geobot 53:1–9

Al-Shehbaz IA (2001) *Cardamine gouldii* (Brassicaceae), a new species from Bhutan. Novon 11:289–291

Al-Shehbaz IA (2002a) New species of *Alyssum*, *Aphragmus*, *Arabis*, and *Sinosophiopsis* (Brassicaceae) from China and India. Novon 12:309–313

Al-Shehbaz IA (2002b) Six new species of *Draba* (Brassicaceae) from the Himalayas. Novon 12:314–318

Al-Shehbaz IA (2002c) New combination in Brassicaceae (Cruciferae): *Draba serpens* is a *Hemilophia* and *D. williamsii* is a *Lepidostemon*. Edinb J Bot 59:443–450

Baruah A, Nath SC (2001) Taxonomic status and composition of stem bark oil of a variant of *Cinnamomum bejolghota* (Lauraceae) from northeast India. Nord J Bot 21:571–576

Dickoré WB (2001) Observations of some *Saussurea* (*Compositae-Cardueae*) of W Kunlun, Karakorum and W Himalaya. Edinb J Bot 58:15–29

Fujihashi H, Akiyama S, Ohba H (2002) Origin and relationships of the Sino-Himalayan *Impatiens* (Balsaminaceae) based on molecular phylogenetic analysis, chromosome numbers and gross morphology. J Jpn Bot 77:284–295

Fujikawa K, Ohba H (2002) Two new species of *Saussurea* subgenus *Eriocoryne* (*Asteraceae*) from the Nepal Himalaya. Edinb J Bot 59:283–289

Gajurel PR, Rethy P, Kumar Y (2001) A new species of *Piper* (Piperaceae) from Arunachal Pradesh, north-eastern India. Bot J Linn Soc 137:417–419

Iokawa Y, Ohashi H (2002) A taxonomic study of the genus *Campylotropis* (Leguminosae) I, II and III. J Jpn Bot 77:179–222, 251–283, 315–350

Mill RR (2001a) Notes relating to the Flora of Bhutan: XLIII. Scrophulariaceae (*Pedicularis*). Edinb J Bot 58:57–98

Mill RR (2001b) A new sectional combination in *Nageia* Gaertn. (Podocarpaceae). Edinb J Bot 58:499–501

Miyamoto F, Ohba H (2002) Studies of *Juncus* (Juncaceae) in the Sino-Himalayan region II. Taxonomical studies of the *Juncus duthiei* group. J Jpn Bot 77:24–35

Miyamoto F, Akiyama S, Wu S-K, Ohba H (2002) New and noteworthy species of *Bistorta* (Polygonaceae) from the Sino-Himalayan Region. Bull Natl Sci Mus Tokyo Ser B 28:141–148

Noshiro S, Rajbhandari KR (eds) (2002) Himalayan botany in the twentieth and twenty-first centuries. The Society of Himalayan Botany, Tokyo, 212 pp, 8 pls

Ogisu M, Awan MR, Mabuchi T, Mikanagi Y (2002) Morphology, phenology and cytology of *Hepatica falconeri* in Pakistan. Kew Bull 57:943–953

Pearce N, Cribb PJ, Renz J (2001) Notes relating to the Flora of Bhutan: XLIV. Taxonomic notes, new taxa and additions to the Orchidaceae of Bhutan and Sikkim (India). Edinb J Bot 58:99–122

Pimenov MG, Kljuykov EV (2002a) Identity of *Archangelica roylei* Lindl. and its consequences for the nomenclature of some West Himalayan Umbelliferae. Rep Spec Nov Regni Veg 113:335–341

Pimenov MG, Kljuykov EV (2002b) A new species of *Lalldhwojia* Farille (Umbelliferae) from Nepal. Willdenowia 32:93–97

Pimenov MG, Kljuykov EV, Ostroumova TA (2001) Towards a clarification in the taxonomy of Sino-Himalayan species of *Selinum* L. s. l. (Umbelliferae). The genus *Oreocome* Edgew. Willdenowia 31:101–124

Rashid MH, Jong K, Mendum M (2001) Cytotaxonomic observations in the genus *Aeschynanthus* (Gesneriaceae). Edinb J Bot 58:31–43

Robson NKB (2001) Studies in the genus *Hypericum* L. (Guttiferae) 4(1). Sections 7. *Roscyna* to 9. *Hypericum* sensu lato (part 1). Bull Nat Hist Mus Lond (Bot) 31:37–88

Robson NKB (2002) Studies in the genus *Hypericum* L. (Guttiferae) 4(2). Section 9. *Hypericum* sensu lato (part 2): subsection 1. *Hypericum* series 1. *Hypericum*. Bull Nat Hist Mus Lond (Bot) 31:37–88

Singh AP (compil) (2001) Flowering plants of Nepal *(Phanerogams)*. Department of Plant Resources, National Herbarium and Plant Laboratories, Godavary, 399 pp

Staudt G, Dickoré WB (2001) Notes on Asiatic *Fragaria* species: *Fragaria pentaphylla* Losinsk. and *Fragaria tibetica* spec. nov. Bot Jahrb Syst 123:341–354

Snogerup S, Zika PF, Kirschner J (2002) Taxonomic and nomenclatural notes on *Juncus*. Preslia 74:247–266

Tateishi Y, Maxted N (2002) New species and combinations in *Vigna* subgenus *Ceratotropis* (Piper) Verdc. (Leguminosae, Phaseoleae). Kew Bull 57:625–633

Thapa N (2002) Pteridophytes of Nepal. Department of Plant Resources, National Herbarium and Plant Laboratories, Godavary, 175 pp

Tomooka N, Maxted N, Thavarasook C, Jayasuriya AHM (2002) Two new species, sectional designations and new combinations in *Vigna* subgenus *Ceratotropis* (Piper) Verdc. (Leguminosae, Phaseoleae). Kew Bull 57:613–624

Tripathi S (2001) *Curcuma prakasha* sp. nov. (Zingiberaceae) from north-eastern India. Nord J Bot 21:549–550

Yonekura K, Ohashi H (2001) Taxonomic studies of *Bistorta* (Polygonaceae) in the Himalayas and adjacent regions (I). J Jpn Bot 76:344–353

Yonekura K, Ohashi H (2002) Taxonomic studies of *Bistorta* (Polygonaceae) in the Himalayas and adjacent regions (2) – *Bistorta amplexicaulis* (D. Don) Greene and its allies, with special reference to the ochreae and leaf sheathes of cauline leaves. J Jpn Bot 77:61–81

Yoo K-O, Malla KJ, Wen J (2001) Chloroplast DNA variation of *Panax* (Araliaceae) in Nepal and its taxonomic implications. Brittonia 53:447–453

Shinobu Akiyama
Department of Botany
National Science Museum
Amakubo 4-1-1
Tsukuba 305-0005, Japan
e-mail: akiyama@kahaku.go.jp

Hideaki Ohba
Department of Botany
University Museum
University of Tokyo
Hongo 7-3-1
Tokyo 113-0033, Japan

Ecology

Hydraulic Redistribution

Ronald J. Ryel

1 Introduction

Hydraulic redistribution is the passive movement of water via roots from regions of wetter soil to regions of dryer soil. While the phenomenon has been the subject of numerous papers and recent reviews (Caldwell et al. 1998; Horton and Hart 1998; Jackson et al. 2000), it is absent or minimally discussed in recent textbooks on plant ecology, physiology, ecophysiology and soil water processes (e.g., Larcher 1995; Marshall et al. 1996; Hillel and Hillel 1998; Lambers et al. 1998; Fitter and Hay 2002; Gurevitch et al. 2002; Taiz and Ziegler 2002). Minimal consideration of hydraulic redistribution as an important process is due in part to the relatively recent discovery of its occurrence in the field and to a lack of a full understanding of the ecological consequences and significance of this water movement. In the past few years, however, more has been learned about hydraulic redistribution and its importance to plants and movements of water in the soil. In this review, the ecological consequences of this phenomenon will be examined to the extent of current understanding.

2 The Phenomenon

Various forces affect water flow through soils. The driving force for water movement differs depending on whether soils are saturated or unsaturated with water. Water movement in soils saturated with water results primarily from gradients in pressure and gravity. In unsaturated soils, differences in water potential (matric potential) and gravity are the forces responsible for water movement. Once the larger pore spaces in unsaturated soil become drained via gravity and preferential flow, subsequent water movements occur almost entirely through gradients in water potential. In soils at and below field capacity, water moves via infiltration from wetter to dryer regions, with the rate and direction of water movement depending on the water potential gradients. Movement can be vertical (up or down) or lateral

Progress in Botany, Vol. 65

with the rate depending on the water potential gradient, water potential of the wetter region and soil characteristics. Water movement by infiltration has often been incorrectly considered to be from capillary processes but water actually moves via film flow (Marshall et al. 1996). When transpiration by plants and evaporation from the soil surface are considered, water movement via infiltration can be enhanced by increasing the potential gradients through soil water loss. These processes, however, do not change the method of water movement, but simply enhance infiltration.

Plants, on the other hand, through their roots, provide an additional mechanism for water movement in unsaturated soil. With roots acting as passive conduits for water, plants can redistribute water from soil regions of high water potential to low water potential. The process involves roots taking up water as during transpiration in wetter soil regions, but also effuxing water into dryer soil. It is important to differentiate hydraulic redistribution from preferential flow associated with roots where water moves along root channels or root surfaces (Johnson 1987; Branswijk 1988). Hydraulic redistribution typically occurs during periods when stomatal aperture is minimal as the atmospheric draw on water during periods of transpiration is sufficiently stronger than that provided by potential gradients in soil. For most plants, hydraulic redistribution would occur at night (Richards and Caldwell 1987; Caldwell and Richards 1989), but CAM species have been observed to redistribute water during the day as would be expected with the inverse pattern of stomatal activity (Yoder and Nowak 1999).

The redistribution of water through roots was originally termed "hydraulic lift" as roots of the shrub *Artemisia tridentata* were observed to "lift" water from deep, wet soil regions and release this water into shallower, dryer soil regions during nighttime periods when transpiration had ceased (Richards and Caldwell 1987). The term was extended to "hydraulic redistribution" (Burgess et al. 1998) as water movements via roots were more recently found to occur laterally (DR Smart, personal communication) and downward (Schulze et al. 1998; Smith et al. 1999; Burgess et al. 2001a; Scholz et al. 2002; Hultine et al. 2003a; Ryel et al. 2003) in addition to upward.

To date, hydraulic redistribution has been observed in upwards of 50 plant species (Jackson et al. 2000) from arid to relatively mesic climates and includes trees, shrubs, grasses and herbs (Caldwell et al. 1998). The phenomenon is likely to occur for most plants where sizable soil moisture gradients occur in the active rooting zone (Caldwell and Richards 1989). Although hydraulic redistribution was first hypothesized to occur in the field by Mooney et al. (1980) and the first strong evidence was provided by Richards and Caldwell (1987), movement through roots was observed many years prior in laboratory settings. Caldwell et al. (1998) provide a comprehensive history of the experiments which indicated water movement by roots between separated soil compartments of differing water content. The earliest of these date to Magistad and Breazeale (1929) and Breazeale (1930) which looked at the rehydration of roots below the wilting point that could be moistened if water was applied to another part of the root system. In addition, some simulation models of water uptake by plants (e.g., Landsberg and Fowkes 1978; Kirkham 1983; Campbell 1985) would predict hydraulic redistribution of soil water during periods of no plant transpiration unless the hydraulic conductivity between the root and soil were reduced during nighttime periods of no transpiration.

If water moves via roots from wetter to drier regions of the soil, this implies that while roots act as conduits for redistributed water, they must be able to exhibit reverse flow of water (i.e., be "leaky"). Caldwell et al. (1998) provide a critical review of the rectifier properties of roots; that is the potential for water to more readily flow into than out of roots, a property that would greatly reduce the potential for hydraulic redistribution. They conclude there are no observed mechanisms in a diel or shorter time scale that would result in true rectification and prevent reverse flow. Also, they point out that the limited degree of rectification considered to be a property of the Casparian bands and suberin lamellae which prevent reverse flow along the symplastic and apoplastic pathways (Peterson et al. 1992) does not preclude water loss via additional pathways as considered in the composite transport model (Steudle 1994).

3 Detecting and Measuring Hydraulic Redistribution

Evidence for the redistribution of water within the soil via passive movement through roots comes from three types of measurements. These include measurements of bulk soil moisture, xylem root sap flow and movements of labeled water. All methods for documenting hydraulic redistribution are labor intensive and difficult to conduct on large scales.

The first measurements of hydraulic redistribution in the field were conducted with screen-cage thermocouple psychrometers (Richards and Caldwell 1987). These psychrometers measure bulk soil water potential with a sensitivity that remains relatively constant for water potentials below about 0.3 MPa, but are relatively insensitive in wetter soils. The sensitivity of the psychrometers can permit observation of diurnal fluctuations in soil water potential (Fig. 1, above) that result from transpiration losses followed by water recharge during periods of minimal transpiration. Psychrometers have been used for a variety of species and communities to document and validate hydraulic redistribution (e.g., Williams et al. 1993; Dawson 1993,

1996; Yoder and Nowak 1999; Ryel et al. 2002; Ludwig et al. 2003). Manipulation of the diurnal light cycle (Williams et al. 1993; Yoder and Nowak 1999) or suppressing transpiration by enclosing plants in plastic bags (Caldwell and Richards 1987) has been used to verify that plant activity is responsible for the water movement. Psychrometers can effectively be used to calculate rates of soil water movement due to hydraulic redistribution when water potential can be converted to volumetric soil water content (Hanks and Ashcroft 1980; Campbell 1985). This is usually done by comparing water potential measured by psychrometers with simultaneous direct measurements of volumetric soil moisture (e.g., gravimetric soil water measurements or time-domain reflectrometry). Because of temperature sensitivity, psychrometers are typically installed at depths greater than 30 cm (Richards and Caldwell 1987; Ludwig et al. 2003) such that temperature fluctuations are small enough to be corrected with a calibration model (Brown and Bartos 1982). Even when installed at sufficient depth, diurnal fluctuations in temperature can still create sizable cyclic patterns in measured soil water potential which do not reflect water movement at deep soil depths if the data acquisition system is not sufficiently buffered from temperature fluctuations (R.J. Ryel, unpubl. data). Millikin Ishikawa and Bledsoe (2000) also indicate that for water potentials below –3 MPa, the corrections of Brown and Bartos (1982) may not effectively correct for temperature effects.

Instruments which directly measure bulk soil water content (e.g., time-domain reflectrometry and neutron probe; Pearcy et al. 1989) often do not have the sensitivity to effectively show diurnal fluctuations in soil water content that result from hydraulic redistribution, especially when the soil is quite dry (Fig. 1, below). In addition, low rates of hydraulic redistribution can result in relatively small diurnal changes in soil water potential which can be masked by signal variability. However, hydraulic redistribution can still be documented in the soil during periods of rapid soil water recharge following rainfall events (Ryel et al. 2003) and when the upper soil layers do not dry as fast as would be predicted by the fraction of roots in these layers (Ryel et al. 2002). Such assessment can be greatly enhanced by using simulation models (see Sect. 5).

A second technique for assessing hydraulic redistribution by roots is through measurements of root xylem sap flow (Burgess et al. 2000a). During periods of transpiration, water moves from the soil, through roots and is transpired by foliage. However, when transpiration is effectively ceased, reverse sap flow in roots can occur if water is released into the soil. Sap flow sensors have been utilized to document and quantify hydraulic redistribution in several woody species (Burgess et al. 1998, 2001a; Smith et al. 1999; Scholz et al. 2002; Hultine et al. 2003a,b).

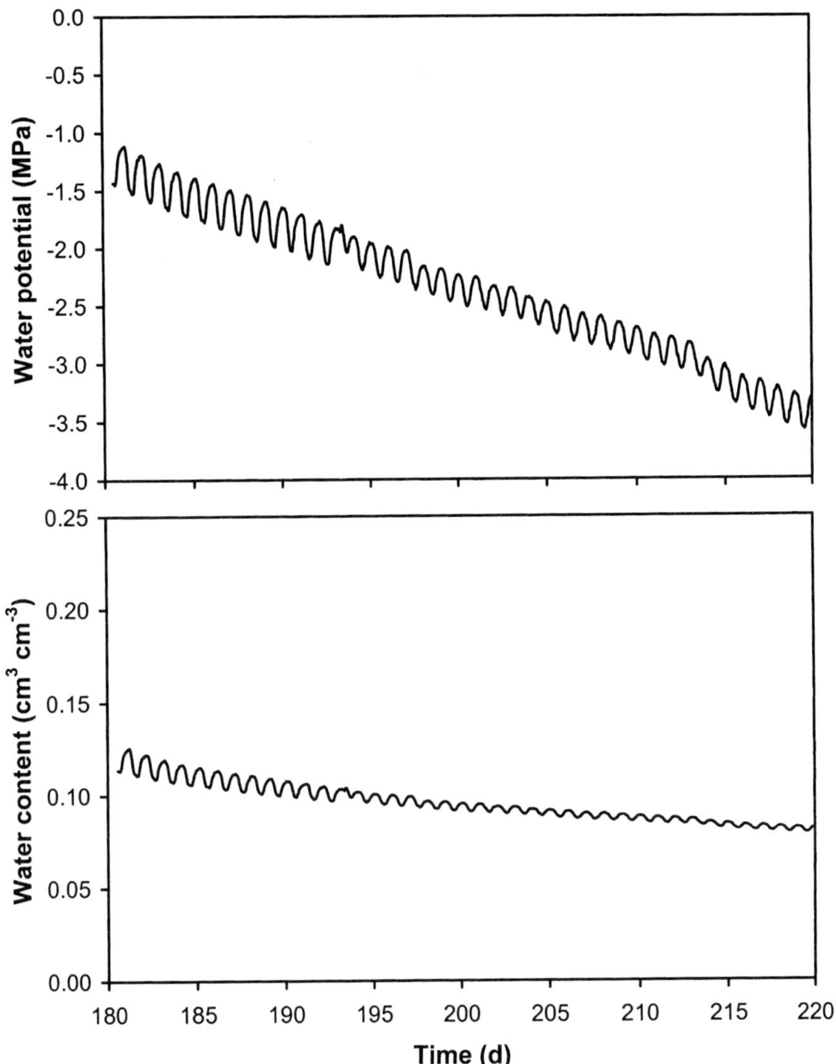

Fig. 1. *Above* Water potential measured hourly at 35-cm depth in a stand of *Artemisia tridentata* in northern Utah, USA, showing classic cyclic pattern of water uptake for transpiration during the day followed by nighttime recharge (M.M. Caldwell, J.H. Richards, K. Williams, J.H. Manwaring, unpubl.). *Below* Water content from same data calculated using a relationship from van Genuchten (1980), parameterized as in Ryel et al. (2002)

Sap flow rates in roots resulting from hydraulic redistribution are typically low due to the small size of roots and the relatively low quantities of water that are often redistributed. Because of these low flow rates, many methods of measuring sap flow are not sensitive enough for detecting the reverse flow (Becker 1998; Burgess et al. 2001b). Two measurement procedures have been successfully used to detect and measure hydraulically redistributed water by roots. One procedure uses constant-power heat-balance sap flow gauges. A heater provides constant heat to a short section of the root, and the temperature gradient across the heater is measured with temperature sensors. From the heat applied and the temperature gradient, the direction and velocity of sap flow can be calculated (Smith and Allen 1996; Smith et al. 1999). A second promising procedure uses a thermometric technique where a short pulse of heat is followed in the sap stream and sap velocity calculated using the heat ratio method (Burgess et al. 2000a, 2001b,c). Both techniques permit determination of sap flow direction and an estimation of sap flow volume given a cross-sectional area of xylem vessels. The pulse method has the advantage of lower power consumption and greater sensitivity to low sap flow rates (Burgess et al. 2001b).

A final method used to detect hydraulic redistribution is through the use of stable isotopes. Variations in the oxygen and hydrogen isotopes in water can be used as markers for water use and redistribution by roots. Deuterated water can be used as a natural marker (Dawson 1993, 1996), applied to exposed roots (Caldwell and Richards 1989) or added to the soil water (Schulze et al. 1998; Brooks et al. 2002). Differences in ^{18}O by water source have also been used to trace water use and movement in soils as affected by roots (Ludwig et al. 2003). Isotope tracers are primarily used to verify water movement by hydraulic redistribution, but if carefully utilized can be effectively used to quantify the water redistributed. These isotope labels are most effective in verifying hydraulically redistributed water when this water movement greatly exceeds the potential for water to move via infiltration. Burgess et al. (2000b) suggested that relatively uniform soil water isotope profiles may indicate that hydraulic redistribution has moved deeper water into shallow soil layers; typically water in shallow and deep soil regions would have different isotope signatures (Dawson and Ehleringer 1991; Thorburn and Walker 1993). Isotope techniques have also been used to document water parasitism by shallow-rooted plants associated with other deeper-rooted plants which are hydraulically lifting water (Caldwell and Richards 1989; Caldwell 1990; Dawson 1993; Brooks et al. 2002).

4 Ecological Implications

Hydraulic redistribution is a process that occurs at the individual plant level. However, the effects of this water movement by roots extend beyond the individual to community and ecosystem levels (Caldwell et al. 1998; Horton and Hart 1998; Jackson et al. 2000). The ecological implications of hydraulic redistribution as understood today are reviewed below. Future research will likely broaden this understanding, and may identify yet unknown effects on individual, community and ecosystem processes.

4.1 Soil Water Dynamics

Until hydraulic redistribution was identified as occurring in field settings, water movement through the soil matrix was assumed to be driven by gravity, infiltration processes and preferential flow. Certainly, most of the water movement in soils can quite easily be explained by these processes and much has been learned about soil water movement without consideration of hydraulic redistribution (e.g., Campbell 1985; Marshal et al. 1996; Hillel and Hillel 1998). However, some phenomenon could not be explained such as the relatively moist upper soil layers coinciding with a mat of dense roots under *Prosopis tamarugo* in the Atacama Desert (Mooney et al. 1980) and the diurnal fluctuations in moisture observed in soil underlying canopies of *Artemisia tridentata* (Richards and Caldwell 1987). Once hydraulic redistribution was identified as the process behind these observations, research has broadened the perspective of this water movement. Studies have indicated that hydraulic redistribution is a necessary process to fully account for observed water dynamics in soils, particularly when active roots link soil regions of differing water potentials.

The root distributions in most plant communities are heavily skewed to shallow soil depths (Schenk and Jackson 2002), principally for acquisition of soil nutrients where they are most plentiful and perhaps for quick usage of rainfall which infiltrates these shallow layers. During daily transpiration, such skewed root distributions will quickly dry out the upper soil layers where roots are concentrated, and in many cases, infiltration from deeper, wetter soil layers is insufficient to replenish this water on a daily basis. Water redistributed upward to these dry, shallow soil layers by roots from deeper, wetter soil layers, however, has been implicated in helping to maintain observed relatively uniform soil moisture conditions (Fig. 2; Breazeale and Crider 1934; de Kroon et al. 1996; Ryel et al. 2002).

Hydraulic redistribution has also been implicated in helping to rapidly recharge deep soil layers after precipitation events at rates far exceeding that possible via infiltration (Smith et al. 1999; Burgess et al. 2001a; Ryel et al. 2003). Such movements likely occur in arid or semi-arid environments (Smith et al. 1999) in situations where groundwater is far below the rooting zone of plants and recharge occurs from surface only. Ryel et al. (2003) indicated that up to all of the water from small rain events (<8 mm) infiltrating below 0.3 m likely moves to depth via hydraulic redistribution

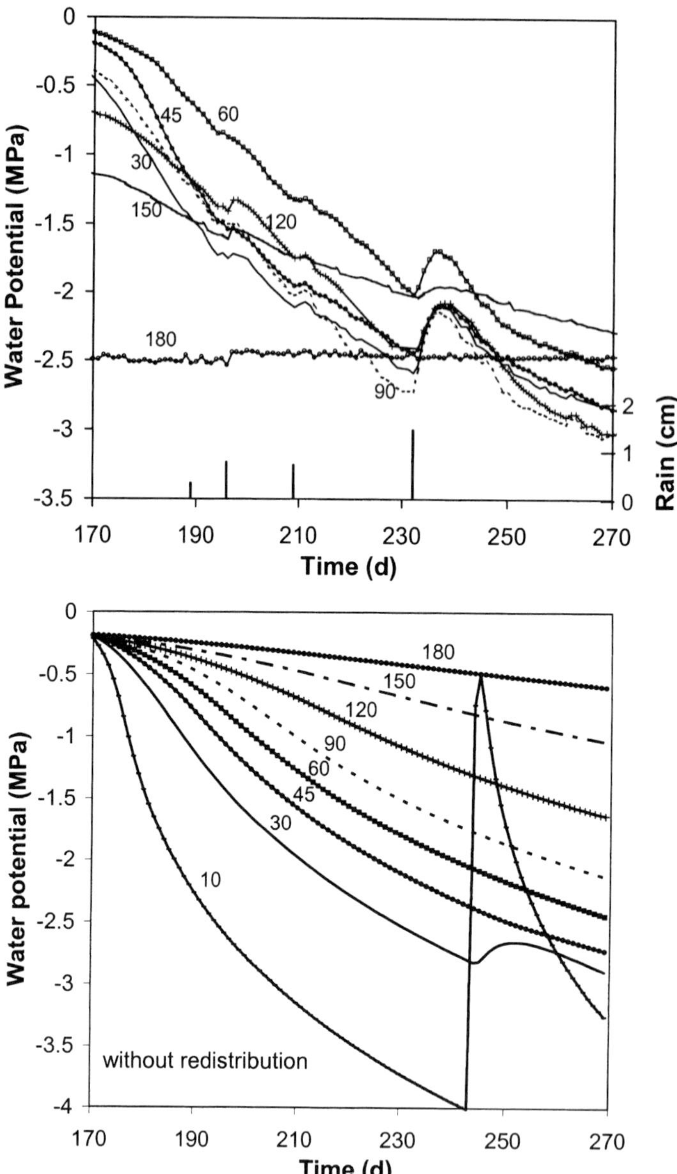

Fig. 2. *Above* Daily average water potential measured at seven depths (m) in a stand of *Artemisia tridentata* in central Utah, USA, showing homogenization of soil water potential. *Below* Simulated water potential for a stand of *A. tridentata* without hydraulic redistribution showing divergent water potentials through a similar period. Note that this simulation began with soil water at field capacity and uniform throughout the soil column as opposed to a more varied range of water potentials as shown in the *upper panel*. (Both figures modified from Ryel et al. 2002)

and up to 74% for a 36-mm rain event. They also provide evidence that during the fall-spring recharge period, 67% of all water moved downward below 0.1 m and 87% of the water moved below 0.3 m was likely via roots. In addition to rapidly recharging water to depth following precipitation events, hydraulic redistribution may permit water to move to layers well below those possible through infiltration (Schulze et al. 1998) or across soil layers of different textures (Ryel et al. 2003) which could interrupt infiltration (Clothier et al. 1977; Campbell 1985). Differences in quantity and timing of hydraulic redistribution, by species however, may result in different soil moisture dynamics (Yoder and Nowak 1999; Hultine et al. 2003b).

4.2 Plant Water Status

Transpiration rates of plants can be limited by soil moisture content where roots are actively taking up water. As the soil dries and water potentials become lower, reduced hydraulic conductivity in the soil reduces the rate at which water can move from the soil to roots and subsequently the rate of transpiration. Where roots are concentrated, water uptake for transpiration can rapidly reduce soil water to potentials where water extraction begins to limit transpiration. For rooting distributions where roots are concentrated in shallow soil, these layers become dryer more rapidly than in deeper layers where roots are fewer. As the upper soil layers become drier, the lower rooting densities in the deeper soil where more water remains can be too low to provide sufficient water for high transpiration rates. Lowered transpiration rates reflect lower stomatal conductance (Mott and Parkhurst 1991) and lowered potential for carbon fixation (Sperry 2000).

Several studies indicate that decreases in transpiration due to lower soil water potential in root-laden upper soil layers can be partially offset by water hydraulically redistributed from deeper, wet soil to shallow dry soil at night. While the upper soil layers are depleted of water during the day, nighttime recharge through hydraulic redistribution can partially replace the water lost in the upper soil layers. With higher water content in the upper soil layers the following morning, transpiration rates would be higher as the plant would have this reserved water to use where root densities were greatest (Caldwell et al. 1998; Horton and Hart 1998). Measurements of hydraulically redistributed water indicate that the quantity of water redistributed to shallow soil layers can be a sizable portion of the daily transpiration rate. Measurements include 14% for the shrub *Gutierrezia sarothrae* (Wan et al. 1993), 25% for the tree *Acer saccharum* (Emerman and Dawson 1996), and approximately 33% for the shrub *Artemisia tridentata* (Richards and Caldwell 1987). Brooks et al. (2002) indicated that hydraulic redistribution in forests of *Pinus ponderosa* and *Pseudotsuga menziesii* would respectively replace 28 and 35% of the water transpired daily and result in 16 and 21 additional days of stored water in the upper soil horizons after a 60-day drought.

Translating hydraulically lifted water to increased transpiration has been indicated by some research. Caldwell and Richards (1989) found that by experimentally circumventing hydraulic redistribution by illuminating the shrub *Artemisia tridentata* throughout the night, transpiration in the subsequent day were reduced by 25–50%. They suggest that much of this reduction in transpiration was due to reduced water availability in upper soil layers. Simulation modeling of soil water dynamics in the same species indicated that transpiration could be increased up to 20% with hydraulic redistribution (Ryel et al. 2002). Dawson (1996) has indicated that *Acer saccharum* trees had higher transpiration rates in deep-rooted individuals that could perform hydraulic redistribution than individuals with shallower roots. Jackson et al. (2000) indicated that annual water use by these trees is 19–40% greater than in a forest where hydraulic redistribution is limited by minimal groundwater. Mendel et al. (2002) showed with simulations for a canopy similar to the *Acer saccharum* in Emerman and Dawson (1996) that higher transpiration rates could occur over a period of a few weeks with hydraulic redistribution once soil water limited transpiration rates. Dawson (1998) in experiments with seedlings of *Acer saccharum* showed that maintaining higher transpiration rates throughout the day through hydraulic redistribution translated to greater carbon fixation which translated to greater rates of root (24–30%) and shoot (8–14%) growth than found for seedlings where hydraulic redistribution was suppressed.

Model simulations by Ryel et al. (2002) indicate that transpiration rates are enhanced most when sizable gradients occur between shallow and deep soil layers that promote maximum rates of hydraulic redistribution. Their model indicated that the period of enhanced transpiration can be relatively short if deeper soil depths are not recharged from below or laterally as the recharging soil region becomes too dry to supply large quantities of water to shallow soil layers. However, if deep soil layers are steadily recharged, higher transpiration rates can be maintained for prolonged periods of time. Thus, for species like *Acer saccharum* in systems with relatively permanent groundwater as described by Emerman and Dawson (1996), transpiration may be enhanced through much of the summer period.

Plant water status also often reflects soil water conditions (Slatyer 1967; Schulze 1991; Leffler et al. 2002; but see Donovan et al. 2001, 2003), and higher transpiration rates are not the only indication of improved physiological function resulting from hydraulic redistribution. If plant water potential is largely tied to soil water potential in the rooting zone, increasing the soil water potential where roots are densest could increase plant water potential. Tyree and Sperry (1989) link plant water potential to the potential for cavitation and higher water potential resulting from hydraulic redistribution may reduce cavitation. Reduced cavitation in roots may keep more roots alive and available to respond to resource pulses resulting from rain events (Noy-Meir 1973). Soil water potential is also linked to hydraulic conductivity between the soil and roots. Lowered water potential increases air spaces in the soil (Campbell 1985; Nobel 1991) and can cause roots to shrink and lose contact with the soil (Huck et al. 1970; Caldwell et al. 1998). With hydraulic redistribution, more roots may maintain hydraulic connectivity due to higher water potential than without hydraulic redistribution.

Plants may also conserve water through hydraulic redistribution and maintain higher plant water potential during drought periods (Ryel et al. 2004). Using the model of Ryel et al. (2002) they show that when soil water recharge occurs primarily from the surface, precipitation redistributed more uniformly in the soil column by roots slows the rate at which this water can subsequently be taken up by plants. This prolongs water availability during periods of drought. By spreading out water more uniformly in the soil column at lower water potentials following precipitation events, water use is reduced due to lower soil conductivity. The greater remaining soil water and more uniform distribution result in higher plant predawn water potentials and higher transpiration rates later in the drought period. This phenomenon could occur both following summer precipitation events and overwinter soil-water recharge.

Effects on plant water status related to hydraulic redistribution can have effects at scales above the individual plant. By increasing transpiration rates due to hydraulically redistributed water, plants can affect water and carbon balance at the community and ecosystem scales (Horton and Hart 1998; Jackson et al. 2000). This may include greater water usage and carbon fixation at the community level, and altered soil water and groundwater dynamics at the ecosystem level. Jackson et al. (2000) indicate that in a forest of *Acer saccharum*, 3–6% less water is available per month for stream flow due to the higher transpiration rates resulting from hydraulically lifted water. At the individual level, greater quantities of water available for transpiration may also reduce the potential for xylem cavitation as plants often operate close to their hydraulic limits (Sperry et al. 1998). Hydraulic failure at the individual level can result in the subsequent death of branch units or whole plants and can limit the distribution of plants (Tyree and Ewers 1991; Portwood et al. 1997; Davis et al. 2002; Sperry and Hacke 2002) and ultimately affect community structure. Finally, downward redistribution of water may permit roots of some species to reach deep water sources (Burgess et al. 1998; Schulze et al. 1998; Hultine et al. 2003b), permitting establishment and persistence in soils where these species may otherwise not survive.

4.3 Nutrient Availability

The availability of nutrients to plants are functions of both nutrient pools and diffusion rates of nutrients through the soil matrix. Nutrient pools are affected by soil chemical properties and the activity of microbial populations. Microbial populations are affected by temperature, labile carbon availability and soil moisture (Stark 1994). Diffusion rates are primarily affected by the quantity of water in the soil, with rates declining rapidly with decreasing soil moisture (Nye and Tinker 1977; Jackson and Caldwell 1996; Ryel and Caldwell 1998). Plant uptake of nutrients is often through diffusion processes as opposed to mass flow through the transpiration stream (Barber 1995).

Hydraulic redistribution increases the water content in shallow soil layers where most of the soil nutrients and microbes occur. This has been hypothesized to increase the availability of nutrients for plants and to increase the activity of soil microbes (Richards and Caldwell 1987; Caldwell et al. 1998). While a plausible idea, field studies to assess these hypotheses are still lacking. Caldwell and Manwaring (1994) tested the hypothesis whether nutrient heterogeneity decreased with hydraulic redistribution and found no change in heterogeneity in the spatial distribution of nutrients. Simulations by Ryel et al. (2002), however, indicated that the period soil water was sufficient for nutrient diffusion increased under hydraulic redistribution. Their result suggests that plants are more able to acquire nutrients with hydraulic redistribution than without.

Caldwell et al. (1998) and Horton and Hart (1998) have hypothesized that hydraulic redistribution may also help maintain the integrity of the external mycorrhizal mycelium and possibly improve nutrient status of mycorrhizal-infected plants. The hypothesis was tested by Querejeta et al. (2003) in a microcosm experiment with seedlings of *Quercus agrifolia* where water applied to tap roots was labeled with tracer dyes. Their results indicated water hydraulically redistributed at night from the tap roots was translocated from lateral roots to mycorrhizal fungi, but not to saprotrophic/parasitic fungi. Hyphae of the symbiotic mycorrhizal fungi were found to persist during the experiment during prolonged periods of drought and with water potential values as low as –20 MPa. They suggested that the maintained activity of mycorrhizal fungi could potentially improve the nutrient status of species which are able to hydraulically redistribute water from depth during periods of drought.

4.4 Plant–Plant Interactions

Hydraulically redistributed water has been implicated to affect plant-plant interactions for soil water (Caldwell et al. 1998; Horton and Hart 1998). When water is lifted by deeply rooted plants to shallow soil layers, the potential occurs for shallow rooted plants to access this water (water parasitism). Uptake of hydraulically lifted water was first documented by Caldwell and Richards (1989) where deuterated water taken up by deep roots of *Artemisia tridentata* was subsequently found in the stems of a neighboring tussock grass, *Agropyron desertorum*. A subsequent study (Caldwell 1990), however, could not show reductions in the transpiration rate of the grass when hydraulic redistribution by the shrub was prevented. Ludwig et al. (2003) indicated limited importance of hydraulic redistribution on an East African savannah by finding soil water potential under trees crowns of *Acacia tortilis* trees that hydraulically lifted water with deep roots was less than soil water potential outside the tree canopy where only herbaceous vegetation occurred. In a *Quercus douglasii* woodland Millikin Ishikawa and Bledsoe (2000) found that hydraulic redistribution by the oaks occurred too late in the season to provide water for use by annual forage grasses which had senesced. This may be a common phenomenon in arid and semi-arid systems where water potentials in shallow soils are relatively low before hydraulic redistribution becomes significant – water potentials below which many herbaceous plants senesce.

Despite these studies which indicated little interactions between plants for water hydraulically lifted by individuals with deep roots, there is some evidence that shallow-rooted plants can benefit from water hydraulically lifted. Shallow-rooted plants that used a high proportion (up to 60%) of water hydraulically lifted by deep-rooted *Acer saccharum* trees were found by Dawson (1993) to generally maintain higher leaf water potential and stomatal conductance than plants which used little or no redistributed water. Brooks et al. (2002) found that seedlings of *Pinus ponderosa* trees utilized water hydraulically redistributed upward by large, deep-rooted trees. They suggest that this redistributed water may enhance survival of these seedlings during summer drought periods. They calculate that an additional 21 days of stored water is available to seedlings in shallow soil layers during the summer. Measurements of water redistributed upward by roots of the annual grass, *Bromus tectorum*, following senescing of above ground foliage (R.J. Ryel, unpubl. data) may provide some water that can benefit subsequent seedlings of this species. This redistributed water may also provide an effective water bridge between shallow soil layers wetted by late summer rains which induce germination and deeper soil layers which contain greater quantities of water that may be necessary for seedling establishment.

Upward redistribution of water is not the only way plant-plant interactions for water can be affected. The downward redistribution of precipitation events can be a mechanism for deep-rooted plants to store water below the rooting zone of shallow-rooted plants thereby reducing completion for redistributed water (Smith et al. 1999; Ryel et al. 2003). While the effects of downward redistribution on shallow-rooted plants have not been quantified, this movement of water has the potential to be more important than upward redistribution of water because removing water during periods of growth may adversely affect the growth rates of shallow-rooted plants. Also by storing this water in the soil instead of structures as do succulents, plants which redistribute water to depth may be able to allocate more structure to roots and become more effective competitors against shallow-rooted species.

5 Simulation Models for Hydraulic Redistribution

Models for simulating soil water dynamics can be useful tools for helping to determine the occurrence of hydraulic redistribution and quantify the water redistributed through roots. This is particularly important as designing experiments which curtail hydraulic redistribution to assess its importance (e.g., Caldwell and Richards 1989; Yoder and Nowak 1999) can be difficult to implement or interpret (Ryel et al. 2002). In general, models of soil water dynamics and plant water uptake (e.g., Thornley and Johnson 1990; Kropff and van Laar 1993; Timlin et al. 1996; Kemp et al. 1997) have not specifically included hydraulic redistribution. As discussed earlier, models of soil water dynamics that include water uptake by plants can result in water being redistributed by roots unless nighttime hydraulic conductance is reduced (Landsberg and Fowkes 1978; Kirkham 1983; Campbell 1985).

Recently, three different modeling efforts have specifically included hydraulic redistribution. All models assume the redistribution is entirely an hydraulic process with minimal osmotic effects. These models all address the processes of transpiration, water uptake by roots, water movement via roots between soil regions and soil water infiltration.

The first model was a one-dimensional model developed by Williams et al. (1993). This model used the difference in soil water potential among layers and rooting distribution to determine the quantity of water redistributed between layers. Redistribution was assumed to be directly related to the difference in water potential between soil layers and the fraction of roots in the layers. When coupled to measured daytime transpiration rates, the model could simulate the nighttime recharge of water in the upper dry soil layers characteristic of hydraulic redistribution. This model was rather simple in that it assumed soil water infiltration was negligible (a reasonable assumption in relatively dry soils) and water transfer among layers by roots was not reduced by lowered hydraulic conductivity that occurs as soil dries.

A detailed two-dimensional model for soil water dynamics was developed by Mendel et al. (2002). This model couples a model for soil water infiltration with a model for root water extraction that is a function root and shoot hydraulic characteristics. The rate of hydraulic redistribution was assumed to be a function of differences in root water potential between rooting zones, the axial conductivity of roots, and root density. The model was implemented for conditions of relatively high soil moisture where the axial conductivity of roots was assumed to be constant. Transpiration was assumed constant except when the calculated root water potential reached a set minimum, where transpiration was then limited to keep the root water potential from going below this minimum water potential. Groundwater effects on soil moisture were also considered in this model. The model could predict rates of hydraulic redistribution as related to total transpiration that were similar to measured values.

The third model, a one-dimensional model of soil water movement, was developed by Ryel et al. (2002) to simulate vertical changes in soil water content which included hydraulic redistribution. This model assumes redistribution via roots is related to root density and differences in soil water potential, and that transpiration and the potential for hydraulic redistribution are limited by root uptake as the soil dries. Precipitation inputs at the surface and subsurface recharge (shallow groundwater) are also considered. Although it could easily be incorporated, their model did not consider surface evaporation as this process of water loss was assumed to have minimal influences on soil water below 10 cm depth, especially when the soil surface is quite dry. This model contains the essential elements necessary for modeling hydraulic redistribution, and is developed below. Simulated soil water potentials from this model were found to be very close to measured soil water potential for a stand of *Artemisia tridentata*.

The model of Ryel et al. (2002) partitions the soil column into several layers of uniform thickness with water moving between layers (Fig. 3). Changes in water content for layer i are expressed as:

$$\frac{dW_i}{dt} = \frac{dF_i}{dz} + H_i - E_i \qquad (1)$$

where W_i is water storage (cm), F_i is net unsaturated flow of water into the layer (cm h^{-1}), H_i is net water redistributed by roots into the layer (cm h^{-1}),

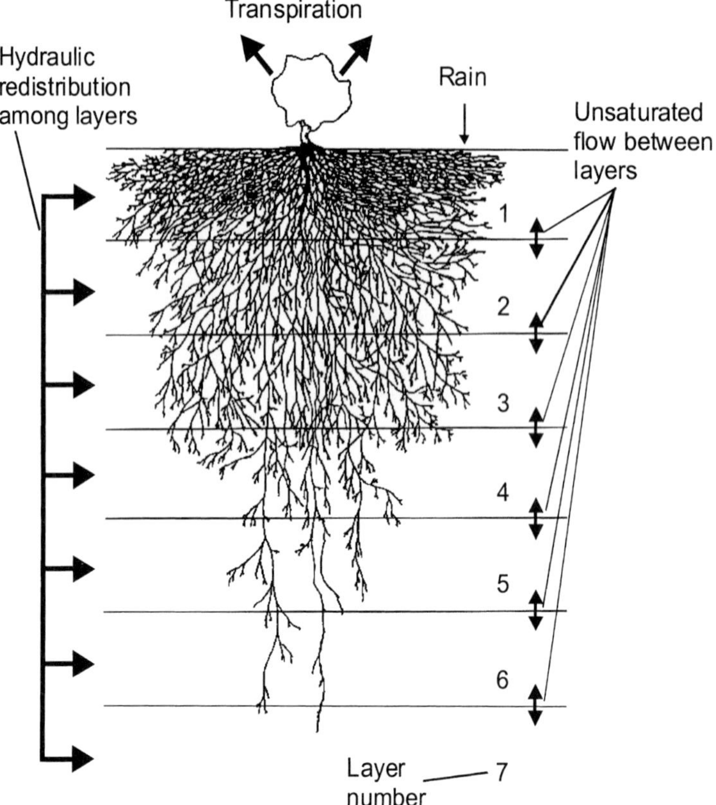

Fig. 3. Structure of model from Ryel et al. (2002) with seven soil layers shown. Unsaturated flow occurs between adjacent layers with different soil water potential, while hydraulic redistribution occurs via roots among all layers where soil water potentials differ. Rates of transpiration losses from each soil layer are based on the root distribution and soil water potential. Rainfall is added to the uppermost soil layer and is redistributed by unsaturated flow and hydraulic redistribution

and E_i is the transpirational water loss from the layer (cm h^{-1}), z is vertical thickness, and t is time. Rainfall (cm h^{-1}) was added to dW_i/dt for the top soil layer.

5.1 Unsaturated Flow

The model assumes that water moves between adjacent layers in response to water potential gradients and gravitational potential according to Buckingham-Darcy's law (Campbell 1985):

$$\frac{dF_i}{dz} = K(\theta_i)\left(\frac{d\Psi_i}{dz} + 1\right) \quad (2)$$

For layer i and vertical thickness z, $K(\theta_i)$ is the unsaturated soil hydraulic conductivity (cm h^{-1}) at volumetric water content θ_i (cm^3 cm^{-3}), and ψ_i is soil-water matric potential (cm; note: 1 MPa=10,200 cm).

Unsaturated soil hydraulic conductivity is related to saturated hydraulic conductivity K_s (cm h^{-1}) and volumetric water content using a relationship from van Genuchten (1980):

$$K_{i,\theta} = K_s S_i^{0.5}\left[1-(1-S_i^{1/m})^m\right]^2 \quad (3)$$

where S_i is relative saturation of the soil:

$$S_i = \frac{\theta_i - \theta_r}{\theta_s - \theta_r} \quad (4)$$

θ_r and θ_s are residual and saturated volumetric water contents (cm^3 cm^{-3}), respectively. The relationship between volumetric water content (θ_i) and soil water potential (Ψ_i) is expressed as in van Genuchten (1980):

$$\theta_i = \frac{\theta_s - \theta_r}{[1+|\alpha\psi_i|^n]^m} \quad (5)$$

where α, and n are parameters related to soil characteristics and m=1–1/n.

5.2 Hydraulic Redistribution

Water movement among soil layers by hydraulic redistribution (H_i) is assumed to move from wetter to dryer layers based on differences in soil water potential (Ψ_i). The rate that water is redistributed is affected by the distribution of active roots, radial conductivity of water between the root-soil interface (rhizosphere conductance), and transpiration activity. Water moving by roots between two layers is assumed to be limited by the layer with the smaller root density regardless of the layer supplying or receiving water, and by the rhizosphere hydraulic conductivity of the supplying layer. This assumes regulation of transpiration rates by stomata is sufficient to prevent limitations in water transport by severe cavitation in either root or shoot xylem (see Sperry et al. 1998, 2002; Kolb and Sperry 1999). Hydraulic redistribution is assumed to occur only during periods when the plant is not transpiring. Ryel et al. (2002) assumed that axial conductivity had minimal effect on water transport, although this could be incorporated into a more complex formulation. Given these assumptions, they model the net water movement into layer i from other layers (j) as:

$$H_i = C_{RT} \sum_j (\Psi_j - \Psi_i) \max (c_i, c_j) \frac{R_i R_j}{1 - R_x} D_{tran} \tag{6}$$

where C_{RT} is the maximum radial soil-root conductance of the entire active root system for water (cm MPa^{-1} h^{-1}), c_i is a factor reducing soil-root conductance based on Ψ_i, R_i is the fraction of active roots in layer i (all R_i must be .5), and D_{tran} is a factor reducing water movement among layers by roots during periods of transpiration. $R_x = R_i$ when $\theta_i > \theta_j$ or $R_x = R_j$ when $\theta_j > \theta_i$. In their implementation, Ryel et al. (2002) assumed D_{tran} was 1.0 during the night when transpiration was minimal and 0.0 during day. This factor could be modified to permit some hydraulic redistribution during the day if the potential gradients between the shoot and atmosphere was less than that for hydraulic redistribution among soil layers. In their implementation, the relative soil-root conductance for water (range 0–1) was modeled using an unpublished empirical relationship from van Genuchten (see Šimůnek et al. 1996):

$$c_i = \frac{1}{1 + \left(\frac{\Psi_i}{\Psi_{50}} \right)^b} \tag{7}$$

Ψ_{50} is the soil water potential (MPa) where soil-root conductance is reduced by 50% and b is an empirical constant, selected by linear (Ryel et al. 2002) or non-linear regression to fit measured data.

5.3 Transpiration

Soil water losses from transpiration are assumed to be primarily limited by the soil-root conductance for water in each layer (Eq. 7). The rate of water extracted by transpiration from a layer is proportional to the quantity of roots in the layer. Transpiration losses from each layer were modeled as:

$$E_i = E_{RT,\,max} c_i R_i \tag{8}$$

where $E_{RT,max}$ is the maximum whole canopy transpiration rate (cm h^{-1}). In their implementation Ryel et al. (2002) assumed that the vapor pressure deficit between the atmosphere and leaf did not limit the rate of transpiration during the day, a more complete representation of the transpiration process could be added to the model. This would be important under conditions when the vapor pressure deficits are low (e.g., when humidity is high, or temperatures low).

6 Perspectives for Future Research

Much of the research concerning hydraulic redistribution has focused on mechanisms and effects at the individual plant level. Minimal work has focused on plant-plant interactions (e.g., Caldwell 1990; Dawson 1993; Brooks et al. 2002; Ludwig et al. 2003) and at the community/ecosystem levels (e.g., Jackson et al. 2000). Certainly documenting and quantifying hydraulic redistribution for other species and plant communities will be valuable in defining the extent of the phenomenon, but the significance of this process will need to be further assessed at the community and ecosystem level to fully understand its importance as an ecosystem process. The issues include studying the effects of hydraulic redistribution on plant survival, plant distribution, species dynamics within plant communities, and ecosystem water, carbon and nutrient cycles. Because manipulative experiments to address hypothesized effects of hydraulic redistribution at larger scales will be difficult to conduct, the use of simulation models will become a significant part of future research efforts on hydraulic redistribution.

7 Conclusions

Hydraulic redistribution is a well documented phenomenon whereby water is passively moved via roots from regions of wetter soil to regions of dryer soil, particularly during periods when transpiration is minimal. Redistribution of soil water by roots has been documented for species occurring from arid to relatively mesic climates and for lifeforms from annuals to trees. Benefits to plants may include increased transpiration, increased nutrient availability, enhanced seedling establishment, water storage, and reduced competition for water. Community and ecosystem level effects of hydraulic redistribution are much less understood, but may include effects on water, carbon and nutrient cycling, plant survival, plant distribution, and species dynamics within plant communities. Simulation models which have recently been developed may become important tools in further understanding the ecological significance of hydraulic redistribution.

Acknowledgement. This work was funded by the National Science Foundation (DEB-9807097) and the Utah Agricultural Experiment Station.

References

Barber SA (1995) Soil nutrient bioavailability: a mechanistic approach. John Wiley and Sons, New York

Becker P (1998) Limitations of a compensation heat pulse velocity system at low sap flow: implications for measurements at night and in shaded trees. Tree Physiol 18:177–184

Branswijk JJB (1988) Modeling of water balance, cracking and subsidence of clay soils. J Hydrol 97:199–212

Breazeale JF (1930) Maintenance of moisture-equilibrium and nutrition of plants at and below the wilting percentage. Ariz Agric Exp Stn Tech Bull 29:137–177

Breazeale JF, Crider FJ (1934) Plant association and survival, and the build-up of moisture in semi-arid soils. Ariz Agric Exp Stn Tech Bull 53:95–123

Brooks JR, Meinzer FC, Coulombe R, Gregg J (2002) Hydraulic redistribution of soil water during summer drought in two contrasting Pacific Northwest coniferous forests. Tree Physiol 22:1107–1117

Brown RW, Bartos DL(1982) A calibration model for screen caged Peltier thermocouple psychrometers. Forest Service Research Paper INT-293. USDA, Ogden, Utah, p 155

Burgess SSO, Adams MA, Turner NC, Ong CK (1998) The redistribution of soil water by tree root systems. Oecologia 115:306–311

Burgess SSO, Adams MA, Bleby TM (2000a) Measurement of sap flow in roots of woody plants: a commentary. Tree Physiol 20:909–913

Burgess SSO, Adams MA, Turner NC, Ward B (2000b) Characterisation of hydrogen isotope profiles in an agroforestry system: implications for tracing water sources of trees. Agric Water Manage 45:229–241

Burgess SSO, Adams MA, Turner NC, White DA, Ong CK (2001a) Tree roots: conduits for deep recharge of soil water. Oecologia 126:158–165

Burgess SSO, Adams MA, Turner NC, Beverly CR, Ong CK, Khan AAH, Bleby TM (2001b) An improved heat pulse method to measure low and reverse rates of sap flow in woody plants Tree Physiol 21:589–598

Burgess SSO, Adams MA, Turner NC, Ong CK, Khan AAH, Beverly CR, Bleby TM (2001c) Correction: an improved heat pulse method to measure low and reverse rates of sap flow in woody plants. Tree Physiol 21:1157

Caldwell MM (1990) Water parasitism stemming from hydraulic lift: a quantitative test in the field. Isr J Bot 39:395–402

Caldwell MM, Manwaring JH (1994) Hydraulic lift and soil nutrient heterogeneity. Isr J Plant Sci 42:321–330

Caldwell MM, Richards JH (1989) Hydraulic lift: water efflux from upper roots improves effectiveness of water uptake by deep roots. Oecologia 79:1–5

Caldwell MM, Dawson TE, Richards JH (1998) Hydraulic lift: consequences of water efflux from the roots of plants. Oecologia 113:151–161

Campbell GS (1985) Soil physics with basic: transport models for soil – plant systems. Elsevier, Amsterdam

Clothier BE, Scotter DR, Kerr JP (1977) Water retention in soil underlain by a coarse-textured layer: theory and a field application. Soil Sci 123:392–399

Davis SD, Ewers FW, Sperry JS, Portwood KA, Crocker MC, Adams GC (2002) Shoot dieback during prolonged drought in *Ceanothus* (Rhamnaceae) chaparral of California: a possible case of hydraulic failure. Am J Bot 89:820–828

Dawson TE (1993) Hydraulic lift and water use by plants: implications for water balance, performance and plant-plant interactions. Oecologia 95:565–574

Dawson TE (1996) Determining water use by trees and forests from isotopic, energy balance and transpiration analyses: the roles of tree size and hydraulic lift. Tree Physiol 16:263–272

Dawson TE (1998) Water loss from tree roots influences soil water and nutrient status and plant performance. In: Flores HE, Lynch JP, Eissenstat DM (eds) Radical biology: advances and perspectives in the function of plant roots. Current topics in plant physiology 18. American Society of Plant Physiologists, Rockville, MD, pp 235–250
Dawson TE, Ehleringer JR (1991) Streamside trees that do not use stream water. Nature 350:335–337
de Kroon H, Fransen B, van Rheenen JWA, van Dijk A, Kreulen R (1996) High levels of inter-ramet water translocation in two rhizomatous *Carex* species, as quantified by deuterium labeling. Oecologia 106:73–84
Donovan LA, Linton MJ, Richards JH (2001) Predawn plant water potential does not necessarily equilibrate with soil water potential under well-watered conditions. Oecologia 129:328–335
Donovan LA, Richards JH, Linton MJ (2003) Magnitude and mechanisms of disequilibrium between predawn plant and soil water potentials. Ecology 84:463–470
Emerman SH, Dawson TE (1996) Hydraulic lift and its influence on the water content of the rhizosphere: an example from sugar maple, *Acer saccharum*. Oecologia 108:273–278
Fitter AH, Hay RKM (2002) Environmental physiology of plants, 3rd edn. Academic Press, San Diego, CA, USA
Gurevitch J, Scheiner SM, Fox GA (2002) The ecology of plants. Sinauer Associates, Sunderland, MA
Hanks RJ, Ashcroft GL (1980) Applied soil physics. Springer, Berlin Heidelberg New York
Hillel D, Hillel D (1998) Environmental soil physics: fundamentals, applications, and environmental considerations. Academic Press, San Diego,
Horton JL, Hart SC (1998) Hydraulic lift: a potentially important ecosystem process. Trends Ecol Evol 13:232–235
Huck MG, Klepper B, Taylor HM (1970) Diurnal variations in root diameter. Plant Physiol 45:529–530
Hultine KR, Cable WL, Burgess SSO, Williams DG (2003a) Hydraulic redistribution by deep roots of a Chihuahuan Desert phreatophyte. Tree Physiol 23:353–360
Hultine KR, Williams DG, Burgess SSO, Keefer TO (2003b) Contrasting patterns of hydraulic redistribution in three desert phreatophytes. Oecologia 135:167-175
Jackson RB, Caldwell MM (1996) Integrating resource heterogeneity and plant plasticity: modeling nitrate and phosphate uptake in a patchy soil environment. J Ecol 84:891–903
Jackson RB, Sperry JS, Dawson TE (2000) Root water uptake and transport: using physiological processes in global predictions. Trends Plant Sci 5:482–488
Johnson CD (1987) Preferred water flow and localised recharge in a variable regolith. J Hydrol 94:129–142
Kemp PR, Reynolds JF, Pachepsky Y, Chen J-L (1997) A comparative modeling study of soil water dynamics in a desert ecosystem. Water Resour Res 33:73–90
Kirkham MB (1983) Physical model of water in a split-root system. Plant Soil 75:153–168
Kolb KJ, Sperry JS (1999) Transport constraints on water use by the Great Basin shrub, *Artemisia tridentata*. Plant Cell Environ 22:925–935
Kropff MJ, van Laar HH (1993) Modelling crop-weed interactions. CAB International, Wallingford, UK
Lambers H, Chapin FS III, Pons TL (1998) Plant physiological ecology. Springer, Berlin Heidelberg New York
Landsberg JJ, Fowkes ND (1978) Water movement through plant roots. Ann Bot 42:493–508
Larcher W (1995) Physiological plant ecology, 3rd edn. Springer, Berlin Heidelberg New York
Leffler AJ, Ryel RJ, Hipps L, Ivans S, Caldwell MM (2002) Carbon acquisition and water use in northern Utah *Juniperus osteosperma* (Utah juniper) population. Tree Physiol 22:1221–1230
Ludwig F, Dawson TE, Kroon H, Berendse F, Prins HHT (2003) Hydraulic lift in Acacia tortilis trees on an East African savanna. Oecologia 134:293–300

Magistad OC, Breazeale JF (1929) Plant and soil relations at and below the wilting percentage. Ariz Agric Exp Stn Tech Bull 25

Marshall TJ, Holmes JW, Rose CW (1996) Soil physics. Cambridge University Press, Cambridge

Mendel M, Hergarten S, Neugebauer HJ (2002) On a better understanding of hydraulic lift: a numerical study. Water Resour Res 38:1183–1192

Millikin Ishikawa C, Bledsoe CS (2000) Seasonal and diurnal patterns of soil water potential in the rhizosphere of blue oaks: evidence for hydraulic lift. Oecologia 125:459–465

Mooney HA, Gulmon SL, Rundel PW, Ehleringer J (1980) Further observations on the water relations of *Prosopis tamarugo* of the northern Atacama Desert. Oecologia 44:177–180

Mott KA, Parkhurst DF (1991) Stomatal responses to humidity in air and helox. Plant Cell Environ 14:509–515

Nobel PS (1991) Physiochemical and environmental plant physiology. Academic Press, San Diego, CA, USA

Noy-Meir I (1973) Desert ecosystems: environment and producers. Annu Rev Ecol Syst 4:25–51

Nye PH, Tinker PB (1977) Solute movement in the soil-root system. Blackwell, Berkeley, CA

Pearcy RW, Ehleringer J, Mooney HA, Rundel PW (eds) (1989) Plant physiological ecology: field methods and instrumentation. Kluwer, Dordrecht

Peterson CA, Myrrmann M, Steudle E (1992) Location of major barrier(s) to movement of water and ions in young roots of *Zea mays* L. Planta 190:127–136

Portwood KA, Ewers FW, Davis SD, Sperry JS, Adams CG (1997) Shoot dieback in *Ceanothus* chaparral during prolonged drought – a possible case of catastrophic xylem cavitation. Bull Ecol Soc Am 78:298

Querejeta JI, Egerton-Warburton LM, Allen MF (2003) Direct nocturnal water transfer from oaks to their mycorrhizal symbionts during severe soil drying. Oecologia 134:55–64

Richards JH, Caldwell MM (1987) Hydraulic lift: substantial nocturnal water transport between soil layers by *Artemisia tridentata* roots. Oecologia 73:486–489

Ryel RJ, Caldwell MM (1998) Nutrient acquisition from soils with patchy nutrient distributions: importance of patch size, degree of variability and root uptake kinetics. Ecology 79:2735–2744

Ryel RJ, Caldwell MM, Yoder CK, Or D, Leffler AJ (2002) Hydraulic redistribution in a stand of *Artemisia tridentata*: evaluation of benefits to transpiration assessed with a simulation model. Oecologia 130:173–184

Ryel RJ, Caldwell MM, Leffler AJ, Yoder CK (2003) Rapid soil moisture recharge to depth by roots in a stand of *Artemisia tridentata*. Ecology 84:757–764

Ryel RJ, Leffler AJ, Peek MS, Ivans CY, Caldwell MM (2004) Water conservation in *Artemisia tridentata* through redistribution of precipitation. Oecologia (in press)

Schenk HJ, Jackson RB (2002) The global biogeography of roots. Ecol Monogr 72:311–328

Scholz F, Bucci SJ, Goldstein G, Meinzer FC, Franco AC (2002) Hydraulic redistribution of soil water by neotropical savannah trees. Tree Physiol 22:603–612

Schulze E-D (1991) Water and nutrient interactions with plant water stress. In: Mooney HA, Winner WE, Pell EJ (eds) Response of plants to multiple stresses. Academic Press, San Diego, pp 89–101

Schulze E-D, Caldwell MM, Canadell J, Mooney HA, Jackson RB, Parson D, Scholes R, Sala OE, Trimborn P (1998) Downward flux of water through roots (i.e. inverse hydraulic lift) in dry Kalahari sands. Oecologia 115:460–462

Šimůnek J, Suarez DL, Šejna (1996) The UNSATCHEM software package for simulating the one-dimensional variably saturated water flow, heat transport, carbon dioxide production and transport, and multicomponent solute transport with major ion equilibrium and kinetic chemistry. Ver 2.0. Res Rep No 141, US Salinity Laboratory, ARS, USDA, Riverside, CA, 186 pp

Slatyer RO (1967) Plant-water relationships. Academic Press, New York

Smith DM, Allen SJ (1996) Measurement of sap flow in plant stems. J Exp Bot 47:1833–1844

Smith DM, Jackson NA, Roberts JM, Ong CK (1999) Reverse flow of sap in tree roots and downward siphoning of water by *Grevillea robusta*. Funct Ecol 13:256–264
Sperry JS (2000) Hydraulic constraints on plant gas exchange. Agric For Meteorol 104:13–23
Sperry JS, Hacke UG (2002) Desert shrub water relations with respect to soil characteristics and plant functional type. Funct Ecol 16:367–378
Sperry JS, Adler FR, Campbell GS, Comstock JP (1998) Limitations of plant water use by rhizosphere and xylem conductance: results from a model. Plant Cell Environ 21: 347–359
Sperry JS, Hacke UG, Oren R, Comstock JP (2002) Water deficits and hydraulic limits to leaf water supply. Plant Cell Environ 25:251–263
Stark JM (1994) Causes soil nutrient heterogeneity at different scales. In: Caldwell MM, Pearcy RW (eds) Exploitation of environmental heterogeneity by plants: ecological processes above- and belowground. Academic Press, San Diego, pp 255–284
Steudle E (1994) Water transport across roots. Plant Soil 167:79–90
Taiz L, Ziegler E (2002) Plant physiology, 3rd edn. Sinauer Associates, Sunderland, MA
Thorburn, PJ, Walker GR (1993) The source of water transpired by Eucalyptus camaldulensis: soil groundwater, or streams? In: Ehleringer JR, Hall AE, Farquhar GD (eds) Stable isotopes and plant carbon-water relations. Academic Press, San Diego, pp 511–527
Thornley JHM, Johnson IR (1990) Plant and crop modelling: a mathematical approach to plant and crop physiology. Clarendon Press, Oxford
Timlin D, Pachepsky YA, Acock B (1996) A design for a modular, generic soil simulator to interface with plant models. Agron J 88:162–169
Tyree MT, Ewers FW (1991) The hydraulic architecture of trees and other woody plants. Tansley Review No. 340. New Phytol 119:345–360
Tyree MT, Sperry JS (1989) Vulnerability of xylem to cavitation and embolism. Annu Rev Plant Physiol Mol Biol 40:19–38
van Genuchten MT (1980) A closed-form equation for predicting the hydraulic conductivity of unsaturated soils. Soil Sci Soc Am J 44:892–898
Wan CG, Sosebee RE, McMichael BL (1993) Does hydraulic lift exist in shallow-rooted species? A quantitative examination with a half-shrub *Gutierrezia sarothrae*. Plant Soil 153:11–17
Williams K, Caldwell MM, Richards JH (1993) The influence of shade and clouds on soil water potential: the buffered behavior of hydraulic lift. Plant Soil 157:83–95
Yoder CK, Nowak RS (1999) Hydraulic lift among native plant species in the Mojave Desert. Plant Soil 215:93–102

Ronald J. Ryel
Department of Forest
Range, and Wildlife Resources
Utah State University
Logan, Utah 84322–5230, USA
Tel.: +1-435-7978119
Fax: +1-435-7973796
e-mail: ron.ryel@usu.edu

New Insights in the Genus *Phytophthora* and Current Diseases These Pathogens Cause in Their Ecosystem

Wolfgang Oßwald, Julia Koehl, Ingrid Heiser, Jan Nechwatal, and Frank Fleischmann

1 *Phytophthora* Species and Their Significance for Diseases on Woody Plants

The genus *Phytophthora* (Greek "plant destroyer") includes many soil-borne species which cause root and collar rot symptoms of herbaceous and woody plants including different forest trees. This review will focus on *Phytophthora* diseases of forest trees, as their significance has increased in recent years. Chlorosis, necrosis and wilt symptoms can be seen on leaves in the whole crown before the tree dies. However, death can take several years. All soil-borne *Phytophthora* species share a common infection cycle which is summarized in Fig. 1.

Oospores germinate in the rhizosphere at low temperatures stimulated by root exudates if the pH lies between 3.9 and 6 (Erwin and Ribeiro 1996). In the presence of free water and attracted by root exudates, zoospores released from zoosporangia will move towards fine roots driven by their two flagella. Most *Phytophthora* species grow from fine roots into coarse roots and from there up into the trunk where they mainly destroy the phloem and cambium tissue. Typical cankers are often found on infected roots and on the trunk.

The phylum Oomycota comprising the order Pythiales with the genus *Phytophthora* has been recently transferred from the Kingdom Mycetae to the Kingdom Chromista (including heterokont algae) due to the fact that the oomycetous organisms differ from true fungi because of their heterokont flagellar apparatus of the zoospores, the lack of chitin and ergosterol in their cell wall and plasmalemma structures and their diploid nature (Cavalier-Smith 1989; Barr 1992; Hawksworth et al. 1995).

The significance and the involvement of different *Phytophthora* species in the decline of woody plants has been shown in forests worldwide. Decline of *Eucalyptus marginata* trees (jarrah) in forests of Western Australia caused by *Phytophthora cinnamomi* is a well-known example of the destructive potential of those root rot pathogens (Old 1979; Weste and Marks 1987; Shearer and Tippett 1989). On specific sites, where water collects

Progress in Botany, Vol. 65

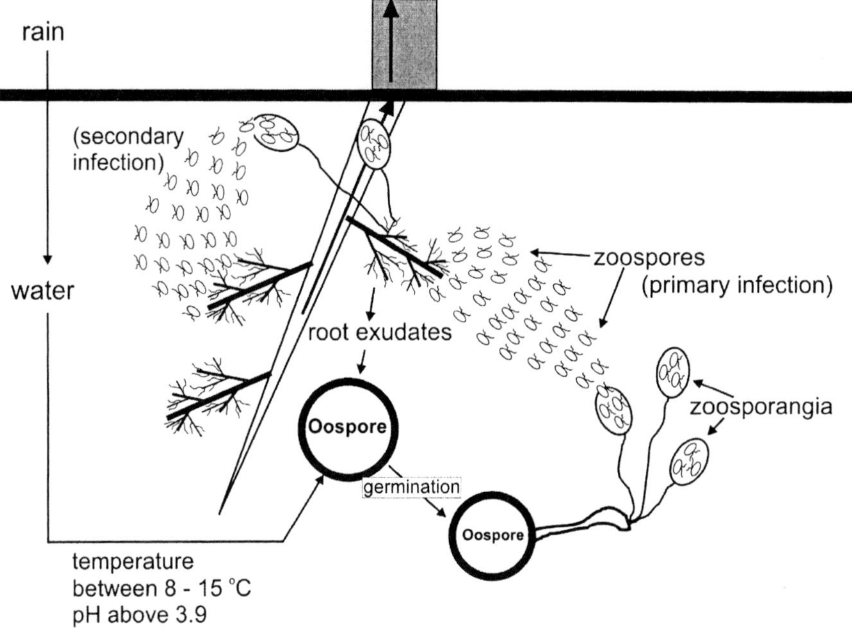

Fig. 1. The infection cycle of soil-borne *Phytophthora* species

above a concreted lateritic layer (hardpan), the soil-borne pathogen produces sporangia and zoospores which infect and kill the vertical roots. This causes the tree to die during the dry summer, because the vertical root system has lost its ability to tap the groundwater at great depths (Shea et al. 1984). *P. cinnamomi* is probably native to Papua New Guinea and was spread from there by man to many parts of the world during the last two centuries. In Western Australia it has been proven that the pathogen was introduced to new areas by road-building equipment and forest machineries. It was also shown that, in the Wilson's Promontory National Park in Victoria, dieback of susceptible plants followed the building of roads with infested gravel obtained from areas where root rot on eucalyptus trees was observed.

P. cinnamomi has a broad host range (more than 1000 species; Erwin and Ribeiro 1996) and was also shown to be associated with mortality of *Quercus suber* (cork oak) and *Q. ilex* (holm oak/holly oak) in the Mediterranean region (Brasier 1992, 1996). In combination with drought, this pathogen is a major predisposing factor in the Iberian oak decline that has recently been confirmed by Sanchez et al. (2002). Computer models indicate that the current activity of *P. cinnamomi* is likely to increase in Europe in the

Mediterranean region and at the coastal area of western Britain up to the year 2050 as a result of global warming (Brasier and Scott 1994).

One attempt to control the spread of *P. cinnamomi* on *Q. ilex* and *Q. suber* in Spain and Portugal in the field was the treatment of declining trees with antifungal materials. Fernandez-Escobar et al. (1999) injected potassium phosphonate, quinosol and carbendazim directly into the trunk using a pressurized injection capsule technique. Only trees treated with potassium phosphonate appeared to stop development of symptoms in the crown during the first and second year, whereas untreated *Q. suber* and *Q. robur* showed an increase in decline. In the third year after treatment, a significant improvement in their general appearance was observed. Recently, Komorek and Shearer (1997) have proved that the fungicide phosphonate (synonym with phosphite) was effective against *P. cinnamomi* in native plant communities in Western Australia by aerial spraying and trunk injection. The active compound, the phosphorous acid (H_3PO_3) is systemically translocated throughout the plant via the xylem and the phloem. It is an environmentally safe, inexpensive chemical and has a very low toxicity to animals. The fungicide is directly toxic to the pathogen and indirectly in stimulating defense responses in the host, which finally inhibit pathogen growth (Guest and Grant 1991; Jackson et al. 2000). Phosphonate cannot eradicate *Phytophthora cinnamomi* from an area once it has established, because protection holds only a few years after application. Therefore, the major strategy for limiting the environmental damage caused by the pathogen remains by means of quarantine and the prevention of the transport of infested soil into uninfested areas.

Phytophthora cambivora is known as the causal agent of ink disease of chestnuts (*Castanea sativa*) in the United States and in Europe (Day 1932; Milburn and Gravatt 1932; Peace 1962). The pathogen infects larger roots and the trunk above the soil line. If the tree is girdled, leaves start to wilt and the tree usually dies within the first year of infection. Occasionally, an inky fluid exudes from dying bark and from infected roots which stains the soil close to the infected tissue. This disease has recently spread dramatically in most chestnut-growing areas in Italy (Anselmi et al. 1996). According to the results of Vettraino et al. (2001), again *P. cambivora* seems to play a major role in the high mortality of *Castanea sativa*.

In the coniferous forests of southwestern Oregon and northern California Port-Orford cedar (POC) (*Chamaecyparis lawsoniana*), also called the Lawson cypress, is killed by *Phytophthora lateralis* (Hansen et al. 2000). The pathogen was reported first in 1923 in nurseries in Seattle growing POC. The disease was so severe that nurserymen were forced to stop selling these plants. At that time *P. lateralis* destroyed the multi-million dollar ornamental cedar industry. The pathogen was reported in the forest in 1952, probably introduced in infested soil with rhododendrons from nurseries. *P. lateralis* has recently been identified for the first time in France on Port-Orford cedar plants from a commercial nursery (Hansen et al. 1999).

The pathogen is not native to the natural range of Port-Orford cedar. The disease is triggered by heavy rains during late winter and early spring. Under these conditions zoospores are released from sporangia which initiate infection of the fine roots. The growing hyphae kill the roots and finally grow up into the trunk where they destroy the phloem and cambium tissue. In the crown the foliage changes colour to bronze and finally to light brown and trees die usually within a year of first crown symptoms.

In cedar country, the road system is largely infested and provides the principal pathway for disease spread. In the forest *P. lateralis* is spread via infested soil distributed by forest machinery or on the feet of humans and animals. It moves downhill in running water and along stream courses. Once introduced, there is no practical means to eradicate the pathogen, and there are fewer and fewer large, disease-free watersheds. As a result of disease, the size and age distribution of cedar have changed. In many areas, large old trees are gone, with only smaller, young trees surviving. *P. lateralis* is now found in most Port-Orford cedar growing areas, including some of the largely protected roadless wilderness areas. Because of this highly aggressive root pathogen, Port-Orford cedar is now considered a threatened species within its native region.

In order to protect Port-Orford cedar, several disease management strategies, known as "POC Management Guidelines" were introduced (Hansen et al. 2000). These include closure of forest roads during the wet season, washing vehicles and harvesting equipment before passing between infested and noninfested areas, or reengineering forest roads to stop or to reduce overland flow of water from infested roads into pathogen-free areas. Besides these efforts to control *P. lateralis* in its ecosystem, a program to select resistant cedar trees from the forest with the aim to breed for enhanced resistance towards *P. lateralis* was started by the local Forest Service in cooperation with the University of Oregon.

2 New *Phytophthora* Species and the Diseases They Cause on Woody Plants

2.1 *Phytophthora ramorum*

Since 1994 a new disease, described as "Sudden Oak Death" (SOD) of *Quercus agrifolia* (coast live oak), *Quercus kelloggii* (California black oak) and *Lithocarpus densiflorus* (tanbark oak) was reported for central and northern California from Monterey to Mendocino County (Garbelotto et al. 2002; Rizzo et al. 2002). Recently, the disease has also been confirmed in southwestern Oregon (Goheen et al. 2002). Symptomatic and dead trees are found in mixed-evergreen forests with and without Douglas fir (*Pseudotsuga menziesii*) and in redwood (*Sequoia sempervirens*) forests. Death of *Q. agrifolia* is also observed in the urban-wild land interface.

This new disease is caused by *Phytophthora ramorum*, a pathogen recently described by Werres et al. (2001). *P. ramorum* was isolated from *Rhododendron* and *Viburnum* in Germany and the Netherlands where it causes a twig blight disease. It kills mature bushes of *Rhododendron* in gardens and nursery plants as well. Morphologically, *P. ramorum* is characterized by the production of many chlamydospores and by ellipsoid, deciduous and semi-papillate sporangia. *P. ramorum* starts growing at 2 °C, the optimum and the maximum temperatures for growth are 20 and 26 °C, respectively. The ITS1 and ITS2 sequences of the rRNA-encoding DNA of *P. ramorum* isolates were closest to those of *P. lateralis*, the causal agent of Port-Orford cedar decline in Oregon and differed by only 11 nucleotides. The ITS sequences of *P. ramorum* from *Rhododendron* and *Viburnum* collected in Europe were identical to those obtained from coast live oak and tanbark oak in California, proving that they belong to the same species, despite the difference in mating type (A1 in Europe, A2 in USA). However, AFLP analyses placed the European and the North American isolates of *P. ramorum* into different genotypes, which grouped into two separate clades. Among the US isolates tested, a single genotype accounted for 82% of all isolates, indicating a largely clonal population in North America. In contrast, each European isolate tested so far in an AFLP study showed a different genotype. These data indicate that the US- and European isolates of *P. ramorum* belong to different populations (Bonants et al. 2002; Ivors et al. 2002).

Infection of oak and tanbark oak by *P. ramorum* starts in the outer bark of the trunk or branches and progresses to the cambium, resulting in the formation of cankers. The infected phloem turns brown and the active growing zone of the pathogen is delimited by a thin black line. This discolouration is mainly found in the bark and to a much lesser extent in the xylem tissue. Cankers normally start on the trunk at the soil line. However, they are also found on branches up to 20 m above the ground. Single cankers can reach 2 m in length. In contrast, *P. ramorum* was never found in roots of oaks or tanbark oaks. Often branches above cankers wilt, while branches below do not show any symptoms. After the pathogen is established in the trunk, the whole foliage turns from a healthy green colour to brown within only a few weeks.

In contrast to oak decline in Mexico or in Europe caused by other *Phytophthora* species (Robin et al. 1992; Brasier et al. 1993; Jung et al. 2000a; Tainter et al. 2000), this pathogen is not necessarily associated with the presence of water in the soil (Garbelotto et al. 2001). According to field observations and seedling experiments, *L. densiflorus* seems to be more susceptible to *P. ramorum* than *Q. agrifolia*. Recent data from Davidson et al. (2002b) indicate that *Umbellularia californica* (California bay laurel) leaves are highly susceptible to *P. ramorum* and produce huge amounts of spores during the rainy season in spring. Thus, infected *U. californica* leaves are thought to be the key factor for the transmission of the disease in California oak woodlands.

Recent investigations by Garbelotto et al. (2002) prove that there are many more hosts in the field for *P. ramorum* besides oaks, tanbark oaks and California bay. The pathogen was isolated from branches and leaves of madrone (*Arbutus menziesii*), huckleberry (*Vaccinium ovatum*), buckeye (*Aesculus californica*) or from leaves of big leaf maple (*Acer macrophyllum*). According to Maloney et al. (2002) and Davidson et al. (2002a), *P. ramorum* was also isolated from dead needles of Douglas fir (*Pseudotsuga menziesii*) and redwood (*Sequoia sempervirens*).

So far as known, *P. ramorum* infects only *Quercus* species that belong to the red oak group. There is great concern that the pathogen could spread to oak forests of the eastern United States. Recent data from Brasier et al. (2002) and Moralejo and Hernandez (2002) have proved that bark of *Quercus ilex*, *Q. cerris*, *Q. rubra* and *Fagus sylvatica* as well as of *Olea europea* are susceptible to *P. ramorum*, whereas the pathogen did not grow in bark of *Tilia cordata*, *Fraxinus excelsior* or *Q. robur*.

The question arises, whether the pathogen was introduced from Europe to the United States by already infected ornamental plants, such as *Rhododendron*. However, due to the molecular population analyses of *P. ramorum* discussed above, it is most probably that the pathogen was separately introduced in both Europe and in North America from a third, still unknown country. In Europe, *P. ramorum* is not yet found in forest ecosystems. It is restricted to nurseries where *Rhododendron* and *Viburnum* plants are infected.

2.2 "Alder-*Phytophthora*"

An unusual mortality of alder (*Alnus glutinosa*) was reported from southern Britain by Brasier et al. (1995), Gibbs (1995) and Gibbs et al. (1999). Diseased trees were found on banks of streams and rivers and also in plantings which were not directly connected to watercourses. Besides *Alnus glutinosa,* disease was also found on gray alder (*A. incana*) and on Italian alder (*A. cordata*). Declining trees show heavy leaf losses and wilted branches in their crowns. Cankers mostly starting from roots are also seen on the trunk which can reach up to 2 m length. Sometimes tarry or rust-coloured exudations seep from the bark.

The disease is caused by an unusual new *Phytophthora* similar to *Phytophthora cambivora*, a known pathogen of hardwood trees in Europe (Brasier et al. 1995). However, it differs from the latter being homothallic, showing high frequency of oospore abortion, having little or no aerial mycelium and lower temperatures for optimum and maximum growth. The analysis of the ITS1 and ITS2 sequences and the results of amplified fragment length polymorphisms (AFLP) experiments demonstrated that the aggressive *Phytophthora* pathogen of alder trees comprises a range of species hybrids involving *Phytophthora cambivora* and a still unknown taxon similar to *Phytophthora fragariae* as parents (Brasier et al. 1999). Besides the "standard" alder *Phytophthora*, which is near tetraploid, other alder *Phytophthora* species, called natural "variants" were described which might be genetic breakdown products of the standard type or products of new hybridization events of both parents. Recently, it has been shown by Delcan and Brasier (2001) that none of more than 4000 oospores of either the "standard" type or other variants investigated germinated, proving that there is a high degree of meiotic irregularities in this hybrid. In consequence, this means that mycelium and zoospores are responsible for the survival and the spread of the pathogen in the field. The latter might infect fine roots or adventitious roots, from where the pathogen grows up into the trunk and destroys the phloem and cambium tissue. Up to now, it is not known how the pathogen impairs water transport from roots to leaves and whether specific toxins produced by the hybrid might be involved in the disease. According to the molecular data, the developmental instabilities of the hybrids and the recombinations in their ITS regions, Brasier et al. (1999) came to the conclusion that the hybrids are of recent origin and that their evolution is still ongoing.

Alder decline and the pathogen were also reported from other European countries such as France where *Alnus glutinosa* was heavily affected along rivers on more than 100 sites in western and northeastern France (Streito et al. 2002), or from Belgium, Sweden, Austria and Germany (Hartmann 1995; Cech 1997; Werres 1998; Jung et al. 2000c; Jung and Blaschke 2001). In all of these countries, the "standard" alder *Phytophthora* was isolated along with other natural variants from declining trees. Recently, Brasier and Kirk (2001) have demonstrated that the standard hybrid and the "Dutch variant" were the most aggressive pathogens towards alder bark, whereas the "German", "UK" and the "Swedish" variants, as well as *P. cambivora,* were only weakly pathogenic. Furthermore, the standard hybrid did not grow in the bark of *Quercus, Castanea, Fagus, Acer, Chamaecyparis* or *Taxus*, proving that it is relatively host-specific.

It is not known where the new hybrid was first generated or how it spread all over Europe. However, there is evidence that infected nursery plants might have been involved in the distribution of the new pathogen.

2.3 *Phytophthora quercina*

Oak decline has been a frequently occurring disease of European oak forests since the beginning of the twentieth century (Delatour 1983; Hartmann et al. 1989) being of local or regional importance in the past. However, since the beginning of the 1980s, oak decline was recorded from all over Europe (Siwecki and Liese 1991; Hartmann 1998; Jung 1998). The main above-ground symptoms include dieback of branches and parts of the crown, yellowing and wilting of leaves, finally resulting in a high transparency of the crown. These symptoms indicate problems with water and nutrient uptake and are found on declining oaks throughout all growth regions. It can take several years until trees die. According to Hartmann et al. (1989), the mortality rate may be up to five trees per hectare and year.

Many investigations proved that declining oaks exhibited severe fine root damage (Vincent 1991; Eichhorn 1992; Blaschke 1994; Jung et al. 1996). Recently, Jung et al. (2000a,b) have carried out a broad survey in Bavaria (Germany) on the occurrence of soil-borne *Phytophthora* species in 35 oak stands, including 217 trees in total on geologically different sites. They found 10 different *Phytophthora* species in 19 stands which were characterized by sandy-loamy, loamy, or clayey soil texture and by a mean pH ($CaCl_2$) between 3.9 and 6.6. No *Phytophthora* species were recovered from soil samples of the other 16 stands on Triassic and Jurassic sandstones, Pleistocene gravels and chalk with mean soil conditions of below pH 3.9. Besides *P. syringae, P. megasperma* and *P. gonapodyides*, which were found at a low frequency, *P. citricola, P. cambivora* and an unknown species were isolated most frequently. This unknown homothallic *Phytophthora*, belonging to Waterhouse's group I of *Phytophthora*, is characterized by paragynous antheridia and papillate sporangia and has recently been described by Jung et al. (1999a) as *P. quercina*. The RAPD and RFLP banding patterns as well as the sequence data of the ITS1 and ITS2 regions confirmed that *P. quercina* is a distinct species (Cooke et al. 1999). *P. quercina* was isolated more frequently from declining than from healthy oaks and proved to be host-specific and highly aggressive to the genus *Quercus* (Jung et al.

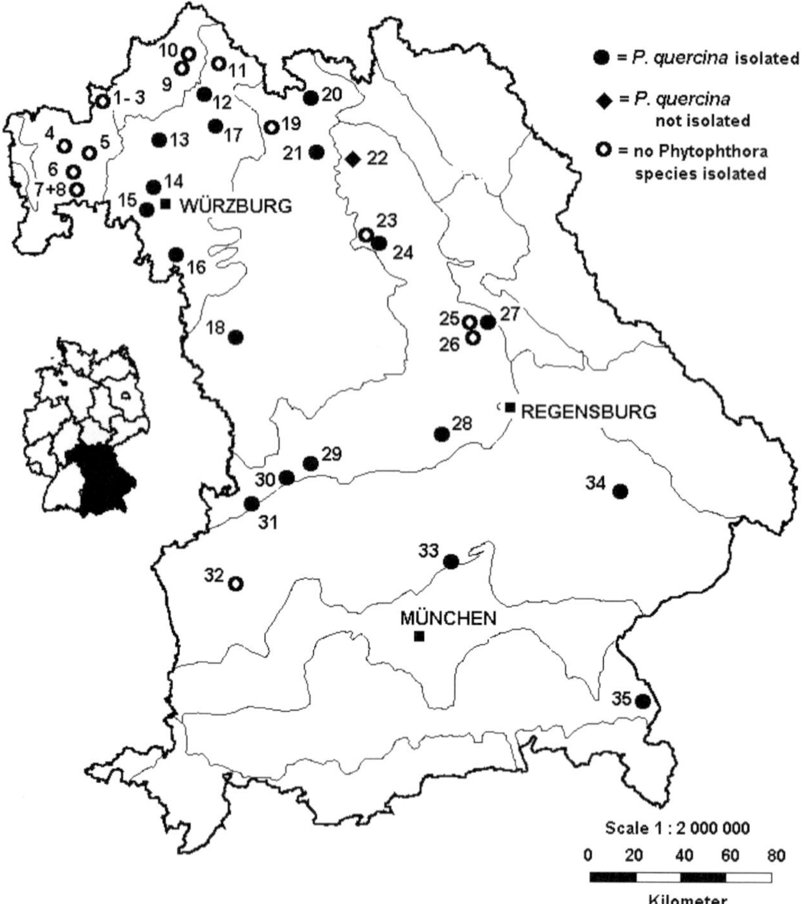

Fig. 2. Distribution map of *Phytophthora quercina* in Bavaria (Jung et al. 2000a)

1999b). Figure 2 shows the distribution of *P. quercina* in oak stands in Bavaria.

Besides the differences in soil pH, about five times higher calcium concentrations were found for the 19 stands infested with *Phytophthora* species as compared to those 16 locations where no *Phytophthora* species were isolated (Jung et al. 2000a). These much higher calcium values might favour or trigger the infection by zoospores as it was shown by Byrt et al. (1982) and Grant et al. (1986) for zoospores of *P. palmivora.*

Statistical analyses clearly showed that crown transparency was negatively correlated with root parameters only in stands where *Phytophthora* pathogens were frequently isolated. In these stands, the degree of crown transparency increased with a decreasing number of root tips, fine roots or

Table 1. Correlation between crown transparency and different root parameters of oak trees

	FR/MR (n m^{-1})	FRL/dwMR (cm g^{-1})	FRT/dwMR (n g^{-1})
19 stands infested with *Phytophthora* species (total 124 trees)			
Spearman correlation	–0.323	–0.396	–0.391
Significance	***	***	***
16 stands not infested with *Phytophthora* species (total 93 trees)			
Spearman correlation	–0.008	–0.057	–0.016
Significance	n.s.	n.s.	n.s.

n, Number; FR, fine roots (diameter mm); MR, mother root (diameter 2–5 mm); FRL, fine root length; FRT, fine root tips; n.s., not significant; *** *p*.<001

decreasing fine root length. No such correlations were found for stands, where these root rot pathogens were not isolated (Table 1; Jung et al. 2000a).

Again, highly significant differences were found for the parameters number of fine roots per meter mother root (FR/MR) or fine root length as well as number of fine root tips per gram mother root (FRL/dwMR or FRT/dwMR), when the root systems of all healthy and declining oaks of *Phytophthora* infested stands were compared (Table 2; Jung et al. 2000a). These data clearly show that healthy-looking trees were characterized by much better developed root systems than declining oaks. However, none of these differences were found for root systems of declining and healthy-looking oaks in *Phytophthora*-free stands.

Surprisingly, the root systems of declining oaks growing in *Phytophthora*-free stands seem to be better developed compared to those of healthy-looking trees of *Phytophthora*-infested stands.

In conclusion, we have to distinguish at least between two different complex diseases which are called "oak decline". At sandy-loamy to clayey sites with a mean soil acidity of higher than pH 3.9 and high calcium concentrations, different *Phytophthora* species, mainly *P. cambivora*, *P. citricola* and *P. quercina*, are strongly involved in oak decline in Bavaria. Recently, Hartmann and Blank (2002) have also reported that the occurrence of *Phytophthora* species was correlated with the amount of calcium saturation in combination with the clay content of the upper rhizosphere soil. These pathogens cause a fine root disease leading to highly significant correlations between crown transparency and important fine root parame-

Table 2. Comparison of root parameters of declining and healthy oaks

	FR/MR (n m^{-1})		FRL/dwMR (cm g^{-1})		FRT/dwMR (n g^{-1})	
	Healthy	Declining	Healthy	Declining	Healthy	Declining
19 stands infested with *Phytophthora* species (total 124 trees)						
Mean	59.2	44.6	121.0	76.1	138.5	88.9
SD	24.8	19.8	66.8	52.2	75.6	68.1
Significance[a]	***		***		***	
16 stands not infested with *Phytophthora* species (total 93 trees)						
Mean	62.8	57.3	174.1	144.6	184.4	156.6
SD	48.3	35.1	120.6	121.5	120.6	123.5
Significance[a]	n.s.		n.s.		n.s.	

n, Number; FR, fine roots (diameter <2 mm); MR, mother root (diameter 2–5 mm); FRL, fine root length; FRT, fine root tips; n.s., not significant; *** *p*.<001 (modified according to Jung et al. 2000a)

[a]Mann-Whitney test

ters. Among all *Phytophthora* species isolated, *P. quercina* seems to play the major role due to its widespread occurrence and its high flexibility concerning site conditions and soil pH (Jung et al. 1996, 1999a). The first visible crown symptoms in mature oaks might be seen when rootlet death begins to exceed rootlet replacement. If the main precondition for infection, the presence of free water in the soil, is fulfilled several times over years or even within 1 year (Erwin and Ribeiro 1996), then the progressive destruction of the fine root system can cause a chronic dieback of the crown, finally weakening the trees and predisposing them to further abiotic and biotic stress factors. Therefore, in combination with summer droughts (Delatour 1983; Hartmann 1998), secondary pathogens (e.g. *Armillaria mellea, Agrilus biguttatus*; Hartmann and Blank 1992) or heavy defoliation (Lobinger 1999), the disease incidence might be higher and rapid mortality of trees can occur.

On sites with sandy or sandy-loamy soil texture and an acidity of lower than pH 3.9, *Phytophthora* species have never been isolated. In addition, at these stands, no significant correlation between root and crown status was found. These site conditions are characteristic for most of the declining oak stands in NW Germany, where the isolation rate for *Phytophthora* species was zero or less than 20%. At these stands these pathogens are not considered to be involved in oak decline (Jung et al. 2000a; Hartmann and Blank 2002). Here, heavy droughts, in combination with defoliation and late winter frost periods (Hartmann et al. 1989) or together with miscellaneous agents such as *Collybia fusipes* (Marcais et al. 1998) might weaken and predispose the trees to the attacks of parasites as mentioned above.

The question arises, whether environmental changes may have occurred that have triggered root infection by natural soil-borne *Phytophthora* species and have unbalanced the host–parasite relationship between oaks and these pathogens. The rise in mean winter temperature of 0.03 °C/year and the seasonal shift of precipitation from summer to the early winter months, reported by Schönwiese et al. (1994), might have favoured root infection by soil-borne *Phytophthora* species which are known to depend on water in the rhizosphere. Furthermore, Mayer (1999) showed, after analyzing forest monitory data from 1984 to 1997, that the probability of oaks showing 25% or more crown transparency is 1.6 times higher after wet winters with average temperatures more than 2 °C above the long-term average values.

Recently, Vettraino et al. (2002) have published their results on the occurrence of different *Phytophthora* species in oak stands in Italy and their association with oak decline. Besides the known species *P. cambivora*, *P. cinnamomi* and *P. cactorum*, which were also isolated from stands in central and southern Italy, *P. quercina* was found only in the northern and central parts of the country. The absence of this species in southern Italy could be explained by its sensitivity to high temperatures. *P. quercina* shows optimum growth near 25 °C and no growth occurs at 30 °C (Jung et al. 1999a). It was the only species significantly associated with declining oaks in the northern and central parts of Italy, proving its pathogenicity to oaks under conditions totally different from those in Germany regarding climate and oak stand composition. In conclusion, the authors consider the significance of *P. quercina* in oak decline in Italy as a predisposing factor weakening the trees and exposing them to final biotic and abiotic factors acting differently at the regional level.

3 Identification and Detection Methods

In order to obtain information on the distribution and spread of different *Phytophthora* pathogens under natural conditions in their ecosystems, specific methods have to be applied which allow the sensitive detection of these pathogens.

3.1 Traditional Isolation and Identification Methods

Baiting with different host plant tissues (e.g. fruits, seedlings or leaflets) to be infected by zoospores is the traditional technique to isolate *Phytophthora* spp. from infected soil samples, since it is often difficult to isolate *Phytophthora* species directly from decayed tissue or from soil. The baiting technique exploits the pathogenicity of *Phytophthora* species towards living bait tissue on which lesions caused by the target species are seen after a certain time. From these appearing lesions the pathogen is transferred to selective agar plates containing a cocktail of different antibiotics and fungicides to inhibit growth of *Pythium* species, bacteria and other fungi (Erwin and Ribeiro 1996). A major problem arising in trying to isolate *Phytophthora* species in the field, is the finding that the season has great influence on the recovery of these soil-borne pathogens. Vettraino et al. (2001) showed that *P. citricola* was isolated from soil around symptomatic *Castanea sativa* trees from March to August and in September, whereas *P. cactorum* and *P. cambivora* were only recovered in spring and autumn and never detected with specific baits in summer and during wintertime. This can be explained by the different demands of *Phytophthora* species on growth conditions in their environment (e.g. soil temperature, moisture, soil pH; see Fig. 1) and in general by their limited saprophytic ability and their restricted capability to grow and to compete in soil with other microorganisms.

Once growing on selective agar, identification of the *Phytophthora* species can be achieved using traditional taxonomy of this genus which is mainly based on morphological structures (Waterhouse 1963; Stamps et al. 1990). These include characteristics of sporangia, which are either papillate (Waterhouse groups I and II), semi-papillate (Waterhouse groups III and IV) or non-papillate (Waterhouse groups V and VI). In addition, the form of the antheridium being amphigynous or paragynous is used for taxonomical purposes. Finally, it is distinguished whether the taxon is inbreeding (homothallic) or outbreeding (heterothallic). Due to a high level of variation of these characters within and between species, it is difficult to identify species and to determine their natural relationship. For this reason the species boundaries of, e.g., *P. cryptogea*, *P. drechsleri* and *P. megasperma* species complexes, were uncertain for a long time (Brasier 1991; Hansen 1991). In summary, these conventional isolation and identification methods are time-consuming and not suitable for screening large numbers of samples.

3.2 Identification and Detection of *Phytophthora* Species Using rDNA Gene Analyses

In order to improve the accuracy and efficiency of the identification of *Phytophthora* species and to obtain information on their evolutionary relationship, different molecular methods have been developed during the last 10 years. They are often based on the ribosomal DNA (rDNA) which encodes the different pieces of RNA that form the backbone of ribosomes (Fig. 3).

Each nucleus of eukaryotic organisms contains the ribosomal DNA genes as a multigene family in which the copies are arranged in tandem repeats. Each rDNA gene repeat consists of a single transcription unit which includes the 5.8S small subunit and the 18S and 28S large subunit genes, separated by two internal transcribed spacers ITS1 and ITS2. The rDNA genes, like the RNA pieces they code for, are highly conserved across different life forms. However, the ITS1 and ITS2 spacers that do not contribute to any part of the ribosomes have developed more rapidly, therefore being highly variable in their sequences. Differences within the ITS1 and

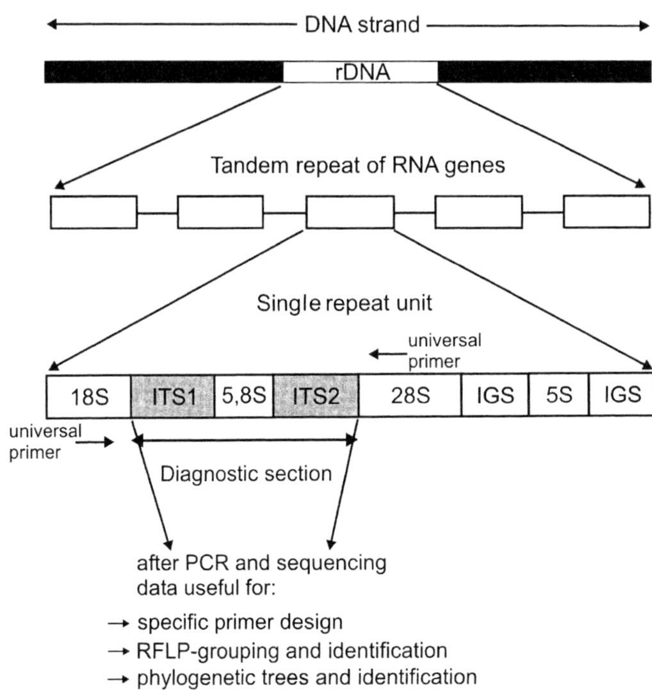

Fig. 3. The organization of the *Phytophthora* rDNA repeats

ITS2 sequences can be used to discriminate between species within the same genus and are very useful for primer design.

The evolutionary conservation of the 18S RNA, the 5.8S RNA and the 28S RNA allows the amplification of the ITS-regions of *Phytophthora* and most other eukaryotes using the PCR technique in combination with the universal primers of White et al. (1990).

As the ITS1 and ITS2 sequences differ from one *Phytophthora* species to another in most cases, primers can be generated that will detect a single species or small groups of species. This approach was used besides others by Schubert et al. (1999) to design an ITS primer for the specific PCR-based detection of *P. citricola*, which is associated with root-rot diseases in European deciduous forests. The diagnostic PCR system was able to detect this pathogen in beech seedlings, artificially infected under controlled conditions. Recently, Nechwatal et al. (2001) have proved that *P. citricola* and *P. quercina* could be detected in soil samples collected under declining oaks by using a combination of PCR and a baiting technique. Nested PCR allowed the detection of as few as 5 zoospores of *P. citricola* and 300 zoospores of *P. quercina*. Further, ITS primers were developed to identify and to detect *P. lateralis* (Winton and Hansen (2001), *P. fragariae* (Bonants et al. 1997), *P. nicotianae* (Grote et al. 2002) or *P. infestans* (Tooley et al. 1997). In order to quantify fungal infection, methods such as image analysis of the amplicon after separation on agarose gels or quantitative competitive PCR (Nicholson et al. 2002) have been developed which are all highly time-consuming and often make laborious post-PCR processing steps necessary.

These problems were solved by the invention of real-time quantitative PCR methods which allow the online monitoring of the amplification of diagnostic target sequences with a fluorescence detection system during the PCR steps (Orlando et al. 1998; Bustin 2000; Walker 2002; Bell and Ranford-Cartwright 2002; Fig. 4).

This technique uses an additional dual-labelled fluorogenic probe of about 25 bp that hybridizes to the target DNA within the region defined by the two specific PCR primers during the annealing phase (A). The specific probe carries two fluorescent dyes, a reporter (R) at the 5' end which can be excited with an argon-ion-laser at 488 nm and a quencher (Q) at the 3' end. The proximity of the reporter to the quencher dye results in suppression of the reporter fluorescence. During the extension phase of PCR, the 5' nuclease activity of the *Taq* polymerase cleaves the reporter dye from the fluorogenic probe (B), allowing the fluorophore to emit light after excitation. Further cleavage of the probe by the *Taq* polymerase from the target strand allows the primer extension to continue to the end of the template strand. Additional reporter dye molecules are cleaved from their corresponding probe with each cycle, causing an increase in fluorescence inten-

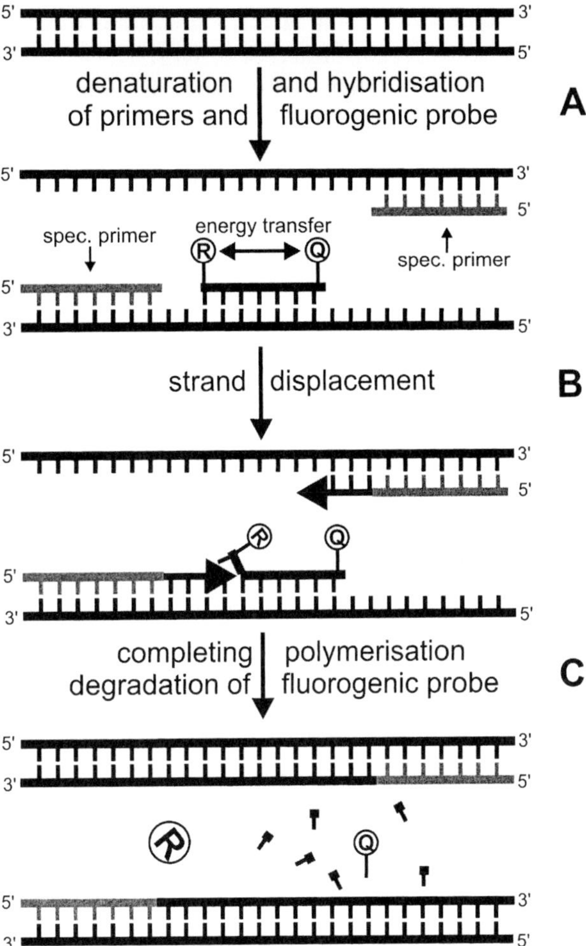

Fig. 4. The TaqMan assay

sity that is proportional to the amount of amplicon produced (C). For each sample a specific Ct value (threshold cycle) is calculated. The Ct value is defined as the cycle number at which a statistically significant increase in the reporter fluorescence is detected first. The Ct values are used to calculate the starting copy number of the target for each sample. The advantage of this technique is that besides precision a high reproducibility over a linear range of six orders of magnitude or more can be achieved.

The TaqMan assay in combination with dual-labelled fluorogenic probes has been recently used by Böhm et al. (1999) to quantify *P. infestans* and *P. citricola* in infected plant tissue. Winton et al. (2001) successfully used the TaqMan method to detect host and pathogen DNA simultaneously in a one-tube assay analyzing Douglas-fir needles infected with *Phaerocryp-*

topus gaeumannii using species-specific primers and TaqMan probes labelled with different fluorescent dyes. These results demonstrate that the novel, highly sensitive real-time PCR techniques are powerful tools in modern phytopathological research in order to detect and quantify invading microorganisms even in the absence of visible host plant symptoms.

After sequencing the PCR products and aligning the sequence data of different isolates (species), phylogenetic analyses can be performed in order to obtain detailed phylograms of genera and species. The most widely used software packages for this purpose are PHYLIP (Felsenstein 1993), PAUP* (Swofford 2002), or MEGA (Kumar et al. 2001). Lee and Taylor (1992) were the first to study phylogenetic relationships of different *Phytophthora* species comparing ITS1 and ITS2 sequences. Crawford et al. (1996) used the same approach and were successful in confirming the reclassification of *P. medicaginis*, *P. trifolii* and *P. sojae* as biological species which were formerly classified as *formae speciales* of *P. megasperma*. Recently, Förster et al. (2000) and Cooke et al. (2000) have investigated phylogenetic relationships of different *Phytophthora* species. The latter generated a phylogram of genera and species of the *Saprolegniales* and *Peronosporales* including Pythiaceae on the basis of the ITS regions of the rDNA. The vast majority of the 50 *Phytophthora* species investigated formed a closely related group and clustered with *Pythium*, *Peronospora* and *Halophytophthora*, but was distant from the genera of the *Saprolegniales*. One of the main conclusions of this comprehensive study on evolutionary relationships in *Phytophthora* is that the six taxonomic groups created by Waterhouse (1963), which are mainly based on presence and form of sporangial papilla, do not represent natural assemblages. Consequently, Förster et al. (2000) drew the conclusion that the characteristics used for taxonomic purposes in the genus *Phytophthora* are of limited value for deducing phylogenetic relationships. However, the Waterhouse groups will undoubtedly remain of value for the identification of *Phytophthora* species until molecular identification methods are more widely used. In this respect it should be mentioned that ITS sequences will not always distinguish between species. For example, *P. inflata* shares identical ITS sequences with *P. citricola*, and *P. melonis* with *P. sinensis*. In these cases, however, the findings are consistent with the proposed conspecificity of the taxa (Cooke et al. 2000).

Despite such minor limitations, the ITS sequences published in numerous studies and made available through the World Wide Web in large databases (GenBank, EMBL, DDBJ), will greatly facilitate and speed up the identification of *Phytophthora* species by providing the possibility to compare the ITS sequences of unknown species with those published in the databases. Thus, besides assessing phylogenetic relationships, sequence data can be used to confirm the specific status of undescribed species, and to support the erection of new taxa by molecular evidence, as it has been shown recently for *P. quercina* (Cooke et al. 1999; Jung et al. 1999a), *P. psychrophila, P. europaea, P. uliginosa* (Jung et al. 2002), *P. ramorum* (Werres et al. 2001), *P. pistacia* (Mirabolfathy et al. 2001) and *P. ipomoeae* (Flier et al. 2002). The recent transfer of *Pythium undulatum* to *Phytophthora undulata* (Dick 1990), however, was not supported by molecular data (Cooke et al. 2000). The ITS sequences have also been recently used to reveal phylogenetic relationships among other oomycete taxa, e.g. the Saprolegniaceae (Daugherty et al. 1998; Leclerc et al. 2000), *Pythium* (Matsumoto et al. 1999) or other fungal genera, such as *Trichoderma, Nephroma* and *Alternaria* species (Kullnig-Gradinger et al. 2002; Lohtander et al. 2002).

Alternatively, when sequencing tools are not available, quick and simple identification of unknown *Phytophthora* isolates can be obtained using PCR-RFLP techniques. After PCR amplification, the ITS spacer regions can be digested into fragments with different restriction enzymes, finally providing species-specific digest patterns on agarose gels (Cooke and Duncan 1997; Vettraino et al. 2001; Duncan and Cooke 2002). Details of this technique, as well as a summary of the banding patterns of a large set of *Phytophthora* species, are summarized on a website (http://www.cabi.org/-bioscience/phytophthora/background.htm), allowing scientists to identify unknown isolates even in only basically equipped laboratories. However, closely related species, such as *P. infestans, P. phaseoli*, and *P. mirabilis*, will not be distinguished by this technique (Tooley et al. 1996). In that case amplified fragment length polymorphism (AFLP) fingerprinting can be used to discriminate between species (Vos et al. 1995; Mirabolfathy et al. 2001) and even between individuals or strains of the same species (Majer et al. 1996; Duncan and Cooke 2002).

4 Impact of *Phytophthora* Root Pathogens on Physiological Parameters of Infected Trees

In order to understand how soil-borne *Phytophthora* pathogens interfere and finally kill their hosts in their ecosystems, one has to study the effect of those pathogens on the physiology of infected host plants. Unfortunately, fewer results on physiological parameters are available for woody compared to herbaceous plants, because it is much more difficult and time-consuming working with woody plants in the field or even under controlled greenhouse conditions. However, some important results will be discussed here.

Most plants infected with different *Phytophthora* pathogens show typical wilt symptoms on aboveground organs before they die. Sterne et al. (1978) investigated the influence of *P. cinnamomi* on 8- to 10-year-old avocado trees under field conditions. They showed that transpirational flux density in healthy trees reached a maximum between noon and early afternoon with a mean value of 1.78 μg cm^{-2} s^{-1} for leaves exposed to full sunlight. However, diseased trees exhibited much lower values with means between 0.2–0.5 μg cm^{-2} s^{-1} and, although they stayed fairly constant through the whole day, they peaked about 2 h earlier compared to uninfected avocado trees. In the same way as transpiration was impaired, also conductance of water vapour of *P. cinnamomi* infected avocado was significantly reduced compared to control trees. Furthermore, Sterne et al. (1978) reported that the leaf xylem water potential (Ψ_{xylem}) was greater in healthy avocado trees than in trees infected with *P. cinnamomi*. In addition, diseased trees showed values for leaf water potential of –8 bar at night and in the morning, whereas those of control trees reached values of –1.5 bar for Ψ_{xylem} measured at the same time. The main result of the studies of Sterne et al. (1978) was that the xylem pressure potential of infected plants was not coupled with transpirational flux over the day, because in diseased trees transpiration and xylem pressure decreased in parallel after 10:00 a.m. until late afternoon. The same behaviour was found for water-stressed avocado trees. The authors concluded that infection with *P. cinnamomi* caused such major changes in the water transport system between roots and leaves that typical control mechanisms (e.g. stomatal closure) could not correct the water deficits.

Similar experiments were done by Dawson and Weste (1984). They studied water relation changes of *Eucalyptus sieberi* (susceptible) and *E. maculata* (field-resistant) infected with the same pathogen under controlled conditions. A major reduction in the net hydraulic conductivity of the root system was found along with lesion extension in roots. This reduction preceded all the effects on leaves, e.g., the reduction of leaf conductance to water vapour loss, of leaf water potential or the reduction of leaf relative water content by at least 24 h or more. These changes in water relations were only present in *E. sieberi*, susceptible to *P. cinnamomi*. Recently, Robin et al. (2001) have shown that the most noticeable effect of the root infection by *P. cinnamomi* on *Quercus ilex* (holm oak) and *Q. suber* (cork oak) was the reduction in stomatal conductance and in leaf water potential. In addition to changes in water relations, Ploetz and Schaffer (1989) investigated the impact of *P. cinnamomi* on net CO_2 assimilation of avocado plants. A significant reduction in assimilation, transpiration and stomatal conductance was measured as early as 3 days after infection and these parameters declined to non-detectable levels within 1 week. Recently, Fleischmann et al. (2002) have compared the influence of different *Phyto-*

phthora species on root development, net CO_2 assimilation and transpiration of beech (*Fagus sylvatica*) saplings. A significant reduction in net CO_2 assimilation and in transpiration of *P. citricola* and *P. cambivora* infected plants was visible only after bud break in the second year of the experiments, possibly indicating that newly developed fine roots are the primary target for these root pathogens. This was proved, when the root systems were analyzed at the end of the experiment. Interestingly, photosynthesis and transpiration were reduced to almost zero some days before plants started to wilt. Water use efficiency data clearly indicated that infected plants suffered from severe drought. Similar investigations were done by Oßwald et al. (1999) working with *Quercus robur* seedlings infected with the oak pathogen *P. quercina*. In parallel with the loss of fine roots and root tips, again CO_2 assimilation, transpiration and stomatal conductance were decreased some weeks after infection started. During the last 2 months of the experiment, the pathogen was no longer detectable in roots. At that time, strong regeneration of the root system was visible, and photosynthesis reached levels comparable with those of control plants. Cahill et al. (1986) reported on changes of cytokinin concentrations in xylem exudates of *Eucalyptus* seedlings following infection with *P. cinnamomi*. They found a significant reduction of the zeatin-type and isopentyladenine-type cytokinin in the xylem exudates of susceptible *E. marginata*, whereas no reduction was measured in the exudates of resistant *E. calophylla*. The authors suggest that failure of cytokinin transport from the root system to leaves may be responsible for the failure in water transport and for the symptoms of *P. cinnamomi* infection observed in infected susceptible *Eucalyptus* plants.

Only a few reports have been published within the last 20 years dealing with biochemical defense reactions of woody plants infected with different *Phytophthora* root rot pathogens. Okey et al. (1997) showed that *P. palmivora*, the causal agent of *Phytophthora* canker on cocoa, caused a strong increase in phenylalanine ammonium lyase (PAL) and peroxidase activity only in the roots of the resistant clone. No change in the activity of either enzymes was measured in roots of the susceptible cultivar. The most comprehensive study in terms of lignin induction and resistance of woody plants to *Phytophthora* root rot pathogens was done by Cahill and McComb (1992). They compared defense reactions of *Eucalyptus marginata* (susceptible) and *E. calophylla* (resistant) infected with *P. cinnamomi*. The authors proved that restriction of the pathogen growth in *E. calophylla* was due to a strong induction of PAL, followed by lignin synthesis and the accumulation of phenolics in infected tissue. Important biochemical defense reactions of woody plants infected with different *Phytophthora* pathogens have been recently summarized by Oßwald (2002). All these results indicate that cell wall reinforcement by lignin and the accumulation of phenolic compounds are involved in resistance of woody plants to *Phytophthora* pathogens.

Such defense responses are often triggered in resistant host tissue by elicitors released by the invading pathogen. A possible candidate for such elicitor structures of *Phytophthora* pathogens might be the elicitins (Tyler 2002). Elicitin elicitors are small proteins consisting of 98 amino acids, abundantly released into liquid culture medium by all *Phytophthora* species examined so far and some *Pythium* species. As most virulent tobacco isolates of *P. parasitica,* the causal agent of tobacco black shank, do not produce elicitins (Bonnet et al. 1994; Kamoun et al. 1994), these proteins have been proposed to mediate the action of avirulence genes in the resistant *P. parasitica*–tobacco interaction. Members of this new protein family trigger a defense response in tobacco and their role as elicitors has been well investigated in tobacco plants and cell suspension cultures by many groups (for review see Tyler 2002).

In contrast to the resistant *Phytophthora*–tobacco interaction, high amounts of the elicitin quercinin were found in roots of *Quercus robur* being susceptible towards *P. quercina* (Brummer et al. 2002). They localized quercinin, the elicitin of *P. quercina,* in infected oak roots. Immunogold- and immunofluorescence-labelled antibodies raised against cryptogein proved the existence of quercinin around the hyphae and within the apoplastic space. Moreover, immunofluorescence staining of quercinin supported the theory that elicitins are also released into the invaded host cell (Brummer et al. 2002; Fig. 5)

These results indicate that elicitins might also act as toxins or as pathogenicity factors in order to weaken the host tissue.

Destruction of chloroplasts and especially of chloroplast membranes by elicitins was shown earlier by Milat et al. (1991) and Heiser et al. (1999). Treatment with culture filtrate from *P. quercina* or cryptogein, an elicitin from *P. cryptogea,* led to severe structural changes in tobacco cells. Loss of turgidity, disorganization of parenchyma cells and cell wall alterations were observed. Elicitins also triggered a massive membrane degradation on thylakoids and the envelope membrane of the chloroplasts (Fig. 6).

Destruction of chloroplasts was also detected by measuring chlorophyll fluorescence of cryptogein-treated tobacco leaves, an elicitin from *P. cryptogea* (Milat et al. 1991). Those leaves revealed a strong decrease in maximum fluorescence, indicating a reduced efficiency of photosystem II that was correlated with the unstacking of grana, observed with electron microscopy (Milat et al. 1991).

In summary, it turns out that *Phytophthora* elicitins act as elicitors inducing defense responses in resistant host tissue or as toxins weakening

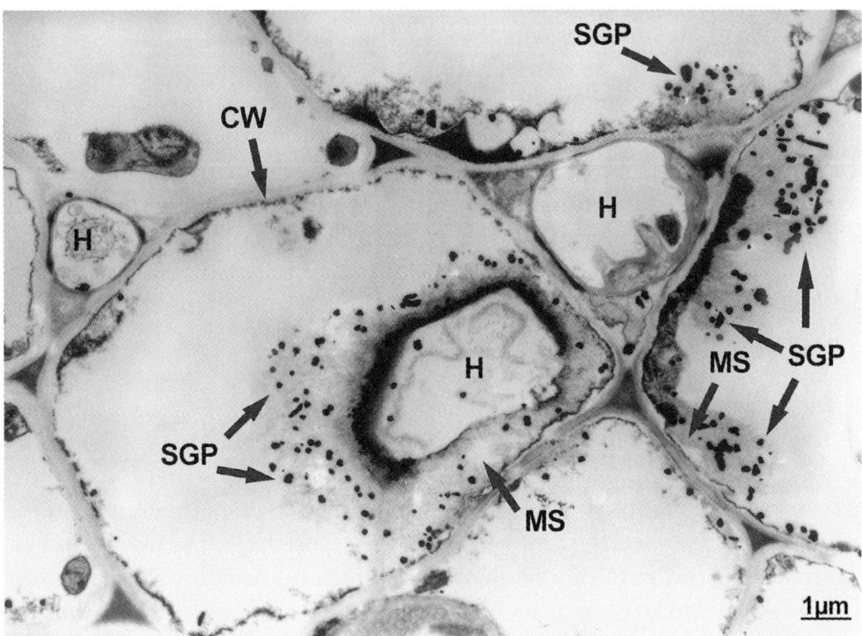

Fig. 5. Transmission electron microscopy in combination with silver-enhanced immunogold labelling to localize the *P. quercina* peptide quercinin in infected oak roots. Accumulation of silver-enhanced gold particles (*SGP*) is found around hyphae (*H*) within invaded cells, embedded in a matrix-like structure (*MS*) (Brummer et al. 2002)

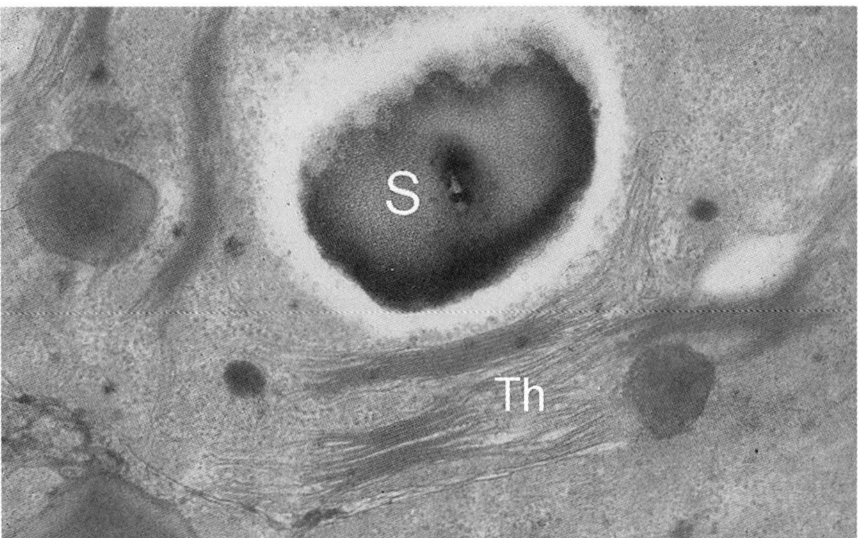

Fig. 6. Separation of chloroplast grana thylakoid membranes after treatment of tobacco leaves with the elicitin quercinin from *P. quercina* (Oßwald et al. 1999)

susceptible tissue which is invaded by the pathogen afterwards. Elicitin peptides can also be transported from infected root systems into leaves via sap flow, where they might interact with chloroplasts impairing photosynthesis or with membranes finally causing wilt symptoms. Due to the finding that net hydraulic conductivity of *Eucalyptus* root systems infected with *P. cinnamomi* was blocked almost completely, although the pathogen had infected only 8–15% of the total root system, Dawson and Weste (1984) also concluded that *Phytophthora* enzymes or toxins released by the pathogen have to be involved in symptom development.

5 Summary and Outlook

The genus *Phytophthora* comprises more than 60 species that are responsible for major plant diseases all over the world. These pathogens produce mobile spores (zoospores) in the presence of free water that mainly attack the fine roots or the collar region of trees. From there, many *Phytophthora* species grow into larger roots or up into the trunk and destroy the cambium and phloem tissue. Finally, infected trees suffer from water stress due to root rot or stem girdling.

Between 1900 and 1950, *Phytophthora cinnamomi* and *P. cambivora* were the primary cause for the high mortality of native chestnut (*Castanea sativa*) in southern Europe (Peace 1962). A similar epidemic on *Castanea sativa* was also reported in the southeastern USA, again caused by *P. cinnamomi* (Crandall et al. 1945). The same pathogen was also responsible for the death of jarrah (*Eucalyptus marginata*) forest and the corresponding understorey in Western Australia (Shearer and Tippett 1989) as well as for severe stem canker of American red oak (*Quercus rubra*) in southwestern France (Moreau and Moreau 1952). Both *P. cinnamomi* and *P. cambivora* are believed to have been introduced to Europe. *P. cinnamomi* is probably native to Papua New Guinea–Celebes area of the southwest Pacific. From there, it was spread to many parts of the world by man.

Since 1990, several new diseases have been reported on woody plants associated with well-known *Phytophthora* pathogens or with newly described ones. In 1993, Brasier et al. showed the involvement of *P. cinnamomi* in the decline and mortality of *Quercus ilex* and *Q. suber* in southwestern Iberia. In the same year, a new disease on *Alnus glutinosa* was diagnosed first in Britain. Brasier et al. (1999) proved that the disease is caused by a new *Phytophthora* species hybrid, involving *P. cambivora* as one of the parents, which induces typical collar rot symptoms and bark lesions. Current results of Jung et al. (2000a) suggest that decline of deciduous oaks, *Q. robur* and *Q. petraea*, across central Europe on non-acidic soils is associated with the recently described, highly virulent *P. quercina* in combination with other *Phytophthora* species such as *P. cambivora* and *P. citricola* causing severe loss of feeder roots. Most recently, Rizzo et al. (2002) have reported that the newly described *P. ramorum* (Werres et al. 2001) causes extensive mortality of several *Quercus species* and *Lithocarpus densiflorus* in California woodlands.

The invention of molecular methods, such as PCR in combination with sequencing of amplified spacer rDNA and RFLP as well as AFLP, has revolutionized mycology and plant pathology. These techniques allow scientists working with the genus *Phytophthora* to detect these pathogens highly sensitively, to distinguish between closely related species and to determine phylogenetic relationships among species. It is now possible to obtain the specific ITS sequence in a few days and to compare it with the database of all described species, thus greatly speeding up identification of isolated species. Due to the high sensitivity of PCR, one can detect *Phytophthora* species in plants that are only weakly infected even if the plants are not showing symptoms. The great advantage of the PCR techniques compared to baiting is that PCR allows the detection and identification of the pathogens, even if they are no longer alive in infected tissue. Under these conditions baiting would give negative and PCR positive results. A big step forward was the invention of quantitative real-time PCR methods which allow the online monitoring of the amplification of diagnostic target sequences.

There are only few results in the literature on biochemical and physiological investigations of woody plants infested with root rot pathogens of the genus *Phytophthora*, compared to what is known for herbaceous plants (Cahill and McComb 1992; Oßwald et al. 1999; Robin et al. 2001; Fleischmann et al. 2002). It is not fully understood which mechanisms impair and finally block water transport in trees infested with *Phytophthora* species. In this field much more work has to be done, in particular to understand the role of *Phytophthora* elicitins as pathogenicity factors or toxins in compatible interactions, or as elicitors for the induction of defense reactions in resistant woody plants.

References

Anselmi N, Giordano E, Vannini A, Troiani I, Napoli G, Crivelli L (1996) Il mal dell'inchiostro del castano in Italia:una vecchia malattia ritorna attuale. Linea Ecol 28:39–44

Barr DJS (1992) Evolution and kingdom of organisms from the perspective of a mycologist. Mycologia 84:1–11

Bell AS, Ranford-Cartwright LC (2002) Real-time quantitative PCR in parasitology. Trends Parasitol 18 (8):337–342

Blaschke H (1994) Decline symptoms on roots of *Quercus robur*. Eur J For Pathol 24:386–398
Böhm J, Hahn A, Schubert R, Bahnweg G, Adler N, Nechwatal J, Oehlmann R, Oßwald W (1999) Real-time quantitative PCR:DNA determination in isolated spores of the mycorrhizal fungus *Glomus mosseae* and monitoring of *Phytophthora infestans* and *Phytophthora citricola* in their respective host plants. J Phytopathol 147:409–416
Bonants P, Hagenaar-de Weerdt M, Van Gent-Pelzer M, Lacourt I, Cooke D, Duncan J (1997) Detection and identification of *Phytophthora fragariae* Hickmann by the polymerase chain reaction. Eur J For Pathol 103:345–355
Bonants P, de Weerdt M, Baayen R, de Gruyter H, Veld WM, Kroon L (2002) Molecular identification and detection of *Phytophthora* species and populations of *P. ramorum*. Sudden Oak Death, the state of our knowledge. Science Symposium, 17 – 18 Dec 2002, Monterey, CA, p 16
Bonnet P, Lacourt I, Venard P, Ricci P (1994) Diversity in pathogenicity to tobacco and in elicitin production among isolates of *Phytophthora parasitica*. J Phytopathol 141:25–37
Brasier CM (1991) Current questions in *Phytophthora* systematics: the role of the population approach. In:Lucas JA, Shattock RC, Shaw DS, Cooke IR (eds) Phytophthora. Cambridge University Press, Cambridge, pp 104–128
Brasier CM (1992) Oak tree mortality in Iberia. Nature 360:539
Brasier CM (1996) *Phytophthora cinnamomi* and oak decline in southern Europe. Environmental constraints including climate change. Ann Sci For 53:347–358
Brasier CM, Kirk SA (2001) Comparative aggressiveness of standard and variant hybrid alder phytophthoras, *Phytophthora cambivora* and other *Phytophthora* species on bark of *Alnus*, *Quercus* and other woody hosts. Plant Pathol 50:218–229
Brasier CM, Scott J (1994) European oak declines and global warming: a theoretical assessment with special reference to the activity of *Phytophthora cinnamomi*. EPPO Bull 24:221–232
Brasier CM, Robredo F, Ferraz JFP (1993) Evidence for *Phytophthora cinnamomi* involvement in Iberian oak decline. Plant Pathol 42:140–145
Brasier CM, Rose J, Gibbs JN (1995) An unusual *Phytophthora* associated with widespread alder mortality in Britain. Plant Pathol 44:999–1007
Brasier CM, Cooke DEL, Duncan JM (1999) Origin of a new *Phytophthora* pathogen trough interspecific hybridization. Proc Natl Acad Sci USA 96:5878–5883
Brasier CM, Rose J, Kirk SA, Webber JF (2002) Pathogenicity of *Phytophthora ramorum* isolates from North America and Europe to bark of European Fagaceae, American *Quercus rubra*, and other forest trees. Sudden Oak Death, the state of our knowledge. Science Symposium, 17 – 18 Dec 2002, Monterey, CA, p 30
Brummer M, Arend M, Fromm J, Schlenzig A, Oßwald W (2002) Ultrastructural changes and immunocytochemical localisation of the elicitin quercinin in *Quercus robur* L. roots infected with *Phytophthora quercina*. Physiol Mol Plant Pathol 61:109–120
Bustin SA (2000) Absolute quantification of mRNA using real-time reverse transcription polymerase chain reaction assays. J Mol Endocrinol 25:169–193
Byrt PN, Irving HR, Grant BR (1982) The effect of cations on zoospores of the fungus *Phytophthora cinnamomi*. J Gen Microbiol 128:1189–1198
Cavalier-Smith T (1989) The kingdom Chromista. In: Green JC, Leadbeater BSC, Diver WL (eds) The chromophyte algae: problems and perspectives. Syst Assoc Spec, vol 38. Clarendon Press, Oxford, pp 381–407
Cech TL (1997) *Phytophthora* – Krankheit der Erle in Österreich. Forstschutz Aktuell 19/20:14–16
Cahill DM, McComb JA (1992) A comparison of changes in phenylalanine ammoniumlyase activity, lignin and phenolic synthesis in the roots of *Eucalyptus calophylla* (field resistant) and *E. marginata* (susceptible) when infected with *Phytophthora cinnamomi*. Physiol Mol Plant Pathol 40:315–332

Cahill DM, Weste GM, Grant BR (1986) Changes in cytokinin concentrations in xylem extrudate following infection of *Euclyptus marginata* Donn ex Sm with *Phytophthora cinnamomi* Rands. Plant Physiol 81:1103–1109

Cooke DEL, Duncan JM (1997) Phylogenetic analysis of *Phytophthora* species based on ITS1 and ITS2 sequences of the ribosomal RNA gene repeat. Mycol Res 101 (6):667–677

Cooke DEL, Jung T, Williams NA, Schubert R, Bahnweg G, Oßwald W, Duncan JM (1999) Molecular evidence supports *Phytophthora quercina* as a distinct species. Mycol Res 103 (7):799–804

Cooke DEL, Drenth A, Duncan JM, Wagels G, Brasier CM (2000) A molecular phylogeny of *Phytophthora* and related Oomycetes. Fungal Genet Biol 30:17–32

Crandall BS, Gravatt GF, Ryan M (1945) Root disease of *Castanea* species and some coniferous and broadleaf nursery stocks, caused by *Phytophthora cinnamomi*. Phytopathology 3:162–180

Crawford AR, Bassam BJ, Drenth A, MacLean DJ, Irwin JAG (1996) Evolutionary relationships among *Phytophthora* species deduced from rDNA sequence analysis. Mycol Res 100 (4):437–443

Daugherty J, Evans TM, Skillom T, Watson LE, Money NP (1998) Evolution of spore release mechanisms in the Saprolegniaceae (Oomycetes): evidence from a phylogenetic analysis of internal transcribed spacer sequences. Fungal Genet Biol 24:354–363

Davidson JM, Garbelotto M, Koike ST, Rizzo DM (2002a) First report of *Phytophthora ramorum* on Douglas-Fir in California. Plant Dis 86:1274

Davidson JM, Maloney PE, Wickland AC, Morse AC, Jensen CE, Rizzo DM (2002b) Transmission of *Phytophthora ramorum* via *Umbellularia californica* (California bay) leaves in California oak woodlands. Sudden Oak Death, the state of our knowledge. Science Symposium, 17 – 18 Dec 2002, Monterey, CA, p 9

Dawson P, Weste G (1984) Impact of root infection by *Phytophthora cinnamomi* on the water relations of two *Eucalyptus* species that differ in susceptibility. Phytopathology 74:486–490

Day WR (1932) The ink disease in England. Forestry 6:182

Delatour C (1983) Les deperissements de chenes en Europe. Rev For Fr 15:265–282

Delcan J, Brasier CM (2001) Oospore viability and variation in zoospore and hyphal tip derivatives of the hybrid alder Phytophthoras. For Pathol 31:65–83

Dick WM (1990) *Phytophthora undulata* comb. nov. Mycotaxon 35:449–453

Duncan J, Cooke D (2002) Identifying, diagnosing and detecting *Phytophthora* by molecular methods. Mycologist 16(2):59–66

Eichhorn (1992) Wurzeluntersuchungen an sturmgeworfenen Bäumen in Hessen. Forst Holz 47:555–559

Erwin DC, Ribeiro OK (1996) *Phytophthora* diseases worldwide. APS Press, St Paul, MN

Felsenstein J (1993) PHYLIP: phylogeny inference package (version 3.5c). Distributed by the author, Department of Genetics, Univ Washington, Seattle

Fernandez-Escobar R, Gallego FJ, Benlloch M, Membrillo J, Infante J, Perez de Algaba A (1999) Treatment of oak decline using pressurized injection capsules of antifungal materials. Eur J For Pathol 29:29–38

Fleischmann F, Schneider D, Matyssek R, Oßwald W (2002) Investigations on net CO_2 assimilation transpiration and root growth of *Fagus sylvatica* infested with four different *Phytophthora* species. Plant Biol 4:144–152

Flier WG, Grünwald NJ, Kroon LPNM, van den Bosch TBM, Garay-Serrano E, Lozoya-Saldana H, Bonants PM, Turkensteen LJ (2002) *Phytophthora ipomoeae* sp. nov. a new homothallic species causing leaf blight on *Ipomoea longipedunculata* in the Toluca Valley of central Mexico. Mycol Res 106(7):848–856

Förster H, Cummings MP, Coffey MD (2000) Phylogenetic relationships of *Phytophthora* species based on ribosomal ITS I DNA sequence analysis with emphasis on Waterhouse groups V and VI. Mycol Res 104(9):1055–1061

Garbelotto M, Svihra P, Rizzo DM (2001) Sudden oak death syndrome fells three oak species. Calif Agric 55 (1):9–19

Garbelotto M, Rizzo DM, Davidson JM, Frankel SJ (2002) How to recognize symptoms of diseases caused by *Phytophthora ramorum* causal agent of Sudden Oak Death. http://nature.Berkeley.EDU/comtf/pdf/EducationalMaterial/SODDiagnosis2002.pdf

Gibbs JN (1995) *Phytophthora* root rot disease in Britain. EPPO Bull 25:661–664

Gibbs JN, Lipscombe MA, Peace AJ (1999) The impact of *Phytophthora* disease on riparian populations of common alder (*Alnus glutinosa*) in southern Britain. Eur J For Pathol 29:39–50

Goheen EM, Hansen EM, Kanaskie A, McWilliams MG, Osterbauer N, Sutton W (2002) Sudden oak death caused by *Phytophthora ramorum* in Oregon. Plant Dis 86:441

Grant BR, Griffith JM, Irving HR (1986) A model to explain ion-induced differentiation in zoospores of *Phytophthora palmivora*. Exp Mycol 10:89–98

Grote D, Olmos A, Kofoet A, Tuset JJ, Bertolini E, Cambra M (2002) Specific and sensitive detection of *Phytophthora nicotianae* by simple and nested-PCR. Eur J Plant Pathol 108:197–207

Guest DJ, Grant BR (1991) The complex action of phosphonates as antifungal agents. Biol Rev 66:159–187

Hansen EM (1991) Variation in the species of the *Phytophthora megasperma* complex. In: Lucas JA, Shattock RC, Shaw DS, Cooke IR (eds) Phytophthora. Cambridge University Press, Cambridge, pp 148–163

Hansen EM, Streito JC, Delatour C (1999) First confirmation of *Phytophthora lateralis* in Europe. Plant Dis 83:587

Hansen EM, Goheen DJ, Jules ES, Ullian B (2000) Managing Port-Orford cedar and the introduced pathogen *Phytophthora lateralis*. Plant Dis 84 (1):4–14

Hartmann G (1995) Wurzelfäule der Schwarzerle (*Alnus glutinosa*) – eine bisher unbekannte Pilzkrankheit durch *Phytophthora cambivora*. Forst Holz 50:555–557

Hartmann G (1998) Aktuelles Eichensterben in Niedersachsen – Ursachen und Gegenmaßnahmen. Forst Holz 53:733–735

Hartmann G, Blank R (1992) Winterfrost, Kahlfraß und Prachtkäferbefall als Faktoren im Ursachenkomplex des Eichensterbens in Norddeutschland. Forst Holz 47:443–452

Hartmann G, Blank R (2002) Vorkommen und Standortbezüge von *Phytophthora*-Arten in geschädigten Eichenbeständen in Nordwestdeutschland (Niedersachsen, Nordrhein-Westfalen und Schleswig-Holstein) (occurrence and site relations of *Phytophthora* spp. in declining oak stands in northwestern Germany). Forst Holz 57 (18):539–545

Hartmann G, Blank R, Lewark S (1989) Eichensterben in Norddeutschland:Verbreitung, Schadbilder, mögliche Ursachen. Forst Holz 44:475–487

Hawksworth DL, Kirk PM, Sutton BC, Pegler DN (1995) Ainsworth and Bisby's dictionary of the fungi, 8th edn. CAB International, Wallingford,

Heiser I, Fromm J, Giefing M, Koehl J, Oßwald W (1999) Investigations on the action of *Phytophthora quercina*, *P. citricola* and *P. gonapodyides* toxins on tobacco plants. Plant Physiol Biochem 37:73–81

Ivors K, Hayden K, Garbelotto M, Rizzo D (2002) Molecular population analyses of *Phytophthora ramorum*. Sudden oak death, the state of our knowledge. Science Symposium, 17 - 18 Dec 2002, Monterey, CA, p 17

Jackson TJ, Burgess T, Colquhoun I, Hardy GE (2000) Action of the fungicide phosphite on *Eucalyptus marginata* inoculated with *Phytophthora cinnamomi*. Plant Pathol 49:147–154

Jung T (1998) Die *Phytophthora*-Erkrankung der Europäischen Eichenarten – Wurzel zerstörende Pilze als Ursachen des Eichensterbens. Lincom Europe, Munich, Germany

Jung T, Blaschke M (2001) *Phytophthora* – Wurzelfäule der Erlen. LWF Merkblatt 6:1–3

Jung T, Blaschke H, Neumann P (1996) Isolation, identification and pathogenicity of *Phytophthora* species from declining oak stands. Eur J For Pathol 26:253.272

Jung T, Cooke DEL, Blaschke H, Duncan JM, Oßwald W (1999a) *Phytophthora quercina* sp. nov., causing root rot of European oaks. Mycol Res 103(7):785–798

Jung T, Schlenzig A, Blaschke H, Oßwald W (1999b) *Phytophthora*-Kartierung für Stiel- und Traubeneiche in Bayern auf der Grundlage der Waldzustandserfassung 1994: Versuch einer Korrelation mit Kronentransparenz, Wurzelschäden und Standortfaktoren. Final report of the project F44, Technical University Munich, Germany

Jung T, Blaschke H, Oßwald W (2000a) Involvement of soilborne *Phytophthora* species in Central European oak decline and the effect of site factors on the disease. Plant Pathol 49:706–718

Jung T, Blaschke H, Oßwald W (2000b) Involvement of *Phytophthora* species in central and western European oak decline and the influence of site factors and nitrogen input on the disease. In: Hansen EM, Sutton W (eds) *Phytophthora* diseases of forest trees. Proceedings of the 1st International Meeting of Phytophthoras in Forest and Wildland Ecosystems, 1999, Grant Pass, Oregon, USA, pp 28–33

Jung T, Schlenzig A, Blaschke M, Adolf B, Oßwald W (2000c) Erlensterben durch *Phytophthora* – droht Bayerns Erlen eine Epidemie? LWF Aktuell 24:22–25

Jung T, Hansen EM, Winton L, Oßwald W, Delatour C (2002) Three new species of *Phytophthora* from European oak forests. Mycol Res 106 (4):397–411

Kamoun S, Young M, Förster H, Coffey MD (1994) Potential role of elicitins in the interaction between *Phytophthora* species and tobacco. Appl Environ Microbiol 60 (5):1593–1598

Komorek BM, Shearer BL (1997) Application techniques and phosphonate movement in the host. In: Murray D (ed) Control of *Phytophthora* and *Diplodina* canker in WA. Western Australia Department of Conservation and Land Management, Perth, pp 1–33

Kullnig-Gradinger CM, Szakacs G, Kubicek CP (2002) Phylogeny and evolution of the genus *Trichoderma*:a multigene approach. Mycol Res 106 (7):757–767

Kumar S, Tamura K, Jakobsen IB, Nei M (2001) MEGA2: molecular evolutionary genetics analysis software. Bioinformatics 17:1244–1245

Leclerc MC, Guillot J, Deville M (2000) Taxonomic and phylogenetic analysis of *Saprolegniaceae* (Oomycetes) inferred from LSU rDNA and ITS sequence comparison. Antonie van Leeuwenhoek 77:369–377

Lee SB, Taylor JW (1992) Phylogeny of five fungus-like protoctistan *Phytophthora* species, inferred from the internal transcribed spacers of ribosomal DNA. Mol Biol Evol 9 (4):636–653

Lobinger G (1999) Zusammenhänge zwischen Insektenfraß, Witterungsfaktoren und Eichenschäden. Berichte aus der Bayerischen Landesanstalt für Wald- und Forstwirtschaft No 19. Bayerische Landesanstalt für Wald- und Forstwirtschaft, Freising, Germany

Lohtander K, Oksanen I, Rikkinen J (2002) A phylogenetic study of *Nephroma* (lichen-forming Ascomycota). Mycol Res 106 (7):777–787

Majer D, Mithen R, Lewis BG, Vos P, Oliver RP (1996) The use of AFLP fingerprinting for the detection of genetic variation in fungi. Res 100 (9):1107–1111

Maloney PE, Rizzo DM, Koike ST, Harnik TY, Garbelotto M (2002) First report of *Phytophthora ramorum* on Coast Redwood in California. Plant Dis 86:1274

Marcais B, Martin F, Delatour C (1998) Structure of *Collybia fusipes* populations in two infected oak stands. Mycol Res 102:361–367

Matsumoto C, Kageyama K, Haruhisa S, Hyakumachi M (1999) Phylogenetic relationships of *Pythium* species based on ITS and 5.8S sequences of the ribosomal DNA. Mycoscience 40:321–331

Mayer FJ (1999) Beziehungen zwischen der Belaubungsdichte der Waldbäume und Standortfaktoren – Auswertung der bayerischen Waldzustandsinventuren. Forstliche Forschungsberichte München 177, pp 1–195

Milat M, Ducruet J, Ricci P, Marty F, Blein J (1991) Physiological and structural changes in tobacco leaves treated with cryptogein, a proteinaceous elicitor from *Phytophthora cryptogea*. Phytopathology 81(11):1364–1368

Milburn M, Gravatt GF (1932) Preliminary note on a *Phytophthora* root disease of chestnut. Phytopathology 22:977–978

Mirabolfathy M, Cooke DEL, Duncan JM, Williams NA, Ershad D, Alizadeh A (2001) *Phytophthora pistaciae* sp. nov. and *P. melonis*:the principal causes of pistachio gummosis in Iran. Mycol Res 105(10):1166–1175

Moralejo E, Hernandez L (2002) Inoculation trials of *Phytophthora ramorum* on detached Mediterranean sclerophyll leaves. Sudden oak death, the state of our knowledge. Science Symposium, 17 - 18 Dec 2002, Monterey, CA, p 32

Moreau C, Moreau M (1952) Etude mycologique de la maladie de l'encre du chene. Rev Pathol Veg Entomol Agric Fr 31:201–231

Nechwatal J, Schlenzig A, Jung T, Cooke DEL, Duncan JM, Oßwald W (2001) A combination of baiting and PCR techniques for the detection of *Phytophthora quercina* and *P. citricola* in soil samples from oak stands. For Pathol 31:85–97

Nicholson P, Turner AS, Edwards SG, Bateman GL, Morgan LW, Parry DW, Marshall J, Nuttall M (2002) Development of stem-base pathogens on different cultivars of winter wheat determined by quantitative PCR. Eur J Plant Pathol 108:163–177

Okey EN, Duncan EJ, Sirju-Charran G, Sreenivasan TN (1997) *Phytophthora* canker resistance in cacao: role of peroxidase and phenylalanine ammonium lyase. J Phytopathol 145:195–299

Old KM (ed) (1979) *Phytophthora* and forest management in Australia. CASIRO, Melbourne, Australia, 114 pp

Orlando C, Pinzani P, Pazzagli M (1998) Development in quantitative PCR. Clinical chemistry and laboratory. Medicine 36:255–269

Oßwald W (2002) Biochemical defence reactions of woody plants infected with *Phytophthora* or *Pythium* fungi. Proceedings of the Cost Action G4, Thessaloniki, Greece, pp 1–15

Oßwald W, Brummer M, Fromm J, Schlenzig A, Koehl J, Jung T, Heiser I, Matyssek R (1999) Investigations on photosynthesis of oak seedlings infected with *Phytophthora quercina* and characterization of the *P. quercina* toxin quercinin. In: Hansen EM, Sutton W (eds) *Phytophthora* diseases of forest trees. Proceedings from the first international meeting on Phytophthoras in forest and wildland ecosystems, 1999, Grants Pass, Oregon, pp 67–70

Peace TR (1962) Pathology of trees and shrubs. Oxford University Press, Oxford

Ploetz RC, Schaffer B (1989) Effects of flooding and *Phytophthora* root rot on net gas exchange and growth of avocado. Phytopathology 79:204–208

Ponchet M, Panabieres F, Milat ML, Mikes V, Montillet JL (1999) Are elicitins crytograms in plant-oomycete communications? Cell Mol Life Sci 56:1020–1047

Rizzo DM, Garbelotto M, Davidson JM, Slaughter GW, Koike ST (2002) *Phytophthora ramorum* as the cause of extensive mortality of *Quercus* spp. and *Lithocarpus densiflorus* in California. Plant Dis 86:205–214

Robin C, Desprez-Loustau ML, Delatour C (1992) Factors influencing the enlargement of trunk cankers of *Phytophthora cinnamomi* in red oak. Can J For Res 22:367–374

Robin C, Capron G, Desprez-Loustau L (2001) Root infection by *Phytophthora cinnamomi* in seedlings of three oak species. Plant Pathol 50 (6):708–716

Sanchez ME, Caetano P, Ferraz J, Trapero A (2002) *Phytophthora* disease of *Quercus ilex* in south-western Spain. For Pathol 32:5–18

Schönwiese CD, Rapp J, Fuchs T, Denhard M (1994) Observed climate trends in Europe 1891–1990. Meteorol Z 3:22–28

Schubert R, Bahnweg G, Nechwatal J, Jung T, Cooke DEL, Duncan JM, Müller-Starck G, Langebartels C, Sandermann H Jr, Oßwald W (1999) Detection and quantification of *Phytophthora* species which are associated with root-rot diseases in European deciduous forests by species-specific polymerase chain reaction. Eur J For Pathol 27:1–19

Shea SR, Shearer BL, Tippett JT, Deegan PM (1984) A new perspective on Jarrah dieback. For Focus 31:3–11

Shearer BL, Tippett JT (1989) Jarrah Dieback: the dynamics and management of *Phytophthora cinnamomi* in the jarrah (*Eucalyptus marginata*) forest of South-Western Australia. Department of Conservation and Land Management. Res Bull No 3:1–76

Siwecki R, Liese W (eds) (1991) Oak decline in Europe. Proceedings of an International Symposium, 1990, Polish Academy of Sciences Kornik, Poland

Stamps DJ, Waterhouse GM, Newhook FJ, Hall GS (1990) Revised tabular key to the species of *Phytophthora*. CABI Mycology Institute Mycology Paper 162, Cambridge, MA, 28 pp

Sterne RE, Kaufmann MR, Zentmyer GA (1978) Effect of *Phytophthora* root rot on water relations of avocado: interpretation with a water transport model. Phytopathology 68:595–602

Streito J-C, Legrand PH, Tabary F, Jarnouen de Villartay G (2002) *Phytophthora* disease of alder (*Alnus glutinosa*) in France: investigations between 1995 and 1999. For Pathol 32(3):179–185

Swofford DL (2002) PAUP*. Phylogenetic analysis using parsimony (*and other methods). Version 4. Sinauer Associates, Sunderland, MA

Tainter FH, O'Brien JG, Hernandez A, Orozco F, Rebolledo O (2000) *Phytophthora cinnamomi* as cause of oak mortality in the State of Colima, Mexico. Plant Dis 84:394–398

Tooley PW, Carras MM, Falkenstein KF (1996) Relationships among group IV *Phytophthora* species inferred by restriction analysis of the ITS2 region. J Phytopathol 144:363–369

Tooley PW, Bunyard BA, Carras MM, Hatziloukas E (1997) Development of PCR primers from internal transcribed spacer region 2 for detection of *Phytophthora* species infecting potatoes. Appl Environ Microbiol 63:1467–1475

Tyler BM (2002) Molecular basis of recognition between *Phytophthora* pathogens and their hosts. Annu Rev Phytopathol 40:137–167

Vettraino AM, Natili G, Anselmi N, Vannini A (2001) Recovery and pathogenicity of *Phytophthora* species associated with a resurgence of ink disease in *Castanea sativa* in Italy. Plant Pathol 50:90–96

Vettraino AM, Barzanti GP, Bianco MC, Ragazzi A, Capretti P, Paoletti E, Luisi N, Anselmi N, Vannini A (2002) Occurrence of *Phytophthora* species in oak stands in Italy and their association with declining oak trees. For Pathol 32:19–28

Vincent JM (1991) Oak decline: alteration of the fine root biomass with the progress of the disease. In: Siwecki R, Liese E (eds) Oak decline in Europe. Proceedings of an International Symposium, 1990, Polish Academy of Sciences, Kornik, Poland, pp 173–175

Vos P, Hogers R, Bleeker M, Reijans M, van de Lee T, Hornes M, Frijters A, Pot J, Peleman J, Kuiper M, Zabeau M (1995) AFLP:a new technique for DNA fingerprinting. Nucleic Acids Res 23(21):4407–4414

Walker N J (2002) A technique whose time has come. Science 296:557–559

Waterhouse GM (1963) Key to the species of *Phytophthora* de Bary. Mycology Paper No 92, CMI, Kew, 22 pp

Werres S (1998) Erlensterben. AFZ / Der Wald 10:548–549

Werres S, Marwitz R, Man in´t Veld WA, de Cock WAM, Bonants PJM, de Weerdt M, Themann K, Ilieva E, Baayen RP (2001) *Phytophthora ramorum* sp. nov., a new pathogen on *Rhododendron* and *Viburnum*. Mycol Res 105(101):1155–1165

Weste G, Marks GC (1987) The biology of *Phytophthora cinnamomi* in Australasian forests. Annu Rev Phytopathol 25:207–229

White TJ, Bruns T, Lee S, Taylor J (1990) Amplification and direct sequencing of fungal ribosomal RNA genes for phylogenetics. In: Innis MA, Gelfand DH, Sninsky JJ, White TJ (eds) PCR protocols: a guide to methods and applications. Academic Press, San Diego, pp 315–322

Winton LM, Hansen EM (2001) Molecular diagnosis of *Phytophthora lateralis* in trees, water, and foliage baits using multiplex polymerase chain reaction. For Pathol 31:275–283

Winton LM, Stone JK, Watrud LS, Hansen EM (2001) Simultaneous one-tube quantification of host and pathogen DNA with real-time polymerase chain reaction. Phytopathology 92:112–116

Xue B, Goodwin PH, Annis SL (1992) Pathotype identification of *Leptosphaeria maculans* with PCR and oligonucleotide primers from ribosomal internal transcribed spacer sequences. Physiol Mol Plant Pathol 41:179–188

Wolfgang Oßwald
Frank Fleischmann
Lehrbereich Krankheiten der Waldbäume/
Phytopathologie
Technische Universität München
Life Science Center Weihenstephan
Am Hochanger 13
85354 Freising/Weihenstephan, Germany
e-mail: Osswald@wzw.tum.de
e-mail: Fleischmann@wzw.tum.de

Jan Nechwatal
Lehrstuhl für Phytopathologie
Universität Konstanz
78457 Konstanz, Germany
e-mail: jan.nechwatal@uni-konstanz.de

Julia Koehl
Ingrid Heiser
Lehrstuhl für Phytopathologie
Technische Universität München
Life Science Center Weihenstephan
Am Hochanger 2
85350 Freising/Weihenstephan, Germany
e-mail: J.Koehl@lrz.tum.de
e-mail: Heiser@lrz.tum.de

Carbon and Water Fluxes in Mediterranean-Type Ecosystems – Constraints and Adaptations

J.S. Pereira, J.S. David, T.S. David, M.C. Caldeira, and M.M. Chaves

1 Introduction

The regions with a climate of the Mediterranean-type have a rainy winter and a dry and hot summer. They are located on the western side of continents between latitudes 30 to 43° north or south (Walter 1973). The dry summer climate of Mediterranean regions arises from the seasonal change in position of the semi-permanent high-pressure zones that are centred over the tropical deserts at about 20° latitude north and south of the equator. The persistent flow of stable air out of these high-pressure centres during summer brings several months of hot, dry weather (Archibold 1995). This type of climate occurs in the Mediterranean Basin, California, on the western foot of the High Andes (Chile), the southwestern tip of Africa and southwestern Australia. In all these regions evergreen sclerophyllous woody plants are the dominant type of vegetation (Walter 1973), but high diversity of species and life forms is the rule. All the areas with Mediterranean endemics belong to the so-called hot-spots for biodiversity. Although the Mediterranean biome represents less than 2% of the world's surface, it houses 20% of the world's total floristic richness (Médail and Quézel 1997).

Carbon and water fluxes between the vegetation and the atmosphere in Mediterranean regions are severely limited by the lack of rainfall in the summer when high potential evapotranspiration and high solar irradiance combine as stress factors. The precipitation regime is most distinctive. Annual precipitation varies between 275 and 900 mm with at least 65% falling during the winter months (Aschmann 1973). The evaporative potential of the atmosphere in the hot and dry summer far exceeds the then scarce or absent rainfall and the ratio of precipitation to annual potential evapotranspiration (P/PET) may reach values below 0.2. The diminished water-holding capacity of the soils due to sheet and rill erosion and low organic matte contents often exacerbates this climatic drought.

Another characteristic of Mediterranean-type ecosystems is the great variability in total annual amounts and monthly distribution of rainfall. For example, from 1982 to 1992, annual rainfall varied from a maximum of

Progress in Botany, Vol. 65

1068 mm to a minimum of 513 mm in a catchment in central Portugal (David et al. 1994). The evaporative potential of the atmosphere is much more conservative between years (David et al. 1994). The variability in water balance increases with increasing aridity. For example, in a study using a drought index, Miranda et al. (2002) found that the southern regions of Portugal (more arid) had a much higher frequency of extreme droughts during the 20th century than the northern part of the country (more humid). The occurrence of severe droughts may be critical as they may lead to severe plant water deficits and high tree mortality as documented by Lloret and Siscart (1995), for example, with heavy *Quercus ilex* tree mortality in 1994 in northeastern Spain.

In the Mediterranean summer, super-optimum temperatures and high irradiance from clear skies usually exacerbate the physiological constraints resulting from water deficits. A gradual increase in soil water deficits will result in progressive stomatal closure towards the middle of the day, reducing both transpiration and carbon assimilation. When those environment constraints are prolonged, down-regulation of photosynthesis or even photoinhibition may occur, resulting in a depression in ecosystem carbon uptake (Werner et al. 1999, 2002; Chaves et al. 2002). Plants, however, have a variety of adaptations that minimize the effects of hot and dry weather conditions, ranging from morphological characteristics as small leaf size (assisting in an efficient heat dissipation) to an increased protection against excessive radiation and high temperatures at the molecular level (Demmig-Adams and Adams 1996; Faria et al. 1996, 1998). During these high-temperature episodes in the summer, plant respiration is also likely to increase, further diminishing the net carbon gain by the canopy.

The winter season is moderately cold with the number of chilling events augmenting steadily with altitude and distance from the ocean. In trees and shrubs the evergreen habit brings the possibility of carbon assimilation in winter, when temperatures are mild, but cold in winter is a major limiting factor for species distribution (Larcher 2000). However, when low temperatures and high irradiances are combined, large depressions of photosynthesis often occur in the morning, leading to inhibition of leaf photochemistry. Annual and biennial plants native to the Mediterranean are in general well adapted to the relatively cold winter months, taking profit of the good water availability of the rainy season and complete their life cycle before the summer stress.

Environmental stress may increase in severity with climate change. Most future climate scenarios suggest that warming will tend to increase the length and severity of summer drought. For example, the future scenarios for the Iberian Peninsula, according to the Hadley Centre Regional Climate Model, suggest that the climate under a CO_2 concentration twice the present, may become substantially warmer in winter with a longer dry season (Miranda et al. 2002). Indeed, the climate in Portugal, has been warming during the last quarter of the 20th century with an increased frequency of dry years (Miranda et al. 2002). In the next 80 years spring rainfall may decline 13% and air temperature is bound to increase by 1.7 to 5.5 °C in southern Portugal (Hulme et al. 1999). More important, however, will be the putative increase in the length of the dry season. We may therefore expect an increase in the summer stress, a lessening of the winter cold limitation and, eventually, more soil erosion due to concentration of rainfall in the winter months. It is possible that the positive effect of CO_2 fertilization may be limited to fertile soils (Maroco et al. 2002). In any case, water deficits will remain a major limiting factor to carbon assimilation and transpiration in the Mediterranean areas.

In addition to drought, plants in Mediterranean-type environments suffer from limitations due to poor nutrition as nutrients have often been leached away by heavy rains in winter and, in some cases, human interference (agriculture, grazing, deforestation for domestic and industrial energy, and bush fires). The situation seems to be more acute in the strongly leached soils derived from nutrient-poor substrates of the Australian heathlands and South African fynbos than in the sclerophyllous vegetation of Chile or the Mediterranean garigue (Rambal 2001). The basic pH of calcium-rich soils of some regions may be limiting to calcifuge species, such as *Q. suber* (Blanco et al. 1997) restricting the availability of phosphorus and micronutrients such as iron.

The hot and dry summers and the fairly high primary productivities that allow the accumulation of dry necromass, make most Mediterranean regions prone to bush fires. These have certainly played an important role in the evolution of Mediterranean vegetation (Archibold 1995; Blanco et al. 1997). In recent years, many Mediterranean areas have experienced the abandonment of agriculture, the reduction in domestic animals and the expansion of forests and shrublands, with the consequent fuel accumulation that allow frequent high-intensity fires. In some cases the fire return interval is short enough to inhibit the development of forests (Walter 1973; Archibold 1995). In future climate scenarios the climatic risk of forest fires may increase several-fold (Pereira et al. 2002).

Covering about 2.75 Mkm^2 worldwide (Rambal 2001) Mediterranean-type ecosystems harbour an important human population, have a high biodiversity and are vulnerable to water shortages and desertification processes. Several authors have reviewed the productivity and physiology of Mediterranean vegetation recently, e.g., Rambal (2001) and Roda et al. (1999). In this work, we will try to update the progress in the understanding of the main constraints that determine the carbon and water fluxes between the vegetation and the atmosphere, with special emphasis in water shortage. Together with soil nutrient limitations, fire and anthropogenic perturbations (which will not be reviewed here), water deficits determine the facies of Mediterranean landscapes. Understanding how plants cope with drought is essential to model the biome responses to climate change and its role in the terrestrial carbon cycle.

2 The Main Plant Functional Groups – An Overview

The structure of the Mediterranean vegetation varies from low shrubs to close canopy forests and park-like woodlands of the savannah type to grasslands (Walter 1973; Archibold 1995; Rambal 2001). This complexity reflects a variety of adaptations in resource acquisition and utilization by plants.

The diversity both in species and in life forms results from spatial variation in resource availability (e.g., topographical gradients in water and nutrient availability), disturbance (e.g., fire and droughts), high topographical and climatic variability and heterogeneity (e.g., latitudinal and altitudinal gradients in which sclerophyllous vegetation merges into temperate-zone or alpine types of vegetation) and, of course, the ancient and long-lasting anthropogenic perturbation (di Castri 1981). In many areas the present vegetation composition and structure (e.g., evergreen oak open woodlands of the savannah type, or dense coppiced forest stands, steppes) reflect human land use.

Species that perform similar roles in an ecosystem can be regarded as belonging to the same functional group (Hobbs et al. 1995). Functional groups in this review are defined in relation to their contribution to carbon and water cycling. The main traits of interest are plant growth form (above- and belowground), and related aspects of physiological function as well as time of residence in the ecosystem (annual versus perennial). Variability in the spatial and temporal availability of soil moisture (Fuentes et al. 1995) and in the opportunity for net carbon uptake provide important dimensions for the niche diversification in Mediterranean ecosystems. Even within the three major groups – trees, shrubs and herbaceous species – there is high functional diversity.

The dominant tree species are evergreen sclerophylls that often rely on deep rooting and access to water in deep soil horizons. At extremes, this is a strategy that is most likely to succeed when the number of large trees per unit of surface area is low (e.g., less than 40 trees/ha) so that single individuals may profit from a water reservoir not used by competing plants. This is the case of the savannah-like systems such as the *dehesas* and *montados* in

Spain and Portugal, respectively, and similar systems that occur in areas of California and Chile (Joffre and Rambal 1993; Archibold 1995). However, the same species (e.g., *Q. ilex*) may occur either in open woodlands or in closed canopy dense stands with leaf area indexes that may reach up to 5 (Sala 1999) and all the trees have access to the same reservoir. The role of isolated trees in resource capture and survival deserves further study, especially in contrast with closed canopy stands. Winter deciduous trees that display a strategy of tolerance to drought and an efficient protection against high irradiance can also co-exist in the transition to colder winters (Damesin and Rambal 1995). However, it is in the riparian habitats that winter deciduous trees most often occur taking advantage of abundant water and nutrients.

The shrubs are typical components of the Mediterranean vegetation and consist of a great diversity of genera, habit (evergreen vs. drought deciduous), and size as well as strategies of resource capture. Whereas some of the sclerophyllous evergreens belong to the same hydrostable functional group as trees, others, as the malakophyllous shrubs, display a strategy of enduring the summer drought reducing transpiration and carbon assimilation to a minimum by enhanced leaf abscission in summer, but having a high activity in winter and early spring (Walter 1973; Harley et al. 1987; Werner et al. 1999).

The Mediterranean seasonality favours C_3 annual species and drought-resistant perennials in the herbaceous component. The longer and drier the summer, the stronger the seasonality and greater will be the predominance of annual species, which take advantage of the rainy winter season and avoid the drought in the summer. Where summer conditions are milder, perennial grasses and forbs are more prominent (Jackson 1985). However, wide fluctuations in species composition and productivity are often observed within a site due to changes in the amount and distribution of rainfall (Pitt and Heady 1978; Figueroa and Davy 1991; Hobbs and Mooney 1991; Espigares and Peco 1995).

3 Seasonality of Carbon Assimilation at the Leaf Level

In Mediterranean woody evergreens growth is usually limited in winter because it requires higher temperatures for photosynthesis (Rambal 2001). In a survey by Rambal (2001) the optimum temperature for photosynthesis for woody plants was ca. 20 °C and the capacity for acclimation in the course of the year was limited in apparent contrast with the reasonable capacity for acclimation of leaf photosynthesis that seems to occur in herbaceous plants. The latter can grow at lower temperatures than woody plants and

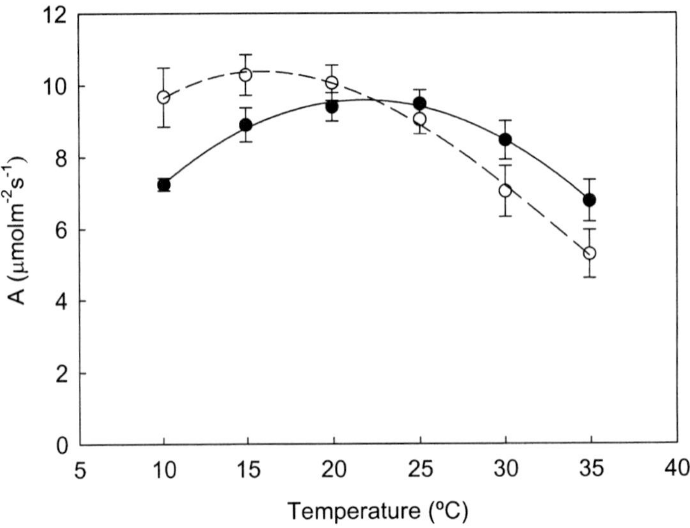

Fig. 1. Photosynthesis (*A*) response to temperature of *Cynara cardunculus* leaves grown under two temperature regimes (*filled symbols* day/night temperatures of 25/20 °C; *open symbols* 15/10 °C) measured at ambient CO_2 and photosynthetic photon flux density of 800 μmol quanta m^{-2} s^{-1}. (P.F. Santos, M.M. Chaves and J.S. Pereira, unpubl.)

they are able to complete their life cycles before water is depleted from the upper soil horizons colonised by their roots. Photosynthetic acclimation to low temperatures was observed, for example, in the case of a common Mediterranean biennial, *Cynara cardunculus*. Maximal rates of photosynthesis were observed between December and March and temperatures that maximize photosynthetic rates were shown to decrease from 22.7 °C in plants grown under a thermal regime of 25/20 °C (day/night) to 13.9 °C in plants grown under a regime of 15/10 °C (Fig. 1).

In general, minimum rates of carbon assimilation occur in the summer, especially in the afternoon, because stomata close not only in response to soil water deficits but also in response to dry air as well as to high temperature (Correia et al. 1990; Chaves et al. 2003). In addition, a high degree of co-ordination between stomatal response and photosynthetic capacity is observed in species surviving the summer stress, which leads the photosynthetic machinery to a down-regulated state of functioning (Werner et al. 1999, 2002). This type of response has been recognized to have a protective role.

Although photochemical reactions are known to be very resistant to dehydration (Genty et al. 1987), it is now well established that superimposition of drought with high irradiance and temperature predisposes the photosynthetic apparatus to "down-regulation" or to photoinhibition, because leaves with closed stomata are unable to efficiently dissipate absorbed solar radiation as latent heat (Chaves 1991). Photoinhibition is a very common phenomenon under Mediterranean climatic conditions (Werner et al. 2002). This phenomenon is usually reversed overnight, as shown in different species (Demmig-Adams and Adams 1992; Epron et al. 1992; Quick et al. 1992). In the Mediterranean tree *Q. suber*, for example, morning values of the quantum efficiency of photosystem II (PSII), expressed by the ratio of variable to maximal fluorescence, were always the highest of the whole day (Faria et al. 1998). The afternoon depression of PSII efficiency is paralleled by an increase in the zeaxanthin content in the leaves, which is formed in response to the high levels of incident sunlight by the de-epoxidation of violaxanthin. The existence of a correlation between the ability to dissipate energy and the concentration of zeaxanthin in the leaves was shown in several species (Demmig-Adams and Adams 1992), indicating that xantophylls (e.g. zeaxanthin) play a role in the non-radiative energy dissipation by the leaves (Bilger and Bjorkman 1994). The impact of photoinhibition on whole canopy production under mild photoinhibitory conditions is small in evergreen Mediterranean sclerophyllous plants (Werner et al. 2001). Under such conditions the canopy photoinhibition model (CANO- PI) developed by Werner et al. (2001), predicts a daily loss of 7.5–8.5% of potential carbon gain in the upper sunlit canopy layers and a loss of ca. 3% in the bottom canopy. However, it is expected that under additional environmental stresses (e.g. more severe water deficits), photoinhibition can be enhanced, producing a higher impact in the whole plant carbon gain.

It is known that temperature has differential effects on various plant processes. Photosynthesis and related metabolic processes like photochemistry, carboxylation and oxygenation by Rubisco, synthesis and export of carbohydrates, all have different temperature sensitivities and optima for temperature. On the other hand, dependence of photosynthesis on temperature varies with changes in environmental conditions, such as light, CO_2 concentration and water status of the plant (Leegood 1995). As temperature rises, net carbon assimilation increases up to an optimum that varies with genotype and preconditioning of the plants. The attenuation of the rate of increase of photosynthesis with temperature corresponds to an increase in the rate of oxygenation relative to carboxylation, in other words, an increase in the photorespiratory rates (Berry and Björkman 1980). Besides the observed increase in photorespiration, a decrease in the absorptive capacity of PSII due to the migration of the core PSII away from the LHCII has also been reported under these conditions (Sundby et al. 1986), corresponding to the so-called state-transition quenching of chlorophyll fluorescence.

Mediterranean plants possess several mechanisms of protection against the effects of high temperatures and among them the emission of volatile isoprenoid compounds (Silva et al. 1999; Llusia and Penuelas 2000; Staudt et al. 2001). These emissions are light dependent and temperature sensitive (Hanson and Sharkey 2001) and have been reported to increase thermotolerance during summer stress (Loreto et al. 1998; Logan et al. 2000), presumably by increasing the stability of thylakoid membranes (Sharkey et al. 2001). In evergreen oaks, such as *Quercus coccifera* and *Q. ilex* or *Q. suber* that do not have anatomical structures specialized for monoterpene accumulation as conifers, the rates of emission depend on temperature and solar radiation (Silva et al. 1999; Llusia and Penuelas 2000). Emissions increase exponentially with temperature up to an optimum and decline at higher temperatures (Singsaas and Sharkey 2000). This explains partly the strong seasonality with maximum values attained in the spring and early summer (Llusia and Penuelas 2000). During the day, the highest levels of emissions occur in the afternoon (Silva et al. 1999). This is coherent with the postulated function of monoterpene emission as a short-term thermotolerance mechanism that functions best when short periods of high temperature occur, as in the spring and early summer in the Mediterranean. This means that during this period Mediterranean-type ecosystems function like temperate forests, but in the summer the continuous exposition to high temperatures make more long-term thermotolerance mechanisms, such as those described above, necessary (Singsaas and Sharkey 2000).

The respiratory costs of growth and maintenance of plant organs is of pivotal importance to the overall carbon balance of plants, since they represent a major source of CO_2 released by the plants. In fact, up to 35% of the daily fixed carbon can be respired by the leaves alone. Several studies indicate that, in Mediterranean-type ecosystems, evergreen species tend to exhibit greater tissue construction costs and lower maintenance costs than deciduous species (Merino et al. 1982; 1984; Merino 1987; Villar and Merino 2001). This may be the result of evergreens being normally more abundant in areas of limited resources (water and nutrients), with a consequent higher investment in defensive molecules against biotic and abiotic stresses (Martinez et al. 2002). Plant adaptation to low resources is also based on high flexibility in metabolic rates, including the reduction of respiration rates when resources are low. As pointed out by Mooney et al. (1975), leaf abscission following a drought spell may be equally important in cutting down transpiration as in reducing respiration during periods when carbon assimilation is very low.

Optimization of respiration rates in relation to temperature has also been observed in evergreen trees, allowing a dramatic reduction in the costs of tissue maintenance (Pereira et al. 1986). In winter, plants tend to exhibit higher Q_{10} (the proportional increase for each 10 °C rise) for the dark respiration than in the summer. This is normally associated with higher concentrations of carbohydrates in leaves of cold-acclimated plants. On the other hand, it was shown that leaf respiration is partially inhibited by light, when measured at moderate temperatures, and also that the temperature sensitivity of respiration is greatest in darkness, decreasing as irradiance increases (Atkin et al. 2000). This means that at high irradiance leaf respiration is almost insensitive to temperature and this has implications on the magnitude of respiration at high temperatures in the light.

3.1 Surviving the Drought: Drought Avoidance and Drought Tolerance

3.1.1 Water Acquisition

During winter, because rainfall is usually abundant, water is not a limiting factor for plant growth and metabolism. Transpiration rates are then mainly determined by external climatic conditions, i.e., radiation and vapour pressure deficit. Winter transpiration is normally low as both solar radiation and vapour pressure deficit are then at their lowest. Carbon assimilation will then be mainly determined by solar radiation, air temperature and photosynthetic capacity.

The seasonal trend of decreasing transpiration results from progressive stomatal closure, i.e., stomata remain close for longer periods in the day as drought progresses. In three Mediterranean "macchia" sites the daytime average canopy conductance declined by up to 90% during the dry season (Reichstein et al. 2002b). Similar decreases were found, for example, in stands of several other species in Mediterranean regions (e.g. *Pinus pinaster*, *Eucalyptus globulus* and *Q. ilex*) (Loustau et al. 1996; David et al. 1997; Infante et al. 1997; Tognetti et al. 1998).

In some cases, however, transpiration does not decline in summer in spite of the atmospheric drought. Jiménez et al. (1996) and David (2000) reported that, in a laurel forest in Tenerife and in a holm-oak (*Q. ilex* ssp. *ballota*, syn. *Q. rotundifolia*) *montado* in Portugal, respectively, transpiration was not reduced during the summer. On the contrary, it increased during that period, following closely the patterns of radiation. The maintenance of high rates of transpiration in the summer may be explained by root access to deep soil water reservoirs. Consistently predawn leaf water potentials remained high throughout (David 2000). In the Portuguese holm-oak site the groundwater table was approximately 13 m below soil surface in summer that would be easily reachable by this species' roots. Correia et al.

(2001) and White et al. (2000) also found that *Ceratonia siliqua* trees in southern Portugal and *Eucalyptus camaldulensis* and *Eucalyptus saligna* in Australia, respectively, were able to maintain high predawn leaf water potentials during the summer due to deep rooting.

Similar situations in other Mediterranean areas may be more frequent than expected since it is known that plants that grow well into the summer drought frequently depend on the ability to tap water from permanent water tables (Canadell et al. 1996). The proportion of biomass allocated to below-ground tissues is usually increased as the environment becomes more severe (Canadell et al. 1999). Canadell et al. (1996) estimated an average rooting depth of 12.6 m for sclerophyllous Mediterranean trees (mainly *Eucalyptus* and *Quercus* spp.). This drought avoidance strategy of trees relies not only on deep rooting but also on the spread of horizontal shallow roots in the surface soil far beyond the crown width (Breman and Kessler 1995; Canadell et al. 1999; Verdaguer et al. 2000). The importance of these drought-avoidance processes does not depend solely on the species but also on the local hydrogeological conditions. When access to deep water is not possible, trees must rely on the shallow soil water supply, suffering severe water stress in summer. Other drought-tolerance features will then play the most important role.

In a dry soil, plants may also enhance their capability to extract water through osmotic adjustment (Morgan 1984). Increased rigidity of cell walls may also enhance tolerance to water deficits at the leaf level (Wilson et al. 1980). Although these processes are meaningful under the Mediterranean type of conditions (Rodrigues et al. 1993), they may not be determinant in deep-rooted trees. Sala and Tenhunen (1994) observed that osmotic adjustment and changes in tissue elasticity did not significantly increase drought resistance in the *Q. ilex* trees.

3.1.2 Stomatal Control

Stomatal control of gas exchange tends to avoid desiccation that might result in cell dehydration and runaway cavitation of the xylem, leading to plant death. The already mentioned tendency for midday stomatal closure in the conditions of the Mediterranean summer seems to agree with the hypothesis that plant evolution brought about a short-term optimisation of carbon assimilation versus water loss (Cowan 1982; Raven 2002). If daily carbon assimilation is to be maximised at a given level of daily transpiration, stomatal conductance must decrease in order to counterbalance the increasing evaporative demand of the atmosphere towards the middle of the day that would inflate the water cost of CO_2 assimilation (Schulze and Hall 1982).

There is evidence that stomatal closure is likely to be mediated by chemical signals travelling from the dehydrating roots to the shoots, with abscisic acid (ABA) as the most likely chemical in the regulation of stomatal behaviour (Davies and Zhang 1991; Chaves et al. 2003). However this regulation is complex and variable according to species, being more important in isohydric species, e.g., some deciduous trees that tend to maintain leaf water potential during the day nearly constant at a value that does not depend on soil water status until plants are close to death (Tardieu and Simonneau 1998; Wilkinson and Davies 2002). Indeed the regulation of stomatal aperture involves long-distance transport of ABA as well as the modulation of ABA concentration by xylem and leaf pH, which may increase in conditions of high evaporative demand. Increased pH could partly explain stomatal closure in the afternoon in well-watered plants (Chaves et al. 2003). However, this straightforward process may be even more complex or irrelevant in large trees as suggested by Perks et al. (2002) due to the long time needed for a direct chemical signal to travel from roots to leaves. Stomata may also respond to changes in the rate of water supply through changes in xylem conductance and this suggests that the long-term stomatal behaviour evolved to prevent the risk of runaway cavitation and the loss of water transport capability in the plant (Jones and Sutherland 1991; Sperry 2000; Cruiziat et al. 2002).

3.1.3 Xylem Vulnerability

When severe summer water stress cannot be avoided, plant survival can only be guaranteed by the maintenance of the integrity of the xylem conducting system (Tyree and Sperry 1989; Jones and Sutherland 1991). This can be achieved if leaf and xylem water potential does not fall below a xylem cavitation threshold (Tyree and Sperry 1989; Jones and Sutherland 1991; Sperry 2000; Buckley and Mott 2002). Actual minimum values of leaf water potential experienced in field conditions are usually within a safety margin of catastrophic xylem embolism (Cochard et al. 1996; Sperry 2000). The water potential cavitation threshold is a species-specific value (Tyree and Sperry 1989) being much lower in drought-adapted than in well-watered species (Sperry 2000; Lemoine et al. 2001). According to Cruiziat et al. (2002) the resistance to cavitation is probably the most important parameter determining the drought tolerance of a tree. Mediterranean-adapted evergreen oaks (*Q. ilex* and *Q. suber*) are much less vulnerable to drought than mesic-zone oaks, since they have much lower xylem cavitation thresholds (Tyree and Cochard 1996). Within the Mediterranean oaks, *Q. ilex* has a lower xylem water potential cavitation threshold when compared to *Q. suber* (Tyree and Cochard 1996) and to *Q. pubescens* (Tognetti et al. 1998). Other Mediterranean species, namely *Phillyrea latifolia*, are even less vulnerable to cavitation than *Q. ilex* (Martinéz-Vilalta et al. 2002).

The main way plants prevent leaf and xylem water potential from falling below the cavitation threshold is clearly through stomatal regulation of leaf transpiration (Jones and Sutherland 1991; Sperry 2000; Buckley and Mott 2002; Cruiziat et al. 2002). Whatever the mechanisms, stomata can then be envisaged as the plant pressure regulators (Jackson et al. 2000; Buckley and Mott 2002). Results of Salleo et al. (2000) in *Laurus nobilis* further suggest that stomatal closure may not be related to a negative feedback response to leaf water potential but rather triggered when the cavitation threshold is reached. It is not clear how this hydraulic based explanation of stomatal closure works and how it links with other mechanisms of stomatal aperture control (Sperry 2000; Buckley and Mott 2002).

When the leaf water potential threshold is reached an internal hydraulic limit is imposed to the root-leaf flow path that ultimately determines the maximum transpiration rate that can be sustained by the plant (Hogg and Hurdle 1997; David 2000; Cruiziat et al. 2002). The actual value of this maximum boundary of hydraulic flow (F) is not only determined by the leaf water potential threshold (Ψc) but also by the water potential at the supply sources to the roots (Ψs) and by the total plant hydraulic conductance (k), as can be seen from the Darcy's law for steady-state conditions (Sperry 2000; Cruiziat et al. 2002):

$$F = k\,(\Psi s - \Psi c)$$

As both (Ψs and k usually decrease as summer drought progresses, maximum hydraulic limits of sap flow and transpiration accordingly also decrease. This may not be the case, however, when tree roots have direct access to large groundwater reservoirs, being then both k, (Ψs, (Ψc and, consequently, the maximum transpiration rate is almost constant during the summer drought (David 2000).

The values of whole plant hydraulic conductance (k) may also be indicative of the plant resistance to drought. Hydraulic conductance has been found to be usually lower in drought-adapted species (Nardini et al. 1999; Lemoine et al. 2001). Nardini et al. (1999) found that root and shoot leaf-specific conductances were lower in drought-adapted oak species (*Q. suber*, *Q. pubescens*, *Q. petraea*) than in water-demanding oak species. Tognetti et al. (1998) found lower xylem leaf-specific conductivity in *Q. ilex* than in *Q. pubescens* in Italy. It is suggested that a decrease in hydraulic conductance helps to limit the transpiration flux, which in itself is an important feature of drought resistance. Nardini et al. (1999) found that leaf blade hydraulic conductance tended to be lower in drought-adapted *Quercus* species. They argued that this might cause an advantageous substantial drop in leaf water potential, promoting a more efficient stomatal closure before xylem cavitation threshold is reached.

3.1.4 Leaf and Fine Root Shedding (Hydraulic Fuses)

In addition to stomatal control of leaf water potential and transpiration, hydraulic segmentation may also be important for preventing catastrophic xylem cavitation (Zimmermann 1983; Jackson et al. 2000; Cruiziat et al. 2002). In some situations, leaves, petioles, minor branches or small roots can be more vulnerable to cavitation than larger branches, stems and tap roots (review in (Jackson et al. 2000; Cruiziat et al. 2002)). Under severe drought cavitation may preferentially occur localized on those more vulnerable and replaceable units (hydraulic fuses) ultimately protecting the main and non-replaceable xylem conducting system. The underlying processes of leaf abscission and fine root shedding are not yet totally clarified (Cruiziat et al. 2002). Martinéz-Vilalta et al. (2002) found that vulnerability to xylem embolism was always higher in roots than in stems for all studied tree and Mediterranean shrub species. The observation that average leaf longevity may increase in dry years in deep-rooted evergreen trees (Sabaté et al. 1999), may reflect a greater vulnerability of fine roots and the continued use of moisture through the deep root system in these plants, following the death of fine roots.

4 Water Use Efficiency

Where water is scarce the efficiency of its use, measured as biomass synthesized per unit of transpiratory water loss (WUE), is a critical parameter to understand plant performance. At the leaf level, the intrinsic water use efficiency (WUE_i), i.e., carbon assimilation divided by stomatal conductance, is a function of the ratio between intercellular and atmospheric CO_2 (c_i/c_a) and tends to increase with better nutrition (higher photosynthetic capacity) and greater leaf mass to area. Under mild water deficits WUE_i may increase as a result of the non-linear relationship between carbon assimilation and stomatal conductance, i.e., water loss is more restricted than the inhibition of photosynthesis (Chaves et al. 2003). However, in the Mediterranean summer, the high evaporative demand of the atmosphere is the main determinant of the seasonal course of transpiration, prevailing over the short term stomatal control of water loss. As a consequence, the amount of carbon fixed per unit of water transpired (WUE) tends to decrease drastically in summer in spite of stomatal closure with water deficits and the short-term optimization of carbon assimilation. As water stress becomes more severe, a decline in photosynthetic capacity while respiration and cuticular transpiration are maintained or even increase will bring about further decreases in intrinsic WUE. Patchy stomatal closure, which is common in Mediterranean evergreens under drought (Beyschlag et al. 1992) may also contribute to a further decrease in WUE (Beyschlag and Eckstein 2001).

Decreases in WUE during the summer have been observed consistently. For example, this was the case in *Q. ilex* stands (Gracia et al. 1999) where WUE dropped from more than 2.5 mmol CO_2 mol^{-1} H_2O in March, to negative values in the dry summer. At the ecosystem level, Reichstein et al. (2002b) also reported a decrease in water use efficiency of gross carbon uptake during the summer drought in several Mediterranean ecosystems.

A question that now arises is, why do Mediterranean woody sclerophyllous plants have leaves in the summer when only low WUE is allowed? One hypothesis is that they sacrifice WUE to gain the possibility to assimilate carbon as soon as the rains start in autumn. As shown in Fig. 2, transpiration in a *Q. ilex* stand in Portugal began (stomata opened) after a few days of rain in autumn. Intense stand carbon assimilation began as soon as stomata opened (eddy covariance data, not shown) when the understorey was completely dry. This rapid response of the whole-canopy gas exchange to soil re-wetting is important because the beginning of the rainy season may be erratic and a green canopy would have opportunistic advantages for carbon gain. Deciduous plants might waste part of the mild autumn to reconstruct the canopy at expenses of carbon reserves. The early start in carbon gain may be one advantage for evergreen trees because it takes place at a time when active organic matter decomposition occurs (see below) and abundant nutrients may be available. Adding to the winter assimilation this may extend the growth season and allow evergreen trees and shrubs in the Mediterranean to accumulate more biomass and produce greater root systems than other life forms.

5 Carbon Balance at Canopy Level

5.1 Changes in Leaf Area Index

Net primary productivity of Mediterranean ecosystems is quite variable, ranging from ca. 0.2 to ca. 1.8 kg m^{-2} $year^{-1}$ (Rambal 2001). This variability

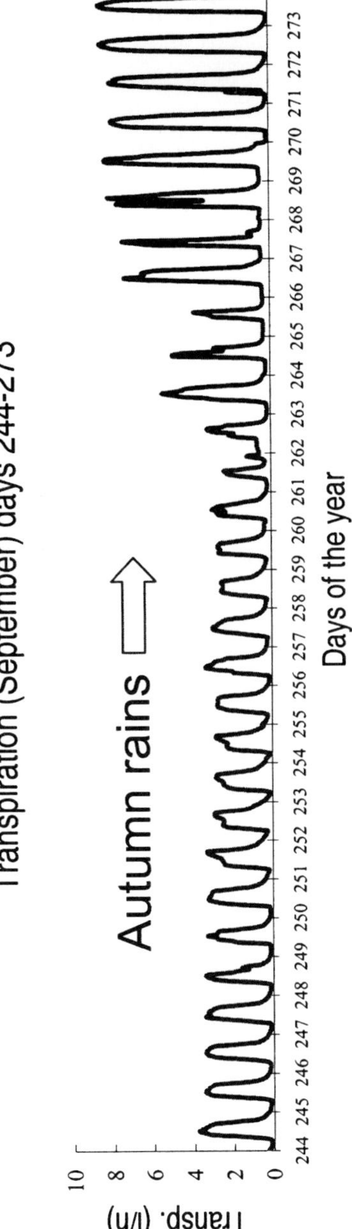

Fig. 2. Transpiration rates of *Q. ilex* ssp. *ballota* in Évora, Portugal, during the month of September. The units are litres per hour (l/h). Stomata opened and transpiration increased dramatically after the autumn rains, which occurred mainly between day 261 and day 267 and amounted to 80 mm. (J.S. David and T.S. David, unpubl.)

represents different degrees of environmental constraints on leaf area index (LAI), photosynthetic rate and plant respiration losses. At the community level, LAI is a determinant of primary productivity and transpiration and it varies with soil water and nutrient availability (Sala 1999). Leaf area index may range from very low values in open woodlands and shrublands (Rambal 2001) to more than 5, as in *Q. ilex* forests (Sabaté et al. 1999). Net primary production and LAI normally decrease in the less fertile hill tops (Sabaté et al. 1999). The relative importance of nutrient and water availability on LAI can be studied by manipulative experiments and is illustrated in Table 1.

In simple monospecific *Eucalyptus globulus* canopies under the mild Mediterranean climate of coastal Portugal, the fraction of light intercepted by the canopy (**f**) increased substantially with abundant water and nutrients (treatment IL), whereas both f and the quotient biomass/intercepted radiation ((ε) were lower in the rain-fed control (C). The addition of nutrients (F) increased LAI to values slightly below those attained by irrigating in the summer (I). Nevertheless, drought in the summer was a major limiting factor because ε in treatment F changed little when compared to the control (non-fertilized).

In the case of open woodlands as *montado/dehesa*, there is large seasonal and spatial variations in LAI as the grass cover is ephemeral and trees evergreen. In a cork oak stand in Portugal, Sá (2001) found that maximum LAI reached by the herbaceous layer of vegetation was ca. 4 under tree cover and 4.5 in the open, in a wet year (precipitation 42% above the average). In an average rainfall year, the LAI of the herbaceous layer was ca. 0.7 and 1.5 under the tree canopy and in the open, respectively. However, this green cover lasts only from October to the end of the spring. The estimated tree

Table 1. Average estimated leaf area index (LAI); aboveground biomass productivity (NPP); mean fraction of solar radiation intercepted by the foliage (**f**, with incident photosynthetically active radiation of 4.9 GJ m^{-2} $year^{-1}$) and the quotient of biomass production:radiation intercepted (ε) in the period June 1988 to July 1989 in an *Eucalyptus globulus* experimental plantation near Obidos, central Portugal (J.S. Pereira, C. Araujo and S. Linder, unpubl.; see text for treatment definition)

Treatments	LAI	NPP ($t\ ha^{-1}\ year^{-1}$)	f	ε ($kg\ GJ^{-1}$)
IL	3.8	35.7	0.92	0.8
I	3.0	29.5	0.78	0.8
F	2.8	26.9	0.77	0.7
C	2.3	23.7	0.71	0.7

cover LAI was 0.55, in a nearby open oak woodland (*Q. ilex* ssp. *ballota* and *Q. suber*) with a percent tree cover of 32% (unpubl. data).

The spring growth flush pattern is the most frequent in the Mediterranean woody vegetation with the exception of a few summer-growing species (Rambal 2001). Leaf area often reaches a maximum after the spring growth flush, when new leaves are added to pre-existing ones (Sabaté et al. 1999) as in *Q. ilex*, whose leaves may last for several years (Arco et al. 1991; Sabaté et al. 1999). In some cases there are substantial seasonal changes in LAI of evergreen canopies. For example, in cork oak (*Q. suber*) the new foliage growth occurs late (April–May) as compared to other Mediterranean evergreens, but previous year's leaves fall at the time of spring growth flush (Pereira et al. 1987), which means an almost deciduous behaviour. In general, the amount of leaves in the canopy reaches a minimum in spring before the current year leaves have fully elongated. In Mediterranean oaks, a second growth flush may occur in the autumn if high temperatures and rainfall would allow it (Pereira et al. 1987), but is normally less intense than the first growth flush and contributes little to a seasonal change in LAI.

Phenology is very sensitive to climate change (Myneni et al. 1997; Osborne et al. 2000). There are not many studies on changes in phenology of the Mediterranean-type of ecosystems in relation to climate change. Nevertheless, there are indications that phenology is changing with warming of climate. In *Pinus pinaster* the main growth flush is in the spring, but a second (polycyclic) flush may occur (Kremer and Roussel 1982). In Portugal, the probability of occurrence of polycyclism in *P. pinaster* populations has increased several-fold in recent years (Paulo and Tavares 1996), when the warming of winters occurred in the country (Miranda et al. 2002). Osborne et al. (2000) related *Olea europaea* flowering phenology to warming and predicted a significant advancement in olive flowering dates with global warming. What this would mean, if generalized to vegetative growth, may be a better utilization of the winter-spring period in the scenario of longer summers, because new leaves will be in general more productive than older overwintering leaves. The effect of climate change in the phenology of Mediterranean vegetation is an issue that needs to be further studied.

Leaf production and the balance between leaf production and leaf shedding may vary from wet to dry years. Leaf production increased in wet years compared with dry years as illustrated by the studies in *Q. ilex* forests of Catalonia (Sabaté et al. 1999). Leaf turnover, however, was slower in dry years than in wet years, leading to a relatively sluggish response of leaf area to climate. In cork oak the annual litterfall can be equated to total leaf production because there the full 1-year leaf crop is substituted by the time of the spring growth flush (Pereira et al. 1987; Escudero et al. 1992; Sá 2001): In a cork oak stand in Portugal the maximum litterfall in a series of 5 years

(1994–1998) occurred in a wet and warm year (precipitation 60% above the average) when leaf biomass was more than twice the biomass produced in a dry year with precipitation 54% below the average (Sá 2001). However, a simple relationship between litterfall and precipitation could not be obtained because other factors such as cold in winter, the number of dry months irrespective of annual rainfall or the occurrence of water deficits in the spring may have influenced leaf production and longevity.

5.2 Rainfall Interception and Spatial Redistribution of Soil Moisture

Rainfall interception losses are important in dense canopies depending on rainfall distribution patterns, canopy storage and the rate of evaporation from wet surfaces. In the sclerophyllous evergreen shrub canopies of the Californian chaparral, rainfall interception may reach 19 to 12% of annual precipitation (Poole et al. 1981), but higher estimates were given for dense *Q. ilex* stands in Spain (Bellot et al. 1999).

In widely spaced woodlands, rainfall interception by isolated trees promotes the aerial redistribution of throughfall underneath the tree crowns. Throughfall below the upwind part of the crown is frequently higher than open area rainfall (King and Harrison 1998). This upwind throughfall concentration has been found under *Olea europeae* (Gómez et al. 2002), *Q. ilex* (Calabuig 1992) and *Q. ilex* ssp. *ballota* trees (David 2000). This results from the accumulation of crown dripping and direct fall of wind-driven rain and may be essential to explain the ecology of the open woodlands.

Neutron probe measurements carried out in *dehesas* in Spain have shown that more water is stored and available in soils underneath tree crowns (Joffre and Rambal 1988, 1993). The authors ascribe this effect to the improved soil properties below canopy. However, this may also be a consequence of the throughfall accumulation below parts of the crown projected area (David 2000). Whatever the cause, this promotes the well-documented difference of herbs and shrubs inside and outside the crown projected area, which overall increases the biodiversity of pasture in open oak woodlands (González-Bernaldez et al. 1969; Marañon 1985, 1986, 1991; Barrantes et al. 1986; Joffre and los Llanos 1986).

Interception loss, i.e., evaporation from wet tree crowns during and immediately after rain, has seldom been measured with accuracy in Mediterranean woodlands such as *dehesas/montados*. Difficulties arise from the fact that below crown throughfall does not only result from canopy dripping but also from wind driven rainfall. With a new measuring procedure, David (2000) estimated the annual interception loss from an isolated *Q. ilex* ssp. *ballota* tree as 21.7% of gross rainfall and 28% of tree evapotranspiration, being the remainder 72% due to transpiration. These interception loss estimates are similar to those obtained in dense *Q. ilex* stands in Spain (Comín et al. 1987; Bellot et al. 1999) but higher than in pine and eucalypt forests under a Mediterranean climate in Portugal (Valente et al. 1997).

One process that contributes to the heterogeneity of soil moisture distribution is hydraulic lift, i.e., water movement from deep to shallow roots when surface soil dries, which may play an important role in the ecology and water relations of dry woodlands (Caldwell and Richards 1989). This night-time water transfer process has been well documented for the California blue oak (*Quercus douglasii*) woodlands (Ishikawa and Bledsoe 2000), as well as for other species as *Acer saccharum* (Dawson 1993; 1996), some species of the Brazilian cerrado (Jackson et al. 1999) and for *Eucalyptus viminalis* (Phillips and Riha 1994). In an old-growth *Pinus ponderosa* stand of a summer drought climate of Oregon (USA) Brooks et al. (2002) estimated that about 35% of total daily water utilization from the upper 2 m of soil were replaced by hydraulic lift during the early part of the dry summer. They concluded that this hydraulic redistribution could slow the rate of soil water depletion for ca. 3 weeks after a 2-month drought and that part of the redistributed water was used for pine transpiration. Another source of heterogeneity in soil moisture is the concentration of water near the trees in Mediterranean open woodlands through the downward flux through roots (i.e. the inverse hydraulic lift) as demonstrated in the Kalahari sands by Schulze et al. (1998).

6 Water Constraints at the Community Level and the Role of Biodiversity

Mediterranean ecosystems seem to maintain rather higher levels of structural, morphological, and functional diversity. There is a wide range of structural types from more or less dense forests to pure grasslands. Functional diversity as related to water use is, for example, expressed by shallow or deep-rooted, perennial or annuals. Recently, it has been recognized that biodiversity can have a role in community resource use and in water and carbon fluxes (e.g., Naeem et al. 1994; Tilman 1997; Hector et al. 1999).

It has been argued that biodiversity can affect various ecosystem processes, e.g. primary productivity or nutrient retention (e.g. Naeem et al. 1994; Tilman 1997). In several experiments, where herbaceous species diversity has been controlled it was observed a positive relationship between species richness and productivity (e.g., Naeem et al. 1994; Tilman 1997; Hector et al. 1999). The same results were found in a Mediterranean grassland in Portugal where species richness was experimentally manipulated (Hector

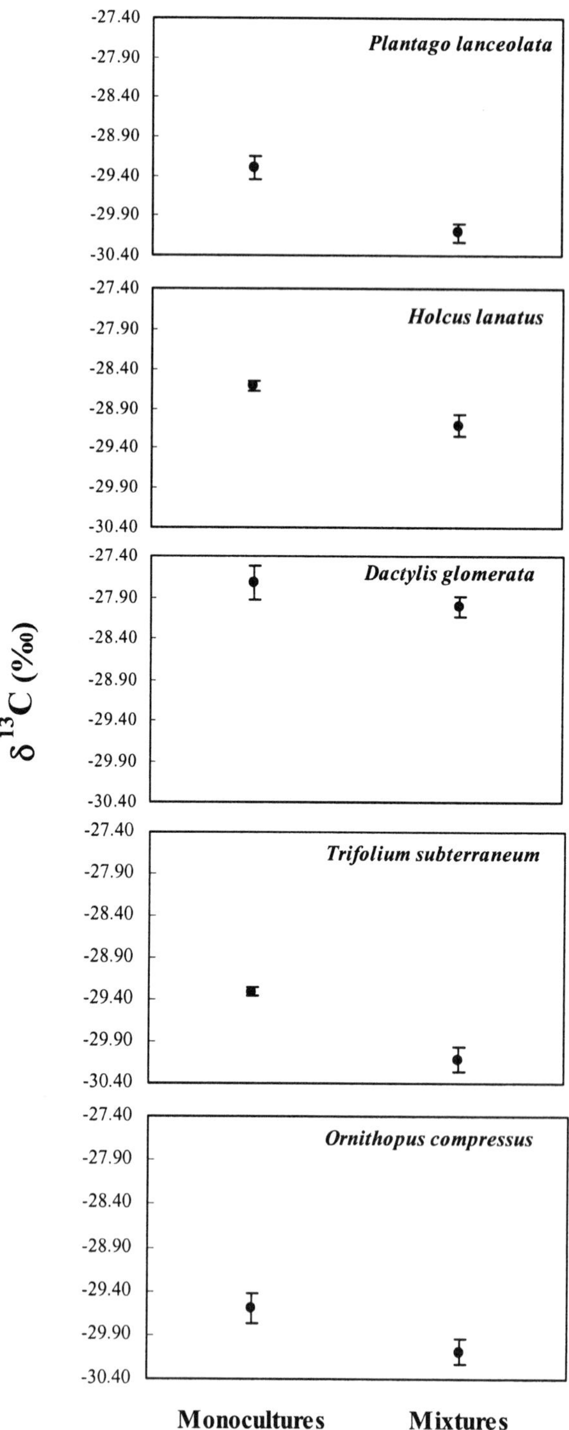

Fig. 3. Leaf carbon isotope ratio ((δ^{13}C) of *Plantago lanceolata, Holcus lanatus, Dactylis glomerata, Trifolium subterraneum*, and *Ornithopus compressus* in monocultures and in species-rich mixtures (8 and 14 sown species). *Points* are means of 1997 and 1998 measurements and *bars* are 1 SE. (Caldeira et al. 2001)

et al. 1999; Caldeira et al. 2001). In this experiment, using the stable carbon isotope composition ($\delta^{13}C$), it was shown that plants growing in species-rich communities had decreased stomatal diffusion limitations to photosynthesis than plants of the same species growing in monocultures (Fig. 3). As water uptake depends on root activity where moisture is available (Lauenroth and Aguilera 1998), leaf ($\delta^{13}C$ was consistent with plants in the species-rich communities having more water available in the upper soil where roots were concentrated. The co-occurring species had different structural and functional attributes (e.g. annuals and perennials) that probably related to complementarity in water use. This can explain the early season differences, with higher total cover in species-rich mixtures than in monocultures. Moreover, the resulting higher biomass and cover in species-rich communities could also have been important in reducing evaporation and percolation loss. Biodiversity effects partly arise from complementarity in resource use and/or in positive interactions.

When considering other growth forms (or functional groups) as trees or shrubs that can have access to other sources of water the overall balance of water in the ecosystem is different from a situation of pure grassland. Trees have access to a greater proportion of the soil volume than grasses. In Spanish Mediterranean woodlands, it was demonstrated that the water balance of grass and tree-grass components can differ by as much as 50% (Joffre and Rambal 1988, 1993). Similar results were observed in oak woodland-annual grasslands of central California (Lewis 1968).

Although tree cover in the tree-grass component averaged between 15 and 20% of total cover (40–50 trees/ha), evapotranspiration was 45–60% greater than for grasses alone (Joffre and Rambal 1988, 1993). At the same time, soil water storage was much greater for the tree-grass component of the ecosystem. This may be due to greater amounts of soil organic matter beneath the trees, which would lead to a greater capacity for soil water storage, or higher water infiltration into the soil. Other direct or indirect effects could be attributed to trees as for example, hydraulic lift (Dawson 1993), even though Ishikawa and Bledsoe (2000) refer that in the California *Q. douglasii* woodlands, hydraulic lift appeared to occur too late in the summer to benefit the annual herbaceous plants.

In many Mediterranean woodlands there are only one or two species of dominant trees, indicating that functional group identity or even species identity can be functionally more important than species number (Roy 2001). The introduction of new life form such as trees and possibly shrubs can significantly ameliorate the effects of water stress for all the vegetation of Mediterranean woodlands (Dawson 1993). Complementarity in water use as well as positive interactions can allow a more efficient use of the limited water resources by plants, which inhabit this ecosystem. Nevertheless, Bertness and Callaway (1994) suggest that neighbour effects vary from positive to negative as productivity increases. This is supported by experiments in Wisconsin old fields, where the effects of trees on grasses shifted from positive to negative as resource limitations shifted from belowground to aboveground (Ko and Reich 1993).

7 The Other Component of Carbon Balance – Heterotrophic Respiration

As happens with the net primary productivity (NPP) in Mediterranean ecosystems, the net ecosystem exchange rate (NEE), i.e., the difference between photosynthesis and whole ecosystem carbon loss through respiration of Mediterranean may be quite variable and much lower than the NPP due to heterotrophic respiration (Valentini et al. 2000). The number of cases where annual NEE was measured using the eddy co-variance method is restricted. In Italy, Valentini et al. (2000) reported NEE of 660 g C m^{-2} $year^{-1}$ for a broadleaf evergreen canopy (*macchia*), whereas we measured 80 g C m^{-2} $year^{-1}$ in a evergreen oak woodland in the Mitra site near Évora, Portugal, during 1999 (unpubl. data). Ecosystem respiration in forest stands and woodlands may be quite high (Valentini et al. 2000). For example, the eddy covariance measurements in the Mitra site near Évora, revealed that about 220 g m^{-2} $year^{-1}$ of carbon, i.e., ca. 70% of the gross net productivity (GPP), were lost by heterotrophic and autotrophic respiration.

In forests and shrublands, ecosystem respiration is dominated by root and microbial soil respiration (Valentini et al. 2000). In addition to the response to temperature, a wealth of evidence is accumulating on the role of drought in limiting heterotrophic respiration, especially in the Mediterranean (Reichstein et al. 2002a,b). In a study involving three Mediterranean woody plant sites, Reichstein et al. (2002b) concluded that soil temperature and moisture content explained 70 to 80% of the variability in soil respiration. In spite of the common assumption that temperature is the leading factor determining soil respiration, in the same study in an evergreen oak "montado" in the district Cambisol in Évora (Portugal), this was only true when soil moisture was above 15% v/v (J. Banza, M.Rayment, J.S. Pereira, unpubl. data). This means that soil respiration revealed a decrease in temperature sensitivity (Q_{10}) in response to drought. In this case, the apparent Q_{10} of soil respiration measured with a field chamber, increased from less than 1 in dry soil to 2 at 20% v/v soil moisture content. Reichstein et al. (2002a) found similar results for other Mediterranean forests and shrublands. As the topsoil was below 15% v/v for most of the year, we may conclude that water deficits are the major limiting factor for soil heterotrophic respiration in Mediterranean-type of ecosystems (Reichstein et al. 2002a).z

A large part of such losses of CO_2 (more than 40% of the GPP) occurred as heterotrophic respiration immediately after rain events at the end of the summer. In one case the amount of rainfall was too small to have any effect on plant metabolism but was enough to induce a large CO_2 evolution from soil respiration (M. Rayment, J. Banza, J.S. Pereira, J.S. David, unpubl. data). It is hypothesised that this occurs as re-wetting the soil surface determined a massive increase in soil microbial activity and a fast decomposition of stored litter and dead fine roots. There are indications of a dramatic increase in microbial activity and a large pulse in N mineralization in the soil under similar conditions, after end-of-summer rains in the same ecosystem in Portugal (C. Azevedo and M. Madeira, pers. comm.). Analysing field chamber measurements of soil respiration from 17 sites worldwide, most in Mediterranean-type ecosystems, Reichstein et al. (submitted) explained most of the inter-site variability using a simple model that included temperature, rainfall and LAI, the latter being a measure of plant productivity and the source of decomposable organic matter.

8 Conclusions

Carbon and water fluxes between the vegetation and the atmosphere in Mediterranean regions are severely limited by the lack of rainfall in summer when high potential evapotranspiration and high solar irradiance combine as stress factors. Most climate change scenarios suggest a future exacerbation of the severity of summer stress in these regions. The high diversity of taxa and life forms of the Mediterranean ecosystems implies a variety of adaptations to environmental constraints and resource capture. Plant functional types range from annual drought avoiders to drought-enduring sclerophyllous evergreen woody plants. Given the seasonal alternation from temperate-humid conditions in winter to a dry summer, the foliage sacrifices productivity and water use efficiency in summer in exchange for the opportunistic use of autumn rains which have an erratic start. On the other hand, winter growth is one of the characteristics of the herbaceous annuals, well exploited by farmers and agronomists for cereal and pasture cultivation and breeding.

In the dominant sclerophyllous woody plants, morphological adaptations to summer drought such as small leaf size combine with protection at the cellular and molecular level. When water deficits start to build up, a progressive stomatal closure towards the middle of the day is observed, reducing both transpiration and carbon assimilation. When summer constraints are prolonged, down-regulation of photosynthesis or even photoinhibition may occur, resulting in a depression in total ecosystem carbon uptake. On the other hand, to cope with severe water deficits, the maintenance of the integrity of the xylem conducting system is essential. Stomatal closure and the shedding of plant parts (leaves and fine roots) seem to be triggered when the runaway cavitation vulnerability threshold is approached, which varies with species and is lower (more negative water potentials) in Mediterranean trees than in mesophytic plants.

Maximum spatial and temporal heterogeneity in leaf area index and resource capture occur in the evergreen savannah-type woodlands where a temporary herbaceous stratum combines with a permanent but sparse tree cover. These scattered trees may capture an additional share of water (throughfall plus wind driven rainfall) as a result of rain interception under the upwind part of the crown and, possibly, channelling of water to deeper soil horizons through their root system. The access to ground water through deep rooting may allow trees to survive periods of drought when the physiological limits of tolerance would be reached without access to those stored water resources. This may also explain the unsuccessful reforestation attempts in years of average rainfall. On the other hand, complementarity in the use of water resources between strata (e.g. trees vs. shrubs) or within the herbaceous component may permit a high biological diversity.

Net ecosystem productivity (NEE), i.e. net primary productivity minus heterotrophic respiration, is a key parameter for carbon sequestration. In Mediterranean-type ecosystems NEE may be quite variable with vegetation (e.g., from 660 to 80 g C m^{-2} $year^{-1}$) and inter-annually as a function of precipitation and temperature variation. In addition to its dependence from net primary productivity, NEE is also quite dependent from soil respiration. Contrary to net primary productivity, spatial and temporal variation in heterotrophic (soil) respiration is not well known. In Mediterranean environments drought not only reduces canopy carbon assimilation but also strongly inhibits soil respiration, overruling temperature, as a major determinant for of soil respiration.

References

Archibold OW (1995) Ecology of world vegetation. Chapman and Hall, London

Arco JM, Escudero A, Vega Garrido M (1991) Effects of site characteristics on nitrogen retranslocation from senescing leaves. Ecology 72:701–708

Aschmann H (1973) Distribution and peculiarity of Mediterranean ecosystems. In: di Castri F, Mooney HA (eds) Mediterranean type ecosystems: origin and structure. Springer, Berlin Heidelberg New York, pp 11–19

Atkin OK, Evans JR, Ball MC, Lambers H, Pons TL (2000) Leaf respiration of snow gum in the light and dark. Interactions between temperature and irradiance. Plant Physiol 122:915–923

Barrantes O, Fernández R, Joffre R, Ortega F (1986) Influencia del arbolado sobre el pasto en las dehesas. I Congresso Florestal Nacional, 2 a 6 de Dezembro 1986; Fundação Calouste Gulbenkian, Lisboa, Portugal, pp 280–285

Bellot J, Àvila A, Rodrigo A (1999) Throughfall and stemflow. In: Roda F, Retana J, Gracia CA, Bellot J (eds) Ecology of mediterranean evergreen oak forests. Springer, Berlin Heidelberg New York, pp 209–222

Berry J, Björkman O (1980) Photosynthetic response and adaptation to temperature in higher plants. Annu Rev Plant Physiol 31:491–543

Bertness MD, Callaway R (1994) Positive interactions in communities. TREE 9:191–193

Beyschlag W, Eckstein J (2001) Towards a causal analysis of stomatal patchiness: the role of stomatal size variability and hydrological heterogeneity. Act Oecol 22:161–173

Beyschlag W, Pfanz H, Ryel RJ (1992) Stomatal patchiness in Mediterranean evergreen sclerophylls. Phenomenology and consequences for the interpretation of the midday depression in photosynthesis and transpiration. Planta 187:546–553

Bilger W, Bjorkman O (1994) Relationships among violaxanthin deepoxidation, thylakoid membrane conformation, and nonphotochemical chlorophyll fluorescence quenching in leaves of cotton (*Gossypium-Hirsutum* L). Planta 193:238–246

Blanco EM, González MAC, Tenorio MC, Bombín RE, Antón MG, Fuster MG, Manzaneque AG, Manzaneque FG, Saiz JCM, Juaristi CM, Pajares PR, Ollero HS (1997) Los bosques Ibéricos. Una interpretación geobotánica. Planeta, Barcelona

Breman H, Kessler JJ (1995) Woody plants in agro-ecosystems of semi-arid regions. Springer, Berlin Heidelberg New York

Brooks JR, Meinzer FC, Coulombe R, Gregg J (2002) Hydraulic redistribution of soil water during summer drought in two contrasting Pacific Northwest coniferous forests. Tree Physiol 22:1107–1117

Buckley TN, Mott KA (2002) Stomatal water relations and the control of hydraulic supply and demand. Progress in botany 63. Springer, Berlin Heidelberg New York, pp 309–325

Calabuig EL (1992) Bioclima. El Libro de las Dehesas Salamantinas. Junta de Castilla y León, Gráficas Ortega, Salamanca, Spain, pp 125–178

Caldeira MC, Ryel RJ, Lawton JH, Pereira JS (2001) Mechanisms of positive biodiversity-production relationships: insights provided by ^{13}C analysis in experimental Mediterranean grassland plots. Ecol Lett 4:439–443

Caldwell MM, Richards JH (1989) Hydraulic lift – water efflux from upper roots improves effectiveness of water uptake by deep roots. Oecologia 79:1–5

Canadell J, Jackson RB, Ehleringer JR, Mooney HA, Sala OE, Schulze ED (1996) Maximum rooting depth of vegetation types at the global scale. Oecologia 108:583–595

Canadell J, Djema A, López B, Lloret F, Sabaté S, Siscart D, Gracia CA (1999) Structure and dynamics of the root system. In: Roda F, Retana J, Gracia CA, Bellot J (eds) Ecology of mediterranean evergreen oak forests. Springer, Berlin Heidelberg New York, pp 47–59

Chaves MM (1991) Effects of water deficits on carbon assimilation. J Exp Bot 42:1–16

Chaves MM, Pereira JS, Maroco JP, Rodrigues ML, Ricardo CPP, Osorio ML, Carvalho I, Faria T, Pinheiro C (2002) How plants cope with water stress in the field. photosynthesis and growth. Ann Bot 89:907–916

Chaves MM, Maroco JP, Pereira JS (2003) Carbon assimilation and growth under drought – from genes to the whole plant. Funct Plant Biol 30:1–26

Cochard H, Breda N, Granier A (1996) Whole tree hydraulic conductance and water loss regulation in *Quercus* during drought: evidence for stomatal control of embolism? Ann Sci For 53:197–206

Comín MP, Escarré A, Gracia CA, Lledó MJ, Rabella R, Savé R, Terradas J (1987) Water use by *Quercus ilex* L. in forests near Barcelona, Spain. In: Tenhunen JD, Catarino F, Lange

OL, Oechel W (eds) Plant response to stress. Functional analysis in mediterranean ecosystems. Springer, Berlin Heidelberg New York, pp 259–266

Correia MJ, Chaves MM, Pereira JS (1990) Afternoon depression in photosynthesis in grapevine leaves-evidence for a high light stress effect. J Exp Bot 41:417–426

Correia MJ, Coelho D, David MM (2001) Response to seasonal drought in three cultivars of *Ceratonia siliqua*: leaf growth and water relations. Tree Physiol 21:645–653

Cowan IR (1982) Regulation of water use in relation to carbon gain in higher plants. In: Lange OL, Nobel PS, Osmond CB, Ziegler H (eds) Encyclopedia of plant physiology – physiological plant ecology II. Springer, Berlin Heidelberg New York, pp 589–613

Cruiziat P, Cochard H, Ameglio T (2002) Hydraulic architecture of trees: main concepts and results. Ann For Sci 59:723–752

Damesin C, Rambal S (1995) Field study of leaf photosynthetic performance by a Mediterranean deciduous oak tree (*Quercus pubescens*) during a severe summer drought. New Phytol 131:159–167

David JS, Henriques MO, David TS, Tomé J, Ledger DC (1994) Clearcutting effects on streamflow in coppiced *Eucalyptus globulus* stands in Portugal. J Hydrol 162:143–154

David TS (2000) Intercepção da precipitação e transpiração em árvores isoladas de *Quercus rotundifolia* Lam. (Rainfall interception and transpiration in isolated *Quercus rotundifolia* trees). Ph D, Universidade Técnica de Lisboa, Lisboa, Portugal

David TS, Ferreira MI, David JS, Pereira JS (1997) Transpiration from a mature *Eucalyptus globulus* plantation in Portugal during a spring-summer period of progressively higher water deficit. Oecologia 110:153–159

Davies WJ, Zhang J (1991) Root signals and the regulation of growth and development of plants in drying soil. Annu Rev Plant Physiol Plant Mol Biol 42:55–76

Dawson TE (1993) Hydraulic lift and plant water use: implications for water balance, performance and plant-plant interactions. Oecologia 95:565–574

Dawson TE (1996) Determining water use by trees and forests from isotopic, energy balance and transpiration analyses: the roles of tree size and hydraulic lift. Tree Physiol 16:263–272

Demmig-Adams B, Adams WW III (1992) Photoprotection and other responses of plants to high light stress. Annu Rev Plant Physiol Plant Mol Biol 43:599–626

Demmig-Adams B, Adams WW III (1996) The role of xanthophyll cycle carotenoids in the protection of photosynthesis. Trends Plant Sci 1:21–26

di Castri F (1981) Mediterranean-type shrublands of the world. In: di Castri F, Goodall DW, Specht RL (eds) Mediterranean-type shrublands. Elsevier, Amsterdam, pp 1–52

Epron D, Dreyer E, Breda N (1992) Photosynthesis of oak trees [*Quercus petraea* (Matt.) Liebl.] during drought under field conditions: diurnal course of net CO_2 assimilation and photochemical efficiency of photosystem II. Plant Cell Environ 15:809–820

Escudero A, Arco JM, Sanz IC, Ayala J (1992) Effects of leaf longevity and retranslocation efficiency on the retention time of nutrients in the leaf biomass of different woody species. Oecologia 90:80–87

Espigares T, Peco B (1995) Mediterranean annual pasture dynamics: impact of autumn drought. J Ecol 83:135–142

Faria T, Garcia-Plazaola JI, Abadia A, Cerasoli S, Pereira JS, Chaves MM (1996) Diurnal changes in photoprotective mechanisms in leaves of cork oak (*Quercus suber*) during summer. Tree Physiol 16:115–123

Faria T, Silvério D, Breia E, Cabral R, Abadia A, Abadia J, Pereira JS, Chaves MM (1998) Differences in the response of carbon assimilation to summer stress (water deficits, high light and temperature) in four Mediterranean tree species. Physiol Plant 102:419–428

Figueroa ME, Davy AJ (1991) Response of Mediterranean grassland species to changing rainfall. J Ecol 79:925–941

Fuentes ER, Montenegro G, Rundel PW, Arroyo MTK, Ginocchio R, Jaksic FM (1995) Functional approaches to biodiversity in the Mediterranean-type ecosystems of central

Chile. In: Davis GW, Richardson DM (eds) Mediterranean-type ecosystems-the function of biodiversity. Springer, Berlin Heidelberg New York, pp 185–232

Genty B, Briantais JM, Da Silva JBV (1987) Effects of drought on primary photosynthetic processes of cotton leaves. Plant Physiol 83:360–364

Gómez JA, Vanderliden K, Giráldez JV, Fereres E (2002) Rainfall concentration under olive trees. Agric Water Manage 55:53–70

González-Bernaldez F, Morey M, Velasco F (1969) Influences of *Quercus ilex* on the herb layer at the El Prado forest (Madrid). A multivariate approach to community structure, diversity and environmental factors. Bol Real Soc Española de Hist Nat (Biol) 67:265–284

Gracia CA, Sabaté S, Martinez JM, Albeza E (1999) Functional responses to thinning. In: Roda F, Retana J, Gracia CA, Bellot J (eds) Ecology of mediterranean evergreen oak forests. Springer, Berlin Heidelberg New York, pp 329–338

Hanson DT, Sharkey TD (2001) Effect of growth conditions on isoprene emission and other thermotolerance-enhancing compounds. Plant Cell Environ 24:929–936

Harley PC, Tenhunen JD, Beyschlag W, Lange OL (1987) Seasonal changes in net photosynthesis rates and photosynthetic capacity in leaves of *Cistus salvifolius*, a European Mediterranean semi-deciduous shrub. Oecologia 74:380–388

Hector A, Schmid B, Beierkuhnlein C, Caldeira MC, Diemer M, Dimitrakopoulos P, Finn JA, Freitas H, Giller PS, Good J, Harris R, Hochberg P, Huss-Danell K, Joshi J, Jumpponen A, Korner C, Leadley P, Loreau M, Minns A, Mulder CPH, O'Donovan G, Otway SJ, Pereira JS, Prinz A, Read DJ, Scherer-Lorenzen M, Schulze E-D, Siamantziouras A-SD, Spehn EM, Terry AC, Troumbis AY, Woodward FI, Yachi S, Lawton JH (1999) Plant diversity and productivity experiments in European grasslands. Science 286:1123–1127

Hobbs RJ, Mooney HA (1991) Effects of rainfall variability and gopher disturbance on serpentine annual grassland dynamics. Ecology 72:59–68

Hobbs RJ, Richardson DM, Davis GW (1995) Mediterranean-type ecosystems: opportunities and constraints for studying the function of biodiversity. In: Davis GW, Richardson DM (eds) Mediterranean-type ecosystems-the function of biodiversity. Springer, Berlin Heidelberg New York, pp 1–42

Hogg EH, Hurdle PA (1997) Sap flow in trembling aspen: implications for stomatal responses to vapor pressure deficit. Tree Physiol 17:501–509

Hulme M, Sheard N, Markham A (1999) Escenarios de cambio climático para la Península Ibérica. Climatic Research Unit, Norwich

Infante JM, Rambal S, Joffre R (1997) Modelling transpiration in holm-oak savannah trees: scaling up from the leaf to the canopy. Agric For Meteorol 87:273–289

Ishikawa CM, Bledsoe CS (2000) Seasonal and diurnal patterns of soil water potential in the rhizosphere of blue oaks: evidence of hydraulic lift. Oecologia 125:459–465

Jackson LE (1985) Ecological origins of California´s mediterranean grasses. J Biogeogr 12:349–361

Jackson PC, Meinzer FC, Bustamante M, Goldstein G, Franco A, Rundel PW, Caldas L, Igler E, Causin F (1999) Partitioning of soil water among tree species in a Brazilian Cerrado ecosystem. Tree Physiol 19:717–724

Jackson RB, Sperry JS, Dawson TE (2000) Root water uptake and transport: using physiological processes in global predictions. Trends Plant Sci 5:482–488

Jiménez MS, Cermák J, Kucera J, Morales D (1996) Laurel forests in Tenerife, Canary islands: the annual course of sap flow in *Laurus* trees and stand. J Hydrol 183:307–321

Joffre R, los Llanos C (1986) Systèmes d'élevage et mise en valeur du milieu: étude des dehesas de la Sierra Norte de Seville. 10 Encontro sobre montados de sobro e azinho; 1986; Évora, Portugal, pp 193–214

Joffre R, Rambal S (1988) Soil water improvement by trees in the rangelands of southern Spain. Act Oecol 9:405–422

Joffre R, Rambal S (1993) How tree cover influences the water balance of Mediterranean rangelands. Ecology 74:570–582

Jones HG, Sutherland RA (1991) Stomatal control of xylem embolism. Plant Cell Environ 14:607–612
King BP, Harrison SJ (1998) Throughfall patterns under an isolated oak tree. Weather 53:111–121
Ko LJ, Reich PB (1993) Oak trees effects on soil and herbaceous vegetation in savannas and pastures in Wisconsin. Am Midl Nat 130:31–42
Kremer A, Roussel G (1982) Composantes de la croissance en hauteur chez le Pin maritime (*Pinus pinaster* Ait.). Ann Sci For 39:77–97
Larcher W (2000) Temperature stress and survival ability of Mediterranean sclerophyllous plants. Plant Biosyst 134:279–295
Lauenroth WK, Aguilera MO (1998) Plant-plant interactions in grasses and grasslands. In: Cheplick GP (ed) Population biology of grasses. Cambridge University Press, Cambridge, pp 209–230
Leegood RC (1995) Effects of temperature on photosynthesis and photorespiration. In: Smirnoff N (ed) Environment and plant metabolism, flexibility and acclimation. Bioscientific, Oxford, pp 45–62
Lemoine D, Peltier J-PGM (2001) Comparative studies of the water relations and the hydraulic characteristics in *Fraxinus excelsior*, *Acer pseudoplatanus* and *A. opalus* trees under soil water contrasted conditions. Ann For Sci 58:723–731
Lewis DC (1968) Annual hydrologic response to watershed conversion from oak woodland to annual grassland. Water Resource Res 4:59–72
Lloret F, Siscart D (1995) Los efectos demográficos de la sequía en poblaciones de encina. Cuad Soc Española Cienc For 2:77–81
Llusia J, Penuelas J (2000) Seasonal patterns of terpene content and emission from seven Mediterranean woody species in field conditions. Am J Bot 87:133–140
Logan BA, Monson RK, Potosnak MJ (2000) Biochemistry and physiology of foliar isoprene production. Trends Plant Sci 5:477–481
Loreto F, Forster A, Durr M, Csiky O, Seufert G (1998) On the monoterpene emission under heat stress and on the increased thermotolerance of leaves of *Quercus ilex* L. fumigated with selected monoterpenes. Plant Cell Environ 21:101–107
Loustau D, Berbigier P, Roumagnac P, Arruda-Pacheco C, David JS, Ferreira MI, Pereira JS, Tavares R (1996) Transpiration of a 64-year-old maritime pine stand in Portugal. 1. Seasonal course of water flux through maritime pine. Oecologia 107:33–42
Marañon T (1985) Diversidad floristica y heterogeneidad ambiental en una dehesa de Sierra Morena. Anales de Edafologia y Agrobiologia, II. Biol Veg Agrobiol 44:1183–1197
Marañon T (1986) Plant species richness and canopy effect in the savanna-like dehesa of SW Spain. Ecol Mediterranea 12:131–141
Marañon T (1991) Diversidad en comunidades de pasto Mediterráneo: modelos y mecanismos de coexistencia. Ecologia 5:149–157
Maroco JP, Breia E, Faria T, Pereira JS, Chaves MM (2002) Effects of long-term exposure to elevated CO2 sand N fertilization on the development of photosynthetic capacity and biomass accumulation in *Quercus suber* L. Plant Cell Environ 25:105–113
Martinez F, Lazo YO, Fernandez-Galiano JM, Merino J (2002) Root respiration and associated costs in deciduous and evergreen species of *Quercus*. Plant Cell Environ 25:1271–1278
Martinéz-Vilalta J, Prat E, Oliveras I, Piñol J (2002) Xylem hydraulic properties of roots and stems of nine Mediterranean woody species. Oecologia 133:19–29
Médail F, Quézel P (1997) Hot-spots analysis for conservation of plant biodiversity in the Mediterranean Basin. Ann Mo Bot Gard 84:112–127
Merino J (1987) The costs of growing leaves of Mediterranean plants. In: Tenhunen JD, Catarino F, Lange OL, Oechel W (eds) Plant response to stress. Functional analysis in mediterranean ecosystems. Springer, Berlin Heidelberg New York, pp 553–564

Merino J, Field C, Mooney HA (1982) Construction and maintenance costs of Mediterranean-climate evergreen and deciduous leaves. I. Growth and CO_2 carbon dioxide exchange analysis. Oecologia 53:208–213

Merino J, Field C, Mooney HA (1984) Construction and maintenance costs of Mediterranean-climate evergreen and deciduous leaves. II. Biochemical pathway analysis. Act Oecol 5:211–229

Miranda P, Coelho FES, Tomé AR, Valente MA (2002) 20th century Portuguese climate and climate scenarios. In: Santos FD, Forbes K, Moita R (eds) Climate change in Portugal. Scenarios, impacts, and adaptation measures. Gradiva, Lisboa, Portugal, pp 25–83

Mooney HA, Harrison AT, Morrow PA (1975) Environmental limitations of photosynthesis on a California evergreen shrub (*Heteromeles arbutifolia*). Oecologia 19:293–301

Morgan JM (1984) Osmoregulation and water stress in higher plants. Annu Rev Plant Physiol Plant Mol Biol 35:299–319

Myneni RB, Keeling CD, Tucker CJ, Asrar G, Nemani RR (1997) Increased plant growth in the northern high latitudes from 1981 to 1991. Nature 386:698–701

Naeem S, Thompson LJ, Lawler SP, Lawton JH, Woodfin RM (1994) Declining biodiversity can alter the performance of ecosystems. Nature 368:734–737

Nardini A, Tyree MT (1999) Root and shoot hydraulic conductance in *Quercus* species. Ann For Sci 56:371–377

Nardini A, Lo Gullo MA, Salleo S (1999) Competitive strategies for water availability in two Mediterranean *Quercus* species. Plant Cell Environ 22:109–116

Osborne CP, Chuine I, Viner D, Woodward FI (2000) Olive phenology as a sensitive indicator of future climatic warming in the Mediterranean. Plant Cell Environ 23:701–710

Paulo M, Tavares M (1996) Policiclismo do crescimento do pinheiro bravo das dunas do litoral Português. Silva Lusitana 4:25–38

Pereira JS, Tenhunen JD, Lange OL, Beyschlag W, Meyer A, David MM (1986) Seasonal and diurnal Patterns in leaf gas-exchange of *Eucalyptus-globulus* trees growing in Portugal. Can J For Res 16:177–184

Pereira JS, Beyschlag G, Lange OL, Beyschlag W, Tenhunen JD (1987) Comparative phenology of four Mediterranean shrub species growing in Portugal. In: Tenhunen JD, Catarino FM, Lange OL, Oechel WC (eds) Plant response to stress. Springer, Berlin Heidelberg New York, pp 503–514

Pereira JS, Correia AV, Correia AP, Branco M, Bugalho M, Caldeira MC, Souto-Cruz C, Freitas H, Oliveira AC, Pereira JMC, Reis RM, Vasconcelos MJ (2002) Forests and biodiversity. In: Santos FD, Forbes K, Moita R (eds) Climate change in Portugal. Scenarios, impacts and adaptation measures. Gradiva, Lisboa, Portugal, pp 363–414

Perks MP, Irvine J, Grace J (2002) Canopy stomatal conductance and xylem sap abscisic acid (ABA) in mature Scots pine during a gradually imposed drought. Tree Physiol 22:877–884

Phillips JG, Riha SJ (1994) Root growth, water uptake and canopy development in *Eucalyptus viminalis* seedlings. Aust J Plant Physiol 21:69–78

Pitt MD, Heady HF (1978) Responses of annual vegetation to temperature and rainfall patterns in northern California. Ecology 59:336–350

Poole DK, Roberts SW, Miller PC (1981) Water utilization. In: Miller PC (ed) Resource use by chaparral and matorral. Springer, Berlin Heidelberg New York, pp 123–149

Quick WP, Chaves MM, Wendler R, David M, Rodrigues ML, Passaharinho JA, Pereira JS, Adcock MD, Leegood RC, Stitt M (1992) The effect of water stress on photosynthetic carbon metabolism in four species grown under field conditions. Plant Cell Environ 15:25–35

Rambal S (2001) Hierarchy and productivity of Mediterranean-type ecosystems. In: Roy J, Saugier B, Mooney HA (eds) Terrestrial global productivity. Academic Press, New York, pp 315–344

Raven JA (2002) Selection pressures on stomatal evolution. New Phytol 153:371–386

Reichstein M, Tenhunen JD, Roupsard O, Ourcival JM, Rambal S, Dore S, Valentini R (2002a) Ecosystem respiration in two Mediterranean evergreen Holm Oak forests: drought effects and decomposition dynamics. Funct Ecol 16:27–39

Reichstein M, Tenhunen JD, Roupsard O, Ourcival JM, Rambal S, Miglietta F, Peressotti A, Pecchiari M, Tirone G, Valentini R (2002b) Severe drought effects on ecosystem CO_2 and H_2O fluxes at three Mediterranean evergreen sites: revision of current hypotheses? Global Change Biol 8:999–1017

Roda F, Retana J, Gracia CA, Bellot J (1999) Ecology of Mediterranean oak forests. Springer, Berlin Heidelberg New York

Rodrigues ML, Chaves MM, Wendler R, David MM, Quick WP, Leegood RC, Stitt M, Pereira JS (1993) Osmotic adjustment in water stressed grapevine leaves in relation to carbon assimilation. Aust J Plant Physiol 20:309–321

Roy J (2001) How does biodiversity control primary productivity. In: Roy J, Saugier B, Mooney HA (eds) Terrestrial global productivity. Academic Press, San Diego, pp 169–186

Sá CMMSS (2001) Influência do coberto arbóreo (*Quercus suber* L.) em processos ecofisiológicos da vegetação herbácea em áreas de montado. PhD, University of Évora, Évora, Portugal

Sabaté S, Sala A, Gracia CA (1999) Leaf traits and canopy organization. In: Roda F, Retana J, Gracia CA, Bellot J (eds) Ecology of mediterranean oak forests. Springer, Berlin Heidelberg New York, pp 121–133

Sala A (1999) Modelling canopy gas exchange during summer drought. In: Roda F, Retana J, Gracia CA, Bellot J (eds) Ecology of mediterranean oak forests. Springer, Berlin Heidelberg New York, pp 149–161

Sala A, Tenhunen JD (1994) Site-specific water relations and stomatal response of *Quercus ilex* in a Mediterranean watershed. Tree Physiol 14:601–617

Salleo S, Nardini A, Pitt F, Lo Gullo MA (2000) Xylem cavitation and hydraulic control of stomatal conductance in Laurel (*Laurus nobilis* L.). Plant Cell Environ 23:71–79

Schulze ED, Hall AE (1982) Stomatal responses, water loss and CO_2 carbon dioxide assimilation rates of plants in contrasting environments. In: Lange OL, Nobel PS, Osmond CB, Ziegler H (eds) Encyclopedia of plant physiology – physiological plant ecology II. Springer, Berlin Heidelberg New York, pp 181–230

Schulze ED, Caldwell MM, Canadell J, Mooney HA, Jackson RB, Parson D, Scholes R, Sala OE, Trimborn P (1998) Downward flux of water through roots (i.e. inverse hydraulic lift) in dry Kalahari sands. Oecologia 115:460–462

Sharkey TD, Chen X, Yeh S (2001) Isoprene increases thermotolerance of fosmidomycin-fed leaves. Plant Physiol 125:2001–2006

Silva P, Nunes T, Campos C, Mariz M, Pio C (1999) Emissões de compostos orgânicos voláteis pela floresta de sobreiro em Portugal. 6th Conferência Nacional sobre a Qualidade do Ambiente; 1999 20 a 22 de Outubro; Lisboa, Portugal, pp 627–637

Singsaas EL, Sharkey TD (2000) The effects of high temperature on isoprene synthesis in oak leaves. Plant Cell Environ 23:751–757

Sperry JS (2000) Hydraulic constraints on plant gas exchange. Agric For Meteorol 104:13–23

Staudt M, Mandl N, Joffre R, Rambal S (2001) Intraspecific variability of monoterpene composition emitted by *Quercus ilex* leaves. Can J For Res 31:174–180

Sundby C, Melis A, Maenpaa P, Andersson B (1986) Temperature-dependent changes in the antenna size of photosystem-II – reversible conversion of photosystem-II-alpha to photosystem-II-Beta. Biochim Biophys Act 851:475–483

Tardieu F, Simonneau T (1998) Variability among species of stomatal control under fluctuating soil water status and evaporative demand: modelling isohydric and anisohydric behaviours. J Exp Bot 49:419–432

Tilman D (1997) Distinguishing between the effects of species diversity and species composition. Oikos 80:185

Tognetti R, Longobucco A, Raschi A (1998) Vulnerability of xylem to embolism in relation to plant hydraulic resistance in *Quercus pubescens* and *Quercus ilex* co-occurring in a Mediterranean coppice stand in central Italy. New Phytol 139:437–447

Tyree MT, Cochard H (1996) Summer and winter embolism in oak: impact on water relations. Ann Sci For 53:173–180

Tyree MT, Sperry JS (1989) Vulnerability of xylem to cavitation and embolism. Annu Rev Plant Physiol Mol Biol 40:19–38

Valente F, David JS, Gash JHC (1997) Modelling interception loss for two sparse eucalypt and pine forests in central Portugal using the reformulated Rutter and Gash analytical models. J Hydrol 190:141–162

Valentini R, Matteucci G, Dolman AJ, Schulze ED, Rebmann C, Moors EJ, Granier A, Gross P, Jensen NO, Pilegaard K, Lindroth A, Grelle A, Bernhofer C, Grunwald T, Aubinet M, Ceulemans R, Kowalski AS, Vesala T, Rannik U, Berbigier P, Loustau D, Gudmundsson J, Thorgeirsson H, Ibrom A, Morgenstern K, Clement R, Moncrieff J, Montagnani L, Minerbi S, Jarvis PG (2000) Respiration as the main determinant of carbon balance in European forests. Nature 404:861–865

Verdaguer D, Casero P, Molinas M (2000) Lateral root development in a woody plant, *Quercus suber* L (cork oak). Can J Bot 78: 1125–1135

Villar R, Merino J (2001) Comparison of leaf construction costs in woody species with differing leaf life-spans in contrasting ecosystems. New Phytol 151:213–226

Walter H (1973) Vegetation of the earth. Springer, Berlin Heidelberg New York

Werner C, Correia O, Beyschlag W (1999) Two different strategies of Mediterranean macchia plants to avoid photoinhibitory damage by excessive radiation levels during summer drought. Act Oecol 20:15–23

Werner C, Ryel RJ, Correia O, Beyschlag W (2001) Effects of photoinhibition on whole-plant carbon gain assessed with a photosynthesis model. Plant Cell Environ 24:27–40

Werner C, Correia O, Beyschlag W (2002) Characteristic patterns of chronic and dynamic photoinhibition of different functional groups in a Mediterranean ecosystem. Funct Plant Biol 29:999–1011

White DA, Turner NC, Galbraith JH (2000) Leaf water relations and stomatal behavior of four allopatric *Eucalyptus* species planted in Mediterranean southwestern Australia. Tree Physiol 20:1157–1165

Wilkinson S, Davies WJ (2002) ABA-based chemical signalling: the co-ordination of responses to stress in plants. Plant Cell Environ 25:195–210

Wilson JR, Ludlow MM, Fisher MJ, Schulze ED (1980) Adaptation to water stress of the leaf water relations of four tropical forage species *Panicum maximum* var. trichoglume, *Heteropogon contortus, Cenchrus ciliaris, Macroptilium atropurpureum.* Aust J Plant Physiol 7:207–220

Zimmermann MH (1983) Xylem structure and the ascent of sap. Springer, Berlin Heidelberg New York

J.S. Pereira
J.S. David
M.C. Caldeira
M.M. Chaves
Instituto Superior de Agronomia
Tapada da Ajuda
1349-017 Lisboa, Portugal
Tel.:+351-21-3653483
Fax: +351-21-3645000
e-mail: jspereira@isa.utl.pt

T.S. David
Estação Florestal Nacional (INIAP)
Quinta do Marquês
2780-159 Oeiras, Portugal

Plants and Geothermal CO_2 Exhalations – Survival in and Adaptation to a High CO_2 Environment

Hardy Pfanz, Dominik Vodnik, Christiane Wittmann, Guido Aschan, and Antonio Raschi

1 Introduction – Analyzing Plant Life at the Extremes of CO_2

Modern plants live in a rather "low CO_2"- world when compared to CO_2 concentrations in the atmosphere during prehistoric evolution (Petit et al. 1999). Similar to Mars and Venus, CO_2 on planet Earth might have been as high as 90–98% during the early days of photosynthetic evolution (Emiliani 1992; Raven 1995; Grace and van Gardingen 1997). But concentrations decreased gradually during epochs to reach only a few hundred ppm, although [CO_2][1] is again steadily increasing since the last 200 years (Bowes 1993). Values nowadays are dramatically lower by a factor of 3,000 than in those ancient times. Nevertheless, a further increase from presently 360 ppm (0.036% w/v) to ca. 700 ppm (0.07% w/v) is thought to take place within the current century (WMO 1990; Bowes 1993; IPCC 1996). Predictions for the second half of the present century range from 415 to 575 ppm depending on a CO_2 emission rate of ±2% (Houghton et al. 1990; Cook et al. 1997). During evolution, plants had to cope with and adapt to a slowly but permanently changing CO_2 environment including periods with increasing and others with decreasing CO_2. Although ambient CO_2 does nowadays not saturate C_3 photosynthesis, plants have evolved mechanisms to rather effectively capture and photo-reduce the oxidised carbon to the level of carbohydrates. Yet, short-term fumigation at elevated CO_2 increases leaf net photosynthesis rates (at least transiently) of nearly all plants studied so far (Constable et al. 1992; Bowes 1993; Drake et al. 1997; for poikilohydric plants see Tuba et al. 1998, 1999) and CO_2 fertilisation consequently has been used for a long time by horticulturists in greenhouse production to increase productivity and yield (Schwiebert 1985; Enoch and Kimball 1986). But even in natural environments plants are confronted with carbon dioxide levels well above ambient. Forest floor plants growing on humus-rich soils are thought to be transiently confronted with [CO_2] up to 1.5%

[1] [CO_2] means concentration of carbon dioxide

Progress in Botany, Vol. 65

(Larcher 2001). In the roots and rhizomes of (semi-)aquatic swamp plants intercellular CO_2 concentrations may even reach 4% (Brix 1988a,b; Constable et al. 1992). Elevated carbon dioxide concentrations above 600 ppm have been found at night in fruit tree plantations (Blanke and Holthe 1997), in alfalfa crops (Raschi unpubl.) and only marginal lower values were found within natural forests (e.g. Fan et al. 1990; Bazzaz and Wayne 1994). Within tree stems and branches CO_2 concentrations vary between 1–26.3% (for recent reviews see Pfanz and Aschan 2001; Pfanz et al. 2002).

Predictions on the plant's behaviour to an increasing CO_2 environment are difficult to verify. Fumigation of plants in controlled-environment facilities (climatic chambers), OTCs (open top chambers) or FACE (free-air carbon dioxide enrichment) systems are only short termed with a maximum length of not more than one decade. Areas with naturally elevated carbon dioxide are therefore elegant model ecosystems (although they are not ideal in any respect; see below) where long-term adaptation (acclimation) to CO_2 enrichment can be studied. Ecophysiological work in mofette[1] areas thus simulates atmospheric changes in the forthcoming decades and therefore allows us to get a glimpse of the probable future life on our planet (Miglietta et al. 1993; Cook et al.1997). Depending on the site and the cultivation state of the CO_2 springs, plants may have had the chance to adapt to the prevailing CO_2-enriched atmospheric gas regime over decades, centuries or even millennia.

2 Mofettes – Natural CO_2 Springs

Similar to solfatares (sulphur-emitting vents) and fumaroles (water vapour emitting vents), mofettes (carbon dioxide springs) occur at sites of former volcanic action. Several regions have been identified where the soil CO_2 concentration or the CO_2 efflux from the soil is anomalous. The normal, ambient CO_2 concentration within the different soil layers does not exceed 0.2–1% (v/v) (Brook et al. 1983; Kiefer and Amey 1992), whereas in mofette areas the CO_2 concentration within the different soil horizons may be as high as 100% (Pfanz et al., unpublished results; Vodnik et al. 2001, 2002a).

In Europe, mofettes have been reported to exist in W Germany, Iceland, NE Slovenia, E Czechia, Hungary, Romania, Spain, Greece, Portugal, Aus-

[1] Depending on the literature studied (biological, geological), several termini are synonymously used for an enhanced CO_2 gas flux from the soil to the atmosphere: mofette, CO_2 spring, CO_2 vent, CO_2 anomaly, mineral CO_2 spring, geothermal CO_2 exhalations, natural CO_2 spring

tria and central France and in large numbers in central (Panichi and Tongiorgi 1975; Raschi et al. 1997, 1999) and S Italy (Baubron et al. 1990). In those regions CO_2 of deep mantle origin is released into the atmosphere by mostly small gas vents (hole size ca. 100–400 cm^2) although even bigger ones exist on active volcanoes.

The distribution of the vents and the gas-piping cracks within the soil and the parent rock are spread irregularly (see Sect. 2.4). During wet periods and/or directly after rain events, the gas escaping from the smaller holes can directly be spotted by the bubbles which are created at the soil surface, whereas during dry periods only a faint whistling noise may be heard. Although, when compared to other potentially toxic gases, the toxicity threshold of CO_2 is rather high (HCl 0.1%; H_2S 0.1%; SO_2 0.04%; CO 0.4%; HF 0.015%; Baxter and Kapila 1989; Martini 1997), carbon dioxide can clearly cause fatal harm to organisms. From 5% onward CO_2 can cause irritations. Humans can not breathe air containing more than 10% CO_2 without loosing consciousness. From 30% onward CO_2 leads to hypoxia which is accompanied by rapid breathing, dizziness, visual disturbances, rapid heart rate and sometimes death (Stupfel and Le Guern 1989 and references therein). Dead bodies of birds and mammals, reptiles, and insects in the direct neighbourhood of mofettes therefore hint to extremes in carbon dioxide. At distinct sites human life is endangered by the escaping gas, when volcanic sweet water lakes suddenly release huge amounts of carbon dioxide after a year-long over-saturation of the water by submerged gas vents. In Cameroon nearly 2000 local people and their live-stock died in 1984 and 1986 within several hours by a limnic CO_2 eruption (lake Nyos, lake Monoun: Stager 1987; Sigurdsson et al. 1986).

2.1 Different Types of Mofettes

Natural CO_2 springs can occur in flat regions like arable farmlands or meadows (Stavešinci, Slovenia), in swamps and bogs (Soos, Czech Republic), in hilly terrain (Eifel Mountains) or in valley-like carstic dolinas (Il Bossoleto, Italy). Principally they occur along geological disturbances (see Weinlich et al. 1999; Rogie et al. 2001; Weise et al. 2001). Besides forming dry mofettes, the escaping CO_2 gas may also pass aqueous solutions (water tables, creeks, rivers or lakes, see Laacher See, Eifel, FRG; river Ščavnica, SLO or river Ambra, and lake Monticchio, southern Italy) leading to acidic springs (mineral water). In an aqueous environment water plants and the water fauna may be affected.

2.2 Where Does the CO_2 Originate?

Carbonic rocks of deep mantle origin melt because of high temperatures and high pressure within the earth's crust. Carbon dioxide and several other gases (e.g., H_2S, CH_4, Rn) are freed and then follow the path of the least resistance according the physical law of diffusion along pre-existing concentration gradients. In the well-known mofette field of Stavešinci (NE Slovenia) $\delta^{13}C$ values ranged around –3.9‰ (Vodnik et al. 2002a, Pezdiĉ et al. 1998), while $\delta^{13}C$ of the normal air is reported to range from –7 to –9‰ (Mook 1986). Values of $\delta^{13}C$ measured for the natural carbon dioxide spring of Stavešinci are comparable to similar sources of CO_2 measured in some other places in Slovenia, e. g., Rogaška Slatina with –5.2‰ (Pezdiĉ et al. 1995). A similar $\delta^{13}C$ enrichment was found also in many Italian CO_2-rich mineral springs (Panichi and Tongiorgi 1975; Chiodini et al. 1995; Etiope 1997; Raschi et al. 1997; Miglietta et al. 1998a; Badiani et al. 2000; Rogie et al. 2000).

2.3 How Pure Is the Gas and How High Are Exhalation Fluxes?

In mofettes, the gas escapes from the vents generally at ambient temperature conditions and it is normally composed of CO_2 (50–99.9%), nitrogen (5–10%), methane (up to 3.2%) and eventually traces of sulphurous compounds (H_2S and/or SO_2) (see Duchi et al.1985; Pfanz and Vodnik, unpubl. data). Once above-ground, the highly concentrated gas is either rapidly diluted in turbulent air or it transiently forms CO_2 gradients or even gaseous CO_2 lakes persisting on the ground at least for several hours (Rogie et al. 2000).

It is still very hard to estimate the non-volcanic lithospheric degassing into the atmosphere, as, so far, only volcanic emissions have been studied in detail and have been taken in account in global C budget analyses. According to the most recent inventories, the lithosphere may emit at least 600 Mt CO_2/y (Mörner and Etiope 2002), while subaerial volcanoes may release about 300 Mt CO_2/y. In mofettes of North America daily fluxes of CO_2 from the soil range from 100 to 30,000 g CO_2 d^{-1} m^{-2}. The normal efflux is given as 25 g CO_2 d^{-1} m^{-2} (Farrar et al. 1999). For the whole region of the Mammoth Mountains, Farrar et al. (1999) calculate 530 Mg d^{-1} (=530×10^6 g d^{-1}). For Mount Aetna (Sicily) 10^{12} mol CO_2 y^{-1} are given and for Mount Kilauea and Mount St. Helens still $0.3–0.4\times10^{12}$ mol CO_2 y^{-1} have been published (Gerlach 1991). Pre-industrial CO_2 degassing was calculated to be $6–7\times10^{12}$ mol CO_2 y^{-1} but has nowadays nearly doubled (11×10^{12} mol CO_2 y^{-1}; Berner 1990; Gerlach 1991).

2.4 Spatial Distribution of CO_2 Within a Mofette Field

The geogenic CO_2 is not equally released from the soil. Resistance to penetrating gas is dependent on disturbances and irregularities within the parent rock and the overlaying soil layers. In some cases the gas just follows cracks in the soil. Figure 1 shows the distribution of soil CO_2 within the rooting zones of the herbaceous plants of a meadow as measured at 20 cm depth (Strmec, Stavešinci, Slo). At this site clearly three main vents can be distinguished with CO_2 concentrations gradually fading with increasing distance from the vents. CO_2 concentrations in the bulk air surrounding the plant tufts were difficult to determine because of the prevailing winds blowing with different speeds and rapidly changing directions. Therefore measurements were performed only during calm days. Measurements using a porometer, a portable gas analyser (landfill gas analyser), or Dräger diffusion tubes (cf. Sestak et al. 1971) indicated that directly at the soil surface CO_2 concentrations can be in the percentage range (see also Meister et al. 1999; Turk et al. 2002; Raschi unpublished). At a height of 10–50 cm, CO_2 concentrations higher than 10.000 ppm may still occur within the plant

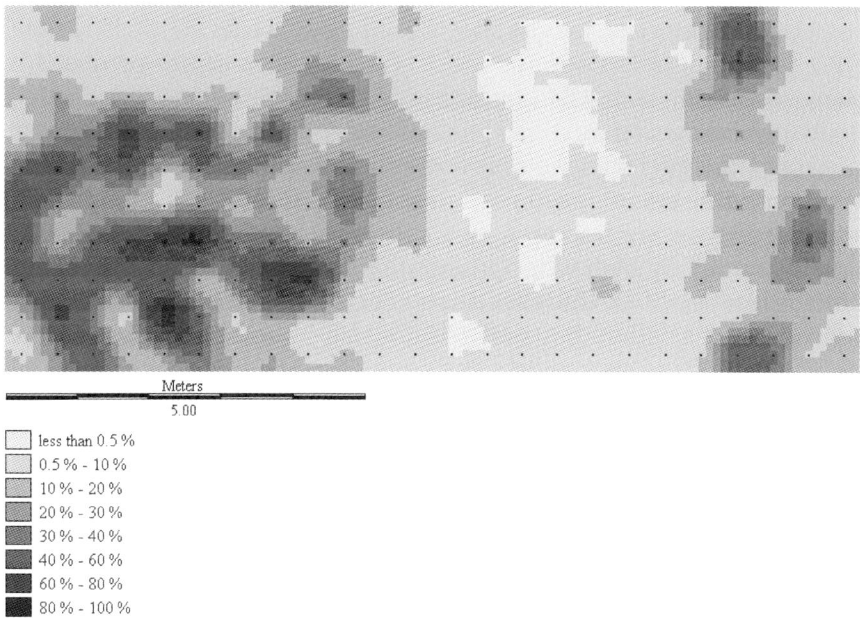

Fig. 1. Spatial pattern of soil CO_2 concentration measured at 20 cm depth by using the GA 2000 (Ansyco) infrared gas analyser. Permanent research plot in Stavešinci, density of sampling was four samples per square metre

canopy. At 50 cm, 3,000–4,000 ppm CO_2 were measured during a calm morning. In no case concentrations higher than 360–500 ppm were measured at a height of 1.5–2 m. Measurements of daily fluctuations and vertical distributions of atmospheric CO_2 have been performed in Italy at Solfatara (Miglietta and Raschi 1993) and at Il Bossoleto (van Gardingen et al. 1995). The vents were either located along a stream, the river bed being about 4 m lower than the surrounding fields (Solfatara) or in a 6-m-deep circular dolina (Il Bossoleto). During calm days large differences among the spots were evidenced at Solfatara; above the bigger emission vents CO_2 concentrations higher than 3,000 ppm were measured even at a height of 4.5 m. At Bossoleto, concentrations as high as 70% were recorded at the soil level early in the morning, decreasing to about 2,000 ppm during daytime; at 3–4 m height, daytime concentrations ranged between 500 and 800 ppm (Chaves et al. 1995).

2.5 Microclimatic Change at a Mofette Site

To estimate the microscalic influence of the degassing CO_2 on the local micro-environment we simultaneously compared the microclimatic factors in a centre of two main vents with the conditions at the edge of the respective mofette site. In both sites alterations of the temperature regime, VPD and evaporation were assessed (Table 1). The centre of *vent 1* is characterized by a drastic increase of air and soil temperatures and a doubling of evaporation, while the alterations in *vent 2* are less significant. Generally, the microclimate near the ground is influenced by the canopy structure of the predominant vegetation (e.g., Jones 1992). In both mofette sites the centres are indicated by a clear reduction in plant height and density, caused by high CO_2 concentrations in the air and in the rooting horizon (see Sect. 5.2). Thus the centres of both main vents are covered with sparse, low vegetation, but *vent 1* is situated in an abundantly covered ruderalized area consisting of some grass species and herbaceous plants including perennial neophytes. In contrast, *vent 2* is localised in a sparse meadow mainly composed of different Poaceae. Due to the higher biomass density around the edge of *vent 1* the microclimatic change between the centre and the edge is more pronounced there. Elevated [CO_2] in the soil and air of a mofette centre alters plant allocation and growth, which affects community structure and therefore modifies microclimate as well as local soil properties. The complex interactions make it necessary to take into account these combinations of single factors in order to make consequential statements about the microscalic effects of elevated CO_2 concentrations in the vicinity of a mofette.

Table 1. Alterations of microclimatic factors close to two different main vents at the geothermal mofette field Strmec near Stavešinci (vent 1: 19–21 May 2002; vent 2: 22–24 May 2002). Comparative measurements were performed in the centre (high CO_2) and in the edge (low CO_2) of a mofette, distance about 5 m. Values are expressed as percentage difference of the respective low CO_2 – site (100%)

		Air temperature (0.5 m)	Air temperature (0.1 m)	Temperature at soil surface	Soil temperature (−0.1 m)	VPD	Evaporation
Vent 1	Daily mean	+6.3	+4.7	+26.6	+12.2	+6.5	
	Daytime mean	+7.8	+7.0	+35.2	+13.2	+7.2	+50
	Maximum	+1.5	+4.0	+59.0	+12.5	−11.9	
	Minimum	+3.4	−11.6	+5.1	+8.0	+33.9	
Vent 2	Daily mean	+0.2	+5.3	+6.4	+2.8	−6.6	
	Daytime mean	+0.8	+6.4	+8.9	+0.6	−5.8	+14
	Maximum	−0.8	+7.7	+2.7	−2.7	−4.2	
	Minimum	−16.7	+1.7	−33.0	+8.0	+48.0	

3 Effects on Vegetation

3.1 Do Plant Populations at Mofettes Consist of Highly Adapted "High CO_2" Species?

The question of whether specialised "mofette plants" exist in the regions of natural carbon dioxide springs has often been put forward (von Faber 1925; Poli 1970). Although the answer is still open, it seems to be clear that some species/subspecies exclusively exist near genuine mofettes or mofettes/solfatares (CO_2 springs with impurities in gaseous sulphur compounds). In such a mofette system in a river bed near Viterbo (Solfatara, Italy) specialised sub-species of *Agrostis* and *Scirpus* occur. The river valley is nearly devoid of other plants, only the 4-m higher rims of the adjacent banks carry trees. Within the valley, *Scirpus lacustris* (or *Schoenoplectus tabernaemontanum* ssp. *lacustris)* and *Agrostis canina* ssp. *monteluccii* (cf. Fordham et al. 1997a,b) occur in large stands.

In principal, mofette areas are characterised by paucispecific plant communities, which occur in typical patterns around the gas vents (Selvi and Bettarini 1999). The growth pattern of those distinct communities is mainly determined by the distance to the gas vents and thus by the pre-existing [CO_2] gradient (in the soil and/or in the air). Studying the composition of grass communities close to natural mofette areas in central Italy, Selvi (1997) reports on an azonal endemic association that surrounds the gas vents, by *Agrostidetum caninae* ssp. *monteluccii.* This association belongs to the geothermal acidophilic vegetation types and is characterised by few plant species and a high ecological specialisation. Cook et al. (1997) made an inventory of vascular plants at a mofette on Iceland (near Olafsvik) and found 49 different herbaceous species (and six mosses). Mostly the plants belonged to communities of perennial pasture vegetation, some of them indicated disturbance by man and livestock. Most abundant was *Nardus stricta*, where distinct ecotypes of reproductively isolated genotypes, probably adapted to a CO_2-enriched environment, were described.

Also Selvi and Bettarini (1999) give a list of plant species growing on several mofette sites in central Italy. They compare the floristic composition of solfataras (springs with a relatively high hydrogen sulphide emission), mofettes on non-acid soils (parent rock of calcareous nature) and so-called soffioni ("soffioni boraciferi"; thermally hot emissions of H_2S, CO_2 and boric acid, but primarily water vapour). Also here several trees, shrubs and herbaceous species were found. The interesting point of all these studies was the finding that the genus *Agrostis* plays a major role in colonising geothermal biotopes in Mediterranean regions. Three different species were found from which *A. canina ssp. monteluccii* was found on hyperacid soils (with a high aluminium content), whereas *A. castellana* grew mainly on acidic soils and tolerated strong soil heating. The third species, *A. stolonifera* was found to be especially resistant to soil anoxia, to the presence of inorganic salts and to leaf heating (Selvi and Bettarini 1999).

As hypoxia and even anoxia may occur in soils near CO_2 vents also many "flooding-tolerant" plant species exist in mofette areas. Rather common in mofette fields are *Phragmites australis, Schoenoplectus lacustris, Scirpus* spec., *Juncus effusus,* and *Agrostis stolonifera* (Selvi and Bettarini 1999); these species are, according to Crawford and Braendle (1996), medium or highly tolerant to oxygen deprivation.

3.2 Phenological Aspects

Reported phenological responses concerning a potential doubling of the [CO_2] are earlier flowering, an accelerated fruit ripening and senescence and a shortened growth period (Krupa and Kickert 1993; Manning and von Tiedemann 1995; for a recent analysis see Edwards et al. 2003). Miglietta et al. (1998b) demonstrated that potato flowering was progressively shifted when 460, 560 and 660 μmol mol^{-1} CO_2 were used instead of ambient air. Yet, several studies revealed, that the phenological responses to [CO_2] are much weaker than expected (Ewert and Pleijel 1999; Wagner et al. 2001). In a FACE experiment on a semi-natural grassland (Raschi unpublished), an accelerated flowering was evidenced only in annual species, while fumigation had no effect on the phenology of perennial species.

Cook et al. (1998) monitored leaf senescence of *Nardus stricta* growing near a cold CO_2 spring in Iceland. High CO_2 exposed plants (estimated mean [CO_2] ca. 790 μmol mol^{-1}) revealed an accelerated senescence and became dormant much earlier than plants growing apart from the spring. Recent data on *Plantago lanceolata* hint in the same direction (Edwards et al. 2003). On the other hand, no acceleration of flowering by elevated CO_2 occurred in different natural CO_2 spring species studied by Körner and Miglietta (1994) and in the Stavešinci mofette (SLO) a delayed flowering or even no flowering is frequently observed in plants growing at [CO_2] extremes (Kaligariĉ 2001; Pfanz and Vodnik unpublished). Yet, according to Cook's results, leaf senescence is accelerated in different plant species growing under high [CO_2] at the Stavešinci mofette. Several studies showed that CO_2 effects on phenology are related to nutrient availability (Sigurdsson 2001) and that an accelerated senescence can be overcome by fertilisation (see below). Since as a rule, a decrease in leaf nitrogen and an increase of C:N ratio can be found in mofette plants (see Sect. 4.2), nutrient imbalance could be an important signal for an altered phenological behaviour.

Not only senescence but also other developmental processes are influenced by an increased C:N ratio. It was shown that seed germination decreases under CO_2 exposure in response to a C:N increase (Andalo et al. 1996). In other species germination could be directly stimulated by elevated CO_2 (for *E. crus-galli* see Yoshioka et al. 1998). This could also explain the presence of germinating and growing *Echinochloa* plants at the sites with extreme CO_2 concentrations in the natural CO_2 spring Stavešinci (Kaligarič 2001).

4 Influences of Enhanced CO_2 on Below-Ground Plant Parts

In natural CO_2 springs, CO_2 may not solely influence above-ground plant parts (see Sect. 5), but may also affect below-ground roots, rhizomes and other storage organs. Aside from direct effects of gaseous and dissolved CO_2 on roots, the lowered O_2 partial pressure may have deleterious effects on root growth and metabolism.

4.1 Hypoxia and Anoxia – A Special Threat for Mofette Plants?

Oxygen depletion in soil is a rather common threat to plant roots. In waterlogged soils oxygen diffusion is too slow and often plant roots get hypoxic or even anaerobic (anoxic) for a certain time. Under these circumstances the roots cannot obtain enough oxygen for respiration and react with fermentative processes (Crawford and Braendle 1996). If the O_2 partial pressure decreases below the threshold of 1–5 kPa (hypoxia) the respiratory quotient exceeds 1 (Larcher 2001). Thus, ethanol and lactic acid will be formed, finally destroying root compartmentation and root function (Ponnamperuma 1984; Sorrell 1999). In parallel, the formation of ethylene and abscisic acid leads to epinastic leaves. But also in mofette areas oxygen concentrations in the rooting horizon of plants growing in the vicinity of CO_2 vents are lower than on control sites. This creates transient hypoxic or even anoxic conditions and therefore root respiration could be affected.

4.1.1 Direct Influences on the Root System

Compared to extensive data on the shoot/root relationship in plants as influenced by artificially elevated CO_2 (Gregory et al. 1996; Rogers et al.

1996; Pritchard and Rogers 2000) little is known on the allocation of carbon in plants growing in natural carbon dioxide springs (but see Edwards et al. 2003).

Šajna et al. (2002) studied below-ground phytomass on a semi-natural wet meadow at the natural CO_2 spring in Rihtarevci near Radenci (NE Slovenia). Pooled samples were taken along the CO_2 transect which was set by estimating air (from 392 to 1,074 ppm) and soil CO_2 (from 0.12 to 4.58%) concentrations using diffusion tubes (Dräger, Germany). There was no correlation between root biomass and CO_2 concentration and it was suggested that the soil humidity had a stronger influence on roots than existing CO_2 gradients.

In fumigation experiments, root growth at elevated CO_2 occurred closer to the upper soil horizon (Van Vuuren et al. 1997; Arnone et al. 2000). Also in natural CO_2 springs it is frequently observed that roots are growing mainly in a thin upper layer of the soil. The driving forces for a changed root distribution and root longevity are most probably increasing hypoxia and the increase in soil moisture. Studies on different populations of *Agrostis canina*, *Plantago major* and *P. lanceolata* revealed that long-term adaptation to elevated CO_2 is associated with increased growth and higher carbon allocation to the roots (Fordham et al. 1997a,b; Fordham and Barnes 1999; Edwards et al. 2003). This would help to prevent the feedback inhibition of photosynthetic capacity often found in plants growing in natural carbon dioxide springs (Badiani et al. 2000).

4.1.2 Root Respiration

Two different responses of plant respiration to elevated CO_2 have been distinguished by Amthor (1995). Direct (short-term) effects are observed when CO_2 is rapidly increased, resulting in a partial or total inhibition of respiration (reduced enzyme activity; Gonzalez-Meler et al. 1996; Gonzalez-Meler and Siedow 1999; Baker et al. 2000). Indirect (long-term) effects on respiration were shown to occur during growth at elevated CO_2. Effects are mediated by growth rate, by non-structural carbohydrate concentrations and tissue composition (Hamilton et al. 2001).

Compared to the number of studies dealing with the effect of elevated CO_2 on the respiration of green tissues (see the review by Drake et al. 1999) information on root respiration is rather scarce (but see Cheng 1999). Soil $[CO_2]$ can significantly influence the respiratory activity of roots (Clinton and Vose 1999). For some plants inhibition of root respiration was found at concentrations normally occurring in soil (Qi et al. 1994; Clinton and Vose 1999; Nobel and Palta 1989) but not in other species (e.g., Bouma et al. 1997a, b; Lambers et al. 1996a,b, 2002). They presume that most plants are less sensitive to high [CO_2] in the root environment. Yet, recent studies suggest that the direct effects of CO_2 on respiration might have been overestimated (Jahnke 2001; Burton and Pregitzer 2002; Jahnke and Krewitt 2002).

Direct effects of CO_2 on respiration occur in environments where the [CO_2] are not stable. In this respect natural carbon dioxide springs are extremes, known for dramatic short-term CO_2 fluctuations (Raschi et al. 1997; Badiani et al. 2000). Soil CO_2 measurements at the Stavešinci mofette area showed non-homogeneously distributed CO_2 emissions with concentrations ranging from 0.3 to 100% (see Fig. 1). Beside this spatial variability, temporal changes in local soil gaseous conditions can be expected as a consequence of rapid changes in the atmosphere (wind) and lateral diffusion under changing soil conditions (e. g., soil drying and re-wetting). Maĉek et al. (2002) measured the root respiratory potential of *Echinochloa crus-galli*, *Setaria pumila* and *Zea mays* growing in a natural CO_2 spring area. The plants were selected according to their height and the preliminary measured soil CO_2 concentration in the rooting zone. Root respiratory potential (measured as electron transport - ETS - activity) was determined on root tip segments (1-cm length) using the iodo-nitrotetrazolium salt (INT) method described by Kenner and Ahmed (1975). No significant effects of high rhizospheric CO_2 concentration on root respiratory potential were found for *Echinochloa* and *Setaria* (Table 2). The low sensitivity of *Echinochloa* root respiration to an enhanced [CO_2] was confirmed by fumigation experiments where the offspring of high- and low-CO_2-grown mofette plants were exposed to near ambient (ca. 370 μmol mol^{-1}) and elevated CO_2 (ca. 1,900 μmol mol^{-1}; Maĉek et al. 2002). Neither effect of high CO_2 on root respiratory potential nor on the root respiration as measured using oxygen electrodes was found (Maĉek et al. unpublished). Thus, *Echinochloa* and *Setaria* are relatively insensitive to high CO_2 and direct effects of CO_2 at the natural spring are not expected. In *Zea mays* a significantly lower root respiratory potential of mofette-grown plant roots was measured (Table 2). On the one side, it is suggested that efficient root respiration has to be maintained under hypoxic conditions to ensure the development of adaptive responses (Bragina and Grinieva 1998), whereas, on the other hand, tolerance to hypoxia is often accompanied with low oxygen consumption by roots (Huang and Johnson 1995). Yet, as a significant decrease in soil oxygen partial pressure has been observed concomitantly with the increased [CO_2], root respiration could be decreased by the limited oxygen supply (Pfanz and Vodnik, unpublished). A decreased root respiration has been observed under hypoxic conditions (Huang and Johnson 1995; Gibberd et al. 2001) and was shown to be species (or cultivar) specific and was

Table 2. Root respiratory potential of different plant species from the geothermal mofette Stavešinci. (Data from Maĉek et al. 2002; with permission)

Plant species	CO_2 exposure[a]	Mean height (cm)[b]	ETS[c]
E. crus-galli var. *crus-galli*	Low (0.4%)	62.3±11.7	1.52±0.19
	High (26%)	15.9±1.8	1.55±0.18
Setaria pumila	Low (0.4%)	51.0±6.8	0.34±0.05
	High (26%)	28.0±5.6	0.32±0.06
Zea mays	Low (0.1–0.4%)	239.0±21.0	1.12±0.13
	High (over 10%)	114.2±7.9	0.95±0.13

[a] Measured as soil CO_2 concentration (25 cm depth) by a gas analyzer GA 2000 (Ansyco, FRG).
[b] By ANOVA, n=10.
[c] Electron transport activity (ETS) given as µg O_2 g^{-1} fresh wt. h^{-1}, by ANOVA, n=12.

dependent on the tolerance mechanisms (Vartapetian and Jackson 1997). Ethanol formation was shown to increase in *Phacelia* growing in the vicinity of CO_2 vents in an arable field in the Eifel Mountains (FRG). The ethanol contents in the roots increased from 114 µg kg^{-1} dm in control plants, via 149 µg kg^{-1} dm to reach 185 µg kg^{-1} dm close to the CO_2 vents (Stubbe and Pfanz, unpublished results).

4.2 Nutrient Availability and Mineral Nutrition

When grown under elevated CO_2 the mineral content in plant tissues is mostly found to be reduced. A decrease in nitrogen and increased C/N ratios were observed both in artificial (Coleman et al. 1993; Rogers et al. 1996; Cotrufo et al. 1998; Geiger et al. 1999; Matt et al. 2001) and natural CO_2 enrichment studies (Körner and Miglietta 1994; Bettarini et al. 1995; Vodnik et al. submitted; but see Bettarini et al. 1997; Pritchard and Rogers 2000; Penuelas et al. 2001). Very often decreasing nutrient concentrations in plants are the result of a dilution effect due to accumulation of non-structural carbohydrates (Kühny et al. 1991), of a shortage of these nutrients in the root environment and of down-regulated root activity.

Yet, mineral nutrition under elevated CO_2 is also strongly affected by the nutrient availability in the soil (Joel et al. 2001). Although negligible effects of doubled atmospheric [CO_2] on nutrient (nitrogen) availability in fumigation experiments were published (Arnone 1997) it may not be generalised to the conditions in natural CO_2 springs. At these sites mineral availability is limited by a retarded mineralisation (Cotrufo et al. 1998) and by redox reactions (i.e., reduction of mineral ions, denitrification, Mn^{2+}, Fe^{2+}, H_2S formation) occurring during hypoxia (Marschner 1995). In addition, nutrient uptake will also be reduced by the limited oxygen supply of the roots.

When different species (*Phleum pratense, Dactylis glomerata, Solidago gigantea, Zea mays, Juncus effusus*) were studied at the Stavešinci mofette, a general decrease of leaf nitrogen and an increase of C/N ratio was found in plants growing in the vicinity of the CO_2 vents (Vodnik et al. 2003). Interestingly, in all these species a photosynthetic down-regulation was measured (Cook et al. 1998; see also Sect. 5.3.2). In timothy grass grown at different [CO_2] at the Stavešinci mofette, the content of S, P, K, and Zn decreased the closer the plants grew to the emitting vents. At a constant C content, the decrease was around 40% in P, and 20% in K and S for the most exposed plants (Pfanz et al., unpublished results). Similar results were obtained for *Erica, Myrtus, Juniperus* (Penuelas et al. 2001) and *Dactylis glomerata* and *Zea mays* (Vodnik et al. unpublished).

5 Influences of High CO_2 on Above-Ground Plants Parts

Although the CO_2 concentrations within the canopy do not reach values as high as those measured in the soil, the air surrounding the plants is generally enriched with carbon dioxide. Depending on the microclimatic situation (mainly wind speed and direction) CO_2 concentrations may reach up to 3,000–10,000 μmol mol^{-1} (ppm) which is 8 to 28 times ambient. Concentrations also vary transiently with extremes measured only during calm early morning hours and in geo-morphological depressions. Plant habitus, as well as organ morphology, tissue anatomy, but also physiology and biochemistry are affected.

5.1 Anatomical and Cyto-Chemical Aspects

Only few studies have been made on anatomical changes of plants growing in mofette areas, although the number of stomata has been counted frequently (see below). Turk et al. (2002) examined bog rush (*Juncus effusus*), a rush grass commonly found in wet mofette fields of Germany, Slovenia, and Czechia. The plants grew in a wet ditch where due to the artificial shape of the ditch and the slope within the ditch, a nearly linear CO_2 gradient was usually formed during calm days. Plant height and stem diameter decreased with an increasing vicinity to the vents; thus stem slenderness (ratio of stem

Table 3. Morphological features of *Juncus effusus* plants growing at seven locations along the ditch (n=15 per growing site). Shoot height and stem diameter increase with the distance from the vent. Plants near the vent are more stout and far from it more slender. As shown in columns 5 and 6, the pith represents the largest part of the stem cross-sectional area. Following columns represent some features of the bog rush stems based on microscope examination of stem cross sections made 2 cm above the highest base leaf. Diameter of epidermis was determined using eyepiece micrometer; sclerenchyma bundles were counted by Optimas 5.0 Image analysing system under polarised light microscopy; the same method was used to determine the total area of sclerenchyma bundles; number of vascular bundles was counted under bright field microscopy. (Data represent the mean of 15 measurements ±SD.) Data were rearranged according to shoot height; from Turk et al. (2002 with kind permission)

Distance from the vents (m)	Average plant height (cm)	Stem diameter (mm)	Slenderness (W/H×100)	Stem section area (mm^2)	Pith area (mm^2)	Thickness of epidermis (μm)	Number of sclerenchymal bundles
0.2	15	1.80±0.20	11.97	2.53	2.52	2.19±0.39	25.1±6.2
2	25	2.03±0.33	8.11	3.23	3.22	2.13±0.16	25.9±7.2
0.9	40	2.75±0.08	6.86	5.92	5.90	1.64±0.08	51.1±6.4
7	60	2.85±0.08	4.76	6.40	6.38	1.64±0.08	59.4±7.2
18	80	2.74±0.09	3.43	5.91	5.89	1.64±0.08	47.1±7.3
12	90	2.76±0.26	3.07	5.99	5.98	1.49±0.16	45.7±8.3
23	100	3.52±0.20	3.52	9.75	9.73	1.59±0.08	75.9±16.7

width to stem height) clearly responded. The epidermal thickness was higher in plants close to the vents (2.19 µm as compared to 1.59 µm), whereas the pith area was four times larger in control plants (9.73 mm^2) than in plants close to the emission (2.52 mm^2). Furthermore, the number of vascular bundles and the total area of sclerenchymal bundles clearly followed the CO_2 gradient being larger at control levels than in the close vicinity (Table 3).

5.1.1 Intercellular Air Space

Leaves of mofette plants appear more xeromorphic than control plants and also contain more sclerenchymatous tissues (Turk et al. 2002). To clarify whether also the intercellular air volume is influenced by the prevailing CO_2 environment, measurements were done with *Dactylis glomerata* (cocksfoot) growing in a [CO_2] gradient. Using the infiltration method of Pfanz and Dietz (1987), mature leaves of *Dactylis* growing at different distances to a CO_2 vent were investigated. Close to the vent (41% CO_2 in the rooting zone) the leaves revealed an intercellular airspace of 27.2±2 µl cm^{-2} leaf surface, whereas in plants growing at the control sites (0.2% CO_2) 35.87±1.8 µl cm^{-2} were measured. Results performed with *Alopecurus pratensis* (meadow foxtail) showed a similar tendency and underline the more xeric character of high CO_2-grown plants.

5.1.2 Chlorophyll Contents

Photosynthetic pigments are not necessarily influenced by an elevated CO_2 regime, but if in excess, the plant's appearance rarely is chlorotic at the highly concentrated sites. Yet, in all species examined so far, a more or less pronounced decrease in chlorophyll was found (Table 4). In *Phleum pratense* the differences in the chlorophyll content were minor (Pfanz et al., unpublished results), while in other grasses (e.g., *Alopecurus pratensis*) and herbaceous plants (*Taraxacum officinale*) chlorophyll was more strongly affected (Table 4). Even in young woody species (*Populus tremula*) a slight decrease in chlorophyll was found when trees grew in the close vicinity of vents.

Table 4. Chlorophyll contents of leaves of selected plant species as grown at different locations within a mofette (Strmec, Slovenia). Chlorophyll was determined in the field on leaves still attached to the plant using a Chlorophyll Meter (SPAD 502, Minolta). The CO_2 concentrations are not directly comparable as all plants grew at different loci. Yet, "low" means a [CO_2] range from 0.1–1.9%, "medium"=2.5–5.7%, and "high"=19.0–27% (41% in *Dactylis*) as measured at 20-cm depth within the rooting zone of the individual plants

Species	Low CO_2	Medium CO_2	High CO_2
Phleum pratense	31.6±3.7	28.9±2.5	27.1±3.3
Dactylis glomerata	35.9±1.8	25.3±4.5	27.2±2.0
Alopecurus pratensis	38.9±1.2	31.3±3.0	24.4±4.2
Taraxacum officinale	40.5±1.8	32.8±1.2	30.9±2.7
Populus tremula	35.1±2.01	32.9±1.82	29.4±3.44

5.2 Growth of Mofette Plants

According to theory and assuming permanent unlimited nutrient availability, plant growth should be stimulated in a high-CO_2 environment. In artificial fumigation systems using climatic chambers, OTCs, or FACE systems this assumption mostly proves true, although sometimes only transiently. Many authors report an increase in growth or relative growth rate (RGR) of C_3, C_4 and even CAM plants (Badger 1992; Weigel et al. 1994; Amthor 1995; Jongen et al. 1995; Badiani et al. 2000 and references therein). In field crops grown under elevated CO_2, a stimulation of seasonal plant growth of up to 20% was reported (Rosenzweig and Perry 1994; Kimball et al. 1995). As an example of long-term tree response, forest patches of *Quercus ilex* growing for 30 years in the vicinity of two natural CO_2 springs (Rapolano, Lajatico) in Italy show a CO_2-induced increase in stem biomass of about 12% (Hättenschwiler et al. 1997).

The picture of herbaceous plants growing at natural CO_2 springs is clearly different. Depending on the mean CO_2 concentrations in the air and – probably more importantly – in the rooting horizon, plant growth is strongly inhibited and has so far never been shown to be stimulated. Growth reduction or retardation was shown in (semi-)natural stands (probably nutrient limited) as well as in agriculturally fertilised sites (principally not nutrient limited). In our studies growth always strictly followed the CO_2 concentrations measured in the soil at 20 cm depth. When soil gas concentrations were determined directly within the rooting horizon of the plants, plant height directly corresponded to the [CO_2] (and concomitantly the

Table 5. Morphometrical parameters (total height, leaf area, dry matter of leaves, flowers, stems, and roots) of *Phacelia tanacetifolia* as grown in a mofette field in the Eifel Mountains. Plants were chosen according to different CO_2 concentrations within the rooting zone at 20 cm depth. At each plot ten plants were harvested. (Stubbe and Pfanz, unpubl. results)

	2.5% CO_2	41.4% CO_2	88.8% CO_2
Total plant height (cm)	70	58	31
Total leaf area (cm^2)	39.7	16.8	2.1
Total leaf matter (g dm[a])	69.1	5.4	2.9
Total flower matter (g dm)	65.6	6.7	3.9
Total stem matter (g dm)	189.9	14.7	5.3
Total root matter (g dm)	1.9	1.5	0.7
Total plant matter (g dm)	326.5	28.2	12.8

[a] *dm*, Dry matter.

[O_2]) in the rooting horizon (Pfanz et al., unpubl. results). As soil [CO_2] are much more stable and as measurements are therefore more reliable than those performed in air, a prerequisite for correlating growth and [CO_2] in natural CO_2 springs is the parallel measurement of the gaseous soil milieu.

Phleum pratense plants were smallest (mean height: 22.8±4.6 cm) in the close vicinity to the vents and highest (ca. 134.7±14.5 cm) at a maximum distance, i.e. at the lowest soil CO_2 concentrations; mean CO_2 concentration resulted in medium growth (46.6±7.3 cm). This behaviour is meanwhile shown for a number of C_3 (*Alopecurus pratensis, Poa pratensis, Solidago gigantea, Plantago lanceolata and P. major, Taraxacum officinale*) and C_4 species (*Echinochloa crus-galli, Setaria pumila, Zea mays*). A negative correlation between total plant height and soil CO_2 concentrations was also found in *Phacelia tanacetifolia* (Table 5) and Triticale (Fig. 2) growing in an arable field in the Eifel Mountains (FRG). The two respective fields were ploughed and fertilised yearly. In *Phacelia* not only plant height, but also dry matter, number and size of leaves, number of flowers and size of fruits or seeds were influenced by elevated CO_2 (Table 5).

In experiments where the high-CO_2 adapted grassland community within the Bossoleto spring was submitted to CO_2 depletion there was no significant effects at the community level, while in individual species growth was significantly decreased (Vaccari et al. 2001). Patches of the same grassland were moved to CO_2 fumigation rings, and submitted to ambient (350 ppm) and elevated (700 ppm) carbon dioxide concentrations, as well

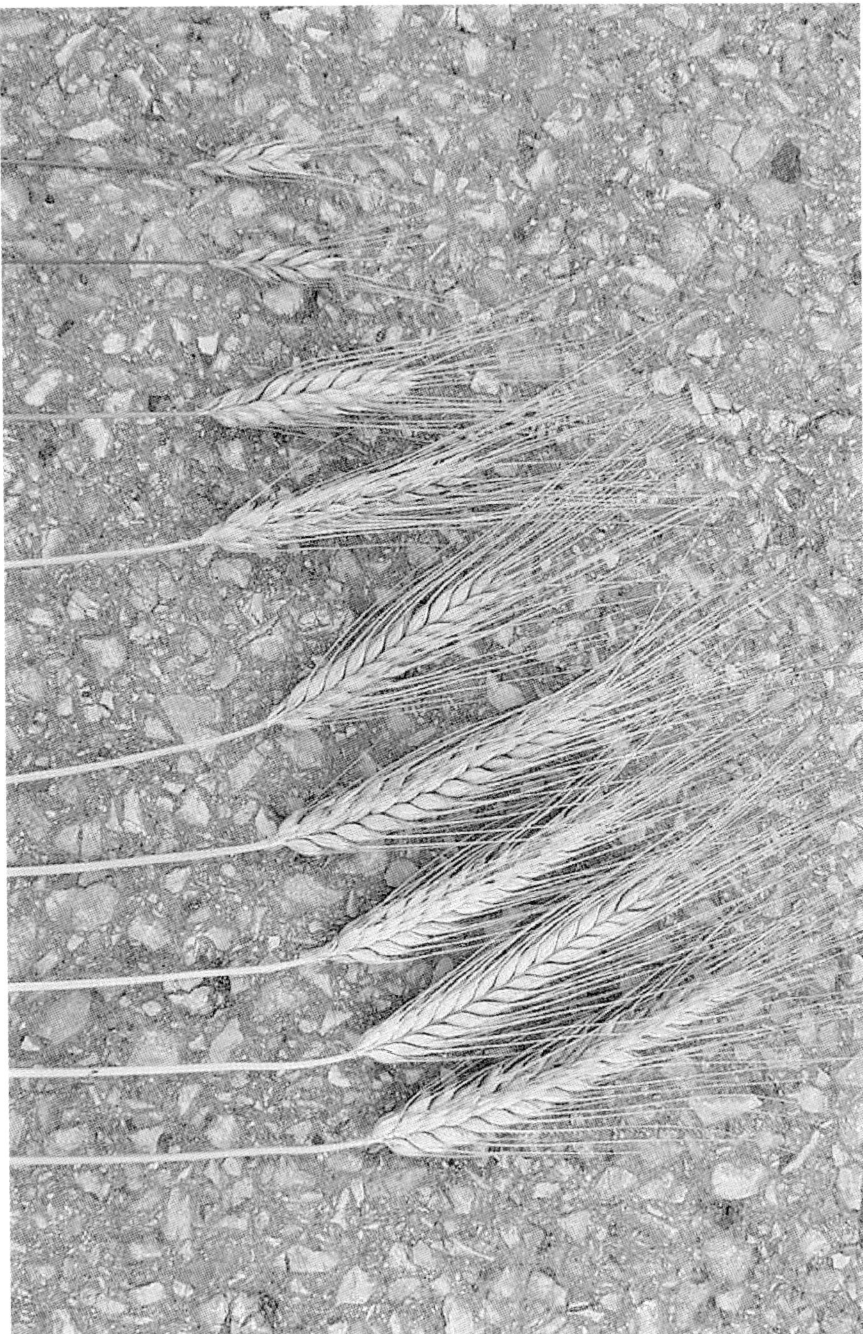

Fig. 2. Ears of *Triticale* sp. as grown in a field with CO_2 vents. The field is ploughed and fertilised every year. From *left* to *right* plants grew an increasing distance from the vent. The closer the plants grew to the CO_2-emitting vent, the smaller the plant and the smaller the respective ear. (Photo Pfanz 2000)

as to fertilisation. The effect of fertilisation was significant for both fumigated and non-fumigated patches. In all cases, elevated CO_2 resulted in an increase in non-structural carbohydrates, suggesting that the lack of response of nutrient-poor ecosystems is not a short-term experimental artefact, and may be maintained over the long term (Navas et al. 1995; Leadly and Körner 1996; Leadly and Stöcklin 1996). In contrast, other studies found no evidence that growing under nutrient-poor conditions suppresses the CO_2 response of plants (Lloyd and Farquhar 1996; Curtis and Wang 1998). As already shown above, most authors describe an enhanced root growth and therefore a higher root/shoot ratio in plants grown at elevated CO_2 (Rogers et al. 1994, Badiani et al. 2000; see also Edwards et al. 2003). In mofette plants shoot growth is decreased, but root growth seems to be nearly constant, also leading to an increased root/shoot ratio.

5.3 Gas Exchange of Mofette Plants

It is commonly assumed that normal plants live in a world of increasing, but still rather low [CO_2] in the atmosphere. As the present-day [CO_2] is roughly 360 ppm (which is still in the linear section of the ACi curve), an increase in CO_2 will in the short run lead to an increase in photosynthesis. Nevertheless, there are several plants that are exposed to (or create themselves) an atmosphere clearly higher than ambient. Herbaceous plants growing close to or even in very humus soils sometimes realise [CO_2] up to 1% (=10,000 ppm!). The composition of lacunal air in *Phragmites australis* also greatly differs from outside conditions. Depending on the organ standing in water or emerging from the water and depending on the light conditions, the lacunal gas system contains 0.2 to 1.8% CO_2 (3.2%–4% CO_2 after a prolonged darkness; Brix 1988a,b; for *Typha latifolia* see Constable et al. 1992). In woody plants respiratory CO_2 can accumulate in the wood and inner bark of stems and branches to give 1–26% (Mc Dougal and Working 1933: Pfanz and Aschan 2001; Pfanz et al. 2002). These plants or plant parts have adapted to [CO_2] which are clearly not limiting photosynthesis during normal growth. Furthermore, when ice covers single leaves or plants during wintertime (ice encasement), [CO_2] can rise up to 44% in the small air phase around the frozen tissue (Rakitina 1970; Andrews and Pomeroy 1991).

5.3.1 Physico-Chemical Considerations of CO_2 Action

The substrate of photosynthesis is carbon dioxide (MW=44.01 g mol^{-1}), a fully colourless, invisible, non-smelling gas, with a specific density of 1.85 kg m^{-3} (at 15 °C and 1 bar) and thus being heavier than air. The latter fact is responsible for the accumulation and formation of carbon dioxide lakes if mofettes occur in hilly terrain. CO_2 is the anhydride of carbonic acid and due to its high solubility in water it dissolves in aqueous solutions to form carbonic acid (H_2CO_3 or chemically more exact $CO_2 \cdot H_2O$). As carbonic acid is a weak acid with pK values of 6.25 and 10.48 (Hocking and Hocking 1977), it readily dissociates to liberate protons and the anions (bi-)hydrogen carbonate (HCO_3^-) and carbonate (CO_3^{2-}) according to:

$$CO_2 + H_2O \leftrightarrow H_2CO_3 \leftrightarrow HCO_3^- + H^+ \leftrightarrow CO_3^{2-} + 2H^+ \quad (1)$$

Using the equation of Henderson Hasselbalch (see Stryer 1988) the distribution of the different species at a given pH is easily determined. Furthermore, the concomitant proton stress within the different tissues can be determined by calculating the liberated protons (Pfanz and Heber 1986, 1989).

Yet, compared to well-known potentially acidic air pollutants like SO_2, extremely high CO_2 concentrations are needed to directly inhibit photosynthesis by acidification (see Table 3 in Pfanz and Heber 1986). Under ambient [CO_2] (350 ppm) and even at doubled [CO_2] (700 ppm), CO_2 per se should never affect the plant's pH-stat homeostasis (cf. Pfanz and Heber 1986, 1989; Wagner 1990). Yet, in the close vicinity to CO_2 vents extremely high concentrations occur that would affect intracellular pH values (see Fig. 3 below).

5.3.2 Photosynthetic Carbon Assimilation at CO_2 Extremes

5.3.2.1 In Vitro Studies

Enhanced carbon dioxide can directly affect photosynthesis of plants. If concentrations exceed a certain threshold, a reduction of photosynthesis is found within several minutes after application. In Fig. 3, young barley leaves (freed from epidermal layers to allow free gas diffusion into the leaf mesophyll) were kept in highly buffered solutions within the chambers of Clark-type oxygen electrodes. Increasing amounts of carbon dioxide were added and the resulting photosynthetic response was measured. The response proved to be highly pH-dependent. When the experiments were run at pH 8, photosynthesis followed a clear CO_2 saturation kinetic as expected

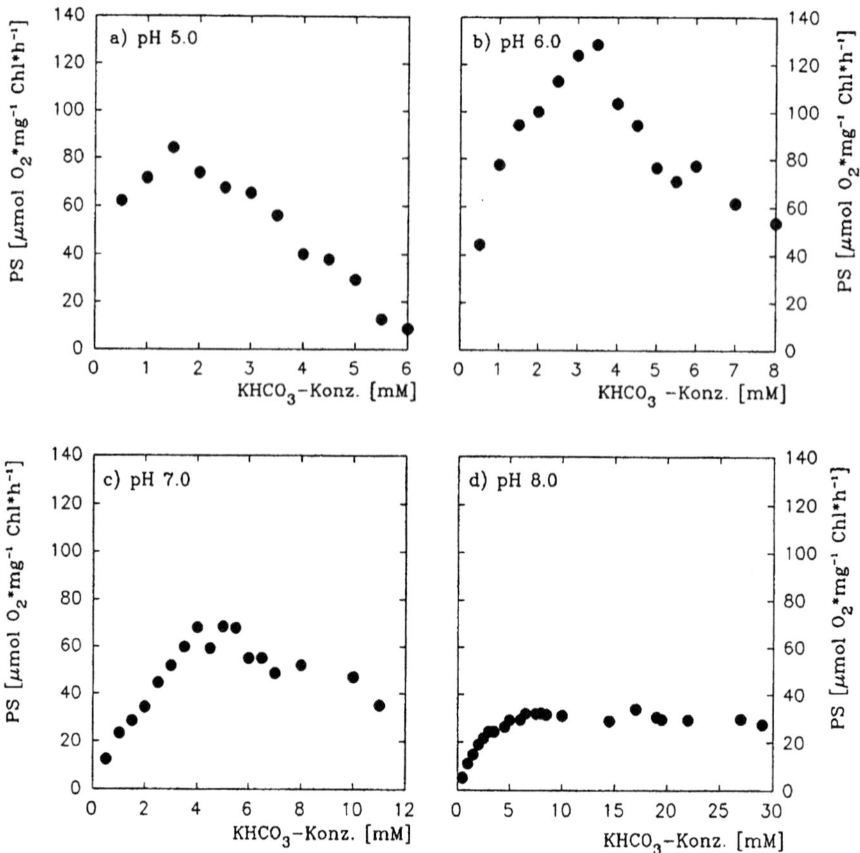

Fig. 3. pH-dependent CO_2 curves of barley leaf photosynthesis as measured in a Clark-type oxygen electrode at 20 °C and a PFD of 1,200 μmol photons m^{-2} s^{-1}. The mesophyll of 10-day-old barley leaves was freed by epidermal peeling and after infiltration with the assay buffer (50 mM MES, 2 mM $CaCl_2$, 1 mM KNO_3, 2 mM $MgSO_4$) oxygen evolution was measured after another 5 min incubation in the dark (Pfanz and Scheewe, unpublished)

from the known behaviour of the AC_o curves (assimilation versus external "outside air" CO_2 concentration, in contrast to AC_i). Only at the highest CO_2 concentration (30 mM $KHCO_3$), the obtained photosynthetic rates tended to decrease. Decreasing pH of the external bathing medium of the leaf led to increased (absolute) rates of photosynthesis, but CO_2-induced inhibition occurred at lower [$KHCO_3$]. Whereas at pH 7 still 6 mM $KHCO_3$ was needed to start inhibition of photosynthesis, only 4 mM was applied at pH 6, and at pH 5 inhibition started already at a [$KHCO_3$] of only 1.5 mM. Similar data were obtained using needle leaves of spruce (*Picea abies*; Pfanz and Rumpel, unpubl.).

The reason for this behaviour is the reaction of CO_2 and its dissociation products at different pH values (see Eq. 1 above). The lower the pH, the higher the portion of non-dissociated "neutral" carbon dioxide freed from bicarbonate. The "neutral" carbon dioxide readily diffuses into the protoplasts (Gimmler, pers. comm.); inside the cytoplasm it hydrates, forming the extremely unstable carbonic acid (H_2CO_3) which rapidly dissociates due to the prevailing pH values to liberate protons. The proton production (which is accompanied by the formation of HCO_3^- and CO_3^{2-}) will then lower the pH in cellular compartments (e.g. the chloroplasts) thus reducing or completely inhibiting photosynthesis (for details see Raven 1985; Pfanz and Heber 1986, 1989; Pfanz 1994)

5.3.2.2 In Situ Studies

Under ambient [CO_2] photosynthesis of most higher plants is not saturated. Up to a certain extent, increasing [CO_2] leads to an increasing rate of photosynthesis. In a broad range of C_3 species, a doubling of current [CO_2] causes a 23–58% increase in leaf photosynthetic rate (Drake et al. 1997). This fact is used by horticulturists in greenhouses to increase yield (Schwiebert 1985). Plants growing in areas of elevated CO_2 could therefore show enhanced carbon acquisition rates and a higher photosynthetic carbon flow. Yet, experiments with double ambient CO_2 clearly showed that the effects on plants can be rather inhomogeneous. Species-specific reactions on enhanced CO_2 were found with cabbage characterised by a lower carboxylation efficiency (CE) and maximum photosynthetic capacity (A_{max}) (Sage et al. 1989); with cotton no change was recorded (Radin et al. 1987) and with soybean, where A_{max} and CE were enhanced due to an increase in CO_2 (Campbell et al. 1988; see Bowes 1993). A decrease in maximal activity of Rubisco towards CO_2 (Pearson and Brooks 1996, Curtis et al. 1995 – for woody species) in carboxylation efficiency (McKee et al. 1995) and in the down-regulation of photosynthetic genes is discussed (Stitt 1991; Garcia et al. 1994; Van Oosten and Besford 1995; Webber et al. 1994). Down-regulation of net photosynthesis may also occur directly by carbohydrate accumulation in sink-limited plants (Rogers et al. 1994; Barnes et al. 1995).

Published data on photosynthetic performance of mofette plants seem to be as inconsistent as results from experiments with artificially doubled CO_2 (Miglietta et al. 1995; Miglietta 1997; Raschi et al. 1997, 1999; Badiani et al. 2000). It seems to be clear that an increased [CO_2] is not *per se* desirable for plant performance. A reduction in maximum photosynthesis was published for *Phragmites australis* (Miglietta et al. 1995) and for different calcareous grassland species (Miglietta et al. 1998a) growing in an Italian dolina mofette (Il Bossoleto); but reduction was not found in *Scirpus lacustris* (Miglietta 1997). According to the recent review of Badiani et al. (2000) no consistent data exist up to now on the photosynthetic behaviour of mofette plants nor on adaptive photosynthetic strategies of autotrophic plants growing under extreme CO_2 conditions in natural CO_2 spring areas.

Yet, our studies at the Slovenian mofette field Strmec clearly showed some common features of the photosynthetic reactions of several plant species. Up to now several C_3 grasses (*Phleum pratense, Alopecurus pratensis, Juncus effusus)*, C_3 dicots and woody species (*Plantago major, Carpinus betulus)* and the C_4 grasses *Zea mays* (Vodnik et al. 2003), *Setaria viridis, Echinochloa crus-galli*)(Vodnik et al. 2002b) were examined and the following observations made (see Table 6):

1. In all species studied so far maximum photosynthesis (A_{350}, A_{700}, and A_{max}, measured at 2,000 ppm CO_2) decreased with increasing vicinity to the CO_2 gas vents. An increase in photosynthesis (as expected from CO_2 fertilisation experiments) was never observed in the field.
2. A_{max} could not be measured as CO_2 concentrations necessary to saturate photosynthesis of "high-CO_2"-grown plants, because those concentrations could not be supplied with the existing equipment. Calculation revealed that [CO_2] would be in the range of several 10,000 ppm.
3. The CO_2 compensation point clearly correlates with the growth distance to the vents. The closer the growth vicinity to the vent and thus the higher the actual CO_2 load, the lower was the CO_2 compensation point determined.
4. The carboxylation efficiency was decreased in plants growing close to the vents.

In the following, representative results for timothy grass (*Phleum pratense*) are given.

When measured using 360 ppm CO_2 (which is thought to be the CO_2 control level at unpolluted rural sites) net photosynthesis was twice as high in plants from control plots (16.9 µmol CO_2 m^{-2} s^{-1}) than in plants grown at elevated soil CO_2 (ca. 8.93–7.79 µmol CO_2 m^{-2} s^{-1}; see Table 6). Net photosynthesis of *Phleum* under "normal" ambient CO_2 conditions is thus dramatically reduced in high-CO_2 plants. Nevertheless, the term "ambient CO_2" must be considered carefully in a region where CO_2 concentrations are never really stable and may fluctuate in air within minutes (or even seconds) from 0.036 to 1%. Similar results were held when A_{700} was determined (see Table 6). Also, at concentrations doubling ambient, photosynthesis of *Phleum* grown at elevated CO_2 was decreased by 50%. Experiments with herbaceous plants fumigated with 360 or 700 ppm CO_2 in OTCs or FACE systems normally resulted in an (at least transient) increase in photosynthesis (Eamus and Jarvis 1989; Allen 1990; Kimball et al. 1995;

Table 6. Photosynthetic parameters of the leaves of *Phleum pratense* growing at different CO_2 concentrations in the soil (see Table 1) at the geothermal mofette field Strmec near Stavešinci; n=5–6 ±SE

Parameter	Soil CO_2 (plot 4b)	Soil CO_2 (plot 7)	Soil CO_2 (plot 2)
A_{350}[a]	16.89	8.93	7.79
A_{360}[b]	17.15	9.19	8.08
A_{700}[c]	21.63	15.24	14.88
A_{2000}[d]	23.04±3.2	20.38±1.72	20.32±3.43
SEC (ppm)[e]	1,500$_m$	ca. 20,000$_c$	ca. 30,000$_c$
CE[f]	0.0042	0.0022	0.0023
CO_2 compensation point[g]	36	93	144

[a–d]Net assimilation rates measured at 350, 360, 700, or 2000 ppm CO_2 in (μmol CO_2 fixed m^{-2} s^{-1}), respectively.

[e]Saturating external CO_2 concentration (SEC) for maximal photosynthesis (measured *m* or calculated *c* in (μmol mol^{-1}); data are given with standard error.

[f]Carboxylation efficiency (mol m^{-2} s^{-1}).

[g]CO_2 compensation point (μmol mol^{-1}).

Fordham and Barnes 1999). However, at prolonged fumigation, the enhanced growth or stimulated photosynthesis was often revised (Stitt 1991; Bowes 1993; Webber et al. 1994; Drake et al. 1997). An adaptive down-regulation of net photosynthesis has been published by numerous authors considering plants from artificial fumigation experiments; but also opposite reactions have been observed (for review see Bowes 1993; Drake et al. 1997).

Maximum photosynthetic activities in *Phleum* varied between plants growing at sites with different CO_2 concentrations (A_{2000}; however, proper photosynthetic capacities, PSC, could not be determined with the portable photosynthesis system LI-6400, LI-COR, USA, which does not allow measurement of CO_2 concentrations higher than 2,000–2,200 ppm). Maximum CO_2 assimilation (23 μmol CO_2 m^{-2} s^{-1}) was obtained at the low-CO_2 plot (0.3% CO_2). With values around 20 μmol CO_2 m^{-2} s^{-1} fixed, maximum assimilation rates at the elevated CO_2 plots were slightly lower. Interestingly, CO_2 saturation of photosynthesis was reached at 1,500 ppm CO_2 only with the "low CO_2" plants, whereas the plants growing at the "medium-" and "high CO_2" plots (3.6 and 26%, respectively) did not reach saturation under the conditions applied. Mathematical approximation of the data led to probable (but theoretical) saturation concentrations of around 20,000–30,000 ppm CO_2 (Table 6).

In *Phleum*, not only net photosynthesis measured at 350, 700, or 2,000 ppm varied between the plants grown at different CO_2 regimes, but also the initial slope of the CO_2 curves, the carboxylation efficiency (CE as a measure of activity and efficiency of Rubisco) greatly differed between the growth variants. CE had values of 0.0042 mol m^{-2} s^{-1} when grown under normal CO_2 conditions, but values decreased to ca. 0.0022 mol m^{-2} s^{-1} with increasing CO_2 (for comparison see Bauer and Martha 1981; Bauer et al. 1983). These findings are in contrast to those of Fordham and Barnes (1999) who stated that long-term adaptation at elevated CO_2 in *Agrostis* and *Plantago* does not appear to be linked in the intrinsic capacity for photosynthesis. Analysing A/Ci curves of *Scirpus lacustris* populations grown in mofettes and in control sites, Miglietta (1997) and van Gardingen et al. (1997) found no differences in indicative photosynthetic parameters (see also Jacob et al. 1995; Miglietta et al. 1995; Badiani et al. 2000). Yet, Cook et al. (1998) found such differences in *Nardus stricta*.

Most interestingly the CO_2 compensation point of the ACi curves of timothy grass revealed big differences. For the "low CO_2" grass tuft, the concentration of the applied CO_2 necessary to compensate for the leaves' respirational CO_2 loss was close to 36 ppm CO_2. When grown under an increased CO_2 regime, the compensation point increased via 93 ppm to reach 144 ppm CO_2 under the maximal CO_2-regime (Table 6). *Phleum* had adjusted respiration and photosynthetic carbon assimilation according to the prevailing CO_2 regime during germination and growth.

5.3.3 Respiration

A great number of publications deals with the probable effects of an enhanced CO_2 concentration on mitochondrial respiration of green organs or roots. Yet, results are contradictory (Jahnke 2001; see also Sect. 4.1.3). Only very few experiments were done quantifying the effects of CO_2 extremes in the vicinity of CO_2 springs on root respiration or even temporary root fermentation. Basic considerations on enzyme kinetics would lead to the assumption that short-term effects of an increased [CO_2] necessarily would reduce or even inhibit mitochondrial respiration by pure feedback inhibition.

5.4 Transpiration

Increasing the CO_2 concentration of the air phase surrounding a leaf normally leads to a concomitant decrease in stomatal aperture (Heath 1948). For many different plant species grown under elevated CO_2 an average decrease in stomatal conductance of 20% is reported (Field et al. 1995). The likely impacts of an increased [CO_2] to plants growing in carbon dioxide springs are therefore a reduced transpiration (presumably affecting plant surface temperatures) and a probable increase in water use efficiency. As will be discussed below, a decrease in leaf conductance can be achieved by a reduction in stomatal density and stomatal apertures. In a synopsis of the changes observed in plant ecosystems permanently exposed to natural CO_2-enrichment, Badiani et al. (2000) compared the findings for the water status and water use efficiency (WUE) of several species. In mature trees of *Quercus ilex, Q. pubescens* and *Salix herbacea* growing in mofettes highly enriched in CO_2 (e.g. Il Bossoleto, Italy) WUE was increased. In the same species a decrease in stomatal conductance and in transpiration was found. An increased water potential and an increased osmotic potential as well as a decreased tendency for xylem embolisms were stated. It is worth noticing that in dry Mediterranean summers the differences between enriched and not-enriched plants tend to disappear, or are limited to morning hours, as stomata are closed throughout most of the daytime. The response of stomata to increasing vapour pressure deficit is usually steeper in control than in fumigated leaves. Yet, considering a larger scale, high CO_2 might exert little or no effect on regional evapotranspiration (Melillo et al. 1990).

5.4.1 Stomatal Patchiness

Many observations hint to the fact that in heterobaric leaves stomatal aperture is not homogeneous and the stomata in intercostal areas differ in aperture from the neighbouring zones (Beyschlag and Eckstein 1997). Furthermore, under constant conditions stomata tend to close with increasing and to open with decreasing [CO_2]. As a consequence, stomatal conductance decreases at higher [CO_2] positively affecting water use efficiency (WUE). Stomatal patchiness was tested with *Solidago gigantea* in a mofette field. On a sunny day in May during late morning representative leaves of *Solidago* growing at three different loci varying in soil [CO_2] were harvested and infiltrated with water (Pfanz 1994; Pfanz and Dietz 1987). Intercostal areas with open stomata were easily infiltrated, whereas closed or semi-closed areas were not. Figure 4 shows gradual differences in the amount of water-infiltrated leaf area and thus in stomatal apertures of the leaves (cf.

Fig. 4. Stomatal patchiness of *Solidago gigantea* as grown at different loci on a mofette field. The leaves of the plants were harvested at 11.00 a.m. on a sunny day in May 2002. The leaves were immediately infiltrated according to Pfanz and Dietz (1987). Images obtained with an Olympus digital camera; photograph of the backlit lower leaf surface. Plants grew at a soil $[CO_2]$ of 0.4% (*upper panel*), 3.4% (*middle panel*), and 26% (*lower panel*). *Darker colours* indicate no infiltration (closed stomata), whereas *brighter colours* show good infiltration (open stomata) (Pfanz unpublished)

Beyschlag and Pfanz 1990). A direct influence of CO_2 on stomatal aperture of leaves of mofette plants can be assumed.

5.4.2 Stomatal Densities

Stomatal densities seem to respond to changes in atmospheric [CO_2]. Beerling and Chaloner (1993) clearly showed that within the last 3,000 years the number of stomata on the leaves of olive trees (*Olea europaea*) decreased from 700 mm^{-2} (1327 b.c. in pharaoh's time) to about 500 mm^{-2} in recent trees. They plot a more or less linear relationship between the atmospheric [CO_2] and the percent change in stomatal densities (see Fig. 4 in Beerling and Chaloner 1993). Also, Heath and Mansfield (1999) show a nearly linear correlation between the stomatal conductance of *Salix herbacea* and the atmospheric [CO_2]. It is therefore to be expected that also plants growing in areas of naturally enhanced CO_2 could anatomically react and adjust the number of stomata to the necessary level. Yet, many hidden traps exist that make evaluation not an easy task. The number of stomata of a given leaf area may vary even within one species. Shade and sun leaves differ in stomatal number; leaf age influences densities and the water regime during growth plays a predominant role in determining the number of stomata per given area. Therefore, results obtained in plants growing near mofettes are rather heterogeneous (see Bettarini et al. 1998; Paoletti et al. 1998). Bettarini et al. (1998) examined 17 plant species from three different plant communities growing in regions of naturally enhanced [CO_2] and found clear changes in stomatal conductance (see above) but not in leaf anatomy. Only in three species (*Fraxinus ornus, Conyza canadensis*, and *Stachys recta*) a reduction in stomatal densities was observed.

On the other hand, Tognetti et al. (2000), working on Mediterranean shrubs growing at the CO_2 spring of Lajatico, found a significant reduction in stomatal density in *Myrtus communis*, but a negligible one in *Juniperus communis*. Interestingly, in the first species the dimension of the stomatal apparatus and the pore length were increased, while in the second one these changes were much less evident. Reductions in stomatal density along a carbon dioxide gradient were evidenced in *Scirpus lacustris* (Bettarini et al. 1997), while on *Quercus ilex*, the reduction in stomatal density was much more limited in CO_2 spring-grown trees than in growth cabinet experiments (Paoletti et al. 1998). On the basis of the results from carbon dioxide springs, the response of stomatal density to elevated carbon dioxide seems to be limited to some species, while the presence of sulphur pollutants does not seem to affect the patterns of stomatal density (Bettarini et al.1997).

6 Conclusive Remarks and Coda

Plants growing in natural CO_2 springs are influenced by an atmosphere strongly enriched in CO_2. This would make research within mofettes an ideal tool to predict probable changes of individual plant behaviour under elevated CO_2 and to estimate response on a larger scale. Yet, there are several facts that raise the question whether mofettes are really ideal places for studying CO_2 enrichment. Beside high fluctuations of air CO_2 and unstable gaseous conditions, there are many other environmental factors (availability of mineral nutrients and water, an increase in temperature) that can easily modify plant response under elevated CO_2 (see van Gardingen et al. 1997). In addition, measured effects may be caused not only by the increased $[CO_2]$ in air but also by the decreased $[O_2]$ in the soil. Very low partial pressures of oxygen that have been measured in the soil at the most exposed spring sites indicate that the effects of soil hypoxia on root functioning and rhizosphere processes are likely and these could be of crucial importance for plant performance. Due to the direct and indirect impact of elevated CO_2 on below- and above-ground plant organs, the causal analysis of the effects is not simple. Moreover, below-ground processes at natural CO_2 springs have been much less intensively studied when compared to shoot processes (photosynthesis, stomatal response). Studies of root growth and functioning, root symbioses (Rillig et al. 2000), microbial functioning (O'Neill 1994) and soil chemical processes (Zak et al. 1993; Gahrooee 1998) are therefore needed.

So far, research at natural carbon dioxide springs has been mainly run in order to improve predictions on long-term plant responses to a doubled atmospheric CO_2 concentration. For this purpose studies were focused on plants growing in a gaseous environment with CO_2 concentration close to 700 $\mu mol\ CO_2\ mol^{-1}$. Indeed, these studies resulted in important information for understanding CO_2 effects (Raschi et al. 1997; Badiani et al. 2000) but at the same time many aspects of life under naturally elevated CO_2 have been neglected (functioning under extreme CO_2 concentrations, dynamics of the response to changing gaseous environment, etc.). Some studies revealed that the quality of information on physiological response can be improved when measurements are performed on plants growing under various CO_2 conditions (CO_2 gradient).

Mofettes are the only known natural areas where life under naturally elevated CO_2 can be closely followed. Despite the inherent difficulties we believe that there are still many research opportunities which have not been exploited.

Acknowledgement. We are grateful to Dr. Horst Kämpf (GFZ Potsdam) and Dr. Diethard Meyer (Institute of Geology, University of Duisburg-Essen) for literature, guided field trips and helpful discussions. We gratefully acknowledge the technical help of Gudrun Friesewinkel, Christa Kosch, Sabine Kühr, Silke Lisiecki-Bracht and Alfred Lenk. Special thanks are extended to Dipl. Umweltwiss. Kim Stubbe for her work on *Phacelia*. The work was supported by the Research-Pool 2001 (09112000) of the University of Duisburg-Essen (H.P. and G.A.) and by a grant J4-2186-0486 from the Ministry of Science, Education and Sport of Slovenia (DV). We express our deepest thanks to the families Kurbus and Sisko who hosted us with great hospitality and interest in our work. This work was performed within the framework of the COST Action 627 of the European Commission.

References

Allen JH (1990) Plant response to rising carbon dioxide and potential interactions with air pollutants. J Environ Qual 19:15–34

Amthor JS (1995) Terrestrial higher-plant response to increasing atmospheric CO_2 in relation to the global carbon cycle. Global Change Biol 1:243–274

Andalo C, Godelle B, Lefranc M, Mousseau M, Bottraud T (1996) Elevated CO_2 decreases seed germination in *Arabidopsis thaliana*. Global Change Biol 2:129–135

Andrews CJ, Pomeroy MK (1991) Low temperature anaerobiosis in ice encasement damage to winter cereals. In: Jackson MB, Davies DD, Lambers H (eds) Plant life under oxygen deprivation. SBP Acad Publishing, The Hague, pp 85–99

Arnone JA (1997) Indices of plant N availability in an alpine grassland under elevated atmospheric CO_2. Plant Soil 190:61–66

Arnone JA, Zaller JG, Spehn EM, Niklaus PA, Wells CE, Körner C (2000) Dynamics of root systems in native grasslands: effects of elevated atmospheric CO_2. New Phytol 147:73–86

Badger M (1992) Manipulating agricultural plants for a future high CO_2 environment. Aust J Bot 40:421–426

Badiani M, Raschi A, Paolacci AR, Miglietta F (2000) Plants responses to elevated CO_2; a perspective from natural CO_2 springs. In: Agrawal SB, Agrawal M (eds) Environmental pollution and plant response. Lewis, Boca Raton, pp 45–81

Baker JT, Allen LH Jr, Boote KJ, Pickering NB (2000) Direct effects of atmospheric carbon dioxide concentration on whole-canopy dark respiration of rice. Global Change Biol 6:275–286

Barnes JD, Ollerenshaw JH, Whitfield CP (1995) Effects of elevated CO_2 and/or O_3 on growth, development and physiology of wheat (*Triticum aestivum* L.) Global Change Biol 1:129–142

Baubron JC, Allard P, Toutain JP (1990) Diffuse volcanic emissions of carbon dioxide from Vulcano Island, Italy. Nature 344:51–53

Bauer H, Martha P (1981) The CO_2 compensation point of C_3 plants – a re-examination. I. Interspecific variability. Z Pflanzenphysiol 103:445–450

Bauer H, Martha P, Kirchner-Heiss B, Maierhofer J (1983) The CO_2 compensation point of C_3 plants – a re-examination. II. Intraspecific variability. Z Pflanzenphysiol 109:143–154

Baxter PJ, Kapila M (1989) Acute health impact of the gas-release at Lake Nyos, Cameroon. J Volcanol Geotherm Res 39:265–275

Bazzaz FA, Wayne PM (1994) Coping with environmental heterogenity: the physiological ecology of tree seedlings regeneration across the gap-understory continuum. In: Caldwell MM, Pearcy RW (eds) Exploitation of environmental heterogeneity by plants. Academic Press, San Diego, pp 349–390

Beerling DJ, Chaloner WG (1993) Stomatal density responses of Egyptian *Olea europaea* leaves to CO_2 change since 1327 b.c. Ann Bot 71:431–435

Berner RA (1990) Global CO_2 degassing and the carbon cycle. Geochim Cosmochim Acta 54:2889–2894

Bettarini I, Calderoni G, Miglietta F, Raschi A, Ehleringer J (1995) Isotopic discrimination and leaf nitrogen content of *Erica arborea* L. along a CO_2 concentration gradient in a CO_2 spring in Italy. Tree Physiol 15:327–332

Bettarini I, Miglietta F, Raschi A (1997) Studying morpho-physiological responses of *Scirpus lacustris* from naturally CO_2-enriched environments. In: Raschi A, Miglietta F, Tognetti R, van Gardingen PR (eds) Plant responses to elevated CO_2. Evidence from natural springs. Cambridge University Press, Cambridge, pp 134–147

Bettarini I, Vaccari FP, Miglietta F (1998) Elevated CO_2 concentration and stomatal density: observations from 17 plant species growing in a CO_2 spring in central Italy. Global Change Biol 4:17–22

Beyschlag W, Eckstein J (1997) Stomatal patchiness. In: Behnke H-D, Esser K, Kadereit JW, Lüttge U, Runge M (eds) Progress in botany, vol 59. Springer, Berlin Heidelberg New York, pp 283–298

Beyschlag W, Pfanz H (1990) A fast method to detect the occurrence of non-homogeneous distribution of stomatal aperture in heterobaric plant leaves. Experiments with *Arbutus unedo* L. during the diurnal course. Oecologia 82:52–55

Blanke MM, Holthe PA (1997):Bioenergetics, maintenance respiration and transpiration of pepper fruits. J Plant Physiol 150:247–250

Bouma TJ, Nielsen KL, Eissenstat DM, Lynch JP (1997a) Estimating respiration of roots in soil:Interactions with soil CO_2, soil temperature and soil water content. Plant Soil 195:221–232

Bouma TJ, Nielsen KL, Eissenstat DM, Lynch JP (1997b) Soil CO_2 concentration does not affect growth or root respiration in bean or citrus. Plant Cell Environ 20:1495–1505

Bowes G (1993) Facing the inevitable: plants and increasing atmospheric CO_2. Annu Rev Plant Physiol Plant Mol Biol 44:309–332

Bragina TV, Grinieva GM (1998) Gas exchange and respiration in waterlogged maize. Russian J Plant Physiol 45:582–585

Brix H (1988a) Light-dependent variations in the composition of the internal atmosphere of *Phragmites australis* (Cav.) Trin. Ex Steudel. Aquat Bot 30:319–329

Brix H (1988b) Uptake and photosynthetic utilisation of sediment derived carbon by *Phragmites communis* (Cav.) Trin. Ex Steudel. Aquatic Bot 38:377–389

Brook GA, Folkoff ME, Box EO (1983) A world model of soil carbon dioxide. Earth Surface Processes Landforms 8:79–88

Burton AJ, Pregitzer KS (2002) Measurement carbon dioxide concentration does not affect root respiration of nine tree species in the field. Tree Physiol 22:67–72

Campbell WJ, Allen LH, Bowes G (1988) Effects of CO_2 concentration on Rubisco activity, amount, and photosynthesis in soybean leaves. Plant Physiol 88:1310–1316

Chaves MM, Pereira JS, Cerasoli S, Clifton-Brown J, Miglietta F, Raschi A (1995) Leaf metabolism during summer drought in *Quercus ilex* trees with lifetime exposure to elevated CO_2. J Biogeogr 22:255–259

Cheng W (1999) Rhizosphere feedbacks in elevated CO_2. Tree Physiol 19:313–320

Chiodini G, Frondini F, Ponziani F (1995) Deep structures and carbon dioxide degassing in central Italy. Geothermics 24:81–94

Clinton BD, Vose JM (1999) Fine root respiration in mature eastern white pine (*Pinus strobus*) in situ:the importance of CO_2 in controlled environments. Tree Physiol 19:475–479

Coleman JS, McConnaughay KDM, Bazzaz FA (1993) Elevated CO_2 and plant nitrogen use – is reduced tissue nitrogen concentration size-dependent. Oecologia 93 195–200

Constable JVH, Grace JB, Longstreth DJ (1992) High carbon dioxide concentrations in aerenchyma of *Typha latifolia*. Am J Bot 79:415–418

Cook AC, Oechel WC, Sveinbjornsson B (1997) Using Icelandic CO_2 springs to understand the long-term effects of elevated CO_2. In: Raschi A, Miglietta F, Tognetti R, van Gardingen

PR (eds) Plant responses to elevated CO_2. Cambridge University Press, Cambridge, pp 87–102

Cook AC, Tissue DT, Roberts SW, Oechel WC (1998) Effects of long-term elevated [CO_2] from natural CO_2 springs on *Nardus stricta*: photosynthesis, biochemistry, growth and phenology. Plant Cell Environ 21:417–425

Cotrufo MF, Ineson P, Scott A (1998) Elevated CO_2 reduces the nitrogen concentration of plant tissues. Global Change Biol 4:43–54

Crawford RMM, Braendle R (1996) Oxygen deprivation stress in a changing environment. J Exp Bot 47:145–159

Curtis PS, Vogel CS, Pregitzer KS, Zak DR, Teeri JA (1995) Interacting effects of soil fertility and atmospheric CO_2 on leaf area and carbon gain physiology in *Populus x euramericana* (Dode) Guinier. New Phytol 129:253–263

Curtis PS, Wang XZ (1998) A meta-analysis of elevated CO_2 effects on woody plant mass, form and physiology. Oecologia 113:299–313

Drake BG, Gonzalez-Meler MA, Long SP (1997) More efficient plants:a consequence of rising atmospheric CO_2. Annu Rev Plant Physiol Plant Mol Biol 48:609–639

Drake BG, Azcon-Bieto J, Berry J, Bunce J, Dijkstra P, Farrar J, Gifford RM, Gonzalez-Meler MA, Koch G, Lambers H, Siedow J, Wulschleger S (1999) Does elevated atmospheric concentration inhibit mitochondrial respiration in green plants? Plant Cell Environ 22:649–657

Duchi V, Minissale A, Romani L (1985) Studio geochimico su acque e gas dellàrea geotermica lago di Vico-M.Cimini (Viterbo). Atti Soc Toscana Sci Nat Ser A 92:237–254

Eamus D, Jarvis PG (1989) The direct effects of increase in the global atmospheric CO_2 concentrations on natural and commercial temperate trees and forests. Adv Ecol Res 19:1–55

Edwards GR, Clark H, Newton PCD (2003) Soil development under elevated CO_2 affects plant growth response to CO_2 enrichment. Basic Appl Ecol 4:185–195

Emiliani C (1992) Planet earth: cosmology, geology, and the evolution of life and environment. Cambridge University Press, Cambridge

Enoch HZ, Kimball BA (eds) (1986) Carbon dioxide enrichment of greenhouse crops, vol I and II. CRC Press, Boca Raton, FL

Etiope G (1997) Migration in the ground of CO_2 and other volatile contaminants. Theory and survey. In: Raschi A, Miglietta F, Tognetti R, van Gardingen PR (eds) Plant responses to elevated CO_2. Cambridge University Press, Cambridge, pp 7–20

Ewert F, Pleijel H (1999) Phenological development, leaf emergence, tillering and leaf area index, and duration of spring wheat across Europe in response to CO_2 and ozone. Eur J Agron 10:171–184

Fan SM, Wofsky SC, Bakwin PS, Jacob DJ (1990) Atmosphere-biosphere exchange of CO_2 and O_3 in the central amazon forest. J Geophys Res 95:16851–16864

Farrar CD, Neil JM, Howle JF (1999) Magmatic carbon dioxide emissions at Mammoth Mountain, California. US Geological Survey Water-Resources Investigations Report, pp 98–4217

Field CB, Jackson RB, Mooney HA (1995) Stomatal response to increased CO_2: implications from the plant to the global scale. Plant Cell Environ 18:1214–1225

Fordham M, Barnes J (1999) Growth and photosynthetic capacity in *Agrostis canina* and *Plantago major* adapted to contrasting long-term atmospheric CO_2 concentrations. In: Raschi A, Vaccari FP, Miglietta F (eds) Ecosystem response to CO_2: the MAPLE project results. RDG European Commission, Brussels, pp 143–157

Fordham MC, Barnes JD, Bettarini I, Polle A, Slee N, Raines C, Miglietta F, Raschi A (1997a) The impact of elevated CO_2 on the growth of *Agrostis canina* and *Plantago major* adapted to contrasting CO_2 concentrations. In: Raschi A, Miglietta F, Tognetti R, van Gardingen PR (eds) Plant responses to elevated CO_2. Cambridge University Press, Cambridge, pp 174–196

Fordham MC, Barnes JD, Bettarini I, Polle A, Slee N Raines C, Miglietta F, Raschi A (1997b) The impact of elevated CO_2 on the growth and photosynthesis in *Agrostis canina subsp. monteluccii* adapted to contrasting atmospheric CO_2 concentrations. Oecologia 110:169–176

Gahrooee FR (1998) Impacts of elevated atmospheric CO_2 on litter quality, litter decomposability and nitrogen turnover rate of two oak species in a Mediterranean forest ecosystem. Global Change Biol 4:667–677

Garcia RL, Idso SB, Wall GW, Kimball BA (1994) Changes in net photosynthesis and growth of *Pinus eldarica* seedlings in response to atmospheric CO_2 enrichment. Plant Cell Environ 17:971–978

Geiger M, Haake V, Ludewig F, Sonnewald U, Stitt M (1999) The nitrate and ammonium nitrate supply have a major influence on the response of photosynthesis, carbon metabolism, nitrogen metabolism and growth to elevated carbon dioxide in tobacco. Plant Cell Environ 22:1177–1199

Gerlach TM (1991) Present-day CO_2 emissions from volcanoes. Eos, Transactions, American Geophysical Union 72(23):249–255

Gibberd MR, Gray JD, Cocks PS, Colmer TD (2001) Waterlogging tolerance among a diverse range of *Trifolium* accessions is related to root porosity, lateral root formation and 'aerotropic rooting'. Ann Bot 88:579–589

Gonzalez-Meler MA, Siedow JN (1999) Direct inhibition of mitochondrial respiratory enzymes by elevated CO_2: does it matter at the tissue or whole-plant level? Tree Physiol 19:253–259

Gonzalez-Meler MA, Ribas-Carbo M, Siedow JN, Drake BG (1996) Direct inhibition of plant mitochondrial respiration by elevated CO_2. Plant Physiol 112:1349–1355

Grace J, van Gardingen PR (1997) Sites of naturally elevated carbon dioxide. In: Raschi A, Miglietta F, Tognetti R, van Gardingen PR (eds) Plant responses to elevated CO_2. Cambridge University Press, Cambridge, pp 1–6

Gregory PJ, Palta JA, Batts GR (1996) Root systems and root:mass ratio – carbon allocation under current and projected atmospheric conditions in arable crops. Plant Soil 187:221–228

Hamilton JG, Thomas RB, DeLucia EH (2001) Direct and indirect effects of elevated CO_2 on leaf respiration in a forest ecosystem. Plant Cell Environ 24:975–982

Hättenschwiler S, Miglietta F, Raschi A, Körner C (1997) 30 years' in situ tree growth under elevated CO_2: a model for future forest responses? Global Change Biol 3:463–471

Heath OVS (1948) Control of stomatal movement by a reduction in the normal [CO_2] of the air. Nature 161:179–180

Heath J, Mansfield T (1999) CO_2 enrichment of the atmosphere and the water economy of plants. In: Agrawal SB, Agrawal M (eds) Environmental pollution and plant response. Lewis, Boca Raton, pp 33–42

Hocking D, Hocking MB (1977) Equilibrium solubility of trace atmospheric sulfur dioxide in water and its bearing on air pollution injury to plants. Environ Pollut 13:57–64

Houghton JT, Jenkins GJ, Ephraums JJ (1990) (eds) Climate change: the IPCC Scientific Assessment. Cambridge University Press, Cambridge

Huang BR, Johnson JW (1995) Root respiration and carbohydrate status of 2 wheat genotypes in response to hypoxia. Ann Bot 75:427–432

IPCC (Intergovernmental Panel for Climatic Change) (1996) Climate change. The second assessment report of IPCC working group I. Cambridge University Press, Cambridge, UK

Jacob J, Greitner C, Drake BG (1995) Acclimation of photosynthesis in relation to Rubisco and non-structural carbohydrate contents and in situ carboxylase activity in *Scirpus olneyi* grown at elevated CO_2 in the field. Plant Cell Environ 18:875–884

Jahnke S (2001) Atmospheric CO_2 concentration does not directly affect leaf respiration in bean or poplar. Plant Cell Environ 24:1139–1151

Jahnke S, Krewitt M (2002) Atmospheric CO_2 concentration may directly affect leaf respiration measurement in tobacco, but not respiration itself. Plant Cell Environ 25:641–651

Joel G, Chapin FS, Chiariello NR, Thayer SS, Field CB (2001) Species specific responses of plant communities to altered carbon and nutrient availability. Global Change Biol 7:435–450

Jones HG (1992) Plants and microclimate, 2nd edn. Cambridge University Press, Cambridge, UK

Jongen M, Jones MB, Hebeisen T, Blum H, Hendrey G (1995) The effect of elevated CO_2 concentrations on the root growth of *Lolium perenne* and *Trifolium repens* in a FACE system. Global Change Biol 1:361–370

Kaligariĉ M (2001) Vegetation patterns and responses to elevated CO_2 from natural CO_2 springs at Strmec (Radenci, Slovenia). Acta Biol Slovenica 44:31–38

Kenner AA, Ahmed SI (1975) Measurements of electron transport activities in marine phyto-plankton. Mar Biol 33:117–120

Kiefer RH, Amey RG (1992) Concentrations and controls of soil Carbon dioxide in sandy soil in the North Carolina coastal plain. Catena 19:539–559

Kimball BA, Pinter PJ, Garcia RL, Lamorte RL, Wall GW, Hunsaker DJ (1995) Productivity and water-use of wheat under free-air CO_2 enrichment. Global Change Biol 1:429–442

Krupa SV, Kickert RN (1993) The greenhouse effect – the impacts of carbon dioxide (CO_2), ultraviolet B (UV-B) radiation and 0zone (O_3) on vegetation (crops). Vegetatio 104:223–238

Körner C, Miglietta F (1994) Long term effects of naturally elevated CO_2 on Mediterranean grassland and forest trees. Oecologia 99:343–351

Kühny JS, Peet MM, Nelson PV, Willits DH (1991) Nutrient dilution by starch in CO_2-enriched *Chrysanthemum*. J Exp Bot 42:711–716

Lambers H, Atkin OK, Scheurwater I (1996a) Respiratory patterns in roots in relation to their functioning. In: Waisel Y, Eshel A, Kafkafi U (eds) Plant roots: the hidden half. Marcel Decker, New York, pp 232–362

Lambers H, Stulen I., van der Werf A (1996b) Carbon use in root respiration as affected by elevated atmospheric CO_2. Plant Soil 187:251–263

Lambers H, Atkin OK, Millenaar, FF (2002) Respiratory patterns in roots in relation to their functioning. In: Waisel Y, Eshel A, Kafkafi U (eds) Plant roots: the hidden half, 3rd edn. Marcel Dekker, New York, pp 521–552

Larcher W (2001) Ökophysiologie der Pflanzen, 6th edn. Ulmer, Stuttgart

Leadley PW, Körner C (1996) Effect of elevated CO_2 on plant species in a highly diverse calcareous grassland. In: Körner C, Bazzaz FA (eds) Carbon dioxide, populations and communities. Academic Press, San Diego, pp 158–175

Leadley PW, Stöcklin J (1996) Effects of elevated CO_2 on model calcareous grassland:community, species, and genotype level responses. Global Change Biol 2:389–397

Lloyd J, Farquhar GD (1996) The CO_2 dependence of photosynthesis, plant-growth responses to elevated atmospheric CO_2 concentrations and their interaction with soil nutrient status. I. General principles and forest ecosystems. Funct Ecol 10:4–32

Maĉek I, Pfanz H, Vodnik D, Batiĉ F (2002) Growth and root respiration of C_4 plants under CO_2 enrichment. Acta Biol Slovenica (in press)

Manning WJ, von Tiedemann A (1995) Climate change potential effects of increased atmospheric carbon dioxide (CO_2), ozone (O_3), and ultraviolet-B (UV-B) radiation on plant diseases. Environ Pollut 88:219–245

Marschner H (1995) Mineral nutrition of higher plants, 2nd edn. Academic Press, London, 889 pp

Martini M (1997) CO_2 emission in volcanic areas: case histories and hazards. In: Raschi A, Miglietta F, Tognetti R, van Gardingen PR (eds) Plant responses to elevated CO_2. Cambridge University Press, Cambridge, pp 35–44

Matt P, Geiger M, Walch-Liu P, Engels C, Krapp A, Stitt M (2001) Elevated carbon dioxide increases nitrate uptake and nitrate reductase activity when tobacco is growing on nitrate, but increases ammonium uptake and inhibits nitrate reductase activity when tobacco is growing on ammonium nitrate. Plant Cell Environ 24:1119–1137

McDougal DT, Working EB (1933) The pneumatic system of plants, especially trees. Carn Inst Wash Publ 441

McKee IF, Farage PK, Long SP (1995) The interactive effects of elevated CO_2 and O_3 concentration on photosynthesis in spring wheat. Photosynth Res 45:11–16

Meister MH, Bolhar-Nordenkampf HR, Kropf PJ (1999) Plant growth modified by the CO_2 gradient along the road ditch at mofeta Strmec. In: Kaligariĉ M (ed) SNACE, FACE and OTCs CO_2 enrichment at the leaf/air interface and/or at the root/soil interface; results in growth and development of plants. Abstracts of International Workshop, Maribor, pp 30–31

Melillo J, Callaghan TV, Woodward FI, Salati E, Sinha SK (1990) Effects on ecosystems. In: Houghton JT, Jenkins GJ, Ephraums JJ (eds) Climate change: the IPCC Scientific Assessment. Cambridge University Press, Cambridge, pp 283–310

Miglietta F (1997) Non-traumatic responses of natural vegetation to long-term carbon dioxide enrichment. In: Allen LH, Kirkham MB, Olszyk DM, Whitman CE (eds) Advances in carbon dioxide effects research. ASA Special Publ 61. Amer Soc Agronomy, Madison, Wisconsin, pp 101–112

Miglietta F, Raschi A (1993) Studying the effect of elevated CO_2 in the open in a naturally enriched environment in central Italy. Vegetatio 104:391–400

Miglietta F, Raschi A, Bettarini I, Resti R, Selvi F (1993) Natural CO_2 springs in Italy: a resource for examining the long-term response of vegetation to rising atmospheric CO_2 concentration. Plant Cell Environ 16:873–878

Miglietta F, Badiani M, Bettarini I, van Gadingen PR, Selvi F, Raschi A (1995) Preliminary studies of a long-term CO_2 response of Mediterranean vegetation around natural CO_2 vents. In: Moreno JM, Dechela EC (eds) Global changes and mediterranean-type ecosystems. Springer, Berlin Heidelberg New York, pp 102–117

Miglietta F, Bettarini I, Raschi A, Körner C, Vaccari FP (1998a) Isotope discrimination and photosynthesis of vegetation growing in the Il Bossoleto CO_2 spring. Chemosphere 36:771–776

Miglietta F, Magliulo V, Bindi M, Cerio L, Vaccari FP, Loduca V, Peressotti A (1998b) Free air CO_2 enrichment of potato (*Solanum tuberosum* L.): development, growth and yield. Global Change Biol 4:163–172

Mörner NA, Etiope G (2002) Carbon degassing from the lithosphere. Global Planet Change 33:185–203

Mook WG (1986) ^{13}C in atmospheric CO_2. Neth J Sea Res 20:211–223

Navas ML, Guillerm JL, Fabreguettes J, Roy J (1995) The influence of elevated CO_2 on community structure, biomass and carbon balance of Mediterranean old-field microcosms. Global Change Biol 1:325–335

Nobel PS, Palta JA (1989) Soil O_2 and CO_2 effects on root respiration of cacti. Plant Soil 120:263–271

O'Neill EG (1994) Responses of soil biota to elevated atmospheric carbon dioxide. Plant Soil 165:55–65

Panichi C, Tongiorgi E (1975) Carbon isotopic composition of CO_2 from springs, fumaroles, mofettes, and travertines of central and southern Italy: a preliminary prospection method of geothermal area. In: Proceedings of the Second UN Symposium on the development and use of Geothermal Resources. San Francisco, CA, USA, 20–29 May 1975

Paoletti E, Nourrisson G, Garrec JP, Raschi A (1998) Modifications of the leaf surface structures of *Quercus ilex* L. in open, naturally CO_2-enriched environments. Plant Cell Environ 21:1071–1075

Pearson M, Brooks GL (1996) The effect of elevated CO_2 and grazing by *Gastrophysa viridula* on the physiology and re-growth of *Rumex obtusifolius*. New Phytol 133:605–611

Penuelas J, Filella I, Tognetti R (2001) Leaf mineral concentrations of *Erica arborea, Juniperus communis* and *Myrtus communis* growing in the proximity of a natural CO_2 spring. Global Change Biol 7:291–301

Petit JR, Jouzel J, Raynuad D, Barkov NI, Barnola J-M, Basile I, Bender M, Chappellaz J, Davis M, Delaygue G, Delmotte M, Kotlyakov VM, Legrand M, Lipenkov VY, Lorius C, Pepin L, Ritz C, Saltzman E, Stievenard M (1999) Climate and atmospheric history of the past 420,000 years from the Vostok ice core, Antarctica. Nature 399:429–436

Pezdiĉ J, Dolenec T, Pirc S, i ek D (1995) Hydrogeochemical properties and activity of the fluids in the Pomurje Region of the Pannonian Sedimentary Basin. Acta Geol Hung 39:319-340

Pezdiĉ J, i ek D, Wolf M, Trettin R, Stichler W, Fritz P (1998) Influence of geogenic carbon dioxide on the plant's growth. Rud-Metal Zb 45:154-157

Pfanz H (1994) Apoplastic and symplastic proton concentrations and their significance for metabolism. In: Schulze E-D, Caldwell MM (eds) Ecophysiology of photosynthesis. Ecological studies 100. Springer, Berlin Heidelberg New York, pp 103-122

Pfanz H, Aschan G (2001) The existence of bark and stem photosynthesis in woody plants and its significance for the overall carbon gain. An eco-physiolocigal and ecological approach. In: Esser K, Lüttge U, Kadereit JW, Beyschlag W (eds) Progress in botany 62. Springer, Berlin Heidelberg New York, pp 477-510

Pfanz H, Dietz KJ (1987) A fluorescence method for the determination of the apoplastic proton concentration in intact leaf tissues. J Plant Physiol 129:41-48

Pfanz H, Heber U (1986) Buffer capacities of leaves, leaf cells, and leaf cell organelles in relation to fluxes of potentially acidic gases. Plant Physiol 81:597-602

Pfanz H, Heber U (1989) Determination of extra- and intracellular pH values in relation to the action of acidic gases on cells. In: Linskens HF, Jackson JF (eds) Modern methods of plant analysis NS, vol 9. Gases in plant and microbial cells. Springer, Berlin Heidelberg New York, pp 322-343

Pfanz H, Aschan G, Langenfeld-Heyser R, Wittmann C, Loose M (2002) Tree stem photosynthesis. Naturwissenschaften 89:147-162

Poli E (1970) Aspetti della vita vegetale in ambienti vulcanici. Ann Bot (Roma) 30:59-79

Ponnamperuma FN (1984) Effects of flooding on soils. In: Kozlowski TT (eds) Flooding and plant growth. Academic Press, London, pp 9-45

Pritchard SG, Rogers HH (2000) Spatial and temporal deployment of crop roots in CO_2-enriched environments. New Phytol 147:55–71

Qi J, Marshall JD, Mattson KG (1994) High soil carbon dioxide concentrations inhibit root respiration of Douglas fir. New Phytol 128:435–442

Radin JW, Kimball BA, Hendrix DL, Mauney JR (1987) Photosynthesis of cotton plants exposed to elevated levels of carbon dioxide in the field. Photosynth Res 12:191–203

Rakitina ZG (1970) Effect of an ice crust on gas composition of the internal atmosphere of winter wheat. Sov Plant Physiol 17:907–912

Raschi A, Miglietta F, Tognetti R, van Gardingen PR (eds) (1997) Plant responses to elevated CO_2. Cambridge University Press, Cambridge

Raschi A, Vaccari FP, Miglietta F (eds) (1999) Ecosystem response to CO_2: the MAPLE project results. RDG European Commission, Brussels

Raven JA (1985) pH-regulation in plants. Sci Progr 49:495–509

Raven JA (1995) The early evolution of land plants: aquatic ancestors and atmospheric interactions. Bot J Scotland 47:151–175

Rillig MC, Hernandez GY, Newton PCD (2000) Arbuscular mycorrhiza respond to elevated atmospheric CO_2 after long-term exposure: evidence from a CO_2 spring in New Zealand supports the resource balance model. Ecol Lett 3:475–478

Rogers HH, Brett Runion G, Krupa SV (1994) Plant responses to atmospheric CO_2 enrichment with emphasis on roots and the rhizosphere. Environ Pollut 83:155–160

Rogers HH, Prior SA, Runion GB, Mitchell RJ (1996) Root to shoot ratio of crops as influenced by CO_2. Plant Soil 187:229–248

Rogie JD, Kerrick DM, Chiodini G, Frondini F (2000) Flux measurements of non-volcanic CO_2 emission from some vents in central Italy. J Geophys Res 105:8435–8445

Rogie JD, Kerrick DM, Sorey ML, Chiodini G, Galloway DL (2001) Dynamics of carbon dioxide emission at Mammoth Mountain, California. Earth Planetary Sci Lett 188:535–541

Rosenzweig C, Parry ML (1994) Potential impact of climate change on world food supply. Nature 367:133–138

Sage RF, Sharkey TD, Seemann JR (1989) Acclimation of photosynthesis to elevated CO_2 in 5 C_3 species. Plant Physiol 89:590–596

Šajna N, Meister M, Bolhar-Nordenkampf HR, Kaligarič M (2002) Possible influence of naturally elevated CO_2 concentration on a semi-natural wet meadow. In: Kaligarič M, Škornik S (eds) Symposium "Flora and Vegetation in Changing Environment" 6 Abstracts. Maribor, p 46

Schwiebert G (1985) CO_2-Düngung im Gartenbau. Zierpflanzenbau 20:27–33

Selvi F (1997) Acidophilic grass communities in central Italy: composition, structure and ecology. In: Raschi A, Miglietta F, Tognetti R, van Gardingen PR (eds) Plant responses to elevated CO_2. Cambridge University Press, Cambridge, pp114–133

Selvi F, Bettarini I (1999) Geothermal biotopes in central-western Italy from a botanical view point. In: Raschi A, Vaccari FP, Miglietta F (eds) Ecosystem response to CO_2: the MAPLE project results. Research Directorate general Unit D.1.1. European Commission EUR 19100, pp 1–12

Sestak Z, Catsky J, Jarvis PG (1971) Plant photosynthesis production. Manual of methods. W Junk Publ, The Hague

Sigurdsson BD (2001) Elevated [CO_2] and nutrient status modified leaf phenology and growth rhythm of young *Populus trichocarpa* trees in a 3-year field study. Trees Struct Funct 15:403–413

Sigurdsson H, Devine JD, Tchoua FM, Presser TS, Pringle MKW, Evans WC (1986) Origin of the lethal gas burst from lake Monoun, Cameroun. J Volcanol Geothermal Res 31:1–16

Sorrell BK (1999) Effeccts of external oxygen demand on radial oxygen loss by *Juncus* roots in titanium citrate solutions. Plant Cell Environ 22:1587–1593

Stager C (1987) Silent death from Cameroon's Killer Lake. Natl Geogr Mag 172:404–420

Stitt M (1991) Rising CO_2 levels and their potential significance for carbon flow in photosynthetic cells. Plant Cell Environ 14:741–762

Stryer L (1988) Biochemistry, 3rd edn. Freeman, New York

Stupfel M, Le Guern F (1989) Are there biomedical criteria to assess an acute carbon dioxide intoxication by a volcanic emission? J Volcanol Geothermal Res 39:247–264

Tognetti R, Minnocci A, Penuelas J, Raschi A, Jones MB (2000) Comparative field water relations of three Mediterranean shrub species co-occurring at a natural CO_2 vent. J Exp Bot 51:1135–1146

Tuba Z, Csintalan Z, Szente K, Nagy Z, Grace J (1998) Carbon gains by desiccation-tolerant plants at elevated CO_2. Funct Ecol 12:39–44

Tuba Z, Proctor MCF, Takacs Z (1999) Desiccation-tolerant plants under elevated air CO_2: a review. Z Naturwiss 54:788–796

Turk B, Pfanz H, Vodnik D, Batič F, Šinkovič T (2002) The effects of elevated CO_2 in natural CO_2 springs on bog rush (*Juncus effusus* L.) plants. I. Effects on shoot anatomy. Phyton 42:13–23

Vaccari FP, Bettarini I, Giuntoli A, Miglietta F, Raschi A (2001) Mediterranean grassland community under elevated [CO_2]:observations from a CO_2 spring. J Medit Ecol 2:41–50

Van Gardingen PR, Grace J, Harkness DD, Miglietta F, Raschi A (1995) Carbon-dioxide emissions at an italian mineral spring – measurements of average CO_2 concentration and air-temperature. Agric For Meteorol 73:17–27

Van Gardingen PR, Grace J, Jeffree CE, Byari SH, Miglietta F, Raschi A, Bettarini I (1997) Long-term effects of enhanced CO_2-concentrations of leaf gas exchange:research opportunities using CO_2 springs. In: Raschi A, Miglietta F, Tognetti R, van Gardingen PR (eds) Plant responses to elevated CO_2. Cambridge University Press, Cambridge, pp 69–86

Van Oosten JJ, Besford RT (1995) Some relationships between the gas-exchange, biochemistry and molecular biology of photosynthesis during leaf development of tomato plants after transfer to different carbon-dioxide concentrations. Plant Cell Environ 18:1253–1266

Van Vuuren MMI, Robinson D, Fitter AH, Chasalow SD, Williamson L, Raven JA (1997) Effects of elevated atmospheric CO_2 and soil water availability on root biomass, root length, and N, P and K uptake by wheat. New Phytol 135:455–465

Vartapetian BB, Jackson MB (1997) Plant adaptations to anaerobic stress. Ann Bot 79:3–20

Vodnik D, Turk B, Pfanz H, Batiĉ F, Wittmann C, Kaligariĉ M, Zupan G (2001): Rast in delovanje rastlin pri povecanih koncentracijah oglikovega dioksida ob naravnih virih CO_2. (Plant growth and functioning at elevated CO_2 concentrations near natural CO_2 spring). Proceedings of the Slovenian Congress on elevated CO_2. In: Komac M (ed) Varstvo zraka v Sloveniji. Ljubljana, pp 205–211

Vodnik D, Pfanz H, Wittmann C, Maĉek I, Kastelec D, Turk B, Batiĉ F (2002a) Photosynthetic acclimation in plants growing near a carbon dioxide spring. Phyton 42:239–244

Vodnik D, Pfanz H, Mâcek I, Kastelec D, Lojen S, Batiĉ F (2002b) Photosynthesis of cockspur [*Echinochloa crus-galli* (L.) Beauv.] at sites of naturally elevated CO2 concentration. Photosynthetica 40:575–579

Vodnik D, Šircelj H, Kastelec D, Maĉek I, Pfanz H, Batiĉ F (2003) The effects of natural CO2 enrichment on the growth of maize. J Crop Prod (in press)

Von Faber A (1925) Untersuchungen über die Physiologie der javanischen Solfatarenpflanzen. Flora 118:89–110

Wagner U (1990) Kinetik und Mechanismus der pH-Stabilisierung in grünen Blättern höherer Pflanzen. Doctoral Thesis, University of Würzburg

Wagner J, Lüscher A, Hillebrand C, Kobald B, Spitaler N, Larcher W (2001) Sexual reproduction of *Lolium perenne L.* and *Trifolium repens* L. under free air CO_2 enrichment (FACE) at two levels of nitrogen application. Plant Cell Environ 24:957–965

Webber AN, Nie G-Y, Long SP (1994) Acclimation of photosynthetic proteins to rising atmospheric CO_2. Photosynth Res 39:413–419

Weigel H, Manderscheid R, Jäger H-J, Mejer GJ (1994) Effects of season-long CO_2 enrichment on cereals. I. Growth performance and yield. Agric Ecosyst Environ 48:231–236

Weinlich FH, Bräuer K, Kämpf H, Strauch G, Tesar J, Weise SM (1999) An active sub-continental mantle volatile system in the western Eger rift, Central Europe: gas flux, isotopic (He, C, and N) and compositional fingerprints. Geochim Cosmochim Acta 63:3653–3671

Weise SM, Bräuer K, Kämpf H, Strauch G, Koch U (2001) Transport of mantle volatiles through the crust traced by seismically released fluids:a natural experiment in the earthquake swarm area Vogtland/NW Bohemia, central Europe. Tectonophysics 336:137–150

Yoshioka T, Satoh S, Yamasue Y (1998) Effect of increased concentration of soil CO_2 on intermittent flushes of seed germination in *Echinochloa crus-galli* var. *crus-galli*. Plant Cell Environ 21:1301–1306

Zak DR, Pregitzer KS, Curtis PS, Teeri JA, Fogel R, Randlett DI (1993) Elevated atmospheric CO_2 and feedback between carbon and nitrogen cycles. Plant Soil 151:105–117

Prof. Dr. Hardy Pfanz
Dr. Christiane Wittmann
Dr. Guido Aschan
Institut für Angewandte Botanik
Universität Duisburg-Essen
Universitätsstrasse 5
45117 Essen, Germany
e-mail: hardy.pfanz@uni-essen.de

Doz. Dr. Dominik Vodnik
Oddelek za agronomijo
Biotehniška fakulteta
Univerza v Ljubljani
1000 Ljubljana, Slovenia

Dr. Antonio Raschi
IATA, CNRS
50144 Firenze, Italia

Recent Advances in Understanding Plant Invasions

Hansjörg Dietz and Tom Steinlein

1 Introduction

In the last two decades biological invasions have drawn increasing attention (see e.g., Drake et al. 1989; Lodge 1993; Williamson 1996; Lonsdale 1999; Alpert et al. 2000; Kolar and Lodge 2001). Compared to other subjects of ecological investigation, studies related to biological invasions have almost exploded in number during the last years (Fig. 1). The fact that an increasing number of ecologists gets caught up in the problem of biological invasions can be attributed to several reasons. First, biological invasions

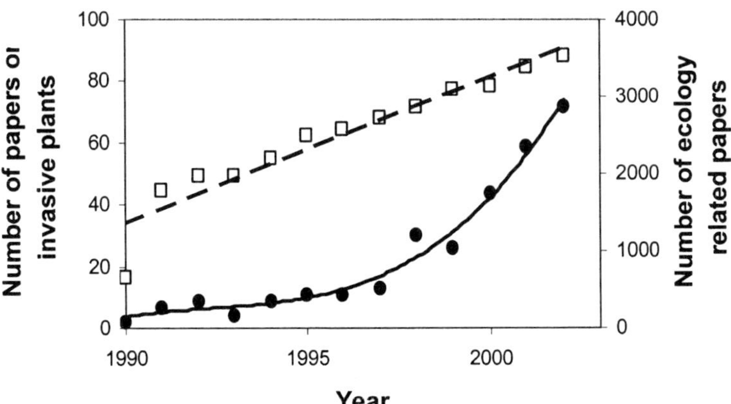

Fig. 1. Increase in the number of published papers on invasive plants between 1990 and 2002 (*filled circles*, plotted against the left *y*-axis). The data were obtained by a year-specific search in the ISI WebOfScience database using *invasive plant(s)* or *plant invasion(s)* as search terms (for the title and the abstract). Note that not all papers on plant invasions were retrieved by these search terms, nor are papers on other invasive organisms included. To account for the constantly increasing journal coverage by ISI, the development in the number of invasion-related studies is shown calibrated against the development in the number of ecology-related studies in the same period of time (*open squares*, plotted against the right *y*-axis). While the number of ecology-related studies increased fairly linearly there has been an exponential increase in the number of studies related to invasive plants in the last years

Progress in Botany, Vol. 65

are increasingly recognized as one of the most important threats to biodiversity (D'Antonio and Vitousek 1992; Vitousek et al. 1997a; Walker and Steffen 1997; Mooney 1999), i.e. biological invasions tend to homogenize the earth's biota (Lodge 1993; Vitousek et al. 1997b; Mooney 1999; Mack et al. 2000). Second, biological invasions can also pose severe environmental, economic and sometimes even health threats (Vitousek et al. 1997a,b; Mack et al. 2000 and references therein, Pimentel et al. 2000). Invasive species affect native species and ecosystems by competing directly for resources that native species require, by altering ecosystem functions and processes such as nutrient and hydrologic cycles, and fire frequency and/or intensity. There are virtually no natural areas left that have not felt the impact of non-native invaders (Usher 1988). Third, biological invasions represent great natural experiments for the ecologist whose investigation is extremely valuable for the understanding of population spread (Sakai et al. 2001) and community- and landscape-level processes affecting the patterns and abundance of species at large spatial and temporal scales, i.e. scales which are otherwise hardly accessible for experimental ecologists. Furthermore, research is increasingly focusing on the problems related to global change, which include biological invasions because species invasions do not only respond to but are also an integral part of global change (Vitousek et al. 1996). In addition, biological invasions are still increasing due to increasing transport, commerce and changing land use regimes (cf. di Castri 1989; Cohen and Carlton 1995; Walker and Steffen 1997; Mack et al. 2000).

As a consequence of this strong interest in biological invasions and the large amount of published studies available on this topic, several authors have published review articles summarizing diverse aspects of invasions such as species invasiveness (Crawley et al. 1996; Kolar and Lodge 2001), habitat invasibility (Levine and D'Antonio 1999; Stohlgren et al. 1999) biological invasions from a population biological perspective (Sakai et al. 2001), the role of enemy release in plant invasions (Keane and Crawley 2002), biological invasions and global change (Dukes and Mooney 1999), the ecological impact of invaders (Parker et al. 1999), the concepts and definitions of plant invasions (Richardson et al. 2000b), approaches to study biological invasions (Mack 1996; Vermeij 1996) and more general reviews (Alpert et al. 2000; Mack et al. 2000). Our intention with this progress report is to draw on these reviews and on original work to give a broad summary on all relevant aspects that have been investigated so far in the search for the causes of plant invasions.

Invasion ecologists now seem to agree that three factor complexes largely determine (plant) invasions: species invasiveness, habitat invasibility and propagule pressure (Lonsdale 1999; Davis et al. 2000). Correspondingly, this review deals with (1) potential species traits conferring increased invasiveness (including growth and reproductive strategies, resistance against herbivory and genetic aspects (e.g. Meekins and McCarthy 1999; Chen et al. 2002; Lee 2002), (2) characteristics of the habitat or community favouring plant invasions (e.g. changing resource availability, disturbance regime, herbivore effects and native species diversity and competitiveness; cf. Lonsdale 1999; Prieur-Richard et al. 2000, 2002) and (3) factors determining propagule pressure (i.e., spatio-temporal processes underlying high propagule availability for increased invasion success). However, due to limited space, we will not review the distinct impacts of invaders in terms of economics and nature conservation, neither will we present biological control mechanisms of invaders.

It was our objective not only to briefly discuss the diverse array of (potential) factors influencing plant invasions but also to provide a concluding discussion that attempts to evaluate the different factors as to their explanatory potential or their importance for invasion success and to analyze current generalizations on the factors of plant invasions with regard to their robustness. Our review covers mainly the most recent literature and, due to limited space, cannot claim to be exhaustive in literature coverage.

There are many different terms and definitions for non-indigenous species ("aliens", "alien invaders", "neophytes", "xenophytes"). Cronk and Fuller's (1995) definition implies an impact on the system that is invaded: "a plant invader is an alien plant spreading naturally (without direct assistance of people) in natural or semi-natural habitats, to produce a significant change in terms of composition, structure or ecosystem process." We will use the more unbiased definition given by Daehler (2001) who argued that the primary criterion for a species to be considered an "invader" (other than being new to a certain region) should be that the new species is "spreading" in a new environment.

2 Invader Traits (Invasiveness)

In the last decades there have been many attempts to find traits that make a specific species invasive. Already in 1965 Baker (1974) found several plant traits ("weedy traits") correlated with the weediness of specific plant species and plants with many of these characteristics were more likely to be highly weedy than plants with only a few of these traits. However, up to now there have been no indications for the presence of a suite of universally successful invader's plant traits (Rejmánek and Richardson 1996; Williamson 1999). The success of invaders is always dependent on specific climatic conditions, the invasibility of the habitat where they arrive (see below) and on interactions between the invader and the organisms of the invaded (plant) community. In the following, we will briefly discuss the plant traits that are thought to be most conducive to the spread and persistence of plant invaders. Certain species biological characteristics have been investigated to predict their invasiveness under specific environmental conditions. In general, it has been found that the sets of plant traits contributing to successful invasion into human-made habitats differ from those supporting invasion into relatively undisturbed vegetation (large and few vs. numerous and small seeds, clonal vs. sexual reproduction, high vs. low competitive strength, tolerance to harsh environmental conditions, initial growth rates of seedlings, resource allocation patterns, etc.). The task is very complex as investigations in the United Kingdom illustrate: when analyzing native and alien plants which are expanding their ranges, in the great majority of cases, expanding aliens and natives are functionally indistinguishable (Thompson et al. 1995).

2.1 Biogeographical Aspects

It has been suggested that species populations which successfully invade new geographic areas originate in more competitive genotypes of their native areas of occurrence. This is more likely to be the case when a species has a wide geographical distribution, which may give rise to a wide variety of ecotypes. Most non-indigenous plants invade latitudes similar to their native occurrences. However, the invaders' function in their new habitats may be fundamentally different from that in the native communities. (Rejmánek and Richardson 1996; Callaway and Aschehoug 2000). Prinzing et al. (2002) mentioned that native growth ranges may be indicative for the areas of successful spread of aliens. They investigated central European aliens in two Argentine provinces characterized by warm and dry climatic conditions (Buenos Aires and Mendoza), and found that these alien species also have a preference for warm, dry and nitrogen-rich conditions in central Europe. Woody plants native to one part of North America were unlikely to become invasive in regions characterized by different climatic conditions elsewhere in North America (with some notable exceptions such as *Robinia pseudoacacia* in California, Oregon, and Washington; Reichard 1997). In California 21% of the invaders and none of the non-invaders were from the Mediterranean area. This region of investigation has Mediterranean-like climate, with cool moist winters and hot dry summers. This precipitation pattern might be a problem for the many species from summer-rain regions (e.g. Europe and temperate Asia) but not for those from the Mediterranean (Reichard 1997).

All these examples suggest that the characteristics of the native range of a species provide some clues to the (climatic) suitability of the introduced range.

2.2 Vegetative Growth

Rapid growth rates (e.g. Huxman et al. 1998; Pattison et al. 1998), or the ability to respond plastic to changing environmental conditions (e.g. Steinlein et al. 1996; Dietz and Ullmann 1997) are often discussed as crucial traits of invaders. An extreme example of the possible consequences of rapid growth rates is salvinia (*Salvinia molesta*), a floating aquatic fern, which is capable of doubling its population size every 2–3 days under ideal conditions, quickly choking out water bodies that it infests (Cronk and Fuller 1995). Invasion success can also be related to tolerance of adverse environmental conditions or the ability to thrive in more diverse habitats (e.g. Glenn et al. 1998; Chen et al. 2002; Uveges et al. 2002).

One of the few general attributes of competitive invasive plant species is that they tend to allocate more biomass to growth, and less to reproduction or defense, than less competitive species. This shift and/or difference in allocation patterns can occur not only at the species level; many studies show differences in growth dynamics and patterns between ecotypes, or even populations. Baruch and Goldstein (1999) compared 83 populations of 34 native and 30 invasive species in Hawaii and found that specific leaf area (SLA) and net CO_2 assimilation was higher and leaf construction cost lower in invasive species. Bastlova and Kvet (2002) examined native and non-native populations of *Lythrum salicaria*. The more vigorous growth of non-native invasive *L. salicaria* plants was the result of differences in dry weight partitioning (higher allocation to shoots) between native and non-native plants.

2.3 Clonal Growth and Reproduction

Vegetative reproduction, such as root sprouting and layering, may facilitate rapid population increase and contribute to population recovery following disturbance. Clonal growth appears to be significantly more common in invasive woody plants than in non-invasive woody plants (Reichard 1997). On the other hand, Pyšek (1997) found that, in general, clonal growth is somewhat underrepresented among invasive alien plants. This may be explained by the rather poor generative reproduction of clonal plants compared to non-clonal species. Many invasive plant species can reproduce both by seed and vegetative growth. But the role of clonality in plant invasions is context-dependent. Clonal invaders are more prevalent in wetter and colder than in drier and warmer climatic areas, and in natural, less disturbed rather than in man-made habitats. Compared to non-clonals, clonal invaders appear to be at a disadvantage in the dispersal phase of invasion (Pyšek 1997). As efficient long-distance seed dispersal is essential for the spread of alien plant species and as clonal plants often lack efficient sexual reproduction mechanisms, these species rely on anthropogenic transport or other transport media. They can be seriously invasive if fragments of (usually subterraneous) plant parts are dispersed by water flow, animals or anthropogenic disturbance. Their ability to regenerate rapidly may help them spread vegetatively at new sites. Examples of those clonal weeds are *Cyperus esculentus* L. which spreads by tubers and rhizomatous growth (see ter Borg et al. 1998 and references therein), *Fallopia japonica* which seems to spread exclusively by regeneration of rhizome fragments in introduced areas (Seiger 1997 and references therein) and *Lepidium latifolium* and *Rorippa austriaca* spreading by translocated root fragments and clonal growth by lateral roots (Young et al. 1997; Dietz et al. 2002). Once established, clonal plants seem to be more persistent and competitive, which leads to an effective occupation of the available space (Pyšek 1997). Anthropogenic spread via soil transport may particularly favour invaders with clonal growth by lateral roots that are able to regenerate from root or rhizome fragments (Dietz et al. 2002). Some invaders even propagate by regeneration from shoot fragments. For example, in the noxious invader *Lythrum salicaria* approximately 80% of the shoots survived fragmentation and produced adventitious roots and lateral shoots (Brown and Wickstrom 1997).

2.4 Competitive Ability

Competitive superiority of invaders in resource uptake (e.g. light and space) is often supposed to help them outcompete native species. Furthermore, some invasive plant species may possess competitive mechanisms that are not present in the communities that they invade. This way they may disrupt inherent, coevolved interactions among long-associated native species.

Competition between co-occurring native plant species and invasive species has not been frequently addressed so far, i.e. controlled growth experiments are rare and it is currently difficult to know to what extent this kind of interaction is involved in the invasion process. In a multiple deWit replacement series Meekins and McCarthy (1999) demonstrated that only one (*Quercus prinus*) out of three native species (one herbaceous annual, two woody perennials) was negatively affected by the invasive species *Alliaria petiolata*. Bakker and Wilson (2001) compared the invasive C_3 grass *Agropyron cristatum* and the native C_4 grass *Bouteloua gracilis* in a mixed grass prairie in Canada. In a transplant experiment in the field they observed strong competitive effects of *A. cristatum* on other plants and this may prevent other species from establishing in fields dominated by this invader. In pot and field experiments Dietz et al. (1998, 1999a) have demonstrated that the invasive species *Bunias orientalis* (Brassicaceae) is a weak competitor compared to co-occurring native ruderal forbs and other traits may account for its rapid spread in Franconia (Germany).

Increased competitive ability can emerge not only from increased allocation to growth, but also from weak co-adaptation between native and invasive species. Furthermore, a species' relative competitive ability may be more important for its invasion success, depending on environmental and community conditions (e.g. presence or absence of mycorrhizae, herbivores, fire, drought or resources) and on the starting conditions (Woitke and Dietz 2002).

Competition between native and alien invasive species for pollinators can be an important factor in reproduction and success of invasive species (see Grabas and Laverty 1999). The increasing spread of invasive plant species raises the possibility that pollination of natives might suffer when they are sympatric with one of the invasives. In contrast to competition for light, water and nutrients, which requires close contact of competitors, competition for pollinators may act over large distances.

In a 2-year study Brown et al. (2002) observed that the invasive *Lythrum salicaria* significantly reduced both pollinator visitation and seed set in the native *Lytrum alatum*. Furthermore, pollinators moved frequently between the two plant species, which may cause heterospecific pollen transfer. They concluded that if similar patterns (limitations in pollinator visitation, seed set reduction and pollen transfer between native and non-native species) occur in the field, invasive plants may be an even greater threat to natives and the community they live in than previously thought. On the other hand, lack of pollinators may limit further spread of non-indigenous invasive plant species as shown for *Lonicera japonica* (Larson et al. 2002). Barthell et al. (2001) stressed the importance of another factor: mutalistic effects between invasive animals (here honey bees) and invasive plants (here yellow star-thistle) may occur. Their results suggest that these "alien" alliances between invasive animals and plants may extend to other invasive plant and pollinator species (see section on 'facilitation' below).

2.5 Morphological Plasticity

Plasticity in general is reported to be one of the most important traits of successful invaders. Plastic responses to specific environments can be either morphological or physiological or there may be wide-ranging responses in the plants' whole life cycle. Gibson et al. (2002), for example, showed that a highly plastic morphological response to local microhabitat conditions is likely to ensure the persistence of *Microstegium vimneum*, an invasive grass in southern Illinois. For *Phalaris arundinacea* Maurer and Zedler (2002) demonstrated that its rapid expansion into a variety of wetlands is a function of morphological plasticity (changes in root:shoot ratios and lateral expansion rates), nutrient availability and clonal subsidy. In *Vaccinium myrtilloides* persistence in deep shade was related to significant morphological and biomass allocation plasticity (Moola and Mallik 1998). In a field study in Lower Franconia (Germany) Dietz et al. (1999a) found that high morphological plasticity of rosette growth is advantageous for persistence of the alien invader *Bunias orientalis* (Brassicaceae) in situations of high competition and in mown communities. Lockhart (1996) found that the invasiveness of the melaleuca tree (*Melaleuca quinquenervia*), an emergent, semi-aquatic tree, is mainly caused by the morphological plasticity of leaf forms that allow it to invade wetland and terrestrial habitats. Schweitzer and Larson (1999) tested the degree of morphological plasticity (here, internode length of climbers, internode number and shoot biomass) of the invasive climber species *Lonicera japonica* and its native congener *L. sempervirens*. *Lonicera japonica* responded to climbing supports with a 15.3% decrease in internode length, a doubling of internode number and a 43% increase in shoot biomass. In contrast, climbing supports did not influence internode length or shoot biomass for *L. sempervirens*, and only resulted in a 25% increase in internode number. This plasticity may allow *L. japonica* to actively place plant modules in favourable microhabitats and ultimately increase its fitness.

2.6 Reproductive Traits

High reproductive effort and success are very common in successful invaders (Steinlein et al. 1996; Meyer 1998; Dietz et al. 1999a; Mandák and Pyšek 1999; Radford and Cousens 2000; Bastlova and Kvet 2002). Harper (1977) pointed out that colonizing plants allocate more resources to reproductive than to vegetative growth. High reproductive output can stabilize new populations and help them persist at a new site, but efficient means of seed dispersal (see below) are also required for spreading of the species. In addition, distinct flower phenologies may favour invasives over native plants (e.g. Bastlova and Kvet 2002). Other common features of invasive species are prolonged flowering and fruiting periods (see below).

Arenas et al. (2002) found that the massive reproductive output of the invasive seaweed *Sargassum muticum* and its limited dispersal range accounts for local and dense recruitment patterns. High seed output, the flowering phenology and the breeding system of *Miconia calvescens* (Melastomataceae), a dominant invasive species in the tropical oceanic island of Tahiti, enables this plant to build up populations rapidly from even a single propagule (Meyer 1998). For the colonization of new sites, long-distance seed dispersal is a prerequisite for successful invasive spread, as shown for *Solidago altissima*, an aggressive invader in central Europe (Meyer and Schmid 1999).

In many cases high reproduction leads to the build-up of large seed banks, that can respond efficiently to frequent disturbance events and give rise to strong seedling recruitment at different times in the year (e.g. Dietz and Steinlein 1998). If reproduction and therefore seed output is high, local dense recruitment in the new habitat is likely to occur (e.g. Steinlein et al. 1996; Arenas et al. 2002; McDowell and Turner 2002).

2.7 Genetic Variation

In the last decades much research has concentrated on the molecular or genetic background of increased invasiveness (e.g. Baker and Stebbins 1965; Ellstrand and Schierenbeck 2000; Reznick and Ghalambor 2001; Lee 2002). Invasive species can often respond quickly and efficiently to new environmental conditions. Changes in the selection regime acting on the invader include both increased selection for adapted genotypes and increase in genetic differentiation among populations. Lee (2002) stressed the importance of genetic attributes such as additive genetic variance, epistasis, hybridization, genetic tradeoffs and the action of small numbers of genes for the success of invaders. Because invaders often have low initial genetic diversity when they arrive in a new location, a high degree of plasticity might be an alternative way of coping with new environments. Alien species adjust in due course to the novel and diverse selective regimes that they encounter as they expand in their new locations (Lambrinos 2002). Sexual species may have a greater ability than asexual species to adjust landscapes with diverse selection pressures. For example, in the invasive plant species *Sapium sebiferum*, the Chinese tallow tree, there were significant post-invasion genetic differences (Siemann and Rogers 2001). Post-introduction adaptation of alien plants may contribute to their invasive success. Adams et al. (1998) argued that new mutations and/or allelic combinations are responsible for the recent invasion of abandoned farmlands by *Juniperus ashei* in the United States.

For evolutionary adaptation to environmental changes, e.g., in the new habitats, additive genetic variance (AGV, i.e. the proportion of genetic variance in a character that is due to the additive component of allelic effects) is necessary, providing the main substrate for selection. High levels of AGV have been found in source populations for traits facilitating invasions (Dietz et al. 1999b; Pappert et al. 2000). High levels of AGV could be lost during founder events. In *Rubus alceifolius*, an invasive weed, successive nested founder events appear to have resulted in cumulative reduction in genetic diversity (Amsellem et al. 2000). This loss in AGV could be alleviated by inter- or intraspecific hybridization of invasive populations with native or non-native populations (Lee 2002).

There are many studies demonstrating positive effects of hybridization on invasiveness (e.g. Ellstrand and Schierenbeck 2000; Milne and Abbott 2000). Hybridization and changes in ploidy level, especially genetic variation from multiple origins of polyploidy within allopolyploid "species", might be a strong determinant of fitness in invasive species. In their survey Ellstrand and Schierenbeck (2000) found 28 plant taxa out of 12 plant families where invasiveness emerged after hybridization and about 24 examples of invasive lineages that are supposed to have a hybrid origin. Evidence for hybridization as a driver for biological invasions was also shown for *Rhododendron ponticum*, where the invasive biotypes seem to be hybrids with 27 North American species (Milne and Abbott 2000). Hybridization can (but does not have to) lead to adaptive evolution by (1) creating novel genotypes with better adaptations to the new environment, (2) creating genetic variation, generating fixed heterosis and (3) dumping genetic load.

The role of a small number of genes as having profound impacts on the invasiveness of species is discussed by Lee (2002). In *Sorghum halepenese*, an allopolyploid grass, a small number of genes associated with invader traits like fast growth, efficient dispersal and high persistence distinguishes it from closely related crop plants (Paterson et al. 1995). Linde et al. (2001) revealed similar results for the flowering time of *Capsella bursa-pastoris*, an invasive weed on a world-wide scale.

3 Environmental Traits (Habitat Invasibility)

Habitat invasibility, i.e. the susceptibility of a specific habitat to invasions by exotic plants is a very complex attribute. Factors contributing to differences in habitat invasibility include disturbance, resource availability, habitat fragmentation and accessibility, evolutionary history, propagule pressure, predation, mutualism and competition (e.g. D'Antonio et al. 1999; Alpert et al. 2000; Richardson et al. 2000a). Furthermore, these factors interact with each other and with specific species traits (Richardson et al. 2000b). Therefore, a causal reasoning of differences in habitat invasibility is difficult to attain (Kolb et al. 2002). Invasibility is an emergent property of invaded ecosystems and their established species that may be affected by extrinsic factors like climatic variation and that affects only the extinction rates of the invaders, and not their immigration rates (Lonsdale 1999). In the following the main mechanisms determining habitat invasibility will be discussed.

3.1 General Resource Availability and Global Change

There are several studies that examined the influence of elevated CO_2 on individually grown invasives or plants grown in monoculture and revealed that positive correlations between growth and elevated CO_2 exist [*Bromus tectorum* invading America (Smith et al. 1987), *Pueraria lobata* (Sasek and Strain 1988) and *Lonicera japonica* (Sasek and Strain 1991) invading parts of Europe and New Zealand]. However, the impact of elevated CO_2 on the behaviour of invasive species in a community context is rather poorly understood. Dukes (2002a) compared the effects of CO_2 enrichment on the invasive *Centaurea solstitialis* grown in monoculture and in serpentine grassland. Aboveground biomass increased by about 70% in monoculture but in competition with the grassland species there was no significant increase compared to the native species. Smith et al. (2000) observed increases in productivity and success of the invasive grass *Bromus madritensis* in the Mojave Desert in North America with increasing CO_2 but they found high interannual differences in the reaction to increased CO_2.

Davis et al. (2000) and Davis and Pelsor (2001) presented a theory of invasibility being dependent on fluctuations (increases) in resource availability. They argued that fluctuations in resource availability are the key factor controlling invasibility, the susceptibility of an environment to invasion by non-resident species. They conclude that the elusive nature of the invasion process arises from the fact that it depends upon conditions of resource enrichment or release that have a variety of causes but which occur only intermittently and, to result in invasion, must coincide with availability of invading propagules.

3.2 Habitat Fragmentation and Patch Size

Although habitat loss and fragmentation are expected to enhance invasive spread, it is not clear at what level of landscape disturbance and patch size this might occur (With 2002). Especially the complex interplay of spatial configuration of fragments, the changes in physical, chemical and biotic fluxes between the fragments and the change in the fragments itself must be further elucidated (see also the discussion in Sect. 4 on 'propagule pressure'). For example, Kiviniemi and Eriksson (2002) found no relationship between the size of grassland fragments and invader spread. Honnay et al. (2002) stressed the importance of edges as accompaniments of fragmentation functioning as pathways for spread of invaders because many alien species move primarily between habitat types. Conversely, in some (rare) cases, if invasive species with limited dispersal abilities occur, habitat fragmentation may prevent the spread of these species (With 2002). Fragmentation is often accompanied by other disturbance effects. For example, Ross et al. (2002) demonstrated for a eucalyptus forest that fragmentation of the habitat alone had a lower effect on invasibility than disturbance combined with fragmentation.

3.3 Diversity and Richness

That there is a relationship between (native) species diversity and habitat invasibility is one classic tenet in invasion ecology (for a review see Levine and D'Antonio 1999). Elton (1958) suggested that communities develop increasing invasion resistance as increasing diversity leads to more tightly connected food webs. This proposed relationship has been supported by many studies, e.g., community assembly models (e.g. Case 1990, 1991; Drake 1990) and experimental analyses (Tilman 1997; Knops et al. 1999; Naeem et al. 2000; Prieur-Richard et al. 2000; Dukes 2001; Kennedy et al. 2002); see also Loope and Mueller-Dombois (1989); Lodge (1993); Pysek and Pysek (1995); Lavorel et al. (1999); Tilman (1999). In the experimental studies it was found that species-rich communities were more resistant to plant invasions than were species-poor communities. In addition, there are indications that higher evenness reduces the invasibility of habitats (Wilsey and Polley 2002).

Different mechanisms may be involved in higher resistance of diverse communities to invasions. Many different species may draw on resources more efficiently, i.e. increasing diversity may lower available resources for potential invaders by a combination of selection processes and niche-complementarity effects (Hector et al. 2001, see also Crawley 1987; Tilman et al. 1996; Hooper and Vitousek 1997; Tilman 1999). However, even highly diverse communities may not be saturated and species richness and niche saturation may not be positively correlated (Troumbis et al. 2002). Higher diversity resisting invasions appears to act on a small scale, for in experimental grassland plots Kennedy et al. (2002) demonstrated that resistance to invasion increases with higher diversity by increased plant density and species richness in the local neighbourhoods, a mechanism that was also found to reduce the success of the invasive *Centaurea solstitialis* in grassland microcosms (Dukes 2002a).

On the other hand, with increasing species diversity the probability increases that the community includes one or several species that are highly competitive compared to possible invaders (Dukes 2002b). The importance of this 'sampling' effect is under current debate in the literature that reports the effects of experimentally manipulated diversity on community invasibility (e.g. Huston 1997; van der Heijden et al. 1999; Wardle 1999; Loreau 2000). Conversely, even the rarer species in a community may add considerably to its diversity-driven resistance against invasions. In an experiment where less abundant species were removed and where this biomass removal was matched by an equivalent biomass removal of abundant species in control plots the exotic grass *Lolium temulentum* established better in plots with reduced species diversity (Lyons and Schwartz 2001). As in this particular case, with limiting water availability less common species might fill niches of resource uptake that are not occupied by more dominant species, lowering resource availability to levels that cannot be tolerated by potential invaders. A greater plant diversity may also increase herbivory of exotic plants (see section on herbivory above).

Positive relationships between the richness of the native flora and plant invasions have also been reported (Lonsdale 1999, see also Knops et al. 1995; Robinson et al. 1995; Case 1996; Planty-Tabacchi et al. 1996, Palmer and Maurer 1997; Stohlgren et al. 1999; Pyšek et al. 2002). These positive relationships may simply reflect greater habitat diversity that is positively correlated with richness of the plant community (Levine and D'Antonio 1999; Lonsdale 1999), i.e. the exotic species are furthered by greater habitat diversity just as the natives are. In addition, in more diverse communities invasions may be facilitated by a more diverse array of pollinators, dispersers, fungi and bacteria (Richardson et al. 2000b).

If observational and experimental studies are compared there is a tendency for conflicting results in that invasibility appears to be negatively related to diversity in experimental studies but increases with diversity in observational study (Naeem et al. 2000). This inconsistency was attributed to extrinsic factors that co-vary with plant diversity and invasion in observational studies. This discrepancy may be also due to the crucial role of native competitive dominants reducing invasion success: in observational studies the occurrence of competitive dominants is often associated with a generally low diversity at productive sites, whereas in recent experimental setups competitively dominant species occurred more frequently in high diversity plots (Wardle 2001 and references therein). Part of the conflict in the results on the relationship between invasibility and species diversity may be resolved if scale is explicitly taken into account (but see Stohlgren et al. 2003). At regional or even larger scales, species diversity is highly indicative of habitat diversity, resource heterogeneity and higher propagule supply and likely explains the positive relationship between habitat invasibility and species diversity at that scale (see Planty-Tabachy et al. 1996; Knops et al. 1999; Levine and D'Antonio 1999; Stohlgren et al. 1999, Levine 2000). In fact, native species richness and the discrimination between island and mainland habitats and reserve vs. non-reserve sites explained about 70% of the variation in exotic plant richness worldwide (Lonsdale 1999).

To merely relate habitat invasibility to species diversity may be inconclusive because species diversity is a rather coarse attribute of a habitat and may not be very informative in relation to ecosystem function. Thus, there have been various recent attempts to look at the role of specific plant functional types and plant functional group richness as potentially better predictors of habitat invasibility. For example, by decreasing resource availability, high functional diversity reduced the success of the invasive annual *Centaurea solstitialis* in grassland community microcosms, whereas species richness alone did not reduce the invasibility of the community (Dukes 2001). Other studies have also reported differences between native functional plant types in resisting plant invaders (Larson and McInnis 1989, D'Antonio 1993; Roché et al. 1994, Tilman 1997, Ferrell et al. 1998, Crawley et al. 1999, Symstad 2000; Woitke and Dietz 2002). In fact, community composition has been often found to have a higher impact on invasibility than species richness itself (Hector et al. 2001, and references therein). This is very apparent for the presence or absence of legumes (Fabaceae) that may strongly influence nitrogen availability in the habitat with possible consequences for habitat invasibility (Maron and Connors 1996; Bishop 2002). Changes in the community composition may also change the competitive situation, which is discussed as one of the most important determinants of the success of plant invasions, particularly including interactions of seedlings of the invasive species with established native plants (Crawley et al. 1999).

The explicatory power of species diversity for habitat invasibility may also increase if community maturity is considered simultaneously. Resident species in more recent communities may have had less time to adapt to local conditions and may have lower competitive abilities than those in older communities; i.e. less mature communities tend to be more invasible (Shea and Chesson 2002).

3.4 Substrate, Habitat Disturbance, Resource Availability and Their Interactions (with Diversity)

Habitat invasibility is not only affected directly or indirectly by biotic factors such as species diversity and herbivore pressure, but also by abiotic factors like the type of the substrate, habitat disturbance and resource availability, i.e. factors that often co-vary with species diversity (Naeem et al. 2000 and references therein). As one of the few generally accepted causal factors disturbance promotes most plant invasions. Disturbance often leads to a disruption of species interactions. It may cause empty niches invaders can occupy and disturbance may be the starting point for new invasions. There are many examples for the increased spread of invasive species along with disturbance (e.g. Burke and Grime 1996; Dietz and Steinlein 1998; Stapanian et al. 1998; Downey and Brown 2000; Buckland et al. 2001). With increasing rates of disturbance, not only the number of invasive species is likely to increase, but also their fraction of total biomass in the vegetation (Crawley 1987; Burke and Grime 1996). Widespread disturbance events that may promote the spread of invasive plant species include overgrazing (e.g. Olson and Wallander 1999; Johnston and Pickering 2001; Scott et al. 2001), mowing (e.g. Steinlein et al. 1996; Dietz and Steinlein 1998), soil disturbance (e.g. Mazia et al. 2001; Woitke and Dietz 2002) and fire (e.g., Rouget et al. 2001; Brooks 2002; Litton and Santelices 2002).

As a result of disturbance or in addition to it, elevated resource availability also increases habitat invasibility (Horvitz et al. 1998; Davis et al. 2000 and references therein). For example, invasibility tends to be higher at sites with deeper soils: the invasion success of *Centaurea solstitialis* was greatest on slopes with deep soils in eastern Washington grasslands (Roché et al. 1994). In a survey of invasions in gullies and on shallow patches in the coastal prairie at Bodega Head, Kolb et al. (2002) found that soils deeper than 90 cm are more susceptible to invasion than shallower soils. In their study, deeper soils tended to be less acidic but to have higher water content so that invasibility appeared to be negatively related to soil acidity but positively related to water availability. As a further example, consider increased light availability following disturbance, e.g., in forest gaps. Plant invaders may benefit from such situations because, in general, as mostly early successional species they show a higher degree of plasticity to light variability than late successional, shade-tolerant species (cf. Chazdon 1996).

The abiotic resistance of habitats to invasions appears to increase with more limiting conditions (see Alpert et al. 2000), particularly in 'harsh' habitats, as a result of the increasing need of special adaptations to successfully colonize under adverse conditions. In this situation the success of an invader may presuppose that its maintenance requirement does not increase as much as that of a resident with environmental harshness, or that it responds more efficiently to increased resources (Chesson and Huntley 1997; Shea and Chesson 2002). For example, the spread of recently introduced exotic species was found to be slower in harsh serpentine soils than in more productive oak woodlands (Williamson and Harrison 2002). Within serpentine meadows, the proportion of exotic species in the vegetation decreased with increasing calcium-magnesium to phosphorus ratio (Harrison 1999). Likewise, in a study of mixed-grass prairie in North Dakota (USA), Larson et al. (2001) found that mesic communities had both higher numbers and a greater abundance of exotic plants than did drier communities. In Ohio (USA), experimentally sown *Alliaria petiolata* performed better in lowland and at forest edges than in the upland and in the interior of forests (Meekins and McCarthy 2001). Fire has also been found to reduce habitat invasibility and therefore, has been suggested as a management tool to control and prevent exotic species invasions in some grasslands (Leach and Givnish 1996).

While variations in a single factor may sometimes largely determine habitat invasibility, it is more common – as usual in ecology – that the interactions of several factors simultaneously control differences in the patterns. For example, in a field experiment, invasion of *Dactylis glomerata* into oak woodland was negatively related to species richness, whereas its invasion success in serpentine meadows was positively correlated to species richness (Williamson and Harrison 2002). Within serpentine habitats ungrazed verges had a higher proportion of exotic species than the grazed central areas, whereas on non-serpentine soils the pattern was reversed (Safford and Harrison 2001). An interaction effect between disturbance and native species composition on community invasibility was observed by Symstad (2000). Following disturbance, bare ground in plots that contained rapidly spreading native C_3 graminoids was quickly filled where plots with a dominant native C_4 graminoid were not readily colonized. As a result, invasibility of the plots with C_3 graminoids was significantly lower. In a similar vein, invasibility of disturbed, productive herbaceous communities by the Brassicaceae forbs *Bunias orientalis* and *Rorippa austriaca* in the mid-Main valley in northern Bavaria (Germany) appeared to be strongly dependent on the composition of the resident species and interaction effects with type of disturbance (Woitke and Dietz 2002). The performance of both exotics profited from disturbance in association with the later-successional semi-shrub *Rubus caesius*, whereas, in association with the ruderal competitor grass *Arrhenatherum elatius*, their growth was strongly reduced under unmanaged or mown conditions. A third treatment, ,as soil disturbance, however, furthered growth of *R. austriaca* strongly and limited the growth of the grass.

3.5 Herbivory

Introduced plants may support considerable animal communities that live and/or feed on them. About fifty species of enemies were found on *Carthamus lanatus* in Europe, with three of them attacking only this species (Sheppard and Vitou 2000). In Kings Park near Perth, Western Australia, native and exotic eucalypts did not differ consistently in leaf damage or their arthropod community (Radho-Toly et al. 2001). The authors mentioned the possibility that a trend for higher nutrient richness of the leaves of the exotic species may compensate for their alien nature in their attractiveness for native herbivores. Hence, in this case, at least for the sapling stage that was investigated, there was no release from herbivory that could have been a factor contributing to the invasiveness of the exotic eucalypts.

The amount of herbivory or the diversity of herbivores on a given invasive plant species may also be influenced by characteristics of the invaded plant community. In a study on the two annual invasives *Conyza bonariensis* and *C. canadensis* in Mediterranean communities of annual plants, herbivory of the exotics increased with increasing plant species richness and was also dependent on the presence of specific functional plant groups. It appears that a richer plant community offers more herbivore niches and may thus support a higher diversity of herbivores on exotic plants (Prieur-Richard et al. 2002).

Some exotic plants may be superior to native plant species in tolerating or compensating for herbivore damage which could contribute to their invasiveness. For example, moderate herbivory on the widespread and very abundant invasive weed *Centaurea maculosa* in North America appears to have weak effects on the plant and may rather stimulate compensatory growth and even increase its competitive ability by reducing the resources available for neighbouring species (Müller-Schärer and Schroeder 1991; Steinger and Müller-Schärer 1992; Callaway et al. 1999). Following defoliation, compensatory growth in *Centaurea melitensis*, an invasive annual from Eurasia, was promoted by the simultaneous presence of the native Californian bunchgrass *Nassella pulchra* and soil fungi (Callaway et al. 2001).

Compared to its native congener, *Lonicera sempervirens*, the invasive *Lonicera japonica* does not only experience lower herbivory in its new range but also shows higher allocation to stems and leaves, suggesting a compensatory response to herbivory (Schierenbeck et al. 1994). Likewise, in a study in the Venezuelan savannah the native C_4 grass *Trachypogon plumosus* was more affected by defoliation than the invasive C_4 grass *Hyparrhenia rufa*, which is likely due to high allocation to leaf and culm production in *H. rufa*, while *T. plumosus* was unable to compensate for the lost biomass (Baruch and Bilbao 1999). Another mechanism conferring tolerance to grazing was observed for the clonal forb *Euphorbia esula*. This species allocated resources mainly to its extensive lateral root system, even after defoliation, thereby maintaining its high competitiveness for belowground resources (Olson and Wallander 1999).

It is widely accepted that introduced plants often lack many or all of their native herbivores or, more generally, do suffer less from enemies than their native counterparts in the new area (e.g. DeBach and Rosen 1991; Keane and Crawley 2002). This can have several important consequences for the invasion potential of the plant species. The introduced plants may lose adaptations and resistance to herbivory over time (Janzen 1975; Painter et al. 1989; Blossey and Nötzold 1995; Daehler and Strong 1997). There may be a shift from a defense-oriented growth strategy to a more competitive growth strategy that results in higher plant vigour (e.g., Fowler et al. 1996) or poorly defended but rapidly growing genotypes may be selectively favoured in the absence of herbivores (EICA hypothesis, Blossey and Nötzold 1995). Indeed, invasive plant species are often reported to grow or reproduce more vigorously in the new area (Blossey and Nötzold 1995; Blossey and Kamil 1996; Fowler et al. 1996; Rees and Paynter 1997; Willis and Blossey 1999; Willis et al. 2000, but see Thébaud and Simberloff 2001). While there are studies that support the EICA hypothesis (Blossey and Kamil 1996; Zangerl and Berenbaum 1997) most studies do not draw a clear picture or even contradict the EICA hypothesis. For example, Daehler and Strong (1997) found that *Spartina alterniflora* plants that were resistant to herbivory by *Prokelisia marginata*, a specialist leaf-hopper, had a higher intrinsic growth rate (i.e. were more 'vigorous') than less resistant forms. Willis et al. (1999) found that the phenolic content of *Lythrum salicaria* leaves was significantly higher in indigenous genotypes, as predicted by the EICA hypothesis. However, the phenolic content was generally low, and therefore probably played only a minor role in anti-herbivore defense. In a review Bergelson and Purrington (1996) concluded that costs were unequivocally involved in resistance to enemies in only about 50% of the studies and that costs were least often associated with herbivore resistance. Furthermore, doubts may be raised that the EICA mechanism has developed in the short period of time since most exotics were introduced (<200 years) (Daehler and Strong 1997; Willis et al. 1999, 2000).

Introduced plants may profit from decreased enemy attacks in the new area in more general terms. Keane and Crawley (2002) discuss the different possibilities under the collective term 'enemy release hypothesis' (ERH). Their arguments for ERH include that (1) natural enemies do have important effects on plant populations; (2) their impact on native species is greater than on exotic species; and (3) plants are able to take advantage of a reduction in enemy regulation, resulting in increased population growth. This is strongly indicated by the often rapid reduction of population size of an invasive plant once a specialist enemy from the native range of the plant is introduced to its new area (Fowler et al. 1996; Keane and Crawley 2002).

Assumptions of the enemy release hypothesis for introduced plants are that (1) specialist enemies of the study species are absent from the new region; (2) host switching by specialist enemies of native congeners is rare (although it has been reported, see below) and (3) generalists have a greater impact on the native competitors (Keane and Crawley 2002). The authors argue that the introduced species will suffer less from enemies in the new area and the relative loss to enemies is shifted onto the species' competitors, resulting in competitive release for the invaders. Furthermore, an invader lacking most specialist natural enemies could rise to a high density, maintaining generalist natural enemies that strongly regulate the native community (Shea and Chesson 2002). Invaders that show high fecundity or individual growth in the native range and thereby compensate for losses to natural enemies (e.g. grazed plants) could gain a strong advantage in an introduced area without their specialist natural enemies (Shea and Chesson 2002).

Support for the ERH comes from a comparative study on *Clidemia hirta* (Melastomataceae) in its native range in Costa Rica and in invasive populations in Hawaii (DeWalt et al. 2001). In the native range, *C. hirta* suffered from a significantly higher rate of mortality due to insects and pathogens than in the introduced range, linking the invasion success of *C. hirta* in Hawaii to increased survivorship of enemy-free seedlings. In fact, Greiling and Kichanan (2002) reported that insect herbivores most strongly influenced regeneration of the native *Monarda fistulosa* in old-field seedling regeneration indicating that their influence may decrease the proportion of native plants and increase non-native invasive species in old-field communities. In addition, a prolonged fruiting period and more or less continuous seed rain which is typical for many invasive plants may attract a greater variety and number of generalist seed predators or seedling herbivores which could have a proportionately greater impact on natives during their peak period of seed production (Ghazoul 2002).

However, there are several circumstances under which introduced plants suffer from herbivory in the introduced area. Specialist enemies of native congeners may switch hosts (Connor et al. 1980; Jobin et al. 1996); the plant and its enemy may have both been introduced to the same region (da Ros et al. 1993) and the plants may be attacked by native or introduced generalist herbivores, such as slugs (Buschmann et al. 2002) or grasshoppers (Joern 1989). Generalist herbivores may prefer native species over invasives, however, because they might be better adapted to native species. Conversely, invasive plants of Eurasian or African origin are probably better adapted to domestic livestock grazing than native plants in invaded communities without a recent history of domestic grazing pressure (Caldwell et al. 1981). Native herbivores that feed on related native plants may more readily attack plant invaders. The fruits of *Mahonia aquifolium*, a North American shrub invasive in central Europe, are infested more by the fruit fly *Rhagoletis meigenii* than those of its native host *Berberis vulgaris* of the same plant family. Still, seed predation was relatively low and the impact on the invasion process was also low (Soldaat and Auge 1998).

As with other factors involved in invasion processes, herbivory often interacts in a complex way with other biotic or abiotic factors. For example, the relative invasive potential of the perennial grasses *Cortaderia selloana* and *C. jubata* in California was strongly influenced by the presence of generalist herbivores. In addition, herbivore effects were dependent on the community type. Without herbivory, *C. jubata* transplants had higher mortality than *C. selloana* transplants due to physiological stress. However, in the presence of mammalian herbivores, herbivory became the dominant source of transplant mortality, and the survivorship of both species was indistinguishable in all habitats (Lambrinos 2002).

3.6 Facilitation

While variations in habitat invasibility are mostly discussed in relation to resistance of the habitat (by competitive effects, harshness of the conditions etc.) there may be biotic interactions facilitating the invasion of (particular) exotics as, for example, in the California coastal prairie where the native, nitrogen-fixing shrub *Lupinus arboreus* facilitates invasion of exotic weeds (Maron and Connors 1996).

Invasion success may depend on pollinator and dispersal service as well as on fungal or other microorganismal symbionts. In most cases these services or suitable mutualists appear to be readily available in the introduced area. This is, in part, due to the widespread introduction of important symbionts themselves (such as mycorrhizal fungi and the honey bee *Apis mellifera* as a generalist pollinator) (see the discussion in Richardson et al. 2000b and references therein). In those cases where these symbionts are missing, the invasion process might be strongly hampered. For example, the absence of ectomycorrhizal fungi restricts the invasion of conifers into meadows (Terwilliger and Pastor 1999). Low visitation by pollinators to the exotic shrub *Cytisus scoparius* impairs its invasive spread (Parker 1997).

Existing aliens in a community often facilitate further invasions by direct mutualisms or indirect effects such as modified disturbance regimes. Synergistic processes in which exotic species facilitate each other and which may result in accelerated invasion have been referred to as 'invasional meltdown' by Simberloff and von Holle (1999). For example, invaders that profit from elevated nitrogen availability in the soil may be facilitated by previous invasions of nitrogen-fixing plants such as *Myrica faya* (Wall and Moore 1999). Some invaders increase disturbance intensity and might, thereby, facilitate the invasion of other species. Furthermore, invasive plants may generally alter nutrient cycling, nutrient availability, moisture relations, and fire cycles such that successional trajectories, trophic interactions, and food webs may be irrevocably altered (Asner and Beatty 1996; Chapin et al. 1997). This adds to the complexity of habitat invasibility because habitat invasibility might change in unpredictable directions due to these changes in ecosystem function.

4 Propagule Pressure

Besides invader traits and the invasibility of a plant community, increased immigration rates or propagules (seeds, fruits or vegetative dispergules) availability (propagule pressure) are crucial for the onset and process of invasions (Ewel 1986; Newsome and Noble 1986; Williamson 1989; Pimm 1991; Davis et al. 2000; Williamson and Harrison 2002). Figure 2 summarizes the main processes affecting plant invasions. Regardless of site conditions, successful colonization requires sufficient availability of viable propagules of potential invaders or repeated introductions into the appropriate habitat (see Rejmánek 1989; DeFerrari and Naiman 1994; Williamson 1996; Richardson et al. 2000b). Historic data testify the importance of propagule pressure for plant invasions: for introduced plants that were offered to the public for only one year in Florida in the period from 1881–1937 just 1.9% became invasive, whereas plants that were sold for three decades had a 69% chance of success (Pemberton, in Enserink 1999). Likewise, most successfully invading trees were commonly planted ornamentals or stem from forestry plantations suggesting that the ability of

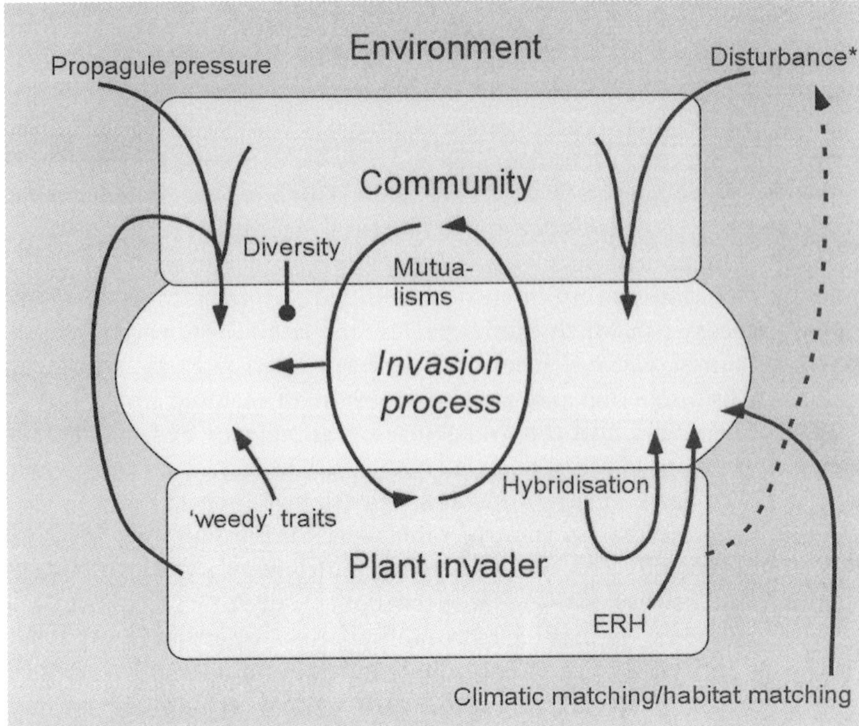

Fig. 2. Scheme of important interactions between plant invaders, the invasion process and the invaded community. *Arrows* indicate promoting influences on invasion success, *capped lines* influences that diminish invader success. (Disturbances in the widest sense include global change, altered resource availabilities, habitat fragmentation, fire, erosion, mowing and soil disturbance.) *ER* Enemy release *List of disturbances

introduced trees to invade neighbouring forests presupposes that they were introduced in large numbers (Fine 2002).

Increasing propagule pressure rises the chance for establishment of introduced plants, because if there are too few individuals they may not meet and reproduce. With more individuals there is a higher probability that suitable habitats are reached and there is risk-spreading, i.e. the probability that the whole founder population succumbs to hazards or pathogens diminishes. In fact, high propagule pressure seems to be exceptional in that it is consistently predictive of invasion success as is the extent of human disturbance (Fine 2002). In a recent review (Lonsdale 1999) concluded that propagule pressure might account for up to 56% of the variance in exotic plant richness among invaded sites (worldwide comparison).

Noble (1989) reported a striking example of increased propagule pressure following introduction into new areas. In both *Acacia longifolia* that was introduced to the Mediterranean fynbos of South Africa and in *Chrysanthemoides monilifera* that was introduced to Mediterranean Australia the number of viable seed increased tremendously in the invaded as compared to the native area (from 50 to 2000 m^{-2} for *C. monilifera* and from 10 to 7400 m^{-2} for *A. longifolia*). The reason for the buildup of this high propagule pressure appeared to be release from specialized seed predators.

According to model results, even introduced plants with low competitive ability can invade communities that have low levels of disturbance, provided that a sufficient number of propagules is introduced (Rejmánek 1989). Actually, even highly diverse rain forests can be invaded, suggesting that high functional diversity may be overpowered by sufficient exotic propagule pressure (Fine 2002). In addition, shade-tolerant exotic plants that are able to disperse their seeds in large quantities can invade even undisturbed tropical forest communities (Fine 2002). Levine (2000) manipulated species diversity/richness and propagule pressure in tussock communities in a riparian area in California and found that the number of seeds added to a tussock was a better predictor of exotic species establishment than species richness of a tussock.

High propagule pressure might operate at two different levels: high propagule pressure of a particular species increases this species' probability to invade, whereas a large exotic species pool increases the probability that one or a few species invade even if the single species does not show high abundance of propagules initially. For example, even frequent fires that may prevent plant invasions in grasslands (Leach and Givnish 1996) may fail to control plant invasions into these communities if the exotic species pool is large, e.g., as a consequence of increasing habitat fragmentation and increasing invasion of neighbouring habitats (Smith and Knapp 2001).

Propagule pressure adds an explicit spatio-temporal dimension to the invasion process, i.e. propagule pressure is a stochastic and spatially and temporally variable factor that has a great influence on whether and how an invasion will happen in a certain area or habitat (e.g., Dietz 2002; Pauchard et al. 2003). For example, higher propagule pressure can build up when suitable habitats are spatially clumped, promoting local establishment and colonization. On the other hand, dispersal through the landscape may be facilitated by a scattered pattern of suitable habitat (fragments) that allow exotic species to 'percolate' through the landscape using these fragments as 'stepping stones' (With 2002). At a larger scale, propagule pressure (i.e. the transport and frequency of arrivals) may be the crucial factor in island invasions (Williamson 1996). Islands are more invaded than comparable mainlands, indicating that propagule pressure may be higher on islands (Lonsdale 1999).

5 Conclusions

The discussion so far should have made it clear that (plant) invasions are very complex phenomena that defy simple causes and explanations in most cases. While it is possible to stress generalizations such that disturbances further invasions, higher diversity at larger scales increases invasibility, and that many plant invaders profit from enemy release in the introduced range, there is the caveat that these generalizations are rather coarse-level and may have low predictive power for particular plant invasions. For example, type of disturbance × species invader trait interactions may be crucial for whether invasions are considerably furthered by disturbance regimes or not (Woitke and Dietz 2002); the presence or absence of particular functional types or species in the native community may determine invasion success of some aliens more than community diversity and the amount of native herbivores on exotic plants 'compensating' for the possible lack of their specialized herbivores in the new area may be highly variable as may be the capability of the alien plants to compensate for herbivore damage (i.e. the degree to which the alien plant is able to tolerate damage without considerable loss of fitness).

Grime (2001) introduced a hierarchical scheme of plant strategies that assigns these strategies to various levels that differ along two gradients: generality vs. accuracy and the number of different strategies present. For example, the r/K concept has the highest generality as a (plant) strategy but is rather inaccurate, is based on very few generally available organism traits (life expectancy, reproductive strategy) and therefore is unlikely to explain a high percentage of a particular species' whole functional ecological type. Conversely, different ecotypes of a species are particular strategies that represent this species' functional type accurately. However, there may be myriads of different strategies among plants at that level and so we have very little generality. For the causal factors underlying (plant) invasion success, a similar scheme may be adopted (Fig. 3). For example, environments may be classified as to their resource pulse regimes, communities as to their diversity and alien plants as to their possession of 'weedy traits'. These classifications are generally applicable and may be evaluated singly or in combination to predict average patterns in invasion phenomena with some success. However, in the case by case quest to understand and predict the processes related to invasion in a particular habitat or for a certain species, reliance on these few general findings is likely an insufficient or perhaps even the wrong approach.

Thus, we agree with Lambrinos (2002) that every invasion seems to be a particular case and the need to investigate the specific ecological aspects of these particular cases will remain in the future. Still, reviews of such case

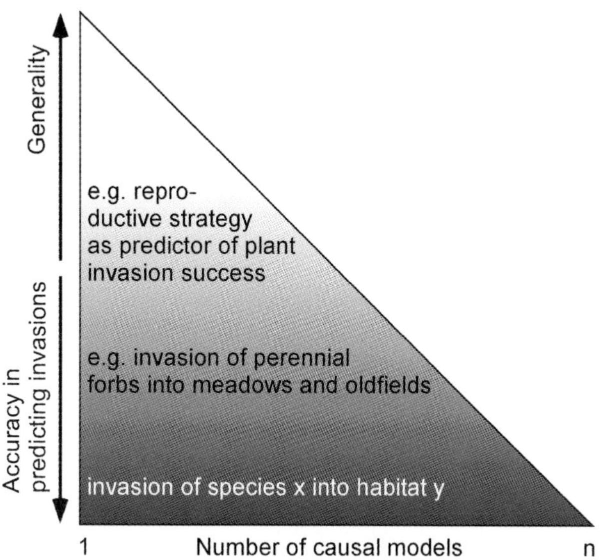

Fig. 3. Trade-off between generality and accuracy of prediction in causal models of plant invasions (examples of invasion systems that increase in specificity from top to bottom are given within the *triangle*)

studies and meta-analysis of them will facilitate the evaluation of newly investigated invasion systems that bear similarities with previously studied ones.

References

Adams RP, Flournoy LE, Singh RL, Johnson H, Mayeux H (1998) Invasion of grasslands by *Juniperus ashei*: a new theory based on random amplified polymorphic DNAs (RAPDs). Biochem Syst Ecol 26:371–377

Alpert P, Bone E, Holzapfel C (2000) Invasiveness, invasibility and the role of environmental stress in the spread of non-native plants. Perspectives in plant ecology. Evol Syst 3:52–66

Amsellem L, Noycr JL, Le Bourgeois T, Hossaert-McKey M (2000) Comparison of genetic diversity of the invasive weed *Rubus alceifolius* Poir. (Rosaceae) in its native range and in areas of introduction, using amplified fragment length polymorphism (AFLP) markers. Mol Ecol 9:443–455

Arenas F, Viejo RM, Fernandez C (2002) Density-dependent regulation in an invasive seaweed: responses at plant and modular levels. J Ecol 90:820–829

Asner GP, Beatty SW (1996) Effects of an African grass invasion on Hawaiian shrubland nitrogen biogeochemistry. Plant Soil 186:205–211

Baker HG (1965) Characteristics and modes of origin of weeds. In: Baker HG, Stebbins GL (eds) The genetics of colonizing species. Academic Press, New York

Baker HG (1974) The evolution of weeds. Annu Rev Ecol Syst 5:1–24

Baker HG, Stebbins GL (1965) The genetics of colonizing species. Academic Press, New York

Bakker J, Wilson S (2001) Competitive abilities of introduced and native grasses. Plant Ecol 157:117–125

Barthell JF, Randall JM, Thorp RW, Wenner AM (2001) Promotion of seed set in yellow star-thistle by honey bees: evidence of an invasive mutualism. Ecol Appl 11:1870–1883

Baruch Z, Bilbao B (1999) Effects of fire and defoliation on the life history of native and invader C_4 grasses in a Neotropical savanna. Oecologia 119:510–520

Baruch Z, Goldstein G (1999) Leaf construction cost, nutrient concentration, and net CO2 assimilation of native and invasive species in Hawaii. Oecologia 121:183–192

Bastlova D, Kvet J (2002). Differences in dry weight partitioning and flowering phenology between native and non-native plants of purple loosestrife (*Lythrum salicaria* L.). Flora 197:332–340

Bergelson J, Purrington CB (1996) Surveying patterns in the cost of resistance in plants. Am Nat 148:536–558

Bishop JG (2002) Early primary succession on Mount St. Helens: impact of insect herbivores on colonizing lupines. Ecology 83:191–202

Blossey B, Kamil J (1996) What determines the increased competitive ability of non-indigenous plants? In: Moran VC, Hofmann JH (eds) Proceedings of the IX International Symposium on Biological control of weeds Stellenbosch, South Africa. University of Cape Town, Cape Town, pp 3–9

Blossey B, Notzöld R (1995) Evolution of increased competitive ability in invasive nonindigenous plants: a hypothesis. J Ecol 83:887–889

Brooks ML (2002) Peak fire temperatures and effects on annual plants in the Mojave Desert. Ecol Appl 12:1088–1102

Brown BJ, Wickstrom CE (1997) Adventitious root production and survival of purple loosestrife (*Lythrum salicaria*) shoot sections. Ohio J Sci 97:2–4

Brown BJ, Mitchell RJ, Graham SA (2002) Competition for pollination between an invasive species (purple loosestrife) and a native congener. Ecology 83 (8):2328–2336

Buckland SM, Thompson K, Hodgson JG, Grime JP (2001) Grassland invasions: effects of manipulations of climate and management. J Appl Ecol 38:301–309

Burke MJW, Grime JP (1996) An experimental study of plant community invasibility. Ecology 77:776–790

Buschmann H, Edwards PJ, Dietz H (2002) Research project: does herbivory by slugs influence the invasiveness of perennial Brassicaceae? Bull Geobot Inst ETH 68:73–81

Caldwell MM, Richards JH, Johnson DA, Nowak RS, Dzurek RS (1981) Coping with herbivory: photosynthetic capacity and resource allocation in two semiarid *Agropyron* bunchgrasses. Oecologia 50:14–24

Callaway RM, Aschehoug ET (2000) Invasive plants versus their new and old neighbors: a mechanism for exotic invasion. Science 290:521–523

Callaway RM, DeLuca T, Belliveau WM (1999) Herbivores used for biological control may increase the competitive ability of the noxious weed *Centaurea maculosa*. Ecology 80:1196–1201

Callaway RM, Newingham B, Zabinski CA, Mahall BE (2001) Compensatory growth and competitive ability of an invasive weed are enhanced by soil fungi and native neighbours. Ecol Lett 4:429–433

Case TJ (1990) Invasion resistance arises in strongly interacting species-rich model competition communities. Proc Natl Acad Sci USA 87:9610–9614

Case TJ (1991) Invasion resistance, species build-up and community collapse in model competition communities. In: Gilpin ME, Hanski I (eds) Metapopulation dynamics. Academic Press, London, pp 239–266

Case TJ (1996) Global patterns in the establishment of exotic birds. Biol Conserv 78:69–96

Chapin FS III, Walker BH, Hobbs RJ, Hooper DU, Lawton JH, Sala OE, Tilman D (1997) Biotic control over the functioning of ecosystems. Science 277:500–503

Chazdon RL (1996) Spatial heterogeneity in tropical forest structure: canopy palms as landscape mosaics. Trends Ecol Evol 11:8–9

Chen HJ, Qualls RG, Miller GC (2002) Adaptive responses of *Lepidium latifolium* to soil flooding: biomass allocation, adventitious rooting, aerenchyma formation and ethylene production. Environ Exp Bot 48:119– 128

Chesson P, Huntly N (1997) The roles of harsh and fluctuating conditions in the dynamics of ecological communities. Am Nat 150:519– 553

Cohen AN, Carlton JT (1995) Biological study: nonindigenous aquatic species in a United States estuary: a case study of the biological invasions of the San Francisco bay and delta. US Fish and Wildlife Service, Washington DC

Connor EG, Faeth SH, Simberloff D, Opler PA (1980) Taxonomic isolation and the accumulation of herbivorous insects: a comparison of introduced and native trees. Ecol Entomol 5:205– 211

Crawley MJ (1987) What makes a community invasible? In: Gray AJ, Crawley MJ, Edwards PJ (eds) Colonization, succession and stability. Blackwell, Oxford, pp 429– 453

Crawley MJ, Harvey PH, Purvis A (1996) Comparative ecology of the native and alien floras of the British Isles. Philos Trans R Soc Lond B 351:1251– 1259

Crawley MJ, Brown SL, Heard MS, Edwards GR (1999) Invasion-resistance in experimental grassland communities: species richness or species identity? Ecol Lett 2:140– 148

Cronk QCB, Fuller JL (1995) Plant invaders. Chapman and Hall, London

D'Antonio CM (1993) Mechanisms controlling invasion of coastal plant communities by the alien succulent *Carpobrotus edulis*. Ecology 74:83– 95

D'Antonio CM, Vitousek PM (1992) Biological invasions by exotic grasses, the fire/grass cycle, and global change. Annu Rev Ecol Syst 23: 63– 87

D'Antonio CM, Dudley TL, Mack M (1999) Disturbance and biological invasions: direct effects and feedbacks. In: Walker LH (ed) Ecosystems of disturbed ground. Elsevier, New York, pp 413– 451

da Ros N, Ostermeyer R, Roques A, Raimbault JP (1993) Insect damage to cones of exotic conifer species introduced in arboreta. 1. Interspecific variations within the genus *Picea*. J Appl Entomol 115:113– 133

Daehler CC (2001) Two ways to be an invader, but one is more suitable for ecology. ESA Bull 82:101– 102

Daehler CC, Strong DR (1997) Reduced herbivore resistance in introduced smooth cordgrass (*Spartina alterniflora*) after a century of herbivore-free growth. Oecologia 110:99– 108

Davis MA, Pelsor M (2001) Experimental support for a resource-based mechanistic model of invasibility. Ecol Lett 4:421– 428

Davis MA, Grime JP, Thompson K (2000) Fluctuating resources in plant communities: a general theory of invasibility. J Ecol 88:528– 534

DeBach P, Rosen D (1991) Biological control by natural enemies. Cambridge University Press, Cambridge

DeFerrari CM, Naiman RJ (1994) A multi-scale assessment of the occurrence of exotic plants on the Olympic Peninsula, Washington. J Veg Sci 5:247– 258

DeWalt SJ, Ickes KL, Denslow JS (2001) Test of the release from natural enemies hypothesis using the invasive tropical shrub *Clidemia hirta*. In: Abstracts of the Ecological Society of America, 86th Annual Meeting. Ecological Society of America, Washington, DC, p 80

di Castri F (1989) History of biological invasions with emphasis on the Old World. In: Drake J, di Castri F, Groves R, Kruger F, Mooney HA, Rejmánek M, Williamson M (eds) Biological invasions: a global perspective. Wiley, New York, pp 1– 30

Dietz H (2002) Plant invasion patches – reconstructing pattern and process by means of herb-chronology. Biol Invasions 4: 211– 222

Dietz H, Steinlein T (1998) The impact of anthropogenic disturbance on life stage transitions and stand regeneration of the invasive alien plant *Bunias orientalis* L. In: Strafinger U, Edwards K, Kowarik I, Williamson M (eds) Plant invasions: ecological mechanisms and human responses. Backhuys Publishers, Leiden, pp 169– 184

Dietz H, Ullmann I (1997) Phenological shifts of the alien colonizer *Bunias orientalis* L.: an image-based analysis of temporal niche separation. J Veg Sci 8:839– 846

Dietz H, Steinlein T, Ullmann I (1998) The role of growth form and correlated traits in competitive ranking of six perennial ruderal plant species grown in unbalanced mixtures. Acta Oecol 19:25–36

Dietz H, Steinlein T, Ullmann I (1999a) Establishment of the invasive perennial herb *Bunias orientalis* L.: an experimental approach. Acta Oecol 20:621–632

Dietz H, Fischer M, Schmid B (1999b) Demographic and genetic invasion history of a 9-year-old roadside population of *Bunias orientalis* L. (Brassicaceae). Oecologia 120:225–234

Dietz H, Köhler A, Ullmann I (2002) Regeneration growth of the invasive clonal forb *Rorippa austriaca* (Brassicaceae) in relation to fertilization and interspecific competition. Plant Ecol 158: 171–182

Downey PO, Brown JMB (2000) Demography of the invasive shrub Scotch broom (*Cytisus scoparius*) at Barrington Tops, New South Wales: insights for management. Aust Ecol 25:477–485

Drake JA (1990) Communities as assembled structures – do rules govern pattern. Trends Ecol Evol 5:159–164

Drake JA, Mooney HA, di Castri F, Groves RH, Kruger FJ, Rejmánek M, Williamson M (1989) Biological invasions. A global perspective. Wiley, Chichester, 525 pp

Dukes JS (2001) Biodiversity and invasibility in grassland microcosms. Oecologia 126:563–568

Dukes JS (2002a) Comparison of the effect of elevated CO_2 on an invasive species (*Centaurea solstitialis*) in monoculture and community settings. Plant Ecol 160:225–234

Dukes JS (2002b) Species composition and diversity affect grassland susceptibility and response to invasion. Ecol Appl 12:602–617

Dukes JS, Mooney HA (1999) Does global change increase the success of biological invaders? Trends Ecol Evol 14:135–139

Ellstrand NC, Schierenbeck KA (2000) Hybridization as a stimulus for the evolution of invasiveness in plants? Proc Natl Acad Sci USA 97:7043–7050

Elton C (1958) The ecology of invasions by plants and animals. Methuen, London

Enserink M (1999) Biological invaders sweep in. Science 285:1834–1836

Ewel JJ (1986) Invasibility: lessons from South Florida. In: Mooney HA, Drake JA (ed) Ecology of biological invasion of North America and Hawaii. Springer, Berlin Heidelberg New York, pp 214–230

Ferrell MA, Whitson TD, Koch DW, Gade AE (1998) Leafy spurge (*Euphorbia esula*) control with several grass species. Weed Technol 12:374–380

Fine PVA (2002) The invasibility of tropical forests by exotic plants. J Tropical Ecol 18:687–705

Fowler SV, Harman HM, Memmott J, Paynter Q, Shaw R, Sheppard AW, Syrett P (1996) Comparing the population dynamics of broom, *Cytisus scoparius*, as a native plant in the United Kingdom and France, and as an invasive alien weed in Australia and New Zealand. In: Moran VC, Hoffman JH (eds) Proceedings of the IX International Symposium on Biological control of weeds. University of Cape Town, Cape Town, pp 19–26

Ghazoul J (2002) Seed predators and the enemy release hypothesis. Trends Ecol Evol 17:308

Gibson DJ, Spyreas G, Benedict J (2002) Life history of *Microstegium vimineum* (Poaceae), an invasive grass in southern Illinois. J Torrey Bot Soc 129:207–219

Glenn ER, Tanner S, Mendez T, Kehret D, Moore J, Garcia, Valdes C (1998) Growth rates, salt tolerance and water use characteristics of native and invasive riparian plants from the delta of the Colorado River, Mexico. J Arid Environ 40 (3):281–294

Grabas GP, Laverty TM (1999) The effect of purple loosestrife (*Lytrum salicaria* L., Lythraceae) on the pollination and reproductive success of sympatric co-flowering wetland plants. Ecoscience 6:230–242

Greiling DA, Kichanan N (2002) Old-field seedling responses to insecticide, seed addition, and competition. Plant Ecol 159:175–183

Grime JP (2001) Plant strategies, vegetation processes and ecosystem properties. Wiley, Chichester
Harper J (1977) Population biology of plants. Academic Press, London
Harrison S (1999) Native and alien species diversity at the local and regional scales in a grazed California grassland. Oecologia 121:99–106
Hector A, Dobson K, Minns A, Bazeley-White E, Lawton JH (2001) Community diversity and invasion resistance: an experimental test in a grassland ecosystem and a review of comparable studies. Ecol Res 16:819–831
Honnay O, Verheyen K, Hermy M (2002) Permeability of ancient forest edges for weedy plant species invasion. For Ecol Manage 161:109–122
Hooper DU, Vitousek PM (1997) The effects of plant composition and diversity on ecosystem processes. Science 277:1302–1305
Horvitz CC, Pascarella JB, McMann S, Freedman A, Hofstetter RH (1998) Functional roles of invasive non-indigenous plants in hurricane-affected subtropical hardwood forests. Ecol Appl 8:947–974
Huston MA (1997) Hidden treatments in ecological experiments: re-evaluating the ecosystem function of biodiversity. Oecologia 110:449–460
Huxman TE, Hamerlynck EP, Jordan DN, Salsman KJ, Smith SD (1998) The effects of parental CO2 environment on seed quality and subsequent seedling performance in *Bromus rubens*. Oecologia 114:202–208
Janzen DH (1975) Behavior of *Hymenaea courbaril* when its predispersal seed predator is absent. Science 189:145–147
Jobin A, Schaffner U, Nentwig W (1996) The structure of the phytophagous insect fauna on the introduced weed *Solidago altissima* in Switzerland. Entomol Exp Appl 79:33–42
Joern A (1989) Insect herbivory in the transition to California annual grasslands: did grasshoppers deliver the coup de grass? In: Huenneke LF, Mooney HA (eds) Grassland structure and function: California annual grasslands. Kluwer, Dordrecht, pp 117–134
Johnston FM, Pickering CM (2001) Alien plants in the Australian Alps. Mountain Res Dev 21:284–291
Keane RM, Crawley MJ (2002) Exotic plant invasions and the enemy release hypothesis. Trends Ecol Evol 17:164–170
Kennedy TA, Naeem S, Howe KM, Knops JMH, Tilman D, Reich P (2002) Biodiversity as a barrier to ecological invasion. Nature 417:636–638
Kiviniemi K, Eriksson O (2002) Size-related deterioration of semi-natural grassland fragments in Sweden. Div Distrib 8:21–29
Knops JMH, Griffin JR, Royalty AC (1995) Introduced and native plants of the Hastings Reservation, central coastal California: a comparison. Biol Conserv 71:115–123
Knops JMH, Griffin JR, Royalty AC (1999) Effects of plant species richness on invasion dynamics, disease outbreaks, insect abundances and diversity. Ecol Lett 2:286–293
Kolar CS, Lodge TS (2001) Progress in invasion biology: predicting invaders. Trends Ecol Evol 16:199–204
Kolb A, Alpert P, Enters D, Holzapfel C (2002) Patterns of invasion within a grassland community. J Ecol 90:871–881
Lambrinos JG (2001) The expansion history of a sexual and asexual species of *Cortaderia* in California, USA. J Ecol 89 (1):88–98
Lambrinos JG (2002) The variable success of *Cortaderia* species in a complex landscape. Ecology 83:518–529
Larson DL, Anderson PJ, Newton W (2001) Alien plant invasion in mixed-grass prairie: effects of vegetation type and anthropogenic disturbance. Ecol Appl 11:128–141
Larson KC, Fowler SP, Walker JC (2002) Lack of pollinators limits fruit set in the exotic *Lonicera japonica*. Am Midland Nat 148:54–60
Larson LL, McInnis ML (1989) Impact of grass seedings on establishment and density of diffuse knapweed and yellow starthistle. Northwest Sci 63:162–166

Lavorel S, Prieur-Richard AH, Grigulis K (1999) Invasibility and diversity of plant communities: from patterns to processes. Div Distrib 5:41– 49

Leach MK, Givnish TJ (1996) Ecological determinants of species loss in remnant prairies. Science 273:1555– 1561

Lee CE (2002) Evolutionary genetics of invasive species. Trends Ecol Evol 17:386– 391

Levine JM (2000) Species diversity and biological invasions: relating local process to community pattern. Science 288:853– 854

Levine JM, D'Antonio CM (1999) Elton revisited: a review of evidence linking diversity and invasibility. OIKOS 87:15– 26

Linde M, Diel S, Neuffer B (2001) Flowering ecotypes of *Capsella bursa-pastoris* (L.) Medik. (Brassicaceae) analysed by a cosegregation of phenotypic characters (QTL) and molecular markers. Ann Bot 87:91– 99

Litton CM, Santelices R (2002) Early post-fire succession in a *Nothofagus glauca* forest in the Coastal Cordillera of south-central Chile. Int J Wildl Fire 11:115– 125

Lockhart CS (1996) Aquatic heterophylly as a survival strategy in *Melaleuca quinquenervia* (Myrtaceae). Can J Bot Rev Can Bot 74:243– 246

Lodge DM (1993) Biological invasions: lessons for ecology. Trends Ecol Evol 8:133– 137

Lonsdale WM (1999) Global patterns of plant invasions and the concept of invasibility. Ecology 80:1522– 1536

Loope LL, Mueller-Dombois D (1989) Characteristics of invaded islands, with special reference to Hawaii. In: Drake JA, Mooney HA, di Castri F, Groves RH, Kruger FJ, Rejmánek M, Williamson M (eds) Biological invasions: a global perspective. Wiley, Chichester, pp 257– 280

Loreau M (2000) Biodiversity and ecosystem functioning: recent theoretical advances. Oikos 91:3– 17

Lyons KG, Schwartz MW (2001) Rare species loss alters ecosystem function – invasion resistance. Ecol Lett 4:358– 365

Mack RN (1996) Predicting the identity and fate of plant invaders: emergent and emerging approaches. Biol Conserv 78:107– 121

Mack RN, Simberloff D, Lonsdale WM, Evans H, Clout M, Bazzaz FA (2000) Biotic invasions: causes, epidemiology, global consequences, and control. Ecol Appl 10:689– 710

Mandák B, Pyšek P (1999): Effects of plant density and nutrient levels on fruit polymorphism in *Atriplex sagittata*. Oecologia 119:63– 72

Maron JL, Connors PG (1996) A native nitrogen-fixing shrub facilitates weed invasion. Oecologia 105:302– 312

Maurer DA, Zedler JB (2002). Differential invasion of a wetland grass explained by tests of nutrients and light availability on establishment and clonal growth. Oecologia 131:279– 288

Mazia CN, Chaneton EJ, Ghersa CM, Leon RJC (2001) Limits to tree species invasion in pampean grassland and forest plant communities. Oecologia 128:594– 602

McDowell SCL,Turner DP (2002) Reproductive effort in invasive and non-invasive *Rubus*. Oecologia 133:102– 111

Meekins JF, McCarthy BC (1999) Competititve ability of *Alliaria petiolata* (garlic mustard, Brassicaceae), an invasive, nonindigenous forest herb. Int J Plant Sci 160:743– 752

Meekins JF, McCarthy BC (2001) Effect of environmental variation on the invasive success of a nonindigenous forest herb. Ecol Appl 11:1336– 1348

Meyer AH, Schmid B (1999) Seed dynamics and seedling establishment in the invading perennial *Solidago altissima* under different experimental treatments. J Ecol 87:28– 41

Meyer JY (1998) Observations on the reproductive biology of *Miconia calvescens* DC (Melastomataceae), an alien invasive tree on the island of tahiti (South Pacific Ocean). Biotropica 30:609– 624

Milne RI, Abbott RJ (2000) Origin and evolution of invasive naturalized material of *Rhododendron ponticum* L. in the British Isles. Mol Ecol 9:541– 556

Moola FM, Mallik AU (1998) Morphological plasticity and regeneration strategies of velvet leaf blueberry (*Vaccinium myrtilloides* Michx.) following canopy disturbance in boreal mixed wood forests. For Ecol Manage 111:35- 50

Mooney HA (1999) Species without frontiers. Nature 397:665- 666

Müller-Schärer H, Schroeder D (1991) The impact of root herbivory as a function of plant density and competition: survival, growth, and fecundity of *Centaurea maculosa* in field plots. J Appl Ecol 28:759- 776

Naeem S, Knops JMH, Tilman D, Howe KM, Kennedy T, Gale S (2000) Plant diversity increases resistance to invasion in the absence of covarying extrinsic factors. Oikos 91:97- 108

Newsome A, Noble IR (1986) Ecological and physiological characters of invading species. In: Groves RH, Burden JJ (eds) Ecology of biological invasions: an Australian perspective. Australian Academy of Science, Canberra, pp 1- 20

Noble IR (1989) Attributes of invaders and the invading process: terrestrial and vascular plants. In: Drake JA, Mooney HA, di Castri F, Groves R, Kruger FJ, Rejmánek M, Williamson M (eds) Biological invasions a global perspective. SCOPE 37. Wiley, Chichester, pp 301- 314

Olson BE, Wallander RT (1999) Carbon allocation in *Euphorbia esula* and neighbours after defoliation. Can J Bot 77:1641- 1647

Painter EL, Detling JK, Steingraeber DA (1989) Grazing history, defoliation, and frequency-dependent competition: effects on two North American grasses. Am J Bot 76:1368- 1379

Palmer MW, Maurer T (1997) Does diversity beget diversity? A case study of crops and weeds. J Veg Sci 8:235- 240

Pappert RA, Hamrick JL, Donovan LA (2000) Genetic variation in *Pueraria lobata* (Fabaceae), an introduced, clonal, invasive plant of the southeastern United States. Am J Bot 87:1240- 1245

Parker IM (1997) Pollinator limitation of *Cytisus scoparius*, an invasive exotic shrub. Ecology 788:1457- 70

Parker IM, Simberloff D, Lonsdale WM, Goodell K, Wonham M, Kareiva PM, Williamson MH, Von Holle B, Moyle PB, Byers JE, Goldwasser L (1999) Impact: toward a framework for understanding the ecological effects of invaders. Biol Invasions 1:3- 19

Parker M, Thompson JN, Weller SG (2001) The population biology of invasive species. Annu Rev Ecol Syst 32:305- 332

Paterson AH, Schertz KF, Lin YR, Liu SC, Chang YL (1995) The weediness of wild plants: molecular analysis of genes influencing dispersal and persistence of johnsongrass *Sorghum halepense* (L.). Proc Natl Acad Sci USA 92:6127- 6131

Pattison RR, Goldstein G, Ares A (1998) Growth, biomass allocation and photosynthesis of invasive and native Hawaiian rainforest species. Oecologia 117:449- 459

Pauchard A, Alaback PB, Edlund EG (2003) Studying plant invasions in protected areas at multiple scales: *Linaria vulgaris* (Scrophulariaceae) in the West Yellowstone area. Western North Am Nat (in press)

Pimentel D, Lach L, Zuniga R, Morrison D (2000) Environmental and economic costs of nonindigenous species in the United States. BioScience 50:53- 65

Pimm SL (1991) The balance of nature? University of Chicago Press, Chicago

Planty-Tabacchi AM, Tabacchi E, Naiman RJ, Deferrari C, Decamps H (1996) Invasibility of species-rich communities in riparian zones. Conserv Biol 10:598- 607

Prieur-Richard AH, Lavorel S, Grigulis K, Dos Santos A (2000) Plant community diversity and invasibility by exotics: the example of *Conyza bonariensis* and *C canadensis* invasion in Mediterranean annual old fields. Ecol Lett 3:412- 422

Prieur-Richard AH, Lavorel S, Linhart YB, Dos Santos A (2002) Plant diversity, herbivory and resistance of a plant community to invasion in Mediterranean annual communities. Oecologia 130:96- 104

Prinzing A, Durka W, Klotz S, Brandl R (2002) Which species become aliens? Evol Ecol Res 4:385- 405

Pyšek P (1997) Clonality and plant invasions: can a trait make a difference? In: de Kroon H, van Groenendael J (eds) The ecology and evolution of clonal plants. Backhuys Publishers, Leiden, pp 405–427

Pyšek P, Pyšek P (1995) Invasion by *Heracleum mantegazzianum* in different habitats in the Czech Republic. J Veg Sci 6:711–718

Pyšek P, Vojtech J, Kucera T (2002) Patterns of invasion in temperate nature reserves. Biol Conserv 104:13–24

Radford IJ, Cousens RD (2000) Invasiveness and comparative life-history traits of exotic and indigenous *Senecio* species in Australia. Oecologia 125:531–542

Radho-Toly S, Majer JD, Yates C (2001) Impact of fire on leaf nutrients, arthropod fauna and herbivory of native and exotic eucalypts in Kings Park, Perth, Western Australia. Aust Ecol 26:500–506

Rees M, Paynter Q (1997) Biological control of Scotch broom: modelling the determinants of abundance and the potential impact of introduced insect herbivores. J Appl Ecol 34:1203–1221

Reichard SH (1997) What traits distinguish invasive plants from non-invasive plants? In: California Exotic Pest Plant Council, 1996 Symposium Proceedings, pp 1–9

Rejmánek M (1989) Invasibility of plant communities. In: Drake JA, Mooney HA, di Castri F, Groves RH, Kruger FJ, Rejmánek M, Williams M (eds) Biological invasions: a global perspective. Wiley, Chichester, pp 369–388

Rejmánek M, Richardson DM (1996) What attributes make some plant species more invasive? Ecology 77:1655–1661

Reznick DN, Ghalambor CK (2001) The population ecology of contemporary adaptations: what empirical studies reveal about the conditions that promote adaptive evolution. Genetica 112:183–198

Richardson DM, Allsopp N, D'Antonio CM, Milton SJ, Rejmánek M (2000a) Plant invasions – the role of mutualisms. Biol Rev Cambridge Philos Soc 75:65–93

Richardson DM, Pyšek P, Rejmánek M, Barbour MG, Panettta FD, West CJ (2000b) Naturalization and invasion of alien plants: concepts and definitions. Div Distrib 6:93–107

Robinson GR, Quinn JF, Stanton ML (1995) Invasibility of experimental habitat islands in a California winter annual grassland. Ecology 76:786–794

Roché BF Jr, Roché CT, Chapman RC (1994) Impacts of grassland habitat on yellow starthistle (*Centaurea solstitialis* L.) invasion. Northwest Sci 68:86–96

Ross KA, Fox BJ, Fox MD (2002) Changes to plant species richness in forest fragments: fragment age, disturbance and fire history may be as important as area. J Biogeogr 29:749–765

Rouget M, Richardson DM, Milton SJ, Polakow D (2001) Predicting invasion dynamics of four alien *Pinus* species in a highly fragmented semi-arid shrubland in South Africa. Plant Ecol 152:79–92

Safford HD, Harrison SP (2001) Grazing and substrate interact to affect native vs exotic diversity in roadside grasslands. Ecol Appl 11:1112–1122

Sakai AK, Allendorf FW, Holt JS, Lodge DM, Molofsky J, With KA, Baughman S, Cabin RJ, Cohen JE, Ellstrand NC, McCauley DE, O'Neil P, Parker IM, Thompson JN, Weller SG (2001) The population biology of invasive species. Annu Rev Ecol Syst 32:305–332

Sasek TW, Strain BW (1988) Effects of carbon dioxide enrichment on the growth and morphology of kudzu (*Pueraria lobata*). Weed Sci 36:28–36

Sasek TW, Strain BW (1991) Effects of CO_2 enrichment on the growth and morphology of a native and an introduced honeysuckle vine. Am J Bot 78:69–75

Schierenbeck KA, Mack RN, Sharitz RR (1994) Effects of herbivory on growth and biomass allocation in native and introduced species of *Lonicera*. Ecology 75:1661–1672

Schweitzer JA, Larson KC (1999) Greater morphological plasticity of exotic honeysuckle species may make them better invaders than native species. J Torrey Bot Soc 126:15–23

Scott NA, Saggar S, McIntosh PD (2001) Biogeochemical impact of *Hieracium* invasion in New Zealand's grazed tussock grasslands: sustainability implications. Ecol Appl 11:1311–1322

Seiger LA (1997) The status of *Fallopia japonica* (*Reynoutria japonica*; *Polygonum cuspidatum*) in North America. In: Brock JH, Wade M, Pyšek P, Green D (eds) Plant invasions – studies from North America and Europe. Backhuys, Leiden, pp 95–102

Shea K, Chesson P (2002) Community ecology theory as a framework for biological invasions. Trends Ecol Evol 17:170–176

Sheppard AW, Vitou J (2000) The effect of a rosette-crown fly, *Botanophila turcica*, on growth, biomass allocation and reproduction of the thistle *Carthamus lanatus*. Acta Oecol 21:337–347

Siemann E, Rogers WE (2001) Genetic differences in growth of an invasive tree species. Ecol Lett 4:514–518

Simberloff D, von Holle B (1999) Positive interactions of nonindigenous species: invasional meltdown? Biol Invasions 1:21–32

Smith MD, Knapp AK (2001) Size of the local species pool determines invasibility of a C4-dominated grassland. Oikos 92:55–61

Smith SD, Strain BR, Sharkey TD (1987) Effects of carbon dioxide enrichment on four Great Bassin [USA] grasses. Funct Ecol 1:139–144

Smith SD, Huxmann TE, Zitzer SF, Charlet TN, Housman DC, Coleman JS, Fenstermaker LK, Seemann JR, Nowak RS (2000) Elevated CO2 increases productivity and invasive species success in an arid ecosystem. Nature 408:79–82

Soldaat LL, Auge H (1998) Interactions between an invasive plant, *Mahonia aquifolium*, and a native phytophagous insect, *Rhagoletis meigenii*. In: Starfinger U, Edwards K, Kowarik I, Williamson M (eds) Plant invasions – ecological mechanisms and human responses. Backhuys Publishers, Leiden, pp 347–360

Stapanian MA, Sundberg SD, Baumgardner GA, Liston A (1998) Alien plant species composition and associations with anthropogenic disturbance in North American forests. Plant Ecol 139:49–62

Steinger T, Müller-Schärer H (1992) Physiological and growth responses of *Centaurea maculosa* (Asteraceae) to root herbivory under varying levels of interspecific plant competition and soil nitrogen availability. Oecologia 91:141–149

Steinlein T, Dietz H, Ullmann I (1996) Growth patterns of the alien perennial *Bunias orientalis* L. (Brassicaceae) underlying its rising dominance in some native plant assemblages. Vegetatio 125:73–82

Stohlgren TJ, Binkley D, Chong GW, Kalkhan MA, Schell LD, Bull KA, Otsuki Y, Newman G, Bashkin M, Son Y (1999) Exotic plant species invade hot spots of native plant diversity. Ecol Monogr 69:25–46

Stohlgren TJ, Barnett DT, Kartesz JT (2003) The rich get richer: patterns of plant invasions in the United States. Front Ecol Environ 1:11–14

Symstad AJ (2000) A test of the effects of functional group richness and composition on grassland invasibility. Ecology 81:99–109

Ter Borg SJ, Schippers P, Van Groenendael JM, Rotteveel TJW (1998) *Cyperus esculentus* (yellow nutsedge) in NW Europe: invasions on a local, regional and global scale. In: Starfinger U, Edwards K, Kowarik I, Williamson M (eds) Plant invasions – ecological mechanisms and human responses. Backhuys, Leiden, pp 261–273

Terwilliger J, Pastor J (1999) Small mammals, ectomycorrhizae, and conifer succession in beaver meadows. Oikos 85:83–94

Thébaud C, Simberloff D (2001) Are plants really larger in their introduced ranges? Am Nat 157:231–236

Thompson K, Hodgson JG, Rich TCG (1995) Native and alien invasive plants: more of the same? Ecography 18:390–402

Tilman D (1997) Community invasibility, recruitment limitation, and grassland biodiversity. Ecology 78:81–92

Tilman D (1999) The ecological consequences of changes in biodiversity: a search for general principles. Ecology 80:1455–1474
Tilman D, Wedin D, Knops J (1996) Productivity and sustainability influenced by biodiversity in grassland ecosystems. Nature 379:718–720
Troumbis AY, Galanidis A, Kokkoris GD (2002) Components of short-term invasibility in experimental Mediterranean grasslands. Oikos 98:239–250
Usher MB (1988) Biological invasions of nature reserves: a search for generalisations. Biol Conserva 44:119–135
Uveges JL, Corbett AL, Mal TK (2002) Effects of lead contamination on the growth of *Lythrum salicaria* (purple loosestrife). Environ Pollut 120:319–323
van der Heijden MGA, Klironomos JN, Ursic M, Moutoglis P, Boller T, Wiemken A, Sanders IR (1999) "Sampling effect," a problem in biodiversity manipulation? A reply to David A. Wardle. Oikos 87:408–410
Vermeij GJ (1996) An agenda for invasion biology. Biol Conserv 78:3–9
Vitousek PM, D'Antonio CM, Loope LL, Westbrooks R (1996) Biological invasions as global environmental change. Am Sci 84:218–228
Vitousek PM, D'Antonio CM, Loope LL, Rejmánek M, Westbrooks R (1997a) Introduced species: a significant component of human caused global change. N Z J Ecol 21:1–16
Vitousek PM, Mooney HA, Lubchenco J, Melillo JM (1997b) Human domination of earth's ecosystems. Science 277:494–499
Walker B, Steffen W (1997) An overview of the implications of global change for natural and managed terrestrial ecosystems. Conserv Ecol [online]1(2) http://www.consecol.org/vol1/iss2/
Wall DH, Moore JC (1999) Interactions underground. Soil biodiversity, mutualism, and ecosystem processes. BioScience 49:109–117
Wardle DA (1999) Is "sampling effect" a problem for experiments investigating biodiversity-ecosystem function relationship? Oikos 87:403–407
Wardle DA (2001) Experimental demonstration that plant diversity reduces invasibility – evidence of a biological mechanism or a consequence of sampling effect? Oikos 95:161–170
Williamson J, Harrison S (2002) Biotic and abiotic limits to the spread of exotic revegetation species. Ecol Appl 12:40–51
Williamson M (1989) Mathematical models of invasion. In: Drake JA, Mooney HA, di Castri F, Groves RH, Kruger FJ, Rejmánek M, Williamson M (eds) Biological invasions: a global perspective. Wiley, Chichester, pp 329–350
Williamson M (1996) Biological invasions. Chapman and Hall, London
Williamson M (1999) Invasions. Ecography 22:5–12
Willis AJ, Blossey B (1999) Benign climates don't explain the increased vigour of non-indigenous plants: a cross-continental transplant experiment. Biocontrol Sci Technol 9:567–577
Willis AJ, Thomas MB, Lawton JH (1999) Is the increased vigour of invasive weeds explained by a trade-off between growth and herbivore resistance? Oecologia 120:632–640
Willis AJ, Memmott J, Forrester RI (2000) Is there evidence for the post-invasion evolution of increased size among invasive plant species? Ecol Lett 3:275–283
Wilsey BJ, Polley HW (2002) Reductions in grassland species evenness increase dicot seedling invasion and spittle bug infestation. Ecol Lett 5:676–684
With KA (2002) The landscape ecology of invasive spread. Conserv Biol 16:1192–1203
Woitke M, Dietz H (2002) Shifts in dominance of native and invasive plants in experimental patches of vegetation. Perspectives in plant ecology. Evol Syst 5:165–184
Young JA, Palmquist DE, Wotring SO (1997) The invasive nature of *Lepidium latifolium*: a review. In: Brock JH, Wade M, Pyšek P, Green D (eds) Plant invasions – studies from North America and Europe. Backhuys, Leiden, pp 59–68
Zangerl AR, Berenbaum MR (1997) Cost of chemically defending seeds: furanocoumarins and *Pastinaca sativa*. Am Nat 150:491–504

Dr. Hansjörg Dietz
Geobotanisches Institut der ETH
Zürichbergstrasse 38
8044 Zürich, Switzerland
e-mail: dietz@geobot.umnw.ethz.ch

Dr. Thomas Steinlein
Lehrstuhl für experimentelle Ökologie und Ökosystembiologie
Universität Bielefeld
Universitätsstrasse 25
33615 Bielefeld, Germany

History of Flora and Vegetation During the Quaternary

Burkhard Frenzel

1 The Problem of Timing and Extent of Last-Glacial Inland Ice Masses in Northern Eurasia

Grosval'd (1983, 1988) had speculated about the existence of a huge last-glacial inlandice, which was said to have covered vast areas of Northern Eurasia, including Middle-Siberia and which might have joined the Tibetan inlandice. This in its turn is said to have existed during the first part of the last glaciation, triggering the later ice advances of the Northern Hemisphere, i.e. those of the so-called Last Glacial Maximum (LGM; Kuhle 1991, 2000). If these inland ice masses should have actually existed, the plants and animals of the Northern Hemisphere would have been seriously affected. Concerning the Tibetan Plateau, it meanwhile could be shown that an inland ice had never existed there (Frenzel and Liu 2001; Kamp and Haserodt 2002; also see the vast literature cited in these papers). Instead, there had only been various types of mountain glaciations occurring in this area. On the other hand, the question as to the extent and timing of the last-glacial North-Eurasiatic glaciations (approximately 115,000 to 11,560 b.p.) has meanwhile been investigated intensively by several research groups. It turned out that Grosval'ds hypothetical north-Eurasiatic inland ice had never existed. This has long been evident; new results on the timing and extent of last-glacial ice masses in these regions are most interesting.

In the centre of the Arctic Ocean there had existed during imprecisely dated phases of the last glaciation very thick ice masses, which had polished and striated the bedrock at the bottom of this ocean (Belyaeva et al. 1994; Jakobsson et al. 2001). Yet, these ice masses evidently had not moved on to the surrounding continents. For instance, the Wrangel Island to the north of Chukotka had experienced at least after about 64,000 b.p. small mountain glaciations, only (Karhu et al. 2001). The same seems to have held true for the southern part of the Bol'shoy Lyakhovskiy Isle, since about 200,000 b.p. (Schirrmeister et al. 2002), and in southwestern Alaska during the last glaciation – even in the proximity of the Pacific Ocean – there had been a small piedmont glaciation at about 60,300 b.p., followed by a much smaller mountain-glaciation between about 30,000 and 17,000 b.p. (Briner et al. 2001). This may help to understand why the northeastern Siberian moose genetically resembles much more the American moose than its north Eurasian relatives (Boeskorov et al. 1993), since an exchange of animal taxa via the Bering Sea area seems always to have been possible. Evidently the high mountains of northeastern Siberia during the last glaciation had merely experienced a relatively small mountain glaciation (P.M. Anderson et al. 2002), which seems to have happened before the LGM. Middle Siberia along the middle Lena River had not been covered during the last and the last-but-one glaciations by an inland ice (Kamaletdinov and Minyuk 1991). Moreover, the surroundings of Lake Baykal seem to have experienced at least during the last three glacial periods small mountain glaciations, only (Trofimov et al. 1994). Thus, there should have existed during the last glaciation a wealth of different habitats for plants and animals within the lowland and mountain areas of central and eastern Siberia.

The Taymyr Peninsula has meanwhile repeatedly been investigated. Here, like in other regions of northern Eurasia, during the last glaciation, there seem to have been at least three inland ice advances, which had come from the north or the northwest. Yet the endmoraine systems cover only the lowlands and hill regions to the north of the Byrranga Mountains. They date from 81,000 or 78,000 b.p., 66,000 to 52,000 and from about 20,000 b.p. (Alexanderson et al. 2001). These glaciers do not seem to have moved into the lowlands to the south of this mountain system (Ebel et al. 1999; Hahne and Melles 1999; Niessen et al. 1999; Siegert et al. 1999; Kienast et al. 2001; Andreev et al. 2002).

It is in general held that in Siberia the last glaciation had consisted of the – older – Zyryanka Glaciation, the Karga-Interstadial and the Sartan Glaciation, which is thought to have corresponded to the LGM. It could be shown now that the ice masses of the Sartan Glaciation, which had originated in the Putorana Mountains, had blocked for some time the runoff of the river Yenisey, causing ice-dammed lakes. This had happened to the north of approximately 61°N (Goncharov 1991; Sukhorukova et al. 1991), yet these ice masses did not move far to the west. Moreover, Sukhorukova et al. (1991) state that the preceding Ermakovo (Zyryanka-) Glaciation, too, had not extended remarkably beyond the Sartan endmoraines.

With regard to the Yamal Peninsula, it is said that during the last glaciation either no glaciers had invaded this region (Gataullin 1991), or an inland-ice coming from the southwest had reached the west coast of the Yamal Peninsula at about 80,000 to 60,000 b.p., only, but not during the LGM (Forman et al. 2002). Farther to the west the situation becomes even more complicated. It seems to be relatively well documented now that during the last glaciation the Barents and Kara Seas repeatedly had experienced the forming and waning of huge ice masses. During the earlier parts of the last glaciation they also had moved onto the continent. Epshteyn et al. (1999) had already stressed that the southern part of the Barents Sea and the adjacent regions of northernmost Russia had seen during the last glaciation three important ice-advances. Yet at that time, these advances could not be dated accurately. Moreover, it was said that Novaya Zemlya had contributed much to the provenances of the drift material. Later, these ice-advances were dated, though sometimes only provisionally. The strongest ice advances seem to have occurred between 110,000 and 90,000 b.p., 75,000 to 65,000 b.p. and 60,000 to 50,000 b.p. (Gataullin et al. 2001; Henriksen et al. 2001; Houmark-Nielsen et al. 2001; Kjaer et al. 2001; Knies et al. 2001; Mangerud et al. 2001; Manley et al. 2001). On the other hand, ice masses dating from the LGM (about 22,000 to 17,000 b.p.) do not seem to have entered the north European continent from the north. The ice-advances dating from the beginning of the last glaciation had dammed up a huge lake, the so-called Komi Lake (approximately 100,000 to 80,000 b.p.; Henriksen et al. 2001), which evidently had drained to the southwest in direction to the Baltic Sea (Maslenikova and Mangerud 2001). The significance of Novaya Zemlya as a center of an important inland-ice glaciation is questionable. This island seems to have experienced a strong glaciation during the first part of the last glaciation, only (before 40,000 b.p.; Zeeberg et al. 2001), but during the LGM the glaciation of Novaya Zemlya has been small only (Zeeberg et al. 2001). Shcherbakov et al. (1999) discussed the different types of glacigene sediments within the Barents Sea Basin without dating them exactly.

In contrast to these strong ice advances during the earlier parts of the last glaciation, originating in the Barents and Kara Seas, traces of contemporaneous Scandinavian ice advances, directed to the east, are rare. Arnold et al. (2002) stressed that the Scandinavian glaciers of deep-sea stage 4 had rapidly retreated to the Scandinavian Fjäll during the phases of climate improvement of deep-sea stage 3, but that they had begun to advance again at about 30,000 b.p., only. This ice-advance had never reached the Peza region to the south of the Timan-Peninsula, but in the Arkhangel'sk region it had arrived at about 17,000 b.p. (Lyså et al. 2001) or somewhat earlier (Kjaer et al. 2001). Further to the south the endmoraine-systems surrounding Lake Kubenskoe (to the north of Vologda) are dated at about 18,000 b.p. (Lunkka et al. 2001). On the other hand, Lake Onega became free of ice already at about 14,250 to 12,750 (calibrated – cal. – b.p.; Saarnisto and Saarinen 2001). Winograd (2001) pointed out that the huge water masses, which had formed the inland ice masses of the LGM had originated in the subtropical oceans. This had already been mapped by Frenzel et al. (1992) as concerns the Northern Hemisphere. The development of the inland ice masses at about 90,000, 60,000 and 20,000 b.p. has been modelled by Siegert et al. (2001), those of Scandinavia at about 50,000, 38,000 and 36,000 b.p. by Arnold et al. (2002).

To sum up, all what could be stated here shows that northern Eurasia had experienced during the last glaciation at least three, probably four, strong ice advances, the most important of which were those of the first part of the last glaciation. Yet these advances of inland ice masses never had reached those dimensions as were hypothetically discussed by Grosval'd (1983, 1988). Thus, there had existed ample room and a huge variety of ecological conditions, which could be used by different plant and animal taxa to survive even times of utmost severeness of climate in the higher latitudes.

2 Methods and Their Handicaps for Reconstructing Past Migrations of Plant Taxa and Vegetation History

It is a well-known methodological approach for reconstructing past changes in flora and vegetation to compare past pollen spectra with those of present-day surface samples, in order to better understand what the past flora and vegetation might have looked like. This method already meets with considerable problems within the temperate zones and under relatively moist climates. The difficulties increase considerably when approaching subpolar and polar regions or steppes and deserts (see already Frenzel 1969). Some new investigations focus on these problems again concerning northern Eurasia. Kozhevnikov (1995b) has stressed that the interpretation of pollen samples in terms of vegetation types to some extent always will be ambiguous, due to a wealth of external factors, which influence the past and present pollen sedimentation and due to the poor precision in determining palynologically the ecologically relevant taxa. Thus it is recommended to use as many other indicators as possible. These objections and recommendations are exemplified by a comparison between the present-day vegetation and the modern pollen flora of surface samples in the catchment area of river Chantalveergyn in Chukotka (Kozhevnikov 1995a). Blagoveshchenskaya (1995) has comparable objections. Thus she uses several statistical methods and correcting factors for describing relatively correctly the recent vegetation of the Volga Hills, as being reflected in surface samples. Kvavadze (2001), on the other hand, studied the annually changing modern pollen sedimentation in forests of eastern Georgia, Caucasia. Comparable investigations were done by Parshall and Calcote (2001) in Michigan and Wisconsin taking into consideration the different pollen production of the taxa involved. Considering the always changing effects of wind and insect activities in transporting pollen and spores, Zhilyaev (1996) relatively cautiously interprets surface samples in the Carpathian Mountains in terms of the prevailing vegetation. In contrast to what has been stated here, Simakova (1991) thinks that the recent vegetation patterns found on hills, mountains, alpine summits and steppic sites in the western Sayan Mountains, southern Siberia, are represented quite well by pollen surface samples. Subalpine meadows there seem to show an excessively high share of tree pollen types. Yu et al. (2001) arrive at more or less the same conclusion as Simakova. These authors investigated the modern pollen rain reflected in surface samples of the Tibetan Plateau. Yet, it seems that the site quality, most of all the moisture conditions of the micro-sites studied, were not duely taken into consideration, though precisely this is a well-known prerequisite for correctly interpreting former distribution patterns of various taxa and of the prevailing vegetation using the pollen flora of surface samples (Frenzel 1969). When studying the northeast Siberian vegetation of the last glaciation, Berman and Alfimov (1992) stress that the last-glacial "tundra-steppe" there is more or less an artefact, only, caused by wind and river transport from various habitats to the sites investigated. To avoid these difficulties, Nepomilueva and Duryagina (1990) had analyzed the recent pollen flora of a great variety of sites of the Middle Timan Mountains, North Europe, when studying the past vegetation patterns of these mountains. Yet, the results obtained are not very convincing. On the other hand, Bezus'ko et al. (1992) used recent surface samples to investigate the Holocene history of the Chenopodiaceae-steppe vegetation along the coasts of the Black Sea. Thereby the authors rely on the palynological criteria of the Chenopodiaceae family elaborated by Monoszon (1952) to palynologically differentiate between various species of this family. Yet, these "characteristics" are not reliable (Frenzel 1968).

Of course, sporomorphs can be transported by wind, rivers and marine currents far into the oceans. This problem is studied comprehensively by Cremer (1999) and by Naidina and Bauch (1999) as concerns the Laptev Sea, in northeastern Siberia. Here, the influence of river waters is remarkable (Are 1999; Ivanov and Piskun 1999). Modern diatom assemblages at the bottom of a lake on the Kunashir Island have been investigated by Lupikina (1989), those in the Kara Sea by Djinoridze et al. (1999), taking into consideration there 91 modern diatom species. Kunz-Pirrung (1999), on the other hand, studied the dependance of the distribution patterns of dinoflagellates and of Chlorococcales on the salinity of the marine waters in the Laptev Sea, and Sarmaja-Korjonen (2001) dealt with the meaning of cladoceran communities for reconstructing past lake-level changes in southern Finland. It seems that these taxa are quite useful for reconstructing the timing of lake-level changes but not for evaluating the amounts of this process in the past.

Taking all these data together, it is quite evident that a wealth of indicators for studying past phytogeographical and palaeoecological conditions in various regions of northern Eurasia is available, yet that a comprehensively and carefully done criticism will always be indispensable. Just this very topic will be dealt with immediately.

3 Holocene History of the Polar Tree-Line in Northern Eurasia

After the retreat of the last-glacial ice masses, vegetation immigrated astonishingly fast into these new areas. In general, it is unknown where the glacial refugia of all the relevant plant and animal taxa might have been situated. Kullman (2000) discussed intensively the problem of last-glacial *Picea*-refugia in westernmost Norway, thus strongly contributing to solve an old problem, which has already been discussed intensively in plant geography and taxonomy (Lindqvist 1948). It can be shown now that *Picea abies* had immigrated into Jämtland probably from some till now unknown regions of westernmost Norway at about 11,000/10,500 ^{14}C-years b.p. Comparably interesting observations were discussed by Anderson et al. (2002) in easternmost Yakoutia. Here, it seems that *Larix dahurica* had immigrated the catchment area of the Indigirka River at about 9600 cal. b.p. from the north, whereas the same tree species seems to have arrived in the upper Kolyma river region some 2300 years later, here coming from the south.

At any rate the tree-taxa seem to have migrated appreciably fast at the Late-glacial to Holocene transition and in general they are thought to have expanded their distribution areas during the Early Holocene much more to the north than they are thriving at present. Several papers deal with this interesting question when discussing the Late-glacial and Holocene vegetation history of various parts of northern Eurasia: Northeastern Yakoutia (Anderson PM et al. 2002); the coastal areas of the Laptev Sea (Naidina and Bauch 2001; Ukrajntseva et al. 1989); the Taimyr Peninsula (Kozhevnikov et al. 1993; Ebel et al. 1999; Hahne and Melles 1999; Kienel 1999; Siegert et al. 1999; Kienast et al. 2001; Andreev et al. 2002); southern surroundings of the Putorana Mountains at the Lower Yenisey river (Kozhevnikov and Ukrajntseva 1992); southern surroundings of the Kara Sea (Ukrajntseva 1990; Khitun 1994; Andreev et al. 2001; Forman et al. 2002); northern parts of European Russia (Yurkovskaya et al. 1989; Nepomilueva and Duryagina 1990; Oksanen et al. 2001; Pitkänen et al. 2002); northernmost Scandinavia, including the Kola Peninsula (Kullman 2000; Luoto and Seppälä 2000;

MacDonald et al. 2000; Mäkelä and Hyvärinen 2000; Corner et al. 2001; Hammarlund et al. 2002). Kremenetskiy et al. (1996) have produced most interesting maps of the position of the polar timberline during various phases of the Holocene for *Picea obovata* (modern times, 9000 b.p., 6000 b.p.), *Larix* spp. (modern times, 10,000 b.p., 8000 b.p.), *Betula* Sect. *Albae*, i.e. *Betula pubescens*, *Betula pendula* (modern times, 9000 b.p.), *Duschekia* (*Alnaster*) *fruticosa* (modern times, 9000 b.p.). The maps are given for the region between the Kanin Peninsula in the west to the basin of river Kolyma in the east, together with additional data concerning the distribution history of *Pinus pumila*, *Ribes*, *Rubus idaeus*, *Vaccinium uliginosum* and *Oxycoccus palustris*. The maps mentioned are essentially based on a wealth of pollen-analytical studies by several Russian research teams. Yet it is a well-known fact that the significance of long-distance transport of sporomorphs in the tundra region must not be neglected, and that this source of error becomes even more important if the bio-production of surrounding vegetation decreases, as in the Arctic tundra. Here even macrofossils may cause problems, since they can be transported by wind over large distances, most of all in winter-time; this is well established for the European tundras (Thomson 1933, 1956) and the Tibetan Plateau (own observations). Some time ago it could be shown, using the international literature (Frenzel 1969), that the percentages of tree pollen in relation to the total sum of sporomorphs counted has in the tundras of northern Siberia and of northern America, at about 400 to 500 km to the north of the polar tree line, values of between 5 and 30 to 40%, the exact percentage always depending on the pollen-binding capacity of the sediments studied and on the intensity of pollen production of the surrounding vegetation. Thus, for reconstructing the former position of the tree line under these harsh climatic conditions pollen analysis alone may be very misleading. In this respect, the dating of tree trunks and stumps repeatedly found within the modern tundra zones becomes very important. P.M. Anderson et al. (2002) reported that within the catchment area of the upper Indigirka River, northeasternmost Yakoutia, larch never seems to have passed its modern northern distribution limit, but that it had immigrated this region by about 9000 cal. b.p. as has already been stated. Yet, tree-like *Betula* and *Alnus* had invaded the present-day tundra zone there between 9000 and 8500 cal. b.p. On the Taymyr Peninsula repeatedly trunks and stumps of various tree species have been found in the modern tundra vegetation. In the surroundings of the Taymyr Lake stumps of larch have been observed, which date from about 10,500±500 ^{14}C-years b.p. (Ukrajntseva 1990; Hahne and Melles 1999). Other observations date from 8390±70 cal. b.p. (Andreev et al. 2002), yet larch does not seem to have passed its present-day timberline to the east of Lake Taymyr (Kozhevnikov et al. 1993). According to Ukrajntseva (1990) larch could be found at about 9730±300 ^{14}C-years b.p. on the Gyda Peninsula at about 70°N. Siegert et al. (1999) reported on several larch trees dating from ca. 8000 to 7000 b.p. to the south of Lake Taymyr, where at present the transition from the southern to the northern tundras is situated. Here, the last larch-stumps are said to date from 2880±60 cal. b.p. At the west coast of the Yamal Peninsula trunks of *Betula*, dating from 10,000 to 9000 cal. b.p. were found more than 200 km farther to the north than the modern tree-line is situated (Forman et al. 2002). Several finds of birch trees and of *Alnaster fruticosa*, dating from about 12,000 ^{14}C-years b.p., are also reported by Kremenetskiy et al. (1996) from the Taymyr Peninsula and from north-eastern Yakoutia. In the Bol'shezemel'skaya Tundra, to the west of the Polar Ural Mountains, Oksanen et al. (2001) reported several findings of leaves of birches (species names not mentioned) and of conifer bud scales at about 67°N, 62°E, dating from ca. 9250±60 or ca. 9420 b.p. to about 1920 b.p. Yet, here, again, one is confronted with the problem, whether these leaves and bud scales actually point to the former existence of the taxa mentioned at the sites studied or whether they merely might have been transported there by winds. Finally, on the Kola Peninsula at about 68°43' to 68°25'N, 35°10' to 19'E *Pinus sylvestris* had moved between 6680 and 3820 b.p. some 20 km farther to the north than it occurs at present and between 5890 and 3450 b.p. some 40 m higher up into the mountains than it can be found there today. Here, it is said that the tree density was highest between 7000 and 5000 b.p. These positions were reached by pine trees evidently some centuries later than in the Fennoscandian Mountains (MacDonald et al. 2000).

Taking all this together, it seems to be proven that some tree species, though not all of those which can be found at the modern polar tree-line, had already arrived at the beginning of the Holocene at the position of the modern tree-line and in several sites they had passed it moving far to the north. This might be interpreted as indicating a generally warmer climate there than is met with at present. The quite different configuration of the coast line needs to be taken into account, however, since at that time it lay in various regions of the study area some hundred kilometers farther to the north. Thus, climate should have been much more continental than it is at present. This would mean that given the solar radiation was constant, summer temperatures should have been higher than they are nowadays. Moreover, it has to be taken into consideration that man not only on the Taz Peninsula has driven the polar tree-line back since unknown times (Khitun 1994).

4 Holocene Palaeoecology

4.1 Eastern Asia

Zhou et al. (2002) intensively studied the past 20,000 years of palaeoecological development of the transition zone between the recent desert and loess-steppe areas of northern China. Ten sites were investigated, five lying in the modern desert, and five on the loess plateau. The profiles were dated using thermoluminescence and ^{14}C. It turned out that between ca. 21,200 and more than 11,300 b.p. the entire region was situated within the desert zone. Yet, during the Late-Glacial and the Holocene nowadays fossil soils were repeatedly formed: At about 11,300 b.p., ca. 10,600 to ca. 10,000 b.p., 9900 to 8400 b.p., ca. 7800 to 7000 b.p. and 5900 to 4200 b.p. At present, the boundary between desert and steppe is situated at nearly the same position as it was during the LGM, but the position of the summer monsoon-front is situated nowadays nearly where it was during the Holocene climatic optimum. Thus, it seems that the climatologically not explainable positioning of the eastern border of the desert might have been caused by man's activities. In this respect it is interesting to realize that from about 10,000 years b.p. onwards many archaeological sites existed in the present-day desert. Already at about 3000 to 2000 ^{14}C-years b.p. an increase in the population density had caused a strong destruction of the natural vegetation, thus triggering an intensive activity of wind and water and a remarkable desertification. Comparable facts were reported by Huang et al. (2000) for the southern part of the loess plateau immediately to the north of the town of Xi An. This area also has been strongly influenced by neolithic and bronze-age man. Here, several fossil soils can be found, covered by loess layers. It was shown that the influence of neolithic agriculture has been very strong, interestingly during times of soil formation as well as of loess sedimentation. It seems that the Holocene climatic optimum has been interrupted by a phase of loess sedimentation, from ca. 6000 to 5000 b.p., without an interruption in agriculture. These observations are so remarkable, because several other research groups interprete past changes in pedogenesis, vegetation and human activities in East Asia in terms of climatic change, only. For instance, Liu et al. (2002) studied the history of the steppe to forest transition immediately to the north of the Great Wall on the western slopes of the Qingan Mountains. Here, at about 10,000 b.p. steppe vegetation and some coppices of birch seem to have been widely spread, followed by the immigration of various forest types, into which since about 5900 b.p. dune-sands had moved from the west and vegetation changed considerably: *Quercus* and *Betula* forests were replaced by pine stands and by the spread of a steppe vegetation, poor in taxa. All this is explained by the authors in terms of climate change, only, because an intensive impact of man is said to have begun at about the 17th century though much older traces of human activities are known here quite well, including the Great Wall. It seems that the influence of climate has been overemphasized considerably, as will be shown later.

Maxwell (2001) pollen analytically investigated the vegetation history and changes of the palaeoecological conditions of northeastern Cambodia, since about 9300 b.p. The history of the natural vegetation consisting at first of semi-evergreen forests with a strong share of subtropical taxa is well documented. At about 5300 b.p. vegetation had changed relatively suddenly by an increasing amount of taxa, which characterize secondary vegetation types, together with strongly increasing amounts of charcoal in the sediment. The ensuing changes in this vegetation are interpreted as indicating climate change, only, leading to drier conditions. On the other hand, it is not mentioned that the history of human activities is very long here (see already Eickstedt 1944). Moreover, it could be shown that on the Tibetan Plateau a wealth of peat bogs began to be formed since about 5800 b.p. and that since then in general these bogs grew more or less at the same velocity without showing phases of a climatically induced retardation in their growth rates (Frenzel 2002). Thus the influence of climate does not seem to have been very important. Guo et al. (2000) dealt with Holocene non-orbital climatic changes studied in the Sahara and – less accurately dated – in China. Concerning China, the authors state that climate had changed repeatedly between warmer and moister viz. colder and drier conditions; yet they also state that it is difficult in China to accurately differentiate between adverse environmental phases caused by climate or by man. This difficulty becomes evident when the observations by Korotky et al. (2000) and by Razjigaeva et al. (2002) on the Holocene vegetation history of the isles Kunashir and of Iturup, Kuriles, are compared with those of Verkhovskaya and Esipenko (1993) and of Verkhovskaya (1990) on the surroundings of Vladivostok. In the latter region cultural layers from final neolithic times up to the Middle Ages in various sites show a forest-steppe-like vegetation, rich in ecologically different taxa. It seems to have been replaced more recently by a vegetation, which is much richer in forest species than it was before. This is interpreted in terms of an initially drier climate (in the immediate vicinity of the Sea of Okhotsk!), which has been replaced by a moister and cooler climate. Yet, this transition cannot be traced on the two Kurile Islands mentioned, though here in several sites vegetation history, together with the history of marine trans- and regressions were studied. It can be shown, that the exacting broadleaf-forest types on these two islands seem to have been replaced at about 3400 to 2700 b.p. by more cold-resistant species, including various conifers. Indications of a neolithic to Medieval dry and warm climate, which had become cooler and moister during the last centuries are lacking here. Again one has to answer the question, whether the "forest-steppe" in the surroundings of Vladivostok is a reliable indicator of a warmer and drier climate or of a strong human impact, only. These difficulties in interpreting palaeoecological processes correctly become evident in view of the investigations of Hong et al. (2000) on a 6000-years-long $\delta^{18}O$ sequence done on the cellulose of *Phragmites* and of Cyperaceae in a north-Chinese peat bog. It is missed that *Phragmites* and Cyperaceae do not belong to the same botanical family (authors: Gramineae) and that the region studied has a well-pronounced monsoon climate (authors: annual air humidity 70±3%!). The data were compared with those of Greenland, Scandinavia etc., though the dating quality is poor. Oldfield (2001) gave a good criticism of this way of climate research. Chu et al. (2002) used a multidisciplinary approach for studying the medieval warm period on the Leizhou Peninsula, southern China, in the immediate vicinity of the Isle of Hainan. Here, the lake sediments are analyzed in terms of climatic change, only. The amounts of total carbon, total nitrogen, anorganic carbon, total organic carbon, and biogene silicates in lake sediments were analyzed. The dating quality is not very strong. The data are being compared with those of old Chinese phenological reports. So, phases of relatively cold or warm climates are differentiated from one another and they are compared with those of the Tibetan Guliya ice cap, with that of Quelccaya Mt. in South America, and with Mono Lake in North America. Though the dating quality of the sediments in the lake studied is not good, "convincing" correlations with the other three sites are found; yet the agricultural activities in the surroundings of the small lake, which might have influenced the amounts of the chemical elements studied are not taken into consideration. The same historical period is analyzed by Qian and Zhu (2002) in the surroundings of Beijing. Here, historical data since 1450 A.D. and stalagmite growth-rates are analyzed. The general tendency of climatic change of this time, which is known from other regions, can be followed quite well there, though it has not been studied how intensively human impact on the surroundings of the cave might have influenced the water-budget of the cave. These data can be improved or corroborated by palaeoclimatological investigations using written reports of the last 2000 years, together with tree-ring data and with data obtained from Tibetan ice-caps (Yang et al. 2002), or relying on dendroclimatology, only: northwestern Karakorum: Esper et al. (2002); western Himalaya, last 200 years: Yadav and Singh 2002); Karakorum, the last 500 years: Esper (2000); eastern Tibet: Bräuning (2001a,b, 2002); increasing activity of the Southwest Monsoon during the last 400 years, as calculated from the foraminifera fauna of the Arabian Sea: D.M. Anderson et al. (2002).

All these data mentioned point to the fact that at least the Middle and Upper Holocene changes of climate have not been very strong, though they had happened repeatedly. Thus, one should think much more intensively about human impact on the vegetation and on the ecological setting in the regions studied than is being done repeatedly. This scepticism or even criticism is in line with that of Oldfield (2001).

For reconstructing past climates the correspondence of modern distribution patterns of some taxa with certain characteristics of present-day climate are very important. One plant species which in Europe is looked upon as indicating warm interglacial climates is *Brasenia schreberi*. Yet, Shibneva (1991) reports on the modern occurrence of this plant species within the catchment area of the lower Amur river in the immediate or at least very close vicinity to the southern boundary of the area of permafrost soil. Again, it becomes evident that much more reliable data are needed for realistically reconstructing past climatic and ecological conditions than are repeatedly used today.

4.2 Northern Eurasia

The Siberian coastal seas are generally shallow. Thus, the timing of the Late-Glacial to Holocene transgression becomes important even for a good understanding of the development of climate on the nearby continent. Bauch et al. (2001) studied this process in the north Siberian shelf area. The dramatic sea-level rise had begun there at about 14,200 cal. b.p.; at about 11,100 cal. b.p. the coastline stood at –50 m; 9800 b.p. at ca. –43 m and by about 8900 b.p. at –31 m. At ca. 5000 b.p. the highest Holocene sea-level stand had been reached. Naidina and Bauch (2001) studied the pollen flora of a marine core situated some 300 km to the north of the East Yakoutian coast, to the southwest of the Newsiberian Islands. Since about 9400 cal. b.p. the appreciably rich pollen flora was evidently brought there by river waters. It reflects quite well the development of vegetation on the continent, beginning with the establishment of a shrub-tundra. The long-distance transport of tree pollen had reached its maximum between 5500 and 2700 cal. b.p.; then it had declined remarkably. Ukrajntseva et al. (1989) reported on the Late-Glacial to Holocene development of vegetation on the Bol'shoy Lyakhovskiy Island. Yet most of these data were obtained by pollen analysis, which can be strongly influenced by long-distance transport under these arctic conditions. It is a well-known fact that seeds and fruits can be transported easily over long distances by birds. So observations of Solovieva (1999) on the travel-routes of *Somateria spectabilis* are welcome. *Somateria* spends winter-time in the Bering Sea area and stays some time in April in polynyas near the Novosibirskie Islands, then moving farther to the west. The bird exemplifies the similarly long-distances traveled by various others.

Kienel (1999) investigated in Lama Lake, Taymyr Peninsula, palaeoecological changes, which had occurred during the final parts of the last glaciation and during the Holocene, as being reflected in the diatom floras of various ages. The results are compared with much older pollen-analytical data. The diatom floras reflect quite well changes in the alkalinity of the surrounding soils. It is interesting to note that even here man has to be taken into consideration when discussing palaeoecological changes. Yet, Pitul'ko (1999) could show that a previously held view of human activities at 74°08'N, 99°06'E, dating from between 9000 and 5000 b.p. is wrong. It evidently merely dates from 1800±75 b.p. Pitkänen et al. (2002) investigated very critically the vegetation history in the Salym-Yugan mire, part of the huge Vasyugan mire (60°10'N, 72°50'E). It is interesting to note that the authors think that a tree pollen percentage of less than 25% of the total sum is indicative of a tree-less vegetation (also see Sects. 2 and 3). The pollen diagrams seem to begin during Younger Dryas times, when some types of tundra or of a cold-steppe vegetation had been very characteristic here. It is suggested that *Betula* forests began to thrive there since late Boreal times, with an immigration of spruce at about 9900 cal. b.p. Ukrajntseva (1990) studied in the southeastern Taymyr Peninsula the Late-Glacial and Holocene vegetation history, to some extent using botanical macrofossils, yet most of all pollen analysis.

According to stumps and trunks of larch, which could be dated to about 10,500 b.p., it is suggested that warmth had reached its maximum at about 10,000 to 9000 b.p. Yet, of course here as well, the quite other configuration of the coast-line has to be taken into consideration. Using pollen percentages and pollen-influx data, Andreev et al. (2001) followed the episodes of vegetation history on the Yugorski Peninsula, Kara Sea, since full-glacial times (26,200 b.p.), when vegetation seems to have been extremely scarce, here. Even during the Bølling-Allerød-interstadial, vegetation was very sparse. It is suggested that the warmest climates of the Holocene had existed here at about 9500 cal. b.p. Yet, at that time the site was situated far away from the coast-line with its strong influence of cold winds. On the other hand, the maximal pollen concentration was observed from about 8500 to 8300 b.p. Since about 4,000 b.p. climate had become colder, paralleled by a remarkable increase in the tundra area. These data are interesting, because they can also shed new light on the migration processes of flora and vegetation in the Bol'shezemel'skaya Tundra, which were studied long ago by Schäfer and Frenzel (1959). It has been stated that during Late-Glacial and Holocene times the north Siberian coasts had experienced very strong changes in their horizontal positions. This does not hold for the Kola Peninsula (Corner et al. 2001), though the ice load had been very strong there.

Elina and Antipin (1992), Elina and Lebedeva (1992), Elina and Yurkovskaya (1992), Elina et al. (1994a,b, 1995, 1996), and Yurkovskaya et al. (1989) were strongly interested in the Late-Glacial and Holocene history of the ecological conditions and of vegetation in the area between the Lake Onega and the Arkhangel'sk region. Problems of the development of peat bogs are discussed equally intensively as those of the immigration history of various tree taxa. Repeatedly these pioneering investigations suffer from the poor dating quality. On the other hand, in one investigation (Elina et al. 1996) 28 out of 52 ^{14}C-datings are discarded, since they do not "fit into the general picture". This seems to be a very dangerous approach. Some of these investigations are used for reconstructing the history of climate quantitatively (Elina et al. 1995, 1996). A very detailed history of changing temperatures and of moisture (precipitation) is given, which seems to be overinterpreted. This becomes evident when the data are compared with those obtained by Bigler et al. (2002; also see Seppä and Birks 2002) in a small lake of the Abisko area in northern Sweden. In this study, sporomorphs, diatoms and cladocerans are used together with determination of the loss of ignition. Approximately the last 10,600 cal. years are taken into consideration and it can be shown that the time between 10,600 and about 6000 b,p. repeatedly is extremely difficult to be discussed in terms of climate history, because the indicators of climate used do not match with modern schemes. On the other hand, just this very time can be relatively well studied by glacier history in the Jostedalsbreen region, western Norway (Dahl et al. 2002), where two glacial advances took place between 10,100 and 9700 cal. B.P. The general trends in vegetation history of the Abisko area, northern Sweden, are discussed for the Holocene by Hammarlund et al. (2002), together with an analysis of climate history, using the $^{18}O/^{16}O$ quotients of lacustrine carbonates. It is interesting to note that the authors are sceptical against too simple a way of interpreting the $\delta^{18}O$ of precipitation in terms of temperatures, only. Luoto and Seppälä (2000) show that in Finnish Lapland remains of former peat bogs, which date from atmost ca. 4000 b.p. are at present destroyed by wind, needle-ice and reindeer browsing (also see van Vliet-Lanoë and Seppälä 2002). These bogs had begun to be formed in an *Empetrum* heathland. According to Mäkelä and Hyvärinen (2000), *Pinus* had arrived in the Inari-Lapland area at about 7500 b.p.; faint traces of spruce pollen are recorded since ca. 3500 b.p. Between both these data the influx-rate of the Ericaceae was appreciably high. It is interesting to see that according to size-statistics of the pollen grains, *Betula tortuosa* could not be traced there uninterruptedly. Much more important has been *Betula pubescens*, and only at the very beginning and at the end of this sequence *Betula nana* could be traced. Nesje and Dahl (2003) stress that the "Little Ice Age" of the 14th to the mid-20th century evidently has been caused in Scandinavia by an increase in winter precipitation, but not by a mere decrease in temperatures. Comparable palaeoclimatic enigmata are dealt with by Oksanen et al. (2001) concerning the beginning of permafrost actions in peat bogs of the Bol'shezemel'skaya Tundra. Here, according to the very critically discussed results, permafrost seems to have begun to act in some of the bogs studied at about 3100 b.p., a second phase in permafrost formation is said to have begun at ca. 2200 to 1900 b.p., and the third one at about 600 to 100 b.p. When permafrost began to act, the growth rates of the peat layers were strong.These observations seem to corroborate the earlier view, that the beginning of permafrost in peat bogs depends to some extent on the thickness of the peat layers formed, not only on the winter-temperatures (Frenzel 1983).

Seppä and Birks (2002) have studied the Holocene climate history in northwesternmost Finnish Lapland using transfer functions elaborated from 113 sites in Finland. The basic palaeobiological data are obtained from the pollen flora, beginning at about 10,400 cal. b.p. It is interesting to learn that the temperature curves and those of the annual precipitation don't show the lot of wiggles as were reconstructed by Elina et al. (1995, 1996). During hypsithermal times (8000 to 6500) mean temperatures of the month of July seem to have been higher than at present by about 1.6 to 1.8 °C with an increase in the annual precipitation rate of about 300 mm against those of today. (As to the synchronous vegetation in northern Sweden see Snowball et al. 2002.) The history of forest fires has been analyzed in eastern Finland (Pitkänen 2000) and in Finnish Lapland (Virkanen 2000), together with the succession after the fires had ended (Pitkänen 2000) and soil erosion (Virkanen 2000). It seems that in Finnish Lapland remarkably strong fires had occurred at about 1600 and 900 cal. b.p., and it might be concluded from the data reported that the fire-periodicity has been ca. 300 years. Vorren (2001) very comprehensively and cautiously has studied the development of eight ombrotrophic peat bogs in the Troms area, northern Norway. These bogs had developed from some types of carr over more or less open Cyperaceae–Amblystegiaceae fens to open Ericaceae–*Sphagnum* bogs. The transition from geogene to ombrogene conditions seems to have happened between 4460 and 2400 cal. b.p. The influence of climate is not very pronounced, yet there was a remarkable change to colder and wetter conditions at about 2750 b.p.

Poska and Saarse (2002) investigated the Holocene history of sea-level changes, of vegetation history and agriculture on the Saaremaa Island (Ösel). The pollen sequences begin asynchronously when the sites emerged from the Baltic Sea. The very early and strong impact of man on vegetation is well documented, beginning at about 6500 to 6000 b.p.

In a very theoretical approach, Bolikhovskaya (1998) described the history of vegetation in central Russia. The aim of the paper was an attempt to characterise the main types of periglacial vegetation in the loess areas of central Russia. Towards these means the author relies on a very elaborate diagnostics in palynology. Yet, it may be doubted that several of the "species" mentioned were determined correctly. On the other hand, Spiridonova (1991) has investigated palynologically the vegetation history in the catchment area of the river Don. The results are given in palaeovegetation maps for an initial interstadial of the last glaciation, an early stadial at about 38,000 b.p., the Bryansk-interstadial (ca. 26,000 b.p.), the LGM (22,000 b.p.), and for five phases of the Holocene. Moreover, it is stressed that during the last interglacial, the so-called Mikulino-interglacial, the forest zone had deeply invaded into the present-day tundra zone, but that during the first part of this interglacial the steppe vegetation had remarkably moved into the modern (natural) forest belt. The interglacial forests are held to have moved onto the steppe during the second part of the interglacial. The dating quality for deciphering these movements is regrettably poor. According to Zernitskaya (1992), the Holocene distribution history of the genus *Quercus* has been relatively complicated in Belarus. The history was deciphered palynologically. It is stated that *Quercus robur* had appeared there already during the Allerød-interstadial, though only in very small quantities. Up to the Atlantic period its share had constantly increased. *Quercus petraea* seems to have appeared there in the southwestern regions of Belarus at the transition from Preboreal to Boreal times. It had reached its maximum values there during the second part of the Atlantic period. The last immigrant is said to have been *Quercus pubescens*, which appeared there during the second part of the Atlantic period. The share of *Quercus petraea* declined since Subboreal times and *Quercus pubescens* disappeared, but *Quercus robur* constantly strengthened its position. Sycheva and Chichagova (1999) investigated the history of pedogenesis and of soil erosion in the Middle Russian hill region, Kursk Oblast'. Here, repeatedly fossil soils were formed on slope and valley sediments: ca. 7700 to 6600 b.p., 5000 b.p., 3800 to 3300 b.p., 1400 to 1050 b.p. and since about 150 b.p. It seems that there had existed still another phase of pedogenesis, i.e. at about 2500 b.p. It is stressed that soil erosion to some extent has been triggered by climate change, but that the most important factor has been the activity of man, at least since about 2500 b.p.

The history of marine trans- and regressions at the mouth of the river Dnestr was studied by Sadchykova and Chepalyga (1999). Gogichaishvili (1992) has analyzed the most interesting history of forests in the subarid regions of Gruziya. Regrettably the investigated profiles could not be dated accurately. It seems that during the Early Holocene *Carpinus-Quercus* forests were widely spread there, with a small admixture of xerophytic forests, composed of *Juniperus* and *Pinus eldarica.* At the beginning of the Middle Holocene the share of these xerophytic forests increased, but the hornbeam-oak forests seem to have retreated to higher elevations. Finally, during the Late Holocene riverine forests expanded, yet the share of xerophytic forests and of the hornbeam-oak forests declined further on. Though climate might have been of some importance in causing these changes, the most efficient factor has evidently been man.

The Holocene vegetation history of Kazakhstan, especially of the pine forests, has been extensively studied by Kremenetskiy et al. (1994) in several peat bogs and former lakes. The sequences begin with late-glacial *Poaceae-Artemisia* steppes to semi-deserts, with high amounts of *Hippophaë* in the immediate surroundings of lakes. Only at about 9500 b.p. *Betula* seems to have intensively immigrated into the region. It was faintly followed at about 6300 to 6200 b.p. by *Pinus*, and far away by *Ulmus*, and already at ca. 2500 to 2000 b.p. *Pinus* began to retreat. In general it had been held till now that the existence of pine trees there dates from glacial or at least late-glacial times, yet the observations demonstrate that *Pinus* had reached its modern southern boundary of its distribution area during Subatlantic times only; yet, during Boreal times the Turan flora element had spread far to the north. The history of the formation of peat bogs there is described by Sviridenko et al. (1994).

5 Pleistocene Palaeoecology and Vegetation History

In Siberia the last glaciation is divided repeatedly into the initial Zyryanka- or Ermakovo-glaciation, the Karga-interstadial and the ensuing Sartan glaciation, which leads into late-glacial times. The Karga-interstadial is held to have been characterized by nearly interglacial climatic and palaeoecological conditions. Yet, Astakhov (2001) critically analyzed these assumptions. He states that all well-dated facts point to a revision of the scheme mentioned: The Karga-interstadial evidently has been the last interglacial, i.e. the Kazancevo- (Mikulino-, Eemian-) interglacial. Thus, Astakhov confirms what has been stated earlier (Frenzel 1993). On the other hand, it is a well-known fact that the middle part of the last glaciation (i.e., roughly the marine isotope stage 3) in eastern Siberia and in central Asia has experienced strange palaeoecological conditions. An example are observations by Ukrajntseva (1996) on the stomach and intestinal contents of various animal species, which were found in middle and eastern Siberia embedded in permafrost. Six animals were analyzed: a mammoth at the Berezovka died at the end of July or the beginning of August 44,000±3500 b.p.; a horse at Selerikanka on the river El'ga died during the same season in 38,590±1120 b.p.; a bison at Mylakhchinsk on the middle Indigirka River died during the end of June or the beginning of July 29,000±1000 b.p.; a mammoth from the middle Shandrin River, lower Indigirka River, died in early spring 40,350±880 b.p.; a mammoth baby from the middle Kirgillyakh River, 41,900±1000 b.p.; and a mammoth of the river Yuribey, Gyda Peninsula, 9730±300 b.p. All the animals mentioned, with the exception of the Yuribey mammoth, had died during the so-called Karga interstadial. The stomach contents of the Selerikanka horse contained 116 taxa: 96 phanerogams, 20 cryptogams. Among the phanerogams were 12 tree species, 14 species of shrubs and dwarf-shrubs, as well as 72 species of herbs and very small dwarf-shrubs. Most of the taxa mentioned belong today to the boreal and hypo-arctic floras. Here even plants of steppes and of dry meadows were met with. The animal of the Berezovka had 57 taxa in its stomach, that of the Shandrin river 49 taxa, at the Indigirka 61 taxa, at the Kirgillyakh 64 taxa and that of the Yuribey 74 taxa. The taxa could be determined to the species or even subspecies level. It is remarkable that the horse of the Selerikanka had browsed on *Picea ajanensis*, *Picea obovata*, *Pinus sylvestris*, *Betula* Sect. Costatae, which at present can only be found about 1000 km further south. Of course, it is not known which distances these animals had migrated per day. Nevertheless, the very rich floras must have existed relatively nearby. In contrast to these relatively well-dated findings, those described on the Taymyr Peninsula, dating from the Karga interstadial, show infinite ages (Siegert et al. 1999; Andreev et al. 2002), together with a pollen flora which is said to have existed under much warmer and moister conditions than are met with there at present. It may be that here again interstadial and interglacial sediments are confused with one another. Approximately the same period like in northeastern Siberia was dealt with by Markova et al. (2002) in the Russian Plain. Here, 45 sites with mammals and 52 sites with plant remains date from the Bryansk interstadial (33,000 to 24,000 b.p.). Evidently, these findings have no modern analogues. They show a mixture of present-day subarctic and steppe elements. At that time a forest belt had not existed there. The results of this very comprehensive research are given in a phytogeographical map.

The stadial vegetation of the Russian Plain was mapped by Spiridonova (1991), most of all for the catchment area of the river Don. The very open vegetation of the LGM, which seems to have been extremely poor in botanical taxa, was investigated in northeastern Siberia by P.M. Anderson et al. (2002); on the Taymyr Peninsula by Ukrajntseva (1990), Kozhevnikov et al. (1993), Hahne and Melles (1999), Kienel (1999), and Kienast et al. (2001);

on the Yugorski Peninsula by Andreev et al. (2001); in the southern part of the west-Siberian lowlands by Pitkänen et al. (2002); the full-glacial deserts of the Tengger Sha Mo, Mongolia, have been intensively studied by Zhang et al. (2002), those of the northern part of the Chinese loess-plateau by Zhou et al. (2002). Kotlia et al. (1997) have reported on very interesting palaeoecological observations from about 20,000 to 13,000 b.p. in Kashmir and Ladakh; here, at that time C_3-plants seem to have been widely spread, when at the same time soils were formed there, in contrast to desert-like conditions in the adjacent Indian lowlands.

The geobotanical and palaeoecological conditions of the last interglacial and during the transition from this interglacial to the last glaciation were repeatedly analyzed: Yamal Peninsula (Gataullin 1991); northern Norway (Linge and Lauritzen 2001); northwestern Ukraine (Bogutskiy et al. 2000, 2001); the surroundings of Kiew (Gerasimenko 2000); the Vologda area in central Russia (Gey et al. 2001); Belarus (Shalaboda 2001); northwest-Greece (Tzedakis et al. 2002); northern China, the Xining-Lanchou area (Chen et al. 1997); the north-Chinese loess plateau (Rousseau et al. 2000; Rousseau and Kukla 2000); the surroundings of Lake Baykal (Prokopenko et al. 2002). In general these observations confirm previously held views.

6 Palynology and Seed Morphology of Various Botanical Taxa Used for Palaeobotanical Investigations

During the last years a wealth of interesting pollen-analytical and morphological features of various plant taxa has been investigated in Russia, which can contribute to a much better palaeobotanical investigation of botanical remains found in geological samples. These observations and their bibliography are given in Tables 1 and 2.

Table 1. Palynological investigations of various taxa

Species/family	Investigation	Reference
Acer *(Aceraceae)*	50 Species investigated. SEM images	Pozhidaev (1992)
Alliaceae	22 Genera, 74 species investigated. Comprehensive list of characteristic features. SEM images	Kosenko and Kudryashova (1995)
Alnus (Betulaceae)	36 Species investigated; 4 palynological types can be differentiated. Key for the differentiation	Khramova (1996)
Alstroemiaceae	25 Species investigated. List of characteristic features. SEM images	Kosenko (1994b)
Ammolirion	31 Species investigated, SEM images	Kosenko and Sventorzhetskaya (1995)
Anthurium (Araceae)	34 Species studied. 4 morphological types and 7 subtypes described. Concerning palynology the sections of *Anthurium* seems to be heterogeneous. SEM and TEM images	Tarasevich (1989)
Aquilegia (Ranunculaceae)	25 Species studied, together with one *Paraquilegia*-species. Comprehensive morphological list. 7 palynogroups described. SEM images	Vasil'eva and Khramova (1992)
Betonica	Comprehensive list of morphological features. SEM images	Krestovskaya and Vasil'eva (1997)
Blandfordiaceae	9 Genera, 19 species investigated. Comprehensive list and descriptions. SEM images	Kosenko (1994a)
Boraginoideae (Boraginaceae)	16 Genera (24 species) out of 6 tribes investigated. Reliable differentiation between *Arnebia*, *Echium* and *Symphytum* possible. Comprehensive list of characteristic morphological features. SEM images	Popova and Zemskoya (1995)
Calligonum (Polygonaceae)	8 Species studied, together with *Zygophyllum*. 3 Morphological types which do not correspond to taxonomic entities. SEM and TEM images. List describing the morphological features.	Tarasevich and Abdullaeva (1993)

Table 1. *Continued*

Species/family	Investigation	Reference
Carex *(Cyperaceae)*	30 Species studied. Even within one species exine-structures repeatedly differ strongly. SEM images	Tarasevich (1992)
Caryophyllaceae	37 Genera, 192 species investigated. The following form-types can be differentiated: *Spergula marina*, *Spergula morissonii*, *Spergula arvensis*, *Stellaria uliginosa*, *Bufonia tenuifolia*, *Cucubalus baccifer*, *Silene jundzillii*, *Dianthus laevigatus*, *Pleconax subconica*	Romanova 1992
Cercidiphyllum (Cercidiphyllaceae)	*Light microscopy, TEM and SEM*	*Rowley (1992)*
Cichorioideae (Asteraceae)	Three types can be differentiated; type *Scolymus*, type *Cichorium*, type *Scorzonera*	Askerova (1992)
Cistus *(Cistaceae)*	13 Species studied. Two morphological types, which correspond to the subgenera. In certain cases even the determination of fossil species seems to be possible. SEM images	Ukrajntseva (1991)
Corydalis (Fumariaceae)	43 Species investigated. Comprehensive list of characteristic features. SEM images	Kosenko and Mikhailova (1991)
Crambe (Brassicaceae)	26 Species investigated, list of characteristic features. The taxonomic sections can generally be differentiated	Khalilov and Arkhangel'skiy (1991)
Cyananthus (Campanulaceae)	23 *Cyananthus* species studied together with species of the genera *Codonopsis*, *Leptocodon*, *Ostrowskia* and *Platycodon*. List and description, SEM and TEM images	Shrestkha and Tarasevich (1992)
Doryanthaceae	9 Genera, 19 species investigated. Comprehensive list and descriptions. SEM images	Kosenko (1994a)
Dryas (Rosaceae)	Compared with *Cowania* and *Fallugia*. SEM images	Kozhevnikov and Arkhangel'skiy (1996)
Eremurus	31 Species investigated, SEM images	Kosenko and Sventorzhetskaya (1995)

Table 1. *Continued*

Species/family	Investigation	Reference
Eriophorum (Cyperaceae)	12 Species investigated. No determination of different species possible. SEM images.	Novoselova (1994)
Flacourtiaceae	56 Species out of 40 genera investigated, description and SEM images of the pollen morphology	Gavrilova (1993, 1997)
Helianthus (Asteraceae)	7 Species studied. No important differences in the structure of the exine. List of characteristic features. SEM images	Toderich (1992)
Henningia	31 Species investigated, SEM images	Kosenko and Sventorzhetskaya (1995)
Itea (Iteaceae)	3 Fossil and 4 modern species studied. Description of the structures analyzed. SEM images	Rylova (1989)
Lamiaceae	153 Genera, 303 species investigated. Nomenclature and typology of exine-structures. SEM images	Pozhidaev (1989)
Liliaceae s.str.	The genus *Tulipa* can be differentiated into 2 groups palynologically, which in essential represent the sections *Tulipa* and *Eriostemones*: list of characteristic features, SEM images	Kosenko (1991, 1992)
Linaceae	17 Palynogroups belonging to 3 pollen types can be differentiated. Attempt to differentiate the various genera palynologically	Grigor'eva (1990)
Linum (Linaceae)	29 Species investigated. List of descriptions, SEM images	Grigor'eva (1989)
Lysimachia (Primulaceae)	27 Species out of 6 subgenera studied. These subgenera can be palynologically differentiated. SEM images	Sharanina (1994)
Nymphaeales	Sketches of *Cabomba caroliniana*, *Brasenia schreberi*, *Barclaya longifolia*, *Victoria amazonica*, *Nymphaea gigantea*, *Nymphaea candida*, *Nuphar luteum*. SEM and TEM images	Meyer-Melikyan and Diamandopulu (1996)

Table 1. *Continued*

Species/family	Investigation	Reference
Ostrowskia (Campanulaceae)	Together with *Ostrowskia*, species of *Cyananthus*, *Leptocodon*, *Codonopsis* and *Platycodon* were studied. SEM images	Tarasevich and Shrestkha (1992)
Passifloraceae		Gavrilova (1993)
Phormiaceae	9 Genera, 19 species investigated. Comprehensive list and descriptions. SEM images	Kosenko (1994a)
Pteris (Pteridaceae)	15 Fossil form-types described (Caucasian floras of Georgia, Caucasia), 8 species determined. SEM images; palynological descriptions of *Pteris longifolia*, *Pteris grandifolia*, *Pteris vittata*, *Pteris crenata*	Shatilova et al. (1994)
Puccinellia (Poaceae)	Comprehensive list of characteristic features	Ovchinnikova (1990)
Rhopalocarpiaceae		Gavrilova (1993)
Selonia	31 Species investigated, SEM images	Kosenko and Sventorzhetskaya (1995)
Sonchus (Cichoriaceae)	Caucasian species *Sonchus asper*, *Sonchus palustris*, *Sonchus aranticus*	Nazarova (1989)
Stachys (Lamiaceae)	44 Species of *Stachys* and 5 species of *Betonica* investigated	Krestovskaya and Vasil'eva (1997)
Tiliaceae		Gavrilova (1993)
Tulipa (Liliaceae)	16 Species investigated, list of characteristic features. Two evolutionary lines can be differentiated	Daneliya and Kosenko (1990)
Turneraceae		Gavrilova (1993)

Table 2. Seed and fruit types studied concerning the possibility of palaeobotanical determinations

Species/family	Key for the determination of species	Reference
Amaranthus *(Amaranthaceae)*		Tikhomirov and Fedorova (1996)
Amaryllidaceae	Seed structure, including that of *Ixiolirion* and *Doryanthes*	Oganetsova (1990)
Aquilegia (Ranunculaceae)	23 Species studied. SEM images	Vasil'eva (1993)
Argophyllum, Corokia (Argophyllaceae)	Seed morphology and anatomy. SEM images	Lobova (1997)
Comarum palustre, Comarum salesovianum (Rosaceae)	Morphology of the fruits	Tril' and Moskalyuk (1994)
Cyananthus (Campanulaceae)	17 Species out of all sections of this genus. 3 Morphological types can be differentiated: SEM images	Shrestkha and Kravtsova (1992)
Francoa (Francoaceae), Tetilla	3 Species investigated in terms of the seed structure. Seeds quite different from those of the Crassulaceae, Saxifragaceae and Penthoraceae	Nemirovich-Danchenko (1994)
Gypsophila (Caryophyllaceae)	Morphology of the seed-surfaces. Determination of various species seems to be possible. SEM images	Fedotova and Ardzhanova (1992)
Gypsophila (Caryophyllaceae)	10 Species:exine-structure of the seeds. Determinable: *Gypsophila patrinii*, *Gypsophila sambukii*, *Gypsophila altissima*, *Gypsophila cephalodes*. SEM images	Kovtenyuk (1994)
Ixioideae *(Iridaceae)*	Seed structure	Oganetsova (1997)
Nepeta (Lamiaceae)	Morphology of the cuticle of 89 species studied. SEM images	Budantsev (1993a,b)
Parietarieae *(Urticaceae)*	Surface structure of the fruits. SEM images	Kravtsova (1992)
Saxifragaceae	137 Species investigated concerning the surface structure of the seeds. The diagnosis of several species seems to be possible. SEM images	Kul'baeva (1992a,b,c)
Urtica (Urticaceae)	Perikarp structure of 20 species studied. The strong differences correspond well to the taxonomic division of this genus. SEM images	Kravtsova and Gel'tman (1994)

References

In the following list, translations of the Russian titles, which were not given by the authors, are marked by an asterisk.

Alexanderson H, Hjort C, Möller P, Antonov O, Pavlov M (2001) The North Taymyr ice-marginal zone, Arctic Siberia – a preliminary overview and dating. Global Planet Change 31:427–445

Anderson DM, Overpeck JT, Gupta AK (2002) Increase in the Asian southwest monsoon during the past four centuries. Science 297:596–599

Anderson PM, Lozhkin AV, Brubaker LB (2002) Implications of a 24,000-yr palynological record for a Younger Dryas cooling and for boreal forest development in northeastern Siberia. Quat Res 57:325–333

Andreev AA, Manley WF, Ingólfsson Ó, Forman SL (2001) Environmental changes on Yugorski Peninsula, Kara Sea, Russia, during the last 12,800 radiocarbon years. Global Planet Change 31:255–264

Andreev AA, Siegert C, Klimanov VA, Derevyagin AY, Shilova GM, Melles M (2002) Late Pleistocene and Holocene vegetation and climate on the Taymyr Lowland, northern Siberia. Quat Res 57:138–150

Are FE (1999) The role of coastal retreat for sedimentation in the Laptev Sea. In: Kassens H, Bauch HA, Dmitrenko IA, Eicken H, Hubberten HW, Melles M, Thiede J, Timokhov LA (eds) Land-ocean systems in the Siberian Arctic dynamics and history. Springer, Berlin Heidelberg New York, pp 287–295

Arnold NS, van Andel TH, Valen V (2002) Extent and dynamics of the Scandinavian ice sheet during oxygen isotope stage 3 (65,000–25,000 yr b.p.). Quat Res 57:38–48

Askerova RK (1992) Nekotorye rezultaty palinologicheskogo issledovaniya podsemeystva Cichorioideae (Asteraceae) – some results of palynological investigation of the subfamily Cichorioideae (Asteraceae). Bot Zh 77 11:78–81 (no English summary)

Astakhov V (2001) The stratigraphic framework for the Upper Pleistocene of the glaciated Russian Arctic: changing paradigms. Global Planet Change 31:283–295

Bauch HA, Mueller-Lupp T, Taldenkova E, Spielhagen RF, Kassens H, Grootes PM, Thiede J, Heinemeier J, Petryashov VV (2001) Chronology of the Holocene transgression at the North Siberian margin. Global Planet Change 31:125–139

Belyaeva NV, Khusid TA, Chekhovskaya MP (1994) Klimaticheskie sobytiya i izmeneniya cirkulyacii v pleystotsene v tsentral'noy chasti Severnogo Ledovitogo Okeana. (Climatic events and changes in the circulation pattern during the Pleistocene in the centre of the Northern Polar Seas). Byull Kom po izuch chetvert per 61:5–13 (English summary)

Berman DI, Alfimov AV (1992) Kontrastnye ekosystemy v kontinental'nykh rayonakh severo-vostoka Azil. *Contrasting ecosystems within continental regions of northeastern Asia. Doklady Akad Nauk 322, 1:196–199 (no English summary)

Bezus'ko LG, Kostylev AV, Shelyag-Sosenko YP (1992) Marevye stepnoy zony Ukrainy v golotsene. (The Chenopodiaceae of the steppe-zone of the Ukraine in Holocene). Bot Zh 77 11:67–71 (no English summary)

Bigler C, Larocque I, Peglar SM, Birks HJB, Hall RI (2002) Quantitative multiproxy assessment of long-term patterns of Holocene environmental change from a small lake near Abisko, northern Sweden. Holocene 12:481–496

Blagoveshchenskaya NV (1995) Subretsentnye sporovo-pyl'tsevye spektry i ikh sopostavlenie s sovremennoy rastitel'nosti tsentral'noy chasti Privolzhskoy Vozvyshennosti. (Subrecent spore-pollen spectra and their comparison with modern vegetation of the central Volga Upland). Bot Zh 80 10:66–73 (English summary)

Boeskorov GG, Shchelchkova MV, Revin YV (1993) Kariotip losya (*Alces alces* L.) iz severo-vostochnoy Azii. *The karyotype of elk (*Alces alces* L.) from northeastern Asia. Doklady Akad Nauk 329:506–508 (no English summary)

Bogucki A, Cyrek K, Konecka-Betley K, Lanczont M, Madeyska T, Nawrocki J, Sytnyk O (2001) Palaeolithic loess-site Yezupil on Dnister (Ukraine)-stratigraphy, environment and cultures. Studia Quatern 18:25–46

Bogutskiy A, Lanczont M, Racinowski R (2000) Conditions and course of sedimentation of the Middle and Upper Pleistocene loesses in the Halič profile (NW Ukraine). Stud Quat 17:3–17

Bogutskiy A, Cyrek K, Konecka-Betley K, Lanczont M, Madeyska T, Nawrocki J, Sytnyk O (2001) Palaeolithic loess-site Yezupil on Dnister (Ukraine) – stratigraphy, environment and cultures. Stud Quat 18:25–46

Bolikhovskaya NS (1998) Opyt tipizatsii pleystotsenovoy periglyatsial'noy rastitel'nosti lyossovykh oblastey lednikovoy i vnelednikovoy zon Russkoy Ravniny. *An attempt to characterize the Pleistocene periglacial vegetation of the loess-regions within the glacial and periglacial zones of the Russian Plain. Byull Kom po izuch chetvert per 63:20–32 (no English summary)

Bräuning A (2001a) Combined view of various tree ring parameters from different forest habitats in Tibet for the reconstruction of seasonal aspects of Asian monsoon variability. Palaeobotanist 50:1–12

Bräuning A (2001b) Climate history of the Tibetan Plateau during the last 1000 years derived from a network of Juniper chronologies. Dendrochronologia 19:127–137

Bräuning A (2002) Methoden und Probleme paläoökologischer Forschung in Gebirgen Hochasiens. Petermanns Geogr Mitt 146:30–41 (English summary)

Briner JP, Swanson TW, Caffee M (2001) Late Pleistocene cosmogenic ^{36}Cl glacial chronology of the southwestern Ahklun Mountains, Alaska. Quat Res 56:148–154

Budantsev AL (1993a) Osobennosti ul'trastruktury poverkhnosti plodov vidov roda *Nepeta* (Lamiaceae). Ultrastructural features of fruit surface in genus *Nepeta* (Lamiaceae). Bot Zh 78 4:80–87 (no English summary)

Budantsev AL (1993b) Osobennosti ul'trastruktury poverkhnosti plodov nekotorykh rodov triby Nepeteae (Lamiaceae) [The ultrastructural features of fruit surface in some genera of the tribe Nepeteae (Lamiaceae)]. Bot Zh 78 5:100–108 (no English summary)

Chen FH, Bloemendal J, Wang JM, Li JJ, Oldfield F (1997) High-resolution multi-proxy climate records from Chinese loess:evidence for rapid climatic changes over the last 75 kyr. Palaeogeogr Palaeoclimatol Palaeoecol 130:323–335

Chu GQ, Liu JQ, Sun Q, Lu, HY, Gu ZhY, Wang WY, Liu TS (2002) The "Medieval Warm Period" drought recorded in Lake Huguanyan, tropical South China. Holocene 12:511–516

Corner GD, Kolka VV, Yevzerov VY, Møller JJ (2001) Postglacial relative sea-level change and stratigraphy of raised coastal basins on Kola Peninsula, Northwest Russia. Global Planet Change 31:155–177

Cremer H (1999) Spatial distribution of diatom surface sediment assemblages on the Laptev Sea shelf (Russian Arctic). In: Kassens H, Bauch HA, Dmitrenko IA, Eicken H, Hubberten HW, Melles M, Thiede J, Timokhov LA (eds.) Land-ocean systems in the Siberian Arctic dynamics and history. Springer, Berlin Heidelberg New York, pp 533–551

Dahl SO, Nesje A, Lie Ø, Fjordheim K, Matthews JA (2002) Timing, equilibrium-line altitudes and climatic implications of two Early-Holocene glacier readvances during the Erdalen Event at Jostedalsbreen, western Norway. Holocene 12:17–25

Daneliya JM, Kosenko VN (1990) Morfologiya pyl'tsy kavkazskikh vidov roda *Tulipa* (*Liliaceae*). [Pollen morphology in Caucasian species of the genus *Tulipa* (Liliaceae)]. Bot Zh 75 3:293–298 (English summary)

Djinoridze RN, Ivanov GI, Djinoridze EN, Spielhagen RF (1999) Diatoms from surface sediments of the Saint Anna Trough (Kara Sea). In: Kassens H, Bauch HA, Dmitrenko IA, Eicken H, Hubberten HW, Melles M, Thiede J, Timokhov LA (eds.) Land-ocean systems in the Siberian Arctic dynamics and history. Springer, Berlin Heidelberg New York, pp 553–560

Ebel T, Melles M, Niessen F (1999) Laminated sediments from Levinson-Lessing Lake, northern Central Siberia – a 30,000 year record of environmental history? In: Kassens H, Bauch HA, Dmitrenko IA, Eicken H, Hubberten HW, Melles M, Thiede J, Timokhov LA (eds.) Land-ocean systems in the Siberian Arctic dynamics and history. Springer, Berlin Heidelberg New York, pp 425–435

Eickstedt v E (1944) Rassendynamik von Ostasien, China and Japan, Tai und Kmer von der Urzeit bis heute. de Gruyter, Berlin, 747 pp

Elina GA, Antipin VK (1992) Endo- i eksogennye sukcessii rastitel'nosti bolot bassejna Onezhskogo Ozera v golotsene. (Endo- and exogenous successions of mire vegetation of the Onega Lake basin in the Holocene). Bot Zh 77 3:16–30 (English summary)

Elina GA, Lebedeva RM (1992) Dinamika rastitel'nosti i paleogeografiya golotsena Karel'skogo berega pribelomorskoy nizmennosti. (Dynamics of vegetation and Holocene paleogeography of the Karelian coast of the Pribelomorskaya Lowland). Bot Zh 77 5:17–29 (English summary)

Elina GA, Yurkovskaya TK (1992) Metody opredeleniya paleogeograficheskogo rezhima kak osnova ob'ektivizatsii prichin suktsessiy rastitel'nosti bolot. (Methods of paleohydrological regime determination as the basis for the objectivation of the causes of mire vegetation successions). Bot Zh 77 7:120–124 (no English summary)

Elina GA, Filimonova LV, Kuznetsova OL, Lukashov AD, Stoykina NV, Arslanov KhA, Tertichnaya TV (1994a) Vliyanie paleogidrologicheskikh faktorov na dinamiku rastitel'nosti bolot i akkumulyatsiyu torfa. (The relationships between paleohydrology, mire vegetation, dynamics and peat accumulation). Bot Zh 79 1:53–69 (English summary)

Elina GA, Kuznetsova OL, Devyatova RL, Lebedeva RM, Maksimov AL, Stoykina NV (1994b) Sovremennaya i golotsnovaya rastitel'nost' natsional'nogo parka Paanajarvi (Severo-Zapadnaya Kareliya). [The present-day and Holocene vegetation of the Paanajärvi National Park (northwestern Karelia)]. Bot Zh 79 4:13–31 (English summary)

Elina GA, Arslanov KhA, Klimanov VA, Usova LI (1995) Rastitel'nost' i klimatokhronologiya golotsena Lovozerskoy Ravniny Kol'skogo Poluostrova (po sporovo-pyl'tsevym diagrammam bugristo-topyanogo bolota). [Vegetation and climate chronology in the Holocene of the Lovozerskaya Plain in the Kola Peninsula (according to spore-pollen diagrams of a palsa mire)]. Bot Zh 80,3:1–16 (English summary)

Elina GA, Arslanov KA, Klimanov VA (1996) Etapy razvitiya rastitel'nosti golotsena v yuzhnoy i vostochnoy Karelii. (Development stages of Holocene vegetation in southern and eastern Karelia). Bot Zh 81 3:1–17 (English summary)

Epshteyn OG, Romanyuk BF, Gataulin VN (1999) Pleystotsenovye skandinavskiy i novozemel'skiy lednikovye pokrovy v yuzhnoy chasti Barentsovomorskogo shel'fa i na severe Russkoy Ravniny. *The Pleistocene inland ice masses from Scandinavia and Novaya Zemlya in the southern part of the Barents Sea shelf and in the northern part of the Russian Lowlands. Byull Kom po izuch chetvert per 63:126–149 (no English summary)

Esper J (2000) Long-term tree-ring variations in *Juniperus* at the upper timberline in the Karakorum (Pakistan). Holocene 10:253–260

Esper J, Schweingruber FH, Winiger M (2002) 1300 years of climatic history for Western Central Asia inferred from tree-rings. Holocene 12:267–277

Fedotova TA, Ardzhanova RR (1992) Morfologiya semeni roda *Gypsophila* (Caryophyllaceae). [Seed morphology in the genus *Gypsophila* (Caryophyllaceae)]. Bot Zh 77 5:1–16 (English summary)

Forman SL, Ingólfsson Ó, Gataullin V, Manley W, Lokrantz H (2002) Late Quaternary stratigraphy, glacial limits, and paleoenvironments of the Marresale area, western Yamal Peninsula, Russia. Quat Res 57:355–370

Frenzel B (1968) Grundzüge der pleistozänen Vegetationsgeschichte Nord-Eurasiens. Erdwissenschaftliche Forschung 1. Steiner, Wiesbaden, 326 p

Frenzel B (1969) Floren- und Vegetationsgeschichte seit dem Ende des Tertiärs (Historische Geobotanik). Fortschr Bot 31:309–319
Frenzel B (1983) Mires – repositories of climatic information or self-perpetuating ecosystems? In: Gore AJP (ed) Ecosystems of the world 4A. Mires: swamp, bog, fen and moor. Elsevier, Amsterdam, pp 35–66
Frenzel B (1993) The history of flora and vegetation during the Quaternary. Progress in botany 54. Springer, Berlin Heidelberg New Yor, pp 402–427
Frenzel B (2002) History of flora and vegetation during the Quaternary. Progress in botany 63. Springer, Berlin Heidelberg New York, pp 368–385
Frenzel B, Liu SJ (2001) Über die jungpleistozäne Vergletscherung des Tibetischen Plateaus. In: Bussemer S (ed) Das Erbe der Eiszeit (Marcinek-Festschrift). Beier & Beran, Langenweißbach, pp 71–91 (English summary)
Frenzel B, Pécsi M, Velichko AA (1992) Atlas of paleoclimates and paleoenvironments of the Northern Hemisphere; Late Pleistocene–Holocene. Fischer, Stuttgart, 153 p
Gataullin VK (1991) Marresal'skaya svita zapadnogo Yamala – otlozheniya del'ty pra-Obi. *The Marresal series of western Yamal: sediments of the pre-Ob. Byull Kom po izuch chetvert per 610:53–61 (no English summary)
Gataullin V, Mangerud J, Svendsen JI (2001) The extent of the Late Weichselian ice sheet in the southeastern Barents Sea. Global Planet Change 31:453–474
Gavrilova OA (1993) Tipy skul'ptury pyl'tsevykh zeren i ikh znachenie dlya sistematiki semeystva Flacourtiaceae. (Types of pollen grain sculpture and their significance for systematics of the family Flacourtiaceae). Bot Zh 78 12:45–52 (English summary)
Gavrilova OA (1997) Znachenie palinologicheskikh priznakov dlya sistematiki i filogenii triby Flacourtieae (Flacourtiaceae). Bot Zh 82 2:74–79 (no English summary)
Gerasimenko NP (2000) Late Pleistocene vegetational and soil evolution at the Kiev loess plain as recorded in the Stari Bezradychy section, Ukraine. Stud Quat 17:19–28
Gey V, Saarnisto M, Lunkka JP, Demidov I (2001) Mikulino and Valdai paleoenvironments in the Vologda area, NW Russia. Global Planet Change 31:347–366
Gogichaishvili LK (1992) K istorii aridnykh lesov vostochnoy Gruzii v golotsene. On the history of arid forests of eastern Georgia in Holocene. Bot Zh 77 11:63–67 (no English summary)
Goncharov SV (1991) Poslednye lednikovo-podprudnye ozera doliny Yeniseya. *The last ice-dammed lakes in the Yenisey valley. Byull Kom po izuch chetvert per 60:67–27 (no English summary)
Grigor'eva VV (1989) O dimorfizme pyl'tsy nekotorykh distil'nykh vidov roda *Linum* (Linaceae). [On pollen dimorphism in some distylous species of the genus *Linum* (Linaceae)]. Bot Zh 74 1:65–72 (English summary)
Grigor'eva VV (1990) Morfologiya pyl'tsevykh zeren predstaviteley semeystva Linaceae. (The pollen grain morphology in members of the Linaceae family). Bot Zh 75 10:1345–1352 (English summary)
Grosval'd MG (1983) Pokrovnye ledniki kontinental'nykh shelfov. Rezul'taty issledovanii po mezhdunarodnym geofizicheskim proektam. (Ice sheets of continental shelves. Results of research on geophysical projects). Nauka, Moskva, 216 pp (no English summary)
Grosval'd MG (1988) Priznaki pokrovnogo oledeneniya Novosibirskikh Ostrovov i okruzhayushchego shel'fa. *Traces of an inland-ice glaciation of the New-Siberian Isles and of the surrounding shelf area. Doklady Akad Nauk SSSR 302:654–659 (no English summary)
Guo ZT, Petite-Maire N, Kröpelin S (2000) Holocene non-orbital climatic events in present-day arid areas of northern Africa and China. Global Planetary Change 26:97–103
Hahne J, Melles M (1999) Climate and vegetation history of the Taymyr Peninsula since Middle Weichselian time – palynological evidence from lake sediments. In: Kassens H, Bauch HA, Dmitrenko IA, Eicken H, Hubberten HW, Melles M, Thiede J, Timokhov LA

(eds.) Land-ocean systems in the Siberian Arctic dynamics and history. Springer, Berlin Heidelberg New York, pp 407–423

Hammarlund D, Barnekow L, Birks HJB, Buchardt B, Edwards TWD (2002) Holocene changes in atmospheric circulation recorded in the oxygen-isotope stratigraphy of lacustrine carbonates from northern Sweden. Holocene 12:339–351

Henriksen M, Mangerud J, Maslenikova O, Matiouchkov A, Tveranger J (2001) Weichselian stratigraphy and glaciotectonic deformation along the lower Pechora River, Arctic Russia. Global Planet Change 31:297–319

Hong YT, Jian HB, Liu TS, Zhou LP, Beer J, Li HD, Leng XT, Hong B, Qing XG (2000) Response of climate to solar forcing recorded in a 6,000-year $\delta^{18}O$ time-series of Chinese peat cellulose. Holocene 10:1–7

Houmark-Nielsen M, Demidov I, Funder S, Grøsfjeld K, Kjaer KH, Larsen E, Lavrova N, Lyså A, Nielsen JK (2001) Early and Middle Valdaian glaciations, ice-dammed lakes and periglacial interstadials in northwest Russia: new evidence from the Pyoza river area. Global Planet Change 31:215–237

Huang CC, Zhou J, Pang JL, Han YP, Hou CH (2000) A regional aridity phase and its possible cultural impact during the Holocene Megathermal in the Guanzhong basin, China. Holocene 10:135–142

Ivanov VV, Piskun AA (1999) Distribution of river water and suspended sediment loads in the deltas of rivers in the basin of the Laptev and East Siberian Seas. In: Kassens H, Bauch HA, Dmitrenko IA, Eicken H, Hubberten HW, Melles M, Thiede J, Timokhov LA (eds.) Land-ocean systems in the Siberian Arctic dynamics and history. Springer, Berlin Heidelberg New York, pp 239–250

Jakobsson M, Løvlie R, Arnold EM, Backman J, Polyak L, Knutsen J-O, Musatov E (2001) Pleistocene stratigraphy and paleoenvironmental variation from Lomonosov Ridge sediments, central Arctic Ocean. Global Planet Change 31:1–22

Kamaletdinov VA, Minyuk PS (1991) Stroenie i kharakteristika otlozheniy Bestyakhskoy terrasy sredney Leny. Structure and characteristics of the sediments of the Bestyakh Terrace on the middle Lena river. Byull Kom po izuch chetvert per 60:68–78 (no English summary)

Kamp K, Haserodt K (2002) Quartäre Vergletscherungen im Hindukusch, Karakorum und West-Himalaya, Pakistan – ein Überblick. Eiszeitalter Gegenwart 51:93–113

Karhu JA, Tschudi S, Saarnisto M, Kubik P, Schlüchter C (2001) Constraints for the latest glacial advance on Wrangel Island, Arctic Ocean, from rock surface exposure dating. Global Planet Change 31:447–451

Kassens H, Bauch HA, Dmitrenko IA, Eicken H, Hubberten H-W, Melles M, Thiede J, Timokhov LA (eds) (1999) Land-ocean systems in the Siberian Arctic dynamics and history. Springer, Berlin Heidelberg New York, 711 p

Khalilov II, Arkhangel'skiy DB (1991) Palinologicheskoe izuchenie vidov roda *Crambe* (*Brassicaceae*) v svyazi s ego sistematikoy. (Palynological studies in the species of the genus *Crambe* (Brassicaceae) in connection with its systematics). Bot Zh 76 11:1582–1586 (no English summary)

Khitun OB (1994) Severnaya granitsa rasprostraneniya *Larix sibirica* (Pinaceae) na Tazovskom Poluostrove (Zapadnaya Sibir'). [The northern limit of *Larix sibirica* (Pinaceae) on the Tazov Peninsula (West Siberia)]. Bot Zh 79 9:70–72 (English summary)

Khramova EL (1996) Palinomorfologiya roda *Alnus* (Betulaceae). [Palynomorphology of the genus *Alnus* (Betulaceae)]. Bot Zh 81 3:42–53 (English summary)

Kienast F, Siegert C, Dereviagin A, Mai DH (2001) Climatic implications of Late Quaternary plant-macrofossil assemblages from the Taymyr Peninsula, Siberia. Global Planet Change 31:254–281

Kienel U (1999) Late Weichselian to Holocene diatom succession in a sediment-core from Lama-Lake, Siberia and presumed ecological implications. In: Kassens H, Bauch HA, Dmitrenko IA, Eicken H, Hubberten HW, Melles M, Thiede J, Timokhov LA (eds.)

Land-ocean systems in the Siberian Arctic dynamics and history. Springer, Berlin Heidelberg New York, pp 377–405

Kjaer KH, Demidov I, Houmark-Nielsen M, Larsen E (2001) Distinguishing between tills from Valdaian ice sheets in the Arkhangelsk region, Northwest Russia. Global Planet Change 31:201–214

Knies J, Kleiber H-P, Matthiessen J, Müller C, Nowaczyk N (2001) Marine ice-rafted debris records constrain maximum extent of Saalian and Weichselian ice-sheets along the northern Eurasian margin. Global Planet Change 31:45–64

Korotky AM, Razjigaeva NG, Grebennikova TA, Ganzey LA, Mokhova LM, Bazarova VB, Sulerzhitsky LD, Lutaenko KA (2000) Middle- and late-Holocene environments and vegetation history of Kunashir Island, Kurile Islands, northwestern Pacific. Holocene 10:311–331

Kosenko VN (1991) Palinomorfologiya semeystva *Liliaceae* s.str. Palynomorphology of the family Liliaceae s.str. Bot Zh 76 12:1696–1706 (English summary)

Kosenko VN (1992) Morfologiya pyl'tsy i voprosy sistematiki semejstva Liliaceae. Pollen morphology and systematic problems of the Liliaceae family. Bot Zh 77 3:1–15 (English summary)

Kosenko VN (1994a) Morfologiya pyl'tsy semeystva Phormiaceae, Blandfordiaceae i Doryanthaceae. (Pollen morphology of the families Phormiaceae, Blandfordiaceae and Doryanthaceae). Bot Zh 79 7:1–12 (English summary)

Kosenko VN (1994b) Morfologiya pyl'tsy semeystva Alstroemeriaceae. (Pollen morphology of the family Alstroemeriaceae). Bot Zh 79 8:1–8 (English summary)

Kosenko VN, Kudryashova GL (1995) Palinomorfologiya semeystva Alliaceae. (Palynomorphology of the family Alliaceae). Bot Zh 80 6:5–17 (English summary)

Kosenko VN, Mikhailova MA (1991) Morfologiya pyl'tsy roda *Corydalis* (Fumariaceae). [The morphology of pollen in the genus *Corydalis* (Fumariaceae)]. Bot Zh 76 4:556–564 (no English summary)

Kosenko VN, Sventorzhetskaya OYu (1995) Morfologiya pyl'tsy roda *Eremurus* (Asphodelaceae). [Pollen morphology of the genus *Eremurus* (Asphodelaceae)]. Bot Zh 80 8:73–79 (English summary)

Kotlia BS, Bhalla MS, Sharma C, Rajagopalan G, Ramesh R, Chauchan MS, Mathur PD, Bhandari S, Chacko ST (1997) Palaeoclimatic conditions in the Upper Pleistocene and Holocene Bhimtal-Naukuchiatel lake basin in south-central Kumaun, North India. Palaeogeogr Palaeoclimatol Palaeoecol 130:307–322

Kovtenyuk NK (1994) Struktura poverkhnosti semyan sibirskikh *Gypsophila* (Caryophyllaceae) v svyazi s sistematikoy. [The structure of seed surface and the systematics of the Siberian *Gypsophila* species (Caryophyllaceae)]. Bot Zh 79 4:48–51 (English summary)

Kozhevnikov YuP (1995a) O svazi sovremennoy rastitel'nosti i poverkhnostnykh sporo-pyl'tsevykh spektrov na Chukotke (basseyn reki Chantal'veergyn). [On the connection of the extant vegetation and surface spore-pollen spectra on Chukotka (basin of Chantalveergyn river)]. Bot Zh 80 5:74–88 (English summary)

Kozhevnikov YP (1995b) Problemy interpretacii sporovo-pyl'tsevykh spektrov v rekonstruktsii rastitel'nogo pokrova. (The interpretation problems of spore-pollen spectra in vegetational cover reconstruction). Bot Zh 80 9:1–20 (English summary)

Kozhevnikov YP, Arkhangel'skiy DB (1996) Differentsiatsiya pyl'tsy i filogeniya roda *Dryas* (Rosaceae). [Differentiation of pollen and phylogeny of the genus *Dryas* (Rosaceae)]. Bot Zh 81 7:10–22 (English summary)

Kozhevnikov YP, Ukrayntseva VV (1992) Nekotorye osobennosti rastitel'nogo pokrova Evrazii v drevnem golotsene. (Some features of vegetation cover of Eurasia in the earliest Holocene). Bot Zh 77 8:1–9 (English summary)

Kozhevnikov YP, Arslanov KA, Boch MS, Sulerzhitskiy DD, Ukrayntseva VV (1993) Ob informativnosti paleobotanicheskikh materialov s vostochnogo Taymyra. (On the information character of paleobotanical material from the eastern Taymir). Bot Zh 78 3:40–52 (no English summary)

Kravtsova TI (1992) Stroenie ploda v tribe Parietarieae (Urticaceae) v svyazi s ego sistematikoy. [Fruit structure in the tribe Parietarieae (Urticaceae) in relation to its systematics]. Bot Zh 77 7:12–29 (English summary)

Kravtsova TI, Gel'tman DV (1994) Anatomiya i ul'trasculptura perikarpiya i semenoy kozhury u predstaviteley roda *Urtica* (Urticaceae). [Pericarp and seed coat anatomy and ultrastructure in representatives of *Urtica* (Urticaceae)]. Bot Zh 79 2:27–44 (English summary)

Kremenetskiy KV, Tarasov PE, Cherkinskiy AE (1994) Istoriya ostrovnykh borov Kazakhstana v golotsene. (Holocene history of the Kazakhastan "island" pine forests). Bot Zh 79 3:13–29 (English summary)

Kremenetskiy KV, McDonald GM, Galabala RO, Lavrov AS, Chichagova OA, Pustovoydov KE (1996) Ob izmeneniy severnoy granitsy arealov nekotorykh vidov derev'ev i kustarnikov v golotsene. (On the Holocene shift of the northern limit of some species of trees and shrubs]. Bot Zh 81 4:10–25 (English summary)

Krestovskaya TV, Vasil'eva IM (1997) Palinologicheskoe izuchenie vidov roda *Stachys* i *Betonica* (Lamiaceae). [Palynological studies in the genera *Stachys* and *Betonica* (Lamiaceae)]. Bot Zh 82 11:30–41 (English summary)

Kul'baeva BZ (1992a) Anatomiya semyan predstaviteley semeystva Saxifragaceae. (Seed anatomy in representatives of the Saxifragaceae family). Bot Zh 77 3:36–49 (no English summary)

Kul'baeva BZ (1992b) Poverkhnostnaya struktura predstaviteley semeystva Saxifragaceae. (The seed structure surface in representatives of the family Saxifragaceae). Bot Zh 77 4:61–68 (no English summary)

Kul'baeva BZ (1992c) Poverkhnost' semyan predstaviteley semeystva Saxifragaceae:Tipologiya i znachenie dlya sistematiki i filogenii. (Seed surface in representatives of the family Saxifragaceae: typology and significance for systematics and phylogeny). Bot Zh 77 8:98–106 (no English summary)

Kullman L (2000) The geoecological history of *Picea abies* in northern Sweden and adjacent parts of Norway. A contrarian hypothesis of postglacial tree immigration. Geoöko 21:141–172

Kunz-Pirrung M (1999) Distribution of aquatic palynomorphs in surface sediments from the Laptev Sea, eastern Arctic Ocean. In: Kassens H, Bauch HA, Dmitrenko IA, Eicken H, Hubberten HW, Melles M, Thiede J, Timokhov LA (eds) Land-ocean systems in the Siberian Arctic dynamics and history. Springer, Berlin Heidelberg New York, pp 561–575

Kvavadze E (2001) Annual modern pollen deposition in the foothills of the Lagodekhi Reservation (Caucasus, East Georgia), related to vegetation and climate. Acta Palaeobot 41:355–364

Lindqvist B (1948) The main varieties of *Picea Abies* (L.) Karst. in Europe, with a contribution to the theory of a forest vegetation in Scandinavia during the last Pleistocene glaciation. Acta Horti Bergiani XIV:249–342

Linge H, Lauritzen S-E (2001) Stable isotope stratigraphy of a late Last Interglacial speleothem from Rana, northern Norway. Quat Res 56:155–164

Liu HY, Xu LH, Cui HT (2002) Holocene history of desertification along the woodland-steppe border in northern China. Quat Res 57:259–270

Lobova TA (1997) Morfologiya i anatomiya semyan rodov *Argophyllum* i *Corokia* (Argophyllaceae). [Seed morphology and anatomy in the genus *Argophyllum* and *Corokia* (Argophyllaceae)]. Bot Zh 82 9:68–78 (English summary)

Lunkka JP, Saarnisto M, Gey V, Demidov I, Kiselova V (2001) Extent and age of the Last Glacial Maximum in the southeastern sector of the Scandinavian Ice Sheet. Global Planet Change 31:407–425

Luoto M, Seppälä M (2000) Summit peats ("peat cakes") on the fells of Finnish Lapland:continental fragments of blanket mires? Holocene 10:229–241

Lupikina EG (1989) Vidovoy sostav diatomovykh vodorosley donnykh otlozheniy ozera Goryachego (ostrov Kunashir). [Species composition of diatoms in sediments of Lake Goryachy (Kunashir Island)]. Bot Zh 74 6:852–855 (no English summary)

Lyså A, Demidov I, Houmark-Nielsen M, Larsen E (2001) Late Pleistocene stratigraphy and sedimentary environment of the Arkhangelsk area, northwest Russia. Global Planet Change 31:179–199

MacDonald GM, Gervais BR, Snyder JA, Tarasov GA, Borisova OK (2000) Radiocarbon dated *Pinus sylvestris* L. wood from beyond tree-line on the Kola Peninsula, Russia. Holocene 10:143–147

Mäkelä E, Hyvärinen H (2000) Holocene vegetation history at Vätsäri, Inari Lapland, northeastern Finland, with special reference to *Betula*. Holocene 10:75–85

Mangerud J, Astakhov VI, Murray A, Svendsen JI (2001) The chronology of a large ice-dammed lake and the Barents-Kara ice sheet advances, Northern Russia. Global Planet Change 31:321–336

Manley WF, Lokrantz H, Gataullin V, Ingólfsson Ó, Forman SL, Andersson T (2001) Late Quaternary stratigraphy, radiocarbon chronology, and glacial history at Cape Shpindler, southern Kara Sea, Arctic Russia. Global Planetary Change 31:239–254

Markova AK, Simakova AN, Puzachenko AY, Kitaev LM (2002) Environments of the Russian Plain during the Middle Valdai Briansk interstade (33,000 – 24,000 yr b.p.) indicated by fossil mammals and plants. Quat Res 57:391–400

Maslenikova O, Mangerud J (2001) Where was the outlet of the ice-dammed Lake Komi, northern Russia? Global Planet Change 31:337–345

Maxwell AL (2001) Holocene monsoon changes inferred from lake sediment pollen and carbonate records, northeastern Cambodia. Quat Res 56:390–400

Meyer-Melikyan NR, Diamandopulu N (1996) Ul'trastruktura pyl'tsevykh zeren predstaviteley poryadka Nymphaeales. (Ultrastructure of pollen grains of the order Nymphaeales). Bot Zh 81 7:1–9 (English summary)

Monoszon MK (1952) Opisanie pyl'tsy vidov Chenopodiaceae rastushchikh na territorii SSSR (dlya tseley pyl'tsevogo analiza). *Description of the pollen grains of those species of the Chenopodiaceae family which grow in the USSR (for pollen analysis). Trudy In-ta Geogr Akad Nauk SSSR 52:127–196

Naidina OD, Bauch HA (1999) Distribution of pollen and spores in surface sediments of the Laptev Sea. In: Kassens H, Bauch HA, Dmitrenko IA, Eicken H, Hubberten HW, Melles M, Thiede J, Timokhov LA (eds) Land-ocean systems in the Siberian Arctic dynamics and history. Springer, Berlin Heidelberg New York, pp 577–585

Naidina OD, Bauch HA (2001) A Holocene pollen record from the Laptev Sea shelf, northern Yakutia. Global Planet Change 31:141–153

Nazarova EA (1989) Karyological and palynological study of representatives of the genus *Sonchus* from the Caucasus. Bot Zh 74 1:53–59 (in Russian; no English summary)

Nemirovich-Danchenko EN (1994) Stroenie semyan vidov rodov *Francoa* i *Tetilla* (Francoaceae). [The seed-structure of the species of the genera *Francoa* and *Tetilla* (Francoaceae)]. Bot Zh 79 7:21–27 (English summary)

Nepomilueva NJ, Duryagina DA (1990) K istorii listvennichnikov srednego Timana v golotsene (Komi ASSR). On the history of larch forests in the Middle Timan in Holocene (The Komi Autonomous Soviet Socialist Republik). Bot Zh 75 3:326–334 (English summary)

Nesje A, Dahl SO (2003) The "Little Ice Age" – only temperature? Holocene 13:139–145

Niessen F, Ebel T, Kopsch C, Fedorov GB (1999) High-resolution seismic stratigraphy of lake sediments on the Taymyr Peninsula, central Siberia. In: Kassens H, Bauch HA, Dmitrenko IA, Eicken H, Hubberten HW, Melles M, Thiede J, Timokhov LA (eds) Land-ocean systems in the Siberian Arctic dynamics and history. Springer, Berlin Heidelberg New York, pp 437–456

Novoselova MS (1994) Palinologicheskoe izuchenie roda *Eriophorum* (Cyperaceae). [Palynological study of the genus *Eriophorum* (Cyperaceae)]. Bot Zh 79 6:62–64 (English summary)

Oganetsova GG (1990) Osobennosti anatomicheskoy struktury semeni nekotorykh amarillidovykh (Amaryllidaceae) v svyazi s ikh sistematikoy i filogeniey. (Seed and fruit anatomy of some Amaryllidaceae in connection with their systematics and phylogeny). Bot Zh 75 5:615–630 (English summary)

Oganetsova GG (1997) Struktura semeni nekotorykh irisovykh v svyazi s sistematikoy, geografiey i filogeniey semeystva Iridaceae. II. Podsemeystvo Ixioideae. (The seed-structure of some Iridaceae in connection with their systematics, geography and phylogeny. II. Subfamily Ixioideae). Bot Zh 82 3:79621 (no English summary)

Oksanen PO, Kuhry P, Alekseeva RN (2001) Holocene development of the Rogovaya river peat plateau, European Russian Arctic. Holocene 11:25–40

Oldfield F (2001) A question of timing:a comment on Hong, Jiang, Lui, Zhou, Beer, Li, Leng, Hong and Qin. Holocene 11:123–124

Ovchinnikova SV (1990) Izuchenie pyl'tsy sibirskikh vidov *Puccinellia* (Poaceae) dlya tseley sistematiki. Study of pollen of *Puccinellia* (Poaceae) species from Siberia for the purposes of systematics. Bot Zh 75 11:1522–1534 (no English summary)

Parshall T, Calcote R (2001) Effect of pollen from regional vegetation on stand-scale forest reconstruction. Holocene 11:81–87

Pitkänen A (2000) Fire frequency and forest structure at a dry site between AD 440 and 1110 based on charcoal and pollen records from a laminated lake sediment in eastern Finland. Holocene 10. 221–228

Pitkänen A, Turunen J, Tahvanainen T, Tolonen K (2002) Holocene vegetation history from the Salym-Yugan Mire area, West Siberia. Holocene 12:353–362

Pitul'ko VV (1999) Archaeological survey in Central Taymyr. In: Kassens H, Bauch HA, Dmitrenko IA, Eicken H, Hubberten HW, Melles M, Thiede J, Timokhov LA (eds) Land-ocean systems in the Siberian Arctic dynamics and history. Springer, Berlin Heidelberg New York, pp 457–467

Popova TN, Zemskaya EA (1995) Palinomorfologicheskoye izuchenie nekotorykh vidov semeystva Boraginaceae (podsemeystva Boraginoideae). [Palynomorphology study of some species of Boraginaceae (subfamily Boraginoideae)]. Bot Zh 80 10:1–13 (English summary)

Poska A, Saarse L (2002) Vegetation development and introduction of agriculture to Saaremaa Island, Estonia. Holocene 12:555–565

Pozhidaev AE (1989) Exine structure in pollen grains of the Lamiaceae family. Bot Zh 74 10:1410–1422 (in Russian; English summary)

Pozhidaev AE (1992) Pyl'tsa roda *Acer* (Aceraceae) i izomorfizm otklonyayushchikh form pyl'tsa dvudol'nykh. (The pollen of the genus *Acer* (Aceraceae) and isomorphism of deviating forms of pollen in Dicotyledons). Bot Zh 77 11:84–88 (no English summary)

Prokopenko AA, Karabanov EB, Williams DF, Khursevich GK (2002) The stability and the abrupt ending of the last interglaciation in southeastern Siberia. Quat Res 58:56–59

Qian WH, Zhu YF (2002) Little Ice Age climate near Beijing, China, inferred from historical and stalagmite records. Quat Res 57:109–119

Razjigaeva NG, Korotky AM, Grebennikova TA, Ganzey LA, Mokhova LM, Bazarova VB, Sulerzhitsky LD, Lutaenko KA (2002) Holocene climatic change and environmental history of Iturup Island, Kurile Islands, northwestern Pacific. Holocene 12:469–480

Romanova LS (1992) Palinomorfostruktura gvozdichnykh (Caryophyllaceae). Structure of palynomorphs of carnations (Caryophyllaceae). Bot Zh 77 11:81–84 (no English summary)

Rousseau DD, Kukla G (2000) Abrupt retreat of summer monsoon at the S1/L1 boundary in China. Global Planet Change 26:189–198

Rousseau DD, Wu NQ, Guo ZT (2000) The terrestrial mollusks as new indices of the Asian paleomonsoons in the Chinese loess plateau. Global Planet Change 26:199–206

Rowley JR (1992) Pollen of *Cercidiphyllum* (Cercidiphyllaceae). Bot Zh 77 11:1–3 (Russian summary)

Rylova TB (1989) Morfologicheskie osobennosti pyl'tsy nekotorykh iskopaemykh i sovremennykh vidov *Itea* (Iteaceae). [Morphological features of pollen in some fossil and extant species *Itea* (Iteaceae)]. Bot Zh 74 5:694–699 (no English summary)

Saarnisto M, Saarinen T (2001) Deglaciation chronology of the Scandinavian ice sheet from the Lake Onega basin to the Salpausselkä end-moraines. Global Planet Change 31:387–405

Sadchikova TA, Chepalyga AL (1999) Pozdnechetvertichnye istoriya limanov severozapadnogo Prichernomor'ya. *Upper Quaternary history of the marine embayments of the northwestern Black Sea coast. Byull Kom po izuch chetvert per 63:76–85 (no English summary)

Sarmaja-Korjonen K (2001) Correlation of fluctuations in cladoceran planktonic:littoral ratio between three cores from a small lake in southern Finland: Holocene water-level changes. Holocene 11:53–63

Schaefer H, Frenzel B (1959) Beiträge zur Kenntnis der Flora des Ostteiles der Großen Samojedentundra. Bot Jahrb 78:367–434

Schirrmeister L, Oezen D, Geyh MA (2002) ^{230}Th/U dating of frozen peat, Bol'shoy Lyakhovsky Island (northern Siberia). Quat Res 57:253–258

Seppä H, Birks HJB (2002) Holocene climate reconstructions from the Fennoscandian tree-line area based on pollen data from Toskaljavri. Quat Res 57:191–199

Shalaboda VL (2001) Characteristic features of Muravian (Eemian) pollen succession from various regions of Belarus. Acta Palaeobot 41:27–41

Sharanina EA (1994) Raznoobrazie ul'trastruktury poverkhnosti semyan v rode *Lysimachia* (Primulaceae). Ultrastructure diversity of seed surface in the genus *Lysimachia* (Primulaceae). Bot Zh 79 7:13–20 (English summary)

Shatilova II, Meyer-Melikyan NR, Mchedlishvili NSh (1994) Spory *Pteris* (Pteridaceae) v neogenovykh otlozheniyakh zapadnoy Gruzii. [The spores of *Pteris* (Pteridaceae) in Neogene deposits of western Georgia]. Bot Zh 79 4:34–38 (English summary)

Shcherbakov FA, Pavlidis YA, Ionin AS, Dunaev NN (1999) Osnonvnye tipy ledovoy sedimentatsii na glyatsial'nom shel'fe v pozdnechetvertichnoe vremya. *Main types of glacigene sedimentation on the glacial shelf during late-quaternary times. Byull Kom po izuch chetvert per 63:114–118 (no English summary)

Shibneva IV (1991) O nakhodkakh *Brasenia schreberi* (Cabombaceae) na severe Primorskogo Kraya. [On the findings of *Brasenia schreberi* (Cabombaceae) on the north of the Primorye Territory]. Bot Zh 76 4:619–624 (no English summary)

Shrestkha KK, Kravtsova TI (1992) Anatomiya i ul'trastruktura semennoy kozhury v rode *Cyananthus* (Campanulaceae) v svyazi s ego sistematikoy. [Anatomy and ultrastructure of the seed-surface of the genus *Cyananthus* (Campanulaceae) in connections with its systematic position]. Bot Zh 77 6:18–29 (English summary)

Shrestkha KK, Tarasevich VF (1992) Sravnitel'no-morfologicheskoe issledovanie pyl'tsy roda *Cyananthus* v svyazi s sistematikoy i polozheniem ego v semeystve Campanulaceae. (Comparative pollen morphology of genus *Cyananthus* in relation to its systematics and its position within the family Campanulaceae). Bot Zh 77 10:1–13 (English summary)

Siegert C, Derevyagin AY, Shilova GN, Hermichen W-D, Hiller A (1999) Paleoclimatic indicators from permafrost sequences in the eastern Taymyr Lowland. In: Kassens H, Bauch HA, Dmitrenko IA, Eicken H, Hubberten HW, Melles M, Thiede J, Timokhov LA (eds) Land-ocean systems in the Siberian Arctic dynamics and history. Springer, Berlin Heidelberg New York, pp 477–499

Siegert MI, Dowdeswell JA, Hald M, Svendsen JI (2001) Modelling the Eurasian ice sheet through a full (Weichselian) glacial cycle. Global Planetary Change 31:367–385

Simakova AN (1991) Subfossil'nye spektry razlichnykh rastitel'nykh zon Zapadnogo Sayana. *Subfossil pollen spectra of different vegetation zones in the Western Sayan Mts. Byull Kom po izuch chetvert per 60:108–110 (no English summary)

Snowball I, Zillén L, Gaillard M-J (2002) Rapid early Holocene environmental changes in northern Sweden based on studies of two varved lake-sediment sequences. Holocene 12:7–16

Solovieva DV (1999) Spring stopover of birds on the Laptev Sea polynya. In: Kassens H, Bauch HA, Dmitrenko IA, Eicken H, Hubberten HW, Melles M, Thiede J, Timokhov LA (eds.) Land-ocean systems in the Siberian Arctic dynamics and history. Springer, Berlin Heidelberg New York, pp 189–195

Spiridonova EA (1991) The development of vegetation in the catchment area of the river Don during the Upper Pleistocene and the Holocene (in Russian). Nauka, Moskva, 221 p (no English summary)

Sukhorukova SS, Shevka AY, Krivonogov SK, Bakhareva VA, Panychev VA, Orlova LA (1991) Novye materialy k stratigrafii pokrovnykh otlozheniy pravoberezh'ya srednego Eniseya v svyazi s problemoy vozrasta oledeneniy. *New data concerning the stratigraphy of near-surface sediments on the right-hand bank of the middle Yeniseiy river in connection with the age-problem of the glaciations. Byull Kom po izuch chetvert per 60:116–122 (no English summary)

Sviridenko BF, Zaripov RG, Litovchenko OG (1994) Rastitel'nost' i stratigrafiya dvukh bolot severnogo Kazakhstana. (The vegetation and stratigraphy of two mires of northern Kazakhstan). Bot Zh 79 11:66–76 (English summary)

Sycheva SA, Chichagova OA (1999) Radiouglerodnaya khronostratigrafiya golotsenovykh otlozheniy Srednerusskoy Vozvyshennosti. *Radiocarbon chronostratigraphy of Holocene sediments of the central-Russian Hills. Byull Kom po izuch chetvert per 63:104–113 (no English summary)

Tarasevich VF (1989) Ul'trastruktura pyl'tsevykh zeren roda *Anthurium* (Araceae) v svyazi s sistematikoy. [Pollen grain ultrastructure in the genus *Anthurium* (Araceae) in connection with its systematics]. Bot Zh 74 3:314–324 (English summary)

Tarasevich VF (1992) Palinologicheskoe izuchenie roda *Carex* (Cyperaceae). [Palynological study of the genus *Carex* (Cyperaceae)]. Bot Zh 77 11:4–15 (English summary)

Tarasevich VF, Abdullaeva AT (1993) Morfologiya pyl'tsy nekotorykh vidov roda *Calligonum* (Polygonaceae). [Pollen morphology in some species of the genus *Calligonum* (Polygonaceae)]. Bot Zh 78 12:111–117 (English summary)

Tarasevich VF, Shrestkha KK (1992) Palinologicheskie dannye o polozheniy roda *Ostrowskia* v semeystve Campanulaceae.(Palynological data on the position of the genus *Ostrowskia* within the Campanulaceae family). Bot Zh 77 9:27–30 (no English summary)

Thomson PW (1933) Zur Genese der Dryastone. Beitr Kenntnis Estlands 18 3:11–12

Thomson PW (1956) Beitrag zur Kenntnis arktischer Sedimente. Eiszeitalter Gegenwart 7:176–178

Tikhomirov VN, Fedorova TA (1996) Morfologicheskoe issledovanie semyan predstaviteley roda *Amaranthus* (Amaranthaceae). [Morphological study of seeds in members of the genus *Amaranthus* (Amaranthaceae)]. Bot Zh 81 11:54–62 (English summary)

Toderich KN (1992) Morfologiya pyl'tsy nekotorykh vidov roda *Helianthus* (*Asteraceae*). [Pollen morphology in some species of the genus *Helianthus* (Asteraceae)]. Bot Zh 77 10:24–31 (no English summary)

Tril' VM, Moskalyuk NV (1994) Morfologo-biologicheskoe izuchenie plodov vidov *Comarum palustre* i *Comarum salesovianum* (Rosaceae). [Morphological and biological study of fruits in *Comarum palustre* and *Comarum salesovianum* (Rosaceae)]. Bot Zh 79 10:59–64 (English summary)

Trofimov AG, Kulagina NV, Kulikov OA (1994) Geokhronologicheskie rubezhi pleystotsena severnogo Pribaykal'ya (po dannym radiotermolyuminestsentnogo analiza). *Geochronological boundaries within the northern Baikal-area (according to radio-thermoluminescence data). Doklady Akad Nauk 334 1:80–82 (no English summary)

Tzedakis PC, Frogley MR, Heaton THE (2002) Duration of last interglacial conditions in northwestern Greece. Quat Res 58:53–55

Ukrayntseva VV (1990) Novye paleobotanicheskie i palinologicheskie svidetel'stva ranne-golotsenovogo poteplenia klimata v vysokikh shirotakh Arktiki. (New paleobotanical and palynological evidence that during the Early Holocene the climate in the Arctic high latitudes was getting warmer). Bot Zh 75 1:70–74 (no English summary)
Ukrayntseva VV (1991) Palinologicheskoe izuchenie roda *Cistus* (Cistaceae). [Palynological study of the genus *Cistus* (Cistaceae)]. Bot Zh 76 7:979–985 (no English summary)
Ukrayntseva VV (1996) Flory pozdnego pleystotsena i golotsena Sibiri. (Late Pleistocene and Holocene floras of Siberia). Bot Zh 81 12:37–48 (English summary)
Ukrayntseva VV, Arslanov KA, Belorusova ZM, Ustinov VN (1989) Pervye dannye o ranne-golotsenovoy flore i rastitel'nosti ostrova Bol'shoy Lyakhovskiy (Novosibirskiy Arkhipelag). [The first data on the Early Holocene flora and vegetation of the Big Lyakhovsky Island (the Novosibirsk Archipelago)]. Bot Zh 74 6:782–793 (English summary)
Van Vliet-Lanoë B, Seppälä M (2002) Stratigraphy, age and formation of peaty earth hummocks (pounus), Finnish Lapland. Holocene 12:187–199
Vasil'eva IM (1993) Morfologiya semyan i anatomicheskaya kharakteristika semenoy kozhury nekotorykh vidov roda *Aquilegia* (Ranunculaceae). [The morphology of seeds and anatomical characteristics of seed coat in some species of the genus *Aquilegia* (Ranunculaceae)]. Bot Zh 78 4:67–80 (no English summary)
Vasil'eva IM, Khramova EL (1992) Morfologiya pyl'tsevykh zeren roda *Aquilegia* (Ranunculaceae) flory SSSR v svyazi s ego sistematikoy. [Pollen-grain morphology in the genus *Aquilegia* (Ranunculaceae) of the USSR flora in connection with its systematics]. Bot Zh 77 4:32–39 (no English summary)
Verkhovskaya NB (1990) O rastitel'nosti yuzhnogo Sikhote-Alinya v srednevekov'e. (On the vegetation of the southern parts of the Sikhote-Alin in the Middle Ages). Bot Zh 75:11:1555–1564 (no English summary)
Verkhovskaya NB, Esipenko LP (1993) O vremeni poyavleniya *Ambrosia artemisiifolia* (Asteraceae) na yuge Rossiyskogo Dal'nego Vostoka. [The time of the *Ambrosia artemissifolia* (Asteraceae) appearance in the south of Russian Far East]. Bot Zh 78 2:94–101 (English summary)
Virkanen J (2000) The effect of natural environmental changes on sedimentation in Lake Kuttanen, a small closed lake in Finnish Lapland. Holocene 10:377–386
Vorren K-D (2001) Development of bogs in a coast-inland transect in northern Norway. Acta Paleobot 41:43–67
Winograd IJ (2001) The magnitude and proximate cause of ice-sheet growth since 35,000 yr b.p. Quat Res 56:299–307
Yadav RR, Singh J (2002) Tree-ring-based spring temperature patterns over the past four centuries in Western Himalaya. Quat Res 57:299–305
Yang B, Bräuning A, Johnson KR, Shi YF (2002) General characteristics of temperature variation in China during the last two millennia. Geophys Res Lett 29, 2:38–1 till 38–4
Yu G, Tang LY, Yang XD, Ke XK, Harrison SP (2001) Modern pollen samples from alpine vegetation on the Tibetan Plateau. Global Ecol Biogeogr 10:503–519
Yurkovskaya TK, Elina GA, Klimanov VA (1989) Rastitel'nost' i paleogeografiya lesnykh i bolotnykh okosistem pravoberezh'ya reki Pinegi (Arkhangel'skaya Oblast'). Vegetation and paleogeography of forest and bog ecosystems on the right-hand beaches of river Pinega (Arkhangel'sk Oblast'). Bot Zhurn 74 12:1711–1723 (in Russian; English summary)
Zeeberg JJ, Lubinski DJ, Forman SL (2001) Holocene relative sea-level history of Novaya Zemlya, Russia, and implications for late-Weichselian ice-sheet loading. Quat Res 56:218–230
Zernitskaya VP (1992) Pyl'tsa *Quercus* (Fagaceae) iz pozdnelednikovykh i golotsenovykh otlozheniy Belarusi. (The pollen of *Quercus* (Fagaceae) from the upper glacial and Holocene of Byelorus). Bot Zh 77 11:71–73 (no English summary)

Zhang HC, Wünnemann B, Ma YZ, Peng JL, Pachur H-J, Li JJ, Qi Y, Chen GJ, Fang HB (2002) Lake level and climate changes between 42,000 and 18,000 ^{14}C yr b.p. in the Tengger Desert, northwestern China. Quat Res 58:62–72

Zhilyaev GG (1996) Rasprostranenie pyl'tsy v populyatsiyakh travyanistykh rasteniy Karpat. Scattering of pollen in populations of herbaceaous plants in the Carpathians. Bot Zh 81 3:53–60 (English summary)

Zhou WJ, Dodson J, Head MJ, Li BS, Hou, YJ, Lu XF, Donahue DJ, Jull AJT (2002) Environmental variability within the Chinese desert-loess transition zone over the last 20,000 years. Holocene 12:107–112

Prof. Dr. Dr. hc. Burkhard Frenzel
Institut für Botanik der Universität Hohenheim - BIO II
Garbenstr. 30
70593 Stuttgart, Germany
e-mail: bfrenzel@uni-hohenheim.de

Subject Index

Druck: Strauss Offsetdruck, Mörlenbach
Verarbeitung: Schäffer, Grünstadt